Workplace Law Handbook 2012:
Health and safety, premises and environment

Edited by Alex Davies

KoganPage

LONDON   PHILADELPHIA   NEW DELHI

ISBN 978-0-7494-6624-4

Published in association with
Workplace Law
110 Hills Road
Cambridge
CB2 1LQ
Tel. 0871 777 8881
Fax. 0871 777 8882
Email info@workplacelaw.net
www.workplacelaw.net

Design and layout by Workplace Law and Amnet Systems Ltd.
Printed and bound by Oriental Press, Dubai.

Cover design by Gary Jobson.

Publisher's note
Every possible effort has been made to ensure that the information contained in this book is
accurate at the time of going to press, and the publishers and authors cannot accept responsibility
for any errors or omissions, however caused. No responsibility for loss or damage occasioned
to any person acting, or refraining from action, as a result of the material in this publication can
be accepted by the editor, the publisher or any of the authors. Readers should be aware that only
Acts of Parliament and Statutory Instruments have the force of law and that only the courts can
authoritatively interpret the law. The views expressed in the *Workplace Law Handbook 2012: Health
and safety, premises and environment* are the contributors' own and not necessarily those of the
publishers.

First published in Great Britain in 2011 by Kogan Page Limited in association with Workplace Law.

120 Pentonville Road	1518 Walnut Street, Suite 1100	4737/23 Ansari Road
London N1 9JN	Philadelphia PA 19102	Daryaganj
United Kingdom	USA	New Delhi 110002
www.koganpage.com		India

ISBN: 978 0 7494 6624 4

British Library Cataloguing-in-Publication Data

A CIP record for this book is available from the British Library.

Contents

Contents

Contents

www.workplacelaw.net

Comment ...

The price of responsibility

Alex Davies, Workplace Law

As news of a double-dip recession dominates headlines, most employers are concentrating on their key business activities to keep their companies afloat. Spending time and money on unnecessary health and safety burdens isn't high on the agenda. Indeed, recent Government initiatives, such as the closure of the HSE's Infoline, and restricting the ways in which accidents and incidents under RIDDOR are reported (including proposals to narrow the scope of reportable injuries – p.618) all point towards a safety culture that continually passes the buck.

People are a company's most vital asset – it's essential we protect their health and safety in a responsible way. Recent weeks have seen mine collapses, explosions and fires, so it's crucial we maintain the UK's excellent safety record in spite of a weak economy and austere cost-cutting measures. The first corporate manslaughter case (p.200) has proved that the Courts won't hesitate to impose fines way beyond a company's financial reach, so having the proper procedures in place is paramount.

On the environmental side, and still on a cost-cutting theme, waste management is increasingly growing in importance. Despite the connotations of waste being something we need to get rid of, what has traditionally been seen as worthless can actually hold value, and so managing waste effectively can reap financial rewards. Indeed, if companies don't abide by the waste hierarchy (p.605), they could expose themselves to potential fines under

Alex Davies is the Project Manager at Workplace Law. She has a degree in Publishing and has worked on a variety of business titles in a range of roles. She has edited the *Workplace Law Handbooks* for the past five editions and enjoys putting together this ever-expanding title whilst cruising the inland waterways of France on her 104-year-old barge.

new legislation (p.707) and increased powers of regulators (p.717).

Knowledge is power, and keeping abreast of the changes is paramount.

The way we consume information has changed radically in the past few years, shifting away from traditional printed words on paper to online newspapers, tailored RSS news feeds, homemade video clips shared on the internet to, at its most radical, condensed 'tweets' in which information of importance is refined to just 140 characters. The result is a transient glut of information, constantly evolving, and instantly forgettable. It's true that once something has been said on the internet, it can't be unsaid – however, I'd argue that due to the colossal size of the internet it's also very hard to find! Useful information can very easily be buried, never to be retrieved.

That's why I think there is still a place for a traditional reference book, to be placed on a shelf or a desk, within easy reach.

The new content in this year's *Handbook* focuses primarily on compliance, and offering solutions. As more responsibility is shifted towards the employer, it's vital to keep informed of the changes occurring in health and safety management and environmental initiatives.

We like to think that when a client chooses Workplace Law, they are putting their trust in us to help them keep up to date with what's important, in an efficient, responsible, and engaging way.

An employer's most valued asset is its people, and in austere times the temptation to cut costs is high on the agenda. But people are what will keep a company afloat during stormy weather, and knowing how to keep them safe is key.

> "People are a company's most vital asset – it's essential we protect their health and safety in a responsible way."

The *Handbook* aims to do just that, in a user-friendly, structured and interesting way. With this in mind, this year's edition contains a brand new 'At a glance' section (p.13), which provides a quick and easy reference point for the key information you need to know over the next 12 months, as well as the usual varied mix of comment pieces, case reviews and chapters to cover all aspects of the modern, dynamic workplace.

I hope you find the content informative and useful and welcome, as ever, any feedback. Please email me on alex.davies@workplacelaw.net with your views, or tweet me at @AlexDaviesWPL. Here's to a safe future.

A. Davies.

@AlexDaviesWPL

About Workplace Law

Workplace Law specialises in health and safety, environmental management and employment law. We help clients achieve excellence by easing the burden of compliance and providing valuable best practice solutions.

Operating nationally from offices in Cambridge and London, we are an established market leader in the provision of information, training, consultancy and support to employers. Through our recruitment arm we place competent professionals in permanent and interim positions with UK employers.

With a 15-year pedigree of excellence, we are trusted by employers throughout the UK to help them get to grips with the law and regulation of managing people and their working environment.

Workplace Law is the specialist advisor to the British Institute of Facilities Management (BIFM) on employment law, health and safety and premises management, and a market leader in the facilities management (FM) and education sectors.

Our specialist services

Workplace Law specialises in supporting employers by providing three core services.

Educating our clients through the provision of an award-winning membership and online information service, and through the delivery of recognised training programmes by classroom, e-learning and blended learning study. Licensed by the CIPD, IEMA, IOSH and NEBOSH, we are one of the UK's leading information and training providers.

Supporting our clients by providing bespoke advice and consultancy, by telephone, online and on-site support throughout the UK. We offer a range of solutions to suit your needs, from one-off projects to indemnified annual support contracts.

Resourcing talent for our clients through our sister company, Workplace Law Career Network, which specialises in the permanent and interim placement of competent professionals in the areas of health, safety and environment (HSE), facilities management, and human resources.

Our specialist subject areas

Workplace Law specialises in three core areas of compliance.

Workplace Law Health and Safety
Workplace Law Health and Safety specialises in health and safety management and compliance. We are an established training provider, licensed by IOSH and NEBOSH to provide accredited health and safety training programmes via classroom and e-learning study modes. We provide advice and support on health and safety strategy and systems such as HSG 65 and OHSAS 18001, in addition to CDM and fire safety compliance.

Workplace Law Environmental

Workplace Law Environmental specialises in environmental management and compliance. We are an established training provider, approved by the Institute of Environmental Management and Assessment (IEMA) to provide certified courses at foundation level. Our specialist consultancy work includes environmental auditing, advice on environmental management systems, including ISO 14001, BREAAM and energy efficiency assessments, and compliance with environmental regulations.

Workplace Law Human Resources

Workplace Law Human Resources specialises in HR and employment law. We are an established training provider, licensed by the Chartered Institute of Personnel and Development (CIPD) to provide accredited training programmes. We provide advice and support for employers on recruitment, performance management, discrimination and dismissal, including representing you in Employment Tribunal proceedings.

Workplace Law Career Network

Workplace Law Career Network is a leading professional services firm specialising in the recruitment of personnel in facilities management, building services, health and safety, and human resources.

We provide permanent recruitment and interim management solutions for employers. Part of Workplace Law, the firm draws on a wide range of resources to offer enhanced support to candidates and clients.

Candidates have access to CIPD-accredited training and updated information through the award-winning Workplace Law service.

Enhanced services for clients include psychometric testing, assessment centres, and advice on employment law, including fair recruitment, TUPE and restructuring and redundancy support.

Workplace Law Career Network provides national coverage, operating from offices in Cambridge and London.

How we do it

Clients trust us to deliver value to their business, challenging the status quo to constantly improve the services we provide.

We do this with energy, enthusiasm and endeavour, to make our community a better place to live and work.

T. +44 (0)871 777 8881
F. +44 (0)871 777 8882

Workplace Law
110 Hills Road
Cambridge
CB2 1LQ

At a glance

Workplace Law Health and Safety

Premises management
Business rates

Area	Rate
England and Scotland	42.6p
Wales	42.8p

Temperature

Guidance	Temperature range
Workplace (Health, Safety and Welfare) Regulations 1992	During working hours, the temperature in all workplaces inside buildings shall be reasonable'
ACoP – Workplace health, safety and welfare. A short guide for managers	In the typical workplace, the temperature should be at least 16°C unless much of the work involves severe physical effort, in which case the temperature should be at least 13°C.
CIBSE Guide A – Environmental design	For air-conditioned buildings in the UK, a dry resultant temperature of between 21°C and 23°C during winter and between 22°C and 24°C in summer for continuous sedentary occupancy. It is recognised that room temperatures in buildings without artificial cooling will exceed the summer values for some of the time but should not exceed 25°C for more than 5% of the annual occupied period (typically 125 hours).
There is currently no maximum temperature for a workplace.	

Welfare facilities

Number of toilets and washbasins for mixed use (or women only)		
No. of people at work	No. of toilets	No. of wash basins
1-5	1	1
6-25	2	2
26-50	3	3
51-75	4	4
76-100	5	5

Toilets used by men only

No. of men at work	No. of toilets	No. of urinals
1-15	1	1
16-30	2	1
31-45	2	2
46-60	3	2
61-75	3	3
76-90	4	3
91-100	4	4

Parking

The BPA, DMUK and BCSC undertook major research in partnership with DfT in 2009, which indicated that the 6% one size fits all approach leads to oversupply in some situations and undersupply in others. It is expected that Inclusive Mobility will be superseded in 2012 with more flexibility in the guidance. The BPA recommends the following allocation.

Size of car park (no. of spaces)	Designated bay provision
1-50	Two + 3% total car park
51-200	Three + 3% of total car park
201-55	Four + 3% of total car park
501-1,000	Five + 3% of total car park
1,000+	Six + 3% of total car park

Environmental management
Energy efficiency

Energy Performance Certificates are required when a commercial building over 50m² is built, sold or rented.
EPCs are valid for ten years but must be renewed if the property is modified.
Display Energy Certificates are required for public buildings over 1,000m².
The DEC is valid for one year and the advisory report is valid for seven years.
For all air conditioning systems first put into service on or after 1 January 2008, the first inspection must have taken place by five years of the date it was first put into service.
Where the effective rated output is more than 250kW the first inspection must have happened by 4 January 2009.
Where the effective rated output is more than 12kW the first inspection must have happened by 4 January 2011.

Waste hierarchy

The generation of waste arising from business activities and its management should be governed by the 'waste hierarchy,' which provides guidelines for increasingly environmentally responsible waste management. It represents a chain of priority for waste management, citing the prevention of waste as the most favourable option. If not possible, waste prevention is succeeded by minimisation, re-use, recycling, energy recovery, then disposal, which is considered the least favourable.

Health and safety
Noise action levels

First action level	A daily personal noise exposure of 80dB(A) and a peak value of 112 pascals.
Second action level	A daily personal noise exposure of 85dB(A) and 140 pascals.
Limit value	A peak sound pressure of 87dB(A) and 200 pascals. The limit value will take into account the reduction afforded by hearing protection.

PAT frequencies

Construction sites 110V equipment	All equipment: every three months
Industrial including commercial kitchens	■ S, IT and M: every 12 months ■ P and H: every six months
Equipment used by the public	■ S and IT: every 12 months ■ M, P and H Class 1: every six months ■ M, P and H Class 2: every 12 months
Schools	■ All Class 1 equipment: every 12 months ■ All Class 2 equipment: every 48 months
Hotels	■ S and IT: every 48 months ■ M and P: every 24 months ■ H: every 12 months
Offices and shops	■ S and IT: every 48 months ■ M and P: every 24 months ■ H: every 12 months

- ■ S = stationary equipment, e.g. vending machine
- ■ IT = IT equipment, e.g. computer
- ■ M = moveable equipment, e.g. extension lead
- ■ P = portable equipment, e.g. fan
- ■ H = handheld equipment, e.g. drill

RIDDOR – reporting accidents and incidents

Type	When to report	How to report	Online form
Deaths	Immediately	By the quickest practical means, usually a telephone call	■ F2508 Report of an injury
Major injury	Immediately	By the quickest practical means, usually a telephone call	■ F2508 Report of an injury
Over-three-day injury	As quickly as practicable and within ten days	Online	■ F2508 Report of an injury ■ OIR9B Report of an Injury Offshore
Injury to third parties or visitors	As quickly as practicable and within ten days	Online	■ F2508 Report of an injury ■ OIR9B Report of an Injury Offshore
Diseases	As quickly as practicable and within ten days	Online	■ F2508A Report of a Case of Disease
Dangerous occurrences	As quickly as practicable and within ten days	Online	■ F2508 Report of a Dangerous Occurrence
Gas incidents	Immediately	By the quickest practical means, usually a telephone call	■ F2508G1 Report of a Flammable Gas Incident ■ F2508G2 Report of a Dangerous Gas Fitting

- ■ Following a three-month consultation with stakeholders in spring 2011, the HSE is seeking a change to the RIDDOR Regulations to extend the reporting threshold.
- ■ The effect of this would be that employers would no longer have to report 'over-three-day injuries' from mid-2012, and would only report 'over-seven-day' accidents in future.
- ■ The changes also include extending the period during which duty-holders must notify a RIDDOR-reportable accident from ten to 15 days after the accident.

Working time limits

Worker	Time
Adults	Adult workers cannot work more than an average of 48 hours per week over a 17-week period.
Young workers	May not work over eight hours per day or 40 hours per week (these hours cannot be averaged).
Child workers (compulsory school age)	■ During term time, a child may not work more than 12 hours per week including a maximum of two hours on a school day or Sunday and a maximum of five hours on a Saturday for 13-14 year olds, or eight hours on a Saturday for 15-16 year olds. ■ Child workers may not work for more than one hour before school. ■ During school holidays, a 13-14 year old may work a maximum of 25 hours per week including up to five hours on a weekday or a Saturday and up to two hours on a Sunday. ■ During school holidays, a 15-16 year old may work up to a maximum of 35 hours per week including up to eight hours on a weekday or a Saturday and up to two hours on a Sunday.

Health and safety statistics

■ In 2009/10, 28.5 million days were lost overall (1.2 days per worker), 23.4 million due to work-related ill health and 5.1 million due to workplace injury.

■ The number of major injuries to employees reported in 2009/10 was 26,061. Slipping and tripping accounted for 41% and falls from height 16%.

■ 152 workers were fatally injured in 2009/10, a rate of 0.5 fatalities per 100,000 workers, down from 180 in 2008/09. Construction and agriculture have the highest number of fatalities, accounting for 42 and 38 fatalities respectively.

Accessible environments

Elspeth Grant, TripleAconsult

Key points

- This chapter examines accessible environments in the wider sense and how best these may be implemented in a manner that provides true benefits for organisations / companies, disabled and non-disabled people alike. Fire safety and evacuation of disabled people is not covered in this chapter and reference should be made to *Disability access and egress* (p.238), *Fire, means of escape* (p.314), and *Personal Emergency Evacuation Plans (PEEPs)* (p.569).
- It is particularly important to ensure that environments are accessible in an economic climate where budgets are tight and every cost unit is likely to be questioned. It is therefore important to remember that an accessible environment is an asset that assists all users, not just disabled people.
- The CBI estimates that disabled people account for a potential £50bn of spend in the country; however, 16 years after the Disability Discrimination Act 1995 (superseded by the Equality Act 2010) was passed, many still cannot access the employment opportunities, services or leisure facilities they require.

Legislation

- Building Regulations – Part B (Fire safety) and Part M (Access to facilities and employment).
- Approved Documents B and M – supporting documents to Building Regulations B and M.
- Special Educational Needs and Disability Act 2001 (SENDA).
- Regulatory Reform (Fire Safety) Order 2005 – fire safety legislation based on risk assessment.
- The Equality Act 2010 – harmonises discrimination law with disability as a protected characteristic.

In brief, Building Regulations Part B and M address the built environment, whilst the Equality Act is focused on the people who use the building, which also includes the built environment. The result is that there is such a thing as a Part M-compliant workplace, yet there is no such thing as anequality-compliant workplace (i.e. although a person can have a Part M-compliant workplace they could still be discriminatory under the Equality Act, as the latter is concerned with the activities and services provided within that area).

British Standards – BS 9999: 2008 (Code of Practice for Fire Safety) and BS 8300: 2009 + A1: 2010 (Design of Buildings: Code of Practice) provide supporting guidance for the legislation.

The Equality Act

The main provisions of the Equality Act came into force in October 2010, with the purpose of harmonising discrimination law and strengthening the law to support progress on equality. As of this point, the Disability Discrimination Act no longer had the force of law and, in April 2011, the integrated public sector Equality Duty, the Socio-economic Duty and Dual Discrimination Protection applied. Subsequent elements of the Act include a ban on discrimination in the provision of goods, facilities, services and public functions, to be introduced in 2012, and the private and voluntary sector gender pay transparency regulations and political parties publishing diversity data will be implemented in 2013.

The major change the Equality Act introducedwas the list regarding how to define disability by reference to day-to-day activities. Thisno longer applies, and someone is considered disabled simply if they have a physical or mental impairment or impairments that have a substantial and long-term adverse effect. The Act also extended the coverage applied to the employment of disabled peopleby introducing a single threshold for the point at which the duty to make adjustments is triggered. This is based on whether a provision or a physical feature of premises places a disabled person at a 'substantial disadvantage' in comparison with a person who is not disabled.

In 2005, the Prime Minister's Strategy Unit stated in *Improving the Life Chances of Disabled People*:

'By 2025, disabled people should have full opportunities and choices to improve their quality of life and be respected and included as equal members of society.'

It must be remembered that, in 2012, not only the Olympics but also the Paralympics will be held in the UK. This will present not only an enormous revenue opportunity but also an enormous challenge in order to make the environment accessible for the tens of thousands of athletes and visitors during this time. Those organisations that ensure that their environments are fully accessible will be those that benefit the most.

It is important to understand the background to disability and the inter-relationship between the legislation that supports accessibility.

There is no longer a register of disabled people, so statistics are gathered from a variety of sources and can vary considerably. The table below identifies the approximate numbers of disabled people in the UK (believed to be over one in five of the population) and will help the focus of attention to be across the wide spectrum of disabilities.

Disability	Number of people in UK
Mental health	The largest group
Arthritis	Over 11 million
Deaf or hard of hearing	8.5 million
Reading difficulties including dyslexia	2.5 – 6 million
Visual impairments	2.5 – 3 million
Wheelchair users	600,000 (of which only 5-7% can never leave their wheelchairs)

Table 1. Disabled people in the UK.

What is discrimination?

Disability has a broad meaning, which has subtly changed from that in the Disability Discrimination Act. It is defined as a physical or mental impairment that has a substantial and long-term adverse effect on the ability to carry out normal day-to-day activities. 'Substantial' means more than minor or trivial. 'Impairment' covers, for example, long-term medical conditions such as asthma and diabetes, and fluctuating or progressive conditions such as rheumatoid arthritis or motor neurone disease.

A mental impairment includes mental health conditions (such as bipolar disorder or depression), learning difficulties (such as dyslexia) and learning disabilities (such as autism and Down's syndrome). Some people, including those with cancer, multiple sclerosis and HIV/AIDS, are automatically protected as disabled people by the Act. People with severe disfigurement will also be protected as disabled without needing to show that it has a substantial adverse effect on day-to-day activities.

The Act protects anyone who has, or has had, a disability. So, for example, if a person has had a mental health condition in the past that met the Act's definition of disability and is harassed because of this, that would be unlawful.

The Act also protects people from being discriminated against and harassed because of a disability they do not personally have. For example, it protects people who are mistakenly perceived to be disabled. It also protects a person from being treated less favourably because they are linked or associated with a disabled person. For example, if the mother of a disabled child was refused service because of this association, that would be unlawful discrimination.

Legal definition of disability

As with the Disability Discrimination Act 2005, the Equality Act 2010 is prescriptive in its approach to who is covered as those with:

- physical impairment – this includes weakening or adverse change of a part of the body caused through illness, by accident or from birth. For example, amongst many other situations, blindness, deafness, heart disease, the paralysis of a limb or severe disfigurement;
- mental impairment – this can include learning disabilities and all recognised mental illnesses – for example, mental illnesses specifically mentioned in the World Health Organisation's International Classification of Diseases are very likely to be included;
- substantial impairment – this does not have to be severe, but is more than minor or trivial; and
- long-term adverse effect – that has lasted or is likely to last more than 12 months, or a normal day-to-day activity – for example one that affects:
  - mobility;
  - manual dexterity;
  - physical coordination;
  - continence;
  - ability to lift, carry or otherwise move everyday objects;
  - speech, hearing or eyesight;
  - memory or ability to concentrate, learn or understand; or
  - perception of the environment and personal safety.

What are reasonable adjustments?

- Making alterations to physical features of premises.
- Changing place of work.
- Assigning duties to others.
- Altering work hours.
- Acquiring or modifying equipment.
- Providing a reader or interpreter.

What determines what is reasonable?

- The effectiveness of the step.
- The practicality of the step.
- Financial and other costs.
- Extent of disruption.
- Employer's financial / other resources.
- The amount of resources already spent.
- Availability of grants / assistance.
- Effect on other employees.
- Adjustments made for other employees.
- The extent to which the disabled person is willing to cooperate.

Building Regulations and British Standards

Building Regulations Part B and Part M and supporting guidance BS 8300: 2009 + A1: 2010 and BS 9999: 2008 (with over 40 separate pieces of guidance regarding means of escape for disabled people), contain the framework for architects and designers to refer to, when implementing a new workplace or changing an existing one. Building Regulations are supported by 'Approved Documents' (AD) which give practical guidance with respect to the Regulations.

If a building is built in accordance with Part M, characteristics that still comply do not have to be altered. Characteristics (such as door handles) not covered by Part M may need to be altered. The Equality Act does not override existing legislation such as Planning Permission or Listed Building Consent.

Building Regulations Part B (Fire safety)
Part B now includes further requirements for means of escape for disabled people, such as emergency voice communications in refuges, management plans for effective evacuation, and clearer signage, whilst as-built plans must include escape routes for those with disabilities.

Building Regulations Part M (Access To and Use of Buildings)
This sets minimum standards for access and use of buildings by all building users, including disabled people. The Regulation avoids specific reference to, and a definition of, disabled people. This inclusive approach means that buildings and their facilities should be accessible and useable by *all* people who use buildings – including parents with children, older people and disabled people.

Previously, Part M covered new buildings and extensions to existing buildings. The 2004 revision brought Part M into line with other parts of the Building Regulations by extending its scope to include alterations to existing buildings and certain changes of use. This covers historic buildings; however, there is recognition of the need to conserve the special characteristics of historic buildings. Approved Document M states that 'the aim should be to improve accessibility where and to the extent to which it is practically possible, always provided that the work does not prejudice the character of the historic building, or increase the risk of long-term deterioration of the building fabric or fittings'.

Access and use
'Reasonable provision shall be made for people to gain access to and use the building and its facilities.'

This does not apply to any part of a building that is used solely to enable the building or any service or fitting within the building to be inspected, repaired or maintained.

Access to extensions to buildings
'Suitable independent access shall be provided to the extension where reasonably practicable.'

This does not apply where suitable access to the extension is provided throughout the building that is extended.

Sanitary conveniences in extensions to buildings
'If sanitary conveniences are provided in any building that is to be extended, reasonable provision shall be made within the extension for sanitary conveniences.'

This does not apply where there is reasonable provision for sanitary conveniences elsewhere in the building that can be accessed by building users.

As the following paragraph, contained in the Approved Document M, indicates, rather than the religious following of Building Regulations, the emphasis should be on solutions that really improve the environment for all disabled people:

'There is no obligation to adopt any particular solution contained in the Approved Document if you prefer to meet the requirement of the regulation in some other way.'

Designers have the option to:

- remove the feature;
- alter the feature so that it no longer has that effect;
- provide a reasonable means of avoiding the feature; and/or
- provide the service by a reasonably alternative means.

BS 8300: 2009 + A1: 2010 Design Guidance

February 2009 saw the launch of BS 8300: 2009 + A1: 2010 Design of buildings and their approaches to meet the needs of disabled people – Code of Practice. The Standard has introduced some flexibility into a number of new areas, including the provision of car parking, the design of stairs, and the provision of accessible bathroom and WC facilities.

Car parking
There is an adjustment in the minimum number of designated Blue Badge parking spaces for some building types, including workplaces, religious buildings and crematoria. The Standard also recommends the provision of a number of additional, enlarged parking bays, designated for general use, and suitable for reclassification as Blue Badge spaces if necessary.

Steps and stairs
The preferred dimensional range for steps and stairs is between 150mm and 180mm for the rise and between 300mm and 450mm for the tread. The Standard also increases the maximum number of risers in a flight to 20 (previously 12, or 16 in small premises with restricted area).

Sanitary facilities
Guidance is given on the provision of an appropriately managed 'Changing Places' facility in many larger buildings and complexes. A Changing Places facility is a combined toilet, bench and changing room with tracked hoist, for use by people with complex and multiple disabilities, who may require the help of up to two assistants.

Residential accommodation
This affects all building types that require permanent sleeping accommodation (i.e. hotels, motels, nursing and residential homes, university and college halls of residence, and relatives' accommodation in hospitals). The minimum of accessible bedrooms as a percentage of the total number of rooms should be:

- 5% with a fixed track hoist system;
- 5% without a fixed track hoist system; and
- 5% capable of being adapted in the future to a standard of accessibility.

The Standard has also made some minor changes to ensure that it is aligned with other Standards, which include BS 9999: 2008 Code of Practice for Fire Safety in the Design, Management and Use of Buildings, and the Approved Document M.

BS 9999: 2008 Code of Practice for Fire Safety in the Design, Management and Use of Buildings

This is probably the least understood area of accessible environments and BS 9999: 2008 provides a great deal of guidance on the subject. There are 33 separate references to fire safety for disabled people, including an entire section providing detailed guidance to underpin the requirements of the Regulatory Reform (Fire Safety) Order 2005 Article 14 (b):

'In the event of danger, it must be possible for persons to evacuate the premises as quickly and safely as possible'.

The table shows the relevant references.

Penalties

The penalties for non-compliance with the legislation vary from case to case but in general will result in:

- *The Equality Act 2010:* A civil court case or Employment Tribunal with the potential for fines and damage to brand through bad publicity; it is also very much driven by what is reasonable. The Equality and Human Rights Commission monitors disability issues and may take up individual cases if appropriate.
- *Building Regulations:* Refusal of planning applications and/or completion when an application has been submitted or criminal proceedings undertaken if Regulations have been found to be broken within the workplace.
- *Regulatory Reform (Fire Safety) Order (RRO):* Criminal prosecution with the potential for fines, imprisonment of the Responsible Person and litigation.

Area	Section
Disabled evacuation	Sections 14 and 46
Horizontal / vertical escape	Section 17 / Clause 18.1, 18.8
Fire doors	Clause 33.1.6
Stairs	Clause 18.5
Lifts	Clauses 46.9, 46.10
Refuges	Clause 46.8 and Annex G
Evacuation strategies	Section 12
PEEPs	Clause 46.7
Test drills / training	Clause 46.11, 46.12
Sheltered housing	Annex Q

Table 2. Disabled references.

- *Corporate Manslaughter and Corporate Homicide Act 2007*: Criminal prosecution of either organisations or individuals, resulting in fines or imprisonment.

SENDA 2001
The Special Educational Needs and Disability Act 2001 (SENDA) amends the DDA and becomes Part 4. There are two separate sections of SENDA:

- Schools; and
- Post 16, which covers nursery schools through to schools for up to 19 year olds.

Schools
Schools have two key duties:

- Not to treat disabled pupils less favourably; and
- To make reasonable adjustments to avoid putting pupils at a substantial disadvantage (which does not necessarily mean the provision of auxiliary aids or altering physical features).

LEAs must draw up access strategies covering:

- access to the curriculum;
- increased access to education; and
- provision of information in a range of formats.

Non-educational services within schools are covered under Part 3.

Higher Education and Further Education
This covers HE and FE and similar designated institutions, including post-16 provision by education authority.

Strategic planning and implementation
A truly accessible environment will only be achieved through a mix of management processes, training and physical changes.

Legally, an employer or service provider must only take 'reasonable' steps to provide access; however, in order to obtain best business benefit and provide the 'best possible' environment for clients and employees, the key is to identify any barriers that may prevent a disabled person in fully participating on the same basis as a non-disabled person.

Before attempting to alter or adapt a building that falls under the requirements of the DDA, consider:

- *Planning*. Think about the holistics of the building.
- *Carry out an access assessment.* This should be appropriate to the building and the applicable legislation.
- *Identify priority of measures.*
- *Develop an action plan.* This should implementthe measures.
- *Review.*In the light of progress and changes in legislation / technical guidance.

Figure 1 (*opposite*) shows the approach for an inclusive strategy.

It is good practice to consult with any appropriate groups of disabled people during the implementation of an accessible environment. As with any market and research activity, this will both add value and ensure that the environment meets the needs of the users.

Fundamental to this exercise is to understand whether there is a requirement to be proactive as in an environment that is open to the public (a bingo hall, supermarket, rail station, school, university) or whether the environment is controlled (an office, staff area).

For example, if the target environment is open to the public and a green field site, then Part M, supported by management processes, should be implemented

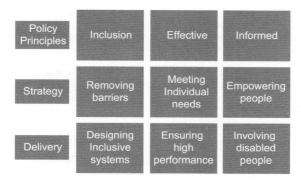

Figure 1. An inclusive strategy.

throughout. However, if the environment is an existing office building and not open to the public then it may be more appropriate to identify potential physical barriers and, if no disabled person is currently employed, then it may be more appropriate to implement impact assessments for all employees to identify whether any alterations to the workplace should be made.

Many organisations undertake expensive access assessments of their environment as a first step to the provision of an accessible environment, without having first identified a strategic approach or clear vision of how changes will be implemented. The Strategic Access Plan will become a blueprint for the implementation of the accessible environment and must include timelines, responsibilities, monitoring procedures and critical success criteria. It is recommended that the access strategy should cover approaches to the following.

Internal communications
- Documentation of an access policy.
- Implementation of a staff communications policy.
- Provision of staff training in assisting disabled people.

- Implementation of a non-discriminatory recruitment procedure.

Health and safety
- Identification of all health and safety issues.
- Creation of overall strategy for means of escape for disabled people.
- Identification of those who may need assistance to evacuate.
- Integration of escape plans with any third party entities.
- Creation of Personal Emergency Evacuation Plans (PEEPs).

External communications
- Identification of the information needs of disabled people.
- Review of existing methods of communication and identify any gaps in information needs; particularly via written media, the internet, websites, call centres, advertisements, etc.
- Documentation of a communication strategy and implementation of service policies for disabled people.
- Provision of accessible booking facilities if appropriate, which must include the same discounts for those unable to book online.
- Provision of information on transportation for both private and public facilities.

Physical environment

- Access assessments of the environment and subsequent tracking of actions by priority and responsibility.
- Resolution of barriers identified during access assessments with a focus on high priority or high risk areas as recommended.
- Maintenance schedules taking into consideration access assessment recommendations.
- Refurbishments and/or new properties.

Products and services

- Develop and implement products and service policies.
- Document store-wide procedures for the provision of service to disabled customers.
- Devise a communication strategy for the public.

Monitoring and evaluation

- Establishment of a baseline for assessing critical success factors.
- Annual reviews and comprehensive reviews of progress.

What if the premises are leased?

If the lease prohibits alterations, a tenant may seek the written consent of the landlord and the landlord cannot unreasonably withhold consent. Consent may be subject to reasonable conditions. It is reasonable for a landlord to withhold consent if:

- the change will result in a substantial permanent reduction in the value of the landlord's interest; and/or
- the change would cause significant disruption or inconvenience to other tenants.

A landlord is required to provide auxiliary aids to assist a disabled tenant to access the property; however, there is no firm definition of what constitutes an auxiliary aid.

Recommendations:

- Access assessments of property portfolio by a qualified access assessor.
- Identify possible improvement works.
- Implement management processes.
- Review and monitor progress to compliance.
- Look at recruitment and HR processes.
- Keep people informed.

Design of the built environment

Like other similar parts of the Regulations the principles under Part M will apply likewise:

- *New work*. Will need to fully comply (e.g. brand new shop front, new internal partitioning).
- *Alteration work*. Should not make any existing situation any worse (e.g. replacing an entrance door).

Access strategy

A strategic approach should be documented for the management of both existing and future employees and providing a service to members of the public with disabilities.

The purpose of this strategy is to both provide business benefit through enhanced services to a new market, whilst reducing the risk of litigation where discrimination is perceived by people with disabilities.

Design appraisals

Examination of proposals for new building works against pre-determined criteria to assess their usability by disabled people at an early stage makes sense. This enables the identification of design changes as early as possible.

Access statements

Access statements document where a refurbishment or works cannot meet the

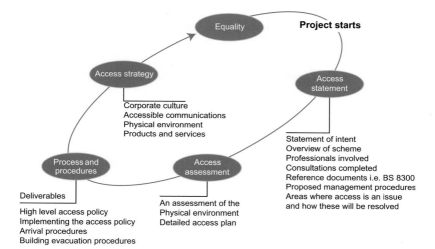

Figure 2. Access strategy.

legal requirements in Part B or Part M of the Building Regulations. The statement explains how the requirement may be met in another way.

Access assessments

An access assessment is a means of:

- examining the accessibility of services and facilities;
- identifying physical barriers to access; and
- measuring how useable facilities are within a building and the services being delivered from it.

Access assessments must be comprehensive and undertaken by qualified, experienced personnel. It often comprises the following elements:

- a walkover survey;
- an assessment of access features; and
- a written report.

The types of physical access that are covered in an access assessment include:

- all disabilities;
- approaches;
- car parking facilities;
- entrances, reception, doorways and corridors;
- lighting / contrast and signage;
- lifts (button height, voice information,etc);
- sanitary facilities;
- communication systems;
- emergency evacuation for the disabled; and
- building management and HR processes.

Action management

Multiple property portfolios should be managed via online databases to ensure that hazards and recommendations can be effectively managed.

Conclusion

It is essential to society and the economy that both disabled and non-disabled people are able to access employment, goods and services. If approached in a logical manner, it is possible to achieve an accessible environment without necessarily increasing cost. Identifying the issues and producing a cohesive plan is key to achieving an accessible environment that meets the needs of disabled people.

See also: Building Regulations, p.76; Disability access and egress, p.238; Fire, means of escape, p.314; Lighting, p.451; Personal Emergency Evacuation Plans (PEEPs), p.569; Toilets, p.669.

Sources of further information

Equality and Human Rights Commission (EHRC): www.equalityhumanrights.com

United Kingdom Disabled People's Council: www.bcodp.org.uk

Centre for Accessible Environments: www.cae.org.uk

Access to Work (Employment support for Disabled People): http://bit.ly/o2C1vF

Accident investigations

Karen Patterson, Kennedys

Key points

- The investigation and analysis of work-related accidents and incidents forms an essential part of managing health and safety.
- HSE guidance on investigating accidents and incidents, which currently takes the form of combined workbook and guidance notes, emphasises that investigation and analysis of work-related accidents and incidents forms an essential part of managing health and safety. Further, it stresses that the learning of lessons from what is uncovered by accident and incident investigation lies at the heart of preventing further events.
- When the guidance was first published, in 2004, the HSE warned: 'To have one accident is bad enough, but to have a further accident because lessons were not learned is inexcusable.'

Legislation

- Health and Safety at Work etc. Act 1974.
- Management of Health and Safety at Work Regulations 1999.

There are legal reasons for investigating accidents. Sections 2 and 3 of the Health and Safety at Work etc. Act 1974 (HSWA) impose general duties on employers to ensure that their employees and those affected by the conduct of their business are, so far as is reasonably practicable, not exposed to risks to their health and safety. Following an accident, in determining whether employers have met this duty, the HSE will often look to see if there have been previous incidents and whether these have been properly investigated as part of the employer's safety management system. A failure to investigate previous incidents or learn lessons from them will be deemed by the HSE to be an aggravating feature in any further incident.

Regulation 5 of the Management of Health and Safety at Work Regulations 1999 also requires employers to plan, organise,

control, monitor and review their health and safety arrangements. Health and safety investigations form an essential part of this process.

In addition to legal reasons for investigating, there are also benefits from doing so – namely the prevention of further similar adverse events, which would include the prevention of business losses due to disruption, stoppage, lost orders, and the impact of criminal and civil legal actions.

It is the potential consequences and the likelihood of the adverse incident recurring that should determine the level of investigation, not simply the injury or ill health suffered on a particular occasion. Thus, for example, near misses can sometimes legitimately warrant more detailed investigation than an accident involving injury.

It is essential that management and the workforce are fully involved in any investigation. Depending on the level of investigation, it may be that employees,

> **Facts**
> - In 2009/10, 28.5 million days were lost overall (1.2 days per worker), 23.4 million due to work-related ill health and 5.1 million due to workplace injury.
> - The number of major injuries to employees reported in 2009/10 was 26,061. Slipping and tripping accounted for 41% and falls from height 16%.
> - 152 workers were fatally injured in 2009/10, a rate of 0.5 fatalities per 100,000 workers, down from 180 in 2008/09. Construction and agriculture have the highest number of fatalities, accounting for 42 and 38 fatalities respectively.
>
> *Source: HSE.*

management and directors of all levels will be involved. It is essential that investigation conclusions are reported directly to someone with the authority to make decisions and act on recommendations.

The urgency of an investigation will depend on the magnitude and immediacy of the risk involved. The general rule is that accidents and incidents should be investigated as soon as possible in order to ensure that evidence is not lost or that memories do not fade. An investigation will involve analysis of all of the information available (i.e. the incident scene, witness evidence, documentary evidence, etc.) to identify what went wrong and determine what steps must be taken to prevent it happening again.

The HSE guidance sets out a step-by-step guide to health and safety investigations.

Step 1. Gathering information
This includes finding out what happened, to establish whether there were any management failings. It is important to capture information as soon as possible. If necessary, work must stop to allow this to take place. If there is any doubt about the safety of a particular type of work, piece of machinery, etc., then this

should be stopped for the duration of the investigation. In the event that this action is not taken, the HSE itself might serve a prohibition notice requiring the work activity to stop immediately where the inspector believes there to be a risk of serious personal injury.

Step 2. Analysing the information
Analysis involves examining all the facts to determine how the accident happened; both the direct and indirect causes. All information gathered should be assembled and examined to identify what information is relevant and what information is missing. The analysis must be carried out in a systematic way so that all possible causes and the consequences and potential consequences of the adverse event are fully reviewed.

Step 3. Identifying suitable risk control measures
Where a proper investigation is undertaken, it should in most cases be possible to identify failings both direct and indirect – frontline and management – which led to the incident or accident taking place. This might include factors such as work pressures on performance, long hours, safety culture, etc. Once the failings have been clearly identified, this should assist in determining the remedial

measures that should be implemented. All possible remedial measures should be risk assessed to ensure that making a change to a work activity, type of machinery, etc., does not in itself import greater risk than was present previously. If several risk control measures are identified, they should be carefully prioritised as a risk control action plan.

Step 4. The action plan and its implementation

At this stage, those people within the organisation who have the authority to make decisions and act on recommendations should be involved. It is good practice for those people to 'accept' the recommendations and to ensure that there is a proper action plan for implementation of the remedial measures identified. This would include a monitoring system to ensure that recommendations are indeed undertaken.

Analysis of major accidents shows that, even where previous incidents have been properly investigated in terms of identifying the failings and the appropriate remedial measures, it is the implementation of the recommendations that has failed to achieve any change and recommendations have become lost in the passage of time. Again, the HSE, when investigating any incident, will look to see whether any previous incidents could and should have been learnt from, including the implementation of recommendations from any previous investigation.

A word of warning

Internal accident investigation reports are generally disclosable to the HSE on request. Although this fact should not impact on an employer's approach to its investigation or its findings, it is often the way in which those findings are expressed that can cause unnecessary difficulties for the employer and individual employees. In the circumstances, care should be taken when drafting investigation reports and those responsible for them should bear in mind that the reports will often have a wider audience than originally anticipated. The involvement of solicitors in the investigation process will sometimes assist in this regard and may also give rise to the possibility of claiming legal privilege for investigation material and reports, which might in turn mean that the HSE and others are not entitled to see them. However, material and reports produced for the dominant purpose of learning safety lessons with a view to identifying and addressing areas for improvement, rather than for the purpose of taking legal advice, will usually be disclosable to the HSE, and potentially to others in the context of civil proceedings.

> *See also*: Health and safety at work, p.361; Health and safety inspections, p.381; RIDDOR, p.618; Safety inspections and audits, p.628; Slips, trips and falls, p.650.

Sources of further information

HSG 245: Investigating accidents and incidents – A workbook for employers, unions, safety representatives and safety professionals: www.hse.gov.uk/pubns/priced/hsg245.pdf

Air conditioning and refrigeration systems

Bob Towse, Heating and Ventilating Contractors' Association (HVCA)

Key points

- Inspections of air conditioning systems are required primarily to improve efficiency and reduce electricity consumption, operating costs and carbon emissions.
- An energy inspection of an air conditioning system must be carried out by an accredited energy assessor.
- Inspections must be a maximum of five years apart.

Legislation

- Energy Performance of Buildings (Certificates and Inspections) (England and Wales) Regulations 2007.
- European Fluorinated Gases Regulations.

Buildings using air conditioning systems are subject to a number of legislative requirements relating to the energy efficiency of the equipment, the control of the refrigerant gases and where the unit has water-cooled condensers or similar the control of Legionella (see p.442).

Energy efficiency

With regard to energy efficiency, all air conditioning systems in a building with a cooling output of more than 12kW – including individual units with a combined output of more than 12kW – are required to be inspected and a report produced that makes recommendations on measures that the building owner can take to improve the energy efficiency of the system.

To comply with the Energy Performance of Buildings (Certificates and Inspections) (England and Wales) Regulations 2007 and similar legislation in Scotland and

Northern Ireland, the building owner, or person with 'control' of the system, shall have an inspection carried out by an accredited air conditioning energy assessor at five-yearly intervals. The first inspection of large installations – more than 250 kW – should have taken place by 4 January 2009, and the remainder (all installations over 12kW) by 4 January 2011.

All completed inspection reports will be lodged on the Landmark EPC register site.

There is not at present any duty on building owners to act on the recommendations made in the assessor's report.

Gases

Many of the gases used in air conditioning systems are Ozone Depleting Substances and their use and destruction at the end of their life must be carefully managed. Many gases that have traditionally been used are now banned within the EU and some installations, which cannot use the more environmentally friendly 'drop in' replacements, will have to be replaced as existing stocks expire.

All installations that use refrigerant gases are required to comply with the European Fluorinated Gases Regulations – known as the F Gas Regulations.

The Regulations place obligations on the 'operator' of stationary refrigeration and air conditioning equipment to ensure that all personnel working on that equipment are appropriately qualified and that the company they are employed by is F Gas certificated.

The Regulations require the operator of stationary refrigeration, air conditioning and heat pump equipment with a refrigerant charge of 3kg or more to maintain records – this requirement also applies to stationary fire protection equipment.

In relation to the requirement to keep records the Regulations state that:

'Operators... shall maintain records on the quantity and type of fluorinated greenhouse gases installed, any quantities added and the quantity recovered during servicing, maintenance and final disposal. They shall also maintain records of other relevant information including the identification of the company or technician who performed the servicing or maintenance, as well as the dates and results of the checks carried out.'

Operators are required to ensure that equipment is regularly checked for leakage by appropriately qualified personnel employed by businesses that are F Gas certificated. Leak checks must be carried out at least annually for small systems and every six months for systems with a refrigerant charge of more than 30kg. These leak checking requirements also apply to stationary fire protection equipment.

Also, operators of equipment that contains 300kg or more of an F Gas should have automatic leakage detection systems installed. The leakage detection systems need to be checked at least once every 12 months. When leak detection systems are installed the leak checking frequency can be halved, although all equipment that contains 3kg or more of an F Gas needs to be leak checked at least once a year.

Inspections

The Energy Performance of Buildings (Certificates and Inspections) (England and Wales) Regulations 2007 were introduced because of the need to cut energy consumption in buildings in the UK. Because of concerns about climate change and to cut down on our energy requirements – we now import a significant amount of our energy and this needs to be reduced – we need to use energy more efficiently in buildings, and air conditioning inspections are part of that drive.

There are somewhere in the region of 50,000 systems in the UK over 250 kW, yet only hundreds of inspections have actually been carried out. So, there is a huge shortfall in the number of inspections that should have been carried out, and that have actually been carried out. It means that there are thousands of air conditioning systems that haven't been inspected.

Air conditioning systems can use as much as 15% of the electricity in a building, so a poorly maintained system could be wasting a lot of electricity and a lot of money.

This lack of compliance stems partly from ignorance, because some building owners don't know about the Regulations, but there is also quite a widespread case of deliberate avoidance. The penalties are not very great, and many people see the requirement as a burden and not a benefit.

Under the Regulations there is potential for a £300 fine for not having a report, and that fine can increase if you don't get a report fairly quickly after conviction.

But there are also safety aspects to it – air conditioning systems can pose a risk of legionella, and a poorly maintained system poses a greater risk. At the moment there is no evidence of trading standards starting active enforcement, but CIBSE has started talking to trading standards bodies and giving them information on what they could usefully do to encourage people to do the right thing. One concern is that there are a lot of air conditioning inspectors who have invested a reasonable amount of time and money in getting themselves accredited to carry out these inspections and they are finding that the work that they were expecting isn't there, because of a lack of compliance.

More importantly, these inspections are there to save carbon emissions, and if they're not being carried out, they're not saving carbon.

> *See also*: Climate change, p.145; Energy performance, p.270; Energy Performance of Buildings Directive, p.275; Environmental Management Systems, p.286; Legionella, p.442.

Sources of further information

Details of accredited air conditioning energy assessors can be located on the Landmark EPC website: www.epcregister.com

Further details on leak check requirements can be found at the DEFRA F-Gas Support website: www.defra.gov.uk/fgas

Alcohol and drugs

Mandy Laurie, Dundas & Wilson

Key points

- There are legal obligations on employers to ensure a safe and healthy workplace for their employees.
- Employees have a right to work in a safe and healthy workplace, and they have responsibilities for their own wellbeing and that of their colleagues.
- Ensuring health and safety in the workplace necessitates the regulation of the use of alcohol and drugs.
- Having clear guidelines regarding the use of alcohol and drugs in the workplace will assist employers in achieving regulation in this area.

Legislation

- Misuse of Drugs Act 1971.
- Health and Safety at Work etc. Act 1974.
- Road Traffic Act 1988.
- Transport and Works Act 1992.
- Disability Discrimination Acts 1995 and 2005.
- Management of Health and Safety at Work Regulations 1999.
- Equality Act 2010.

Legal obligations on employers and employees

The Health and Safety at Work etc. Act 1974 and the Management of Health and Safety at Work Regulations 1999 place a general duty on employers to ensure, so far as is reasonably practicable, the health, safety and welfare of employees.

If an employer knowingly allows an employee under the influence of alcohol or drugs to continue working and this places the employee and/or others at risk, the employer could be prosecuted.

The Road Traffic Act 1988 and the Transport and Works Act 1992 require that drivers of road vehicles must not be under the influence of alcohol or drugs while driving, attempting to drive or when they are in charge of a vehicle.

The Transport and Works Act 1992 makes it a criminal offence for certain workers to be unfit through drink and/or drugs while working on railways, tramways and other guarded transport systems. The operators of the transport systems would also be guilty of an offence unless they had shown all due diligence in trying to prevent such an offence being committed.

The Misuse of Drugs Act 1971 makes it illegal for any person knowingly to permit drug use on their premises except in specified circumstances (e.g. when they have been prescribed by a doctor).

Regulation of alcohol and drugs in the workplace

It is important that employers regulate the use of alcohol and drugs in the workplace because failure to do so can lead to:

- poor performance and reduced productivity;
- lateness and absenteeism;
- accidents and therefore health and safety concerns;
- low morale and poor employee relations; and
- damage to the company's reputation and customer relations.

Case study

Sinclair v. Wandsworth Council (2007)

Mr Sinclair worked for Wandsworth Council. He was caught drinking on duty. The Council agreed to put disciplinary proceedings on hold if he would agree to a referral to the Occupational Health Service. Reluctantly, he agreed to the referral. The Council held a disciplinary hearing and issued Mr Sinclair with a final written warning for his drinking on duty.

Mr Sinclair was caught drunk at work again four weeks later. He was suspended, pending an investigation. He lied during the investigation about drinking and about Occupational Health having referred him for counselling. At the disciplinary hearing, Mr Sinclair said he was cooperating with Occupational Health. He asked that the hearing be suspended. The Council refused. Mr Sinclair had been drunk at his place of work. As he already had a final written warning, he was dismissed.

Mr Sinclair unsuccessfully appealed his dismissal. He then brought a claim before the Employment Tribunal for unfair dismissal.

The Tribunal decided that the dismissal was unfair because the Council had not given its alcohol policy to Mr Sinclair or to his managers, despite the wording of the policy that said it should be circulated to all staff.

The Council had failed to tell Mr Sinclair what he needed to do to avoid disciplinary proceedings, as obliged under the Council's policy. Mr Sinclair had erroneously believed he was doing enough. The Council should have made it clear to Mr Sinclair what he had to do to halt the proceedings against him.

Despite the problems and risks caused by employees being drunk in the workplace, care must be taken when managing these issues. Mr Sinclair was found guilty of being drunk at work twice (and dismissal was found to be a reasonable response), but his dismissal was ultimately found to be unfair because the Council's alcohol policy had not been followed.

When dealing with any instances of alcohol-related misconduct, you should first take into account any relevant policies the employer has in place and follow the appropriate procedure. Although it is a difficult issue for managers and HR to deal with, any issues arising due to an employee's alcohol problems should be addressed in an up-front, but understanding, manner.

The detrimental effects of drug and alcohol misuse in the workplace can be seen in the following statistics:

- A survey carried out by Alcohol Concern found that nearly two-thirds of employers have suffered problems in the workplace as a result of employee alcohol abuse.
- The after effects of alcohol and drug abuse can severely impact employee performance. According to a survey

---

Facts

- In the UK, up to 14.8 million working days are lost each year as a result of alcohol-related absence.
- A survey carried out by Norwich Union Healthcare found that 77% of employers consider that alcohol represents a significant threat to the wellbeing of staff.
- According to a survey by the Chartered Institute of Personnel Development (CIPD), 43% of employers do not have alcohol policies. It also showed that the vast majority of employers (84%) do not offer health awareness programmes for staff.
- In the UK, over 50% of all fatal accidents occurring in the workplace involve alcohol.

Sources: CIPD, HSE, Institute of Alcohol Studies, Medical Council on Alcohol, TUC.

---

conducted by YouGov for PruHealth, every day in the UK around 200,000 workers turn up to work with a hangover.

- In early 2004 the HSE published a report into the scale and impact of illegal drug use by workers and found that 29% of working respondents under 30 admitted to drug use in the previous year.

Alcohol and drug misuse by employees are two problems that increasing numbers of employers face. It is therefore essential that employers have policies and procedures to deal with such problems. What's more, because the topic touches on so many legal issues, such as privacy, conduct, performance, and health and safety, it's important to ensure any policies or procedures are tailored to fit with the working environment.

Policies and procedures

Policies on alcohol and drugs in the workplace should:

- set out the legal obligations behind the policy and summarise the aims of the policy;

- be clear as to whom the policy applies;
- make clear what will be considered to be alcohol and drug misuse and any specific rules / exceptions, e.g. in relation to prescription medicines or a dependency;
- set out the disciplinary action that will be taken following a breach of the policy or cross refer to the disciplinary policy;
- provide advice as to where help can be obtained and details of any support that the employer will provide;
- assure staff that any alcohol or drug problem will be treated in strict confidence; and
- encourage employees to come forward and ask for help.

The ACAS guidelines on 'Health, Work and Wellbeing' provide information on producing alcohol and drug policies (see *Sources of further information*).

In addition to ensuring that there is a drug and alcohol policy in place, employers should consider the following:

- Putting in place procedures to find out if an employee has an alcohol- or drug-related problem, e.g. alcohol / drugs screening or medical examinations. Testing and searching is an invasion of privacy and employers should take into account the nature of their business before imposing such a requirement; routine testing is only likely to be proportionate in safety-critical industries. Further, if employers wish to impose the right to search or test for drugs, this should be referred to in their policy, included in the contract of employment if possible, and consent should be obtained from the individual before carrying out testing / searching in the individual situation. An employer cannot require an employee to take a test / undergo a search, but if they refuse to do so, the policy should state whether disciplinary action will follow.

- Seeking to support the employee in the event that medical advice determines that they have an alcohol- or drug-related problem.

- Ensuring managers receive appropriate training to implement the policy and to ensure that it is applied consistently.

- Considering placing the employee on alternative duties until they have combated their alcohol- or drug-related problem.

- Considering disciplinary action where an employee under the influence of alcohol or drugs places either himself or others at risk in the workplace.

- Considering dismissal on grounds of capability or for some other substantial reason, where there is an ongoing problem and there appears to be no other alternative. A dismissal on capability grounds may be fair where an employee's performance falls below an acceptable standard, or where an employee is absent through ill health because of an alcohol- or drug-related problem. A fair procedure should always be followed.

- Considering reporting any criminal activity to the police. However, there is no obligation for an employer to do so.

See also: Driving at work, p.250; Occupational health, p.519; Violence at work, p.693.

Sources of further information

ACAS – Health, work and wellbeing: www.acas.org.uk/CHttpHandler.ashx?id=854

As an employer, you are committed to providing a safe working environment for your employees. You are also committed to promoting the health and wellbeing of your employees. Alcohol and substance misuse can be detrimental to the health and performance of employees and may pose a potential risk to safety in the workplace and the welfare of other employees. Workplace Law's ***Drug and Alcohol Policy and Management Guide v.2.0*** policy is designed to help protect employees from the dangers of alcohol, drug and other substance misuse and to encourage those with a drug or alcohol problem to seek help. For more information visit www.workplacelaw.net.

Asbestos

Mick Dawson, Dawson Asbestos Consulting Ltd

Key points

- Exposure to asbestos is hazardous to health. The three main conditions – asbestosis, lung cancer and Mesothelioma – can all be fatal. Current HSE estimates put the death rate in the UK from asbestos-induced diseases at almost 4,000 a year, a figure not likely to decrease until 2020 or later.
- Workers in the maintenance and construction trades comprise the largest group of people contracting the diseases, rather than those who worked in the old asbestos manufacturing and installation industry.
- The use of asbestos in buildings and construction materials is now completely prohibited. Building owners and those in control of non-domestic premises have a duty to manage the asbestos-containing materials (ACM) within their buildings by formulating an asbestos management plan.
- This will involve identifying whether buildings have ACM, assessing the risk from each occurrence and having a process to manage the ACM. In addition, any work on ACM or work that is likely to disturb ACM is regulated by the need for additional risk assessments prior to work taking place, control measures to prevent the spread of asbestos, and evidence of training and competency.
- Depending on the risk, work on some ACM may be restricted to HSE-licensed companies who must follow a notification procedure and carry out the work inside segregated work areas, using specially trained operatives who undergo medical surveillance.

Legislation

- Health and Safety at Work etc. Act 1974.
- Management of Health and Safety at Work Regulations 1999.
- Hazardous Waste Regulations 2005.
- The Control of Asbestos Regulations 2006.

The Control of Asbestos Regulations (CAR) 2006 (the Regulations) are the single statutory instrument dealing with asbestos in the UK by implementing the 2003/18/EC amendment to the Asbestos Worker Protection Directive 83/477/EEC. The main parts are as follows:

- Risk assessments to be undertaken and plans of work to be prepared for all work on asbestos.

- Work becomes licensable, notifiable and workers subject to medical surveillance if the Control Limit will be exceeded.
- The Control Limit for all types of asbestos is 0.1 fibres per cubic centimetre (f/cm^3) of air measured over eight hours.
- There is no requirement for licensing, notification or medical surveillance if worker exposure is judged to be 'sporadic and of low intensity'. This is defined as being below the Control Limit and comprised of:
  - short, non-continuous maintenance activities;
  - removal of non-degraded materials firmly linked in a matrix;
  - encapsulation or sealing of ACM in good condition; and
  - air and bulk sampling.

Facts

- Over 4,000 people a year die from asbestos-related deaths, and this figure is due to carry on rising until 2020 at least.
- Around a quarter of these deaths are from workers involved in maintenance and building trades.
- Exposure to asbestos continues to be the greatest single work-related cause of death in the UK.
- It is estimated that from 1900 to 1999 over five million tonnes of raw asbestos fibre was imported into the UK.
- Asbestos-containing materials (ACM) contain an average of 10% asbestos fibre, which means over 50 million tonnes of building products were installed in an estimated one to two million buildings.
- A recent EU ruling means the HSE will be consulting stakeholders regarding a new classification of non-licensed work.

- Awareness training should be given to those whose work may disturb the fabric of the building.
- Selection of Respiratory Protective Equipment (RPE) is based on reducing exposure as low as reasonably practical, not just below the Control Limit.

The provisions fall under the umbrella of the Health and Safety at Work etc. Act 1974. CAR 2006 is supported by two Approved Codes of Practice (ACoPs) and a number of guidance documents.

The primary purpose of the Regulations is to prevent and reduce asbestos exposure to building occupants, visitors and maintenance and construction workers.

Hazardous Waste Regulations 2005

The Hazardous Waste Regulations came into force in England and Wales in July 2005 and superseded the Special Waste Regulations 1996. Asbestos has now been reclassified as a 'hazardous' waste in line with the EU waste classification system.

The changes from the previous Regulations largely affect the registered waste carriers and licensed asbestos removal companies (defined as being 'consignors') who transport the waste to landfill after it has been removed from buildings.

However, building owners or occupiers are defined as being 'producers' of the waste. This means that their buildings need to be notified to the Environment Agency on an annual basis if more than 200 kg of asbestos waste will be removed from their premises. This notification process can either be carried out by the producer or the consignor.

All asbestos products and waste have to be disposed of in accordance with the Regulations by means of a registered carrier (which can be the licensed asbestos removal contractor) under a consignment note procedure.

The Scottish Environmental Protection Agency (SEPA) still recognises the Special Waste Regulations 1996.

ACoPs

The ACoPs give advice on the preferred method of compliance with the

Regulations. This advice has special legal status, which means that if you are prosecuted for a breach of health and safety law and it is proved that you did not follow the relevant provisions of an ACoP, you will need to show that you have complied with the law in some other way or the Court will find you at fault.

Approved Code of Practice, L127: The management of asbestos in non-domestic premises

Regulation 4 of the CAR 2006 imposes a more practical duty on those in control of buildings. The duty-holder will have to:

- take reasonable steps to find ACM and check their condition;
- presume materials contain asbestos unless there is strong evidence they do not;
- make a written record of the location and condition and keep it up to date;
- assess the risk of exposure; and
- prepare a plan to manage that risk.

Regulation 4 is also known as the Duty to Manage, and introduces the concept of an Asbestos Management Plan (or Plan). It applies to managing ACM in non-domestic premises, although the common parts of domestic premises are part of the Regulation. However, landlords have a duty to fulfil the same obligations under the Defective Premises Act 1974 (and also the Management of Health and Safety at Work Regulations 1999), and by taking actions in line with this ACoP they would be demonstrating a satisfactory level of compliance.

The duty-holder is defined as being anyone with a contractual, or tenancy, obligation in relation to the maintenance and repair of buildings, or, where there is no contract, anyone in control of the buildings.

The duty-holder will need to have in place a Plan explaining how the issues of asbestos within their buildings will be addressed. The Plan should consist of:

- details of each ACM and an explanation of the risk assessment (the algorithm);
- a table of priorities and timescales;
- personnel and responsibilities;
- training for employees and contractors;
- procedures for preventing uncontrolled maintenance and building work;
- procedures for ensuring information on ACMs are made available to those that need it;
- arrangements for monitoring ACM; and
- arrangements for updating and reviewing the management plan.

Management options

There are a number of options available to building managers to fulfil their duties under ACoP L127.

Option A

A comprehensive survey and removal programme would ensure full and thorough compliance. This, however, would be disruptive and extremely costly and the HSE does not advise this as the preferred approach.

Option B

Do nothing in the short-term, assume that all unidentified materials contain asbestos, and carry out a survey before any building or maintenance work is carried out. This would protect those coming into contact with asbestos but may lead to greater cost in the long run. In larger premises or multi-building portfolios, it would also be difficult to manage, and hence, prove the ACoP is being complied with.

Option C

Introduce a planned survey programme based on a management system that reviews the property stock and identifies where asbestos is more likely to be present, e.g. in buildings constructed between 1945 and 1985. Surveys would then be carried out in those buildings or areas where ACM are most likely to be present and the risk would be assessed on a case-by-case basis.

If the material is in good condition and unlikely to be disturbed, it can be left in place. An asbestos management system should be maintained and updated when action is taken, as well as ensuring that all information is made available to those who could come into contact with asbestos.

Asbestos surveys

In January 2010, HSG 264 'Asbestos: The survey guide' replaced MDHS 100 as the main reference document for advice on carrying out and using asbestos surveys.

The main features of the guidance are:

- Management surveys – previously known as Type 1 (presumptive) and Type 2 (sampling) surveys.
- Refurbishment / demolition surveys – previously known as Type 3 surveys, requiring a more rigorous inspection for ACM used within the building fabric.
- Better advice on how to use the survey information to manage ACM and develop the asbestos management plan.
- Greater emphasis on the correct planning of surveys.
- Greater clarity on the respective roles of the client and surveyor in the survey and what to expect from each other.
- Strengthening of the section on the competence of the individual surveyor and the survey company and quality control checks that can be used.

- More detailed guidance on refurbishment / demolition surveys.

In the guidance the HSE strongly recommends the use of companies accredited by UKAS to ISO 17020 for carrying out surveys. This is currently the only recognised measure of competence available since the personal certification scheme, ABICS, folded in October 2010.

Approved Code of Practice and Guidance, L143: Control of Asbestos Regulations 2006

In general terms, those ACM posing the highest risk are those containing fibres that are less firmly bound and so have a greater potential for fibre release – sprayed coatings, insulation and asbestos insulating board (AIB). Because it is likely that the Control Limit of 0.1 f/cm^3 (otherwise known as f/ml) would be exceeded during removal, this work would usually need to be carried out by a licensed contractor who notifies the Enforcing Authority 14 days in advance and prepares a bespoke plan of work, or method statement.

The ACoP expands on the concept of 'sporadic and low intensity worker exposure,' (*see above*) a phrase that appears in the EU Directive itself. As well as a comparison with the 0.1 f/cm^3 Control Limit, the Regulation also stipulates that if a peak level of 0.6 f/cm^3 measured over ten minutes is not exceeded then the work is judged to be within this definition.

Other examples of work that fit the definition and would therefore not need a licensed contractor, notification to the Enforcing Authority, or medical surveillance of operatives include:

- materials firmly linked in a matrix, including asbestos cement, textured decorative coatings, bitumen, plastic resin or rubber, cardboard, felt, gaskets and washers; and

■ short, non-continuous, maintenance activities, including any one person carrying out work for less than one hour in a seven-day period, or all workers carrying out work for fewer than two hours in total.

However, all work with ACM, including those listed above, must be undertaken by trained workers, following a risk assessment, and with appropriate controls to prevent exposure.

The ACoP also outlines more defined and explicit training requirements in three areas:

1. Work on licensable ACM.
2. Work on non-licensable ACM.
3. Awareness training for maintenance, installation and construction workers.

Asbestos awareness training

Regulation 10 of CAR 2006 states that awareness training should be given to anyone whose work will disturb the fabric of the building where ACM could be hidden. This would include maintenance staff, electricians, plumbers, painters, HVAC engineers, IT installers and other contracting trades. The only buildings that are exempt would be those that are free from ACM. This does not mean that a building owner or occupier has to train the people themselves, but they will have to ask their suppliers how they intend to comply with the Regulation. This could be achieved by making it a condition of approved supplier status.

Guidance

The Licensed Contractors' Guide, HSG 247

The ACoP is supported by a detailed guidance document published in 2006.

HSG 247 consolidates five previous documents into one single reference point for licensed asbestos removal work.

The Guide embraces subjects such as licences, plans of work, training, PPE, enclosures, controlled removal techniques, waste disposal and decontamination.

Recent EU ruling (non-licensed work)

In February 2011 the EU ruled, by way of a 'reasoned opinion,' that the HSE had not fully transposed the EU Directive with regards to the full definition of 'sporadic and low intensity' which effectively allowed more types of asbestos work to be exempt from notification and medical surveillance obligations than was intended.

The HSE consulted on this towards the end of 2011 and, at the time of writing, the intention is for CAR 2006 and its accompanying ACoP L143 to be amended and introduced by April 2012.

In order to comply with the reasoned opinion a new category of notifiable non-licensed work (NNLW) is being proposed. Employers carrying out this type of low risk, short duration maintenance and repair work will be required to:

■ notify the work to the relevant Enforcing Authority;
■ obtain medical examinations for workers; and
■ maintain a register of each worker of the type and duration of work done.

As a result of the EU ruling the planned new document, the *Non-Licensed Contractors' Guide*, has been delayed.

Minor works (asbestos essentials)

There is additional guidance aimed at works of a minor nature or short duration that the building maintenance and allied trades may carry out. The previous HSG 210 'Asbestos Essentials Task Manual' has been replaced by free download task sheets that describe a safe system of work to complete various repair or maintenance tasks that involve work on ACM. There are around 40 different task sheets available from the HSE asbestos website, describing tasks such as how to carry out minor work on textured coatings or AIB, removing asbestos-containing electrical items and dealing with fly-tipped asbestos waste.

Prohibition and related provisions

The importation, supply and use of ACM containing brown and blue (amosite and crocidolite respectively) asbestos (generally comprising asbestos spray coatings, insulation products and AIB) have been formally prohibited since 1985. The importation, supply and use of chrysotile or white asbestos (such as asbestos cement products) was prohibited in 1999.

Using the prohibition dates can provide an indication of what products could be found in buildings if the date of construction is known.

Checklist of key action points

- Assess the likelihood of ACM in your premises.
- Decide whether a survey is needed.
- Incorporate the survey information into an Asbestos Management Plan.
- Make this information available to anyone on site who may need it, e.g. contractors.
- Keep the information up to date.
- Assess whether removal or repair work will require a licensed or non-licensed contractor.
- Seek specialist advice if unsure of any of the above requirements.

See also: Occupational cancers, p.512; Hazardous waste, p.700.

Sources of further information

HSE – Asbestos: www.hse.gov.uk/asbestos

ATAC – Asbestos Testing and Consulting: www.atac.org.uk

Asbestos Removal Contractors Association: www.arca.org.uk/

Asbestos Building Inspectors Certification Scheme: www.abics.org

HSE – Approved Code of Practice, L127: The management of asbestos in non-domestic premises: www.hse.gov.uk/pubns/priced/l127.pdf

HSG 227: A comprehensive guide to managing asbestos in premises: www.hse.gov.uk/pubns/priced/hsg227.pdf

HSG 264 Asbestos: The survey guide: www.hse.gov.uk/pubns/priced/hsg264.pdf

HSE – Approved Code of Practice and Guidance, L143: Work with materials containing asbestos: www.hse.gov.uk/pubns/priced/l143.pdf

HSG 247 The Licensed Contractors' Guide: www.hse.gov.uk/pubns/priced/hsg247.pdf

HSG 213 Introduction to Asbestos Essentials: www.hse.gov.uk/pubns/priced/hsg213.pdf

The publication of a new HSE guidance document (HSG 264) replaces MDHS 100. It has implications for all duty-holders and organisations throughout the UK regarding the management of asbestos in their buildings.

Workplace Law's *One-day Asbestos Guidance Legal Update* gives an overview of the Guide which is not solely focused on the surveys themselves.

There are several responsibilities for the duty-holder, in particular:

- The selection of an appointed person;
- Responsibility of the duty-holder to check the competency of an asbestos surveyor;
- Responsibility of the duty-holder to develop the survey results into a Management Plan;
- Better planning of the survey and its objectives; and
- Extra guidance on carrying out invasive surveys prior to demolition or refurbishment.

These changes mean that there is more responsibility than ever on the duty-holder; for the first time one must identify an 'appointed person' as well as carrying out quality checks on the surveying company and its work.
Visit www.workplacelaw.net/training/course/id/61 for more information.

Batteries

Phil Conran, 360 Environmental

Key points

- The UK Batteries Regulations 2009 have been transposed from the EU Batteries Directive and affect those that place portable, industrial and automotive batteries on to the UK market. Unlike WEEE, these Regulations only apply to companies based in the UK and *do not* apply to those that supply to the end user in the UK direct from overseas.
- Different requirements are placed on those that place portable batteries on to the market to those that place industrial and automotive batteries. Different Government departments also have responsibility, with DEFRA responsible for portable batteries implementation and BIS responsible for industrial and automotive implementation.
- The Directive has basically been split into two sets of Regulations. The single market provisions that determine markings and banned materials have been in place since September 2008 under the Batteries and Accumulators (Placing on the Market) Regulations 2008. The Waste Batteries and Accumulators Regulations 2009 cover the collection and recycling requirements, which is the main topic covered in this chapter.

Legislation

- Batteries and Accumulators (Placing on the Market) Regulations 2008.
- Waste Batteries and Accumulators Regulations 2009.

Definitions

'Portable batteries' are considered to be those that are sealed, capable of being carried by hand and are *not* automotive or industrial batteries. They include torch batteries of all sizes, button batteries, and mobile phone, computer and any other purpose designed battery.

An 'industrial battery' is designed exclusively for industrial or professional uses, is used as a source of power for propulsion in an electric or hybrid vehicle, is unsealed but is not an automotive battery or accumulator, or is sealed but is not classified as a portable battery.

An 'automotive battery' is a battery of any size or weight that is used for the starting or ignition of the engine of a road-going vehicle or for providing power for any lighting used by such a vehicle.

'Producers' are businesses that manufacture or import batteries (other than for own use), including those that import batteries in products.

'Retailers' are those that sell loose batteries (not in products) to businesses or the public by direct sales or through indirect means such as online or by mail order.

Portable batteries

Producers

All producers, regardless of how many batteries they place on to the UK market, are required to register. Those that supply more than one tonne a year must contribute to the costs of collection and recycling of spent batteries by registering with a compliance scheme and paying towards the costs of collection and recycling, starting from 1 January 2010. Producers that place less than one tonne on the UK market must also register, but through one of the Environmental Agencies rather than a scheme. They have no responsibility for collection and recycling costs. All registered producers will be allocated a unique Producer Registration Number that will remain with them whilst a registered producer.

Compliance schemes must gain approval from the relevant environmental agency – the EA in England and Wales, SEPA in Scotland and NIEA in Northern Ireland.

There is a fee of £17,000 charged to each scheme that submits for approval. In addition, schemes have to pay the Agency an annual registration fee of £90,000, regardless of how many members they have. There is a further annual EA registration fee that must be paid by schemes of £600 per member.

Registration is an annual requirement that must be completed by schemes by 31 October each year. In a similar way to WEEE (see p.737), schemes must also then supply quarterly data to the Agency by the end of the month following a quarter, showing the amount of batteries placed on the market by its members by weight and chemistry and the weight of batteries the scheme has collected over the same period, also by weight and chemistry.

Targets

Targets for collection and recycling have been set in the Regulations to ensure that the UK achieves the 2012 and 2016 targets laid down by the EU Directive. Schemes have to demonstrate how they will meet these targets in their approval process, with the main methods likely to be through relationships with Local Authorities – who will collect batteries at their Civic Amenity sites – and retailers who offer in-store take-back. All costs associated with the collection, transport and recycling of portable batteries will be met by scheme members.

Schemes are expected to meet their targets for each year, but only those in

Year	Directive targets	UK Scheme targets
2010		10%
2011		18%
2012	**25%**	**25%**
2013		30%
2014		35%
2015		40%
2016 onwards	**45%**	**45%**

2012 and 2016 are mandatory. Failure to meet the other targets could, however, lead to a scheme having its approval withdrawn.

Prior to 2010, the UK was currently thought to be recycling 2-4% of portable batteries placed on to the market. The Directive targets below are those that must be achieved by the UK by the end of each target year, whilst the UK targets are those that should be met by schemes by the end of each year.

These targets are applied to the tonnage of batteries placed on to the market by a scheme's members in the previous year in 2010, the average of the previous two years in 2011, then the average of the previous two years *and* the current year in 2012 onwards.

If schemes over-collect, they will be able to trade evidence. However, the EA has made it clear that schemes will *not* get approval if they seek to depend on trading to meet their targets. The Regulations require that all collected batteries must be treated although the recycling targets do not kick in until 2011. Therefore a scheme that collects more batteries than it needs and cannot trade them is stuck with the cost of getting them treated.

Retailers

Retailers that sell more than 32kg a year to both businesses and the public must provide free take-back facilities for their customers as of 1 February 2010. They are then entitled to free collection from the compliance schemes, but batteries must be stored and transported under specific constraints, as mixed portable batteries are considered to be both hazardous and dangerous.

Where a retailer sells less than 32kg a year, they do not have to offer in-store

take-back but may, of course, do so. However, they will not be entitled to the free collection by schemes.

It is expected that retailers will marry up with schemes to arrange a regular collection service, but if a retailer is unable to come to a regular arrangement with a scheme, they are entitled to call any scheme to get their batteries collected free.

32kg in a year equates to approximately 24 AA batteries per week.

Timeline

- By 15 October 2009, existing 'large' producers (more than one tonne on to the market) should have registered with a Compliance Scheme.
- By the end of October 2009, Compliance Schemes should have registered their members with the appropriate environmental agency.
- By 12 November 2009, existing 'small' producers should have registered with an environmental agency.
- Since 1 January 2010, Schemes should have started to pay for any collected batteries to be treated and recycled.
- By the end of January 2010, all producers' data for batteries placed on the UK market in 2009 (from at least 5 May) should have been supplied to the environmental agencies, either direct for small producers or through schemes for large. Data includes chemistry type by weight.
- Since 1 February 2010, any retailer selling more than 32kg a year of batteries to businesses or the public must be offering their customers free take-back of any portable battery, regardless of origin.
- From April 2010, and then quarterly, Schemes must supply data for what their members placed on the market in the previous quarter to the environmental agencies.

- At the end of each year, Schemes must demonstrate to the environmental agencies that they have collected sufficient batteries in the previous year to meet the target set by Government.
- As a condition of approval, Schemes must organise and finance public information and education campaigns.

Costs

For large producers annually:

- A share of a Scheme's £90,000 registration fee;
- An additional agency fee of £600;
- A membership fee; and
- Costs associated with collection, treatment and recycling of their market share of the waste batteries collected.

For small producers:

- An annual £30 registration fee with the agency.

For retailers:

- There should be no costs as Schemes should pay for collection from their premises.

Storage and transport

Mixed waste portable batteries are considered both hazardous and dangerous, thereby requiring storage and transport under the Hazardous Waste Regulations and the Carriage of Dangerous Goods (ADR) requirements.

DEFRA has issued guidance that clarifies the requirements:

- Up to 80kg of batteries may be stored in a location without making a notification or paying a fee.
- A collection vehicle carrying over 333kg of batteries must operate under full ADR. Those carrying less can operate under a reduced ADR,

but must still have drivers that have undergone training, and ensure batteries are transported in closed and marked containers.

- Mixed batteries cannot be transported by air and DEFRA has advised that spent batteries should therefore not be sent through the post. Spent batteries must be collected as Hazardous (or Special in Scotland and NI) Waste. They must therefore be accompanied by consignment notes and the appropriate charges although, in England and Wales, there is only one consignee charge per quarter.
- Where a site has more than 500kg of hazardous waste a year, including mixed batteries, they must be registered as a hazardous waste producer.

Industrial and automotive batteries

Producers

Industrial and automotive producers are required to register with the Department of Business, Innovation and Skills (BIS) each year using the same form of registration as that proscribed for portable battery producer registration. Registration is free.

However, where a producer of industrial and/or automotive (I and A) batteries is also a portable batteries producer, they need not complete a separate registration for the I and A batteries, but must register their I and A data at the same time as the portable data. This will then be submitted to the relevant agency – through a scheme if a large producer or direct if a small producer – who will then pass the data to BIS.

Once registered, they will be allocated a unique Producer Registration Number that will stay with them all the time they are registered, which must be declared when supplying to a customer. Producers are also required, by 1 December each year, to publish how an end user may request take-back from that producer.

I and A producers must then submit data on the amount of batteries they placed on the market in the previous year by 31 March, starting with 2009 data by 31 March 2010. This must be by weight and chemistry. At the same time, they must also report on the weight of I and A batteries collected by them, or taken by them for treatment.

The Regulations ban the disposal of I and A batteries to landfill. Under the Regulations, industrial battery producers have slightly different take-back responsibility to automotive battery producers, but in essence it boils down to the same requirement in that producers must, if requested, take back I and A batteries free of charge from whoever ends up with them, be it an end user, a garage or a collector.

In practice, however, it is unlikely that producers will see many batteries being returned as they have a positive value and therefore are already subject to an efficient national collection infrastructure for recycling. For I and A producers,

therefore, the Regulations are largely a data gathering exercise.

Timeline

- By the end of October each year, all producers (including those that import batteries in vehicles) should have registered with the Department of Business, Innovation and Skills (if also a portable batteries producer, they only need to register with an environmental agency).
- By 1 December each year, producers must publish their take-back plan.
- As of 1 January 2010, producers must provide free take-back.
- By end March 2010 and then annually, producers must supply tonnage and chemistry data to BIS on batteries placed on the market and waste batteries taken back in the previous year.

See also: Recycling, p.605; Hazardous waste, p.700; Waste Electrical and Electronic Equipment: the WEEE and RoHS Regulations, p.737.

Sources of further information

DEFRA: www.defra.gov.uk/environment

BIS: www.bis.gov.uk

Biological hazards

Andy Gillies, Gillies Associates

Key points

- Biological hazards include fungi, moulds, bacteria, viruses and allergens.
- The effects of biological contamination in humans can lead to mild conditions, such as nausea, through to potentially lethal conditions, such as hepatitis.
- Biological agents are classified as hazardous substances under the Control of Substances Hazardous to Health Regulations 2002 (COSHH), which impose a responsibility on employers to carry out a risk assessment for all hazardous substances.
- Effective containment measures, thorough testing and good hygiene practice can help to control the risks of contamination.
- Under COSHH, a biological agent includes any micro-organism, cell culture, or human endoparasite, whether or not genetically modified, which may cause infection, allergy, toxicity or otherwise create a hazard to human health.
- Biological agents are classified into four hazard groups according to their ability to cause infection, the severity of the disease, risk of spread to the community, and availability of effective prophylaxis or treatment.

Legislation

- Notification of Cooling Towers and Evaporative Condensers Regulations 1992.
- Management of Health and Safety at Work Regulations 1999.
- Genetically Modified Organisms (Contained Use) Regulations 2000 (as amended).
- Control of Substances Hazardous to Health Regulations 2002 (as amended).
- The Carriage of Dangerous Goods and Use of Transportable Pressure Equipment Regulations 2004.
- Health and Social Care Act 2008 (Registration of Regulated Activities) Regulations 2009.

Types of hazard

Many workplaces involve a risk of infection to biological agents, either because of intentional work with micro-organisms (e.g. work in a microbiology laboratory) or through incidental exposure (e.g. healthcare, farming). Exposure to micro-organisms may occur in manufacturing processes, in products used at work, or as contaminants in the environment. Examples of typical occupations with a risk of exposure to bioaerosols are people working in waste management (e.g. recycling centres, landfill sites, sewage works), industrial and research laboratories, health care settings, agriculture and food processing sectors. Most micro-organisms are harmless but some can be highly infectious, produce endotoxins, or provoke an immunological response in humans.

Biological agents include harmful micro-organisms such as fungi, bacteria, viruses, internal parasites and biological allergens which may cause asthma and other allergic diseases. Exposure is mainly through inhalation of bioaerosols but in some cases there may be the possibility of skin infection (especially through open wounds) or ingestion from hand-to-mouth transfer. Harm may occur from exposure to the agent itself, exposure to toxins produced by the micro-organism, or an allergic reaction.

1. *Fungi.* These produce spores that can cause allergic reactions when inhaled.
2. *Moulds.* These are a group of small fungi that thrive in damp conditions. They can produce mycotoxins (toxic metabolites) and bring on allergic reactions including athlete's foot and asthma.
3. *Bacteria.* These are very small, single-celled organisms (mean size ~500 nm), which are gradually becoming immune to treatment. The effects of contamination in humans range from mild nausea to potentially lethal conditions such as Legionnaires' Disease and Tuberculosis.
4. *Viruses.* These are minute (20 – 450 nm diameter), non-cellular organisms, smaller than bacteria. As with the common cold, there is no way of controlling viruses other than the body's own natural defences. Extremely dangerous forms of virus include AIDS and hepatitis.
5. *Allergens.* Exposure to bioaerosols may cause allergic respiratory disease. Examples include biological enzymes (e.g. proteases from detergent manufacture), laboratory animal proteins, and shellfish proteins (e.g. crabs, prawns).

The HSE has provided guidance on a range of biohazards including influenza, SARS, legionella, zoonoses (diseases transmitted from animals to humans) and bovine spongiform encephalopathy (BSE). This includes practical advice covering:

■ What is the disease / condition?
■ How do people get it?
■ What are the symptoms?
■ Where does it come from?
■ What measures are there to control the disease / condition?
■ What do you do if you or an employee contracts the disease / condition?

These are all essential areas that should be covered as part of a COSHH assessment.

The Approved List of Biological Agents provides a list of specific organisms covered. Biological agents are classified into one of four hazard groups, according to the risk of infection to a healthy worker. The Hazard Groups have been defined by the Advisory Committee on Dangerous Pathogens (ACDP). The criteria for classifying a biological agent in one of the four hazard groups are:

■ *Group One* – unlikely to cause human disease.
■ *Group Two* – can cause human disease and may be a hazard to employees; it is unlikely to spread to the community and there is usually effective prophylaxis or treatment available.
■ *Group Three* – can cause severe human disease and may be a serious hazard to employees; it may spread to the community, but there is usually effective prophylaxis or treatment available.
■ *Group Four* – causes severe human disease and is a serious hazard to employees; it is likely to spread to the community and there is usually no effective prophylaxis or treatment available.

The required physical containment measures and other exposure controls vary, depending on the classification. Special control measures are stipulated in Schedule 3 of COSHH for Hazard Groups Two, Three and Four biological agents with the most stringent controls applied to HG4.

Working with genetically modified organisms is a specialised field of work which is rapidly expanding. GMOs may be plants, animals, or most commonly micro-organisms like bacteria and fungi. Work with GMOs in contained facilities (in contrast with controlled release trials of genetically modified crops, for example) is normally done in accordance with guidance issued by the Scientific Advisory Committee on Genetic Modification.

Practical scenarios

The risks from biological hazards are increased where a thorough cleaning and maintenance regime is not put in place. The presence of the legionella bacteria in water systems and air conditioning units can result in an outbreak of Legionnaires' Disease; the UK's most high-profile case occurred in Barrow-in-Furness in 2002, where seven people died. (See 'Legionella,' p.442.) More recently, two firms – Eaton Limited (a multinational automotive parts manufacturer) and Aegis Limited (a water treatment services provider) – were found guilty of putting workers and members of the public at risk of exposure to Legionella bacteria and had to pay nearly £250,000 in fines and costs.

Similarly, poor hygiene in the use and handling of food can increase the spread of biological hazards. Employers should put in place a policy for managing food in fridges in order to separate meats, cooked food, raw food and liquids so that the effect of minor spillages is minimised. Many employers also enforce a policy preventing workers from eating at their desks or on the shop floor, and ensure that food waste goes into special bins. The advantage here is that food waste can be dealt with specially (both inside and outside the building) to avoid attracting rodents – a further cause of the spread of disease.

The so-called hospital 'superbug' MRSA is perhaps one of the best-known biological hazards. In 2005, an NHS patient successfully sued a hospital, arguing that MRSA could be classified as a biological agent under the Control of Substances Hazardous to Health Regulations, and that the hospital had breached these Regulations by not ensuring its infection control policy was properly implemented. This approach, using health and safety legislation rather than pursuing more expensive, time-consuming and difficult to prove negligence claims, is likely to be used by an increasing number of claimants.

More recently, the risks from pandemic flu have been to the fore with the spread of swine flu (type A influenza caused by a new H1N1 virus). Guidance to businesses on ways to reduce the risk of spread of infection in the workplace has been issued by the Department of Health and the HPA in collaboration with the HSE.

Risk assessments

A COSHH risk assessment should be undertaken to consider the risks arising from biological hazards. As with other risk assessments, the process should first identify the presence of biological agents in the workplace, and the people who might be affected by them. Special consideration should be given to the susceptibility of potentially exposed groups, for example pregnant women, or those whose immune system is not functioning properly. The assessment should take into account the features of a particular hazard (such as the legionella bacteria) and should evaluate the risk. Additional requirements will apply when handling and using genetically modified organisms (see www.hse.gov.uk/biosafety/gmo/law.htm).

Prevention / control of exposure

Control measures to prevent or adequately control exposure should be defined in the risk assessment. The most important principle underpinning the risk assessment is understanding the chain of infection, from source (the symptomatic individual) via a transmission path to a recipient (the susceptible individual). This recognises the notion that the breaking of any link in the chain will enable the infection to be prevented or controlled. In other words, preventing or controlling the source of the infection and its means of transmission or protecting the potential hosts (recipients) could each provide an effective level of control.

However, as with any hierarchical model, the preferred and most effective method will always be to control as close to the source as possible. Effective containment to prevent escape of micro-organisms from their containers is important and may be achieved through handling in microbiological safety cabinets and good laboratory or industrial practice. Barriers to prevent transmission from the source of micro-organisms to the worker include restrictions on access, medical surveillance and immunisation, and high levels of personal hygiene.

Standard procedures such as autoclaving infected wastes before removal also prevent spread of contamination. The risks from biological hazards will be reduced through regular maintenance and cleaning, and through the exercise of good personal hygiene practice, reinforced through information, instruction, training and supervision. Reliance on protecting the host using PPE should be regarded as a last resort.

In order to conduct a robust risk assessment, it may be necessary to carry out accurate monitoring in the form of environmental testing (such as air and water quality monitoring) where validated monitoring techniques exist. Monitoring

of bioaerosols is complex and usually involves short period sampling on to suitable media (e.g. agar strips or liquid media) followed by rapid transfer to a microbiological laboratory for culturing of viable organisms. Most sampling equipment for bioaerosols is bulky and more suited for area rather than personal sampling. Results are normally reported in numbers of colony-forming units per cubic metre (CFU m^{-3}). Some guideline values have been proposed by authoritative organisations to help with interpretation of measurements, but there are no national exposure limits for bioaerosols in the UK.

Hospital infections

Healthcare-associated infection (HCAI) is an important cause of morbidity and mortality amongst hospital patients. The most well-known are MRSA (Methicillin-Resistant Staphylococcus Aureus) and C_{diff} (Clostridium difficile). Although these are potentially subject to health and safety regulation, the HSE does not generally deal with clinical matters. Extensive guidance and advice is provided by the Department of Health. However, employers should still ensure that HCAIs are addressed as part of the risk assessment protocols performed.

The importance in managing HCAIs is demonstrated in the report by the HSE on the investigation into outbreaks of Clostridium difficile at Stoke Mandeville Hospital, Buckinghamshire Hospitals NHS Trust (see *Sources of further information*). A new Regulation regarding HCAI came into force on 1 April 2009, and requires NHS Trusts to register with the Care Quality Commission and to follow good practices for infection control. A new Code of Practice for the prevention and control of HCAI (the 'Hygiene Code') has been published to help NHS Trusts comply with the Regulation, and can be downloaded from the Department of Health.

See also: COSHH: Control of
Substances Hazardous to Health,
p.207; Legionella, p.442; Water
quality, p.725.

Sources of further information

HSE – COSHH: www.hse.gov.uk/coshh/index.htm

HSE – guidance on infections at work: www.hse.gov.uk/biosafety/infection.htm

The Approved List of Biological Agents: www.hse.gov.uk/pubns/misc208.pdf

Department of Health – healthcare-associated infection:
http://bit.ly/q0kB7v

HSE – Stoke Mandeville Hospital report:
www.hse.gov.uk/healthservices/hospitalinfect/stokemandeville.pdf

Care Quality Commission – healthcare-associated infections:
http://bit.ly/qDabsw

Bomb alerts

Peter Power, Visor Consultants (UK) Limited

Key points

- Bomb threats will always occur in the UK from known terrorist organisations, saboteurs, protestors or other disgruntled individuals or groups. Some will be threats alone with no device; others will be either hoax bombs, or the real thing, varying from explosives to harm people, or to start fires.
- All threats should be considered real until / unless proven otherwise. However, a knee jerk response to evacuate entire buildings whenever any threat is received, irrespective of caller, is seldom advised. A rapid on-site assessment is preferred.
- When action is considered necessary, evacuation is often the second best option compared to staying inside, but moving away from the threat – sometimes referred to as 'invacuation'. In the past many casualties have occurred owing to broken glass and other flying debris outside premises.
- Any suspicious small package or box found must be treated with caution. Suspicious objects must not be handled. They can maim and disfigure for life.

Legislation

The legislation that requires an employer to deal with bomb alerts is the Management of Health and Safety at Work Regulations 1999. Regulation 8 requires employers to put into place procedures for serious and imminent danger and for danger areas.

Introduction

Success of the terrorist depends on two things – the cunning of the terrorist and the reaction to the threat. If we give in and subsequently work and live in fear by assuming that there is nothing we can do to stop them, terrorism has won. The fear of crime would have taken over, despite the reality. Yet, if we ignore the problem we are also exposing ourselves to danger. The answer lies between the two – to avoid the extremes of anxiety and complacency, and to replace fear with vigilance. This advice is aimed at replacing complacency or fear with common sense, vigilance and adhering to best practice.

It is a fact that the greater the number of obstacles between a terrorist and his/her target, the greater the rate of failure to cause damage and/or detection. Therefore, the first response should be to operate a security and general awareness regime to first make it difficult to leave any device in a position where injury or damage could occur. After all, security is something that must concern us all. Such threats will vary according to:

- the intentions of the people / person posing the risk and/or threat;
- their ability to carry out such a threat; and/or
- the perceived activities of the target organisation at any one time, especially if considered by others to be in any way controversial.

You must also consider if you could be the victim of collateral damage from a neighbouring organisation that was the actual target.

Bombs: What do they look like?

A bomb is easily disguised in many ways, but you can be sure the last thing it is likely to look like is a bomb. Very often it is not what a package looks like, but where it is seen that can give it away. In other words, the device can look innocent, but put under a car, or against a door that same object now becomes suspicious. The terrorist will shape, paint, or otherwise disguise a bomb to try and make it inconspicuous. Homemade terrorist bombs are also referred to as Improvised Explosive Devices (IEDs).

Remember that the packaging of a device is left only to the imagination of the terrorist. One of the most serious threats is from bombs carried in vehicles, which can carry the largest quantity of explosives, and from mortars that can be directed on to a target (*see below*). However, much smaller bombs can start serious fires, perhaps out of office hours, and/or be planted to disrupt key systems or utilities.

Apart from blast bombs, there are also incendiary devices designed to create a limited explosion to start a fire. They can be made to fit inside a video cassette case or even a cigarette packet, or other similar small container. Terrorists will carry an incendiary bomb on them until a place is found to conceal it. Places where visitors attend are most at risk from this type of small device, as they can easily be planted in everyday items of furniture.

An interesting idea is to leave your building and then approach it as a potential terrorist might. Ask yourself: 'If I were a bomber, where would I do it:

- To maximise damage?
- To cause the most injury?
- To start the worst fire?
- To give me the best chance of being undetected and make my escape?'

Bombs in the post

Postal bombs, in a variety of sizes, may be sent in envelopes no thicker than a quarter-of-an inch or in larger packages. They can kill. If in any doubt, place the letter or package on a flat surface away from thin walls and windows. Do not move it unnecessarily. Lock the room.

Clear the immediate area and adjacent rooms of all persons and call the police. Meet them with the key to the room where the suspect device is. Do not bend or open any letter or package, or place it in water.

Postal bombs may explode on opening. Other mail may contain alleged infectious and/or contaminated matter. Look for:

- *The postmark* – especially if unusual or foreign. Seek clarification from the addressee (e.g. is it expected?).
- *The writing* – which may be in a 'foreign' style. Do you recognise it?
- *The balance* – is it evenly balanced? If the letter or parcel is lopsided, treat it as suspect.
- *The weight* – if this seems to be excessive for size, treat it as suspect.
- *Any holes* – are there any small holes or pinpoints, which could have been made by wires? Any wires sticking out?
- *Stains* – are there any stains or grease marks?
- *The smell* – some, but certainly not all, explosives have an aroma of marzipan or almonds.
- *The feel* – in the case of letters, it will indicate whether there is only folded paper inside (which will show that it is all right) or if there is stiffening; for example, cardboard or the feel of metal, in which case treat it as suspect.
- *The outline* – are there any unusual outlines if you hold it up to the light?
- *The flap* – is the flap of the envelope stuck down completely? (There is

usually a small gap.) If so, treat it as suspect.

- *Stamps* – Those who send bombs in the post often put too many on so it will always pass through the post office. Bear in mind that most government / official mail is stamped by franking machine without postage stamps being applied.

Everyone must be alert to the danger of postal devices. Do not accept presents or parcels from unknown persons, particularly parcels that are not ordered. Always check deliveries carefully before accepting them. Encourage regular correspondents to write their name on the outside of parcels and bulky letters and give clear instructions to members of your staff.

Vehicle bombs

Vehicle Borne Improvised Explosive Devices (VBIEDs) have been used many times before in the UK and are capable of delivering a large quantity of explosives to cause a great deal of damage. An enterprising terrorist can detonate it from a safe distance using a timer or remote control, or it can be detonated on the spot by a suicide bomber.

When it comes to vehicles, over the past few years it has become apparent that no one type of transport is used, nor one type of bomb. Indeed, to identify a well placed car bomb, for example, you first need to know the underneath of a car quite well to spot something out of place – but how many of us know what the underside of our car looks like? Once again, common sense counts for a great deal, as does trying to spot the unfamiliar.

The police advise that organisations should discourage unchecked vehicles to park in underground service areas, directly below public areas where there will be large numbers of people and where there is a risk of structural collapse. It is good advice to:

- ensure that delivery vehicles arriving at your building are expected by the receiving unit before they are granted access into goods / service areas;
- consider a vehicle search regime at goods / service entrances that is flexible and can be tailored to a change in threat or response level. It may be necessary to carry out a risk assessment for the benefit of security staff who may be involved in vehicle access control;
- create and rehearse bomb threat drills. Bear in mind that, depending on where the suspected VBIED is parked and the design of your building, it may be safer in windowless corridors or basements than outside (invacuation);
- prepare your staff in identifying suspect vehicles, and in receiving and acting upon bomb threats. Key information and telephone numbers should be prominently displayed and readily available; and
- bear in mind that the installation of physical barriers needs to be balanced against the requirements of safety and should not be embarked upon without full consideration of planning regulation and fire safety risk assessment.

Assessing the threat

Assessing the threat from any type of device is probably one of the most difficult aspects when dealing with these incidents. The more accurate the detail passed by all staff, the more likely it is that a correct assessment will be made. Consider:

- the source of the information, including a full description of the object, who found it, and at what time;
- the exact location;
- history of past incidents, including current media portrayal;

- any VIP / prominent or controversial person visiting or otherwise on the premises; and/or
- whether the date is significant, for example, an anniversary, or if part of a campaign / terrorist trial is in progress.

In all cases, prompt common sense actions taken by people on site will mean the difference between either:

- overreaction – disruption to business, loss of important data, panic; and
- over complacency – a blasé approach, hoping it cannot happen.

Anyone who receives a threat by telephone should try to remain calm and obtain as much detail as possible. This will help to assess the threat, and perhaps, ultimately, help in locating any device. A bomb threat, even if there is no actual bomb, is a serious criminal offence.

All rooms, stairways, and halls should be kept clean and tidy, as a matter of routine. When walking around the building all staff should not ignore anything that is out of place.

At times when the threat of bombing is high, a search of handbags and luggage being brought into your workplace may be implemented by your security staff. Apart from actually finding a bomb, the deterrent value of this is enormous – only the most determined of terrorists will attempt to penetrate such a security screen. Also, overt security should reassure staff that proper precautions are being taken.

The police will normally, although not always, leave the decision about evacuation to the occupier of a building. Current Home Office advice recommends a staff 'coordinator' to act as the interface between the police and all occupants.

Invacuate or evacuate?

All staff should know where each assembly point is, and how to go there. There will not be an opportunity during an evacuation to learn about this for the first time. The identification of routes is especially necessary when there are a number of exits from the building.

However, you must try and avoid evacuation as a knee jerk reaction. Remember, people have frequently been badly injured by flying glass and other debris, outside of buildings. Not only that, but the police will want to keep streets clear of people when a vehicle bomb is suspected. Someone, therefore, has to quickly make a balanced decision.

Where invacuation is preferred it may be that areas within the building have previously been identified as providing a safer environment than having to go outside. These may be referred to as Enhanced Protection Areas (EPAs). Such a location(s) might well be safer than evacuation which, if into the street, could easily put people in more danger. In general such an area should be:

- equipped with a means of communication and emergency lighting (these must be tested);
- clear of stairwells or access to lift shafts;
- surrounded by full height masonry concrete walls;
- away from the perimeter structural bay – that part of a floor structure at all levels between the building's perimeter and the first line of support columns; and
- away from windows, external doors and walls.

However, not all buildings have obviously safer areas inside (because of extensive use of glass, etc.) and it may be that directing to the opposite side of the building may be preferable to evacuating and risking glass injuries in the street. Whenever staff are moved, the area where they are to go must be less dangerous than where they have just left. You should always double check that a secondary bomb is not, for example, at the back of a building.

All employees and visitors may be asked to take their personal belongings with them. This will help to avoid unnecessary suspicions being aroused over innocent articles left behind after an evacuation.

Where evacuation is necessary, staff and visitors who have been evacuated may have to remain outside for a long time before the building is declared safe. Consideration should therefore be given to providing shelter, including re-housing elsewhere. Try and avoid using car parks as assembly areas in bomb threat situations.

On their arrival the police will assume control until any object is declared safe. However, they are unlikely to search as only the building occupiers will be familiar with the building layout and thus spot anything out of place.

Where the police have directed an evacuation, they will, as far as possible, remain in control and will declare when the building is safe to re-occupy. Good staff training in advance will help so that every employee will know what to do – and what not to do.

At large scale incidents the emergency services will establish one Joint Emergency Services Control Centre (JESCC) comprising Police, Fire, Ambulance control vehicles, between the inner and outer cordons.

Cordon distances established by the police will vary. In the case of a suspect terrorist bomb that has not exploded (and as a very rough guide depending on wind in some cases) the distances could be:

- Briefcase size object: 100 – 200 metres minimum.
- Larger objects: 200 – 300 metres minimum.
- HGVs etc.: 400 – 500 metres minimum.

Weighing this all up, it will be apparent that evacuation may not always be the most appropriate response to a threat. The purpose of evacuation is to move people away from an actual or potential danger area to a safer place. In all cases the overall priority must be the safety of the public and emergency responders. This must be the focus of the police decision-making process and other factors (e.g. commercial considerations must not be permitted to interfere in achieving this objective).

It should be recognised that many buildings provide significant protection against various hazards, and when deciding whether to evacuate it should therefore be considered if everyone may be safer indoors by invacuating. Examples of where evacuation may be counterproductive include a chemical, biological or radiological release outside a building, the police keeping the streets clear when they suspect a vehicle bomb, or a terrorist bomb threat where internal bomb shelter areas are available and otherwise the best course of action.

Disabled people

Some disabled staff will be unable to make unaided use of the normal alternative means of escape from an office building when the lifts are taken out of action.

If the lifts are taken out of operation during an emergency, volunteer helpers will be required to assist their disabled colleagues. Normally at least two able-bodied people will be required to assist each disabled person, but the form of help needed will vary according to the nature and severity of an individual's handicap.

The drill to be followed by those assisting a disabled person in the event of an emergency will to some extent be dictated by the constraints and layout of the building. It will normally be safer and more practical for anyone who is severely handicapped and using a wheelchair to remain in their chair.

If no means of escape other than by carrying can be arranged, it may be necessary to consider calling for volunteer helpers who are prepared to assist in an emergency.

Confirm, contain and control

Confirm
- Where the incident is – including where it might shortly affect.
- What the threat level is at the time of the incident, and/or current advice from the police.
- What are the most realistic, safest and achievable instructions to give to all people on the premises that you are responsible for.

Contain
- Ensure the emergency services have been called, if required. Have maps ready to give them, along with local knowledge, any witnesses and your identity.
- Stop the incident from spreading. The emphasis, at this stage, is on the response to the incident and not the cause.
- Warn neighbouring premises if necessary.

Control
- If you are in control of the situation, identify yourself clearly. Others may be coming to help and time spent trying to find you will be time wasted.
- If there has been a bomb threat, do not evacuate as a knee-jerk reaction. Very quickly consider (a) who is telling you this information and (b) what the information is. Validate the source of the data and the impact if the truth is being told. Then decide.
- Whatever decisions are to be made must be clear ones. There is no room for ambiguity.
- If the decision is taken to evacuate, make sure staff and visitors are told not to use any staircases, exits, evacuation routes or assembly areas that are dangerous. This includes routes that may take people past suspect bombs.
- In bomb situations, or in other cases where the route to the primary Assembly Point may be jeopardised (e.g. the route goes past the suspect bomb), give route instructions and ensure everyone knows.
- If required, post staff to help direct the emergency services. Make sure they are clearly visible and know what to do.
- Make sure that someone keeps a written note of all decisions made.

See also: Business Continuity
Management, p.92; Emergency
procedures and crisis management,
p.261; Fire, means of escape,
p.314; Personal Emergency
Evacuation Plans (PEEPs), p.569.

Sources of further information

Centre for the Protection of National Infrastructure:
www.cpni.gov.uk/Security-Planning/Business-continuity-plan/Bomb-threats

National Counter Terrorism Security Office: www.nactso.gov.uk/default.aspx

Direct Gov: http://bit.ly/qqcCSQ

NaCTSO – Counter-terrorism protective security advice for hotels and
restaurants: http://bit.ly/od8kJD

Boundaries and party walls

Stuart Wortley, Pinsent Masons Property Group

Key points

- The position of a boundary on a plan is normally only indicative, and its exact position normally has to be plotted on the ground.
- Where possible, boundary disputes should be resolved between neighbours as soon as possible without recourse to litigation.
- Works to a party wall need to conform to the procedures in the Party Wall etc. Act 1996.
- The erection of a wall or fence may require planning permission if it exceeds certain heights.

Legislation

- Town and Country Planning (General Permitted Development) Order 1995.
- Party Wall etc. Act 1996.
- Land Registration Act 2002.
- Antisocial Behaviour Act 2003.
- Land Registration Rules 2003.

Boundary lines

A boundary line is an invisible line between adjoining properties. Its position is ascertained from title deeds and documents, from plans and measurements referred to in these, and from inspection of the site. Frequently, precise positioning is not possible; plans, measurements and descriptions are not necessarily accurate when compared to what appears to be obviously long established on the ground.

There may be a structure along a boundary. This could be a wall or fence on one side of the boundary or a party wall with the boundary along its centre line. The external or (in the case of a multi-let building or terraced house) internal wall of a building may also be a boundary wall or a party wall.

Position of boundary lines

In relation to registered land, the general boundaries rule is that boundary lines on the Land Register are only to be taken as general indications as to their precise position (Land Registration Act 2002, Section 60). This is the starting point for ascertaining the position of a boundary line but is no guarantee on a sale of premises. The commonly used standard conditions of sale reflect this by stating that a seller does not guarantee his property boundaries. An application, by the registered proprietor, may be made to the Chief Land Registrar to determine the exact line of the boundary (Land Registration Rules 2003, Part 10).

If an established boundary structure is reasonably close to this line, it is likely that structure will be immediately adjacent to, or, in the case of a party wall, on the true boundary line. In the case of ancient field boundaries, which can exist in developed areas, where there is or was a hedge and a ditch, the presumption is that the boundary is on the far side of the ditch from the hedge.

Ownership of a boundary

A boundary, as such, does not belong to anyone – it is simply the point at which two properties meet. However, the wall that marks the edge of a building will generally belong to that building and be the responsibility of the owner of the property on that side of the boundary line.

Covenants in title deeds to maintain walls or fences, and 'T' marks on the inside of a boundary line on a deeds plan in the title, are all indicators as to which owner is responsible for or owns the boundary structure. But this may not be conclusive. Where a wall divides two buildings, it will normally be a party wall.

Party walls

A party wall is a wall owned or used by two adjoining owners. Commonly, the boundary runs down the centre line and each owner is deemed to own the wall up to the centre line and to have rights over the other half. Special rules apply under the Party Wall etc. Act 1996. The provisions of this Act are very detailed and before taking any steps under it an owner should consult a surveyor or solicitor with specialist knowledge. The Act also details the different types of party wall. In buildings on several floors in multi-occupation, a floor / ceiling may also be a party structure within the Act.

New walls or fences along the boundary

There are limits on the height of manmade boundary structures that can be erected without obtaining planning legislation (Town and Country Planning (General Permitted Development) Order 1995). There is a maximum height of one metre above ground next to a highway and of two metres on all other boundaries. Anything higher will require planning permission.

The owner of land can put up a wall or fence on his land up to the boundary. It may also be possible to construct a wall or fence across a boundary.

Under the Party Wall etc. Act 1996, an owner may put up a new party wall, after appropriate statutory notices, if his neighbour agrees. If the neighbour does not agree, the new wall must be on the building owner's side, but footings may be placed on the adjoining owner's land. A new wall must not interfere with established rights such as rights of light to windows or rights of way.

Boundary structures next to highways may be subject to restrictions or controls, particularly next to access points and on corners where there are vision splays. While security fencing will be permitted in general, anything that is likely to cause harm to children and deterrents designed to injure trespassers, such as broken glass on top of walls, should be avoided as it is likely to result in civil or criminal liability if someone is in fact injured.

Maintaining the boundary structure

If you own the boundary structure, you will be responsible for maintaining it in a safe condition unless the deeds require the adjoining owner to do so. If it causes damage to the neighbouring property by collapse or by parts of it falling on to that property, you may be liable. You can and should insure against this risk.

Check your policy to see that this risk is covered. The cost of rebuilding in case of damage can also be included in your policy.

Existing structures on boundaries may have rights of support. Any activity on the adjoining property that lessens or removes that support will not generally be permissible. However, if the scheme of work is carried out to a party wall under the Party Wall etc. Act 1996, both building owners should have the comfort of having a scheme approved by independent surveyors.

Entry on to neighbouring land to carry out work

If you need access to the neighbouring property to repair or rebuild the boundary structure (which may actually be the wall of your building), your deeds may give you rights.

You also have limited rights for this under the Access to Neighbouring Land Act 1992 (but this does not allow new development). You need to give notice of your intentions to the owner and occupier of the adjoining property. If he is not willing to give you access, you can apply to the court for an order that he must give necessary access. You will be responsible for any damage caused to his property in the course of exercising the right.

The Party Wall etc. Act 1996 contains alternative arrangements for allowing entry on to adjoining land for carrying out work to party walls (and other party structures) and for excavations below the foundations of the neighbouring property. The Act has a strict regime of notices and counter-notices with detailed timescales and provisions for resolution by independent surveyors if the owners are unable to agree. The Act assumes the involvement of specialist surveyors, whose advice should be sought from the outset.

Hedges and trees

The Antisocial Behaviour Act 2003 contains provisions that enable a Local Authority, on receiving a complaint, to force remedial action in respect of evergreen hedges that are more than two metres high and that adversely affect the reasonable enjoyment of another person's land. In such a case the owner can be required to ensure that the hedge is kept to a height of no more than two metres.

If there are trees or hedges belonging to a neighbouring property that actually overhang your boundary, you can trim the offending overhanging branches. However, you must return the trimmings to the owner of the tree or hedge. Before exercising this right, it is advisable to warn the neighbouring owner that you intend to do so.

Hedges in rural areas are increasingly subject to conservation controls, particularly where they are next to open or agricultural land. For example, hedges more than 20 metres long with open or agricultural land on at least one side may not be removed without the consent of the Countryside Agency.

Boundary disputes

The costs of taking legal action are usually out of all proportion to the value of the property concerned, and only lead to soured relations between adjoining owners. Therefore, try to resolve disputes by agreement.

If you agree terms, record them in writing and place a copy with each owner's deeds.

Where it is not possible to reach agreement, it is possible to apply to the court:

- for a declaration as to the correct position of the boundary; and/or
- for an order for possession of any land on your side of the boundary that is occupied by another party, and damages for any loss you may have suffered. Damages may also be awarded based on the benefit that the party in occupation has enjoyed.

If the dispute is urgent (for example, if your neighbour is carrying out works to a party wall without complying with the proper procedures) you may be able to apply for an order that the works be stopped whilst the dispute is considered by the court. However, if you ultimately lose the argument, damages and costs may be awarded against you.

If the Land Registry makes an error on the title plan as to the location of a boundary line, you may be able to claim to have the title plan corrected or a claim for compensation.

Under the Civil Procedure Rules, there is a duty on both parties to act reasonably in exchanging information and to try to avoid the need for court proceedings.

See also: Buying and selling property, p.107; Landlord and tenant: possession issues, p.438; Property disputes, p.600.

Sources of further information

The Workplace Law website has been one of the UK's leading legal information sites since its launch in 2002. As well as providing free news and forums, our Information Centre provides you with a 'one-stop shop' where you will find all you need to know to manage your workplace and fulfil your legal obligations.

It covers everything from CDM, waste management and redundancy regulations to updates on the Carbon Reduction Commitment, the latest Employment Tribunal cases and the first case to be tried under the Corporate Manslaughter and Corporate Homicide Act, as well as detailed information in key areas such as energy performance, equality and diversity, asbestos and fire safety.

You'll find:

- quick and easy access to all major legislation and official guidance, including clear explanations and advice from our experts;
- case reviews and news analysis, which will keep you fully up to date with the latest legislation proposals and changes, case outcomes and examples of how the law is applied in practice;
- briefings, which include in-depth analysis on major topics; and
- WPL TV – an online TV channel including online seminars, documentaries and legal updates.

Content is added and updated regularly by our editorial team who utilise a wealth of in-house experts and legal consultants. Visit www.workplacelaw.net for more information.

BREEAM

James Tiernan, Bureau Veritas UK Ltd

Key points

- BREEAM stands for Building Research Establishment Environmental Assessment Method.
- BREEAM is a voluntary scheme, although in some cases the achievement of a particular rating may be required to satisfy planning or funding requirements.
- There are dedicated BREEAM schemes for use in different countries, as well as a BREEAM International scheme.
- In the UK there are various BREEAM schemes: New Construction 2011 (which replaces the BREEAM 2008 schemes, and will also be used for refurbishments and fit-outs until the launch of the Non-domestic Refurbishment Scheme in 2012), Communities (for large development projects at the planning stage) and In Use (for existing buildings).
- BREEAM New Construction 2011 now uses a single, consolidated scheme to assess all applicable building types rather than the multiple individual schemes used under BREEAM 2008 (e.g. for offices, retail, industrial, etc).
- The Code for Sustainable Homes, while similar to BREEAM, is a separate scheme for domestic dwellings run by the Department for Communities and Local Government.

What is BREEAM New Construction?

BREEAM New Construction (BREEAM NC) is a single assessment tool that can be used to help improve the sustainability of new developments. Its stated aims are as follows:

- To mitigate the life cycle impacts of buildings on the environment.
- To enable buildings to be recognised according to their environmental benefits.
- To provide a credible environmental label for buildings.
- To stimulate demand for sustainable buildings.

The BREEAM NC scheme groups buildings in four different sectors:

1. Commercial:
   - Offices.
   - Industrial.
   - Retail.
2. Public (non-housing):
   - Education.
   - Healthcare.
   - Prisons.
   - Law courts.
3. Multi-residential accommodation.
4. Other:
   - Residential institutions.
   - Non-residential institutions.
   - Assembly and leisure buildings.
   - Other buildings.

While the BREEAM New Construction scheme is able to cater for speculative (shell and core) constructions, it is not suitable for the following:

- Existing building refurbishment and fit-out (although until the launch of the Refurbishment Scheme the New Construction scheme may be used).
- Existing buildings in operation or existing unoccupied buildings.
- Existing building de-construction.
- Infrastructure projects.

How does a BREEAM NC assessment work?

Only a fully licensed and appropriately qualified BREEAM Assessor can undertake a BREEAM New Construction assessment; a BREEAM Accredited Professional (AP) may also be engaged to assist the Design Team. Note the BREEAM methodology can be used to guide the design team without undertaking a formal BREEAM assessment, although no rating could then be awarded or claimed. A BREEAM NC assessment typically has three stages:

1. Pre-assessment stage.
2. Design Stage (DS) Assessment (leading to Interim or DS Certification), ideally conducted prior to activities commencing on site.
3. Post Construction Stage (PCS) Assessment (leading to Final or PCS Certification).

The pre-assessment stage, while not mandatory, should be considered vital for setting the conditions for the successful achievement of the desired rating by ensuring all key project team members are fully aware of the requirements and their own responsibilities. Under certain circumstances it may be possible to conduct a PCS Only assessment, although further guidance should be sought.

There are nine sections to a BREEAM NC assessment:

1. Management.
2. Health and wellbeing.
3. Energy.
4. Transport.
5. Water.
6. Materials.
7. Waste.
8. Land use and ecology.
9. Pollution.

Each section comprises numerous BREEAM issues relating to a particular facet of the building's performance, for which a different number of credits are available. However, as an Environmental Weighting factor is applied to each section, depending on its perceived importance under the BREEAM scheme, each credit is not of equal value. In addition, for some issues there are minimum mandatory requirements for the number of credits that must be achieved for different ratings to be awarded.

Extensive documentary evidence is required to demonstrate compliance with the requirements of each BREEAM issue. The total weighted credit score awarded by the BREEAM Assessor at the DS and PCS dictates the final BREEAM Rating:

- Outstanding.
- Excellent.
- Very good.
- Good.
- Pass.
- Unclassified.

What is BREEAM In-Use?

BREEAM In-Use (BIU) is an assessment tool that allows property owners and users to monitor and compare their building's current environmental performance against good practice benchmarks in a holistic approach. It is intended to help reduce running costs and improve environmental performance, and can be used on the full range of non-domestic buildings: commercial, industrial, retail, and institutional.

BREEAM-In Use is a cost-effective, efficient and simple assessment tool, which can be added to over time as more information becomes available and improvements occur. It is designed to be user-friendly and allow integration of other building assessment tools such as Energy Performance Certificates and Display Energy Certificates.

BREEAM In-Use Assessment

BIU uses an online questionnaire of 197 questions to assess the building's performance over ten categories:

1. Environment issues.
2. Management.
3. Materials.
4. Transport.
5. Waste.
6. Water.
7. Health.
8. Pollution.
9. Energy.
10. Land use.

The questionnaire can either be completed by a qualified and licensed BIU Auditor or by the building manager who has completed an online training and assessment package; if the latter, then a BIU Auditor must undertake a formal verification of the findings.

A BIU assessment comprises of three separate stages:

1. Part 1: Asset rating.
2. Part 2: Building management rating.
3. Part 3: Organisational rating (currently restricted to offices only).

Asset rating

The inherent performance characteristics of a building based on its built form, construction and services.

Building management rating

Management policies, procedures and practices related to the operation of a building. The consumption of key resources such as energy, water and other consumables; the environmental impacts such as carbon and waste generation.

Organisational effectiveness rating

The understanding and implementation of management policies, procedures and practices, staff engagement and delivery of key outputs.

Declaration of commitment

An organisation can choose to set a commitment to the BRE (e.g. improve its asset rating of all buildings to 'very good' rating by 2012), and this will demonstrate an organisation's commitment to environmental improvement. This assessment is available to portfolio members only (a portfolio member would have more than 30 properties).

Auditing and certification

Following the independent certification by the BIU auditor, BRE Global Quality Assurance undertakes a quality check of the process via three methods:

1. Customer follow up review.
2. Call in of auditor records.
3. Targeted witness assessments.

Following the assessment and audit, BRE Global calculates the building's rating based on the questionnaire and the evidence sent by the auditor. BRE provides the scope to the auditor who produces the certificate and issues it to the client / property owner.

The final certificate produced will show a star rating, depicting the performance of the building against the key performance indicators. There are seven ratings boundaries, ranging from unclassified to outstanding. Each level achieved corresponds to a six star rating scheme:

1. No stars, or Unclassified.
2. *, or Acceptable.
3. **, or Pass.
4. ***, or Good.
5. ****, or Very good.
6. *****, or Excellent.
7. ******, or Outstanding.

Why undertake a BREEAM In-Use Assessment?

The reduction of environmental impact is becoming increasingly important to all companies. Increased building environmental performance can improve environmental efficiency, energy consumption, waste generation and building user wellbeing. This in turn creates savings in operational costs and increased profitability.

Undertaking a BREEAM In-Use Assessment provides a tool to both rate and promote the environmental performance of a building, its management and operation, providing benefits for tenants, property managers and owner occupiers of buildings alike.

Tenants benefit from the visible rating of environmental performance on a given building, thereby adding another tool to aid the selection process when selecting / renewing a tenancy agreement. Furthermore, via the organisational assessment, tenants and owner occupiers can identify means of improvement upon company policy and reduce the impact of their individual operating procedures.

Property managers can use BIU to increase the marketability of a property. Certification of an asset shows potential tenants that environmental impact is of a concern to the manager. Furthermore,

the Declaration of commitment provides property managers with a means of informing potential tenants of their future commitments to achieving given environmental targets across their portfolio.

The output will also give the building's performance against each question category, allowing property managers to assess the areas within their buildings that are lowering the overall performance of a building. Planned improvements can then take place in order to raise the building performance. Inputting these proposed changes into the online questionnaire tool will allow for the potential impact of any amendments to be calculated prior to undertaking works.

Moreover, BREEAM In-Use allows building owners to compare their building's performance against good practice benchmarks. By comparing against good practice, realistic and achievable targets can be set for improvements.

Links to other Environmental Assessment

A BREEAM In-Use assessment allows for the first steps towards implementing and complying with environmental legislation and standards, such as ISO 14001 (see 'Environmental Management Systems', p.286), representing good practice in environmental management. It will also show a business's commitment to Corporate Social Responsibility (see p.202) and can feed directly into an annual CSR statement.

See also: Corporate Social Responsibility, p.202; Environmental Management Systems, p.286; Waste management, p.707.

Sources of further information

Bureau Veritas UK and Ireland – www.bureauveritas.co.uk

BREEAM – www.breeam.org

BRE – www.bre.co.uk

BS 8536: 2010 Facility Management Briefing

Ian R. Fielder, BIFM

Key points

- The publication of a new British Standard in the area of design briefing aims to reinforce current best practices through a focus on the operability of facilities. BS 8536: 2010 Facility Management Briefing – Code of Practice provides owners, operators and designers with guidance on incorporating operational needs and requirements.
- This new British Standard gives owners, operators and designers the opportunity to reinforce best practice when incorporating operational needs and requirements of the design brief.

Beyond buildability

Ensuring that design takes account of operational requirements is a critical factor in the success of a new or refurbished facility. Design decisions have to be based on the correct information and data, and their impact on operations has to be understood before they are committed to construction work and/or installation. Once the facility is operational, it is too late to comment on the suitability or effectiveness of the design. The principle of buildability is widely applied in design; however, the principle of operability is not necessarily practised to the same extent.

Facility management briefing reinforces the importance of adopting a whole life perspective of a facility; not solely its design and refurbishment or construction. For the owner and/or operator of a facility, an objective might be to optimise operational cost over the whole life cycle. In this connection, the facility might need to sustain operations and support use over many decades in an environment in which pressure to reduce energy use and, by implication, carbon emissions, is likely

to increase significantly. The Standard is intended to cover a wide range of facilities whose requirements have to be scrutinised individually. Its target audience is owners, operators and designers, and is as applicable to refurbishment as it is to new build.

Not one brief, but several

The Standard outlines the process of design briefing as opposed to specifying detailed documentation. The rationale is simple. Unless the process is defined and an explicit plan formulated, the need for, and extent of, specific documents cannot be determined reliably. In this regard, documents represent both inputs and outputs of the process. Without that process in place, they may fail to reflect their true purpose, scope and content. Nonetheless, the Standard contains ample guidance on the key outputs (deliverables) in the process – the *statement of needs* and *functional brief* – as well as other matters of a practical nature. It prefaces them with a plan of design briefing – a roadmap of how to initiate briefing – based on the principle of a stage-gated process in which progression to the next stage is conditional upon satisfying predefined criteria.

The term design brief is widely adopted in practice and, for the purpose of the Standard, is an all-embracing concept. The facility management brief can be one of a number of documents collectively referred to as the design brief. Other briefs might be prepared on various aspects of the facility, for example financial and legal matters, procurement and contractual arrangements. Whilst the facility management brief might be prepared as a standalone document, it should nonetheless form an integral part of the design brief.

The Standard is unreservedly concerned with the safe and correct operation of the facility, which it encourages through improved briefing practice, especially in terms of operability and energy-related matters. It also provides an objective basis against which design briefing practices and performance can be measured – that is why it is a Standard. The expectation is that existing plans of work will be adjusted to benefit from the wider range of issues considered.

Provisions to ensure inclusive design and management of the facility are also covered. The treatment is intended to be awareness-raising rather than prescriptive and is typical of the approach taken to the drafting of the standard.

According to Dr Brian Atkin, a member of BSI's Facility Management Technical Committee that oversaw the drafting:

"One of the current weaknesses in practice is the absence of a briefing plan of work that can be tailored to suit different kinds of facilities and to do so without becoming either complicated or prescriptive. There

is a need for something to guide decision-making in design, not to constrain it. The Standard provides a roadmap that draws attention to necessary considerations during the briefing process, particularly those relating to operational needs and requirements. The aim is to strengthen existing best practice by encouraging a balanced approach to briefing."

Current best practices

The Standard reflects current best practices and helps, through its definition of the briefing process, to show where particular practices and procedures fit in. Few, if any, facilities can be designed without wide consultation of one kind or another. The Standard recognises the vital role that stakeholders of many kinds play in the success or otherwise of a facility. By incorporating the requirement to identify and assess stakeholders and then to engage them in the briefing process, the Standard promotes an inclusive approach to design. Examples are given of how an easy, but effective, means for stakeholder impact analysis can be used to determine appropriate response strategies.

Uncertainty and risk abound in construction and so it should not be surprising that these concepts are actively considered. Risk assessment occupies a key position in the briefing process and is broadened to incorporate opportunity assessment. Looking at risks alone ignores potential benefits. Both *downside* and *upside* risks have to be fleshed out during the early stages of design, i.e. during briefing, if the owner and users of the facility are to enjoy the most beneficial choices in design, construction and, importantly, operations.

Feasibility studies occupy a key stage in the design process and the Standard considers the scope and content of them as a means for ensuring that the developing design will be *fit for purpose*. During the development of the design, many events will threaten the integrity of the design and so change control has to be built-in from the outset. Any changes have to be properly evaluated if they are to be approved and that must include their impact on operations and use as well as their cost and time. Too often, in practice, decisions are taken without due consideration of the longer term implications. Looking at the capital cost and construction schedule is short-sighted and runs contrary to best practice project control.

Operations and maintenance might appear to be something to concern facility managers once the facility is operational; but unless there is a properly developed understanding during design of how the facility can be managed safely and correctly, problems will be stored up for the owner and/or operator. The Standard serves as a reminder of needs and requirements in the operational phase. The cornerstone of a push for greater awareness of operational demands is the requirement for a Facility Handbook – an organised collection of documentation covering the operation of the facility. The Handbook brings together, in a single coordinated document, the health and safety file, as-built information, building log-book and other operational data to deliver an information asset for the owner and operator.

The Standard is forward-looking in its scope and treatment by anticipating, for example, tougher measures covering environmental assessment, including energy consumption and carbon emissions. The development of the Common Carbon Metric will play a vital role in future assessments and evaluations of facilities. The particular requirement to consider operational carbon – defined as

the weight of carbon dioxide equivalent attributable to the operational phase – will ensure that a more holistic evaluation of a facility's design over its projected lifetime is undertaken. The Standard complements current and emerging legislation as far as practicable.

There is a danger that Facility Management Briefing will be regarded as simply taking into account the routine, daily chores associated with operating a facility. When all the effort of design, construction, commissioning and handover is done, the owner is at the beginning of the most demanding phase of all and faced with the compelling question of how to gain – financially, socially and environmentally – from the facility over the years and decades to come. For the start-up of operations and beyond to yield no unpleasant surprises, there has to be an explicit plan of how to capture needs and define requirements in a structured manner – briefing cannot simply happen.

In the unstructured environment of the early stages of design, it makes sense to put order into the briefing process. That process should be neither open-ended nor constraining, but should exist in the middle ground to guide the decision-making of the designer, as an individual or as an integrated design team, towards a fully-operable solution that meets requirements. Major facility owners and operators have been working along these lines and those outlined for years, but they are in a minority. Far more owners, operators and designers need something to serve their interests in design briefing to make it a more certain and predictable process.

The value of BS 8536 is significant. The benefit that a true facility management approach offers government, commerce and industry is its contribution during design, whether for new build or refurbishment. Input at this stage will influence operational costs and the efficient use of the asset which is now recognised as representing 80% of whole life cost. When you add the potential to directly influence the carbon footprint, sustainability, working environment and flexibility in use, the gains from such an early input can be invaluable.

In summary

New or refurbished facilities represent opportunities for owners and operators, as well as exposure to numerous risks. Significant amongst those risks will be stakeholders of many kinds, with different powers and degrees of urgency brought to bear on the briefing process. Appropriate engagement of those stakeholders is essential and represents just one example of the competing pressures on designers. By anticipating the hurdles along the way and adopting an inclusive, stage-gated process and plan of work for briefing, the design that ensues is more likely to satisfy operational needs and requirements.

> *See also*: The CRC Energy Efficiency Scheme, p.215; Facilities management contracts, p.304.

Sources of further information

BS 8536: 2010 Facility Management Briefing – Code of Practice was published on 30 October 2010 and is available from BSI at www.bsigroup.com

Building Regulations

Dave Allen and Steve Cooper, BYL Group

Key points

- The Building Regulations are made under powers in the Building Act 1984. They apply in England and Wales, and the majority of building projects have to comply with them. They exist to ensure the health and safety of people in and around all types of building (i.e. domestic, commercial and industrial). They also provide for energy conservation and for access into and around buildings.

- The Building Regulations contain various sections dealing with definitions, procedures, and what is expected in terms of the technical performance of building work. For example, they:

    - define what types of building, plumbing and heating projects amount to 'building work,' and make these subject to control under the Building Regulations;

    - specify what types of building are exempt from control under the Building Regulations;

    - set out the notification procedures to follow when starting, carrying out and completing building work; and

    - set out the 'requirements' with which the individual aspects of building design and construction must comply in the interests of the health and safety of building users, energy conservation, and access into and around buildings.

- The Government recognised that the piecemeal way of reviewing the Regulations made it difficult for the industry and the building control service alike to keep abreast of regulatory changes. The newly introduced Periodic Review Process restricts the review of technical parts of the Regulations in three-yearly cycles, with revisions in 2010, 2013, 2016 and onwards. The periodic review process reduces burdens of change and ensures proportionate, justified and visible regulation that allows stakeholders to plan and adapt for technical changes.

Legislation

- Magistrates' Courts Act 1980.
- Building Act 1984.
- Water Resources Act 1991.
- Workplace (Health, Safety and Welfare) Regulations 1992.
- Regulatory Reform (Fire Safety) Order 2005.
- The Energy Performance of Buildings (Certificates and Inspections) (England and Wales) Regulations 2007.
- The Energy Performance of Buildings (Certificates and Inspections) (England and Wales) (Amendment) Regulations 2008.
- The Energy Performance of Buildings (Certificates and Inspections) (England and Wales) (Amendment No. 2) Regulations 2008.
- The Building (Electronic Communications) Order 2008.

- The Building (Local Authority Charges) Regulations 2010.
- The Building Regulations 2010.
- The Building (Approved Inspectors etc.) Regulations 2010.
- Building and Approved Inspectors (Amendment) Regulations 2010.
- Equality Act 2010.
- The Building (Amendment) Regulations 2011.

What is 'building work'?

'Building work' is defined in Regulation 3 of the Building Regulations. The definition means that the following types of project amount to 'building work':

- The erection or extension of a building;
- The installation or extension of certain building services or fittings;
- An alteration project involving work that will temporarily or permanently affect the ongoing compliance of the building, service or fitting with the requirements relating to structure, fire or access to and use of buildings;
- The insertion of insulation into a cavity wall;
- The underpinning of the foundations of a building; and
- Certain works on a building's thermal elements, works carried out as a result of a change in a building's energy status, and work to upgrade the overall energy performance of larger buildings that are being renovated.

Whenever a project involves 'building work,' it must comply with the Building Regulations. This means that the works themselves must meet the relevant technical requirements and they must not make other fabric, services and fittings worse than they previously were.

The Building Regulations may also apply to certain changes of use of an existing building, even though the work involved might not seem like 'building work'. This is because the change of use may result in the building as a whole no longer complying with the requirements that will apply to its new type of use, and so having to be upgraded to meet additional requirements. It must always be remembered that a change of use under Building Regulations is usually different to a change of use under planning.

Two systems of building control

If the work amounts to 'building work' it will be subject to, and must comply with, the Building Regulations. To help achieve compliance with the Regulations, developers are required to use one of two types of building control service:

1. Local Authority building control service. This can be contacted at the district, borough or city council. For further information see www.labc.uk.com.
2. An approved inspector's building control service. Approved inspectors are private sector companies or practitioners and are approved to carry out the building control service as an alternative to the Local Authority. Most approved inspectors belong to the Association of Consultant Approved Inspectors. For further information see www.approvedinspectors.org.uk.

Competent person schemes

Competent person schemes were introduced by the Government to allow individuals and enterprises to self-certify that their work complies with the Building Regulations as an alternative to submitting a building notice or using an approved inspector. The principles of self-certification are based on giving people who are competent in their field the ability to self-certify that their work complies with the Building Regulations without the need to submit a building notice and thus incurring Local Authority inspections or fees. It is hoped that moving towards self-certification will significantly enhance compliance with the requirements of the Building Regulations, reduce costs for firms joining recognised schemes, and promote training and competence within the industry.

It should also help tackle the problem of 'cowboy builders', and assist local authorities with enforcement of the Building Regulations. Building Control bodies and competent persons work within industry-set performance standards.

Contravening the Building Regulations

The Building Regulations can be contravened in two ways:

1. By not following the correct procedures.
2. By carrying out building work that does not comply with the requirements contained in the Building Regulations.

The Local Authority has a general duty to enforce the Building Regulations in its area and will seek to do so by informal means wherever possible.

Where an approved inspector is providing the building control service, the responsibility for checking that the Building Regulations are complied with during the course of building work will lie with that inspector.

However, approved inspectors do not have enforcement powers. Instead, in a situation where they consider that building work does not comply with the Building Regulations, they will not provide a final certificate and, additionally, will notify the Local Authority that they are unable to continue. If no other approved inspector takes on the work, the Local Authority will take on the building control role. From this point the Local Authority will also have

enforcement powers to require the work to be altered if it considers this necessary.

If a person carrying out building work contravenes the Building Regulations, the Local Authority may decide to take him to the Magistrates' Court, where he could be fined up to £5,000 for the contravention and £50 for each day the contravention continues (Building Act 1984, Section 35).

This action will usually be taken against the owner, although proceedings must be taken within 12 months of the offence (Magistrates' Courts Act 1980, Section 127).

Alternatively, or in addition, the Local Authority may serve an enforcement notice on the owner, requiring alteration or removal of work that contravenes the Regulations (Building Act 1984, Section 36). If the owner does not comply with the notice, the Local Authority has the power to undertake the work itself and recover the costs of doing so from the owner.

Approved Documents

Each Building Regulation is supported by 19 'Approved Documents' that reproduce the requirements contained in the Building Regulations. This is followed by practical and technical guidance, with examples, on how the Regulations can be met in some of the more common building situations. However, there may well be alternative ways of complying with the requirements to those shown.

You are therefore under no obligation to adopt any particular solution shown in an Approved Document if you prefer to meet the requirements in some other way. If an alternative method is used, proof would be required to show that the alternative approach is at least equivalent to that shown in the Approved Documents.

Part A – Structural Stability (2004 edition)

Part A of the Building Regulations is concerned with the strength and stability of a building. It remained almost unchanged between 1992 and 1 December 2004, when a new edition of the Approved Document came into force.

Part A1 seeks to ensure that a building is constructed in a manner and from materials that ensure all loads are transmitted to the ground:

- safely; and
- without causing deflection or deformation of any part of the building, or movement of the ground, that will impair the stability of any part of another building.

Part A2 deals with ground movement and requires a building to be constructed in such a way that ground movement caused by swelling, shrinkage or freezing of the subsoil, or land-slip or subsidence, will not impair the stability of any part of the building.

Structural safety depends on a successful combination of design and construction, particularly:

- loading;
- properties of materials;
- detailed design;
- safety factors; and
- workmanship.

The Approved Document gives detailed guidance on the construction of certain residential buildings no greater than three storeys in height and other small buildings of traditional construction. All other buildings should be designed to the relevant Code of Practice.

Structural design is heavily dependent on the guidance contained in British Standards and Codes of Practice. These give information on:

- loadings;
- structural work in timber;
- structural work in masonry;
- structural work in reinforced, pre-stressed and plain concrete;
- structural work in steel;
- structural work in aluminium; and
- foundations.

Part A3 requires that a building does not suffer from disproportionate collapse as a result of an accident in a part of the building. It seeks to ensure that, should an accident occur in or around a building, such as a gas explosion or vehicle impact, the building is sufficiently tied together to avoid a catastrophic collapse.

Part B – Fire Safety (2006 edition) – Volume 1: Dwelling houses; Volume 2: Buildings other than Dwelling houses

Part B aims to ensure the safety of the occupants and of others who may be affected by a fire in a building, and to provide assistance for firefighters in the saving of lives.

Therefore, buildings must be constructed so that if a fire occurs:

- the occupants are given suitable warning and are able to escape to a safe place away from the effects of the fire;
- fire spread over the internal linings of the walls and ceilings is inhibited;
- the stability is maintained for a sufficient period of time to allow evacuation of the occupants and access for firefighting;
- fire spread within the building and from one building to another is kept to a minimum; and
- satisfactory access and facilities are provided for firefighters.

It is not a function of the Regulations to minimise property damage or insurance losses in the event of fire.

Part B1 applies to all building types (except prisons). In some parts of the country there may be other legislation that imposes additional requirements on the means of escape from a building. In inner London, reference needs to be made to the Building (Inner London) Regulations 1987 and elsewhere Local Acts of Parliament (e.g. the Hampshire Act 1983) may apply.

The Approved Documents give specific guidance on means of escape in Vol. 1: Dwelling Houses and Vol. 2: Buildings other than Dwelling Houses. Volume Two makes reference to general principles that address fire alarm and fire detection systems and horizontal and vertical escape in flats and other building types.

Guidance is also given on:

- lighting of escape routes;
- provision of exit signs;
- fire protection of lift installations;
- performance of mechanical ventilation and air conditioning systems in the event of fire; and
- construction and siting of refuse chutes.

Part B2 gives guidance on the choice of lining materials for walls and ceilings. It concentrates on two properties of linings that influence fire spread:

- The rate of fire spread over the surface of a material when it is subject to intense radiant heating; and
- The rate at which the lining material gives off heat when burning.

It also gives guidance on how these properties can be controlled, mainly by restricting the use of certain materials.

Part B3 deals with Internal Fire Spread and includes measures to ensure:

- stability of the load-bearing elements of the structure of a building for an appropriate time in the event of fire;
- subdivision of the building into compartments by fire-resisting construction such as walls and floors;
- sealing and subdivision of concealed spaces in the construction to inhibit the unseen spread of fire and smoke;
- provision of sprinkler protection to high-rise flats;
- protection of openings and fire-stopping in compartment walls and floors including fire damper guidance; and
- provision of special measures to car parks and shopping complexes.

Part B4 seeks to limit the possible spread of fire between buildings. It does this by:

- making provisions for the fire resistance of external walls and by limiting the susceptibility of their external surfaces to ignition and fire spread;
- limiting the extent of openings and other unprotected areas in external walls in relation to the space separation from the boundary of the site; and
- making provisions for reducing the risk of fire spread between roofs and over roof surfaces.

Part B5 gives guidance on installation of fire mains and hydrants to buildings with further guidance on:

- provision of vehicle access for high reach and pumping appliances;
- access for fire service personnel into and within the building; and
- venting of heat and smoke from basements.

There is interaction between the Building Regulations and other fire safety requirements in England and Wales. It is therefore important for developers and designers to adhere to set procedures to ensure that owners and occupiers do not need to carry out extra building work at the end of a project. The Building Regulations cover means of escape, fire alarms, fire spread, and access and facilities for the fire service, but, for certain buildings, additional requirements are imposed by the Regulatory Reform (Fire Safety) Order 2005. Because of this, there are statutory requirements on the building control body to consult with the fire authority and for fire safety information to be provided to building owners under Regulation 16B. Details of the consultation procedure can be found in *Building Regulations and Fire Safety Procedural Guidance* (4th edition, 2007) published by the CLG.

Part C – Site Preparation and Resistance to Contaminants and Moisture (2004 edition)

Part C of the Building Regulations is concerned with site preparation and resistance to contaminants and moisture. It remained almost unchanged between 1992 and 1 December 2004, when a new edition of the Approved Document came into force.

Part C1 deals with site preparation and the possibility of contaminants.

- Attention needs to be given to the removal of vegetable matter, topsoil and pre-existing foundations.
- Site investigation is recommended as the method for determining how much unsuitable material should be removed.
- Remedial measures are required to deal with land affected by contaminants. These include materials in or on the ground (including faecal or animal matter) and any substance that is, or could become, toxic, corrosive, explosive, flammable or radioactive. Therefore, it includes the naturally occurring radioactive gas, radon, and gases produced by landfill sites, such as carbon dioxide and methane.
- The area of land that is subject to measures to deal with contaminants includes the land around the building.
- Protection from radon includes buildings other than dwellings.
- Guidance is included relating to subsoil drainage and the risk of transportation of water-borne contaminants.

Part C2 deals with resistance to moisture and seeks to ensure that the floors, walls and roof of a building are constructed in such a way that moisture cannot penetrate the building and nor will condensation occur.

- Guidance recommends that in order to reduce the condensation risk to floors, walls and roofs, reference should be made to BS 5250 'Limiting thermal bridging and air leakage: robust details,' and BR 262 'Thermal insulation: avoiding risks'.
- Guidance is now provided on the use of moisture-resistant boards for the flooring in bathrooms, kitchens and other places where water may be spilled from sanitary fittings or fixed appliances.
- Reference is made to BS 8208 for assessing the suitability of cavity walls for filling.
- Where walls interface with doors and windows, checked rebates are now recommended in the most exposed parts of the country.
- Where Part M requires level access, there is a need to pay particular attention to detail in exposed areas to ensure adequate provision is made for resistance to moisture.

Former requirement F2 'Condensation in roofs,' has been transferred to Part C as it deals with effects on the building fabric rather than ventilation for the health of occupants.

Part D – Toxic Substances (1985 edition, amended 1992 and 2000)

Part D is probably the least used of the Approved Documents. It was introduced when urea formaldehyde foam was a popular method of providing cavity insulation in buildings. The foam can give off formaldehyde fumes that can be an irritant to occupants.

The Approved Document gives guidance on:

- the type of materials used for the inner leaf of the cavity together with the suitability of walls for filling;
- details of the foam itself; and
- the credentials of the installer.

In recent years the popularity of urea formaldehyde foam has declined.

Part E – Resistance to the Passage of Sound (2003 edition, amended 2004)

Part E1 deals with the protection against sound from other parts of the building and adjoining buildings. It aims to achieve adequate sound insulation to walls and floors between dwelling houses, flats and rooms for residential purposes.

A room for residential purposes means a room, or suite of rooms, which is not a dwelling house or flat and which is used by one or more persons to live and sleep in, including rooms in hotels, hostels, boarding houses, halls of residence and residential homes, but not including rooms in hospitals, or other similar establishments, used for patient accommodation. The requirements for resistance to airborne and impact sound for floors also include stairs where they form part of the separating element between dwellings.

Site testing of sound insulation is intended on a sampling basis. However, as an alternative, robust details have been available to the industry since 1 July 2004. Robust Details Ltd is a non-profit-distributing company, limited by guarantee, set up by the house-building industry. Its objectives are broadly to identify, arrange testing and, if satisfied, approve and publish design details that, if correctly implemented in separating structures, should achieve compliance with requirement E1. It also carries out checks on the performance achieved in practice. See *Sources of further information*.

Although the design details are in the public domain, their use in building work is not authorised unless the builder has registered the particular use of the relevant design detail or details with Robust Details Ltd to identify a house or flat in which one or more of the design details are being used.

The requirement for appropriate sound insulation testing imposed by the Regulations does not apply to building work consisting of the erection of a new dwelling house (i.e. a semi-detached or terraced house) or a building containing flats where robust details are registered and adhered to.

Part E2 sets standards for the sound insulation of internal walls and floors in dwelling houses, flats and rooms for residential purposes. Site testing is not intended.

Part E3 controls reverberation in the common parts of buildings containing flats or rooms for residential purposes. Site testing is not intended.

All new school buildings are now controlled under the Building Regulations, and Part E4 covers the sound insulation, reverberation time and indoor ambient noise levels. Guidance on meeting the requirement is given in Building Bulletin 93, published by the Department for Education and Skills (DfES).

Part F – Means of Ventilation (2010 edition)

In order to satisfy Part L, buildings should attain a better standard of airtightness than before. This means that there will be less ventilation due to air flowing into and out of the building through gaps and cracks in the structure. To take account of this it has been necessary to improve the ventilation requirements that are described in Approved Document F in order to satisfy requirement F1. The new provisions have been designed to ventilate buildings to an air permeability as good as $3m^3/h/m^2$ at 50 pa. Buildings more airtight than this will require additional ventilation provision.

In the 2010 edition the opportunity has also been taken to improve other aspects of the guidance, and the main changes are listed below:

- A mainly performance-based approach has been adopted to allow designers more flexibility and encourage product innovation.
- Ventilation areas are now described in terms of equivalent area, instead of free area, because equivalent area relates better to air flow.
- More guidance has been given for domestic mechanical and natural ventilation systems.
- Guidance has been given for ventilation of basements in dwellings.
- The recommended air supply rate for offices has been increased from 8l/s per person to 10l/s per person.

- Replacement windows should normally be fitted with trickle ventilators.
- Appendices give guidance on passive stack ventilation; good practice for installing fans in dwellings; and ingress of external pollutants into buildings in urban areas.
- The introduction of a new requirement for the testing of mechanical ventilation air flow rates in new dwellings. The testing must be carried out in accordance with a procedure approved by the Secretary of State and the results and data upon which they are based recorded in a manner so approved. The recorded results and data must be given to the building control body not later than five days after completion of the test.
- The introduction of a requirement for fixed systems for mechanical ventilation to be commissioned where testing and adjustment is possible to ensure that such systems provide adequate ventilation.
- A requirement that the building owner be given sufficient information about a building's ventilation system so that it can be operated and maintained to provide adequate ventilation.

Part G – Sanitation, Hot Water Safety and Water Efficiency (2010 edition)

Part G1 gives guidance on supply of wholesome water for the purposes of drinking, washing or food preparation and the supply of suitable quality water to sanitary conveniences.

Part G2 sets out requirements on water efficiency in dwellings and introduces a new Regulation 17K to implement this. Where there is a material change of use to create a new dwelling in an existing building, either by converting a non-domestic building or by the provision of a flat or flats in a building, the water efficiency requirements in GS and

Regulation 17K will apply, as will the hot water safety requirement in G3.

Part G3 provides enhanced and amended guidance on hot water supply and safety, applying safety provisions to all types of hot water systems and a new provision on the prevention of scalding.

Part G4 provides guidance on sanitary conveniences and hand washing facilities.

Part G5 provides requirements for bathrooms, which apply to dwellings and to buildings containing one or more rooms for residential purposes.

Part G6 provides guidance on kitchens and food preparation areas and includes a new requirement for sinks to be provided in areas where food is prepared.

In addition to the above, a Local Authority may not give a completion certificate until it has received in respect to new dwellings a certificate specifying the calculated potential consumption of wholesome water per person per day. The CLG has published an online document entitled *The Water Efficiency Calculator for New Dwellings – see Sources of further information*.

Part H – Drainage and Waste Disposal (2002 edition)

Part H1 gives guidance on the design of above-ground sanitary pipework and below-ground foul drainage. It stresses the need for the drainage system to be designed and constructed to:

- convey foul water to a suitable outfall (a foul or combined sewer, cesspool, septic tank or holding tank);
- minimise the risk of blockage or leakage;
- prevent the entry of foul air into the building under normal working conditions;
- be ventilated;

- be accessible for clearing blockages; and
- not increase the vulnerability of the building to flooding.

Sewers (i.e. a drain serving more than one property) should normally have a minimum diameter of 100mm when serving no more than ten dwellings. Sewers serving more than ten dwellings should normally have a minimum diameter of 150mm. Access points to sewers should be in places where they are accessible and apparent for use in an emergency.

Part H2 deals with the siting, construction and capacity of waste-water treatment systems and cesspools so that they:

- are not prejudicial to health or a nuisance;
- do not adversely affect water sources or resources;
- do not pollute controlled waters; and
- are not sited in an area where there is a risk of flooding.

They should be adequately ventilated and should be constructed so that the leakage of the contents and the ingress of subsoil water is prevented.

Waste-water treatment systems should be considered only where the nature of the subsoil indicates that the operation of the system and the quality and method of disposal of the effluent will be satisfactory and where connection to mains drainage is not practicable.

The quality of the discharged effluent is not covered by the Building Regulations, but some installations may require consent for discharge from the Environment Agency.

Additionally, the Environment Agency may take action against any person who knowingly permits pollution of a stream, river, lake, etc., or groundwater, by requiring him to carry out works to prevent the pollution (Water Resources Act 1991 (as amended), Section 161a).

Part H2 also contains guidance on:

- the siting and construction of drainage fields; and
- the provision of information regarding the nature and frequency of the maintenance needs of waste-water systems and cesspools.

Part H3 gives guidance on the need for rainwater from roofs and paved areas to be carried away either by a drainage system or by some other means.

Where provided, a rainwater drainage system should:

- carry the flow of rainwater from the roof to a suitable outfall (a surface water or combined sewer, soak-away or watercourse);
- minimise the risk of blockage or leakage; and
- be accessible for clearing blockages.

The Approved Document contains guidance on:

- precautions to be taken where rainwater is permitted to soak into the ground;
- siphonic roof drainage systems;
- eaves drop systems;
- rainwater recovery systems;
- drainage of paved areas;
- the design of soak-aways and other infiltration drainage systems; and
- the use of oil separators.

Part H4 gives guidance on the construction, extension or underpinning of a building over or within three metres of the centreline of an existing drain, sewer or disposal main shown on the sewerage undertaker's sewer records.

Building work should be carried out so that it will not:

- cause overloading or damage to the drain, sewer or disposal main; or
- obstruct reasonable access to any manhole or inspection chamber.

Future maintenance works to the drain, sewer or disposal main must be possible without undue obstruction, and the risk of damage to the building must not be excessive due to failure of the drain, sewer or disposal main. The guidance also explains that precautions should be taken if piles are to be placed close to drains.

Part H5 is designed to ensure that:

- rainwater does not enter the public foul sewer system where it may cause overloading and flooding;
- rainwater does not enter a waste-water treatment system or cesspool not designed to take rainwater where it might cause pollution by overloading the capacity of the system or cesspool; and
- foul water, including run-off from soiled or contaminated paved areas, does not enter the rainwater sewer system or an infiltration drainage system intended only for rainwater.

Part H6 gives guidance on the design, siting and capacity of refuse containers and chutes for domestic developments.

For non-domestic developments it is recommended that the collecting authority is consulted regarding:

- the volume and nature of the waste, and the storage capacity required;
- any requirements for segregation of waste for recycling;
- the method of storage, including any proposals for on-site treatment;
- the location of storage areas, treatment areas and waste collection points; and

the means of access to these, hygiene arrangements, hazards and protection measures.

Part J – Combustion Appliances and Fuel Storage Systems (2010 edition)

Parts J1 to J3 apply only to fixed fuel-burning appliances and incinerators. The Approved Document gives advice on:

- the amount of air supply needed for safe combustion of the fuel;
- the construction of hearths, fireplaces, flues and chimneys;
- the location of boilers; and
- separation of flues and chimneys from structural timbers and thatch.

The 2010 edition introduces a new requirement for the installation of carbon monoxide alarms in dwellings in appropriate circumstances when a combustion appliance is installed.

The guidance also includes ways of demonstrating that the safe performance of combustion installations is not undermined by mechanical extract ventilation systems.

Part J4 calls for a notice providing the performance characteristics of the hearth, fireplace, flue or chimney to be fixed in an appropriate place in the building. The Approved Document gives guidance on the form, content and location of such notices.

Part J5 gives guidance on the protection of oil and LPG fuel storage systems from fire. This includes the positioning and/or shielding so as to protect these systems from fires that might occur in adjacent buildings or on adjacent property.

Part J6 makes provision for protection against leakage from oil storage tanks polluting boreholes and water and drainage courses and for permanent labels containing information on how to respond to oil escapes to be positioned in a prominent position.

Part K – Protection from Falling, Collision and Impact (1998 edition, amended 2000)

Part K1 deals with the design, construction and installation of stairs, ladders and ramps. In a public building, the standard of provision may be higher than in a dwelling to reflect the lesser familiarity and number of users.

It deals with:

- the rise and going of stairs (i.e. their steepness);
- provision of handrails; and
- allowing sufficient headroom over a stair.

Parts K2 and K3 cover:

- provision of guards designed to prevent pedestrians from falling;
- provision of vehicle barriers capable of resisting or deflecting the impact of vehicles; and
- measures to protect people in loading bays from being struck or crushed by vehicles by providing adequate numbers of exits or refuges.

Part K4 gives guidance on the installation of windows so that parts that project when the window is open are kept away from people in and around the building, and on the provision of features that guide people away from open windows, skylights and ventilators.

The Approved Document also describes measures designed to prevent the opening and closing of doors and gates from presenting a safety hazard. These include vision panels in doors and safety features to prevent people being trapped by doors and gates.

Recommendations for the design of stairs for means of escape, included in Approved Document B (Fire Safety), and for the design of stairs and ramps for use by disabled people in Approved Document M

(Access to and Use of Buildings), should also be considered.

Compliance with Part K (and, where appropriate, Part M as it relates to stairs and ramps) would prevent the service of an improvement notice with regard to the relevant requirements of the Workplace (Health, Safety and Welfare) Regulations 1992.

Part L – Conservation of Fuel and Power (2010 edition)

New buildings (Approved Documents L1A and AD L2A)

Compliance involves demonstrating that separate criteria have been met:

1. The CO_2 performance target has been met.
2. No elements of the design fall outside defined limits of design flexibility. This is to ensure that the performance of the building is not critically dependent on a particular feature that might fail to be maintained in proper working order.
3. The building does not suffer from excessive solar gains. This is to ensure that the building has appropriate passive measures to limit any tendency to retro-fit air conditioning.
4. The building as constructed should deliver the calculated CO_2 performance. This includes mandatory air tightness testing and commissioning to be carried out as detailed in the respective Approved Documents.
5. Appropriate information is provided to the user to enable the building to be operated efficiently.

A new requirement is introduced, where Regulation 17C applies, for CO_2 emission rate calculations to be carried out before the start of building work on the erection of a new building and given to the building control body, along with a list of the specifications used in the calculations.

Dwellings (Approved Document L1A)

The CO_2 target is based on achieving a 25% reduction in CO_2 emissions over the 2006 standards for heating, hot water and lighting in a gas-heated dwelling. This equates with a Code for Sustainable Homes Level 3 rating. The next versions of Part L will achieve a 44% reduction in 2013 and 100% reduction by 2016. The target for dwellings heated using fuels other than gas is adjusted by the fuel factor, which relates the CO_2 emission factor of the particular fuel to that of gas. This results in an increased CO_2 target, but a reduced energy consumption target. The energy impact of thermal bridging and secondary heating must be included within the estimate of the performance of the proposed building.

Lighting is a non-tradable item in the CO_2 target (i.e. a fixed amount of low energy lighting is assumed in the calculation, irrespective of the number of low energy fittings that are installed). The guidance also requires that the fittings be installed with an appropriate shade or diffuser. The existing guidance on the minimum number of fittings is now supplemented by a percentage of light fittings criteria, which means that if a large number of less efficient lamps are specified, then an increased number of low energy lights will be needed to compensate.

The limits on design flexibility mean that the average performance of opaque elements should be no worse than the 2006 elemental standards. Guidance on the minimum efficiency of heating and hot water systems is given in a second tier 2010 guidance document. This specifies that all boilers should be rated at least SEDBUK B.

The exemption from the energy efficiency provisions for extensions consisting of a conservatory or porch is amended to grant the exemption only where the existing walls, windows or doors are retained or replaced if removed or where the heating system of the building is not extended into the conservatory or porch.

When extending an existing dwelling, u-values have been tightened, and extra guidance is now provided on the renovation of a thermal element, historical buildings, and swimming pool basins.

New non-domestic buildings (Approved Document L2A)

The CO_2 target is based on achieving an overall reduction in CO_2 emissions of 25% for heating, hot water, ventilation, cooling and lighting; this equates to improvements between 16% for hotels and 40% for air conditioned shallow pan offices. This overall improvement includes the contribution made from a benchmark provision of LZC technology. Credits can be taken for advanced monitoring and control features, and for power factor correction.

The limits on design flexibility mean that the average performance of opaque elements should be no worse than the 2006 elemental standards. Greater clarity has been achieved through the introduction of definitions for such features as display glazing and high usage entrance doors. Guidance on the minimum efficiency of building services systems is given either in the Approved Document or in a second tier guidance document.

The requirement for testing and commissioning has been widened to include leakage testing of ventilation ductwork.

Work in existing buildings (Approved Documents L1B and L2B)

Most of the changes are concerned with bringing more types of work within the control of the Regulations, as described in paragraphs 10 and 11. Particular guidance is given on what would be reasonable provision when renovating a controlled element and the application

of consequential improvements to non-domestic properties.

The standards that the building work is expected to achieve are generally more demanding than in 2006, with the exception of replacement windows, where the standard is unchanged (although for dwellings and similar domestic type buildings, compliance can now be demonstrated using a window energy rating). A separate standard has been introduced for doors.

Part M – Access to and Use of Buildings (2004 edition)

Part M covers access to and use of buildings. It used to deal solely with access and facilities for disabled people but is now universally inclusive to encourage the provision of an accessible environment for all.

Parts M1 to M3 cover:

- the use of access statements;
- access to the main entrances to the building from the edge of the site and from car parking within the curtilage;
- access into and within the building and from one building to another on a site;
- the provision of lifts;
- access to and use of the building's facilities;
- design of the building's elements so that they are not a hazard;
- the provision and design of sanitary accommodation and changing and showering facilities;
- the provision and design of accommodation for disabled people in audience or spectator seating;
- aids for communication for people with impaired hearing or sight;
- accessibility of switches and controls; and
- visually contrasting surfaces and fittings.

Part M4 covers the provision of reasonably accessible sanitary conveniences in dwellings.

There is interaction between the Building Regulations and the Equalities Act 2010. It is therefore important for developers and designers to appreciate how all these pieces of legislation might apply to a particular scheme. This should ensure that owners and occupiers do not need to carry out extra building work at the end of a project.

Part N – Glazing: Safety in Relation to Impact, Opening and Cleaning (1998 edition, amended 2000)

People using buildings may come into contact with glazing in critical locations, such as doors, door side panels and at low level in walls and partitions. Part N1 describes measures to be adopted to reduce the likelihood of cutting and piercing injuries occurring from contact with such glazing by making sure that it will break safely, be robust or be permanently protected.

Part N2 gives guidance on the measures that might be adopted to indicate the presence of large uninterrupted areas of transparent glazing with which people might collide. It does not apply to dwellings.

Part N3 provides guidance on the safe operation of openable windows, skylights and ventilators relating to the location of controls and the prevention of falling. It does not apply to dwellings.

Part N4 gives guidance for the safe means of access for cleaning glazed surfaces where there is danger of falling. Guidance is also given for safe cleaning where the glazed surfaces cannot be reached from the ground, a floor or some other permanent stable surface. Again, it does not apply to dwellings.

Parts N3 and N4 contain similar provisions to those in Regulations 15(1) and 16 of the Workplace (Health, Safety and Welfare) Regulations 1992.

Part P – Electrical Safety (2006 edition)

A new section of the Regulations was introduced on 1 January 2005 to control electrical safety in dwellings. Revisions were introduced on 1 April 2006.

Statutory requirements P1 and P2 have been revoked and replaced with a new requirement P1. This requires only that reasonable provision be made in respect of design and installation. There is no longer a statutory requirement to provide information, although such provision may be needed for those operating, altering or maintaining installations.

Part P applies in England and Wales to fixed electrical installations in dwellings and in:

- dwellings and business premises that have a common supply;
- common access areas in blocks of flats;
- shared amenities in blocks of flats such as laundries and gymnasia; and
- outbuildings, including sheds, garages and greenhouses, supplied from a consumer unit located in any of the above.

Guidance has been altered to make it clear that installations attached to the outside of dwellings are within the scope of Part P.

The Approved Document includes guidance on a long list of circumstances where work is to be regarded as notifiable or non-notifiable.

The main way of complying is to follow the technical rules in BS 7671: 2001 and the guidance given in installation manuals that are consistent with this standard, e.g. IEE On-Site Guide and Guidance Notes Nos 1 to 7.

With certain exceptions, notification of proposals to carry out electrical installation work must be given to a building control body before work begins.

This prior notification is not necessary if:

- the proposed work is to be undertaken by a competent person, i.e. a firm that has been approved by and certificated by an approved competent person scheme.
- the proposed work is minor work. This comprises:
  - work that is not in a kitchen or special location:
    - adding lighting points to an existing circuit;
    - adding socket outlets and fused spurs to an existing circuit; and
    - installation / upgrading of main and supplementary equipotential bonding;

  - work in all locations:
    - replacing accessories such as socket outlets, control switches and ceiling roses;
    - replacement of the cable for a single circuit only;
    - re-fixing or replacing the enclosures of existing components; and
    - providing mechanical protection to existing fixed installations.

Approved Document to support Regulation 7 – Material and Workmanship (1992 edition)

This document covers the requirements for materials and workmanship in construction. It therefore shows ways of establishing the fitness of materials and ways of establishing the adequacy of workmanship.

Building work shall be carried out:

- with adequate and proper materials which are:
  - appropriate for the circumstances in which they are used;

  - adequately mixed or prepared; and
  - applied, used or fixed so as adequately to perform the functions for which they are designed; and

- in a workmanlike manner.

Approved Document – Basements for dwellings

The Approved Document was first published in 1997 and dealt with the 1991 Regulations, and was updated in 2004 to align with the Building Regulations 2000. This publication brings into one document all of the relevant Building Regulations for dwellings that are affected by the inclusion of a basement.

This is a private sector Approved Document, which has been approved by the Secretary of State under Section 6 of the Building Act 1984 as practical guidance on meeting the requirements of relevant paragraphs in Schedule 1 to the Building Regulations 2002 and 2000 (as amended 2001 and 2002) as they apply to the incorporation of basements to dwellings, and has the same standing as HMSO Approved Documents. However, as a private sector Approved Document it also includes good practice guidance on matters such as vehicle access and selecting a construction and water resisting system. The Approved Document can be found on the Basement Information Centre website at www.basements.org.uk.

See also: Accessible environments, p.18; Contaminated land, p.182; Disability access and egress, p.238; Energy performance, p.270; Fire in non-domestic premises, p.317; Temperature and ventilation, p.666; Toilets, p.669.

Sources of further information

Billington, M. J.: *Manual to the Building Regulations* (TSO, 2001).
ISBN: 0 11 753623 7.

Building and planning legislative guidance and downloads of Approved
Documents can be obtained at: www.planningportal.gov.uk

The Government's official website: www.communities.gov.uk

For details of the Future of Building Control Implementation
Plan: www.communities.gov.uk/publications/planningandbuilding/
buildingcontrolimplementation

For a full contact list of Approved Inspectors, visit the Construction Industry
Council at: www.cic.org.uk/services/AIregister.shtml

Robust Details Ltd: PO Box 7289, Milton Keynes MK14 6ZQ; tel. 0870 240 8210;
fax 0870 240 8203; www.robustdetails.com.

CLG: *The Water Efficiency Calculator for New Dwellings*:
www.communities.gov.uk/publications/planningandbuilding/watercalculator

Changes to parts of the Building Regulations in 2010 have health and safety
implications for organisations throughout England and Wales.

Workplace Law's one-day ***Building Regulations Legal Update*** gives an
overview of recent changes to the Approved Documents and will cover known
changes or issues currently under consultation. Emphasis will be placed on the
main changes in 2010, with particular regard to Part L: Conservation of Fuel and
Power, and reference will also be made to Parts F (Ventilation), G (Hygiene) and
J (Combustion Appliances and Fuel Storage Systems). The Legal Update is also
updated to include a session on BS 9999 and how it can be used as alternative to
compliance with Regulation B1 means of warning and escape. Dates confirmed
for 2012 are 14 February 2012 and 6 September 2012.
Visit http://building-regs.workplacelaw.net for more information.

Business Continuity Management

Peter Power, Visor Consultants (UK) Ltd

Key points

- Business Continuity Management (BCM) identifies potential threats to an organisation and the impacts to business operations that those threats, if realised, might cause.
- It provides a framework for building organisational resilience with the capability for an effective response that safeguards the interests of key stakeholders, reputation, brand and value-creating activities.
- In particular, BCM can deliver increased business resilience and drive higher organisational performance.
- BCM is now seen more as an intrinsic strategic tool as opposed to just part of regulatory and compliance requirements.
- Barriers to creating BCM relate to a lack of understanding of the level of resource and commitment required to do the job properly. This is a matter of BCM maturity, and is further reinforcement of the view that the perceived maturity does not match the reality.
- Although the primary role of BCM may be to help organisations recover from an incident, it has many other peripheral benefits. The two largest benefits to businesses are that they feel better prepared for known and unforeseen events and better understand their business. These two benefits taken independently are enough justification for BCM implementation.
- Given that the overall goal of risk management is to manage risks effectively and efficiently, the BCM programme should work within an overall ERM structure. A better understanding of the business and better risk-intelligent decision-making will improve the effectiveness of the overall risk management and resilience strategies. This can potentially lead to a better return from the investment in these areas.
- A recent report by the Business Continuity Institute stated that out of 221 organisations that had implemented Business Continuity Plans (BCPs), 77% were able to recover faster as a result, with a quarter recovering in half the time compared to without the plan. In addition, 55% said their plans had led to substantial cost savings or protected critical revenue streams in the last 12 months.

Legislation

- Civil Contingencies Act 2004.

This Act still exists, but is currently undergoing a review – the explanatory memorandum to the Civil Contingengies Act 2004 (Amendment of List of Responders) Order 2011 states 'The Guidance that has been issued under Section 3 of the Act, *Emergency Preparedness*, will be updated later in 2011 once an ongoing review of the Act has been completed. We will issue interim guidance directly to those affected by the proposed changes highlighted in this Explanatory Memorandum'. At the time of writing, the 2004 Act was still being reviewed.

Case study

At 15:11 hours on Saturday 30 June 2007 (the second busiest day of the year), a Jeep Cherokee 4x4 vehicle was deliberately driven into the main terminal building at Glasgow Airport and set alight. A well rehearsed Emergency Procedure plan (see *'Emergency procedures and crisis management'*, p.261) was put in place to evacuate the building and deal with the fire. It was established quite quickly that this was in fact a terrorist attack and the perpetrators were arrested at the scene.

The airport's integrated emergency plans include a support mechanism whereby off-duty persons were called in to support the frontline staff. The crisis team was initiated and operational within 45 minutes, with a BC recovery team operational an hour later.

In total, around 3,500 passengers were evacuated to the Scottish Exhibition and Conference Centre (SECC) to allow the police to interview them as potential witnesses.

The local BCM strategy served the airport very well on the day of the incident, as everyone knew their business end-to-end processes. They had analysed what could go wrong and how. They had tested plans in place, and rehearsed and fine tuned them.

This enabled the airport to protect its reputation by showing the world how effective its plans were as it reopened the terminal building 23 hours and 59 minutes after the attack.

This was a high profile event for Glasgow Airport in particular, and the aviation industry in the UK in general. It was more challenging as it happened during the busiest period of the year. Through having robust BC plans in place, the airport was able to deal effectively with the incident and return to normality in a staggeringly short period of time.

Gillies Crichton, Head of Assurance, BAA Glasgow Airport, said:

"We were able to demonstrate that Business Continuity Management is an essential part of our ongoing lives… the unthinkable can and does happen, you need to be prepared for it."

The recovery was recognised at the Business Continuity Awards in May 2008, winning the 'Business Continuity Recovery of the Year'.

Overview

Business Continuity (BC) is now a recognised feature in a great many organisations in both public and private sectors. It has evolved considerably over the past few years from a position where it was little more than just IT disaster recovery by another name, to now being a continuing management process that embraces the vertical needs of the entire organisation (between the furthest points in the entire up / down stream supply chain) to more horizontal issues such as reputation and staff welfare.

BC is fundamentally different from disaster recovery, as it focuses on the continuance of key operations, rather than just trying to recover what's left after a disaster. It also differs from insurance, which is chiefly about compensation rather than continuity. It is therefore more realistic to refer to the title of this chapter as BC Management (BCM) rather than just BC or planning, albeit a workable plan is a key deliverable of the BC process.

When properly applied, BCM embraces, or at least works in complete harmony with, the following business and business support functions, irrespective of private or public sector application:

- Facilities management;
- Risk management;
- Physical security;
- IT security;
- Crisis management;
- Staff welfare;
- Operations; and
- Finance.

In the UK, BS: 25999 is the definitive standard for Business Continuity Management (BCM). Readers are encouraged to read this Standard, which is available directly from the BSI website (www.bsi-uk.com). It offers

an accepted framework for incident anticipation and response with a series of recommendations for good practice. Some of the following text is therefore taken directly from the BSI Standard, as well as the British Government UK Resilience website (which contains advice and links on BCM and many other resilience issues) along with up-to-date information from the Business Continuity Institute and guidance from the author of this chapter, who frequently runs BC and Crisis Management workshops. BS: 25999 is in two parts:

- *BS: 25999 Part one*. This Standard establishes the process, principles and terminology of BCM, providing a basis for understanding, developing and implementing business continuity within an organisation and to provide confidence in business-to-business and business-to-customer dealings.
- *BS: 25999 Part two*. BS: 25999-2 specifies requirements for establishing, implementing, operating, monitoring, reviewing, exercising, maintaining and improving a documented BCM System (BCMS) within the context of managing an organisation's overall business risks.

The requirements specified in BS: 25999-2 are generic and intended to be applicable to all organisations (or parts thereof), regardless of type, size and nature of the business. The extent of application of these requirements depends on the organisation's operating environment and complexity. Therefore the design and implementation of a BCMS to meet the requirements of this Standard will be influenced by regulatory, customer and business requirements, the products and services, the processes employed and the size and structure of the organisation. It will not be the intent of this British Standard to imply uniformity in the structure of a BCMS but for an organisation to design a BCMS to be appropriate to its needs and that meets its stakeholder's requirements.

BS: 25999-2 can be used by internal and external parties, including certification bodies, to assess an organisation's ability to meet its own business continuity needs, as well as any customer, legal or regulatory needs.

Business continuity management (BCM) is a process that helps manage risks to the smooth running of an organisation or delivery of a service, ensuring continuity of critical functions in the event of a disruption, and effective recovery afterwards. The Government aims to ensure all organisations have a clear understanding of BCM and its importance, and foster discussion of how best to achieve business continuity. This chapter outlines the importance of BCM and how it should be carried out.

Business Continuity Management

BCM is a generic management framework that is valid across the public, private and voluntary sectors. It is an ongoing process that helps organisations anticipate, prepare for, prevent, respond to and recover from disruptions, whatever their source and whatever aspect of the business they affect. The primary 'business' of private sector organisations is the generation of profit, a process that BCM seeks to protect. Other organisations provide services to the public, and it is equally important that these are protected and resilient.

BCM is a business-owned, business-driven process that establishes a fit-for-purpose strategic and operational framework that:

■ proactively improves an organisation's resilience against the disruption of its ability to achieve its key objectives;

■ provides a rehearsed method of maintaining or restoring an organisation's ability to supply its critical products and services to an agreed level within an agreed time after a disruption; and

■ delivers a proven capability to manage a business' disruption and protect the organisation's reputation and brand.

While the individual processes of BCM can change with an organisation's size, structures and responsibilities, the basic principles remain exactly the same for voluntary, private or public sector organisations, regardless of their size, scope or complexity.

BCM linked to mission and strategy

All organisations, whether large or small, have aims and objectives, such as to grow, to provide services and to acquire other businesses. These aims and objectives are generally met via strategic plans to achieve an organisation's short-, medium- and long-term goals. BCM understanding at an organisation's highest level will ensure that these aims and objectives are not compromised by unexpected disruptions.

The consequences of an incident vary and can be far-reaching. These consequences might involve loss of life, loss of assets or income, or the inability to deliver products and services on which the organisation's strategy, reputation or even survival might depend. BCM needs to recognise the strategic importance of perception, welfare and the interest of all stakeholders.

Examples of stakeholders include, but are not restricted to, internal and 'outsourced' employees, customers, suppliers, distributors, investors, insurers and shareholders. Furthermore, as consequences of a disruption unfold, new stakeholders emerge and have a direct impact on the eventual extent of the damage. Examples of these include competitors, environmentalists, regulators and the media.

BCM is complementary to a risk management framework that sets out to understand the risks to operations or business, and the consequences of those risks. Risk management seeks to manage risk around the critical products and services that enable an organisation to survive. Product and service delivery can be disrupted by a wide variety of incidents, many of which are difficult to predict or analyse by cause. However, risk management relates directly to insurance and where a well assembled and audited BCM structure exists it should be possible to explore a reduction in insurance premiums by virtue of now being able to reduce your exposures and demonstrate acceleration in post-crisis business resumption.

BCM and insurance

- BCM is often seen by insurers as a means to improve the quality of the business they are underwriting, and confirms that BCM helps organisations mitigate impact, recover faster and minimise losses.
- BCM can be used to protect against losses incurred through traditionally non-insurable perils such as supplier insolvency or pandemic influenza.
- BCM can be used to better understand the requirements for Business Interruption (BI) cover (and potentially lower the amount of cover needed).
- BCM can help obtain BI cover where otherwise it would not be available.
- BCM can help to secure optimal terms for cover.

By focusing on the impact of disruption, BCM identifies those products and services on which the organisation depends for its survival, and can identify what is required for the organisation to continue to meet its obligations. Through BCM, an organisation can recognise what needs to be done before an incident occurs to protect its people, premises, technology, information, supply chain, stakeholders and, above all, reputation. Indeed, a crisis well handled can actually increase the value of the organisation and enhance reputation. Moreover, an organisation with appropriate BCM measures in place might be able to take advantage of opportunities that have a high risk.

BCM forms an important element of good business management, service provision and entrepreneurial prudence within which managers and owners have the responsibility to maintain the ability of the organisation to function without disruption.

Organisations constantly make commitments or have a duty to deliver products and services; i.e. they enter into contracts and otherwise raise expectations. All organisations have moral and social responsibilities, particularly where they provide an emergency response or a public or voluntary service. In some cases, organisations have statutory or regulatory duties to undertake BCM.

All business activity is subject to disruptions, such as technology failure, flooding, utility disruption and terrorism. BCM provides the capability to adequately react to operational disruptions while protecting welfare and safety. You should not regard BCM as a costly planning process, but as one that adds value to the organisation. In other words, it should not be thought of as a 'grudge purchase'.

BC plans (BCPs)

Having first established a management framework for BCM, it follows that some form of 'user friendly' plan should be written and be constantly available, plus be regularly updated and exercised.

All organisations should carefully consider the case for BCPs as a systematic basis for managing the continuity of critical functions and recovery of the organisation from disruption. Good BC planning may require both generic and specific plans. A generic plan is a core plan that enables an organisation to respond to a wide range of possible scenarios, setting out the common elements of the response to any disruption (e.g. invocation procedure, command and control, access to financial resources). Within the framework of the generic plan, specific plans may be required in relation to specific risks, sites or services. Specific plans provide a detailed set of arrangements designed to go beyond the generic arrangements when these are unlikely to prove sufficient.

BCPs should be based on systematic identification and assessment of the significant risks of an emergency occurring in an organisation's area. Identifying the risks threatening the performance of critical functions in the event of an emergency will enable organisations to focus resources in the right areas, and develop appropriate plans.

A BCP cannot be considered reliable until it is exercised and has proved to be workable, understood by all involved and not just a bundle of papers that might be designed to protect the author.

Exercising should involve:

- validating plans;
- rehearsing key staff; and
- testing systems that are relied upon to deliver resilience (e.g. uninterrupted power supply).

The frequency of exercises will depend on the organisation, but should take into account the rate of change (to the organisation or risk profile), and outcomes of previous exercises (if particular weaknesses have been identified and changes made).

It is important to ensure that relevant people across the organisation – and in other organisations where appropriate – are confident and competent concerning the plan. A training programme should be developed and run for those directly involved in the execution of the BCP, should it be invoked.

Organisations should not only put BCPs in place, but should ensure they are reviewed regularly and kept up to date. Particular attention may need to be paid to staff changes, changes in the organisation's functions or services, changes to the organisational structure, details of suppliers or contractors, changes to risk assessments and business objectives / processes.

Creating BCM in your organisation
The BCM lifecycle comprises six elements explained below. These can be implemented by organisations of all sizes, in all sectors – public, private, non-profit, educational, manufacturing, etc. The scope and structure of a BCM programme can vary, and the effort expended will be tailored to the needs of the individual organisation, but these essential elements still have to be undertaken.

1. *Understanding the organisation.*
   The activities associated with doing this properly provide information that enables prioritisation of an organisation's products and services, and the urgency of the activities that are required to deliver them. This sets the requirements that will determine the selection of appropriate BCM strategies.

2. *BCM programme management.* This enables the BC capability to be both established and maintained in a manner appropriate to the size and complexity of the organisation – and all its dependencies.

3. *Determining business continuity strategies.* This enables a range of strategies to be evaluated and allows an appropriate response to be chosen for each product or service, such that the organisation can continue to deliver those products and services at an acceptable level of operation and within an acceptable timeframe during and following a disruption. The choice made will take account of the resilience and countermeasure options already present within the organisation.

4. *Developing and implementing a BCM response.* Developing and implementing a BCM response results in the creation of BCPs and crisis management plans that detail the steps to be taken during and after an incident to limit damage to reputation and restore operations.

5. *Exercising, maintaining and reviewing BCM arrangements.* This leads to the organisation being able to demonstrate the extent to which its strategies and plans are complete, current and accurate; and identify opportunities for improvement.

6. *Embedding BCM in the organisation's culture.* This enables BCM to become part of the organisation's core values, and instils confidence in all stakeholders in the ability of the organisation to cope with disruptions.

Creating a BCM policy

The BCM policy should, ideally, define the following processes:

■ The set-up activities for establishing a BC capability.

■ The ongoing management and maintenance of the BC capability.

■ The set-up activities incorporating the specification, end-to-end design, build, implementation and initial exercising of the BC capability.

■ The ongoing maintenance and management activities include embedding BC within the organisation, exercising plans regularly, and updating and communicating them, particularly when there is significant change in premises, personnel, process, market, technology or organisational structure.

You should ensure that the BCM policy is appropriate to the nature, scale, complexity, geography and criticality of the company's activities and that it reflects its culture, dependencies and operating environment. The BCM policy defines the process requirements to ensure that business continuity arrangements continue to meet the needs of the organisation in the event of an incident. This policy should ensure that a business continuity capability is promoted within the organisation's culture and the growth and development of the organisation's products and services. The BCM capability should be integrated into the organisation's change management activity.

You should also develop a BC policy that states the objectives of BCM within the organisation. Initially, this may be a high level statement of intent, which is refined and enhanced as the capability is developed.

The BC policy should provide the organisation with documented principles to which it will aspire and against which its business continuity capability should be measured. The BCM policy should be owned at a high level, e.g. a board director or elected representative. You may consider the following useful when developing a BCM policy:

- Defining the scope of BCM within the organisation.
- BCM resourcing.
- Defining the BCM principles, guidelines and minimum standards for the organisation.
- Referencing any relevant Standards, Regulations or policies that have to be included or can be used as a benchmark.

Any organisation should maintain and regularly review its BCM policy, strategies, plans and solutions on a regular basis in line with the organisation's needs. The scope of the BCM policy should clearly define any limitations or exclusions that apply, e.g. geographical or product exclusions. But however BCM is resourced, there are key activities that should be carried out both initially and on an ongoing basis. These may include:

- Defining the scope, roles and responsibilities for BCM;
- Appointing an appropriate person or team to manage the ongoing BCM capability;
- Keeping the business continuity programme current through good practice;
- Promoting business continuity across the organisation and wider, where appropriate;
- Administering the exercise programme;
- Coordinating the regular review and update of the BC capability, including reviewing or reworking risk assessments and business impact analyses (BIAs – see below);
- Maintaining documentation appropriate to the size and complexity of the organisation;
- Monitoring performance of the BC capability;
- Managing costs associated with the BC capability; and
- Establishing and monitoring change management and succession management regimes.

Preparing for immediate media interest

A well structured BCM process should recognise that how you are portrayed by TV, radio and the press will, to a very large extent, be decided by how well you have prepared for this and fully understand what the media will expect (sometimes demand) of you during any real or perceived crisis. It follows that your media response should be realistic, exercised and documented, which should include:

- An incident communications strategy. A press holding statement might have to be ready in 30 minutes, otherwise speculation will fill in for facts.
- The organisation's preferred interface with the media. Reporters and journalists do not want to speak to spin doctors or third party agencies. They want a senior figure who can appear honest and make sense of what is going on.
- A guideline or template for the drafting of a statement to be provided to the media at the earliest practicable opportunity following the initial incident.
- Appropriate numbers of trained, competent spokespeople nominated and authorised to release information to the media. Ideally, a cadre of suitable trained executives should always be contactable – and a decision taken in advance on what any manager might, for example, say if 'door-stepped' by a reporter. Saying 'no comment' is not a good idea.
- Establishment, where practicable, of a suitable venue to support liaison with the media, or other stakeholder groups.

In some cases, it may be appropriate to provide supporting detail in a separate document, establish an appropriate number of competent, trained people to answer telephone enquiries from the press, prepare background material about the organisation and its operations (this information should be pre-agreed for release) and ensure that all media information is made available without undue delay.

Business impact analysis and determining critical activities

You should determine and document the impact of a disruption to the activities that support critical products and services of your organisation. This process is commonly referred to as a business impact analysis (BIA). For each activity supporting the delivery of critical products and services within the scope of its BCM programme, the organisation should:

- assess over time the impacts, both in financial and operational terms, that the loss or disruption of the activity would have; and
- establish the maximum tolerable period of disruption of each activity by identifying the maximum time period after the start of a disruption within which the activity needs to be resumed, taking into account any business cycle implications, the minimum level at which the activity needs to be performed on its resumption, the length of time within which normal levels of operation need to be resumed and identify any inter-dependent activities, assets, supporting infrastructure or resources that have also to be recovered over time.

When assessing impacts, you should consider those that relate to the business aims of your organisation and objectives and its stakeholders. These may include:

- the impact on staff or public wellbeing;
- damage to, or loss of, premises, technology or information;

- the impact of breaches of statutory duties or regulatory requirements;
- damage to reputation – yours and your stakeholders;
- damage to financial viability;
- deterioration of product or service quality; and
- environmental damage.

The organisation should document its approach to assessing the impact of disruption and its findings and conclusions.

During a disruption, impacts often increase over time and affect each activity differently. Impacts might also vary depending on the day, month or point in the business lifecycle.

You might assess risk according to (a) likelihood and (b) impact and then categorise activities according to their priority for recovery.

Those activities whose loss, as identified during the BIA, would have the greatest impact in the shortest time and which need to be recovered most rapidly may be termed 'critical activities'. You may then wish to focus your planning activities on critical activities, but should recognise that other activities will also need to be recovered within their maximum tolerable period of disruption and might also require advance arrangements to be in place.

The maximum time period for resuming activities can vary between seconds and several months, depending on the nature of the activity. Activities that are time-sensitive might need to be specified with a great degree of accuracy, e.g. to the minute or the hour. Less time-sensitive activities might require less accuracy.

The maximum tolerable period of disruption will influence each activity's recovery time objective when determining BCM strategies.

Exercising

Nothing correlates to how well your organisation will perform in a real drama as much as how accurately you previously exercised your BCM structure in advance. Not only that, but to have learned the lessons from any exercise so that BC becomes a process that can be moved into at the first opportunity to ensure you are ready to not only react, but predict where any crisis might be heading and take avoidance steps without delay.

Exercises or tests can be either desktop or walk through in style where little or no communications are actually put to the test, but a group of key people sit around a table while a facilitator puts an unfolding scenario to them to explore how things might go if it was real. These sessions normally last just a few hours and might be made much more realistic by showing pre-recorded mock TV news broadcasts, etc. However, the aim is not to put people into a gratuitous pressure test, but to explore in a fairly gentle way where any gaps might exist – and to identify them by leaving enough time to hold a thorough and immediate de-brief.

More complex exercises might follow once obvious problems have been resolved following a desktop walk through. These more structured sessions could involve actual working from home, invocation to IT hot sites and your executive layer facing a simulated TV news camera crew.

Any exercise should be realistic, carefully planned, and agreed with stakeholders, so that there is minimum risk of disruption to business processes. An exercise should also be planned such that the risk of an incident occurring as a direct result of the exercise is minimised.

Every exercise should have clearly defined aims and objectives. A post-exercise debriefing and analysis should be undertaken that considers the achievement of the aims and objectives of the exercise. A post-exercise report should be produced that contains recommendations and a timetable for their implementation.

It follows that where exercises show significant deficiencies or inaccuracies in the BCM structure they should be rerun after corrective actions have been completed.

See also: Bomb alerts, p.56; Emergency procedures and crisis management, p.261.

Sources of further information

Business Continuity Institute: www.thebci.org

British Standards Institute: www.bsonline.bsi-global.com

The Institute of Risk Management: www.theirm.org

Business rates

Alex Stevens, GVA

Key points

- Business rates, or non-domestic rates, are a tax on almost all non-domestic property, in addition to some items of rateable plant and machinery. In England and Wales the new rating system came into force on 1 April 1990. Since then we have had five year cycles of rating revaluations.
- The current rating list for England, Wales and Scotland came into force on 1 April 2010 and ceases to exist on 31 March 2015. Northern Ireland deferred its revaluation due from 1 April 2010 and has not had one since 1 April 2003. In all four countries the system is broadly similar, but different statutes exist in Scotland and Northern Ireland and a number of the Welsh Regulations are different to those in England. Case law is relevant to all countries.

Legislation

The most important statutes for legislation for England and Wales are:

- Local Government Finance Act 1988.
- Rating (Valuation) Act 1999.
- Rating (Empty Properties) Act 2007.

The Local Government Finance Act provides the framework for the legislation but the detail is in the Regulations. The Regulations concerning appeals and charging change every five years and since 1988 nearly 500 Regulations have come into force subsequently in England and Wales. A number have been superseded and overridden.

The Rating (Valuation) Act changed the definition of rateable value. Since the Act the assumption is that the landlord puts the property into repair prior to a letting. The tenant then takes a full repairing and insuring lease from year to year but with a prospect of continuance. Prior to the change a ratepayer had successfully argued for a discount on a vacant property in poor repair at Lands Tribunal and the Valuation Officer's attempt to overturn the decision at the House of Lords had failed.

The Rating (Empty Properties) Act 2007 changed the level of charge for vacant properties and void periods after 1 April 2008 and added nearly £1bn of revenue to the Exchequer in 2008/09.

Exemptions

Almost all property and some plant and machinery is liable to be rated. The main exemptions include agricultural land, some places of religious worship, parks and sewers. Other utilities such as electricity, gas and oil pipelines and telecommunication networks are rateable. A full list of exemptions is at Schedule 5 to the Local Government Finance Act.

Rateable Value (RV) and Rates Payable

In all four countries the rateable value is supposed to be the hypothetical rental value two years before each rating list comes into force. For the 2010 rating list we value at 1 April 2008.

Rates payable, in the absence of transition, are: Rateable Value (RV) x Uniform Business Rates (UBR).

The UBR increases each year in line with inflation. At the end of each five-year cycle it is rebased so as to raise the same amount in revenue in real terms. The tax is thus an exceptionally stable one and currently rises in the region of £25bn per annum in England and Wales from 1.8 million properties.

Uniform Business Rate and transition

In recent years Wales and Northern Ireland have not had a transition scheme and in Scotland a decision was made not to have a transition scheme for the 2010 list. The purpose of the transition scheme is to cushion large increases in rate liability between rating lists. This gives certainty to the rate payer. The Governments like to make it self-financing and so there is transition downwards as well as transition upwards. Consequently, if a rate payer has a large fall in rateable value the benefits are considerably smaller than they should be.

For the 2011/12 rate year the UBR in England and Scotland has been set at 42.6p and in Wales 42.8p. Northern Ireland has no Uniform Business Rate and each Local Authority charges its own rate poundage. At present the rate poundages are around 57.5p in the pound.

On top of the Uniform Business Rate a number of supplements may be payable. All 'large' properties in England and Scotland pay a small business rate supplement, which is currently 0.7p in the pound. Other supplements include business improvement district (BIDS) and typically the supplementary rate will be 2p. The most widespread supplementary rate is Crossrail, which was introduced from 1 April 2010 and has been set at 2p in the pound for all rateable values in Greater London exceeding £55,000 and is expected to last over 30 years.

Transition is complex and the increases and decreases in real terms differ depending on the size of the rateable value. A large property is one with a rateable value in excess of RV £18,000 outside London and RV £25,500 in Greater London. The maximum increases and decreases in real terms each year are:

It should be noted that there is no transition on the supplementary rates and this complicates the calculation of rate liability. Between 2009/10 and 2010/11 the Uniform Business Rates fell from 48.1p to 40.7p, due to an increase in rateable values following the revaluation. The fall is approximately 15%. If a rateable value stayed static between the two rating lists

Year	Increases		Decreases	
	Large	Small	Large	Small
2010/11	12.5%	5%	4.6%	20%
2011/12	17.5%	7.5%	6.7%	30%
2012/13	20%	10%	7%	35%
2013/14	25%	15%	13%	55%
2014/15	25%	15%	13%	55%

and it exceeds RV £18,000 outside London it will take until year three, 2012/13, before RV x UBR becomes payable due to the restriction on decreases.

Reliefs

There are a number of reliefs from full liability and they include the following.

Empty property

Non-domestic rates were historically a tax on occupation. From 1 April 2008 the Chancellor decided to raise approximately an additional £1bn and thus changed the rules for Empty Rate Relief in England and Wales. The rules for England and Wales since 2008 are that shops and offices receive a three-month void following vacation of a property and then they pay full rate liability. Industrial and warehouse properties receive a six-month void following vacation and then pay full rates. Prior to the alteration, shops and offices paid at 50% after the three-month void and industrial and warehouses had no empty rate liability. Scotland and Northern Ireland continue to grant the historic reliefs enjoyed in England and Wales, i.e. 50% after a three-month void on a shop or office, etc.

There are exceptions to this rule and the main ones for England and Wales are:

- Listed buildings – pay no rates.
- Small properties with RV of £2,599 or less pay no empty rates.

Since 2008 a new industry has been created to mitigate empty rate liability. This includes occupations of six weeks or more, charitable occupation and statutory exemptions. It is worthwhile seeking advice as to how best mitigate liability. This industry has resulted in relief costing the Government approximately £500m in 2008/09 but £1.2bn in 2009/10.

Small Business Rates Relief

A small business is designated as being one with a rateable value of less then RV £18,000 outside London and RV £25,500 inside Greater London, the same threshold as for transition relief. The aggregate rateable value of each rate payer is considered and so large businesses with many small rateable values do not benefit from the scheme. Rateable values between RV £12,000 and RV £18,000 outside London or RV £25,500 in London do not pay the 0.7p small business rate supplement. Below £6,000 the supplement is removed and 50% relief is given. Between £6,000 and RV £12,000 a sliding scale of reliefs exists.

Charities and Community Amateur Sport Clubs

These receive mandatory relief of 80% and Local Authorities have discretion to give an additional 20% relief or indeed relief where the 80% mandatory relief does not apply. The Community Amateur Sports Club rules are so tightly drawn that very few clubs sign up. Strict guidelines are given to Local Authorities for additional discretionary rate relief and the Local Authority funds most of the relief themselves. Sports clubs are typical examples of where discretionary relief may be applied.

Rural relief

This exists for villages with populations of less then 3,000 for sole village shops, public houses and petrol filling stations.

Checking rateable values and communications

This can be done by logging on the Valuation Office website for England and Wales, the Scottish Assessors website for Scotland and the Lands and Property Services website in Northern Ireland. The addresses are listed in *Sources of further information*.

It should be noted that if the VO or Scottish Assessor change the rateable value they serve notices on the ratepayer. If they wish to seek tenure details they serve notices on the ratepayer. The Scottish questions are brief, whereas in England and Wales the bulk classes (industrial, shops and offices) form of return VO 6003 is very detailed. If new lettings have taken place incentives should be included otherwise the rateable value ascribed to the property may be high. Fines can be charged for late return but rarely are. It is often best to seek professional advice when sending back returns to make sure no prejudicial information is included and all information that will assist in reducing the rateable value is included.

Business rate appeals

Appeals against the 1 April 2010 rating list entry or in response to valuation office notice can be made at any time in England and Wales up until 31 March 2015 but are date-restricted in Scotland to within six months of either event. In Scotland any appeal against the 1 April 2010 entry had to be made by 30 September 2010. A new rate payer also has the right of appeal but only within six months of becoming the rate payer in Scotland. The VO in England and Wales has until the first anniversary of the rating list ending to make amendments and if he does these near the end, or after the rating list has ceased to be in force, the ratepayer has six months in which to lodge an appeal.

Additional appeals can be made for a material change of circumstance. A change needs to be a physical change adversely affecting the value of the property. An example is the demolition and new construction of an adjoining office building or a new shopping centre weakening other pitches in the same town centre. It is advisable to lodge appeals at the outset of an adverse physical change as the date the appeal is lodged is a material consideration and the adverse factor needs to last six months or thereabouts from the date of the appeal if the appeal is to be successful. Substantial and worthwhile savings can be made.

The mechanics of the rating system

The Local Authority collects business rates and passes it on to Central Government, after which it is redistributed according to need. The Local Authority is responsible for telling the Valuation Office, a part of the Inland Revenue, about newly completed buildings, extensions and other changes so that the rating list can be kept up to date at all times.

A rating appeal is to the Valuation Officer (VO) and if unresolved is entered into a programme for discussion. The majority of appeals are then settled by negotiation. The process is slow and typically takes around two years if appeals are lodged near the outset of the rating list. Thereafter, the time period should be swifter. Once all the research has been carried out and initial appeals settled, subsequent appeals take far less time to conclude.

If a settlement cannot be reached the appeal is heard by a Valuation Tribunal for England (VTE), which comprises three members with a clerk who has advised on procedural and other matters. Most hearings are oral and typically the ratepayer's agent has acted as both an Expert Witness and an Advocate against a single VO who has had a dual role. Procedures for the 2010 rating list appeals have changed and Statement of Cases, Experts' Reports and a Statement of Agreed Fact will be required prior to the Tribunal hearing. The clerk's role has changed and Regulations came into force in October 2009 that changed the system.

Professional advice should be sought if a settlement cannot be reached as the rules and formality of Tribunals are becoming more complex to adhere to. Any appeal from the VTE decision is to the Upper Tribunal (Lands Chamber). The Upper Tribunal is the highest valuation court and any appeal is direct to the Court of Appeal on a point of law only.

In Scotland it has always been advisable to have a Barrister and Expert Witness for a hearing and the hearings are far more formal. Scottish appeals can either be referred direct to the Lands Tribunal or heard by the Valuation Appeal Committee, and any appeals beyond either Tribunal are to the Lands Valuation Appeal Court.

What should the ratepayer look out for?

- Always check the rate bill is accurate and if in doubt about transition seek professional advice (some bills have been found to be inaccurate by in excess of £1m due to the complexity of transition and poor Local Authority software).
- Audit historic liability to ensure no overpayments have been made or credits not claimed.
- Ensure reliefs are claimed, including voids when property is vacant.

- Check the floor areas on the Valuation Officer's website are correct or low.
- Check the rates per square metre adopted are fair when compared to nearby comparable property.
- Look out for material changes adversely affecting the property and lodge appeals early.
- Consider how any rent interrelates with the rating assessment.
- If recent improvements have been carried out, check the rateable value was increased from the date they were completed as appeals can lead to increases as well as decreases and the VO does not always pick up on improvements.
- Take trouble in filling out any form of return (VO 6003) as the information can prejudice the ratepayer if it is inaccurate, incomplete or excess information is given to the VO.
- Seek professional advice if in doubt as most agents offer a no win, no fee service.

See also: Dilapidations, p.221; Landlord and tenant: lease issues, p.429; Landlord and tenant: possession issues, p.438.

Sources of further information

Business Link: www.businesslink.gov.uk

The Valuation Office Agency: www.voa.gov.uk

Scottish Assessors Association: www.saa.gov.uk

Land and Property Services: www.lpsni.gov.uk

Buying and selling property

Kevin Boa, Pinsent Masons Property Group

Key points

- A buyer will want to ensure it acquires a property with good and marketable title through due diligence.
- A survey and an environmental audit are recommended.
- Until contracts are exchanged, either party can withdraw without penalty. After contracts are exchanged, both parties are legally bound.
- A buyer is generally responsible for insurance as soon as contracts are exchanged.

Legislation

- Law of Property Act 1925.
- Law of Property (Miscellaneous Provisions) Act 1994.
- Land Registration Act 2002.

This chapter discusses the acquisition and disposal of freehold land and buildings.

A purchase or sale is best described by outlining the various stages involved. This procedure applies in England and Wales. The law and procedure are different in Scotland and are outside the ambit of this chapter.

The pre-contract stage

Firms of chartered surveyors usually deal with the disposal of commercial property. It is not essential to buy or sell through a commercial surveyor, but such firms have knowledge of market conditions.

Surveyors will often agree 'heads of terms' setting out what has been agreed between the seller and the buyer and covering key points including the parties, property, price, timetable and any particular conditions.

A buyer will want to acquire a property that has good and marketable title, as otherwise the buyer might be concerned about its value and the ability to dispose

of the property in the future. Without good and marketable title it might also be difficult to raise finance on the strength of the security of the property. This involves the process of due diligence, i.e. the investigation of the property by the buyer before entering into the contract. There are three possible approaches:

1. A full title investigation is carried out by the buyer's solicitors.
2. A certificate of title is given by the seller's solicitors.
3. Warranties are given by the seller in the contract.

Additionally, there may be a combination of any of these.

Title investigation

This involves the buyer's solicitor investigating the seller's legal title to the property and checking whether there are any restrictive covenants or other matters affecting the property. In addition, pre-contract enquiries will be raised with the seller's solicitors and various searches carried out against the property.

The buyer should also arrange for a survey to be carried out to ensure that the property is sound and that the sale price is not excessive. The buyer will need to consider investigating for environmental contamination, particularly if the property has a history of industrial use.

Certificate of title

Instead of the buyer making a full investigation of the property, the seller's solicitor could produce a certificate of title, confirming that the property has good and marketable title and disclosing all information it has in respect of the property.

Unless the seller gives a warranty as to the accuracy of the certificate, the buyer will in essence be relying on the seller's solicitor's professional indemnity policy.

A certificate is useful if there is insufficient time for the buyer to carry out its investigation, if a large number of properties are involved, and where the seller's solicitor acted on the original purchase and so is already in possession of much of the information needed to complete the certificate.

The terms of the certificate will still need to be negotiated and the statements given may be qualified or disclaimers imposed.

Warranties

The seller can give warranties about the property to the buyer, which provide the latter with a form of insurance. Ideally they should protect the buyer in respect of the presence of factors relating to the property of which it is unaware and which detrimentally affect the property.

Warranties are vital if the property is acquired as part of a share acquisition, to cover the unknown liabilities, actual or contingent, and therefore go beyond matters of pure title.

However, the buyer will need to consider whether the warranties are worth the

paper they are written on as it is obviously essential that the seller is financially able to satisfy any claim under the warranties.

The warranties are usually qualified by a separate disclosure letter, which will be annexed to the contract. If an item is excluded from the warranties, the risk passes to the buyer. It is therefore essential that there is also an express warranty in the contract that the disclosure letter is accurate.

There may also be some general disclaimers, e.g. that the buyer is aware of all matters apparent from a physical inspection of the property (which means that a survey should be commissioned).

The contract

Most disposals or acquisitions are preceded by exchange of contracts. Until contracts are exchanged, neither party is committed to the transaction and either may withdraw at any time.

A contract is a binding agreement between the buyer and the seller, which sets out all the relevant details and the agreed terms. Sometimes completion is conditional, e.g. on planning permission being granted.

Once contracts are exchanged, neither party may withdraw from the transaction without being in breach of contract and facing a possible claim for damages or specific performance. Specific performance is an order by the Court requiring the party in default to complete the transaction. It is a discretionary remedy and will normally be awarded only where damages alone are not regarded as being an adequate remedy.

Contracts should not be exchanged until any pre-contract investigations and surveys have been made and the buyer's funding is in place.

If there is a chain of related sales and purchases (common in residential transactions), then all the parties in the chain need to be ready to exchange at the same time. This can lead to delay.

On an exchange of contracts, a deposit is usually paid by the buyer; normally 10% of the purchase price. It is usual for the deposit to be held by the seller's solicitor pending completion, although in certain circumstances the deposit may be used for an onward purchase by the seller. The buyer may lose its deposit if it does not complete the purchase in accordance with the terms of the contract.

Unless the contract says otherwise, the buyer is responsible for insuring the property as soon as contracts are exchanged.

Completion

Completion is when legal ownership is transferred. It is at this stage that the balance of the purchase price is paid over and possession of the property is given to the buyer. The completion date is fixed in the contract and traditionally is four weeks after the exchange, although it can be much earlier or later (depending on what the parties agree).

The transfer

The transfer is the document that formally transfers the property from the seller to the buyer. The transfer describes the property and any rights, restrictive covenants, shared facilities, rights of way and other matters that affect the property and subject to or with the benefit of which the buyer will take the property.

Where the property is sold subject to covenants (such as restrictions on the use of the property), the buyer will usually agree in the transfer to observe and

perform these covenants and to indemnify the seller for any loss caused to the seller if the buyer fails to comply with them.

Stamp duty land tax

Stamp duty land tax (SDLT) is payable by the buyer on the transaction. A wide range of transactions are subject to SDLT – essentially most dealings with an interest in land. The rate of SDLT varies according to the price paid. Transactions not exceeding a value of £125,000 (for residential property, or £250,000 for first-time buyers) or £150,000 (for commercial property) are exempt from SDLT. The maximum rate is 4% where the price is more than £500,000 (for both residential and commercial property).

On commercial property sales, value added tax at 17.5% (20% from 4 January 2011) may be chargeable on the purchase price, and SDLT in such cases is charged on the VAT-inclusive figure.

After completion

The transfer must be sent to the Land Registry so that the buyer can be registered as the new owner of the property. A registration fee is payable, which is calculated according to the price paid. The maximum fee is currently £920. Once registration is complete, the Land Registry will issue a title information document, which provides that the buyer is the owner of the property and also includes a plan of the property and details of any matters affecting the property, such as rights of way and restrictive covenants. Upon registration the register is deemed conclusive; however, it is recommended that older title deeds are retained to assist in resolution of disputes involving rights of way, boundaries or covenants, and in case rectification of the register is required.

Acquisition of a property as part of a business purchase

Sometimes a property is acquired as part of the purchase of a business. This can happen when a buyer acquires all the shares in a company. The buyer steps into the shoes of the seller so that there is simply a change in control rather than a change in the owner of the business (and as a consequence of any property owned by the company). As there is no change in ownership, the buyer will take on all existing liabilities of the company in respect of the property.

Although searches will be carried out, enquiries raised and a contract entered into, there is no need for a transfer to be completed (because there is no change in the name of the owner) and consequently no need for a Land Registry application to be made. There are various tax and other considerations on a share purchase, including possible stamp duty savings, as stamp duty on a share purchase is currently fixed at 0.5% of the price. Specialist advice on how to structure the acquisition or disposal of the property is recommended.

See also: Business rates, p.102; Planning procedures, p.585; Property disputes, p.600.

Sources of further information

The Workplace Law website has been one of the UK's leading legal information sites since its launch in 2002. As well as providing free news and forums, our Information Centre provides you with a 'one-stop shop' where you will find all you need to know to manage your workplace and fulfil your legal obligations.

It covers everything from CDM, waste management and redundancy regulations to updates on the Carbon Reduction Commitment, the latest Employment Tribunal cases and the first case to be tried under the Corporate Manslaughter and Corporate Homicide Act, as well as detailed information in key areas such as energy performance, equality and diversity, asbestos and fire safety.

You'll find:

- quick and easy access to all major legislation and official guidance, including clear explanations and advice from our experts;
- case reviews and news analysis, which will keep you fully up to date with the latest legislation proposals and changes, case outcomes and examples of how the law is applied in practice;
- briefings, which include in-depth analysis on major topics; and
- WPL TV – an online TV channel including online seminars, documentaries and legal updates.

Content is added and updated regularly by our editorial team who utilise a wealth of in-house experts and legal consultants. Visit www.workplacelaw.net for more information.

Comment ...

A sustainable future

David Sharp, Workplace Law Environmental

It's been reported that environmental management is the fastest growing area of regulation at present in the UK, and next year looks to be no exception. The current consultation on the Site Waste Management Plans Regulations 2008 is likely to have an impact on construction and refurbishment works taking place in the latter part of next year.

It will also be worth looking out for the result of the consultation on the code of practice relating to the waste management 'duty of care', which gives greater clarity to waste-holders to allow them to discharge their obligations.

The Duty of Care requires the holder of the waste, which can be the waste producer, carrier or disposer, to take all reasonable steps to ensure there is no unauthorised deposit, treatment, keeping or disposal.

This includes ensuring it does not escape from your control, is only transferred to an authorised person, and that any transfer of waste is accompanied by a written description.

Waste management will undoubtedly cause employers the most problems over the next few years. It's one of the biggest areas of non-compliance when it comes to undertaking an environmental audit. Our clients include tenants, landlords, building owners and building managers, who have different responsibilities when it comes to waste management regulations. Despite the fact that the legislation has been around for some time, the requirements to

David Sharp is the founder and Managing Director of Workplace Law. David is an experienced commentator on the regulation of the workplace, a regular contributor to industry magazines, and is the co-author of the RIBA Good Practice Guide to Employment. He is a specialist advisor to the Facilities Management Association.

use a registered broker for the transfer of waste still seem to have passed some people by.

The drivers of regulation in the environmental management sector have been carrot and stick. The carrot is resource efficiency (such as carbon reduction, sustainable development) designed to motivate improved business performance and help meet government targets. The stick consists of a combination of 'economic instruments' (trading schemes, taxation) and tougher regulation, which has seen an increase in penalties and the advent of civil sanctions for waste offences.

In the coming year, I can see businesses struggling with the Waste (England and Wales) Regulations 2011, which will require them to demonstrate that the waste hierarchy has been applied on transfer / consignment notes. The Regulations came into force in September 2011 and might slip through the net.

The environmental side of our business has grown incredibly over the past two years, and the most pleasing thing to date has been the praise we received from IEMA for the quality of our submission to teach the IEMA Foundation Certificate.

I'm also extremely proud of being shortlisted for the E-learning Age E-learning Awards 2011 for the development of our IEMA Foundation Certificate course. The course provides a very useful and comprehensive introduction to environmental management, covering a lot of ground in just four days. We put a lot of work into transferring the classroom-based course content into an interactive format that would really make the e-learning experience enjoyable and worthwhile, and the feedback has been tremendous.

Students can study the web-based course over a 24-hour period, where they want, when they want, to fit with their busy lives. We were delighted that our commitment to the e-learning format was recognised by Elearning Age at its awards, given that we were shortlisted from a field of nearly 250 entrants, including major players such as the BBC and the Open University.

In the next year we are introducing the IEMA Associate Certificate course to our environmental training portfolio, which is an exciting development for us. Since we launched the Foundation Certificate at the end of 2010 we have become one of the most popular IEMA-approved training organisations in the UK – we are looking forward to building on this in 2012. We're also broadening our health, safety and environmental operations internationally, with a new office in South Africa, so there should be some great opportunities to develop Workplace Law Environmental there.

> "I wish I had a pound for every time someone asked me...
>
> ... what their responsibilities are for removing waste from their premises."

@Look_Sharp

Carbon and water footprinting

James Tiernan, Bureau Veritas UK Ltd

Key points

- A carbon footprint quantifies the total greenhouse gases (GHGs) attributable to the activities of a person, company, organisation or other entity, and is usually expressed as carbon dioxide equivalent (CO_2e) for simplicity.
- A water footprint quantifies the total freshwater consumption attributable to the activities of a person, company, organisation or other entity.

Legislation

- Climate Change Act 2008.

Carbon footprints

Carbon footprints are now widely used to assess the emissions of the four GHGs – carbon dioxide (CO_2), methane (CH_4), nitrous oxides (N_2O), and sulphur hexafluoride (SF_6) – and two groups of gases – hydrofluorocarbons (HFCs) and perfluorocarbons (PFCs) – identified by the Kyoto Protocol as the most significant GHGs. Since different GHGs can trap different amounts of heat within the atmosphere, they are usually converted into CO_2e; this is the expression of the tonnes of CO_2 that would have the same global warming potential (GWP) as a single tonne of the given GHG over a 100-year period. The GWPs of the four GHGs are:

- CO_2 1
- CH_4 25
- N_2O 298
- SF_6 22,800

Different gases within the HFC and PFC family have varying GWPs and so are calculated according to the GWPs of the actual gases emitted (e.g. 14,800 for HFC-23).

Typically, the main sources of CO_2, CH_4 and N_2O are through the combustion of fossil fuels for energy. This can be direct (such as combustion of gas in a local boiler for heating), or indirect (such as the combustion of coal in a power station to produce the electricity consumed on a site). SF_6 is widely used as an electrically insulating gas in high voltage applications such as transformers and switchgear, and HFCs are widely used as refrigerant gases; emissions from both of these are principally associated with leakage from these applications. PFCs are used in polymer and foam manufacture.

The carbon footprint data may be expressed as an absolute total emissions figure, list of emission from different sources, or expressed as a ratio of other relevant variables such as floor area, number of employees, financial turnover, production output or similar.

Under the Climate Change Act, the Government is to introduce regulations requiring mandatory reporting of GHG emissions by April 2012, or lay a report before Parliament explaining why this has not been done. There will be a further consultation on this guidance before a decision is made to introduce mandatory reporting requirements.

Under the Climate Change Act 2008, DECC / DEFRA released updated guidance on Corporate GHG reporting in September 2009. The key points included:

- Based on the Greenhouse Gas Protocol – the internationally recognised standard for corporate reporting and accounting of GHG emissions – in terms of setting organisational boundaries, Scope 1 (direct), Scope 2 (electricity, heat and steam – indirect) and Scope 3 (other indirect) emissions – and as such contains no major surprises.
- Outlines Standard Practice and Best Practice approaches – Standard includes at least Scope 1 and 2; Best Practice includes Scope 3 emissions (diagram below) – this is in line with GHG Protocol.

- Reporting period should be for a 12-month period, chosen to align with internal reporting needs.
- Use activity data and apply DEFRA / DECC emissions factors (excel sheets available online). Where other emission factors are required (e.g. for international operations), refer to GHG Protocol tools online.
- Report total GHG emissions as a gross tCO_2e figure (i.e. before accounting for any purchased reductions / credits) and for each GHG separately.
- Where applicable, report a net tCO_2e figure (i.e. subtracting purchased carbon credits meeting DEFRA emissions reduction criteria from gross figure).
- Set a reductions target.

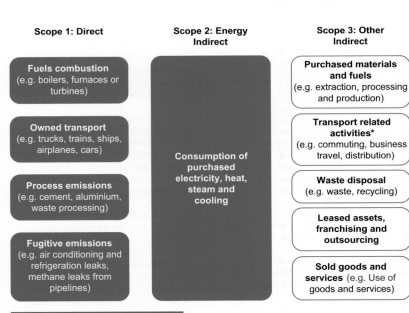

Scope 1: Direct	Scope 2: Energy Indirect	Scope 3: Other Indirect
Fuels combustion (e.g. boilers, furnaces or turbines)		**Purchased materials and fuels** (e.g. extraction, processing and production)
Owned transport (e.g. trucks, trains, ships, airplanes, cars)	**Consumption of purchased electricity, heat, steam and cooling**	**Transport related activities*** (e.g. commuting, business travel, distribution)
Process emissions (e.g. cement, aluminium, waste processing)		**Waste disposal** (e.g. waste, recycling)
Fugitive emissions (e.g. air conditioning and refrigeration leaks, methane leaks from pipelines)		**Leased assets, franchising and outsourcing**
		Sold goods and services (e.g. Use of goods and services)

Key:
Recommended
Discretionary

*From / to point of ownership transfer

Various different methodologies and tools have been developed for calculating carbon footprints, with variations in methodology and approach depending on their origin and intended usage. Reporting schemes, whether voluntary (such as the Carbon Disclosure Project – CDP) – or mandatory (such as the UK's Carbon Reduction Commitment – CDP) may include or exclude different sources or proportions of emissions, and so may not be directly comparable.

A number of proprietary standards and methodologies have been developed, or are under development, such as PAS250 (lifecycle GHG emissions for products and services), CEN TC 350 (in development, concerning construction), and the Greenhouse Gas Protocol, part of which has formed the basis of ISO 14064-I:

- Greenhouse Gas Protocol
  - Developed by the World Business Council for Sustainable Development and World Resources Institute in the late 1990s as guidance for government and business leaders to understand, quantify and manage greenhouse gas emissions. This is guidance only, but is the most widely recognised methodology and the basis for subsequent standards. It is accessible and easy to read.
- ISO 14064 (Part 1)
  - Developed by ISO in 2006 to provide details and requirements for designing, developing, managing and reporting organisation or company level GHG inventories. This is more prescriptive, method based and difficult to read!
- PAS 2050 (Public Available Standard)
  - Released in 2008 by DEFRA and the Carbon Trust and provides a consistent methodology for

the assessment of life cycle GHG emissions from goods and services. The standard is product focused and very detailed (due to complex subject matter).

In October 2009 DEFRA released guidance on Carbon Neutral Claims with the intention to tackle the general lack of transparency about the term 'carbon neutrality'. Previously there was a variety of statements about carbon neutrality based on organisations' own preferred definitions of the term. Some companies might not have sought to reduce emissions, but simply to purchase offset credits. Without a more uniform basis for making statements of carbon neutrality, users of the term are open to accusations of 'greenwash'. Under the Government's definition, achieving carbon neutrality entails the completion of the following three separate stages:

1. Calculate emissions.
2. Reduce them (e.g. through energy efficiency measures).
3. Offset residual emissions. In making a carbon neutral statement, the business or individual should provide clear information on the emissions measured, the reductions made and the offsets purchased.

A claim of carbon neutrality should always be linked to a particular and specified period of time, because doing so will ensure the claim is understandable.

Water footprints

Water footprinting is a relatively new concept (first conceived in 2002) and takes the measurement of water usage beyond the traditional consideration only measuring direct metered supplies. A water footprint calculation seeks to fully quantify, not only this direct consumption, but also the indirect water consumption associated with activities further up the supply chain (such as transport, manufacturing, agriculture and forestry activities). Water footprinting recognises that not all water is the same and broadly categorises water into different types to enable consistent calculations of total water usage by the activities of the person, company, organisation or other entity being measured. The definitions are:

- Blue water: surface and ground water.
- Green water: rainwater stored in soil.
- Grey water: the water required to dilute pollutants to acceptable levels.

Unlike GHG emissions, water footprint data must be carefully interpreted as the wider environmental, social and economic impact of water consumption can vary significantly depending on numerous factors such as location, and the nature of the water consumed such as its abundance and source, and the nature of the usage. For example, a litre of potable water used in the production of a consumer product in an arid location with high population density could have a more detrimental local impact than the same volume of potable water in an area of high rainfall and lower population density. This can be compounded where a footprint calculation includes numerous consumers in different locations at different stages of the supply chain.

As yet there is no global standard methodology for water footprinting; however, various organisations have developed proprietary water footprinting tools and methodologies that can be applied to individuals, products, companies, facilities, regions or whole countries.

See also: Climate change, p.145; Corporate Social Responsibility, p.202; Environmental Management Systems, p.286; Waste management, p.707.

Sources of further information

Bureau Veritas UK and Ireland: www.bureauveritas.co.uk

Greenhouse Gas Protocol: www.ghgprotocol.org

UNESCO Institute for Water Education: www.unesco-ihe.org/

Water footprint: www.waterfootprint.org

Catering: food safety

Jagdeep Tiwana, Bond Pearce LLP

Key points

Food should be prepared in accordance with the Food Safety Act 1990 and other legislation. The Act is the principal food-control measure in the UK. It sets out general duties of caterers in respect of food safety and quality, together with general consumer protection measures. Subordinate legislation made under the Act fleshes out requirements to be followed by caterers and other food operators and businesses in relation to matters such as claims and advertising. European Regulations lay down requirements in respect of hygiene and temperature control.

Legislation

- Food Safety Act 1990.
- Food Labelling Regulations 1996.
- Food Labelling (Amendment) Regulations 1999 and 2003.
- Food Labelling (Amendment) (No. 2) Regulations 2004.
- Food Safety Act 1990 (Amendment) Regulations 2004 implementing Regulation EC 178/2002.
- Genetically Modified Food (England) Regulations 2004 implementing Regulation EC 1829/2003.
- General Food Regulations 2004 implementing Regulation EC 178/2002.
- Regulation (EC) 852/2004 on the Hygiene of Foodstuffs.
- Regulation (EC) 853/2004 on Hygiene Rules for Foods of Animal Origin.
- Food Hygiene (England) Regulations 2006.
- Regulation (EC) 1924/2006 on Nutrition and Health Claims.

Food safety

General duties in relation to safety and quality

It is a criminal offence to render food injurious to health or sell food that is unfit for human consumption. Food can be rendered injurious to health by:

- adding any article or substance to food;
- using any article or substance as an ingredient in the preparation of food – e.g. adding caustic soda instead of baking powder to a product;
- taking out any constituent from a food; and/or
- subjecting the food to any other process or treatment.

In determining whether a food is injurious to health, regard has to be given to three issues:

1. The probable immediate, short-term and/or long-term effects of that food on the health of consumer *and* subsequent generations;
2. The probable cumulative toxic effects; and/or
3. Particular sensitivities of certain consumers.

Unless there has been a deliberate act of sabotage, it is rare for proceedings to be brought under this provision.

Proceedings in relation to food safety are more likely to be brought for selling food that is unsafe. In determining whether food is unsafe, the following issues have to be taken into consideration:

■ The normal conditions of use by the consumer;
■ The normal conditions at each stage of production processing and distribution; and
■ The information provided to consumers.

In addition, a food will be deemed unfit for human consumption if it is unacceptable for human consumption. This could be for reasons of contamination by extraneous matter, through putrefaction or deterioration.

Consumer protection provisions

It is also necessary to protect consumers from food that is safe but of unsatisfactory quality. It is an offence if the food sold to the consumer is not:

■ of the nature demanded (e.g. the food sold is different to that requested, such as reformed fish being sold as scampi);
■ of the substance demanded (this applies where the composition differs from that requested, e.g. glass in a pizza); or
■ of the quality demanded (this includes mouldy, bad and decomposed food as well as food that fails to meet commercial quality standards – e.g. excess sugar in diet cola).

Consumers are also protected from false claims in relation to foods as it is an offence to label, present, display or advertise food in such a way that it is falsely described or is likely to mislead as to the nature, substance or quality of the food. For example, a menu describing a cocktail as Red Bull and vodka would be misleading if the cocktail did not contain Red Bull but a substitute product. For further information on menus, see below.

Other provisions

These are traceability requirements imposed on all food businesses, including caterers, requiring them to keep records of their food suppliers. Internal traceability is not required. This means that it is not necessary to state which specific batch of ingredients went into which meal. The Regulations also impose an obligation on caterers to notify the authorities if they suspect that any food they supply or has been supplied to them is unsafe.

Labelling and menus

The Food Labelling Regulations set out requirements for the labelling of food products. Usually, food sold by caterers does not need to be labelled. However, there are certain exceptions; for example, foods that contain irradiated ingredients.

Any food that is supplied to caterers where there has been an intentional use of GM ingredients at any level must be labelled. This is regardless of whether GM material is detectable in the final product. It is not necessary to label small amounts (e.g. below 0.9% for approved GM ingredients and 0.5% for unapproved varieties that have received a favourable assessment). However, it is less clear whether caterers should continue to provide this information to consumers for foods sold in restaurants, cafés, etc. This is currently being dealt with at EU level and, until clarified, it is unlikely that any action will be taken against caterers that do not pass this information on.

Food that is pre-packed to sell directly to customers does not have to be labelled but where food has been pre-packed by a different supplier, for example where sandwiches have been prepared by another company, the food must be appropriately labelled with an ingredients list and details of allergens such as nuts, milk or fish. In the case of allergens, it may be advisable to provide information to consumers in all circumstances. The Food Standards Agency has issued voluntary Best Practice Guidance for the Provision of Allergen Information for Non Pre-packed Foods.

All this information must be on a label attached to the food, the menu or a notice. Part III of the Regulations requires attention in this context as it applies to claims and descriptions in the labelling and advertising of food products, meaning that it will apply to adverts for food products sold by caterers. It identifies the following:

- *Prohibited claims.* These are claims that a food has a tonic property or that it has the property of treating, preventing or curing disease. This prohibition prevents caterers from making medicinal claims for products on their menus. For example, although it would be permissible to mention the health maintenance properties of calcium, it would not be permissible to say that calcium could help prevent osteoporosis.
- *Restricted claims.* These include claims for reduced or low energy, protein, vitamins and minerals.
- *Misleading descriptions.* Part III sets out when certain wording such as 'ice cream' and 'starch-reduced' can be used.

These prohibitions and restrictions need to be taken into account by caterers when drafting menus. Caterers also need to ensure that all nutrition and health claims comply with European Regulation 1924/2006.

It is essential that the prices of food and drink sold at the premises are clearly displayed. The easiest way to do this is on a menu. Generally, prices should include VAT when appropriate. There are few rules setting out the composition of food; the exceptions are sausages and burgers, which must contain a prescribed amount of meat.

Care has to be taken when drafting menus to ensure that they are not misleading and that they do not falsely describe the foods in the menus. For example, it would not be acceptable to describe an omelette as freshly cooked if the caterer buys it from frozen and then reheats it. Guidance can be found in the report, *Criteria for the Use of the Terms Fresh, Pure, Natural etc. in Food Labelling*, published by the Food Standards Agency, which sets out terms that should be avoided or used only sparingly. These include the terms 'country-style', 'homemade', 'authentic', 'original' and 'traditional'. It also explains when the words 'fresh', 'natural' and 'pure' should be used.

Hygiene

Regulation 852/2004 on the hygiene of foodstuffs identifies the general obligations on proprietors of food businesses in relation to hygiene. Of particular importance is Annex II, which sets out the rules of hygiene. The Annex is divided into 12 chapters that lay down both general and specific legal requirements relating to the hygienic operation of a food business:

- Chapter I lays down general requirements for ensuring hygiene of premises, including cleaning, maintenance, layout, design, construction and size. It also covers welfare amenity provisions (such as sanitary accommodation, washing and clothing), storage facilities and environmental provisions (including temperature, lighting and ventilation).
- Chapter II lays down more specific requirements for food preparation,

treatment areas and rooms. These requirements apply to all rooms where food is prepared, treated or processed, except for dining rooms. It covers matters such as floors, walls, ceilings, overhead fixtures, windows, doors and washing facilities for foods.

- Chapter III covers requirements for moveable or temporary premises such as stalls and delivery vehicles.

- Chapter IV sets out the standards that must be maintained in relation to hygiene when transporting food. This includes caterers that make home deliveries.

- Chapter V requires food businesses to keep clean all articles, fittings and equipment with which food comes into contact and is intended to prevent cross-contamination from one foodstuff to another.

- Chapter VI (food waste) lays down requirements to minimise any food safety risk that inevitably arises from working debris and food waste generally. The new Regulation adds the requirements that all waste is to be eliminated in a hygienic and environmentally-friendly way.

- Chapter VII prescribes the quality of water to be used in a food business. In particular, it requires an adequate supply of 'potable water', which must be used whenever necessary to ensure foodstuffs are not contaminated.

- Chapter VIII (personal hygiene) sets out the legal requirements designed to achieve 'clean person' strategies. Personnel have two general legal obligations. First, every person working in the food-handling area must maintain a high degree of personal cleanliness and must wear suitable, clean and, where appropriate, protective clothing. Second, people working in food-handling areas who are suffering from or carrying a disease likely to be transmitted through food must report

the condition immediately to the food business operator. Such persons must not be permitted to handle food or enter a food handling area if there is any likelihood of indirect contamination.

- Chapter IX sets out provisions applicable to foodstuffs. It requires that food businesses must not accept any raw material or ingredients if they are known to be or suspected to be contaminated. Once accepted, raw materials must be stored in appropriate conditions designed to prevent harmful deterioration and protect them from contamination. Food is to be protected against any contamination and adequate procedures must be in place to control pests.

- Chapter X lays down provisions to prevent contamination when wrapping and packaging foodstuffs.

- Chapter XI sets out requirements applicable to food placed on the market in hermetically-sealed containers.

- Chapter XII stipulates that caterers must ensure that their food handlers are supervised and trained in relation to food hygiene matters. The training must be commensurate with their work activities.

Food safety management

Regulation (EC) No. 852/2004 requires food business operators to identify critical points for food safety, establish and implement monitoring procedures for critical points, establish corrective actions for critical control points and document and record the application of these measures. This is commonly referred to as a Hazard Analysis of Critical Controls Points (HACCP) system, or, if it is applied to the catering industry, it is referred to as Self-Assured Catering.

Article 5 of the Regulation requires food business operators to put in place, implement and maintain a permanent procedure based on the HACCP principles. An HACCP system requires that a catering operation be analysed step by step. In particular, it requires that stages in preparation at which hazards exist be identified, together with the means by which they can be controlled. These may be as simple as introducing temperature controls, e.g. cooking chicken to a minimum temperature of 75°C or ensuring that it is chilled before cooking.

The critical points that need to be adhered to, to ensure food safety, are summarised below:

- Identify any hazards that must be prevented, eliminated or reduced to acceptable levels.
- Identify the critical control points at the step or steps at which control is essential.
- Establish critical limits that separate acceptability from unacceptability for the prevention, elimination and reduction of identified hazards.
- Establish and implement effective monitoring procedures at critical control points.
- Establish corrective actions when monitoring to indicate that a critical control point is not under control.
- Establish procedures to be carried out regularly to verify whether the measures adopted are working effectively.
- Establish records to demonstrate the effective application of these measures.

Food safety checklist
- Are the premises registered with the competent authority?
- Has a liquor licence been obtained where alcohol is to be sold?
- Does the construction and layout of the premises comply with the Hygiene Regulations?
- Are suppliers reputable and reliable?

- Have the critical control points been identified and controls put in place and details of the procedure recorded in writing?
- Have staff been properly trained in all aspects of food safety, hygiene and temperature control?
- Is there a detailed cleaning schedule in place?
- Is waste picked up on a regular basis?
- Are there adequate pest control measures in place?
- Is equipment regularly cleaned and inspected? (This is particularly important for oil filtration systems.)
- Are proper temperature controls in place?
- Are all of the above steps documented?

Temperature controls
Controlling temperatures is one of the most significant ways of ensuring food safety. As of 1 January 2006, this area has been governed by various EC Regulations. The Regulations set out requirements for foods that are likely to support the growth of pathogenic micro-organisms or the formation of toxins at temperatures that would result in a risk to health. In particular, they stipulate when foods must be kept chilled, e.g. dairy products, cooked products containing meat, fish or egg, sandwiches and cooked rice. They also detail temperatures that should be reached during cooking, those that should apply to hot food and cold food displays, and when reheating of food products is acceptable.

Defences
If a breach of food law occurs, it is likely that a prosecution will result. In order to defend proceedings, it is necessary to prove to the court that all reasonable steps had been taken to avoid the offence occurring. This is known as a due diligence defence. What is reasonable will depend on the size of the business and its resources as well as the seriousness of the breach.

If steps have been taken to comply with the legislation and controls have been put in place in relation to hygiene, food safety and management, even if an employee has made a mistake that has caused the system to break down, it is likely that the proceedings can be defended.

Penalties

For a conviction under food law, the penalties can range from a fine of up to £20,000 and/or six months' imprisonment in the lower courts, to an unlimited fine and up to two years' imprisonment in the higher courts.

See also: Catering: health and safety issues, p.123; Legionella, p.442; Packaging, p.541.

Sources of further information

Food Standards Agency: www.food.gov.uk

Chartered Institute of Environmental Health: www.cieh.org.uk

Industry Guide to Good Hygiene Practice: Catering Guide, Chadwick House Group, 1997. ISBN: 0 900 103 00 0.

Catering: health and safety issues

Jagdeep Tiwana, Bond Pearce LLP

Key points

- Caterers must work in a way that protects the health and safety of employees and those people who could be affected by their activities.
- The law imposes general duties on caterers as employers to ensure their businesses are run safely and sets out ways in which this can be achieved.

Legislation

- Health and Safety at Work etc. Act 1974.
- Health and Safety (First Aid) Regulations 1981.
- Electricity at Work Regulations 1989.
- Manual Handling Operations Regulations 1992.
- Workplace (Health, Safety and Welfare) Regulations 1992.
- Reporting of Injuries, Diseases and Dangerous Occurrences Regulations 1995 (RIDDOR).
- Health and Safety (Safety Signs and Signals) Regulations 1996.
- Gas Safety (Installation and Use) Regulations 1998.
- Lifting Operations and Lifting Equipment Regulations 1998.
- Provision and Use of Work Equipment Regulations 1998.
- Management of Health and Safety at Work Regulations 1999.
- Control of Substances Hazardous to Health Regulations 2002.
- Health and Safety (Miscellaneous Amendments) Regulations 2002.
- Management of Health and Safety at Work and Fire Precautions (Workplace) (Amendment) Regulations 2003.
- The Regulatory Reform (Fire Safety) Order 2005.
- The Work at Height Regulations 2005 (as amended).

Health and Safety at Work etc. Act 1974

The Health and Safety at Work etc. Act 1974 imposes a number of general duties on all employers to ensure their business is conducted in a manner that minimises health and safety risks to its employees, customers and third parties such as contractors.

Employers' general duties to employees include:

- devising and maintaining a safe system of work;
- providing and maintaining equipment;
- ensuring that the use, handling, storage and transport of articles and substances are carried out safely;
- providing training and supervision;
- maintaining the workplace (including entrances and exits) in safe condition; and
- providing a safe working environment, which includes adequate facilities.

It is important to note that the general duty to provide safe conditions extends not only to employers but also to people such as landlords who exercise some control over premises or plant and equipment. The way in which an employer fulfils its duties to employees, visitors and third parties is set out in more detail in subordinate legislation.

Risk assessments

The Management of Health and Safety at Work Regulations 1999 set out in broad terms how health and safety should be managed. They provide detailed guidance, and one of the most important issues they address is the requirement to carry out risk assessments. In particular, they require that a suitable and sufficient risk assessment for the activities undertaken by the employer must be carried out.

The purpose of the risk assessment is to identify:

■ hazards caused by the procedures in relation to the business' activities;
■ risks that arise from those hazards; and
■ procedures to eliminate or control those risks.

Where there are more than five employees, significant findings of the risk assessments must be recorded in writing. Guidance issued by the HSE makes it clear that risk assessments should not be over-complicated and should highlight significant hazards that could result in serious harm or affect several people. The assessments should be reviewed from time to time to ensure that the precautions are working effectively. It should be noted that the risk assessment needs only to be suitable and sufficient. This does not mean perfect.

When carrying out risk assessments, specific consideration should be given to vulnerable employees or third parties such as people under the age of 18 or new and expectant mothers. In relation to people under the age of 18, the risk assessment will need to take into account the young person's inexperience and immaturity and lack of awareness of risks within a workplace.

When employing contractors etc., a copy of a risk assessment for the work they undertake should be provided by their employer. The contractor's employer should also be provided with copies of the risk assessments relating to the premises at which the work will be undertaken.

Once hazards and risks have been identified, safe systems of work should be devised to ensure that employees do not come to harm and procedures should be put in place to ensure the safety of third parties. In this respect, it is particularly important that training and information are provided to employees and appropriate third parties to ensure that they are aware of the risks and the necessary procedures to be taken to avoid them. These procedures should be encapsulated in a company health and safety manual. The manual should also set out an organisation chart detailing the health and safety responsibilities of individuals within the company.

It is also necessary to carry out a risk assessment that deals specifically with the risks arising from fire. This risk assessment must identify the steps that need to be taken to meet fire safety requirements and in particular what procedures must be followed in the event of an emergency. It is important that the risk assessment takes into account the needs of disabled people and children, particularly in the case of schools, in the event of a fire. A detailed assessment is necessary and the rules relating to recording significant findings of the risk assessment apply.

Safety signs

The Health and Safety (Safety Signs and Signals) Regulations 1996 provide details of signage that can be used to help control risks that have been identified in risk assessments. This includes signage to identify corrosive, flammable or explosive substances. The Regulations set out the format that signage should take, including details of the shape, colour and pattern of safety signs. It is necessary to ensure that safety signs are maintained in good condition.

Workplaces

The Workplace (Health, Safety and Welfare) Regulations 1992 aim to ensure adequate welfare facilities are provided for employees. In summary, they cover the following areas:

- The maintenance of the workplace, including equipment. In this regard, steps should be taken to ensure that equipment and the workplace in general are maintained in good repair and that any problems that may arise are rectified.
- The provision of ventilation, lighting and temperature control.
- The maintenance of cleanliness in the workplace, including equipment and furniture.
- Protection from falls and falling objects. This is particularly important for the catering industry as one of the greatest causes of accidents is slips, trips and falls. Employers must ensure that floors and traffic routes are of sound construction and do not present tripping or slipping hazards. These traffic routes and floors should be maintained and kept free from waste, clutter and other objects. Where floors are likely to get wet (e.g. around dishwasher areas), procedures should be put in place to minimise the risks arising from this. These could include the use of matting, a wipe-as-you-go procedure, and the provision of non-slip shoes. Physical safeguards should be provided where there is a risk of somebody falling a distance likely to cause personal injury. This calls for the specific consideration of the maintenance and cleaning of

windows and roofs. Windows and doors that are made of glass should contain safety glass and should be marked so they are easily visible.
- The provision of sanitary conveniences and washing facilities. When up to five people are at work at the same time, there should be at least one WC and one wash station, two of each for up to 25 workers, three for up to 50 and so on for every 25 extra workers.
- Changing and rest facilities. A rest-room (a rest area) should be provided with adequate seating for employees. There should be the provision of a warm, dry, well-ventilated area to hang personal clothing, and drinking water must be available. Where employees eat meals at work, then the means of preparing a hot drink must also be provided.

Equipment

The Provision and Use of Work Equipment Regulations 1998 set out the safety requirements for the provision and use of equipment at work. In particular, they require that equipment should be of suitable construction and design for the purpose for which it is intended. Work equipment must be used only in relation to work activities for which it is meant and it should be regularly maintained and inspected. Where appropriate, written instructions and training should be provided to enable employees to use the equipment.

Manual handling

The Manual Handling Operations Regulations 1992 prescribe measures that employers must follow to reduce the risk of injury to their employees from manual handling activities. First, employers are required to establish if the manual handling operation that gives rise to risk can be avoided altogether. This could be done by reorganisation or the introduction of mechanical aids. If the operation cannot be avoided, then it is necessary to assess the risk and to try to implement controls to reduce the risk as far as reasonably practicable.

Once a safe system of work has been devised in relation to manual handling, it is important that it is maintained and that there is adequate monitoring to ensure that there are no deviations from these procedures.

Where automated systems are introduced in place of manual handling, then regard has to be given to the Lifting Operations and Lifting Equipment Regulations 1998. Such equipment must be regularly inspected, and, where an employer does not have the requisite competence to do this, he must retain someone who does. For so long as the equipment is in use, it is necessary to keep a copy of the inspection report for two years or until the next report is made.

Working at height

The Work at Height Regulations 2005, which came into force in April 2006, amended the old Regulations that gave a qualifying height threshold of 2.5m. Regulations now cover all work at height where there is a risk of a fall liable to cause personal injury. Employers must now do all that is reasonably practical to prevent anyone falling.

Their responsibilities as duty-holders include:

- avoiding work at height for their employees where they can;
- using work equipment or other measures to prevent falls where they cannot avoid working at height; and
- where they cannot eliminate the risk of a fall, use work equipment or other measures to minimise the distance and consequence of a fall, should one occur.

Duty-holders must ensure that:

- all work at height is properly planned and organised;
- all work takes account of conditions that could endanger health and safety;
- those involved in work at height are trained and competent;
- the place where work at height is carried out is safe;
- equipment for work at height is appropriately inspected;
- the risks from fragile surfaces are properly controlled; and
- the risk from falling objects is properly controlled.

See 'Industry-specific concerns' for examples of where the new Regulations may affect the industry.

Control of hazardous substances

It is a necessary part of every caterer's business to deal with hazardous substances. Hazardous substances can come in many shapes and sizes such as liquids, gas, dust and micro-organisms and may occur naturally or be artificially produced. If they create a hazard to health, then it is necessary either to prevent exposure to that substance or, if this is not reasonably practicable, to control employees' and third parties' exposure to the substance. It is particularly important to assess the level, type and duration of exposure. Once this has been done, preventative or control measures can then be introduced.

The most common type of hazardous substance found in the catering industry is cleaning materials. In such cases, because the use of such hazardous substances cannot be avoided, information should be provided to employees, stating what action should be taken in the event of an accident or emergency, and employees should be trained on how to use the products safely.

Use of electricity

The design, installation and operation of electrical and gas systems are highly regulated. The Electricity at Work Regulations 1989 and the Gas Safety (Installation and Use) Regulations 1998 set out detailed requirements for the health and safety precautions that must be taken when dealing with gas or electricity.

Accident reporting

Employers are under an obligation to report serious work-related health and safety incidents, diseases or deaths. The Reporting of Injuries, Diseases and Dangerous Occurrences Regulations 1995 (RIDDOR) set out which types of accidents are reportable and how quickly they should be reported. There is now a central reporting system, the Incident Contact Centre at Caerphilly.

There is a distinction between a reportable accident, which is an accident associated with work activities, and reportable diseases, which require to be diagnosed by a doctor. The list of major injuries, dangerous occurrences and diseases that require to be reported are extensive. However, in summary, injuries that should be reported are:

- fractures (except of fingers or toes);
- amputations;
- dislocation of knees, hips, shoulders or spine;
- temporary or permanent loss of sight (including any penetrating injury to the eye);
- injury leading to hypothermia or heat-induced illness;
- unconsciousness;
- admittance to hospital for more than 24 hours; and
- exposure to a harmful substance(s).

Dangerous occurrences that would need to be reported include:

- the collapse, failure or overturning of load-bearing lifting machinery;
- electrical short-circuits leading to a fire explosion;
- the collapse of scaffolding; and
- the escape of any substance that could lead to death or major injury.

Reportable diseases include:

- occupational dermatitis or asthma; and
- repetitive strain-type injuries.

Incidents that must be reported without delay are:

- dangerous occurrences;
- the death of any person due to an employer's activities;
- a major injury suffered by any person at work; and
- injury to any third party (e.g. contractors, visitors, etc.) that requires their admission to a hospital.

Although it is not specified how incidents should be reported, this must be done by the fastest means practicable, which would normally be by telephone.

Other injuries must be notified within ten days. These include a three-day injury suffered by an employee due to a work-related incident. This means that, where a person is unable to carry out work because of an injury for more than three consecutive days, then the incident must be notified. The day of the accident does not count towards this time period, but non-work days are included. There is a specific form, Form F2508, which must be completed to send to the Incident Contact Centre.

Industry-specific concerns

By way of example, a number of activities that take place specifically within the catering industry are considered to be high risk by the HSE and need to be considered in any health and safety compliance or training session.

In particular:

- Boiling water and splashes from hot fat, especially where the cleaning and emptying of deep fat fryers is concerned;
- Slips, trips and falls, particularly where staff may be working on wet or tiled floors;
- Manual handling, particularly in the brewing industry where items such as large barrels of beer may be moved into and around cellars;
- Working in cellars and confined spaces;
- Working with knives and other sharp implements such as meat-slicing machinery;
- Gas safety in kitchens;
- Looking after hands and skin, particularly in relation to preventing contact dermatitis; and
- Food debris on floors.

Offences

Offences under health and safety legislation can be subject to a fine of up to £20,000 in the Magistrates' Courts and an unlimited fine or imprisonment (although this is unlikely in most cases involving a company) in the higher courts.

Defences

There is no true defence to proceedings brought under health and safety legislation. Where a defendant can show that it did everything that was reasonably practicable (i.e. it did all that it could to prevent an accident), then proceedings may be successfully defended. However, it is notoriously difficult to mount such defences.

> *See also*: Catering: food safety, p.117; Manual handling, p.474; Reporting of Injuries, Diseases and Dangerous Occurrences (RIDDOR), p.618; Risk assessments, p.624; Slips, trips and falls, p.650.

Sources of further information

HSE: www.hse.gov.uk

Incident Contact Centre: www.riddor.gov.uk

CCTV monitoring

Lisa Jinks and John Macaulay, Greenwoods Solicitors LLP

Key points

- Employers should note that most CCTV systems will be covered by the Data Protection Act 1998 (the DPA). They should therefore familiarise themselves with the requirements of the DPA and the Information Commissioner's CCTV Code of Practice.
- Before installing a CCTV system, an impact assessment should be carried out to assess whether the use of CCTV is justified or whether another, less intrusive solution (such as improved lighting in a car park), could achieve the same objectives.
- A policy regarding the use of CCTV systems by camera operators and retention of images should be implemented.
- Warning signs should usually be posted, and the organisation must ensure it is able to comply with data subject access requests for images.

Legislation

- Data Protection Act 1998.
- Human Rights Act 1998.

CCTV in the UK

Increasingly, CCTV has become the principal method of carrying out surveillance of areas that may be accessed by the public as well as becoming commonplace in many workplaces in a variety of industries.

While CCTV has an obvious crime-prevention role in the high street and in places such as shops and car parks, its use in the workplace as a means of observation of staff is both less obviously beneficial, and considerably less accepted. Employers who use CCTV in the workplace need to ensure that doing so does not affect overall levels of trust in the employment relationship. Employers are advised to bear in mind the Information Commissioner's CCTV guidance, which recognises the special considerations that apply in the workplace to assist employers to comply with their legal obligations.

Dummy and non-operational CCTV

The Data Protection Act applies to images captured by CCTV. Therefore, if no image is captured, no personal data is processed, so the Data Protection Act would not apply; additionally it would be difficult to argue any infringement of an individual's privacy. Therefore, the guidance in this chapter does not apply to the use of dummy or non-operational cameras.

Data Protection Act 1998

The DPA is the major legal control over CCTV surveillance in the UK, both within and outside the workplace. Most images recorded by CCTV systems will constitute 'personal data' for the purposes of the DPA, for example, where someone is identifiable from the images captured, or where other information relating to a living individual is caught, such as car registration numbers. This is a change from the Information Commissioner's original view that most CCTV images would not be covered by the DPA, and reflects the fact that systems are becoming far more technologically advanced.

> **Facts**
> - It is estimated that there are around 1.85 million CCTV cameras in use in the UK, one for every 32 people.
> - The vast majority of these cameras are operated by private companies on private premises, for example, workplaces.
> - CCTV technology now goes further than simply recording people's movements, to incorporate numberplate and facial recognition, and, in some cases, adding listening functionality.

Furthermore, some of the data recorded by CCTV may constitute 'sensitive personal data' under the DPA, for example, where the images relate to the commission or alleged commission of an offence. In these circumstances, the DPA applies more stringent processing guidelines that must be complied with.

Employers using CCTV systems should also be aware that they must be able to respond to data subject access requests under the DPA. Data subjects – i.e. those individuals about whom the data relates – have the right to see data held about them, and in the CCTV context, this will generally be the right of those recorded by cameras to see a copy of any such recording. Such 'subject access requests' must be complied with within a statutory time period unless one of the limited exemptions in the DPA applies.

To assist employers to comply with the DPA, the Information Commissioner has issued two pieces of guidance. The first relates specifically to CCTV in the workplace, with the second of more general application, relating to all workplace 'monitoring' – see *Sources of further information*.

Human Rights Act 1998
Under Article 8 of the Human Rights Act 1998 (HRA) everyone has 'the right to respect for his private and family life, his home and correspondence'.

It is arguable that workers' right to privacy under Article 8 may be compromised by some use of CCTV, particularly in areas where there is a legitimate expectation of privacy – toilets or private offices, for example. Employers should bear in mind the provisions of the HRA when using CCTV systems, although compliance with the DPA and various Information Commissioner's Codes of Practice will likely lead to compliance with the HRA as well.

Code of Practice for users of CCTV
The Information Commissioner published a revised CCTV Code of Practice in January 2008. It sets out the measures that should be adopted in order to ensure that a CCTV scheme complies with the DPA and provides guidance on good practice.

The Code of Practice applies to most CCTV and other systems that capture images of identifiable individuals, or information relating to individuals, for specific purposes including monitoring their activities or potentially taking action against them (for example, as part of a disciplinary or criminal investigation). Note that the Code does not apply to the covert surveillance activities of law enforcement agencies or the use of conventional cameras (not CCTV) by the media or for artistic purposes such as filmmaking.

Although the Code should be considered in its entirety, Appendix 3 is specifically aimed at employers who use CCTV to monitor their workers, and supplements the Employment Practices Code guidance on monitoring employees (*see below*).

The full text of the Code should be considered by employers, but the main practical requirements are summarised in the following paragraphs.

Impact assessments

Employers should conduct an impact assessment before installing and using CCTV to assess whether the objectives of monitoring can be achieved by a less intrusive means. The guidance lists a number of questions that companies should consider before installing CCTV, such as:

- What are the problems the use of CCTV will address?
- Can CCTV realistically deal with those issues?
- How will the system work in practice, and who will manage it?
- What are the views of those who will be surveyed? Can the impact on them be minimised?

Under the DPA, organisations must notify the Information Commissioner of purposes for which they process data. Such notification should cover data collected through CCTV and so organisations should be clear about the purposes for which they need CCTV and ensure the Information Commissioner is informed of these.

In some cases, it may be appropriate to install CCTV specifically for workforce monitoring, provided this is justified and properly impact-assessed. Workers should normally be made aware that they

are being monitored, but, in exceptional circumstances, covert monitoring may be used as part of a specific investigation, for example, where there is reason to suspect criminal activity or equivalent malpractice, and the decision to use covert recording is taken by senior management having considered the intrusive effects on innocent workers.

Cameras and listening devices should not be installed in private areas such as toilets and private offices, except in the most exceptional circumstances where serious crime is suspected. This should only happen where there is an intention to involve the police, not where it is a purely internal disciplinary matter, and again should be considered properly by senior management prior to authorisation.

Camera positioning and signage

Cameras should be positioned in such a way that they are only able to monitor areas intended to be covered by the CCTV scheme. The operators of the equipment must be aware that they may only use the equipment in order to achieve the purpose as notified to the Information Commissioner. For example, if the aim of the CCTV is to prevent and detect crime, it should not be used for monitoring the amount of work done, or compliance with company procedures.

Clearly visible and legible signs of an appropriate size should be used to inform the workers (and, if appropriate, the public) that they are entering a zone with CCTV coverage. Such signs should detail the identity of the person or organisation responsible for the scheme, its purpose, and details of who to contact regarding the scheme. The contact point should be available during office hours, and workers staffing the contact point should be aware and understand the relevant policies and procedures.

Where CCTV is used to obtain evidence of criminal activity, signage may not be appropriate. However, the Code sets out tight controls over the use of CCTV in these circumstances, and employers should bear in mind the potential preventative effect of CCTV that would be negated where no signage is used.

Image quality

Images captured by CCTV equipment should be as clear as possible to ensure that they are effective for the purposes intended. Cameras should be properly maintained and serviced, and capable of recording with a high resolution. If dates and times are recorded, these should be accurate. Consideration must also be given to the physical conditions in the camera locations (e.g. infrared equipment may need to be used in poorly lit areas). No sound should accompany the images except in limited circumstances. Images should not be retained for longer than necessary and, while they are retained, access to and security of the images must be tightly controlled in accordance with the DPA. Disclosure of images from the CCTV system must be controlled and the reasons for disclosure must be compatible with the purposes notified to the Information Commissioner. All access to or disclosure of the images should be documented.

Use of images

Images from a CCTV camera can be used as evidence in criminal and civil proceedings. However, when a company seeks to rely on CCTV evidence in Court, the Court will have to assess the weight of the CCTV evidence. In many cases, there are problems with the quality of CCTV images, which makes the evidence obtained from CCTV less reliable. For example, the quality of a CCTV image can often be poor, especially where cameras have been placed on tall poles, making identification of individuals difficult. Even

where the camera has recorded a clear image it can be difficult to identify an individual from this image. Concerns are still raised about the reliability of CCTV images as evidence, due to the fact that digital images can be modified. Companies should therefore ensure they can authenticate any digital images presented as evidence.

Employment Practices Code

Part Three of the Information Commissioner's Employment Practices Code on Monitoring at Work sets out the general guidelines for employee monitoring. Employers considering the use of CCTV in the workplace must consider this Code, even where the purpose is not specifically to monitor employees – for example, CCTV systems in shops designed to prevent shoplifting will inevitably also capture workers.

The Code stresses the need for proportionality – any adverse impact of monitoring must be justified by the benefits to the employer and others. Continuous monitoring of particular workers is only likely to be justified where there are particular safety or security concerns that cannot be adequately dealt with in other, less intrusive, ways. All employees and visitors to organisations should be made aware that CCTV is in operation and of the purposes for which the information will be used.

British Standards Institution

The British Standards Institution has issued Code of Practice BS 7958:2009 *Closed circuit television (CCTV): management and operation* to assist CCTV operators' compliance with the DPA (and other applicable legislation) and to ensure that CCTV evidence can be used by the police to investigate crime. The Code is particularly useful where CCTV systems are used in public places, or have a partial view of a public place, and may therefore be of use to workplace managers.

Conclusion

Workplace managers responsible for CCTV use should familiarise themselves with the relevant Information Commissioner's Codes of Practice, as well as the underlying legislation. The main points to consider are:

- An initial assessment should be carried out before installing the CCTV system.
- Clear guidelines for use should be established before the system 'goes live', and these should be communicated to all staff who use, and those who may be caught by, the system.

- Warning signs alerting people to the presence of CCTV, and containing relevant information should normally be put in place.
- CCTV images should be handled in accordance with the requirements of the DPA, in particular those relating to storage and security.
- Employers should ensure they are able to comply with data subject access requests.

See also: Confidential waste, p.167; Loneworking, p.467; Parking, p.547; Violence at work, p.693; Visitor safety, p.697.

Sources of further information

Information Commissioner: www.ico.gov.uk

CCTV Code of Practice: Revised Edition 2008: http://bit.ly/4wioW

Employment Practices Code: June 2005: http://bit.ly/1HLK49

BS 7958: 2009 *Closed circuit television (CCTV): management and operation*: www.bsigroup.co.uk

Case review

What does *Nussbaumer* mean to you?

Association for Project Safety

The *Nussbaumer* case, which has recently been decided by the European Court of Justice (ECJ), could have serious implications for the UK as it goes to the heart of what projects the Temporary and Mobile Construction Sites Directive actually applies to.

This Directive lays down minimum safety and health requirements for temporary or mobile construction sites across Europe. Within the UK it is implemented through the Construction (Design and Management) Regulations 2007 (CDM).

The case, from Italy, determined that where a number of contractors are present on a construction site, EU law requires that a safety coordinator be appointed and that a safety plan be drawn up where there are particular risks, whether or not planning permission is required.

Currently, with the exception of Designer duties, CDM does not apply to domestic projects and is only notifiable when a project exceeds 30 construction days or 500 person days.

On domestic projects, or on projects smaller than the UK's Notification threshold, where there is more than one contractor involved, a simple interpretation of the ECJ decision on an Italian domestic project would seem to suggest that the UK could be disciplined by the EC for not applying the Directive properly. The consequence could be the requirement to

 Amongst APS members are the country's leading architectural, engineering, health and safety, project management and surveying professionals. Our mission is to continuously improve and promote the practice of construction health and safety risk management by:

- setting standards;
- raising performance;
- providing guidance; and
- education and training.

Our vision is that the Association for Project Safety be known by the construction industry as the leading professional body in construction health and safety risk management.

apply the CDM Regulations across the UK to all projects where there is likely to be more than one contractor on site.

The ECJ decision, which challenges the Italian application of the Directive, could in turn present the UK with a need to make significant changes. Should we worry? Well, not really – given that there are excellent safety and (particularly!) health related arguments for having a CDM-C involved in all construction projects – to make sure that risks of all types are eliminated, reduced or controlled and all workers given the potential protection that CDM provides. We are, after all, continually being told that the largest proportion of accidents are on small construction sites and that ill health has basically flat-lined across the board.

It also seems inconceivable (and basically immoral) that many construction workers (there are those, after all, who work on non-notifiable / small / domestic projects much, if not most of the time) should be left beyond the CDM pale – with no one charged with making sure that design, management and construction planning take account of health and safety in the ways envisaged by the CDM Regulations (and the Directive).

The HSE is reluctant to apply the Regulations to domestic projects because of the various 'difficulties' that they foresee. But if CDM-Cs were to be required on all projects then Clients would have the information, advice and assistance that they need and the workforce would have the protection that CDM Regulations proffer – provided that the full CDM regime is applied to each and every project where more than one contractor is likely to be on site at any one time. Clients could also benefit from increased competence at the lower levels of construction activity as well as improved project planning, coordination and management – and, provided that CDM related costs were kept proportionate, there could be a distinct net gain to both clients and industry – as has been demonstrated many times on larger projects.

If CDM has driven down accidents and injuries (health is another story, for another day) then it seems only appropriate that the small projects that are the source of the worst statistics should come into their ambit – and the workforce benefit accordingly. It could easily be said that the real benefits of CDM are still to be gained. Has the UK's non-application of the Directive to small and domestic projects in fact frustrated its impact? Combine this change with a refocusing on health issues – to move those statistics downwards – and the Coalition's Star Chamber could then be shown that the changes in Regulation would result in very significant improvements to both health and safety – and potentially by a lot more than the 5% threshold it has set.

So, perhaps we need to think outside the UK box and inside the EU one – for a change – and the sooner the better!

The *Nussbaumer* case could have serious implications for the UK as it goes to the heart of what projects the Temporary and Mobile Construction Sites Directive actually applies to.

So what would compliance with the directive mean? It would mean that on all projects where there is more than one contractor present there would be coordination to help eliminate and manage risk and a means of ensuring that designers really do take steps to eliminate or minimise health and safety risks.

Will that cost a lot? It needn't – and given current economic conditions it probably wouldn't. Would it save a lot? Almost certainly, both in terms of cost and time as CDM has shown on so many projects. But it could provide the step change that is needed to significantly reduce ill health as well as accidents, injuries and deaths in UK construction.

That could be the most important thing to happen to CDM in a long time, and given the problems that the HSE has identified with smaller construction projects it ought not to be difficult to persuade the Coalition's Star Chamber that these changes to the Regulations could be well worthwhile, especially in terms of potential health gains and potential cost savings to the health and welfare budgets.

If, whilst they are at it, the Regulations extend CDM-C duties to include site health and safety audits – and give CDM-Cs

some teeth – the HSE could significantly reduce reliance on their inspectors' visits to get things to happen on construction sites.

Could it be made to work? Would the Regulations be applied? Well, with CDM-Cs on hand to advise clients and carry out site health and safety audits on projects, they probably would. But the real question is – can we find a better way to improve construction safety and health performance? Can *Nussbaumer* provide the opportunity that we need?

CDM

Simon Toseland, Workplace Law

Key points

- The Construction (Design and Management) Regulations (CDM) 2007 came into force on 6 April 2007.
- They completely replaced the previous CDM 1994 Regulations and the Construction Health, Safety and Welfare Regulations 1996.

Legislation

- Construction (Design and Management) Regulations 2007.

Introduction

The CDM 2007 Regulations give full effect to Council Directive 92/57/EEC, and apply to all construction projects, regardless of the timescale or complexity of the work. The only exception where the CDM 2007 Regulations do not apply is to domestic clients, although any Designer engaged by a domestic Client is still obliged to comply with the CDM 2007 Regulations.

CDM 2007 applies to *all* construction work in the UK, which has a very broad definition. It refers to '*construction, alteration, fitting-out, commissioning, renovation, repair, upkeep, redecoration or other maintenance, decommissioning, demolition or dismantling.*' It also includes site preparation, exploration and surveys.

There are five defined duty-holders under the CDM Regulations, namely the Client, Designer, CDM Coordinator, Principal Contractor and Contractor, all of which have their own statutory responsibilities.

Regulations 2 and 3 provide the criteria as to whether a project is notifiable. If the construction phase is likely to involve more than 30 days or 500 person days of construction work, the project is notifiable

to the Health and Safety Executive (HSE). It is a duty of the CDM Coordinator to ensure that this notification is made.

The Client

CDM 2007 recognises the influence that a Client has over the health and safety of a construction project. After all, they are the ones who appoint the project team and can determine the timescales of the project. Good planning and communication also represent two essential elements of well managed construction projects. In either case, Clients play a significant role. They set the agenda, should make their expectations of high health and safety standards clear to their supply chain, and should also ensure that they and others pass on the necessary information that facilitates effective planning for safe work.

For non-notifiable projects, the Client is directly responsible for implementing health and safety management arrangements, and fulfilling a number of other duties, including assessment of the competency of Contractors and Designers, vetting health and safety documentation and preparing / updating the health and safety file. In these situations the Client is therefore reliant on the competence of the health and safety advice they have in place.

For notifiable projects, the Client has additional duties, which include appointing

Case study

When Hitachi Consulting UK moved into a new central London office, it appointed MMoser Associates as lead Designers and Principal Contractors to oversee the re-fit of its new offices located at the prestigious 2 More, London. Workplace Law advised and assisted on the measures needed to comply with the Construction (Design and Management) (CDM) Regulations, which require the appointment of a Competent CDM Coordinator.

Claire Hollister, UK Operations Manager explains:

"Workplace Law was recommended as a firm who could help me understand my responsibilities and make sure that they were being executed correctly. During our detailed fit-out planning stages, Simon Toseland came to my office and explained all my responsibilities and how Workplace Law could implement procedures and carry out the checks to ensure that any risks were identified and steps implemented to ensure they were significantly mitigated."

Health and safety pre-construction information was gathered and distributed to the construction and design team members, and an 'Information Inventory' was used to identify the specific information required and when it was communicated.

"The day after his appointment, Simon was meeting with my fit-out team and within two weeks our fit-out plan had been updated to reflect his additional requirements," says Hollister. "Although these changes were small, Simon gave me the reassurance that only comes from independent validation that my fit-out team were working to the standards required – this confidence was endorsed when the working site received an audit from the HSE and passed with flying colours."

Health and Safety CDM Workshops were facilitated by Workplace Law to encourage designers to comply with their duties and collectively consider and mitigate hazards. The information was recorded through a design 'CDM' risk register. This register was then used to provide information on design health and safety residual risks, which affected the management of the construction work as well as operational and maintenance issues for which the client would be responsible.

There was, in fact, a high regard for health and safety throughout the lifespan of the project. An unannounced visit by a HSE inspector, during the construction phase, resulted in a clean bill of health for MMoser. There were no reportable accidents and the project was delivered on time and in budget and at the end of the project the health and safety file was prepared and handed over to the client, in person.

a competent CDM Coordinator (CDMC) and Principal Contractor (PC). It should be noted that until such time that the appointment of a CDM-C or PC has been made, the Client retains the role. The longer this appointment process takes, the less of an opportunity there will be to influence design in terms of safety.

	Duties (* additional duties for notifiable projects)
Client (excluding domestic client)	■ Check competence and resources of all appointees. ■ Ensure suitable management arrangements for welfare facilities. ■ Allow sufficient time and resources for all stages. ■ Provide pre-construction information to Designers and Contractors. ■ Appoint CDM Coordinator.* ■ Appoint Principal Contractor.* ■ Provide information relating to the health and safety file to the CDM Coordinator.* ■ Retain and provide access to the health and safety file.*
CDM Coordinator	■ Advise and assist the Client with his/her duties and notify HSE.* ■ Coordinate health and safety aspects of design work and cooperate with others involved with the project.* ■ Facilate good communication between Client, Designers and Contractors.* ■ Liaise with Principal Contractor regarding ongoing design.* ■ Identify, collect and pass on pre-construction information.* ■ Prepare / update the health and safety file.*
Designer	■ Eliminate hazards and reduce risks during design. ■ Provide information about remaining risks. ■ Check Client is aware of duties and CDM Coordinator has been appointed.* ■ Provide any information needed for the health and safety file.*
Principal Contractor	■ Plan, manage and monitor construction phase in liaison with Contractor. ■ Prepare, develop and implement a written plan and site rules. ■ Give Contractors relevant parts of the plan. ■ Make sure suitable welfare facilities are provided from the start and maintained throughout the construction phase. ■ Check competence of all appointees. ■ Ensure all workers have site inductions and any further information and training needed for the work. ■ Consult with the workers. ■ Liaise with CDM Coordinator regarding ongoing design. ■ Secure the site.
Contractor	■ Plan, manage and monitor own work and that of workers. ■ Check competence of all their appointees and workers. ■ Train own employees. ■ Provide information to their workers. ■ Comply with the specific requirements in Part 4 of the Regulations. ■ Ensure there are adequate welfare facilities for their workers. ■ Check Client is aware of duties and a CDM Coordinator has been appointed and HSE notified before starting work.* ■ Cooperate with Principal Contractor in planning and managing work, including reasonable directions and site rules.* ■ Provide details to the Principal Contractor of any Contractor whom he engages in connection with carrying out the work.* ■ Provide any information needed for the health and safety file.* ■ Inform Principal Contractor of problems with the plan.* ■ Inform Principal Contractor of reportable accidents, diseases and dangerous occurrences.*

CDM

The CDM Coordinator

The CDM Coordinator (CDMC) has completely replaced the role of the Planning Supervisor, which was a duty holder under the CDM 1994 Regulations. The CDMC is only appointed on notifiable projects and, once appointed, must advise and assist the Client in complying with their duties.

The CDMC is also required to ensure that arrangements are in place for all aspects of planning and coordinating with health and safety measures primarily before the construction phase starts. This will involve ensuring that Designers are giving due consideration in complying with their duties and that clear channels of communication and cooperation have been established within the project team, particularly around the flow of information.

Initially the CDMC should identify what existing pre-construction information held by the Client exists and that it has been successfully passed on to the Designers and Principal Contractor. Although the CDMC does not have the obligation to prepare a construction phase plan, they are best positioned to advise the Client on its suitability. Another function of the CDMC is to prepare the health and safety file and hand it over to the Client at the end of the project.

The Designer

The Designer has obligations that apply to all construction work in accordance with Regulation 11. The influential role of the Designer on health and safety management in a construction project is underscored by the obligation to satisfy himself that the Client is aware of its duties under CDM 2007.

The CDM 2007 Regulations require the Designer to have regard to a list of specific risks that covers the full lifecycle

of a project, including construction, maintenance, building operational use and eventual demolition.

In the case of a notifiable project, the Designer should not commence work beyond the initial design stage without the confirmation that the CDMC has been appointed as required by Regulation 18.

The need for the Designer to share design information with others, so as to enable them to comply with the CDM 2007 Regulations, is seen as being critical in order to coordinate health and safety hazards and risks.

The Principal Contractor

The role of the Principal Contractor only exists in respect of notifiable projects, as set out in Regulation 22, where they have an obligation to prepare a project-specific Construction Phase Plan before work starts.

The Principal Contractor has the overall responsibility to plan, manage and monitor the construction phase. To assist the Principal Contractor, the CDM 2007 Regulations empower and require the Principal Contractor to manage the activities on site by means of disseminating prescribed information, consulting and giving directions to other contractors.

Contractors other than the Principal Contractor are required to cooperate with the Principal Contractor in accordance with Regulation 19. This includes complying with directions given by the Principal Contractor and providing it with details on the management and prevention of health and safety risks created by the Contractor's work on site. For non-notifiable projects, Contractors have duties to ensure that they satisfy themselves that anyone they employ or engage is competent and adequately resourced. They must also plan and manage the work that they have been appointed to carry out, which can be achieved through consultation of the workforce, cooperating with others, and managing the hazards and risks associated with their working practices.

The Construction Phase Plan

The means by which all the parties to a notifiable project share information and plan for the construction phase is the Construction Phase Plan. The health and safety plan, which was an innovative feature of the CDM 1994 Regulations, has been renamed but maintains its role in linking the duty holders to a project through sharing knowledge, with the objective of improving the exchange and communication of information that affects health and safety.

The Construction Phase Plan is intended to be a dynamic document, subject to continuous review and amendment throughout the construction phase. The Construction Phase Plan also has an important role in fulfilling the general principles of coordination and cooperation that are required by Regulations 5 and 6 of the CDM 2007 Regulations.

The Construction Phase Plan must be project-specific and communicated to others working on the project. Although it is the Principal Contractor who prepares the plan, there is a duty on the Client to ensure that it is both adequate and in place before work starts. The Client can rely on the advice of the CDMC when making this decision.

The suggested contents of the Construction Phase Plan are located within Appendix Three of the CDM 2007 Approved Code of Practice (L144) (see *Sources of further information*).

Approved Code of Practice

The CDM 2007 Regulations should be read together with the Approved Code of Practice (ACoP), which provides practical guidance on compliance with the CDM 2007 Regulations. Failure to comply with the ACoP is not in itself an offence, although such failure may be taken by a court in criminal proceedings as proof that a person has contravened the CDM 2007 Regulations. In those circumstances, it is open to a person to satisfy the court that he has complied with the Regulations in some other way, although that would be very onerous.

The CDM 2007 Regulations are of universal application to construction work in Great Britain. The main reason for the new Regulations was to avoid some of the bureaucracy that became a feature of the CDM 1994 Regulations and to clarify the responsibilities of the parties to a construction project. It would be wrong to assume that the new Regulations create the same duties and obligations.

Any party to a construction project who has responsibility for the design, planning or management of construction work cannot avoid the obligation of becoming familiar with the new CDM 2007 Regulations in making their contribution to the improvement of health and safety in the construction industry.

Enforcement of the CDM 2007 Regulations

Although Clients may be aware that enforcement notices may be served upon Contractors, they may not be aware that notices can also be served upon Designers and themselves. An initiative was undertaken over a three-year period by the HSE's Construction Division. The HSE inspectors conducted a number of in-depth audits of firms of designers and also undertook site visits, with Designers invited to attend. The broad aims of this initiative were to assess Designer performance and to gather intelligence concerning both good and bad design practice. In 2005, a total of 128 design practices were assessed, 14 follow-up visits were made and three improvement notices were served. It should be expected that, in those cases where the HSE Construction Inspector investigates accidents or injuries on a construction site, the investigation trail will be followed back through the Designers to the Client in order to ascertain whether either the Designer or the Client has contributed to any failings in the design, the allocation of resources (including time) or failure to communicate information down the line.

Under CDM 1994, between April 1995 and March 2006, there were, in total, 1,276 successful prosecutions (all of duty holders), under CDM. Of these, 144 were served against the Client.

See also: Construction site health and safety, p.177; Health and safety at work, p.361.

Children at work

Nicola McMahon, Charles Russell LLP

Key points

- A child is a person not over compulsory school age (i.e. up to the last Friday in June in the academic year of his/her 16th birthday). A young person is a person who has ceased to be a child but is under 18.
- The employment of children in the UK is subject to limitations regarding the number of hours that they can work. Children also have a number of special rights and protections in the workplace, justified on health and safety grounds. It is important for all employers of children to be aware of the legal framework.
- The general principle, subject to exceptions, is that a child under 14 may not work.
- Any organisation employing a child of compulsory school age must inform the local education authority in order to obtain an employment permit for that child. Without one, the child may not be covered under the terms of the employer's insurance policy. (This does not apply, however, in respect of children who are carrying out work experience arranged by their school.)

Legislation

- Children and Young Persons Act 1933.
- Children and Young Person Act 1963.
- Education Act 1996 (as amended).
- Employment Rights Act 1996 (as amended).
- Management of Health and Safety at Work Regulations 1999.
- Children (Protection at Work) Regulations 1998-2000 (implementing the provisions of the EC Directive on the Protection of Young People at Work).
- Working Time Regulations 1998.
- Working Time (Amendment) Regulations 2002.

In addition, the United Nations Convention on the Rights of the Child (for these purposes, all persons under 18) provides that member states have an obligation to protect children at work, set minimum ages for employment and regulate conditions of employment. No child may be employed to carry out any work whatsoever (whether paid or unpaid) if s/he is under the age of 14. This is subject to some specific exceptions, e.g. in relation to children working for their parent / guardian, or in sport, television, the theatre or modelling. In the case of a child who is 14 or over, s/he may not:

- do any work other than light work;
- work in any industrial setting, sea-going boat, factory or mine (subject to very specific exceptions);
- work during school hours on a school day;
- work before seven a.m. or after seven p.m.;
- work for more than two hours on any school day;
- work for more than 12 hours in any week during term-time;
- work more than two hours on a Sunday;
- work for more than eight hours on any day s/he is not required to attend school (other than a Sunday) (or five hours if under the age of 15);

- work for more than 35 hours in any week during school holidays (or 25 hours in the case of a child under the age of 15);
- work for more than four hours in any day without a break of at least an hour;
- work without having a two-week break from any work during the school holidays in each calendar year; or
- work without an employment permit, issued by the education department of the local council, and signed by the employer and one of the child's parents.

In addition, each local authority has specific, additional by-laws regulating the employment of children in its area.

Health and safety considerations

The main source of health and safety legislation in relation to the employment of a child or young person is the Management of Health and Safety at Work Regulations 1999. Under these Regulations, the term 'young person' means any person under 18. Before a young person starts work, the employer must carry out a risk assessment, taking into account various issues such as the inexperience and immaturity of the young person, the nature of the workstation, the risks in the workplace, including the equipment which the young person will use, and the extent of health and safety training provided. Young persons cannot be employed to carry out any work:

- beyond their physical or psychological capacity;

- involving harmful exposure to toxic or carcinogenic agents or radiation;
- involving the risk of an accident which a young person is more likely to suffer owing to his insufficient attention to safety or lack of experience; or
- in which there is a risk to health from extreme cold, heat, noise or vibration.

A breach of the Regulations, if sufficiently serious, could lead to criminal prosecution following a complaint to the HSE.

Work experience

Guidelines were issued by the DTI (now BIS) in 2007 regarding work experience placements in the television industry and these are now regarded as being applicable in various industries. The main points are:

- A written document should detail the framework of the placement;
- Placements should be for a fixed period of time (ideally between two weeks and a month);
- Placements should not entail more than 40 hours work per week;
- Specific learning objectives should be agreed between the child on the placement and the provider of the placement; and
- No unpaid volunteer should be under an obligation to comply with an employer's instructions.

See also: Young persons, p.774.

Sources of further information

Worksmart – children's work rights:
www.worksmart.org.uk/rights/childrens_work_rights

Children's Legal Centre – a charity providing advice and information for young people: www.childrenslegalcentre.com/Legal+Advice/Child+law/childemployment/

Climate change

UK Centre for Economic and Environmental Development (UK CEED)

Key points

- Human activity has been dramatically affecting natural cycles by altering natural climate change patterns. In fact, the planet is warming mainly because of the build-up of greenhouse gases, particularly carbon dioxide (CO_2).

- Increasingly, businesses will need to adapt both their strategy and premises in order to assess the risks, and mitigate the effects, of climate change. In applying enterprise and innovation to these challenges there is potential for business opportunities to arise from climate change, along with efficiency savings.

- According to research three-quarters of the largest businesses are at least thinking about climate change, reducing to just over half in smaller companies. However, a new British Gas business report argues that 70% of UK businesses are not currently considering investing in energy-efficiency measures. Ipsos Mori research found 'the progress of most of the organisations we spoke to on climate change adaption is at a fairly early stage'.

Legislation

- The Climate Change Act 2008.
- Climate Change Act 2008 (Credit Limit) Order 2011.
- Climate Change Act 2008 (2020 Target, Credit Limit and Definitions) Order 2009.
- Buildings Regulations 2010.
- Energy Act 2010.
- The Ecodesign for Energy-related Products Regulations 2010.
- The Climate Change Levy (General) (Amendment) Regulations 2011.
- CRC Energy Efficiency Scheme (Amendment) Order 2011.
- Energy Information Regulations 2011.
- A Roadmap for moving to a competitive low carbon economy in 2050 (COM (2011) 112).
- Climate Change Agreement Scheme.
- EU Emission Trading Scheme (EU ETS).
- The National Allocation Plans (NAPs).

How climate change might affect your business

Climate change already affects economic activity and business continuity, and in all business areas as stressed by the UK Climate Impacts Programme (UKCIP), from the more obvious areas of product design or service delivery to markets (changing demand for goods and services); finance (investments, insurance and stakeholder reputation); logistics (vulnerability of supply chain, utilities and transport); premises (impacts on building design, construction, maintenance and facilities management); people (behavioural changes for workforce and customers); and process (production processes and service delivery).

Increasingly, businesses will need to adapt both physically, in terms of premises, and strategically, in order to mitigate, and potentially capitalise on, the effects of climate change upon their business.

Case study

Existing and new-build residential and commercial properties are progressively more exposed to the physical impacts and consequences of climate change in terms of flooding, water scarcity and overheating. Therefore, it has become hugely important to retrofit buildings to ensure their resilience, resistance and long-term sustainability against climate change. Focusing on overheating, a new survey by the Office of National Statistics shows that 92% of office workers lose up to an hour per day due to overheated offices, costing British employers around £19.3m in lost productivity over the summer.

Moreover, The British Safety Council argues that with indoor temperatures above 24°C the propensity for accidents increases and work productivity declines. In order to effectively control indoor environments, a broad spectrum of actions can be undertaken including:

- structural and expensive solutions:
  - external solar control;
  - improved ventilation systems and insulation;
  - energy efficient IT office equipment;
  - replacing carpeted floors with wood and old windows with new, low e-coating windows; and
  - green roof installation;
- more cost-effective measures:
  - better natural ventilation; and
  - blinds; and
- passive actions:
  - not using energy when unnecessary.

These measures can be beneficial at all levels, e.g. by providing cleaner and healthier workplaces while saving money on energy bills (typically between 25 and 30% according to Government estimates) along with cutting CO_2 and other potent greenhouse gas emissions. However, current uptake of these sorts of measures is low due to barriers, e.g. incomplete information and a lack of awareness about available options.

The unprecedented depletion of natural resources, the scale and the speed of environmental degradation, and pollution are unsustainable and worsen climate change. The annual UK production of over 80 million tonnes of waste results in environmental damages and monetary losses for business and households. In 2009, 52% of commercial and industrial waste was recycled or reused in England, but a European Environment Agency report warns that in 2008 UK landfill contributed 30.2% to the total EU15 waste methane emissions.

Managing waste effectively is not only a legal requirement but also can save money by cutting expenses of raw material, packaging, equipment and cutting waste disposal costs. It can even generate a source of revenue in itself,

> **Case study** – *continued*
>
> e.g. waste products can be sold to other businesses for reuse or recycling; it can also improve the image of the company to stakeholders such as customers, suppliers and investors. There are a series of actions helping the UK move towards a zero waste economy. This includes the creation of a waste business policy according to the waste hierarchy (prevent, then minimise, then reuse, then recycle, then energy recovery, and – after all else – dispose) and introducing IT control equipment. Particular care must be taken to dispose of hazardous materials.
>
> As a best practice in waste management, Yorkshire Window Company Ltd, a manufacturer of PVC windows, has managed to make resource efficiency an everyday practice using computer-controlled systems to optimise its processes; establishing separation and recycling routes for materials previously disposed of to landfill; and raising staff awareness and demonstrating commitment from the top of the organisation down. The main resulting benefits have been that waste from profile cutting is down to 1%; automation increased production by 14%; a 68% recycling rate was achieved in 2007; glass deliveries were reduced by 156 per year; and the company achieved ISO 14001 certification in 2006.

In fact, climate change effects on business have the potential to be both positive and negative. UK Trade and Investment report that two-thirds of international businesses consider that a range of opportunities, rather than risks, can arise from climate change. For a given business, these effects are dependent on a range of factors. The factors include:

- business location;
- nature of business activity;
- ability of premises to withstand extreme weather events;
- business customers; and
- length, location and diversity of business supply chain.

Climate change is already having a relatively significant impact upon weather patterns. For example, the growing season in East Anglia is beginning around three weeks earlier than in 1980. A DEFRA survey on climate change adaptation found around four in ten organisations said that they have been significantly affected in the past three years by the kind of extreme weather that climate change can make more likely. The Association of British Insurers has stated:

'Climate change is not a remote issue for future generations to deal with. It is, in various forms, here already, impacting on insurers' businesses now.'

The Association has noted that weather risks for households and property are increasing by 2-4% per annum. Claims for storm and flood damages in the UK doubled to over £6bn over the period 1998-2003, compared to the previous five years. These changes may result in higher levels of business premises damage and disruption. This will result in rising insurance premiums, and the risk that in some areas flood insurance will become unaffordable for some businesses.

These impacts are set to become more accentuated, with increasing temperatures

---

Facts

- Climate change is being caused by increasing concentrations of anthropogenic greenhouse gases (GHG) being released into the atmosphere, leading to an enhanced greenhouse effect and further global warming.
- The atmospheric concentration of CO_2, the most important GHG, is at its highest level for 650,000 years, analyses of gases trapped in ice cores reveal. It has increased by 38% since the Industrial Revolution and because it resides for such a long time in our atmosphere it continues to build up, as we emit more.
- Average global temperature is increasing by 0.2°C every ten years. By 2005 the average global temperature had increased by 0.76°C from pre-Industrial times. Even if global average temperatures rise by only 2°C, 20–30% of species could face extinction. This will have serious effects on our environment, food and water supplies and health.
- The past decade was the warmest since reliable records began in 1880. Globally, nine out of the ten hottest years ever recorded have occurred since 1990. 2010 was the warmest year of the global surface temperature record – the 34th consecutive year with temperatures above the 20th Century average.
- As global temperature increases, sea levels will rise. This is due to both thermal expansion (as water warms it expands), and ice-melt. Significant amounts of water are stored in glaciers, permafrost and ice caps. As warmer weather causes them to melt, water will be released into oceans, raising sea levels.
- Other major effects of climate change are increasing temperatures above the surface and in the depths of the ocean; reduction in Arctic sea-ice; shrinking ice-sheets; changes in rainfall patterns and in eco-systems.
- Total summer rainfall has decreased in most parts of the UK.
- Sea-surface temperatures around the UK have risen by around 0.7°C over the last 30 years.
- In summer 2007, while temperatures in areas of south-east Europe reached 46ºC causing several deaths, forest fires and damages to the agricultural sector, parts of Britain experienced their worst floods for 60 years, resulting in deaths, billions of pounds worth of damage and disruption to water supplies.

Source: Met Office (2011).

and changes to seasonal weather patterns, as well as extreme weather events such as coastal and river flooding, strong winds or high temperatures, with the three more high-risk exposed areas being London, the South-East and the East of England. One principal benefit for the UK will be warmer winters.

Knock-on impacts of these more extreme weather events are also likely to become increasingly important issues, such as water shortages as a result of higher temperatures and increased energy costs and carbon footprint for cooling of premises. For an average global temperature rise of 4°C, models project (although there are a range of computations) the warmest days could be 8°C hotter than at present.

Temperature is of course magnified in urban areas. Knock-on effects will be increased electrical and water use. Employees will be more vulnerable to heat stress. Health and safety legislation states that the workplace should be kept at an optimal temperature of 24°C, with the upper limit previously defined as 30°C.

Under the 2008 Climate Change Act, some companies are now legally obliged to take action on climate risk. This is complemented by the CRC Energy Efficiency Scheme. This is a mandatory scheme for large organisations, which use over 6,000MWh of energy per year, in both the public and private sector. The scheme aims to reduce the energy consumption of these companies, resulting in a net financial gain. There is a league table detailing the comparative performance of participants. With increased awareness of environmental issues amongst the general public, sustainable and 'green' credentials can be financially beneficial for a business' image.

Many processes of the earth underpin the economy. Natural benefits include food, improved air quality, mechanisms for a stable climate, water purification, pollination, prevention of flooding and erosion, as well as enjoyment, recreation and spiritual sustenance. These services are going to be lost or damaged as the climate heats up. Biodiversity and ecosystems could be protected by businesses taking proper account of the real value of natural resources. In the short term, companies can look to source products, or even ideas, from renewable or sustainable sources, or look to offset their impact by participating in projects to improve or sustain biodiversity.

Implementing adaptation and mitigation measures: A new business strategic approach to climate change

Evaluating business strategy has the potential to avoid future disruption and costs, as well as mitigating the current effects of climate change. Integrating current environmental concerns and future legislation into business plans can also open up new business opportunities.

Today, in 30 countries, 11,000 companies including installations such as power stations, combustion plants, oil refineries and iron and steel works, factories making cement, glass, lime, bricks, ceramics, pulp, paper and board are responsible for almost half of the EU's emissions of CO_2 and 40% of its total GHG emissions. The organisations must take into account the EU Emissions Trading System (EU ETS) while setting up their environmental strategy. ETS is the key pillar and tool of the European Union's climate policy for reducing industrial GHG emissions cost-effectively by setting a limit on the total amount of certain GHGs that can be emitted. Companies are given the related emissions allowances which can be traded as needed. The limit on the total number of allowances available makes them valuable.

Although a business may not be directly affected by the growing effects of climate change, by planning ahead it may be possible to limit future effects. In order to successfully integrate climate change into business strategy it is necessary to consider both short- and long-term goals. A strategic business plan should outline these goals and define the steps necessary in order to achieve them. Part of this plan should address the positive impacts of climate change, such as new opportunities, as well as negative impacts; for instance, decreasing purchasing power and effects upon customers and employees, rising costs, changing consumer behaviours and supply chain interruptions.

Sustainability is of paramount importance in updating a strategic business plan in response to climate change. This involves ensuring the future potential of a business' activity will not be inhibited by present modes of operation, taking into account environmental and economic factors. As with all business risks – which according to many authors (e.g. KPMG) can be identified as regulatory, physical, litigation, reputational – those associated with climate change can be managed through taking an approach of risk assessment and mitigation. This may involve using an environmental management system (EMS), business continuity planning, risk management and strategic planning. These processes are effective in assessing the strengths and weaknesses of a business, as well as securing costs and increasing internal efficiency.

The following are some strategic business plan ideas that may assist in mitigating the effects of climate change:

- Green Travel Plans include measures to minimise business trips (e.g. increased use of telephone / video conferencing) and other logistic operations, e.g. car sharing, bike schemes, improved facilities.
- A flexible working hours scheme allows employees to working from home, saving on travel.
- New energy efficient IT office equipment.
- Procurement of more sustainable products.
- Defined waste plan.
- Moving away from the 'just in time' model.
- Regular CSR reporting.

- Implementing regular risk assessment and management plans in order to evaluate potential threats such as flooding, and therefore set insurance plans accordingly. Tools such as the DirectGov website and flood maps on Environment Agency's website can support this.

Insuring against climate change may protect your business against the risks of physical damage of buildings or equipment; however, insurance cannot protect against all possible risks. This is of particular consideration with regards to geographical location. For instance, some areas may be more prone to flooding and the effects of sea-level change than others, making them less insurable.

The Environment Agency reported that the 2007 flooding in the UK led to 180,000 insurance claims, totalling over £3bn, affecting businesses in various ways. For instance, as reported by the CBI in 2009, businesses lost short-term computing capacity and long-held data when network servers were damaged by flood water. Annual flood damages in the UK could total £22bn by 2080.

Practical steps to deal with climate change

A number of practical measures can be taken by businesses to adapt to climate change. Alternative building construction may need to be considered, such as changing roof and window design to withhold increased extreme weather conditions. It may also be necessary to include flood-resilient design into buildings, such as improved drainage systems or physical barriers such as the Flood Ark system.

The extent to which these issues should be taken into account is of course dependent on the location of the business.

For instance, those located closer to the coast are more at risk. This notion is reinforced in a recent paper by AXA Insurance, which predicts that within 200 years, as sea levels rise, cities including London, Edinburgh, Bournemouth and Peterborough could all be under water. Additionally, businesses may be affected by flooding from surface waters such as streams, lakes and rivers. These factors should also be taken into account when considering the location of new premises.

Further emphasis should be placed on improving the energy efficiency of the business' workforce. This will become increasingly necessary as pressure upon utilities grows. Becoming more energy efficient will reduce fuel consumption and, as a consequence, costs. Measures such as natural ventilation and insulation, blinds and energy efficient equipment can contribute to this. Implementation of recycling schemes, energy saving light bulbs and car sharing are also examples of measures that can be taken to mitigate the effects of climate change. There are a host of companies that will produce an energy audit for your business, many of which are free of charge. Audits are also offered by the Carbon Trust, which can also be contacted for general advice, and which even offers low-interest loans to invest in energy saving equipment.

Working from home will decrease electricity and office expenses for the company, whilst saving the carbon emissions of commuting, and saving time for the employee. A further measure could be to extend employee working hours. This can allow hot desk arrangements, which

can offer significant savings over an office relocation and bring the benefit of manning the telephone for longer, although there will be an increase in the hours for heating and ventilation.

Engaging colleagues in raising awareness of climate risks will provide different, valuable perspectives on the impact of climate change upon your business, and is likely to encourage staff to be more receptive of new ways of working. Improving a company's carbon footprint and green credentials will bring tangible financial rewards.

Conclusion
It is apparent that the effects of climate change are likely to grow in the future, creating significant challenges for businesses. With climate change perceived as an inevitability against which adaptation and mitigation measures must be taken, it is the responsibility of businesses to re-evaluate business strategies to reduce risk, and investigate the potential to maximise any opportunities that climate change may present. Sustainability is the key tool to drive innovation and improve business, which all businesses should fully incorporate into their strategic plans.

See also: Business Continuity Management, p.92; The CRC Energy Efficiency Scheme, p.215; Environmental Management Systems, p.286; Environmental risk assessments, p.292; Recycling, p.605.

Sources of further information

Information and data on Climate Change:
www.metoffice.gov.uk/climatechange
http://ec.europa.eu/clima/publications/docs/factsheet-climate-change_en.pdf

Business Link guide to Climate Change and cutting business carbon emissions:
http://bit.ly/nErceh

Assess the environmental compliance of your business: http://bit.ly/rg7un8

KPMG report: http://bit.ly/nHnjPE

DEFRA Climate Adaptation survey: http://bit.ly/aBa5fn

Visit the website for the Government-led initiative ACT ON CO_2, aimed at encouraging and assisting people in reducing their CO_2 emissions:
www.direct.gov.uk/actonco2

The CCR Energy Efficiency Scheme:
www.decc.gov.uk/en/content/cms/what_we_do/lc_uk/crc/crc.aspx

Government guide to flood preparedness:
www.direct.gov.uk/en/HomeAndCommunity/WhereYouLive/FloodingInYourArea/DG_10014599

European Environment Agency publications:
www.eea.europa.eu/publications/#c9=all&c14=&c12=&c7=en&b_start=0

COMAH: Control of Major Accident Hazards

Jan Burgess, CMS Cameron McKenna

Key points

- The growth of industrialisation in the second half of the 20th Century brought with it an increasing number of serious accidents, particularly at sites storing dangerous substances. It soon became apparent that greater legal controls were necessary to prevent and mitigate the effects on people and the environment of those incidents involving dangerous substances.

- The first controls at a European level were a direct result of the Seveso accident in 1976, when dioxin accidentally released from a pesticide-manufacturing plant caused 2,000 people to seek medical attention. Although no immediate fatalities were reported, kilogram quantities of the substance – lethal to human beings even in microgram doses – were widely dispersed, which resulted in immediate contamination of some ten square miles of land and vegetation.

Legislation

- Control of Major Accident Hazards Regulations 1999 (COMAH).
- Control of Major Accident Hazards (Amendment) Regulations 2005.
- Planning (Control of Major Accident Hazards) Regulations 2005.

The 'Seveso' Directives

The first 'Seveso' Directive (Directive 82/501/EEC) was adopted by the European Commission in 1982. Further incidents in Bhopal, India in 1984 (where a leak of methyl isocyanate caused more than 2,500 deaths) and in Basel, Switzerland in 1986 (where firefighting water contaminated with mercury and other chemicals caused massive pollution of the Rhine) led to the Seveso Directive being amended twice to broaden its scope (Directives 87/216/EEC and 88/610/EEC).

The Seveso Directive was superseded in 1996 by Directive 96/82/EC on the control of major accident hazards – the 'Seveso II' Directive. The primary purpose of Seveso II is to prevent major accidents at industrial sites storing or using dangerous substances above certain thresholds, and, if such accidents do occur, to mitigate their impact on humans and the environment. It introduced new requirements for safety management systems, emergency planning and land-use planning. One of its main aims was to address accidents caused by management failures, a significant factor in over 90% of major hazard accidents in the EU since 1982. The Seveso II Directive also gave more rights to the public in terms of access to information and consultation, with operators and public authorities having certain obligations to publish information.

The Seveso II Directive was extended in 2003 by Directive 2003/105/EC, as a result of several serious industrial accidents (at Toulouse, Baia Mare and Enschede), and to take into account new studies on carcinogens and substances dangerous to the environment. Major changes included an extension of the Directive to cover risks arising from storage of explosives and ammonium nitrate, and risks arising from mining operations.

Case studies

On 11 December 2005, a number of large explosions occurred at the Hertfordshire Oil Storage Terminal, generally known as the Buncefield Oil Depot. Fire engulfed a high proportion of the site. The fire burned for several days, destroying most of the site and emitting large clouds of black smoke into the atmosphere. Forty-three people were injured but there were no fatalities. Significant damage occurred to both commercial and residential properties in the vicinity and a large area around the site was evacuated on emergency service advice.

A Government inquiry conducted jointly by the HSE and the EA was initiated to identify the immediate causes of the explosion. The inquiry was fact-finding only and was not intended to attribute fault. The Buncefield Major Incident Investigation Board (the Board) published its Final Report into the incident on 11 December 2008 (see *Sources of further information*). The report summarises the incident and the work of the Board, and brings all of the Board's 78 recommendations together. It also examines the economic impact of the incident.

The Buncefield Standards Task Group (BSTG) (distinct from the Board) issued a report into safety and environmental standards for fuel storage sites on 24 July 2007 (see *Sources of further information*). The BSTG was formed shortly after the incident and consisted of representatives from the COMAH Competent Authority and Industry. The report specifies the minimum expected standards of control that should be in place at all establishments storing large volumes of petroleum, and similar products capable of giving rise to flammable vapour clouds if there is loss of containment.

Following the criminal inquiry into the Buncefield explosion, five companies were fined a total of £9.5m for various breaches of the Health and Safety at Work etc. Act 1974, the Water Resources Act 1991, and COMAH. Two of the five companies were found guilty of breaching Regulation 4 of COMAH, which requires that 'every operator shall take all measures necessary to prevent major accidents and limit their consequences to persons and the environment'.

In February 2008, a much smaller but nevertheless significant blast occurred at Shell's Bacton Gas Terminal in Norfolk. No one was killed or seriously injured. The blast occurred when highly flammable condensate leaked from a corroded vessel into a concrete water storage tank, where it was exposed to an electric heater element. In addition to the blast, there was an unauthorised release into the North Sea of hundreds of tonnes of fire water and firefighting foam. Shell was sentenced in June 2011 after pleading guilty to seven charges under health and safety and environmental legislation, including a charge under Regulation 4 of COMAH. Ipswich Crown Court ordered Shell to pay £1.24m in fines and costs.

At the time of writing, the Police and HSE are continuing their investigations into a fire and explosion at the Chevron Pembroke Refinery on 2 June 2011. Four workers died in the blast.

In December 2010, the European Commission adopted a proposal for a new Directive on the control of major accident hazards involving dangerous substances (see *Sources of further information*).

The proposal primarily addresses the consequences to Seveso II of changes to EU legislation on the classification, labelling and packaging of chemical substances (i.e. Annex I to Regulation (EC) No. 1272/2008 – the 'CLP' Regulation). However, it also recognises that there should be a review of the need for other changes. Amongst other things, it suggests improving access to and the quality of information disseminated to the public about major accident hazard sites.

Implementation of the Seveso II Directive in Great Britain: COMAH

With the exception of the land use planning requirements, the Seveso II Directive was implemented in Great Britain by the Control of Major Accident Hazards Regulations 1999 (COMAH). These Regulations, which came into force on 1 April 1999, replaced the Control of Industrial Major Accident Hazards Regulations 1984 (CIMAH). The HSE, together with the EA (in England and Wales) or the SEPA (in Scotland) act together as the 'Competent Authority' (CA) and are responsible for enforcing the COMAH Regulations.

Companies regulated under COMAH fall into two categories – the lower or upper tier – depending on the quantity of dangerous substances present at a site. Lower tier duties apply to all operators covered by the Regulations, and involve notification to the HSE and EA (or SEPA) of the presence of any dangerous substance. There is also a requirement to prepare a written Major Accident Prevention Policy (MAPP) setting out the overall aims and principles of action for controlling major accident hazards.

If the quantity of a substance held at a particular site is above the threshold laid down in COMAH, the operator producing the substance will fall within the upper tier. Additional duties then apply. They include an obligation to prepare a safety report, which should include information on the safety management system, organisation of the site and its surroundings, and likely causes, probability and consequences of a major accident.

Companies falling within the upper tier must also prepare an on-site emergency plan, and provide information to local authorities to assist them in the preparation of an off-site plan. Emergency plans must be reviewed and (where necessary) revised at least every three years.

The on-site emergency plan

The aim of this plan is to contain and control incidents so as to minimise the effects and limit damage to persons, the environment and property. The plan should also provide for the communication of necessary information to the public and emergency services in the area. It should additionally provide for the restoration and clean-up of the environment following a major accident. The requirements for the on-site emergency plan are contained within Schedule 5 of COMAH.

The off-site emergency plan

The local authority for an area in which a top-tier organisation is located must prepare an emergency plan for dealing with the off-site consequences of a major accident. The objectives and requirements set out in Schedule 5 relating to on-site plans also apply to off-site plans. In preparing the off-site emergency plan, local authorities must consult the operator, emergency services, Competent Authority,

> **Facts**
> - COMAH applies to industrial sites where certain quantities of dangerous substances are kept or used – usually sites that manufacture, process or store dangerous chemicals or explosives, or nuclear sites.
> - The HSE and the Environment Agency (in England and Wales) or the Scottish Environment Protection Agency (in Scotland) are responsible for the enforcement of COMAH.
> - Companies regulated under COMAH will fall under a lower or upper tier.
> - Lower tier duties apply to all operators covered by COMAH and include a duty to notify the relevant authorities of the presence of any dangerous substance.
> - Upper-tier companies (those with particularly high quantities of dangerous substances) must additionally prepare a Safety Report and Emergency Plan.

local health authorities and such members of the public as it deems appropriate.

Changes to COMAH

The COMAH (Amendment) Regulations 2005 (the 2005 Regulations) implemented the health, safety and environment aspects of Directive 2003/105/EC, amending the Seveso II Directive in Great Britain. Since 30 June 2005, operators who come into the scope of the COMAH Regulations, due to an increase in the quantity of dangerous substances present, must submit a notification prior to any construction or operation.

New thresholds and classifications for dangerous substances were introduced, which changed the criteria for triggering the COMAH Regulations, with the result that many businesses were brought within the COMAH Regulations for the first time.

In response to an explosion at a fertiliser plant in Toulouse, some of the main changes introduced by the 2005 Regulations concern ammonium nitrate. The threshold for the nitrogen content derived from ammonium nitrate was reduced. In addition, two new classes of ammonium nitrate were added – one

for fertilisers capable of self-sustaining decomposition and the other for 'off-spec' material. There were also two new entries for potassium nitrate, to reflect the less-combustible nature of certain fertilisers, which would previously have been categorised as oxidising substances and had threshold quantities of 50 and 200 tonnes. In addition, seven carcinogens were added to the existing list.

A category for petroleum products (including naphthas, kerosenes and diesel fuels) was introduced so as to enable the CA to focus its regulatory activity according to the hazards associated with different types of products. The threshold quantities of 2,500 and 25,000 tonnes for petroleum products that trigger the COMAH Regulations are now half those of the former automotive petrol category.

Changes were also introduced in relation to administrative procedures to be followed under COMAH. Those changes included new duties on operators (noted above) to notify the CA when safety reports for upper tier sites are revised or reviewed, and to consult with the EA (or SEPA), employees and members of the public in preparing off-site emergency plans.

Planning regime

Further changes to COMAH were introduced by the Planning (Control of Major Accident Hazards) Regulations 2005 (the Planning Regulations). The Planning Regulations were designed to implement the land use provisions of Directive 2003/105/EC. They are concerned with whether the storage of dangerous substances at a site is an appropriate use of land, taking into account the characteristics of the environment in which it is located. Planning authorities are now obliged to put in place controls on the location of new major accident hazard sites, modifications to existing sites and new developments in the vicinity of existing sites, such as transport links and housing. The overall aim is to ensure that there is an appropriate distance between major accident hazard sites and residential areas.

Implications of the new COMAH regime

As a result of the 2005 amendments, a number of industrial sites previously out of the scope of COMAH came under its umbrella. Similarly, many moved from the lower to the upper tier and thereby became subject to additional duties.

Companies new to COMAH may have to apply for a hazardous substances consent. The cost of obtaining a consent is small, but granting it triggers the formation of a consultation zone around the proposed or existing site. This can result in difficulties when trying to expand activities at a COMAH site and limit future development near the site. Any proposed new developments in the vicinity of existing sites fall under greater scrutiny now than prior to the 2005 amendments.

See also: Accident investigations, p.29; Biological hazards, p.51; COSHH: Control of Substances Hazardous to Health, p.207.

Sources of further information

HSE: Control of Major Accidents Hazards: www.hse.gov.uk/comah/index.htm

Buncefield Major Incident Investigation Board's final report: www.buncefieldinvestigation.gov.uk/reports/index.htm#final

BSTG's final report: www.hse.gov.uk/comah/buncefield/bstgfinalreport.pdf.

European Commission's proposal for a new Directive to replace Seveso II: http://ec.europa.eu/environment/seveso/pdf/com_2010_0781_en.pdf

Competence and the Responsible Person

Kate Gardner, Workplace Law

Key points

- The Corporate Manslaughter and Corporate Homicide Act 2007 and the Health and Safety (Offences) Act 2008 provide the latest legislative impetus for employers to ensure that their employees are competent to undertake what is expected of them.
- Ensuring that employees are competent is the responsibility of senior management. However, competence is not an absolute concept and senior management may wonder how competence can be recognised and assessed.
- The concept of competence in health and safety legislation is so far reaching that the Policy Group of the HSE has published an outline map on competence, training and certification, which is regularly updated and available on the HSE website (see *Sources of further information*).
- The outline map is intended to give an overview of requirements for competence and training in health and safety legislation.
- The concept of competence and its adoption by numerous statutory instruments is such that even the Policy Group has to admit there may be a few omissions.
- Employers or employees have to assess competence simply by relying on the ordinary definition of 'competent' and the HSE's own guidance.

Legislation

- Regulatory Reform (Fire Safety) Order 2005.
- Corporate Manslaughter and Corporate Homicide Act 2007.
- Health and Safety (Offences) Act 2008.

Competency

The *Oxford English Dictionary*'s definition of the relevant aspects of competent is:

- suitable, appropriate;
- sufficient or adequate in amount, or degree;
- legally authorised or qualified, able to take cognisance (of a witness, evidence), eligible, admissible; and

- having adequate skill, properly qualified, effective.

The definition identifies that competence is based on being suitable and appropriate to undertake a task successfully. The need for suitability and appropriateness is only to the extent that the degree of competence is sufficient or adequate.

In considering human factors, competence is defined by the HSE as the ability to undertake responsibilities and perform activities to a recognised standard on a regular basis. Competence is a combination of practical and thinking skills, experience and knowledge.

Case studies

In *Miah v. Thorne Barton Estates* (2007), two companies were ordered to pay damages amounting to £5m to a painter and decorator who suffered brain damage after falling through scaffolding. The High Court heard how Alan Miah was left seriously injured after he fell 20 ft through scaffolding in October 2003. Mr Miah, a self-employed painter from Luton, was subcontracted to work for Thorne Barton Estates Ltd on the conversion of a four storey office building into residential premises.

Karen Morris, HM Inspector for Health and Safety, said:

"This was a terrible accident, but one which was avoidable. An inspection of the scaffold after the incident showed that the scaffold board that broke had several knots in it, which contributed to its lack of strength and subsequent failure. When the scaffold was erected, Thorne Barton Estates Ltd did not receive a handover certificate from Gemini Riteway Scaffolding Ltd. The scaffold was not inspected by a competent person, as required, before it was brought into use for the first time. The scaffold was also not inspected by a competent person at seven-day intervals as required."

In 2004, Thorne Barton Estates was ordered to pay £5,475 in fines and costs following the incident, while Gemini Riteway was fined £17,000. Thorne Barton will pay 30% of the compensation and Gemini Riteway Scaffolding 70%. The settlement consists of a £2.4m lump sum, as well as an index-linked annual payment of £105,000 for the rest of his life – which means that for an average lifespan, he will receive in the region of £5m. Thorne Barton Estates, together with scaffolding firm, Gemini Riteway, admitted liability in the court case.

An external fire risk assessor and a hotel manager were jailed for eight months for fire safety offences in July 2011.

David Liu, who ran The Dial Hotel and Market Inn, pleaded guilty to 15 offences under the Regulatory Reform (Fire Safety) Order 2005, while John O'Rourke, of Mansfield Fire Protection Services, pleaded guilty to two offences under the RRO.

Officers from Nottingham Fire and Rescue Service visited both hotels as part of a routine inspection. They found that both premises were being used to provide sleeping accommodation on the upper floors and that fire precautions, which should have been provided to safeguard the occupants in the event of a fire, were inadequate.

Due to the serious risk to life, they issued prohibition notices preventing any further use of both premises for sleeping accommodation until suitable improvements had been made.

Competence and the Responsible Person

Case studies – *continued*

Mr O'Rourke was prosecuted because he had prepared fire risk assessments for both premises, which failed to identify a number of significant deficiencies that would have placed the occupants at serious risk in the event of a fire.

The offences common to both hotels to which Mr Liu, as the responsible person, pleaded guilty were:

- a lack of a suitable and sufficient fire risk assessment;
- a failure to ensure effective means of escape with doors leading onto corridors not being fire resisting or having self-closers fitted;
- a failure to ensure that emergency routes and exits were provided with emergency lighting;
- a failure to ensure the premises were equipped with appropriate firefighting equipment, detectors and alarms in that there was no fire detection within the bedrooms; and
- a failure to ensure that equipment and devices provided were subject to a suitable system of maintenance in that the fire alarm system, emergency lighting system and firefighting equipment were not tested.

In addition, at the Dial Hotel, officers found both staircases from upper levels terminating in the same ground floor area with no alternative escape routes or separation, a locked fire exit door, and exit routes obstructed by combustible materials. The other offence at the Market Inn related to a missing fire door and a window not being fire resisting.

As well as the prison sentence, Mr Liu was also ordered to pay costs of £15,000, and Mr O'Rourke, as a person other than the responsible person who had some control of the premises, was ordered to pay costs of £5,860, having pleaded guilty to two counts (one for each hotel) of failing to provide a suitable and sufficient fire risk assessment.

The judge said the time had come to send a message to those who conduct fire risk assessments, and to hoteliers who are prepared to put profit before safety.

The requirements identified by the HSE for competence are organised under four headings:

1. General health and safety.
2. Health hazards.
3. Safety hazards.
4. Special hazards.

The HSE further identified 37 general goal-setting requirements where there is no precise detail of how the requirement for competence should be met, including five high risk areas:

1. Construction.
2. Major hazards.
3. Off shore.
4. Railways.
5. Nuclear.

www.workplacelaw.net

Competence has to be assessed with respect to the particular task and associated hazards and risks. The assessment of competence is related to what is necessary to achieve a satisfactory result.

Training and certification are a recognised means of establishing competence in addition to experience. Training and certification alone may not necessarily mean that competence has been achieved.

The wider concept of competence and the importance of experience is highlighted by the judgment in the case of *Maloney v. A Cameron Limited* (1961), where the Court of Appeal was prepared to accept without further comment that a competent person required under the Construction (Working Places) Regulations 1966, who was required to be responsible for inspection and supervision of competent workmen, meant 'possessing adequate experience of such work'.

In a later decision in *Gibson v. Skibs A/S Marina* (1966) the Judge stated:

"Who is a competent person for the purpose of such an inspection? This phrase is not defined. I think that it is obviously to be taken to have its ordinary meaning of a person who is competent for the task. I think that a competent person for this task is the person who is a practical and reasonable man, who knows what to look for and knows how to recognise it when he sees it."

Training

It is of considerable assistance if a person who is required to be competent can demonstrate that they have had the appropriate and relevant training.

Training does not necessarily have to be provided by a recognised training

organisation, although such organisations have the benefit of demonstrating consistency in the training content and delivery, which should be capable of being audited. It is also the case that training organisations can issue certificates of attendance or evidence of having reached an appropriate standard.

Competency should not be considered in isolation and devolved upon one person. All persons involved in activities that affect their own health and safety and the health and safety of others should be competent to undertake the role for which they are engaged. The person at the top of the management structure should be able to demonstrate competence for the role they undertake just as much as the 'front line operative' who has to be competent in their activities.

Competence is about obtaining the best performance from the most important resource any organisation has – its people. To ensure that competence is an integral part of an organisation's culture, incompetence should not be tolerated, and risk assessments and job descriptions should always refer to objective standards for skills and experience and, if necessary, formal training and certification.

Competence and fire safety

As demonstrated by the case study (*above*), fire safety is a particualr area in which competence must be demonstrated.

The one specific piece of legislation that applies to fire safety in most workplaces is the Regulatory Reform (Fire Safety) Order 2005 (RRO). It requires the appointment of a Responsible Person(s) to oversee fire safety in the workplace. The Responsible Person(s) is a person(s) who has a level of control over non-domestic premises and they are required to take reasonable steps to:

- reduce the risk from fire;
- ensure people are able to escape safely if there is a fire; and
- ensure that a suitable and sufficient fire risk assessment is carried out and that measures are taken to address the above risks.

Companies need to have a competent person to conduct this risk assessment, i.e. someone who has the knowledge, training, practical ability, experience and perhaps other qualities to understand the risks and the measures that can be taken to reduce or eliminate them.

For simple workplaces the competence to undertake fire risk assessments might already exist within the business. This could be advantageous because of their understanding of the workplace when compared with an external consultant. On the other hand, if the workplace and the fire precautions are more complex it may be necessary to enlist an external consultant with expert knowledge.

There are no compulsory qualifications, but you might expect a consultant to be a member of the IFE (Institution of Fire Engineers) or hold the NEBOSH Fire Management Certificate as well as have a demonstrable background in fire safety, e.g. ex fire service, fire engineering.

If the nature of the workplace involves varied tasks, external assistance is almost certainly required. It could take between two and three days to complete a fire risk assessment, and since costs will vary from one-man-bands to larger organisations with an array of specialist expertise to call upon, accurate costing is difficult. Including all of the required reports / documentation the cost could be expected to be somewhere between £1,500 and £2,500 at current rates.

In-house versus external consultants

Research by Loughborough University (in 2003) found that the appointment of consultants / contractors is generally price-led, and that competitors are prone to undercut each other. This creates a situation where a client may allow standards to drop by appointing less qualified people.

In some cases a business may choose to train up one of their own employees to act as their 'competent' person.

The advantages of having an internal appointment (as opposed to an external one) are knowledge of the organisation and its key personnel, the likelihood of full-time commitment to the role, and the greater potential to make decisions quickly and decisively.

In all cases it is important that the person undertaking the role is not only competent but is empowered with an adequate level of authority with access to suitable resources.

The nominated person should receive the required level of training in order to fulfil their duties. However they may still need to have access to external technical support in relation to very specialist areas such as fire safety, ergonomics, construction activities or asbestos.

See also: Construction site health and safety, p.177; Health and safety management, p.388, Risk assessments, p.624; Training, p.678.

Sources of further information

Developing and maintaining staff competence, HSE (2002). ISBN: 0 717617 32 7.

A risk assessment is nothing more than a careful examination of what, in your work, could cause harm to people, so that you can weigh up whether you have taken enough precautions or should do more to prevent harm. Risk assessment is the methodology that underpins all decisions about health and safety.

The governing legislation is the Management of Health and Safety at Work Regulations 1999, which require all employers to carry out an assessment of risk by a competent person. Workplace Law's ***Risk Assessment Training*** course examines what is meant by risk assessment, the basic principles of risk assessment, and the HSE's recognised five-step approach to successful risk assessment as part of an overall safe system of work.

The course enables you to identify risks to employees in the workplace and put together assessments to ensure that these do not result in injury. Risk assessment training is essential to ensure your organisation meets its basic commitments to providing a safe work environment for your staff.
Visit www.workplacelaw.net/training/course/id/21 for more information.

Case review

The importance of competence

Simon Toseland, Workplace Law

In the first case of its kind in Northern Ireland, a safety expert and the company he advised were fined for breaching health and safety legislation which led to the death of 52-year-old machinist, Norman McCord.

The case arose out of an incident on 7 September 2009 at Miskelly Brothers Ltd block making yard at Moss Road, Ballygowan, when an employee of the company, Mr Norman McCord, was fatally crushed when making adjustments to a block strapping machine.

Miskelly Brothers Ltd, the owner of the block making yard, was fined a total of £50,000. Steven Jones, trading as Hazron Safety Services, was acting as a health and safety consultant for Miskellys and had been engaged to carry out risk assessments on the equipment involved in the incident. He was fined a total of £4,000.

After the hearing, Mrs Nancy Henry, an Inspector with HSENI's Major Investigation Team said:

"This was a tragic and preventable incident which happened because adjustments that were being made to the machine were poorly planned. The importance of carrying out meaningful risk assessments, which encompass all the activities carried out on a piece of machinery, has yet again been highlighted. This case sends out a clear message to all employers who use health and safety consultants to ensure that they are competent and accredited to undertake

Simon Toseland is Workplace Law's Head of Health and Safety. Simon has over ten years' experience of delivering health and safety consultancy and training, is a Chartered Member of the Institution of Occupational Safety and Health, a Registered Member of the Association for Project Safety, and a Graduate Member of the Institute of Fire Engineers. In January 2011 Simon became approved on the Occupational Safety and Health Consultants Register.

the work required. This is the first time in Northern Ireland that a health and safety consultant has been convicted for a breach of health and safety legislation. HSENI wants to send out a strong message that it will not hesitate to pursue through the courts any consultant who fails in his or her duty of care."

Judge Smyth, who gave Jones time to pay his fine, said while his fine had been reduced, it still represented a significant financial penalty.

In a separate case, Health and safety consultant, Richard Atterby, was fined £1,000 and ordered to pay costs of £700, having pleaded guilty in a hearing at

Bradford Magistrates' Court of an offence under Section 36(1) of the Health and Safety at Work etc. Act 1974 relating to his failure to make a suitable assessment to the risk to health of employees from a substance hazardous to their health.

The Court also heard that an offence committed by his client, George Farrar (Quarries) Ltd, was due to the poor quality of his advice. A number of charges against George Farrar (Quarries) Ltd arose from an incident investigation and site inspection carried out by the HSE.

Richard Atterby had provided health and safety services to George Farrar (Quarries) Ltd for over three years. His work involved carrying out risk assessments on behalf of his client. The court heard that the failure of George Farrar (Quarries) Ltd to make a suitable and sufficient assessment of the risk to the health of their employees from exposure to respirable crystalline silica, arising from the processing of sandstone, was due the poor quality of Mr Atterby's advice.

HSE Principal Inspector, Keith King, said:

"This should serve as a salutary lesson for all employers who rely too readily upon paid advisers. You cannot outsource your responsibilities – the duty of care remains with you as an employer and the selection and use you make of consultants is crucial. Employers have to make absolutely sure that anyone who they commission to carry out assessments on their behalf is fully competent to do so. Consultants should not attempt to give advice on matters unless they have adequate knowledge, training, skills and experience to make the right decisions about risks and precautions that are needed."

The concept of 'competence' is inherent across most safety legislation. The onus is on employers to define competence against the activities managed for health and safety within their own organisation.

The key ingredients that make up an individual's competence are 'qualifications, experience and qualities appropriate to their duties'. A competent individual should be adequately trained, possess the ability to communicate effectively, and have an appreciation of their own limitations and constraints. A client should, therefore, have an understanding of these competency barometers when making appointments.

Before appointing a consultant, the client should have a clear concept of the size, scope and special requirements of the project / task that they will be used for. This will provide an understanding of the competencies required by the consultant they engage. This will also assist the consultant so that they can plan the project and ensure that they utilise only adequately resourced and skilled persons.

The means of assessing competence may not be immediately clear. Where a client chooses to rely solely upon a formal assessment scheme it is essential to have knowledge of its parameters and how it works. There are a number of initiatives that label themselves as a means of demonstrating competency. For example, the Contractors Health and Safety Assessment Scheme (CHAS) accredits organisations that have a proven safety management system, but does not test for experience.

Workplace Law Safety Culture Toolkit

Health and safety culture is simply the attitude about safety that pervades throughout the whole organisation, from top to bottom, and is the norm of behaviour for every member of staff, from directors down to the newest junior. It can be identified, measured, improved and, if neglected, allowed to deteriorate. It is of paramount importance in the prevention of accidents in any organisation.

Workplace Law has developed a safety culture toolkit that will analyse the following areas.

Proceduralisation and compliance culture

- Are safety controls and procedures perceived as being valid and justified?
- How effectively have any new procedures been incorporated into your safety management system?

Awareness and attitude

- How much do individuals take ownership for health and safety in the workplace?
- How often are employees asked for their input into safety issues?

Communication about safety

- How well are your safety messages received and understood by the workforce?
- How open are your employees about safety?
- Is there effective two-way communication about safety?

Management issues

- What is the belief, amongst the workforce, regarding management commitment to health and safety?
- What is the balance between safety, quality and productivity?

Resources and morale

- What strategies are used to encourage safe behaviour? Is this applied over numerous sites?
- Is the workforce pressured to work long hours or take safety shortcuts to improve productivity?

Following the completion of the survey Workplace Law will:

- produce a comprehensive report of our findings;
- identify any key patterns and trends at site and organisation level; and
- prepare and deliver a presentation at board level showing the outcome of our findings.

Visit www.workplacelaw.net for more information.

Confidential waste

David Flint and Valerie Surgenor, MacRoberts LLP

Key points

- Confidential waste includes any record that contains personal information about a particular living individual, or information that is commercially sensitive.
- Examples include correspondence revealing contact details, personnel records, job applications and interview notes, salary records, Income Tax and National Insurance returns, contracts, tenders, purchasing and maintenance records and sensitive industrial relations negotiation material.

Legislation

- Official Secrets Act 1989.
- Data Protection Act 1998 (DPA 1998).
- Freedom of Information Act 2000 (FOIA 2000).
- Freedom of Information (Scotland) Act 2002 (FOISA 2002).
- Waste Electrical and Electronic Equipment (WEEE) Directive (2002/96/EC).
- The Landfill Regulations 2002.
- The Landfill (Scotland) Regulations 2003.
- Waste Electrical and Electronic Equipment Regulations 2004 (WEEE Regulations).
- The Waste Electrical and Electronic Equipment (Amendment) Regulations 2009.
- Data Retention (EC Directive) Regulations 2009.
- The Public Records (Scotland) Act 2011.

Implications of the Data Protection Act 1998

The DPA 1998 does not set out a standard way in which confidential waste should be disposed of; however, businesses must ensure that the steps they are taking meet with the intention of the DPA 1998.

The DPA 1998 covers all computer records, information held in a relevant filing system, discs and CDs (and in the case of public authorities all other information that is not held on computer records or a relevant filing system).

Companies have several responsibilities under the DPA 1998. They must ensure that data is not kept for longer than is necessary and also that when data is finished with it is destroyed in a safe and secure manner. Throwing files away into office bins in the hope that they will be adequately destroyed is not sufficient. Companies must take appropriate technical and organisational measures to prevent against unauthorised or unlawful processing of personal data and against accidental loss or destruction of, or damage to, personal data. The Act specifically states that in deciding the manner in which to destroy data, consideration must be given to the state of technological development at that time, the cost of the measures, the harm that might result from a breach in security, and the nature of the data to be disposed of.

The issues that must be addressed with regards to confidential waste in a paper environment are different to those that must be addressed in an electronic environment. Special care must be taken when destroying electronic records as these can even be reconstructed from deleted information. Erasing or reformatting computer disks (or personal computers with hard drives) that once contained confidential personal information is also insufficient.

Although the DPA 1998 does not prescribe an exact method by which confidential records should be destroyed, employers should consider the following:

- Procedures regarding the storage and disposal of personal data including computer disks, pen drives, USB memory sticks and print-outs should be reviewed.
- Waste paper containing personal data should be placed in a separate 'confidential' waste bin and shredded by a reputable contractor, who meets the new EN15713: 2009 standard on securely destroying confidential waste and is registered and audited to ISO 9001: 2008. It is also advisable to ensure that the contractor's employees are screened in accordance with BS 7858: 2006.
- If sub-contractors are used as data processors, a sub-contractor who gives guarantees about security measures and takes reasonable steps to ensure compliance with those measures should be chosen. Furthermore, a contract should be drawn up with the data controller, and certificates of destruction (and recycling of equipment) of documents by the sub-contractor should be issued as proof that the process has been completed.
- A standard risk assessment should be completed in order to identify threats to the system, the vulnerability of the system, and the procedures that can be put in place in order to manage and reduce the risks.

Penalties for non-compliance

Contravention of the DPA 1998 is a criminal offence and the Information Commissioner has the discretion to impose financial penalties of up to £500,000 for serious breaches of the DPA 1998 on the issuance of a monetary penalty notice.

The ICO has recently published draft guidance on its powers to impose financial penalties. The guidance is not yet in force, as a consultation process is being carried out to allow businesses to comment on the guidance and request any further information which they would find helpful.

The guidance, in its draft form, provides that the ICO will consider the following when deciding the level of fine to be imposed:

- The level of harm individuals have suffered;
- Whether the breach has caused substantial damage or distress;
- Whether the contravention was deliberate; and
- Whether the organisation at fault knew there was a risk but didn't take any steps to prevent the breach.

It is possible for a single breach to attract the maximum penalty of £500,000 if that breach is serious enough, for example, losing customer data due to inadequate security.

Audits

Government departments found guilty of serious breaches of the DPA 1998 may find themselves the recipients of a compulsory audit notice. The purpose of audits is to allow the ICO to assess whether organisations are acting in compliance with the DPA 1998 and identify their strengths and weaknesses. Whilst the ICO hopes that government departments will consent to audits, if they do not, and the ICO takes the view that an audit is necessary, they can issue a compulsory audit notice and thereafter conduct an audit of a department. Whilst compulsory audits only apply to government departments at present, this may be extended to other organisations in the future.

The ICO has issued guidance that deals with both compulsory and voluntary audit notices. The guidance provides that an audit will be deemed necessary where:

- a risk assessment has been conducted and the result indicates that it is probable that personal data is not being processed in compliance with the Act;
- there is likelihood of damage or distress to individuals; and
- the organisation has unjustly refused to consent to an audit.

The risk assessment will take into account factors including, for example, compliance history, business intelligence, and the volume and nature of the personal data being processed.

The ICO has confirmed that a penalty will not be imposed in relation to a breach discovered whilst carrying out an audit. This may offer some comfort and encouragement for organisations to consent to an audit. A further incentive for cooperation is that the ICO has said it will, where appropriate, publicly identify organisations that do not offer a reasonable level of cooperation.

Individuals who suffer damage as a direct result of a contravention of the Data Protection Act by a data controller are entitled to be compensated for that damage. Prosecutions under the DPA 1998 are becoming increasingly common.

There is also a strong commercial incentive for businesses to protect personal data following the publicity of the Sony and Epsilon data breach scandals in 2011. A security hack of Sony's playstation network and user database resulted in the theft of information of what has been estimated to be around 100 million online gaming users. The information taken included names, email addresses, dates of birth, phone numbers, and debit and credit card details. Sony then had to contact all customers to warn them that their information had been taken and that they may receive correspondence, requesting personal information, which claimed to be from Sony. It is considered to be one of the biggest data loss incidents ever to occur.

US company Epsilon suffered data theft in respect of a significant (but undisclosed) number of customers. Epsilon is a marketing company that builds and hosts customer databases for about 2,500 companies, including major retail companies such Marks and Spencers and Play.com. Epsilon refused to disclose the full extent of the breach, but insisted that the data stolen only included email addresses and not credit card details. However, an email address is of significant value to hackers as it allows them to target customers by launching phishing attacks.

The above cases are just a couple of examples of a long line of security breaches in recent years.

Additional responsibilities for Public Authorities

FOISA and FOIA

Section 61 of the FOISA 2002 and Section 46 of the FOIA 2000 place additional responsibilities on Public Authorities regarding the management and disposal of their records. Section 61 and Section 46 respectively state that it is desirable

Confidential waste

for Public Authorities to follow the Code of Practice on Records Management (the Code). The Code sets out various practices regarding the creation, keeping, management and disposition of their records. The implications of the FOISA 2002 and the FOIA 2000 are very far-reaching as the Code is applicable to all records in all formats, including paper, electronic, video and microfilm. It should be noted that the ambit of this definition is wider than that under the DPA 1998, and extends to all personal data.

The issues discussed in the Code affecting the disposal of confidential information can be summarised as follows:

- The disposition of records must be undertaken in accordance with clearly established policies that have been formally adopted by authorities and are enforced by properly authorised staff. Authorities should establish a selection policy that sets out in broad terms the function for which records are likely to be selected for permanent preservation and the periods for which other records should be retained.
- Disposal schedules should be drawn up for each business area. These schedules should indicate the appropriate disposition action for all records within that area.
- A permanent documentation of any records destroyed, showing exactly what records were destroyed, why they were destroyed, when they were destroyed and on whose authority they were destroyed, should be kept. The record should also provide some background information on the records being destroyed, such as legislative provisions, functional context and physical arrangement.
- Records should be destroyed in as secure a manner as is necessary for the level of confidentiality they bear.
- Authorities must have adequate arrangements to ensure that before

a record is destroyed they ascertain whether or not the record is the subject of a request for information under the FOISA 2002 or the FOIA 2000. If a record is known to be the subject of a request under either the FOISA 2002 or the FOIA 2000, the destruction of the record should be delayed until either the information is disclosed or the review and appeal provisions have been exhausted.

Public Records (Scotland) Act 2011

The Public Records Act received Royal Assent on 20 April 2011 and is expected to enter into force towards the end of 2011.

The Act was introduced following a general review of the public records system by the Keeper of Records for Scotland in 2009, which found that individuals were not easily able to access their personal records due to inadequate record management. This review was in response to the Historical Abuse Systemic Review (the Shaw Report) of 2007, which revealed that individuals who had attended residential schools or stayed at children's homes had struggled to trace their personal records for medical, identity or family purposes as a result of poor record keeping.

The Act aims to improve record keeping across a range of public authorities, and prevent confidential records being misplaced or accidently destroyed by requiring public authorities to implement a records management that will create transparency and accountability. The Keeper of Records will have the authority to approve these plans and will issue guidance to assist authorities.

Once approved and implemented, Authorities must ensure that they undertake regular reviews of their records management plan. The Keeper will have the authority to schedule reviews.

If an Authority has failed to comply with the Act, under Section 7 of the Act, the Keeper can issue a notice requiring the Authority to take action within a certain timescale. If the Authority does not comply with the requirements of the action notice, the Keeper may take appropriate steps to publicise the failure of the organisation.

Only public authorities that are named in the Act will be subject to its provisions, including, for example, the Scottish Government, local authorities, the Scottish courts and the NHS. However, if private organisations carry out functions on behalf of any of these authorities, records created by private organisations in respect of those functions will be covered by the Act.

Public authorities, who are subject to the provisions of this legislation, must take action to ensure they implement an appropriate records management plan in order to avoid negative publicity, which could be severely damaging.

Disposal of IT equipment

Companies, when disposing of their IT equipment, must act with extreme care in order to ensure that personal data is completely erased. Despite the fact that contravention of the DPA 1998 is a criminal offence, a study published by the University of Glamorgan's Computer Forensics Team revealed that around half of second-hand computers obtained from various sources contained sufficient information to identify organisations and individuals. This clearly illustrates that businesses are failing properly to delete highly sensitive information stored on their computers before they are sold on, and failing to meet the requirements of the DPA 1998.

There are various ways a business may dispose of its IT equipment responsibly. Shredding disks is considered to be the most effective way to destroy a disk and all the personal data that it contains. Where businesses wish to reuse a disk, unless they can adequately delete the files themselves, refurbishers or recyclers should be used.

The European Waste Electrical and Electronic Equipment (WEEE) Directive seeks to regulate the disposal of electronic equipment by requiring that member states of the European Union ensure that approximately 20% of their WEEE (around four kilograms per person in the UK) is collected and recycled. The Directive received effect in the UK on 1 July 2007. Manufacturers and importers of electrical and electronic equipment (EEE) are responsible for financing the producer compliance regimes for the collection, treatment, recycling and recovery of WEEE. Such schemes are monitored by the relevant environmental agency.

Retailers, on the other hand, must either offer customers a free exchange (or 'take-back') of WEEE for EEE, or help to finance public waste WEEE recycling services. The Regulations are reasonably specific about the sort of equipment that falls under their ambit and the sort of equipment that does not. Replacement peripherals and components, such as hard disk drives, are not considered WEEE under the Regulations except where they are inside equipment that is within their scope at the time of disposal.

Failure to comply with the WEEE Regulations may result in a breach of the DPA 1998, as well as considerable bad publicity.

The Data Retention Regulations

The new Data Retention (EC Directive) Regulations 2009 came into force in April 2009. These Regulations require that public communications providers (defined as providers of a public electronic communications network or service) retain communications data for a specific period, beginning at the date of the communication. This period is 12 months in the UK. The data to be retained includes telephony and internet data stored or logged in the UK, such as who sent the data, where it was sent from and when. For example, telecommunications services will involve retaining data relating to the name and address of the subscriber, the calling telephone number and the date and time of the start and end of the call. This information is subject to data protection and data security. Although the Regulations specify that the data held should be destroyed at the end of the period of retention, there is no specific guidance on how this should be done. The Regulations simply require the data to be deleted 'in such a way as to make access to the data impossible'.

It is therefore important that such public communications providers, and indeed any organisation that processes personal data, create their own successful methods of destruction of confidential waste in order to avoid any loss and any potential security breach.

Official Secrets Act 1989

Under Section 8 of the Official Secrets Act 1989, government contractors and crown servants will be guilty of an offence if they fail to comply with official direction to return or dispose of documents or articles which it would otherwise be an offence under the Act to disclose. Prohibited disclosures include, for example, information in relation to security and intelligence, international relations, and information which could result in the commission of a crime. If Section 8 of the Act is breached then individuals may face a prison sentence of three months, a £5,000 fine, or both.

See also: Waste management, p.707; Waste Electrical and Electronic Equipment: the WEEE and RoHS Regulations, p.737.

Sources of further information

Information Commissioner: www.informationcommissioner.gov.uk

Information Commissioner Guidance on Monetary Penalties:
http://bit.ly/5byF1f

Information Commissioner Guidance on Audit Notices:
http://bit.ly/pkhe6W

British Security Industry Association: www.bsia.co.uk

Confined spaces

Kathryn Gilbertson, Greenwoods Solicitors LLP

Key points

- Entry into confined spaces can be extremely hazardous. An average of 15 people needlessly die each year as a result of lack of oxygen, poisonous gases, fumes or vapours, fire, explosions and excessive heat.
- Tragically, a large number of people who die are attempting a rescue, falling victim to the same conditions that they are attempting to rescue their colleagues from.
- The Confined Spaces Regulations 1997 were introduced to put stricter controls on work in confined spaces. They include a definition of confined spaces and the risks associated with working in them.

Legislation

- Health and Safety at Work etc. Act 1974.
- Personal Protective Equipment at Work Regulations 1992.
- Confined Spaces Regulations 1997.
- Provision and Use of Work Equipment Regulations 1998.
- Management of Health and Safety at Work Regulations 1999.
- Control of Substances Hazardous to Health Regulations 2002.

Meaning of 'confined space'

'Any place, including any chamber, tank, vat, silo, pit, trench, pipe, sewer, flue, well or other similar space in which, by virtue of its enclosed nature, there arises a reasonably foreseeable specified risk.'

Under the Regulations a 'confined space' has two defining features:

1. A place that is substantially (though not always entirely) enclosed. An open-topped tank may be a confined space.
2. There is a reasonably foreseeable risk of serious injury from hazardous substances or conditions within the space or nearby.

The hazards

- Flammable substances and oxygen enrichment.
- Toxic gas, fume or vapour.
- Oxygen deficiency.
- Ingress or presence of liquid.
- Solid materials that can flow.
- Excessive heat.

The above list is not exhaustive. Other hazards may be found that are covered by other Regulations, e.g. noise, dust, asbestos.

Core duties

The three main duties under the Regulations are:

1. Where possible to avoid entering a confined space.
2. If access to a confined space cannot be avoided, a safe system of work must be devised.
3. Prior to starting work you must ensure that adequate arrangements are in place in the event of an emergency.

Risk assessment

The Regulations are quite clear. Can entry be avoided? In some cases you can clean a confined space from the outside using

---

Case study

On 4 July 2011, Scottish Sea Farms was fined £600,000 by Oban Sheriff Court following a double fatality on a barge at Loch Creran. A team of workers were to repair a hydraulic crane on the barge. Having identified the fault, two workers entered a concealed space beneath the decks.

Within a minute of climbing down the ladder, one person passed out whilst another managed to climb out and call for help. Two further workers entered the compartment to rescue their colleague. They both lost consciousness and died.

The HSE identified that the oxygen levels below deck were very low because water in the compartment had cause rust to form, which removed oxygen from the air. Further, the company had failed to survey the barge and identify the confined space. They had also failed to provide training on working within confined spaces. A full risk assessment including air monitoring and testing for oxygen levels should have been carried out.

---

water jetting, steam or chemical cleaning, long-handled tools or in-place cleaning systems. Alternatively, for inspection purposes a CCTV system may be appropriate.

If entry is unavoidable you should assess the general condition of the confined space to identify what may or may not be present. You should consider:

- previous contents;
- residues;
- contamination;
- oxygen deficiency and oxygen enrichment;
- physical dimensions;
- cleaning chemicals;
- sources of ignition; and
- ingress of substances.

It is essential that staff are aware of the type of environment that they are working in and that a permit-to-work system is used to manage the risks. Issues to be addressed should include:

- whether the person responsible for operating the plant is aware of the type of maintenance involved and how long it is likely to take;
- whether staff have been sufficiently informed, instructed, trained and supervised to minimise any incident occurring; and/or
- lack of knowledge by maintenance staff of the environment where work is being carried out (i.e. lack of risk assessments, warning signs, method statements, emergency procedures), leading to ignition of flammable substances (e.g. heat sources such as cigarettes or welding, static and electrical discharge, use of non-spark-resistant tools) or injury / fatality from incorrect personal protective equipment (e.g. respirators) being worn.

Safe system of work

It is imperative that a safe system of work is implemented. To be effective it needs to be in writing. The following factors must be considered.

Supervision

Someone must be in charge and be competent for that role. A supervisor may also need to remain present when work is underway.

Competence for confined spaces working

Training is essential before entering a confined space.

Communications

Operatives inside a confined space must be able to communicate with those outside.

Testing / monitoring the atmosphere

Testing and monitoring equipment is not expensive.

Gas purging

Air or an inert gas, but it is imperative that if inert gas is used the atmosphere is checked before entry.

Ventilation

Fresh air should be drawn from a point where it is not contaminated. Never introduce additional oxygen to 'sweeten' the air!

Removal of residues

Often the purpose of confined space work is to remove residues or rust. Dangerous substances can be released when these are disturbed.

Isolation from gases, liquids and other flowing materials

Disconnect pipes, ducting or insert blanks. If this is not possible valves must be locked shut with no possibility of allowing anything to pass through.

Isolation from mechanical and electrical equipment

The power should be isolated by locking off, unless the risk assessment dictates power must be on for vital services, e.g. firefighting, lighting, communications.

Selection and use of suitable equipment

Must be suitable for purpose, e.g. a lamp certified for use in explosive atmospheres.

Personal protective equipment (PPE) and respiratory protective equipment (RPE)

Always a last resort. May be identified in the risk assessment, in which case it needs to be suitable and used by those entering the confined space. PPE and RPE should be used in addition to engineering controls and safe systems of work.

Portable gas cylinders and internal combustion engines

Do not use a petrol engine within a confined space! Diesel fuelled engines are nearly as dangerous, and are inappropriate unless exceptional precautions are taken.

Gas supplied by pipes and hoses

The use of pipes and hoses for conveying oxygen or flammable gases must be controlled to minimise the risks. For example, at the end of every working period supply valves are securely closed and pipes and hoses are withdrawn from the confined space.

Access and egress

Needs to be adequate and take into account persons wearing PPE and RPE.

Fire prevention

Wherever possible, flammable and combustible materials should not be stored in a confined space.

Lighting

Adequate and suitable lighting, including emergency lighting, should be provided. It may need to be specially protected if used where flammable or potentially explosive atmospheres are likely to occur.

Static electricity
Exclude static discharges and all sources of ignition if there is a risk of a flammable or explosive atmosphere.

Smoking
Prohibited in confined spaces. Exclusion area may be required beyond the confined space.

Emergencies and rescue
Procedures need to be suitable and sufficient and in place before any person enters a confined space.

Limited working time
Under extreme conditions of temperature, humidity or the use of respiratory equipment, time limits on individuals may be required.

Permits-to-work
These will be required where there is a reasonably foreseeable risk of serious injury in entering or working in the confined space. The permit-to-work procedure is an extension of the safe system of work, not a replacement for it!

Health issues
A potential issue for some confined spaces such as sewers – Leptospirosis (Weil's disease) and Hepatitis.

A doctor should be consulted in the event of flu-like illness or fever, particularly where associated with severe headache and skin infections.

Medical advice should be sought if there are persistent chest symptoms, particularly if consistent with asthma or alveolitis (inflammation of the lung).

See also: Construction site health and safety, p.177; Personal Protective Equipment, p.576; Risk assessments, p.624.

Sources of further information

HSE – Confined spaces at work: www.hse.gov.uk/confinedspace/index.htm

INDG 258 Safe work in confined spaces: www.hse.gov.uk/pubns/indg258.pdf

HSE – L101 Safe work in confined spaces: www.hse.gov.uk/pubns/priced/l101.pdf

Construction site health and safety

Kathryn Gilbertson, Greenwoods Solicitors LLP

Key points

The main causes of death and major injury in the construction industry are:

- falling through fragile roofs and roof lights;
- falling from ladders, scaffolds and other workplaces;
- being struck by excavators, lift trucks or dumpers;
- overturning vehicles; and
- being crushed by collapsing structures.

Legislation

- The Construction (Health, Safety and Welfare) Regulations 1996.
- The Lifting Operations and Lifting Equipment Regulations 1998.
- The Management of Health and Safety at Work Regulations 1999.
- The Control of Substances Hazardous to Health Regulations 2002.
- The Control of Vibration at work Regulations 2005.
- The Work at Height Regulations 2005.
- The Construction (Design and Management) Regulations 2007.

Key areas to managing construction safety

Correct planning and implementation for safety

All construction and building work projects have to be managed under the CDM Regulations.

Ladders and towers

- Ladders should not be used without proper justification.
- Where ladders are used they must be used correctly.
- Towers need to be capable of being erected safely.
- Towers need to be erected by properly trained operatives.

Fragile roofs

- No work should be carried out on a fragile roof unless a safe system of work is in place and understood by those carrying out the work.
- It is essential that only trained competent persons are allowed to work on roofs.
- Footwear with a good grip should be worn.
- It is good practice to ensure that a person does not work alone on a roof.

Slips, trips and falls

- Housekeeping needs to be actively managed.
- Walkways must be kept clear.
- Work areas should be tidy and free from obstructions, e.g. trailing leads.

Construction site transport

- All vehicles should be regularly maintained and records kept.
- Only trained drivers should be allowed to drive.
- Vehicles should be fitted with reversing warning systems.
- Traffic routes and loading / storage areas need to be well designed with enforced speed limits.
- Suitable signs should be erected.
- Good visibility and the separation of vehicles and pedestrians must be used.

Facts

- The provisional figures for 2010/11 show that 52 people lost their lives working in construction.
- The construction industry reported the greatest number (811) of major injuries due to falls in 2010/11.
- It also reported a high number (653) of major slip and trip injuries over the same period.

- The safety of members of the public must be considered, particularly where vehicles cross public footpaths.

Lifting operations
- Must be properly planned by a competent person.
- Must be appropriately supervised.
- Must be carried out in a safe manner.
- Lifting equipment that is designed for lifting people should be appropriately and clearly marked.
- Lifting equipment not designed for lifting people but which might be used in error should be clearly marked to show it is not for lifting people.
- Machinery and accessories for lifting loads shall be clearly marked to indicate their safe working loads (SWL).

Occupational health issues
- Musculoskeletal disorders (MSDs) include back pain, usually from poor methods of manual handling such as heavy or awkward lifts. Where MSD trends are increasing, additional training for manual handling should be considered.
- Contact dermatitis can be caused from substances such as cement. Managers should effectively supervise their staff and ensure that gloves are being worn.
- Hand–Arm vibration (HAV) is caused by vibrating or poorly managed hand tools.

- Long-term hearing loss can be caused by excessive noise.
- Asbestos-related illnesses such as Mesothelioma are caused by exposure to asbestos.

Passport schemes
Passport schemes ensure that workers have basic health and safety awareness training. The HSE is encouraging organisations to work together so that one scheme recognises the core training of other schemes.

Passport schemes are not:

- a way of knowing or identifying that a worker is competent;
- a substitute for risk assessment;
- a way of showing 'approval' of a contractor;
- required by law or regulated by law;
- a reason to ignore giving site-specific information; or
- a substitute for effective on-site management.

The HSE and the Environment Agencies do not endorse or approve individual passport training schemes.

Benefits and advantages
- They can help reduce accidents and ill health.
- They can save time and money because workers need less induction training.

- They show a company's commitment to having safe and healthy workers.
- Companies know that workers have been trained to a common, recognised and validated standard.
- Insurance and liability premiums may be reduced if a company can show that all workers have basic health, safety and environmental training.

Recent cases

On 1 July 2011, the CPS decided to charge Lion Steel Ltd with corporate manslaughter. This is the second corporate manslaughter prosecution.

Alison Storey, Reviewing Lawyer in the CPS Special Crime and Counter Terrorism Division, said:

"I have advised Greater Manchester Police to charge Lion Steel Ltd in Manchester with corporate manslaughter following the tragic death of Steven Berry at the Hyde site on Johnson Brook Road when he fell through a fragile roof panel and died as a result of injuries sustained in the fall. I have also decided that three of the company directors – Kevin Palliser, Richard Williams and Graham Coupe – should be charged with gross negligence manslaughter.

"The three men are also charged under Section 37 of the Health and Safety at Work etc. Act 1974 for failing to ensure the safety at work of their employees. Lion Steel is also charged under Sections 2 and 33 of the Health and Safety at Work etc. Act 1974 for failing to ensure the safety at work of its employees.

"I have taken this decision after very carefully reviewing the material gathered in the police investigation and have concluded that there is sufficient evidence for a realistic prospect of conviction and that it is in the public interest to bring these charges."

The preliminary hearing took place at Tameside Magistrates' Court on 2 August 2011, and the trial date has now been set for 12 June 2012.

On 14 January 2011, Leicester Magistrates' Court heard that a construction company endangered the lives of both its workforce and the public while demolishing an old factory in Leicester city centre.

Saleh Properties Ltd was demolishing a disused factory in Orson Street to make way for new homes. Workers had removed structural parts of the building without properly supporting it. Some workers were even spotted standing on the roof, demolishing parts of the building by hand and were working at height without suitable equipment to prevent falls.

The HSE inspector immediately stopped work and served prohibition notices preventing any more activity until a demolition plan was in place and a competent supervisor was on site.

Missing safety signs and fencing were ordered to be installed to ensure members of the public were kept away from the unsafe building.

Saleh Properties Ltd pleaded guilty to breaching Regulations 9 (1)(a) and 28(2) of the Construction (Design and Management) Regulations 2007 and two breaches of Regulation 6(3) of the Work at Height Regulations 2005. The company was fined £4,000 and ordered to pay costs of £1,084.

Inspector, Stephen Farthing, said:

"Saleh Properties showed a horrendous disregard for health and safety, which was not only putting workers at risk, but also passing members of the public going about their daily lives. There was a real danger of this building collapsing.

"The site supervisor had no training in health and safety, no method statements or risk assessments had been carried out before the work started, and there were no welfare facilities for workers.

"To run a construction company in this manner is wholly unacceptable and this prosecution shows that the HSE will clamp down on small construction companies

failing to adhere to basic health and safety regulations."

See also: CDM, p.137; Workplace deaths, p.218; Working at height, p.399; Ladders, p.425; Reporting of Injuries, Diseases and Dangerous Occurrences (RIDDOR), p.618; Slips, trips and falls, p.650; Vehicles at work, p.686.

Sources of further information

HSG 144 – Safe use of vehicles on construction sites:
www.hse.gov.uk/pubns/priced/hsg144.pdf

HSG 150 – Health and safety in construction:
www.hse.gov.uk/pubns/priced/hsg150.pdf

L102 – Construction (Head Protection) Regulations 1989:
www.hse.gov.uk/pubns/priced/l102.pdf

L 144 – Managing health and safety in construction:
www.hse.gov.uk/pubns/priced/l144.pdf

Contaminated land

Guy Willetts, Shoosmiths

Key points

- The contaminated land regime is the name given to the area of environmental law that regulates and aims to prevent pollution of land, as opposed to air and water, which are governed by separate regimes.

- The law relating to contaminated land has grown exponentially over the last 20 years, and there is now a large body of law to draw on (statutes and regulations and case law).

- Fundamentally, though, it can be divided into two parts. Firstly, there is the regulatory side of the law. The key environmental law statutes, the Environmental Protection Act 1990 and the Environment Act 1995 between them set up a regime to deal with contaminated land. The main enforcers of the scheme are either Local Authorities or the Environment Agency, depending on what kind of land we are considering, who owns it and so forth, and are primarily concerned with state intervention in the ownership of land with a view to dealing with contaminated land.

- The second area of law, which must not be forgotten despite not being covered by recent statute, is civil law – that is to say that the rights that one landowner has against a neighbour who might have contaminated his land.

- It is wrong to think that only industrial sites (and therefore those involved in owning industrial sites) are affected by the contaminated land regime. The regime covers any land that might be contaminated land, and as pointed out in the Government report that led to the Environment Act 1995, a document called 'Paying for our Past', significant areas of the country are potentially suffering from historic contamination.

- The country has a long history of industry, going back many centuries, ranging from 20th Century heavy industry, right back to medieval tanning and metal working. All of this has left its mark on the landscape and in the environment, particularly in or around urban areas, some of which may have long since been turned over to less contaminative uses.

Legislation

The main legislation relating to the contaminated land regime is to be found in the Environmental Protection Act 1990, in particular Section 78, which runs to 28 sub-sections. It was brought into being by the 1995 Environment Act. This needs to be read in conjunction with the Statutory Guidance, which goes to several hundred pages, and which sets out the mechanism in detail.

The civil side of the contaminated land regime is based on common law notions of nuisance, negligence, the rule in *Rylands v. Fletcher* as amended, and other creations of the common law.

The contaminated land regime

The Environment Act gives Local Authorities a duty to identify what the Act defines as contaminated land, imposes the obligation to deal with it, and gives them the power to serve remediation notices on a 'appropriate person'. All of these terms are minutely defined in the Act and the statutory guidance. An understanding of how the different definitions fit together is necessary before any view can be taken as to whether any piece of land is a concern to its owner or occupier. The pieces of the jigsaw are as follows.

Contaminated land

Land is defined as contaminated land if as a result of substances in or under it there is a *significant possibility of significant harm*. This is referred to in the jargon as SPOSH. The statutory guidance to the Environment Act explains how SPOSH is to operate. This makes it clear that the mere presence of a pollutant on the land does not of itself make the land contaminated land as defined. If that contamination is properly contained, then there is no significant possibility of any harm occurring and therefore the land is not deemed to be contaminated land. Similarly, even if the pollution is able to escape but causes only some minor insignificant harm, then the land again is not contaminated because there is no significant harm. Arguments can and do revolve around whether a particular type of harm is or is not significant.

In the words of the Statutory Guidance, there must be a 'contaminant', which passes through a 'pathway', and causes harm to a 'receptor'. All three elements are necessary. If any one of these three elements is missing, then the land is not contaminated within the meaning of the statute, and the statutory regime does not apply.

The appropriate person

The idea behind the legislation is that the polluter should pay. Given what we know about the historic nature of a large amount of pollution in the country, it is not always possible to find out who that person is; sometimes the polluter might be identifiable, but has long since either died or (if a company) been dissolved. It might even be that the polluter has no money, so pursuing him would be a waste of time. For that reason a more involved series of rules has been devised to identify who is the 'appropriate person', who is to be made to pay for the clean-up. The appropriate person (and that is the term used in the guidance for the person who will be forced to pay for clean-up) is said by the guidance to be either a Class A or a Class B appropriate person.

A Class A appropriate person is the person who caused or knowingly permitted the pollution. There are complex rules about what exactly these phrases mean, but in simple terms if you are responsible for the pollution, then you are a Class A appropriate person. The Local Authority will, in the first instance, look to you and any others who can be said to be responsible for the pollution, and you are first in their sights. If there are a number of people who have contributed towards the pollution (in other words if there are several Class A appropriate persons), then the Local Authority can determine the proportion of responsibility that each one bears. It is noteworthy that Class A type liability can attach to someone simply because they did not put in place proper procedures to prevent spillages of contaminants. These contaminants can be something as simple as diesel from a filling station, or even cleaning fluids poured down a drain.

If the Local Authority cannot find any Class A appropriate person, it will look for a Class B appropriate person. A Class B appropriate person is one who simply owns or occupies the land but did *not* cause or knowingly permit pollution on it. In the case of historic contamination, this is a fairly common situation. The statutory guidance talks in terms of 'owners or occupiers', without making much of a distinction between the two. 'Owners' includes landlords, and 'occupiers' means, normally, the tenants. Landlords regularly put in leases terms that passes all of the risk of Class B liability on to their tenants, and by and large will be successful in passing on such liability to their tenants in that way. Therefore, when taking out a lease, any business should be careful to understand where its liabilities lie under that lease and to ask appropriate questions at the time that they take on the lease.

Remediation notices

Having identified that the land is contaminated land, as defined, and having identified who the appropriate person is, either Class A or Class B, the Local Authority has a duty to deal with the problem. It has the power, should it so wish, to serve a notice on the appropriate person, requiring that person to deal with the issues. It is interesting to note that since the contaminated land regime was brought into effect only a miniscule number of remediation notices have actually been served. This is because most Local Authorities at that stage speak to the appropriate persons and work out arrangements and schemes of work to deal with the contamination. This can be as straightforward as putting in a cap or a bund to prevent the pollution from going anywhere, thereby stopping the problem at source. On the other hand, it could be something more complex, like having the contaminated land taken away and disposed of by a specialist agency, which is more often than not a particularly expensive solution.

Failure to comply with a remediation notice is an offence, with significant fines and even imprisonment of individuals as

possible punishments, and although this is not the place to discuss personal liability of directions of companies, environmental law is an area where directors and other managers can find themselves personally responsible, in certain circumstances, for the offences of the companies they run.

Civil remedies

The contaminated land regime described so far sets out how Local Authorities are empowered to regulate contaminated land. The high profile cases, however, that tend to follow pollution incidents (whether one-offs or as a remit of pollution over a long period of time) tend to arise because of common law civil remedies.

A recent and well publicised example is the claim brought against Corby Council by the families of a number of children who suffered birth defects as a result of contaminants released into the environment following the clean-up of the former steel works by the Council. The claim was that because of the negligent way in which the Council removed contaminated soil from the former steelworks land, and distributed it around various waste sites and other areas within the borough, children, particularly a number of children born between 1989 and 1999, suffered congenital birth defects including missing fingers and toes and defects to hands and feet. The Judge held, based on expert evidence presented at the trial, that these birth defects could realistically have been caused by the contaminants not being dealt with properly by the Council.

In April 2010 the Council reached a final, binding agreement with 19 young people who suffered personal injury which they allege occurred as a result of the Council's clean-up operation. As a result of the Agreement, the Council agreed to drop its challenge to the High Court ruling, and instead would pay compensation to each of the children, without accepting liability in this case. The financial terms of settlement remain confidential and in the case of the younger children will require approval by the Court. The total compensation bill has been estimated at in excess of £4.5m, and it is thought that up to 60 more families are planning to pursue claims.

Corby Borough Council Chief Executive, Chris Mallender, said:

"The Council recognises that it made mistakes in its clean-up of the former British Steel site years ago and extends its deepest sympathy to the children and their families. Although I accept that money cannot properly compensate these young people for their disabilities and for all that they have suffered to date and their problems in the future, the Council sincerely hope that this apology coupled with today's agreement will mean that they can now put their legal battle behind them and proceed with their lives with a greater degree of financial certainty."

Although the highest profile cases are where individuals suffer as a result of contamination, such as happened at Corby, a large number of cases come to Court because land, as opposed to people, has been contaminated. Where contamination leaches from one piece of land on to a neighbouring piece of land, the neighbour has a right in common law to sue the polluter for damages under the doctrine of Private Nuisance.

He can ask the Court to make a simple order of damages. Alternatively, he can ask the Court to order the polluter to carry out works that would prevent further contamination. There are many cases caused, for example, by faulty underground pipes in petrol and other filling stations, where the contamination has entered into the ground, and leaches on to neighbouring land. Such a situation could well give rise to the kind of issues that are discussed above under the Environment Act, but what is of more concern to the neighbouring landowner is his right to get compensation and/or an order for clean-up through the civil courts against the polluter directly, without involving the Local Authority. Any liability on the polluter as a result of a civil claim against him by his neighbour is in addition to, and not instead of, the liability that the Local Authority might want to impose under the Environment Act.

There have been several debates as to whether the liability for nuisance, or a similar doctrine, known as the rule in *Rylands v. Fletcher,* imposes a strict liability on the polluter (i.e. he is liable even if he did not know the existence of the pollution) but the significant fact is that most pollution inexorably finds a way of spreading. That is in the nature of most forms of pollution. The legal questions therefore in reality play second fiddle to the technical and scientific issues as to whether the pollution being suffered by the claimant genuinely can be said to have emanated from the alleged polluter's land. Most time in cases is spent in dealing with that issue rather than any complex point of law, because the legal aspects of the claim are relatively straightforward: if there has been negligence, or if the polluting land is contaminating a neighbour's land, through seepage or any other mechanism, then whoever can be shown by technical evidence to be the polluter will be liable for clean-up costs and any damages in addition that the neighbouring landowner has suffered.

As an example, the author was involved several years ago in a case where land used for offices had been built on a site which, during the Second World War, had been used as a munitions factory. Arsenic, antimony and other heavy metals had, during the course of the War, found their way into the ground. Due to some heavy rains in the 1990s, they had leached out on to the bed of the local tidal river and with successive high tides were being swept upstream. There, they contaminated the bed on which a waterside development had been planned, significantly increasing the costs of that development. The case did not come to trial, but significant damages had to be paid to the developers, as well as to the owner of the tidal foreshore (The Crown Estate) in addition to an extremely costly clean-up operation both of the polluter's land and the river bed that was affected.

The main issues involved in the case were scientific ones, to do with whether or not the contaminants found upstream at the development site were indeed those emanating from the alleged polluter's land or not. The law showed itself to be capable of providing a suitable remedy to the victims of that pollution.

Conclusion

The long history of industry in this country means that more land is affected by the contaminated land regime than would at first sight appear likely. Hundreds of years of industrial use have led to large areas of land being potentially contaminated in one form or another.

Over the last 20 years, legislation has been brought in to enable Local Authorities and the Environment Agency to deal with the historic effects of contamination, and with continuing pollution. However the existing common law rules, many of which date back to the 19th Century, also from time to time show that they have teeth enough to make the polluter pay, which is the primary intention behind the contaminated land regime.

See also: Batteries, p.46; Environmental Management Systems, p.286; Does the polluter pay?, p.296; Hazardous waste, p.700; Waste management, p.707.

Sources of further information

Environment Agency: www.environment-agency.gov.uk

DEFRA: www.defra.gov.uk/ENVIRONMENT/land/contaminated

Contractors

Sally Cummings, Kennedys

Key points

- An employer's undertaking for the purposes of Section 3 of the Health and Safety at Work etc. Act 1974 may include work undertaken for him by an independent contractor (on the basis that contractors are not employees, but independent third parties).
- Section 3 of HSWA imposes a duty on employers to conduct their undertaking in such a way as to ensure, so far as is reasonably practicable, that persons not in their employment are not exposed to risks to their health or safety.
- This duty may extend to contractors and those employed by them, if it can be said that the risks in question arise as part of the conduct of the employer's 'undertaking'.

Legislation

- Health and Safety at Work etc. Act 1974 (HSWA).
- Management of Health and Safety at Work Regulations 1999 (MHSWR).
- Construction (Design and Management) Regulations 2007 (CDM Regulations).

Overview

Work undertaken by a contractor is usually covered by a separate contract for services. It is good practice for provisions concerning health and safety arrangements and responsibilities to be included within the contract. However, the duties under HSWA and other associated Regulations cannot be delegated from one party to another by contract. Therefore, any contractual provisions that seek to absolve a party of any responsibility will be void; the statutory obligations under HSWA will pertain and will override the contract. The key message to remember is that both parties will have responsibilities under health and safety law: indeed, from an enforcement perspective, in circumstances where contractors are used it is not uncommon for the employer of the injured contractor to be prosecuted under Section 2, as well as the organisation who engaged the contractors under Section 3.

CDM

For most construction activities, the CDM Regulations will apply and the employer (or 'Client' as referred to under the Regulations) and contractor will have specific duties depending upon their role. These are in addition to the duties under HSWA and MHSWR. Certain projects will require a Principal Contractor to be appointed, who is responsible for liaising with any sub-contractors to plan, manage and coordinate the work during the construction phase to make sure that risks are properly controlled. A CDM Coordinator will also be required to advise the Client on health and safety issues during the design, planning and construction phases of the project. Contractors themselves will have other responsibilities regarding their own employees, as well as in planning, managing and monitoring their own role in the project. The duties under the CDM Regulations are quite specific, and should be considered in detail by contractors and employers alike before any construction project is undertaken. Some duties under CDM Regulations will apply to all types of construction project, whereas others will only apply in certain circumstances.

The aim of the Regulations is to ensure that all parties on a construction project work together to ensure they assist each other in complying with their individual duties under health and safety law. For more on the Regulations, see 'CDM' (p.137).

Health and safety management

The arrangements for ensuring proper health and safety management envisaged by the HSE's guidance HSG 65 – 'Successful health and safety management' (see *Sources of further information*) will apply to the work of independent contractors as it does to the employer, if the work forms part of the employer's undertaking. An employer who engages contractors will therefore be expected to have identified all aspects of the work required of the contractor, and all associated risks. The employer should therefore have a health and safety policy in place, together with relevant risk assessments and safe working procedures. He should have systems to ensure that there is proper supply of information, instruction and training to the contractor.

Measures should also be taken to regularly review and audit the contractor's performance to ensure that the contractor's work – which forms part of the employer's undertaking – is being conducted in a way that is legally compliant.

HSG 159 (see *Sources of further information*) gives further guidance to employers on how to manage and work together with contractors to ensure good health and safety management. The guidance offers five practical steps to

assist with matters such as planning jobs involving contractors, offering guidelines for tendering and selecting contractors, and advice as to how to manage and monitor them whilst on site. Whilst it is primarily aimed at small and medium-sized enterprises, it is nevertheless of use to all employers by helping them to assess their current practice and to apply good practice to ensure compliance with the law.

Ensuring competency

The above steps may include undertaking specific checks of the contractor involved to ensure that it is properly licensed, competent and qualified to undertake the role on behalf of the employer (indeed, the CDM Regulations include specific provisions regarding 'competence'). HSE guidance suggests that this would entail looking at the proposed contractor and determining matters such as:

- experience in the type of work concerned;
- qualifications and skills;
- the health and safety training and supervision it provides;
- recent health and safety performance;
- health and safety policies and procedures; and
- its arrangements for consulting its workforce.

It should be noted that the above principles will also apply where sub-contractors are used; the employer must therefore satisfy himself that the contractor has an effective procedure for monitoring and auditing the competence of sub-contractors, and that there are also systems in place for the proper supply of information, instruction and training to the sub-contractor.

Finally, the provisions of MHSWR must also be taken into account in any employer–contractor relationship. To some extent, certain responsibilities overlap with those set out in the CDM Regulations.

Case study

R v. Associated Octel (1996).

The leading case in relation to the 'conduct of one's undertaking,' and the circumstances in which this amounts to a duty under Section 3, is *R v. Associated Octel* (1996). The case considered liability of Octel, the owner of a chemical plant, under Section 3 for injuries sustained by an employee of an independent contractor, who was engaged to carry out repairs and maintenance to the owner's plant. Like all other contractors on the site, the contractor worked subject to Octel's 'permit-to-work' system.

Octel was prosecuted for a breach of Section 3(1) HSWA, but pleaded that the contractor was independent, and did not form part of its undertaking. The case went to the House of Lords, which held that in determining whether the owner of the premises (i.e. Octel) was liable, the question was whether the activity in question – i.e. that gave rise to the risk to the injured contractor – could be described as an activity that was part of the owner's undertaking. The Court held that it was a reasonably foreseeable part of the plant owner's business / undertaking that maintenance and repairs would need to be carried out to its plant by external contractors, and so, to that extent, the owner could be held liable under Section 3 for failing to ensure that the contractor's work was undertaken without risk; just because the maintenance work had been outsourced to a specialist contractor did not mean that Octel divested itself of its ability to control the activities on its site, and therefore of its duty under Section 3(1).

Prior to *Octel* the House of Lords had held in *Austin Rover Group Limited v. HM Inspector of Factories* (1990) that, for the purposes of Sections 2 and 3 of HSWA, the key element of the relationship between the conduct of the employer and liability was the extent to which the employer was able to exercise 'complete control' over the matters to which his duties extend. Therefore, in other words, an employer is only under a duty under Section 3 to exercise control over an activity if it forms part of the conduct of its undertaking.

Where a prosecution is brought and there is a dispute on the facts, it will be for the HSE to prove that a contractor's work formed part of an employer's undertaking. The House of Lords case of *R v. Chargot* (2009) has suggested that if liability under Section 3 is to be proved, "it may be necessary to identify and prove the respects in which the injured person was liable to be affected by the way the defendant conducted his undertaking". This suggests that there must be a clear link between the activities carried out by the duty-holder (employer) and the risk that materialised (to employees of the contractor, for example).

However, this does not mean the MHSWR can be overlooked. The Regulations set out the general requirements for health and safety management systems in all workplaces, and to that extent contain important requirements regarding management, supervisory and monitoring responsibilities.

See also: CDM, p.137; Construction site health and safety, p.177; Health and safety at work, p.361; Risk assessments, p.624.

Sources of further information

HSE: www.hse.gov.uk and information line (0845 345 0055).

L21: Management of health and safety at work: www.hse.gov.uk/pubns/priced/l21.pdf

HSG 159: *Managing contractors – a guide for employers* (1997). ISBN: 07176 1196 5.

HSG 65: Successful health and safety management: www.hse.gov.uk/pubns/priced/hsg65.pdf

INDG 368: Use of Contractors: www.hse.gov.uk/pubns/indg368.pdf

Corporate manslaughter

Daniel McShee and David Wright, Kennedys

Key points

- The passing of the Corporate Manslaughter and Corporate Homicide Act 2007 to introduce a statutory offence of 'Corporate manslaughter' (called 'Corporate Homicide' in Scotland) followed many years of promises by this and previous Governments to reform the law of corporate manslaughter.
- The new offence came into force on 6 April 2008.
- Until then there was a common law offence only, which, in order for a company to be found guilty, required the conviction of an individual person for gross negligence manslaughter and for that person to be so senior within the company that he represented its 'directing mind'.
- This requirement was known as the Identification Principle. The common law offence remains for any management failure resulting in an incident up to 6 April 2008, but not beyond that date.
- It is important to emphasise that the Act is an offence-creating statute rather than a duty-setting one.
- The Act itself imposes no new health and safety duties.

Introduction

It is now more than three years since the Corporate Manslaughter and Corporate Homicide Act 2007 came into force on 6 April 2008. To date, there have been only two reported prosecutions commenced. The defendant in the first of these was Cotswold Geotechnical Holdings Limited, and the case focused on the death of one of the company's geologists, Alexander Wright, on 5 September 2008.

Mr Wright was in the process of taking soil samples from inside an excavated pit when the sides of the pit collapsed, crushing him.

In addition to the charge of corporate manslaughter, the company also faced a charge under Section 2(1) of the Health and Safety at Work etc. Act 1974 for failing to safeguard the health and safety of its employee. This is possible as the Corporate Manslaughter and Corporate Homicide Act specifically allows for

separate charges to be brought under health and safety legislation, where there is evidence of any such offence.

Mr Eaton, the sole Director of the company, was also charged with gross negligence manslaughter, as well as an offence under Section 37 of HSWA.

The case came to a conclusion in February 2011.

However, for reasons explained in the case study overleaf, the case was not at all as ground-breaking in terms of establishing the new law as some had previously predicted.

In July 2011, the second case was brought, against Manchester-based company, Lion Steel Ltd, after one of its employees fell through a fragile roof on 29 May 2008 sustaining fatal injuries. In parallel, three of the company's directors,

Case study

On 23 April 2009, the CPS publicised the first corporate manslaughter charge under the new Act, against Cotswold Geotechnical Holdings Ltd in relation to the death of Alexander Wright in September 2008. Mr Wright was a 27-year-old geologist employee of the company taking soil samples from inside a three-and-a-half-metre-deep trench, which had been excavated as part of a site survey on a development site in Stroud, when the unsupported sides of the trench collapsed about him. Mr Wright had been left working alone in the trench when the company's director, Mr Eaton, left site for the day. A few minutes later, the site owners heard a shout, and by the time they reached the now-collapsing trench Mr Wright was buried up to his head. Tragically, a further collapse resulted in Mr Wright perishing. It transpired that the company's system of work in digging trial pits was highly dangerous and ignored well-known industry practice that prohibited entry into unsupported excavations over 1.2 metres deep.

Cotswold Geotechnical Holdings Ltd was a relatively small company. Its director, Peter Eaton, was also charged with gross negligence manslaughter in April 2009 and with an offence contrary to Section 37 of HSWA 1974 – that offence itself is now imprisonable under the Health and Safety Offences Act 2008, which came into force in January 2009, but not retrospectively for an offence committed in September 2008. Despite suggestions at the time from some pundits that this would be a landmark test case, from the off there was considerable doubt as to whether the case would achieve much, if anything, in terms of interpreting the new law. The fact is that, as a director, it seemed very likely that Mr Eaton would have been a directing mind of the company under the old common law, and consequently were he found guilty then the company would have been guilty of corporate manslaughter under the old common law in any event.

The first hearing occurred in June 2009 but it was not until February 2011 that the three-week trial took place. In the interim, the case against Mr Eaton was dropped, as a result of his suffering terminal illness. The Prosecution, at the direction of trial judge, Mr Justice Field, who was concerned that the jury might become confused, also allowed the HSWA charge against the company to fall away, leaving just the corporate manslaughter charge. The Defence unsuccessfully applied to have the case against the company discontinued.

The company was finally convicted of corporate manslaughter on 15 February 2011 and it was sentenced two days later. Arguably it was the sentencing that gave rise to the only issues of real legal note in the case. Mr Justice Field fined the company £385,000, despite the fact that the company was already understood to be in a 'parlous financial state'. He said the fine marked the gravity of the crime and the deterrent effect it should have on other companies. Although he ordered that the fine could be paid over ten years, he said, "It may well be that the fine in terms of its payment will put this company into liquidation. If that is the case it's unfortunate but unavoidable, it's a consequence of the serious breach."

Kevin Palliser, Richard Williams and Graham Coupe, were charged with gross negligence manslaughter.

So over three years on, *Geotechnical* and *Lion Steel* being the only reported cases to date, is it possible that the Act will fail to have the fundamental effect intended? Or will the impact of the Act slowly build with time?

The 'new' offence of Corporate Manslaughter

The Act was introduced to create a statutory offence of 'Corporate manslaughter' (called Corporate Homicide in Scotland). Prior to that there had been a common law offence only that, in order for a company to be found guilty, required the conviction of an individual person for gross negligence manslaughter and for that person to be so senior within the company that he represented its 'directing mind'. This requirement was known as the Identification Principle. Whilst attempts were made to prosecute large companies under the old law these attempts were all ultimately unsuccessful.

The new Act abolished the common law offence (save for management failures committed prior to 6 April 2008) creating a completely new framework for finding an organisation guilty of corporate manslaughter.

It is important to emphasise that the Act is an offence-creating statute rather than a duty-setting one. The Act itself imposes no new health and safety duties. In other words, the Act is solely designed to make it easier to prosecute organisations where their gross negligence leads to death. The Health and Safety at Work etc. Act 1974 remains as the duty-creating statute in terms of health and safety, and prosecutions under this legislation co-exist with potential prosecutions for corporate

manslaughter under the Act, as seen in the two cases so far.

The wording of the offence of corporate manslaughter is that an organisation is guilty of an offence if the way in which its activities are managed or organised:

- causes a person's death, and
- amounts to a gross breach of a relevant duty of care owed by the organisation to the deceased.

An organisation is only guilty if the way in which its activities are managed or organised by its senior management is a substantial element in the breach. Unhelpfully, the statute itself does not give any clear guidance on what would amount to 'a substantial element of the breach', and the Geotechnical case threw no additional light on this area.

Organisations

Whereas the common law offence only applied to corporate bodies, the new offence applies to a much wider group of organisations including corporations, specified government departments, police forces and partnerships.

Senior management

'Senior Management' is defined in Section 1(4)(c) as the persons who play significant roles in:

- the making of decisions about how the whole or a substantial part of its activities are to be managed or organised; or
- the actual managing or organising of the whole or a substantial part of those activities.

Clearly there will be some people, e.g. members of the Board, who will be involved in substantial parts, or arguably the whole, of the company's decision-making process and fulfil the test in 1(4) (c)(i) above. Under the old law those individuals would be identified as representing the company's 'directing mind' and there does not appear to be any change in respect of people at that level.

However, the definition in Section 1(4)(c)(ii) of a person 'actually managing or organising … a substantial part of those activities' is a major change. The definition includes two strands; the taking of decisions and actually managing those activities. The term 'significant' is intended to capture those whose role in the relevant management activity is decisive or influential, rather than playing a supporting role.

What amounts to a 'substantial' part of an organisation's activities will be important in determining the level of management responsibility engaging the new offence. The scale of the organisation's activities overall will be a factor and it is stated that the offence is intended to cover, for example, management at regional level within a national organisation. In many large companies there will be managers who manage areas or are responsible for a particular discipline.

The definition in Section 1(4)(c)(ii) – 'actually managing a substantial part' – potentially lowers the threshold for these types who might become 'senior management' under the new law but who would not be identified as a directing mind under the old common law.

As such, the definition is bound to catch a layer of management lower in the organisation that would not have been classified as representing the directing mind under the old law. It should also be noted that senior management would cover both those in the direct operational chain of management as well as those

in, for example, strategic or regulatory compliance roles.

The other key point about the definition of 'senior management' is that the new offence allows the prosecutor to aggregate the failures of a number of senior managers rather than relying on the conduct of one single 'directing mind' as required by the identification principle under the old common law.

These two points are the key changes from the old law and will undoubtedly increase the potential for the new offence to be investigated against medium-sized and bigger companies. However, as explained below, there are still hurdles to overcome, particularly as the new legal test requires a significant element of 'gross' failure to be at a senior management level.

Meaning of 'duty of care'

The new offence only applies in circumstances where an organisation owes a duty of care to the victim. This is no different from the old law. Section 2(1) requires the duty of care to arise out of certain specific functions or activities. The effect is that the offence only applies where an organisation owes a duty of care:

- to its employees or other persons working for the organisation;
- as an occupier of premises;
- when the organisation is supplying goods or services;
- when constructing or maintaining buildings, infrastructure or vehicles etc. or when using plant or vehicles etc.; and
- when carrying out other activities on a commercial basis.

Excluded from being a relevant duty of care are public policy decisions made by a public authority.

Gross breach

Section 8 of the new Act requires there to have been a gross breach of the duty of care. The requirement for a breach to be gross is a positive one as it provides a jury with a reminder that the offence should be reserved only for the most serious cases. The notes produced with the draft Bill made it clear that the decision of what is 'gross' is one for the jury. The definition in the Act requires a jury to do the following:

- It must consider whether the evidence shows the organisation failed to comply with any health and safety legislation relating to the alleged breach, and if so:
    - how serious that failure was; and
    - how much of a risk of death it posed.

Section 8(3) goes on to say that in assessing grossness a jury may also:

- consider the extent to which the evidence shows that there were attitudes, policies, systems or accepted practices within the organisation that were likely to have encouraged any such failure or to have produced tolerance of it; and
- have regard to any health and safety guidance relating to the alleged breach.

The first part of this definition causes concern. It leaves open to the jury a very subjective assessment of 'corporate culture' that they may not be in a proper position to judge in context, having never been part of the industry concerned.

Causation

The wording of the offence makes it clear that the gross breach must have 'caused' a person's death. Although this is not explained further in the Act, the guidance that went with the Bill is likely to mean that the usual principles of causation in criminal law will apply, namely that it need not be the sole cause but need only be a cause.

The offence in Section 1 specifies that an organisation is only guilty if the way in which its activities are managed or organised by its senior management is a substantial element in the breach. Whilst the failure need not be the only cause, the difficulty of proving a causal link between failures by someone classified as senior management and the victim's death may provide a significant hurdle to the prosecuting authorities similar to the way it did under the old law, albeit to a lesser degree.

This perhaps explains the reasoning behind the CPS' original decision to bring charges against Geotechnical Holdings for both Corporate Manslaughter and Section 2 of HSWA; if the charge under the Act had failed for want of proof that there was a causal link between the actions of Peter Eaton, as the sole director, and Alexander Wright's death, then the prosecutions under Section 2 and Section 37 might nevertheless still bite, as these did not require proof of causation. In addition, Section 2 of HSWA involves a reverse burden of proof, i.e. it is for the defendant to prove that it did all that was reasonably practicable to avoid the commission of the offence. Whilst the burden of proof remains on the prosecution under Section 37, this only requires evidence of a causal link to the extent that the actions of the individual brought about the offence committed by the company, not that they were causative of the death.

Fines

The offence of corporate manslaughter is an indictable-only offence, meaning that it will automatically be heard by the Crown Court. The sanction upon conviction (as it relates to a corporate entity only and therefore cannot include imprisonment) is a (potentially) unlimited fine (Section 1(6)), although, from a reputational perspective, perhaps the greater sanction will be the stigma of being branded a 'corporate killer'.

The Home Office paper that went with the original Bill stated that the offence would be targeted at the worst cases of management failure causing death. So it was always likely that when considering fines for conviction, the Courts would be looking at figures over and above record fines seen to date under HSWA (such as £15m in the Scottish case of Transco in 2005 and £7.5m for Balfour Beatty's part in the Hatfield rail disaster).

Indeed, Definitive Sentencing Guidelines for Corporate Manslaughter and Health and Safety Offences Causing Death were duly produced by the Sentencing Guidelines Council and came into effect on 15 February 2010 for all sentences passed from that date on (see *Sources of further information*).

For the first time in this area of the law, the Guidelines advocate a quasi-tariff for fines, stating that:

For corporate manslaughter,

'The appropriate fine will seldom be less than £500,000 and may be measured in millions of pounds'.

For health and safety offences causing death,

'...the appropriate fine will seldom be less than £100,000 and may be measured in hundreds of thousands of pounds or more.'

Whilst the Council therefore dispensed with the previous suggestion that it might set the level of fine by linking it to company turnover (originally mooted at between 2.5% and 10% of average turnover), it is clear that the bar has been raised in terms of the starting point for determining the appropriate sentence for cases of corporate manslaughter and health and safety offences causing death.

These figures would be subject to the usual reduction in the event of an early guilty plea, although it is not entirely clear from the Guidelines whether they would be expected to be reduced further in the case of a defendant with limited means, but the assumption would be so.

That being said, the Guidelines also state that whilst the intention is not to put companies out of business by way of the fine, it accepts that in particularly deserving cases, this might be considered an acceptable side-effect of the punishment.

Remedial orders

Section 9 of the new Act gives a Court the power to make a remedial order requiring a defendant to take action to put right health and safety breaches. The Section states that an order can only be made 'on an application by the prosecution specifying the terms of the proposed order'. Further, that any such order must be on such terms (whether those proposed or others) as the court considers appropriate having regard to any representations made and any evidence adduced by the prosecution or the accused.

The power is not a new one. The HSWA (Section 42) contains a similar power for the Court to make remedial orders and has done so for 30 years. It is rarely used. The potential sanction is fraught with difficulties. Any such orders would need to be consistent with the strategy of improving safety across a whole industry.

Publicity orders

Section 10 of the Act will allow a Court sentencing a Corporate Manslaughter case to make a publicity order requiring the convicted organisation to publicise the fact that it has been convicted, particulars of the offence, the level of fine and any remedial order made. The Court will have free reign to decide in what manner the convicted company is to publicise its guilt – advertisements, billboards, etc. This may turn out to be a little used power – given the small number of cases that are expected to be prosecuted, one would expect widespread publicity to be given to the case in any event.

Nevertheless, the Sentencing Guidelines confirm that most prosecutions for corporate manslaughter should be accompanied by a publicity order, so this should be an expected part of future convictions.

DPP consent

Section 17 of the Act requires the consent of the Director of Public Prosecutions before proceedings are instituted. This is a sensible policy decision to reduce the risk of insufficiently well-founded prosecutions, whether brought by the authorities or private individuals, which would ultimately fail but have an unfair, possibly irreparable, effect on an organisation's reputation.

Who will investigate?

The notes to the draft Bill made it clear that only the police and CPS would investigate. Although the Act is silent on this it seems clear that the HSE, although likely to assist in the police investigation, will not be a prosecuting authority for Corporate Manslaughter.

Individual liability

As stated, the Act imposes no new liability on individuals. The proposals in the earlier drafts to disqualify a manager who had

'any influence' on a management failure and potentially to imprison someone who 'contributed substantially' were conceptually unsound. They would have lowered the level of possible criminal sanction to an unacceptable and unreasonable level. If an individual's acts or omissions are judged to be sufficiently serious and causative of a death then, regardless of the new law, they can still be liable for an offence of gross negligence manslaughter. It should also be recalled that the Enforcing Authorities can prosecute senior managers and directors under Section 37 of the HSWA where an offence by the organisation is committed with their consent, connivance or neglect. Similarly, Section 7 creates a duty in the case of all employees to take reasonable care for fellow employees and others. This combination approach to charging appears in both the concluded *Geotechnical* case and the ongoing *Lion Steel* case.

Commencement and extent

The Act applies in England, Wales, Scotland and Northern Ireland to incidents occurring from 6 April 2008. Section 27 makes it clear that the legislation is not retrospective. Any management failings that pre-date April 2008 will have to be continuing at the time of the death to be a cause of death and will then be caught by the new Act. Although controversially the provisions in the Act extending it to cover deaths in custody were not rendered effective when the Act first came into force in April 2008, they were finally implemented on 1 September 2011. From this date on, state and private custody providers have been subject to the Act.

What should organisations do?

Although the Act has brought no new duties, it poses a natural reason and opportunity for organisations to review their safety management approach, their organisational framework and the systems underpinning them. Guidance published by the HSE / IOD, defining what private and public sector directors should do to lead and promote heath and safety, should be examined against organisations' existing safety management procedures to establish how they measure up and also to identify any weaknesses.

Bearing in mind the reach of the Act beyond the boardroom, organisations should not think that the principles set out in the guidance do not apply lower down the management chain. It would also be prudent for organisations, particularly those in high-hazard industries, to review their liability insurance cover to ensure the legal defence costs for the new offence are covered.

Many employers and Public Liability policies will provide such cover but some may not. Dependent on makeup and size, the organisation may wish to explore the possibility of purchasing additional Directors' and Officers' cover or another form of management liability cover. Experienced advice is important in the immediate aftermath of a workplace fatality, particularly as decisions made at this early stage can set the tone for the criminal investigation and can prejudice an organisation's position and that of its directors and employees. In the circumstances, it is sensible to factor this in to the pre-planning of a major accident response.

Conclusion

The new offence created a completely new framework for the prosecution of Corporate Manslaughter. There is no doubt that the new offence has some impact on the potential criminal liability of companies for manslaughter where there have been serious failures since it is no longer necessary to convict one individual alone. As we have seen above, the aggregated failures of a number of senior managers, who form the senior management, is sufficient.

In addition, the second part of the definition of senior management catches people lower in the management chain than those who would have represented the 'directing mind' under the old law, meaning that a much wider part of the workforce could be considered by the prosecuting authorities when looking at whether the offence has been committed.

Having said that, the new offence continues to have a number of safeguards ensuring that the offence is likely to be restricted for the worst cases – this is seen in particular for large companies in the requirements for senior management involvement and for any breach to be gross.

The result of this is that to date the Act has not had the fearsome effect that some predicted and the safeguards that the Act provides is likely to mean that the floodgates will not open in terms of the number of prosecutions brought. In addition, to date no organisation such as a hospital, partnership etc. has been charged with an offence.

What is clear, however, is that there is more focus being placed on considering whether corporate manslaughter might apply in the wake of a workplace fatality. There has been an increase in initial police investigations for corporate manslaughter, albeit the vast majority will not proceed very far, primacy being handed over to the HSE in accordance with the Work Related Deaths Protocol for Liaison, in order for charges under health and safety legislation to be considered. This has implications for organisations as it means more involvement from the police and the greater criminalisation of investigations, which can be very time-consuming and stressful for the organisations and individuals involved. It is also clear from the single case sentenced so far that, in line with the Sentencing Guidelines, the Courts will look to impose very substantial fines for the new offence, fines which might result in small to medium-sized corporate offenders being forced out of business.

See also: Accident investigations, p.29; Construction site health and safety, p.177; Case review: The first corporate manslaughter case, p.200; Workplace deaths, p.218; Directors' responsibilities, p.227; Health and safety at work, p.361; Interviews under caution (PACE), p.421.

Sources of further information

IOD/HSE: Leading Health and Safety at Work: www.hse.gov.uk/pubns/indg417.pdf

Definitive Sentencing Guidelines for Corporate Manslaughter and Health and Safety Offences Causing Death: http://bit.ly/qY5czo

The Corporate Manslaughter and Corporate Homicide Act 2007 is one of the biggest changes to health and safety law since the HSWA. Its influence will affect everyone with responsibility for the health and safety of employees, and will mean greater responsibility is put on the shoulders of managers and directors, to ensure that they are fully compliant, and not at risk of prosecution should an accident occur. Workplace Law's *Corporate Manslaughter and Corporate Homicide Act: Special Report* includes guidance and comment from some of the leading authorities on the subject, and enables directors and managers to identify the risks, put measures in place to safeguard their fortunes and reputations, and to understand the full implications of the Act. For more information visit www.workplacelaw.net.

Workplace Law's one-day *Corporate Health and Safety Briefing* has been specially designed to help directors and senior managers comply with health and safety legislation and to ensure that they discharge their duty of care towards employees, contractors and members of the public. The course covers the syllabus of IOSH Directing Safely, and can be accredited by IOSH subject to completion of a short practical assignment and brief written examination.

In addition, this special briefing session updates delegates on the latest developments in health and safety law and practice, including the latest information on corporate manslaughter legislation and the IoD / HSE guidance on directors and board members involvement in safety leadership.

Visit http://corporate-manslaughter.workplacelaw.net/ for more information.

Case review

The first corporate manslaughter case

Kevin Bridges, Pinsent Masons Employment Group

Alex Wright was working as a geologist for Cotswold Geotechnical Holdings on 5 September 2008, investigating soil conditions in a deep trench on a development plot in Stroud when it collapsed and killed him.

The Court was told that Mr Wright was left working alone in the three-and-a-half-metre deep trench to 'finish up' when the company director left for the day. The two people who owned the development plot decided to stay at the site as they knew Mr Wright was working alone in the trench. About 15 minutes later they heard a muffled noise and then a shout for help.

While one of the plot owners called the emergency services, the other ran to the trench where he saw that a surge of soil had fallen in and buried Mr Wright up to his head. He climbed into the trench and removed some of the soil to enable Mr Wright to breathe. At that point, more earth fell so quickly into the pit that it covered Mr Wright completely and, despite the plot owner's best efforts, Mr Wright died of traumatic asphyxiation.

The prosecution's case was that Mr Wright was working in a dangerous trench because Cotswold Geotechnical Holdings' systems had failed to take all reasonably practicable steps to protect him from working in that way. In convicting the company, the jury found that their system of work in digging trial pits was wholly and unnecessarily dangerous.

Kevin Bridges is a Partner and Regulatory Lawyer at Pinsent Masons, specialising in health and safety and environmental law. He is dual qualified as a solicitor and chartered safety and health practitioner and is a council member of the Institution of Occupational Safety and Health.

Kevin was involved in advising the first company to be charged with the new offence of Corporate Manslaughter (under the Corporate Manslaughter Act 2007), together with its Managing Director who was separately charged with Gross Negligence Manslaughter.

It said the company ignored well-recognised industry guidance that prohibited entry into excavations more than 1.2 metres deep, requiring junior employees to enter into and work in unsupported trial pits, typically from two to three-and-a-half metres deep. Mr Wright was working in just such a pit when he died. There was no person in the dock at Winchester Crown Court during the three-week trial as it is the company, rather than an individual, which is charged with corporate manslaughter.

The judge, in sentencing and handing down this fine of £385,000 to be paid over ten years, has done something really quite unprecedented. The fine represents in excess of 100% of the company's turnover as it was at the time of the incident in 2008. It represents 250% of the company's turnover today. Fines of that magnitude are really unprecedented; I find it quite staggering that the fine has been levied at that amount, particularly when the Sentencing Guidelines Council recommends that you take into account the size and the means of the company, notwithstanding their starting point of half a million pounds, and many millions of pounds. This legislation was really intended for large multinational companies, and this fine I feel is completely disproportionate to the means of this particular company.

Another interesting feature is that the judge ordered the fine to be paid in equal instalments of £38,500 over ten years and in my experience that is also quite a unique feature of this particular case.

The judge made clear that the fine needs to reflect the gravity of the offence, and so sends out a very clear deterrent to those operating in the wider construction industry. He says it wasn't his intention to necessarily put the company out of business – if that was the effect of the fine that he set out then that was an unfortunate but unavoidable consequence. I think he tried to balance the gravity of the offence as he saw it with the means of the company, by setting an extremely high fine but nevertheless spreading that fine out over a very long period of time and that's the way I believe he approached striking that balance.

It's difficult to say whether the fact that this is the first prosecution of its kind has had an impact or not. My view has always been that this legislation was intended to address the difficulty under the pre-existing common law offence of corporate manslaughter that meant large companies avoided and escaped prosecution and conviction for gross, serious breaches that gave rise to a fatal accident. That's what this legislation was intended to achieve. This offence could quite easily have been brought under the old common law; it was never intended for companies the size of Cotswold. There has not been a true test of the corporate manslaughter legislation. Whilst we've seen hints of it, for example, there's been no discussion of what senior management is in the context of this case, so there's been very little of legal interest in this case.

I think this case has been of interest in relation to the approach the judge has adopted in relation to the sentencing, that is absolutely clear. This is the first time we have seen the SGC's guidance apply in the context of a CM case. We will see other cases no doubt in the future where other aspects of the CM legislation will be tested.

Debate has been circling for many years, since this legislation first came in, as to whether it was necessary at all, or whether the Health and Safety at Work etc. Act 1974 was sufficient. This is the most serious offence under which a company can be convicted. What this case shows is that this legislation was never needed and wasn't necessary in order to bring a conviction against a company.

Corporate Social Responsibility

Sally Goodman, Bureau Veritas UK Ltd

Key points

- Corporate Social Responsibility (CSR) is used by leading companies as a strategic framework, through which all company activities are viewed.
- It is also known as Corporate Responsibility (CR) and sometimes as Corporate Citizenship.
- CSR generally concerns all the impacts that a company has upon society and the environment, and the need to deal with those impacts on each group of stakeholders – typically the shareholders, investors, customers, employees, media, suppliers, regulators, communities (global and local) and non-governmental organisations (NGOs).
- Key issues will vary according to industry sector, but include environmental impacts of products and services; impacts of operations on local communities; human rights and labour conditions in the supply chain and companies' own sites; and impacts of products and services upon customers and the public. It can also be considered as a corporate response to sustainable development and builds on environmental / health and safety / quality systems already in place.
- CSR is not a radically new concept, as businesses throughout history have always had to respond to wider societal issues, but increasing concerns about corporate reputation have undermined trust in business and there is a growing belief that effective CSR is a sound investment for companies and a civil society.
- The UK Government believes that mandatory and voluntary disclosures alone are insufficient to generate responses by businesses to legitimate concerns of society. The growing importance of brand and reputation in the corporate sector is acknowledged as being directly related to business success – a 2009 survey of 224 business leaders worldwide showed that 60% believe corporate social responsibility had increased in importance over the previous year (*see Figure 1, opposite*) and 87% say they have focused their CSR efforts to create new efficiencies.
- This is reflected in the investment community, which has increasingly adopted CSR criteria for screening investments with considerable success – the FTSE4 Good Index is the fastest growing investment index in the UK.
- CSR elements affecting Facilities Managers are most like to be around building management, procurement, and travel. These will incorporate carbon reduction / climate change preparedness and occupational health and safety, and in the current economic climate there is an increasing requirement to quantify the financial benefits of such initiatives.

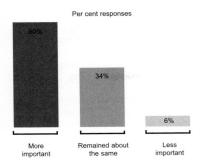

Per cent responses

Figure 1. Change in importance of CSR
to strategic objectives over the past year.
(Source: IBM Institute for Business Value
2009 CSR study.)

Legislation
■ Companies Act 2006.

On 28 November 2005, the UK
Government announced that it would
no longer require quoted companies
to prepare an Operating and Financial
Review (OFR). Instead, through the
Companies Act 2006, companies would
need to include a Business Review in
line with the EU Accounts Modernisation
Directive as part of the Director's Report.

The requirement to include a Business
Review has been enforceable from
October 2007 and requires all quoted
companies (other than small companies),
including subsidiary companies, to
provide more narrative disclosures in the
Directors' Report on the performance of
the business, consistent with its size and
complexity. In addition to information on
risks and uncertainties, and disclosure
of financial key performance indicators,
where appropriate, non-financial key
performance indicators, including
information on employment and
environmental matters, are required.
Clearly, a robust CSR programme will

enable the significance of these
impacts to be understood and any risk
management process to be reported.

Towards the end of 2010, the
Coalition Government instigated
a consultation of how companies
report. This consultation was part of
implementing the Coalition Agreement
commitment to 'reinstate an Operating
and Financial Review to ensure that
directors' social and environmental
duties have to be covered in company
reporting and investigate further ways
of improving corporate accountability
and transparency'.

The objective of the consultation
was to look at ways to drive quality
of company reporting to the level of
the best and thereby enable stronger
and more effective shareholder
engagement. The consultation paper
explored all options – regulatory
and non-regulatory – to achieve the
objectives. It focused in particular
on the business review provisions,
but as part of its exploration of wider
narrative reporting, it also looked
at issues relating to the Directors'
Remuneration Report and a summary
of responses is available on the
BIS website (see Sources of further
information).

First steps
In order to effectively incorporate CSR
into a business model, it is common to
first evaluate the model as a system of
three elements:

1. *Principles* – the core principles
   upon which the business
   operates.
2. *Processes* – mechanisms by
   which principles are implemented.
3. *Outcomes* – the results of
   applying the principles.

By defining the company dynamics in this way, and considering them alongside a generic CSR template, it becomes clearer how a basic CSR framework might be built. For instance, by understanding principles, facilities managers might see how to incorporate a culture of sustainability; by understanding processes, they might outline a plan for ethical procurement; and by understanding outcomes, they might be able to reduce environmental impacts.

Following the adoption of a basic CSR framework, and in order to put the ethos into practice, the next steps would include:

1. Identifying key stakeholders.
2. Identifying key CSR issues (threats and opportunities).
3. Researching best practice in the sector and using indicators as benchmarks.
4. Planning a CSR programme (setting objectives, targets and key performance indicators, or KPIs).
5. Implementing communications and meetings with stakeholders.

The FM would then be well placed to put CSR into practice, and attempt to meet or even exceed the relevant objectives, as measured using the KPIs. An important emphasis under this regime is placed upon dialogue between the FM and important stakeholders (who, for FMs, are likely to be the building owner, its occupants, and suppliers), so as to ensure that appropriate objectives are set and progress is constantly being made, in an interactive and adaptive process.

Putting CSR into practice as an FM
Facilities managers have a key role to play in contributing to CSR, as they can translate the high-level strategic change required by senior decision-makers into day-to-day reality for people in their work or living space. Facilities managers know how buildings work in practice, and the facilities management approach emphasises sustainability, long-term thinking and lifecycle costing.

Reducing greenhouse gas emissions and improving energy performance is a key component of sustainable development – and just one example of where facilities managers are on the 'front line'. Facilities managers can help to achieve organisation and Government targets on energy efficiency and reductions in carbon dioxide emissions as they control heating and cooling systems, lighting and, increasingly, all electronic appliances and information technology in their buildings.

In practice, the adoption of a CSR policy and implementation plan would be made manifest in three main categories:

1. Building management.
2. Procurement.
3. Travel.

In terms of building management, the objective is to fulfil the three pillars of sustainability (economic, social and environmental considerations) by increasing the efficiency with which the building is managed, so as to make financial savings; by creating an amenable and attractive working environment for users; and by reducing the overall environmental impact of the building respectively.

These three strands are closely interrelated, and would be achieved through a number of direct actions; for example, energy and water bills can often be substantially reduced through efficiency gains. Less transparent improvements might be made if, for example, waste streams are carefully organised so that true waste is reduced and reusable waste is sold back to the appropriate sectors (which might reduce waste transport costs besides the clear environmental benefits), or alternatively, if an attractive working environment reduces absenteeism and contributes towards the health of users.

CSR in procurement spans various themes, but largely involves combining ethical and economic considerations in the sourcing of materials and services. Establishing dialogue with suppliers would be essential in ensuring that materials and services are made available without detrimental environmental or social impacts. Despite any moral imperative, in the absence of such considerations it might be the case that various negative externalities are not taken into account, so that true purchasing 'costs' are hidden and poor purchasing choices are made. Consequently, financial gains can often be made in procurement too.

Finally, many environmental gains can be made in fleet management and transport culture. There are increasing numbers of vehicle alternatives available to achieve this end, based upon hybrid, electric and fuel cell technologies, which will probably begin to make long-term economic sense in the face of unstable standard fuel prices and evolving legislation.

Equally, but more simply, the incorporation of CSR into the transport culture so as to encourage public and shared transport, or even to reduce use of vehicles altogether, can have similar benefits to those mentioned above. Many opportunities exist to make gains in each of these three areas that come under sustainability, dependent upon the nature of the business, and which are described in widely available publications (see *Sources of further information*).

Costs and benefits of CSR implementation

The close relationship between CSR and the principle of sustainability demands that, in incorporating CSR into a business model, a long-term view is taken. As such, costs involved in implementation should be considered alongside all potential benefits, which in the long-term would tend to far outweigh costs. These benefits generally originate from four sources:

1. Economic gains.
2. Meeting client demands.
3. Improving public perception.
4. Adapting to legislation.

As previously alluded to, the absorption of CSR into a business strategy tends to make both direct and indirect economic gains possible. However, in the modern business climate, the demands of clients (be they users, businesses or otherwise) increasingly include ethical consideration to some degree, and as such, a clear and transparent CSR framework can provide a business edge.

Equally, the public generally hold a better perception of businesses that demonstrate responsibility to some extent, where this is applicable. Finally, the influence of the sustainability principle upon both European and National policy and legislation continues to increase. In order to strengthen any given business approach for the present and future, the inclusion of a robust CSR framework is considered to be one of the most efficient and necessary solutions.

Standards

The new ISO 26000 Standard, published in November 2010, provides guidance on the underlying principles of social responsibility, the core subjects and issues pertaining to social responsibility and on ways to integrate socially responsible behaviour into existing organisational strategies, systems, practices and processes. By looking at an organisation's behaviour, the ISO 26000 Standard evaluates to what extent it transparently and ethically:

- contributes to sustainable development, including the health and welfare of society;
- takes into account the needs and expectations of stakeholders;

- is compliant with applicable laws and consistent with international norms; and
- integrates and implements these behaviours throughout the organisation.

See also: Building Regulations, p.76; Environmental Management Systems, p.286; International Standards, p.416.

Sources of further information

The EFQM Framework for Corporate Social Responsibility (2004): www.efqm.org

Chartered Management Institute – *Corporate Responsibility: Sustainable Business Practice* (2008): www.managers.org.uk/page/best-practice-corporate-responsibility-sustainable-business-practice

Cranfield University School of Management – *The Doughty Centre 'Guide to Guides': A guide to useful CR / sustainability 'How to embed CR' guides* (October 2009): www.som.cranfield.ac.uk/som/p9346

Business in the Community: www.bitc.org.uk

International Organisation for Standardisation (ISO): www.iso.org/iso/home.htm

BIS Directors' Remuneration Report: The Future of Narrative Reporting – A Consultation: www.bis.gov.uk/assets/biscore/business-law/docs/s/10-1318-summary-of-responses-future-narrative-reporting-consultation.pdf

COSHH: Control of Substances Hazardous to Health

Andy Gillies, Gillies Associates

Key points

- The Control of Substances Hazardous to Health Regulations 2002 implement the requirements of the Chemical Agents Directive (EU no. 98/24/EC) and place a strong emphasis on prevention of exposure to hazardous substances in order to prevent workers suffering ill health. Where prevention is not possible then exposure must be adequately controlled by applying the principles of good control practice. There are many thousands of substances hazardous to health used in industry and other workplaces, but only a few hundred have been assigned exposure limits.

- Workplace Exposure Limits (WELs) have replaced the previous two-tier system of limits. WELs are airborne concentrations of substances averaged over a specified period of time, referred to as a time-weighted average (TWA). Two time periods are used – long-term (eight hours) and short-term (15 minutes).

- The HSE's publication, *Workplace exposure limits* (EH40/2005), now includes the list of substances that have been assigned WELs. A revised list came into force on 1 October 2007 and there are approximately 500 WELs. This follows the implementation of the Second Directive on Indicative Occupational Exposure Limit Values (IOELVs) (2006/15/EC).

- The third IOELV Directive was adopted by the EC in December 2009 and the HSE must implement the new values by December 2011, following a public consultation. Nineteen substances are listed, of which five do not have existing WELs, including bisphenol A, methyl acrylate and sulphuric acid mist.

- Over the next few years, many more substances will be given exposure limits under the Registration, Evaluation and Authorisation of Chemicals Regulation (REACH, EC No. 1907/2006). These new exposure limits will be called DNELs (Derived No. Effect Levels) for airborne concentrations and BMGVs (Biological Monitoring Guidance Values) and will have to be taken into account under COSHH.

- The eight principles of good control practice are explained in Schedule 2A of the COSHH Regulations and will apply regardless of whether a substance has an occupational exposure limit.

Legislation

- Control of Substances Hazardous to Health Regulations 2002 (as amended) (COSHH).
- Registration, Evaluation, Authorisation and Restriction of Chemicals (REACH) Regulation (EC) No. 1907/2006.
- Classification, Labelling and Packaging of substances and mixtures (CLP) Regulation (EC) no. 1272/2008
- Chemicals (Hazard Information and Packaging for Supply) (Amendment Regulations) 2009 (CHIP 4).

Note: CHIP 4 will be amended to meet the requirements of the Classification, Labelling and Packaging Regulations, and will be repealed in 2015 when the CLP Regulations are fully in force.

Case studies

In *Bilton v. Fastnet Highlands Ltd* (1997), the plaintiff, Bilton, worked in a prawn processing factory, and claimed her occupational asthma, caused by certain substances in the workplace, could have been prevented had appropriate measures been taken. While employed in this capacity, she developed occupational asthma as a result of exposure to respirable prawn protein. Her condition was aggravated further by contact with certain substances used and produced in the processing of prawns. The appellant contended that these substances, as well as the prawn protein, fell within the definition supplied by the COSHH Regulations, and that her employer was in breach of this duty.

The employer argued that these were insufficient grounds to claim that a breach of COSHH had occurred, and that the appellant also needed to illustrate what measures should have been taken, but were not, in order to comply with the Regulations. The court referred to the case of *Nimmo v. Alexander Cowan*, where the House of Lords ruled that a plaintiff need only claim that the place in which they had to work was unsafe. It was unnecessary for them to state what they believed was reasonably practicable to actually make and keep it safe.

The Court held that an absolute duty lay with the employer to keep the workplace safe. The appellant need do no more than claim she has suffered injury as a result of the employer failing to discharge their duty under COSHH.

Dugmore v. Swansea NHS Trust and another (2002) was a further important case. It reinforced the authority that:

- the primary duty under COSHH is to prevent exposure altogether where the Regulations apply, unless this is not reasonably practicable;
- if prevention is not reasonably practicable, the secondary duty is 'adequately' to control the exposure. The defence of reasonable practicability qualifies only the primary duty and for purposes of the secondary duty 'adequately' is defined without reference to reasonableness;
- the duties under the Regulations are not subject to foreseeability of risk, nor is it dependent upon what a risk assessment would have revealed; and
- there is no common law duty to dismiss an employee with a particular sensitivity who is willing to take the risk of carrying on working in what for others is a reasonably safe environment.

A more recent case involved exposure to infectious biological agents. The Health Protection Agency (HPA) was prosecuted and fined £25,000 with costs of £20,166 in July 2010 for exposing employees to the risk of infection from E.coli 0157 during waste disposal activities in October 2007 at the Centre for Infections at Colindale. The HPA was found guilty of failing to assess risks during the waste transfer and disposal process, not properly training employees in the use of standard operating procedures, and not remedying defective equipment.

There is now a strong focus in COSHH on good control practice. It is not sufficient to simply demonstrate that exposure is below any applicable exposure limit value. This is a radical change that should encourage improvements in the use of control equipment, ways of working and worker behaviour. This may require a significant update of all COSHH risk assessments. Under the eight principles, employers must:

1. Design and operate processes and activities to minimise emission, release and spread of substances hazardous to health.
2. Take into account all routes of exposure when developing control measures:
    - Inhalation;
    - Skin absorption; and
    - Ingestion.
3. Control exposure by measures that are proportionate to the health risk.
4. Choose the most effective and reliable control options, which minimise the escape and spread of substances hazardous to health.
5. Where adequate control of exposure cannot be achieved by other means, provide, in combination with other control measures, suitable personal protective equipment.
6. Check and review regularly all elements of control measures for their continuing effectiveness.
7. Inform and train all employees on the hazards and risks from the substances with which they work and the use of control measures developed to minimise the risks.
8. Ensure the introduction of control measures does not increase the overall risk to health and safety.

Substances hazardous to health
COSHH applies to a very wide range of substances and preparations (mixtures) with the potential to cause harm if they are inhaled, ingested or come into contact with, or are absorbed through, the skin. COSHH defines a 'substance hazardous to health' as anything that:

1. is listed in Part I of the approved supply list as dangerous for supply within the meaning of the CHIP Regulations, and for which an indication of danger specified for the substance is very toxic, toxic, harmful, corrosive or irritant;
2. the HSE has approved a workplace exposure limit;
3. is a biological agent;
4. is dust of any kind, except dust that is a substance within paragraph (1) or (2) above, when present at a concentration in air equal to or greater than (i) ten mg/m^3, as a time-weighted average over an eight-hour period, of inhalable dust, or (ii) four mg/m^3, as a time-weighted average over an eight-hour period, of respirable dust;
5. not being a substance falling within sub-paragraphs (1) to (4), because of its chemical or toxicological properties and the way it is used or is present at the workplace, creates a risk to health.

Part 5 of the definition is a catch-all for all substances which may be harmful, whether or not they are classified as dangerous or have a WEL assigned to them. The onus is on the employer to identify the hazardous properties of all the substances in use and assess health risks due to the way they are used and the potential exposure of workers and others affected by the work. This broad definition covers substances in the form of solids, liquids, gases, fumes, dusts, fibres, mists, vapours, and biological agents ('germs'). Examples would include:

- chemical substances or preparations such as paints;
- cleaning materials;
- metals;
- asphyxiate gases;
- welding fumes;

- pesticides and insecticides; and
- biological agents such as pathogens or cell cultures.

COSHH does not cover lead, asbestos or radioactive substances, since these are controlled under their own specific Regulations.

Note: The Registration, Evaluation, Authorisation and Evaluation of Chemicals (REACH) Regulation (EC) No 1907/2006 came into effect on 1 June 2007. The Regulations will fundamentally change the format and content of chemical safety information provided by suppliers. For the first time, downstream users of chemicals will have specific duties placed on them that were not previously covered by legislation. (See 'REACH,' p.612.)

COSHH compliance
To comply with COSHH you need to do the following.

Assess the risks to health
- Identify the hazards from substances present in your workplace.
- Consider the risks these substances present to people's health.
- A risk assessment requires consideration of the intrinsic hazard of the substance combined with the degree of exposure and dose received by the worker. There are many determinants of exposure to take into account when carrying out a health risk assessment.
- The amount of detail and information in the risk assessments should reflect the nature and severity of the risks. A RA for a low hazard office environment will be much shorter and simpler than that for a large chemical manufacturing site, for example. The HSE website contains a lot of sensible advice and RA templates that can be used (www.hse.gov.uk/risk/assessment.htm).

Decide what precautions are needed
If you identify significant risks, decide on the action you need to take to remove or reduce them to acceptable levels. If you have five or more employees you must make and keep a record of the main findings of the assessment either in writing or on a computer.

Prevent or adequately control exposure
You are required to prevent exposure to substances hazardous to health, if it is reasonably practicable to do so. You might:

- change the process or activity;
- replace the substance with a safer alternative; or
- use it in a safer form, e.g. pellets instead of powder.
- If prevention is not possible you must adequately control exposure. This may be through use of control equipment such as Local Exhaust Ventilation (LEV), improved ways of working and supervision, and correct worker behaviour and following procedures.
- You should consider and put in place measures appropriate to the activity and consistent with the risk assessment.
- COSHH essentials (www.coshh-essentials.org.uk) can assist in providing advice on the control of chemicals for a range of common tasks, e.g. mixing or drying. It is a practical interactive tool that can assist but not replace the completion of a COSHH assessment. COSHH essentials is supported by a series of general and specific guidance publications aimed at providing practical and informative advice. (www.hse.gov.uk/pubns/guidance/index.htm).

Ensure that control measures are used and maintained
Employees are required to make proper use of control measures. This includes

wearing PPE if necessary, using control equipment provided, and warning supervisors of defects in controls. A common failure with employers is to ignore this point. It is the employer's responsibility to take all reasonable steps to ensure that employees do this.

Some items of equipment will have to be regularly checked to make sure they are still effective. In the case of LEV, a thorough examination and test must be carried out at least once every 14 months. Recent guidance issued by the HSE on controlling airborne contaminants at work (HSG 258 – see *Sources of further information*) gives detailed information on types of LEV, good design principles, and measures for using and testing LEV systems. Users are expected to develop User Manuals and logbooks for each LEV system and perform regular checks to ensure that the system continues to effectively control emissions and exposure.

Respiratory protective equipment (RPE) should be examined and, where appropriate, tested at suitable intervals. Tight-fitting RPE must be fit tested for individuals to ensure a good fit. Fit testing may be carried out using a qualitative or quantitative method and must be performed by a competent person. A voluntary accreditation scheme (Fit2Fit scheme) has been established to accredit individual fit testers. Records of fit tests must be available for all employees who wear tight-fitting RPE. The Regulations set out specific intervals between examinations and you must retain records of examinations for at least five years.

Monitor exposure

You must measure the concentration of hazardous substances to which employees are exposed if your assessment concludes that:

- it is necessary to demonstrate the adequacy of exposure controls, particularly if there could be serious risks to health if control measures failed or deteriorated;
- you need to demonstrate that WEL, BMGV or other exposure limits are not exceeded;
- changes in conditions of work mean that control measures might not be working properly; and/or
- personal exposure monitoring measures the amount of a substance the worker is exposed to. This may be by measuring airborne concentrations, dermal exposures or biological monitoring. Records should be kept for at least five years for general records and for at least 40 years for personal records.

Monitoring exposure to hazardous materials must be done by a competent person using reliable and validated monitoring methods if available. The HSE publishes a list of validated methods in its 'Methods for the Determination of Hazardous Substances' (MDHS) series (see *Sources of further information*).

Carry out appropriate health surveillance

Health surveillance means any activity to obtain information about employees' health, which helps in protecting them against health risks at work. It is necessary when there is a disease associated with the substance, it is possible to detect the disease or adverse changes caused by the exposure to the substance, and where health surveillance information can help to reduce the risk of further harm. You are required to carry out health surveillance in the following circumstances:

- Where an employee is exposed to substances listed in Schedule 6 to COSHH and is working in one of the related processes and there is a reasonable likelihood that an identifiable disease or adverse health effect will result from that exposure.

■ Where employees are exposed to a substance linked to a particular disease or adverse health effect and there is a reasonable likelihood, under the conditions of work, of that disease or effect occurring and it is possible to detect the disease or health effect. Personal records need to be maintained for at least 40 years.

Prepare plans and procedures to deal with foreseeable accidents, incidents and emergencies

This will apply where the work activity goes well beyond the risks associated with normal day-to-day work. If this is the case, you must plan your response to an emergency involving hazardous substances before it happens. This means having the right equipment to deal with an emergency (e.g. appropriate PPE for spill clean-up), and trained staff who understand the health risks and appropriate control measures for dealing with the particular substances.

Ensure that employees are properly informed, trained and supervised

This should include:

■ the names of the substances they work with or could be exposed to, and the risks created by such exposure, and access to any safety data sheets that apply to those substances;
■ the main findings of your risk assessment;
■ the precautions they should take to protect themselves and other employees;
■ how to use personal protective equipment and clothing provided;
■ results of any exposure monitoring and health surveillance (without giving individual employees' names); and
■ emergency procedures that need to be followed.

Note: This requirement is regarded as vital by the HSE. Control measures will not be fully effective if employees do not know their purpose, how to use them properly or the importance of reporting faults.

Safety data sheets, REACH and COSHH assessment

Suppliers of chemicals must provide an up-to-date safety data sheet for all substances classed as 'dangerous for supply'. They provide information on substance hazards, safe handling, storage and emergency measures. This information is helpful to users of chemicals in making their own risk assessments under COSHH.

The classification system for substances and mixtures is changing with the adoption of the CLP Regulation. New pictograms (hazard labels), hazard and precautionary statements, and classification criteria for some physical, human health and environmental hazards will be adopted.

All substances and mixtures on the market at the end of 2010 had to have their classification information notified to the European Chemicals Agency (EChA) to be entered on a Classification and Labelling Inventory. The Approved Supply List has been discontinued and replaced by Annex 6 of the CLP Regulation. Annex 7 of the CLP Regulation contains a translation table to convert classifications made under the Dangerous Substances Directive into new classifications using CLP criteria.

Collecting manufacturers' or suppliers' data sheets and other information does not in itself meet the COSHH requirements to carry out an assessment. The data sheets are a tool to be used in making your assessment.

Note: REACH will result in more extended safety data sheets being produced by

Figure 1. Old and new pictograms.

suppliers. These e-SDS will contain 'Exposure Scenarios' for identified uses of the substance which specify the Risk Management Measures (RMM) to be employed to ensure safe use of the material. It is a legal duty under REACH for Downstream Users to implement the RMM stated in the e-SDS.

Gathering the information is only the first stage; the information must then be used to determine the appropriate control measures needed to protect the health of employees. The assessment record has to include the following information:

- The hazardous properties of the substance.
- Information on health effects provided by the supplier, including information contained in any relevant safety data sheet.
- The level, type and duration of exposure.
- The circumstances of the work, including the amount of the substance involved.

- Activities, such as maintenance, where there is the potential for a high level of exposure.
- Any relevant occupational exposure standard, workplace exposure limit or similar occupational exposure limit.
- The effect of preventive and control measures that have been or will be taken in accordance with Regulation 7.
- The results of relevant health surveillance.
- The results of monitoring of exposure in accordance with Regulation 10.
- In circumstances where the work will involve exposure to more than one substance hazardous to health, the risk presented by exposure to such substances in combination.
- The approved classification of any biological agent.
- Such additional information as the employer may need in order to complete the risk assessment.

See also: Biological hazards, p.51; Local Exhaust Ventilation systems, p.460; Registration, Evaluation and Authorisation of Chemicals (REACH), p.612; Hazardous waste, p.700.

Sources of further information

COSHH Approved Code of Practice and guidance (fifth edition): www.hse.gov.uk/coshh/further/publications.htm

COSHH: Five steps to risk assessment: www.hse.gov.uk/risk/fivesteps.htm

HSE web pages covering risk assessment, COSHH and CLP:

- www.hse.gov.uk/risk/index.htm
- www.hse.gov.uk/coshh/index.htm
- www.hse.gov.uk/ghs/eureg.htm

HSE – Workplace Exposure Limits (EH40/2005) as consolidated with amendments October 2007: www.hse.gov.uk/coshh/basics/exposurelimits.htm

INDG 136: COSHH – Working with substances hazardous to health: www.hse.gov.uk/pubns/indg136.pdf

COSHH essentials: www.coshh-essentials.org.uk

HSE – Controlling airborne contaminants at work: http://news.hse.gov.uk/2008/06/05/hsg-258

HSE – 'Methods for the Determination of Hazardous Substances' (MDHS): www.hse.gov.uk/pubns/mdhs/

The CRC Energy Efficiency Scheme

Dr Anna Willetts, Greenwoods Solicitors LLP

Key points

- The CRC Energy Efficiency Scheme (formerly known as the Carbon Reduction Commitment) is the UK's mandatory climate change and energy saving scheme, which aims to improve energy efficiency and reduce the amount of carbon dioxide emissions in the UK. The scheme is estimated to save 11 million tonnes of CO_2 from the non-traded sector between now and 2022.
- It is central to the UK's strategy for improving energy efficiency and reducing greenhouse gas emissions by 2050 by at least 80% compared with emissions in 1990, as set out in the Climate Change Act 2008.
- It affects large organisations in the private and public sectors, especially at senior level, and aims to encourage changes in behaviour and infrastructure.
- The CRC began in 2010, and the first league table is expected to be published in October 2011. The price mechanism of the scheme will take effect from 2011/12 with the first sale in 2012.

Legislation

- Climate Change Act 2008.
- CRC Energy Efficiency Scheme Order 2010.
- CRC Energy Efficiency Scheme (Amendment) Order 2011.

The relevant legislation that governs the CRC Energy Efficiency Scheme is the Climate Change Act 2008, which was enacted in November 2008 in order to give the UK a legally binding long-term framework to cut carbon emissions. This is implemented through the CRC Energy Efficiency Scheme Order 2010 and the CRC Energy Efficiency Scheme (Amendment) Order 2011.

Who will take part in the CRC?

Organisations that meet the qualification criteria (their 2008 annual electricity supply through all half hourly meters was at least 6,000 MWh) are obliged to participate in the mandatory scheme. Government estimates indicate that around 20,000 public and private sector organisations are required to participate in some way. Failure to comply may result in financial penalties, and the Environment Agency (which is administering the scheme) is likely to publish details of non-complying organisations – 'naming and shaming'.

Around 5,000 organisations will be required to participate fully, which means as well as recording and monitoring their CO_2 emissions, they will also purchase 'allowances', initially sold by the Government, for each tonne of CO_2 they emit. These organisations will include supermarkets, water companies, banks, local authorities and all central Government departments. The more CO_2 an organisation emits, the more allowances it will need to purchase. It is hoped that this will encourage organisations to reduce CO_2 emissions, as the cost saving on energy bills should be a direct incentive.

Following the initial sale period, participant organisations can buy or sell allowances by trading on the secondary market. This enables organisations that have reduced their energy supplies more than they expected to sell some allowances, while those that have higher emissions than anticipated can purchase extra allowances.

A league table will be published by the administrator, and the league table position affects how much of the revenue each organisation receives through a system of bonuses and penalties. It is hoped that this will be an added incentive to reduce CO_2 emissions, as the better a company performs in terms of emissions reduction, the higher it will be represented in the table.

It is anticipated that the savings made should be well in excess of the costs of participating in the scheme. There is a registration fee of £950 for the scheme and organisations were required to be registered before 30 September 2010.

Rule on 'organisation'

The single entity organisation will be responsible for determining qualification and will be the 'primary member' in the CRC. Qualification is also based on supply of electricity to an organisation as a whole, rather than to specific sites.

In general, a public sector entity designated as a 'public authority' under the Freedom of Information (FOI) Act 2000 and the Freedom of Information (FOI(S)) Act (Scotland) 2002 will participate in the CRC on the basis of their individual FOI/FOI(S) listing, or the listing of their organisational type, unless they are legally part of another body, in which case they would participate as part of that parent body.

Subsidiaries

Where an organisation has any subsidiaries that would be eligible to participate in their own right were they not part of a group, these large subsidiaries are known as Significant Group Undertakings (SGUs). Groups with members that are defined as SGUs have some additional administrative requirements. They may also break up the subsidiaries to participate in CRC as individual bodies.

For each of these SGUs, the primary member must:

- provide separate information on the SGU's half hourly electricity supplies, as part of their group registration;
- provide separate information on the SGU's emissions, as part of their group annual report; and
- notify the administrator in the event of a purchase or sale of one of these subsidiaries.

Regulation

The CRC system is regulated by the Environment Agency in England and Wales, the Scottish Environment Protection Agency (SEPA) in Scotland, and in Northern Ireland by the Northern Ireland Environment Agency (NIEA).

What emissions are covered by the CRC?

There are rules covering what emissions count towards CRC emissions that organisations must report to Government. This ensures that organisations do not have to buy allowances for activities or emissions covered by other Government policies.

Emissions for which participants do not have to purchase allowances include:

- domestic accommodation;
- transport emissions;
- emissions from activities covered by a Climate Change Agreement or the EU Emissions Trading System; and
- emissions from consumption outside the UK.

Calculating electricity supply

The administrator should have contacted organisations that were supplied with electricity through at least one settled half hourly meter during 2008. This information includes the 2008 electricity supply data for that meter, which will help calculate your total HHM electricity for 2008.

The basic rule under the CRC is that any electricity supply counts as your responsibility if your organisation has an agreement with another party to supply you with electricity, which you receive via a meter for your own use, and for which you pay on the basis of the meter readings. If you purchase electricity through a third party agent who procures energy services on your behalf and pays the bills, you are responsible as the organisation that contracted the agent.

The only exceptions to these rules are for government departments, which must all participate in the CRC, regardless of whether they meet the qualification threshold or not.

Revenue recycling

The Government Spending Review decision not to proceed with revenue recycling has been criticised by a number of parties, including stakeholders, but the Government says it has had to balance tackling the deficit against a background of high pressure on public finances. The resulting revenue streams were factored into the Government's spending projections for the remainder of the Spending Review period (to the end of FY 2014/15). The first allowance sale for 2011/12 emissions is expected to be a retrospective sale in 2012, which should give participants more time to understand the measuring and monitoring requirements of the scheme.

There are a number of amendments and simplifications to be made to the whole CRC scheme going forward, which are too numerous to go into any great detail here, but the sources of further information listed below provide more detail.

See also: BREEAM, p.67; Climate change, p.145; Corporate Social Responsibility, p.202; Energy performance, p.270; Environmental Management Systems, p.286.

Sources of further information

Department of Energy and Climate Change:
www.decc.gov.uk/en/content/cms/what_we_do/lc_uk/crc/crc.aspx

Environment Agency:
www.environment-agency.gov.uk/business/topics/pollution/98263.aspx

---

Workplace deaths

Kathryn Gilbertson, Greenwoods Solicitors LLP

Key points

- The police and the HSE have an agreed protocol for investigating work-related deaths.
- The HSE has a policy that requires it to investigate individuals for possible proceedings.
- Prosecution (and conviction) rates are increasing under existing laws.
- On 15 February 2011 Cotswold Geotechnical Holdings Ltd became the first company to be convicted under the Corporate Manslaughter and Corporate Homicide Act 2007.
- A second corporate manslaughter trial will commence in June 2012, with proceedings being brought against Lion Steel Ltd.

Legislation
- Corporate Manslaughter and Corporate Homicide Act 2007.

Overview
The police and the HSE jointly investigate workplace deaths, including work-related road traffic accidents. The police will lead the initial investigation and they are often advised in a technical capacity by HSE inspectors. Where either find evidence of culpability, those responsible may be charged:

- with manslaughter, which carries penalties of up to life imprisonment for individuals and an unlimited fine for organisations; or
- under health and safety legislation, which can carry penalties of up to two years' imprisonment for certain specific breaches or an unlimited fine.

Prosecution and conviction rates are increasing.

Cotswold Geotechnical Holdings Ltd was convicted of Corporate Manslaughter following the death of a geologist who was taking soil samples from a pit when it collapsed on to him.

The CPS used guidance published by the HSE on shoring, together with the company's own written policy, signed by its managing director, to prove that activities managed by the senior managers of the business caused the death of Alex Wright.

Manslaughter investigations
Workplace deaths are investigated in accordance with a protocol between the Association of Chief Police Officers (ACPO), the British Transport Police (BTP), the HSE, the Local Government Association (LGA) and the CPS. This also covers work-related road traffic accidents (e.g. involving lorry drivers or managers travelling to meetings).

The protocol gives the police primacy in conducting the investigation and allows them to seek assistance from the HSE, thus utilising the HSE's specialist knowledge and effectively allowing the HSE to conduct its own preliminary investigations. In most cases, the HSE investigates with the police, using its greater knowledge of health and safety management systems, and frequently joint interviews take place. A senior police officer will then review the evidence to decide whether manslaughter charges ought to be brought. The final decision to bring a manslaughter prosecution and/or corporate manslaughter prosecution rests with the CPS. The HSE separately decides on whether to bring proceedings under the HSWA.

Coroner's inquest

The coroner investigates all cases of sudden death. The coroner needs to answer certain questions – to determine who died, when they died, the cause of death, where they died and how the death came about. Workplace fatalities are normally subject to an early preliminary hearing, which is convened in order to allow the deceased to be buried.

The coroner will then adjourn the inquest until after the consideration of proceedings for manslaughter or safety offences (or, in the case of a road-related workplace fatality, possible driving offences) is concluded. This may mean that the inquest will not resume until several years after the accident.

If the CPS decides there is insufficient evidence to bring a prosecution for manslaughter, the coroner's inquest will resume, usually following the coroner receiving the HSE's report into the death.

Inquests into work-related deaths are held before a jury. Several verdicts are available to the jury, including accidental death / misadventure, unlawful killing, or an 'open' verdict. If the jury returns a verdict of unlawful killing, the coroner must refer the case back to the CPS for reconsideration of manslaughter charges. The CPS must then give its reasons if no such charges are brought. At this stage the HSE will consider a prosecution for any breach of health and safety law.

If a criminal trial takes place, which examines the facts and considers the cause of death, then usually only an administrative (paper-based) inquest will be held.

Manslaughter: individuals

An individual may be found guilty of gross negligence manslaughter if a jury is satisfied that:

- the individual owed a duty of care to the deceased;
- the individual breached that duty of care;
- the breach was a substantive cause of death; and
- the individual's conduct was so grossly negligent in all circumstances that the individual deserves criminal sanctions (i.e. imprisonment).

The convicted individual will then face a sentence of up to life imprisonment.

Prosecutions are on the increase. Also, the police are now investigating more instances of work-related deaths.

Those individuals most vulnerable to manslaughter charges are those with day-to-day management of work activities, i.e. directors of small companies, contracts managers, site managers and supervisors.

Manslaughter: organisations

The statutory offence of corporate manslaughter is discussed further in *'Corporate manslaughter'* (p.191).

Other charges

If manslaughter charges are not brought, the HSE is still likely to prosecute culpable individuals and organisations under separate health and safety legislation for health and safety offences.

Minimising liability

The one certain way of avoiding manslaughter charges is to ensure that no one dies as a result of work activities.

Although this may seem obvious, nearly two-thirds of work-related deaths result from just three causes:

1. Falls from height.
2. Being struck by a moving or falling object.
3. Being struck by a moving vehicle.

By focusing their attention on these hazardous areas, employers can minimise the risks and disruption of both deaths and prosecution.

Other simple measures include:

- promptly informing senior managers of any dangerous circumstances or near misses;
- obtaining specialist legal advice on safety systems and protocols, in particular before any incidents arise, and certainly before any police or HSE interviews; and
- dealing sensitively and appropriately with those involved in the accident and relatives of those killed or injured.

Future developments

The prosecution of Lion Steel Ltd and three of its directors – Kevin Palliser, Richard Williams and Graham Coupe – may establish the extent to which senior managers were involved in the decision making that led up to the death of an employee, Steven Berry, whilst working on a fragile roof. The directors are charged with manslaughter by gross negligence and offences under Section 37 of the HSWA. No doubt the CPS will rely on previous warnings and enforcement notices served by the HSE in their efforts to obtain a conviction.

See also: Accident investigations, p.29; Corporate manslaughter, p.191; Health and safety at work, p.361; Interviews under caution (PACE), p.421; Reporting of Injuries, Diseases and Dangerous Occurrences (RIDDOR), p.618.

Sources of further information

Workplace Law's one-day *Corporate Health and Safety Briefing* has been specially designed to help directors and senior managers comply with health and safety legislation and to ensure that they discharge their duty of care towards employees, contractors and members of the public.

The course covers the syllabus of IOSH Directing Safely, and can be accredited by IOSH subject to completion of a short practical assignment and brief written examination.

In addition, this special briefing session updates delegates on the latest developments in health and safety law and practice, including the latest information on corporate manslaughter legislation and the IoD / HSE guidance on directors and board members involvement in safety leadership.

Visit http://corporate-manslaughter.workplacelaw.net/ for more information.

Dilapidations

Andrew Olins, IBB Solicitors

Key points

- Both landlord and tenant should ensure that the scope of repairing obligations in the lease is properly understood.
- The Court's pre-action protocol, or the Property Law Association's protocol, should be used to resolve a claim for dilapidations.
- At an early stage, try and assess the 'true' value of the claim and offer to start negotiations for a settlement of the claim.
- If negotiations for a settlement are unsuccessful, consider making a Part 36 offer to gain protection on costs in any future litigation.
- Be prepared to instruct a building surveyor or a valuer to obtain expert advice.

Legislation

- Landlord and Tenant Act 1927.
- Leasehold Property (Repairs) Act 1938.

Introduction

A claim for dilapidations arises where a landlord considers that its tenant has (wrongly) failed to keep the premises in repair in breach of its repairing obligations under the lease. The landlord can make a claim either during or at the end of the lease. If the claim is made during the lease, the landlord's potential remedies are damages (money compensation), an injunction to force the tenant to comply with its repairing obligations, or forfeiture of the lease. The remedies of an injunction and forfeiture are mutually exclusive. If the claim is made at the end of the lease, the landlord can only seek damages.

The essential issue in any claim for dilapidations is to try and ascertain the value of the claim – in other words, the amount of money, if any, that the tenant will ultimately have to pay to the landlord for failing to keep the premises in repair. Generally speaking, this exercise involves a two-stage process. The first stage is to quantify the cost of carrying out the works necessary to remedy the disrepair. The second stage is to decide whether the cost of these 'remedial works' represents the landlord's recoverable loss.

Stage one

The premises

Looking at the lease is the starting point for quantifying the cost of carrying out the necessary remedial works. Knowing precisely what premises have been let is important as the tenant's repairing obligations normally only relate to the premises actually let. For example, if the tenant occupies a floor in an office block, it is likely that its repairing obligations will only relate to that space, and not the 'common parts' (such as the entrance, reception area, toilets etc.) that the tenant uses in conjunction with other occupants of the block.

Often, the lease will state that 'fixtures and fittings' and plant and machinery are to be treated as part of the premises and, where this occurs, the tenant's repairing obligations apply to these items too.

Tenant's repairing obligations

Once the physical extent of the premises let to the tenant is known, it will be necessary to consider the scope of the tenant's repairing obligations. These obligations or 'covenants' usually fall into five categories:

1. The repairing obligation (properly so-called).
2. The decoration obligation.
3. The reinstatement obligation.
4. The compliance obligation.
5. The yield-up covenant.

The repairing obligation

The repairing obligation invariably imposes an obligation on the tenant to put the premises into a particular standard of repair and to keep it up to that standard throughout the lease. Usually, the standard is 'good and substantial repair' or 'no worse condition than as at the start of the tenancy as evidenced by a schedule of condition'. There is no easy test for determining whether any item of disrepair falls below the standard of repair that the repairing obligation imposes. If a surveyor advising a would-be tenant thinking of taking a new letting of the premises would be concerned by the existence of the item of disrepair, the disrepair is likely to fall below the standard of repair that the repairing obligation imposes. Obviously, whether an item of disrepair troubles the surveyor will depend on the age, character and location of the premises, and the use to which the would-be tenant proposes to make of the premises.

The decoration obligation

The decoration obligation needs little explanation. Usually, the tenant will be under an obligation to decorate the premises at periodic intervals during the lease and at the end of the lease. The standard of decoration that the decoration obligation imposes is to be determined generally by reference to the same criteria applicable to the repairing obligation.

The reinstatement obligation

The reinstatement obligation normally imposes an obligation on the tenant to restore or reinstate the premises at the end of the lease to the condition in which it was at the start of the lease. The reinstatement obligation applies to alterations that the tenant has made to the premises during the lease to make the premises more suitable for its needs. Occasionally, the wording of the reinstatement obligation states that it is only to apply if the landlord, before the end of the lease, requests the tenant to reinstate. If the landlord fails to request reinstatement, or makes a request but leaves the tenant with insufficient time to reinstate the premises before the end of the lease, the tenant may have a good argument for refusing to reinstate.

The compliance obligation

The compliance obligation usually imposes an obligation on the tenant to comply with statutes, regulations or notices served by central or local government or a statutory agency. The compliance obligation is often designed to cover health and safety legislation, the Disability Discrimination Act, and regulations governing the removal of asbestos and other noxious or toxic materials. As legislation tends not to be retrospective, the tenant needs to look carefully to see what, if any, remedial works are actually necessary to comply with the obligation.

The yield-up covenant

The yield-up covenant is usually straightforward. It imposes an obligation on the tenant to hand back the premises to the landlord at the end of the lease in a condition that complies with its obligations.

Landlord's schedule of dilapidations

Once the scope of the tenant's obligations is properly understood, a site inspection and the preparation of a schedule of dilapidations are the next steps in quantifying the cost of the necessary remedial works. To ascertain what remedial work the tenant should carry out, the landlord needs to instruct a suitably-qualified building surveyor to undertake the task. The surveyor should undertake a site inspection a few months before the end of the lease and make a comprehensive note of all items of disrepair found at the premises. The surveyor should be encouraged to take photographs or a video of the condition of the premises, which will be good evidence if, in due course, there is an argument between parties as to the existence of any particular item of disrepair.

The table below represents a typical schedule of dilapidations. The landlord's surveyor will need to complete columns one to five inclusive.

Every item of disrepair that the landlord's surveyor finds during his inspection should be given its own row in the schedule. In column two, the surveyor enters the lease clause that he considers applies. In column three, the surveyor describes the actual disrepair that he found. In column four, the surveyor describes the works that he considers necessary to remedy the disrepair. In column five, the surveyor gives his costing for his proposed remedial works.

If the surveyor exaggerates in the schedule the extent of the disrepair found at the premises, or the cost of necessary remedial works, the landlord could be penalised in costs in any future litigation.

Tenant's response to the schedule of dilapidations

On receiving the landlord's schedule of dilapidations, the tenant will, in turn, wish to seek advice from a building surveyor. The tenant will need advice as to:

- whether the items of disrepair in column three of the schedule fall below the standard of repair imposed by its repairing obligations;
- if so, what remedial works are necessary to make good the disrepair;
- the cost of carrying out the necessary remedial works; and
- whether it should carry out the remedial works or, instead, negotiate a financial settlement with the landlord.

To give this advice, the tenant's surveyor will need to inspect the premises. He, too, should be encouraged to take photographs or a video for future use.

1 Item	2 Lease clause	3 Alleged disrepair	4 Landlord's proposed remedial works	5 Landlord's costings	6 Tenant's comments	7 Tenant's comments on landlord's proposed remedial work	8 Tenant's costings	9 Outcome of without prejudice discussions
1								
2								
3								
4 etc.								

Plant and machinery in particular can throw up particular difficulties for the surveyor. He needs to appreciate that, the mere fact that an item of plant or machinery is old-fashioned, or is stated in 'life span timetables' to be beyond its economic life, or is less efficient than compared with its modern equivalent, or a surveyor acting for a would-be tenant of the premises would not regard it as suitable, does not (of itself) give rise to a breach of the repairing obligation.

Equally, the surveyor needs to understand that, if an item of plant or machinery is in serious disrepair, making replacement the only sensible option, it is likely that the landlord will be able to insist on a modern equivalent, taking advantage of the developments in design and technology that come with it.

After carrying out his inspection, the surveyor will need to complete columns six to eight inclusive of the schedule. In column six, the surveyor states whether the nature and extent of the disrepair in column three is admitted; if not, is any lesser disrepair admitted and, if so, what that lesser disrepair is. In column seven, if any disrepair is admitted, the surveyor gives details of the remedial works that he considers are necessary to make good the disrepair. In completing column seven, the surveyor needs to remember that, as a general rule, if there is more than one method of repair that a reasonable tenant's surveyor would recommend, it is for the tenant to choose which method to adopt. Consequently, the landlord cannot object to the tenant opting for the cheapest method. In column eight, the surveyor puts his costings for his proposed remedial works.

After the tenant's surveyor has completed his part of the schedule, the first stage in valuing the claim – namely, ascertaining

the cost of the necessary remedial works – nears completion. The schedule should now reveal in monetary terms the gap between the parties as to the costs of carrying out the necessary remedial works to make good the disrepair found at the premises.

Discussion between surveyors

The parties' surveyors should be encouraged to hold a meeting on site to go through the schedule with a view to resolving, or at least narrowing, any differences between them as to the nature and extent of any particular item of disrepair, and the cost of making good the disrepair. It is customary for this meeting to be held on a 'without prejudice' basis so that the surveyors can talk candidly and freely without fear that, at a later date, possibly in court, any statements or admissions that are made during their discussions will be held against them.

Where the landlord serves its schedule shortly before the end of the lease, the tenant tends to prefer to negotiate a financial settlement rather than carry out remedial works. The meeting between surveyors affords the tenant a good opportunity to start negotiations for a financial settlement.

Stage two

Calculating diminution

Each party instructing its own valuer is the starting point in determining whether the costs of carrying out the necessary remedial works represents the landlord's recoverable loss. At common law, the landlord's recoverable loss is usually the cost of carrying out the works necessary to remedy the items of disrepair listed in the schedule of dilapidations. However, Section 18(1) of the Landlord and Tenant Act 1927 imposes a statutory cap on the landlord's recoverable loss. The effect of

Section 18(1) is to prevent the landlord's recoverable loss exceeding the amount by which its reversion is diminished owing to the tenant's breach of its repairing obligations.

There is no definitive method for ascertaining whether the statutory cap bites in any particular case. The parties' valuers will endeavour to calculate the value of the landlord's reversion in repair and out of repair, and compare the difference with the cost of carrying out the works necessary to remedy the items of disrepair listed in the schedule. If the difference between the value of the landlord's reversion in repair and out of repair is less than the cost of the necessary remedial works, the statutory cap will bite so as to reduce or even extinguish the landlord's recoverable loss.

The statutory cap tends to bite where the necessary remedial works, if carried out, would be rendered redundant. This occurs where, for example, the landlord intends to demolish or convert the premises (possibly for an alternative use), or to refurbish the premises to improve its existing specification.

Landlord's other recoverable losses

If the landlord is able to establish that it has suffered a loss of rent, or loss of service charge, or had to pay business rates because the tenant failed to hand back the premises in repair at the end of the lease, it may be able to recover these losses. Usually, the landlord will need to show that, if the premises had been handed back in repair, it would have been able to re-let the premises relatively easily. In other words, 'but for' the disrepair, the landlord would not have suffered a 'void'. The period for which loss of rent, rates or business rates can be claimed is linked to the period that it would reasonably take to carry out the necessary remedial works.

If the landlord incurs fees with surveyors, architects, quantity surveyors etc. in connection with the actual carrying out or supervision of the necessary remedial works, these fees can form part of the landlord's recoverable loss. Often, there is a clause in the lease that expressly permits the landlord to recover the fees that it incurs with surveyors and lawyers in preparing and serving the schedule of dilapidations. Where there is such a clause, these fees are recoverable in contract.

Leasehold Property (Repairs) Act 1938

This Act gives the tenant protection against a landlord intent on pursuing a claim for dilapidations for an improper motive. If the lease was granted for a term of seven years or more and there is still three or more years to run, the tenant can serve a notice preventing the landlord from pursuing its claim without the permission of the court. The court is unlikely to give permission unless the works necessary to remedy the disrepair need to be carried out urgently to protect the value of the landlord's reversion or to comply with any legislation or regulations.

Landlord's remedial works

If the lease permits, rather than pursue a claim for dilapidations during the lease, the landlord can enter the premises, carry out the works necessary to remedy the disrepair, and claim reimbursement from the tenant of the cost of those works as a 'debt'. The Leasehold Property (Repairs) Act 1938 (and the protection it affords) has no application to a landlord's claim for reimbursement.

Avoiding litigation

Wherever possible, the landlord and the tenant should seek to avoid the need to litigate a claim for dilapidations. Litigation is invariably expensive and time-consuming. Therefore, the parties should be willing to open negotiations for settlement at an early stage.

The Civil Procedure Rules (the rules of Court) include a pre-action protocol that obliges the landlord and the tenant to participate in a constructive dialogue, involving the exchange of information and documentation, with a view to settling the claim without the need for litigation. The Property Litigation Association has produced its own protocol for handling claims for dilapidations. Whilst the PLA's protocol does not have the force of law, it does offer a clear structure and timetable for promoting the resolution of claims.

There will, of course, be occasions where, despite the parties' best efforts, it is not possible to settle a claim through negotiation. Before the landlord starts litigation, the tenant should give serious thought to making a realistic 'offer to settle' under Part 36 of the Civil Procedure Rules to 'buy' protection on costs.

A landlord who declines an offer to settle and, at trial, fails to achieve a better result than the offer to settle, runs a very serious risk of having to pay the tenant's costs of defending the claim. Accordingly, a landlord who fails to accept a reasonable offer to settle can find that it ends up with a pyrrhic victory. The tenant's offer to settle should be pitched, therefore, at a sum that causes the landlord to reflect on the wisdom of litigating.

Forum for resolving claims

Where the claim for dilapidations is modest and relatively uncomplicated, the appropriate forum for litigating the claim will be the local County Court. However, if the claim is for a substantial sum, say, in excess of £100,000, and there are complicated expert issues to be determined, the claim should be issued in the Technology and Construction Court. The TCC has a specialist panel of judges who have expertise in adjudicating claims for dilapidations. The principal TCC is at St Dunstan's House, Fetter Lane, London, EC4 although there are several district registries in England and Wales where claims can be pursued.

See also: Landlord and tenant: lease issues, p.429; Landlord and tenant: possession issues, p.438.

Sources of further information

Property Litigation Association: www.pla.org.uk

Dilapidations claims can be costly and time-consuming for both the landlord and tenant, and a source of grave contention between the two parties. The legislation surrounding dilapidations has evolved over a long period of time and is notoriously complicated. Workplace Law's *Guide to Dilapidations* provides a better understanding of this complex issue. Written in Workplace Law's jargon-free, plain-English style, this downloadable guide is an indispensable resource for all users and owners of commercial buildings. For more information visit www.workplacelaw.net.

Directors' responsibilities

Rachel Farr and Lorraine Smith, Taylor Wessing

Key points

- Company directors are primarily responsible for the management of their companies and, generally, their duties are owed to the company. However, they also owe duties to the owners as a whole. In addition, they have responsibilities in respect of the company's employees and its trading partners, and under statute these duties may be supplemented or modified by a company's articles of association. Furthermore, many directors will be subject to service agreements (contracts of employment) which may augment these duties.
- Directors are responsible for ensuring that the company complies with the various requirements imposed upon it by law.
- Although generally the company is liable for any failure to comply with legal requirements, in certain circumstances the directors can be held personally liable where the default was due to their neglect or connivance.

Legislation

- Health and Safety at Work etc. Act 1974.
- Company Directors Disqualification Act 1986.
- Insolvency Act 1986.
- Value Added Tax Act 1994.
- Management of Health and Safety at Work Regulations 1999.
- Companies Act 2006.
- Corporate Manslaughter and Corporate Homicide Act 2007.
- Health and Safety (Offences) Act 2008.
- Bribery Act 2010.

Directors' duties under the Companies Act 2006

The Companies Act 2006 (CA 2006) codifies certain key duties of directors. It sets out a statutory statement of seven general duties:

1. A duty to act in accordance with the company's constitution and only to exercise powers for the purposes for which they are conferred.

2. A duty to act in a way which a director considers, in good faith, would be most likely to promote the success of the company for the benefit of its members as a whole.
3. A duty to exercise independent judgement.
4. A duty to exercise reasonable care, skill and diligence.
5. A duty to avoid conflicts of interest.
6. A duty not to accept benefits from third parties.
7. A duty to declare to the other directors an interest in a proposed transaction or arrangement with the company.

These duties are (apart from the duty to exercise reasonable care, skill and diligence) all fiduciary duties. They are expressed to replace the previous common law duties but continue to be interpreted by reference to the body of case law in this area.

These duties are owed to the company and only the company can enforce them, although shareholders can make a claim by statutory derivative action on behalf of the company. CA 2006 has extended the common law derivative action, making it easier for shareholders to bring a claim on behalf of the company against directors and others for negligence, default, breach of duty or breach of trust, where a prima facie case is disclosed and the court gives permission for the claim to continue.

Duty to act in accordance with the company's constitution and only to exercise powers for the purposes for which they are conferred

A director must act in accordance with the company's constitution, which, for this purpose, includes resolutions or decisions made by the company in accordance with its articles of association, as well as the articles of association themselves.

Duty to act in a way which a director considers would be most likely to promote the success of the company for the benefit of its members as a whole

In fulfilling this duty, directors must have regard to a statutory non-exhaustive list of factors, namely:

- the likely consequences of any decision in the long term;
- the interests of the company's employees;
- the need to foster the company's business relationships with suppliers, customers and others;
- the impact of the company's operations on the community and the environment;
- the desirability of the company maintaining a reputation for high standards of business conduct; and
- the need to act fairly as between members of the company.

This duty has extended and replaced the common law duty on directors to act in good faith and in the best interests of the company.

The decision as to what will promote the success of the company, and what constitutes such success, is one for a director's good faith judgement. For a commercial company, 'success' will usually mean a long-term increase in value.

Duty to exercise independent judgement

This duty is likely to be most relevant where a director wishes to bind himself to a future course of action which might be seen as 'fettering the discretion' of the director to make future decisions. It is not infringed by a director acting in a way authorised by the company's constitution or acting in accordance with an agreement duly entered into by the company that restricts the future exercise of discretion by its directors.

Duty to exercise reasonable care, skill and diligence

A director must exercise reasonable care, skill and diligence. The standard expected of him is not only the general knowledge, skill and experience he has (for example, a particular expertise in financial matters), but also the general knowledge, skill and experience that may reasonably be expected given his position and responsibilities to the company.

Duty to avoid conflicts of interest

A director must not use information gained by him as a director to further his own interests (unless he has the consent of the company to do so, as outlined below), nor must he seek to apply company assets for his own gain. For example, a director must not receive commission on a transaction between the company and a third party or offer to take up, on a private basis, work offered to the company.

A director must disclose any direct or indirect personal interests in a contract and will have to account for any profit made unless he complies with the requirement for disclosure before the contract was entered into. These requirements for avoiding conflicts of interest and declaring interests in contracts are dealt with in specific statutory duties, as set out below.

A director must not, without the company's consent, place himself in a position where there is a conflict, or possible conflict, either directly or indirectly, between the duties he owes the company and either his personal interests or other duties he owes to a third party. That applies, in particular, to the exploitation of property, information or opportunities, and whether or not the company could take advantage of the property, information or opportunity. There is no breach if the situation cannot reasonably be regarded as likely to give rise to a conflict of interest.

This duty does not apply to a conflict arising in relation to a transaction or arrangement with the company. In that situation, it is the duty to declare an interest in a proposed or existing transaction or arrangement with the company which applies (*see below*).

The duty to avoid conflicts of interest continues to apply after a person ceases to be a director as regards the exploitation of any property, information or opportunity of which he became aware when he was a director.

The company's consent can be obtained in advance by board authorisation (unless the company's constitution prevents this) but will only be effective if the quorum and voting majority are met without counting the director in question or any other interested director.

For a private company incorporated before 1 October 2008, the board can only do this if the shareholders have previously approved the board giving such authorisations. For a public company, whenever incorporated, the board can only do this if authorised by the articles.

The shareholders themselves can also authorise conflicts of interest that would otherwise be a breach of this duty. To some extent, a company's articles can also contain provisions for dealing with conflicts.

Duty not to accept benefits from third parties

A director has a duty not to accept benefits from third parties, except where the benefit is not likely to give rise to a conflict of interest (in fact, most directors' service agreements will have an express prohibition on accepting benefits or 'kickbacks' from any third party, irrespective of whether or not a conflict of interest might arise).

Benefits conferred by the company, its holding company or subsidiaries, and benefits received from a person who provides the director's services to the company, are excluded.

Duty to declare an interest in a transaction or arrangement with the company

A director has a statutory duty to declare the nature and extent of any direct or indirect interest in a proposed transaction or arrangement with the company. No declaration is needed where:

- the director is not aware of the interest or the transaction in question (unless he ought reasonably to be aware of it);
- it cannot reasonably be regarded as likely to give rise to a conflict of interest;
- the other directors are already aware of it (or ought reasonably to be aware); or
- it concerns terms of his service contract being considered by the board or a board committee.

A director also has a statutory obligation (although not a fiduciary duty) to declare any direct or indirect interest in an existing transaction or arrangement with the company.

Duties to employees

While a director owes no common law duty to consider the interests of the workforce, and a director's duties are owed to the company, CA 2006 recognises the principle that the interests of the workforce fall within the wider picture of the interests of the company, as the interests of employees are one of several matters that directors must have regard to when satisfying their duty to act in the way they consider to be most likely to promote the success of the company.

Directors must comply with employment law in dealings with employees. In some circumstances, directors personally can be sued for unfair work practices such as race, sex, disability and other discrimination. Directors must ensure the company complies with any new employment laws.

Health and safety issues

A company has various obligations to fulfil under the Health and Safety at Work etc. Act 1974 (HSWA) and the Management of Health and Safety at Work Regulations 1992 (MHSWR).

The most important of these are as follows:

- A duty to ensure, so far as is reasonably practicable, the health, safety and welfare of its employees. The size of the company and the activities that are carried on by it will be taken into account when assessing what is reasonably practicable.
- A duty to carry out a risk assessment and implement procedures to minimise any risks that are highlighted.
- A duty to provide (and periodically revise) a written health and safety policy, to implement it and to bring it to the attention of employees.

These are the company's responsibility, but the directors would be breaching their duties to the company by failing to take the appropriate measures. Furthermore, directors can be prosecuted under Section 37 of HSWA where the offence committed by the company occurred with their consent or connivance or through neglect on their part.

If found guilty, directors can be sentenced to up to two years in prison. In addition, the Corporate Manslaughter and Corporate Homicide Act 2007 means the company is liable where a death is caused as a result of a gross breach of a duty of care and the way in which the company's activities were organised or managed by senior management constitutes a substantial element of that breach.

It is wise for directors to ensure they have health and safety systems in place. The HSE provides comprehensive guidance on this issue.

Financial responsibilities

Accounts

CA 2006 requires directors to maintain accounting records that:

- show the company's transactions and its financial position;
- enable the directors to ensure that accounts required under CA 2006 comply with the CA 2006 requirements;
- contain entries of all receipts and payments, including details of sales and purchases of goods, and a record of assets and liabilities; and
- show stock held at the end of each year.

Records must be kept at the company's registered office (unless the directors specify a different location) and be retained for a period of six years for a public company or three years for a private company.

The directors of a public company must lay its annual report and accounts before a general meeting (usually the AGM) and then file them with the Registrar of Companies within six months of the end of the company's financial year. Private companies do not have a statutory requirement to lay their accounts before the company in general meetings but they must file them with the Registrar of Companies within nine months of the end of the company's financial year.

Statutory returns, including the annual report and accounts, the annual return (including a statement of capital) and notice of changes to directors and secretaries, must be filed with the Registrar of Companies on time.

Failure to comply with these requirements renders the company liable to a penalty and directors liable to a fine.

Directors are also responsible for filing tax returns.

Financial management

Directors must exercise prudence in the financial management of the company. In the event of insolvency, directors can find themselves personally liable to creditors where it can be shown that they acted outside of their powers or in breach of their duties or were engaged in wrongful or fraudulent trading. The latter offences will be committed where a director continues to incur liabilities on behalf of the company where he knows or ought to have known that the company was, or inevitably would become, insolvent, or there was no reasonable prospect of repaying debts.

Directors have an obligation not to approve accounts unless they give a true and fair view of the financial position of the company (or of the companies included in the group accounts, to the extent that this concerns members of the company).

Bribery Act 2010

The Bribery Act 2010 (BA 2010) came into effect on 1 July 2011. It is of fundamental importance to all commercial organisations that either operate or are registered in the UK. BA 2010 has reformed the criminal law to provide a modern and comprehensive scheme of bribery offences to enable courts and prosecutors to respond more effectively to bribery, wherever it occurs. It is a far-reaching piece of legislation with some provisions that are more extensive than equivalent laws elsewhere, including the US Foreign Corrupt Practices Act (FCPA). It applies to bribery in both the private and the public sectors and makes illegal the bribery of another person, being bribed, and the bribery of a foreign public official.

Adequate procedures to prevent bribery by a corporate organisation

An offence under BA 2010 is committed by a commercial organisation when a person 'associated' with it (i.e. performs services for it) bribes another person and that bribe is intended to obtain or retain business for the commercial organisation or retain an advantage in the conduct of the organisation's business. It is a defence to this offence if the company can show it had adequate procedures in place to prevent such bribery taking place, which must include:

- *Proportionate procedures* – the procedures an organisation should take must be proportionate to the risks they face. The Adequate Procedures Guidance suggests, for example, that factors such as an organisation's size and the nature and complexity of its business will influence the response required.

- *Top level commitment* – this requires top level management, including directors to ensure that the organisation's staff and those who do business with or for the organisation understand that bribery is never acceptable.

- *Risk assessment* – this requires organisations to assess the nature and the extent of their exposure to external and internal risk of bribery. This assessment needs to be periodic, informed and documented.

- *Due diligence* – this is about knowing who you do business with; the Adequate Procedures Guidance recommends organisations undertake a proportionate and risk based approach to due diligence in respect of persons who will perform services for and on behalf of the organisation.

- *Communication* (including training) – the Government believes the communication of bribery policies and procedures will deter bribery by enhancing awareness of the organisation's procedures and its commitment to their proper application. Training should be used to raise awareness about the threats posed by bribery in general and the sector areas in which the organisation operates.

- *Monitoring and review* – an organisation should monitor the effectiveness of the procedures they put in place and make improvements where necessary.

In order to show adequate procedures, the board of directors will be expected to take responsibility for the matters above.

Individual directors' liability

If a corporate body (of any kind) or a Scottish partnership commits an offence under BA 2010 then a senior officer, including a director, can be personally liable for the same offence if he or she has consented or connived in the commission of the offence. 'Senior officer' is defined widely and includes director, manager, company secretary or similar officer of the corporate body or a partner of the Scottish partnership. Such a senior officer must have a close connection to the UK as defined in Section 12(4) – such as British citizenship, British Overseas citizenship, or being ordinarily resident in the UK.

If found to have committed an offence under BA 2010, a senior officer may be liable to a maximum penalty of ten years' imprisonment and/or an unlimited fine.

Other duties

Directors must maintain various statutory books including:

- a register of members;
- a register of mortgages and charges;
- a register of debenture holders; and
- a register of directors.

The statutory books are to be kept either at the company's registered office, or at a single alternative location (subject to certain criteria being met).

An annual return must be filed with the Registrar of Companies within 28 days of the anniversary of either its incorporation or the made-up date of the company's previous annual return.

Directors must ensure that minutes are taken at board meetings, giving a record of all decisions taken.

Practical steps for directors

All directors should:

- ensure that they are fully aware of their duties under CA 2006 and that their board processes reflect these duties;
- check the company's articles of association to establish the scope of their powers;
- ensure that minutes are maintained;
- be alert to conflicts of interest between the company and themselves as individuals;
- always comply with employment law;
- ensure that the company operates a comprehensive system for assessing and minimising health and safety risks;
- keep informed about the company's financial position – ignorance will not save them from facing personal liability in certain circumstances;
- be very clear about what their service agreements require of them – often their obligations in such agreements will be more onerous than their statutory or common law obligations; and
- make sure that the company obtains insurance to protect them against facing personal liability.

Directors' indemnities

Companies are prohibited from exempting a director from any liability he may incur in connection with any negligence, default, breach of duty or breach of trust by him in relation to the company, unless the provision constitutes a 'qualifying third party indemnity provision' (QTPIP). For the indemnity to be a valid QTPIP the director must not be indemnified against:

- liability the director incurs to the company or a group company;
- fines imposed in criminal proceedings or by regulatory bodies such as the FSA;
- legal costs of criminal proceedings where the director is convicted;
- legal costs of civil proceedings brought by the company or a group company, where judgment is given against the director; or
- liability the director incurs in connection with applications under Sections 661 or 1,157 CA 2006 for which the court refuses to grant the director relief.

Companies can therefore indemnify directors against liabilities to third parties, except for legal costs of an unsuccessful defence of criminal proceedings or fines imposed in criminal proceedings or by regulatory bodies.

Companies can pay a director's defence costs as they are incurred, even if the action is brought by the company itself against the director. The director will, however, be liable to repay all amounts advanced if he is convicted in criminal proceedings or if judgment is given against him in civil proceedings brought by the company or an associated company.

Companies can now also provide slightly wider indemnities to directors of corporate trustees of occupational pension schemes against liability incurred in connection with the company's activities as trustee of the scheme. This is known as a 'qualifying pension scheme indemnity provision' (QPSIP).

A QTPIP or QPSIP must be disclosed in the directors' report in each year that the indemnity is in force and a copy (or summary of its terms) must be available for inspection by shareholders. The QTPIP or QPSIP must also be retained by the company for at least one year after its expiry. Departing directors may also ask for such policies to be retained after their termination as part of exit negotiations.

> *See also*: Corporate manslaughter, p.191; Health and safety at work, p.361; Insurance, p.412.

Sources of further information

IOSH Directing Safely is intended for people with strategic responsibility for determining and implementing effective health and safety management within an organisation. The importance of health and safety in the day-to-day running of a company dictates that senior members of management need to be very aware of their responsibilities. In recent years some high profile cases have clearly demonstrated that the responsibility for managing health and safety starts at board level. The safety message has to be driven from the top down, and directors, although busy people, need to be involved in this process.

The course reflects the principles embodied in:

- The Health and Safety Executive's guidance. 'Successful Health and Safety Management' (HSG65),
- The Turnbull report ('Internal controls: Guidance for Directors on the Combined Code'),
- The DETR/HEC's 'Revitalising Health and Safety' strategy statement.

The course provides directors / owners of UK organisations with an understanding of the moral, legal and business case for proactive health and safety management and to give guidance on effective risk management.
Visit http://iosh.workplacelaw.net/directing-safely for more information.

Comment ...

Health and safety leadership

David Sinclair, Metis Law LLP

The HSE recently ruled out recommending that express legal health and safety duties should be imposed on directors. Instead, the HSE will rely on the implementation of its guidance 'Leading Health and Safety at Work: leadership actions for directors and board members' (INDG 417) and existing provisions in the Health and Safety at Work etc. Act 1974 (HSWA) that impose liability on individual directors and managers.

Despite the HSE's decision, a series of recent events indicate that directors and managers who fail to ensure leadership and effective management of health and safety could come under regulatory scrutiny.

- Since 2008, figures show that 25 people have received prison sentences for breach of health and safety laws, compared with only 14 in the previous 30 years.
- The Court of Appeal has found that the conduct of a company director who took on personal responsibility for health and safety made her personally liable for damages to a seriously injured employee.
- Later this year, three Warwickshire Fire and Rescue Service senior managers will stand trial for gross negligence manslaughter following the death of four firemen. If convicted, the managers face terms of imprisonment and/or substantial fines.

The HSE has also recently updated its guidance to inspectors who visit

Metis Law LLP is a new style of regional law firm with a strong property and construction base. Metis acts across the public and private sector, providing businesses with tailored services and innovative products. The firm has a reputation for delivering solutions to a range of property, facilities and construction organisations.

David Sinclair is one of a few dual-qualified solicitors and safety practitioners in the UK, who have achieved Chartered Health and Safety Practitioner status. Trained originally as a mining engineer, David has a BSc Honours degree in Occupational Health and Safety and a Durham University postgraduate diploma in Environmental Management.

organisations for any reason, requiring them to determine the extent of the organisation's compliance with health and safety law, and assess the effectiveness of health and safety management. A key part of any visit will be inspectors determining the effectiveness of 'leadership' within the organisation.

The updated guidance follows the HSE's evaluation of the implementation of INDG 417, which found that, on the surface,

NHS trusts, charities and not-for-profit organisations score well against other public and private organisations in implementing INDG 417. However, when researchers 'drilled down' they found that many boards appear to be paying only lip service to the guidance.

In any intervention, the HSE will now look for evidence that the Board is leading on shaping the organisation's health and safety ethos, that it is accountable for its decisions that affect health and safety, and that it has visible ownership of health and safety issues.

In determining whether leadership is effective, inspectors will look at the organisation's culture (e.g. by reviewing documents, interviewing staff and inspecting work areas) and assess whether the Board is effective in influencing the attitudes of managers and staff.

> "An increased HSE focus on boards could see more directors and senior managers being prosecuted for health and safety offences, and being ordered to pay damages and costs following a fatality or a serious injury."

According to the guidance, when visiting organisations inspectors should (amongst other things) look for evidence that demonstrates:

- genuine, active leadership from the Board, including a Board Health and Safety Champion, and every member setting an example on health and safety;
- the Board's receipt of regular, appropriate health and safety information;
- the Board has correctly identified and categorised significant health and safety risks, determined what additional control measures are needed, and it has plans in place to implement, monitor and review the effectiveness of those controls;
- Board members regularly discuss health and safety issues with staff and contractors in their workplaces and that issues are listened to and acted upon;
- the Board sets policy and standards and monitors health and safety; and
- that health and safety implications of any business decision are considered and addressed (at the planning / design stage where possible).

The guidance also provides that inspectors should look for a balanced (sensible and proportionate) approach to health and safety management. In determining if there is such an approach, inspectors should, for example, determine that:

- the organisation has effective management arrangements in place, which are ideally externally accredited and that health and safety is integrated into business processes;
- available documentation is current, organised and relevant and that paperwork is not getting in the way of good management;
- the health and safety responsibilities of key people are set out in their job descriptions and staff (and contractors) fully understand the health and safety aspects of their role, including risks and control measures; and

- active and reactive monitoring is undertaken to check controls are working and the root cause of incidents are identified and communication on health and safety issues between the Board and the 'shop floor' is effective.

The courts now have powers to imprison individuals for up to two years for most health and safety offences, and inspectors are required to remind courts of their power to disqualify convicted directors for up to 15 years.

An increased HSE focus on boards could see more directors and senior managers being prosecuted for health and safety offences, and the willingness of the civil courts to look behind the corporate veil and hold directors personally liable could see more directors being ordered to pay damages and costs following a fatality or a serious injury.

It is important for all directors to be fully aware, and comply with the requirements, of INDG 417. In particular, directors must be able to show that they have received appropriate, formal training and that they are competent to undertake the health and safety aspects of their role.

Directors should ensure that documents do not make them personally responsible for health and safety and that they receive regular health and safety updates.

Disability access and egress

Dave Allen and Steve Cooper, Butler & Young Group

Key points

- The DDA and various parts of associated legislation were repealed on 1 October 2010 to be replaced by the Equality Act 2010.

- Although the Disability Discrimination Act 2005 (DDA) had been in place for some time and the various stages of introduction have provided for gradual change to the built environment, there has always been a need to consider health and safety issues and, in particular, fire safety and evacuation provision. Unfortunately, guidance on the subject has been slow to keep pace with the need for all buildings to become accessible.

- Central to the principle of accessibility to buildings and facilities, and the extent of provision or adaptation necessary, is the tenet of 'reasonableness'. This still applies in principle to fire safety provisions for disabled people, such as the creation of refuge areas (*see below*) or installation of evacuation lifts. Clearly the extent and/or nature of use of a building will have a bearing upon what can or should be expected, but this will also impact upon the evacuation strategy and any consequential limitation on access.

- Safety has occasionally been stated as a reason for not allowing some disabled people full access to all parts of a building on the basis that the risks cannot be overcome or managed effectively. As a result, there is a potential conflict between the Equality Act requiring a disabled person to have equal rights and benefits that a non-disabled person enjoys, and the need to assess and prevent any safety risks.

- Great emphasis has been placed in the past on management responsibilities and the principles to be adopted. However, practical information on how to go about planning for the safe evacuation of disabled people has, until recent years, been lacking.

Legislation

- Regulatory Reform (Fire Safety) Order 2005.
- Equality Act 2010.

Guidance

Guidance can effectively be divided into two elements:

1. Physical provision within the building environment to aid the evacuation process (principally Building Regulations Approved Document B and BS 9999).
2. Management and procedural guidance (Section 9 of BS 9999 and *Fire Safety*

Risk Assessment – Supplementary guide on Means of escape for Disabled People).

Approved Document B

Being restricted to new buildings, changes of use, and relevant alterations to buildings, Approved Document B has limited impact. The guidance in paragraphs 4.7 – 4.14 of Approved Document B is essentially an extract of the most relevant sections of BS 5588 Part 8 with respect to the provision of refuge areas and associated communication provision. There is also reference to warnings for people

Case study

Shaban v. Wharfe's Restaurant and Tea Rooms (2007)

A disabled woman who was subjected to 'overt and aggressive' behaviour when she tried to visit a tea shop was awarded £4,500 in compensation by Yeovil County Court.

Jazz Shaban won a disability discrimination claim against Wharfe's Restaurant and Tea rooms after District Judge Brian Smith said that the tea-shop owner had presented a "wholly unmeritorious defence, which exhibited a lack of sensitivity".

The court heard that Shaban, who has brittle bones and uses a wheelchair, visited Wharfe's Restaurant and Tea room, in Shaftesbury, Dorset with some friends for lunch, in May 2006.

Shaban, whose case was supported by the Disability Rights Commission (DRC), attempted to access the downstairs seating area, which was down two steps, when the restaurant owner, Rosemary Wharfe shouted, "Not in here, we cannot accommodate that," referring either to the wheelchair or to Shaban, or both.

Shaban and her friends left the tea room immediately, but Wharfe continued to shout at them. In a subsequent letter the defendant admitted refusing access to the party, but said that wheelchairs could be 'pulled' upstairs.

Wharfe chose not to appear in court, and in her defence papers referred to the case as 'scurrilous' and 'tawdry'.

However, District Judge Smith said, "This is the clearest possible case of discrimination," and that the defendant's attempt to justify that discrimination "fails utterly".

He also found statements made by Wharfe and given in correspondence to the court, to have been "deeply unfeeling and patronising". Wharfe, he said, had behaved in an overt and aggressive way in public.

with impaired hearing where others may not be available to make a person aware of an emergency in a room. This could include, for example, hotel bedrooms and sanitary accommodation.

BS 9999 – Code of Practice for Fire Safety in the Design, Management and Use of Buildings

The Code provides guidance on provision to assist the escape of disabled people, including more comprehensive information on the design, location and associated provisions within a refuge area within the appendices. Commentary on fire alarms is also provided.

Refuges

A refuge is a specified area relatively safe from the effects of a fire where a disabled person (not just a wheelchair user) may travel with or without assistance and await

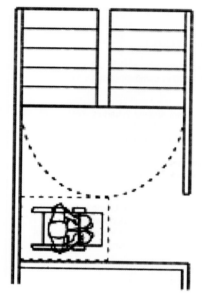

Figure 1.

rescue by designated helpers without impeding the escape route of others. In multiple floor buildings it is also beneficial in allowing a place of temporary rest during an evacuation process (*see Figure 1*).

Within the refuge area there should be a means of achieving two-way communication with a manned control point. The importance of this communication, properly used, cannot be over-emphasised. The risk that a person needing assistance may be overlooked is real, and the fire safety consequence and psychological impact on the individual of feeling isolated must be avoided. Most alarms are likely to be false and it is essential that disabled people are made aware of this as soon as an emergency condition is determined. This may then avoid potentially stressful, risky and

painful evacuation, which to most would be nothing more than a minor inconvenience.

Managing fire safety
Section 5 of the document regarding the evacuation of disabled people provides guidance on management procedures and practices that should be implemented and how they interact with the physical provisions in place.

Reference is made to disabled people having a Personal Emergency Evacuation Plan (PEEP) and differentiates between plans for people who regularly attend the site, those who do not but make themselves known on arrival, and those who do not identify themselves to staff; however no guidance is provided of the detail of a plan or any practical information on processes and evacuation techniques. See *'Personal Emergency Evacuation Plans (PEEPs)'* (p.569).

Fire Safety Risk Assessment Supplementary Guide on Means of escape for Disabled People (DCLG)
An important development with respect to the responsibility and process of evacuation came in 2006 with the introduction of the Regulatory Reform (Fire Safety) Order. The Order, made under the Regulatory Reform Act, places a duty on a person within a building or organisation to take reasonable steps to reduce the risk from fire and ensure occupants, including disabled people, can safely escape if a fire does occur.

In March 2007, a supplementary guide in association with the Department of Communities and Local Government (DCLG) was made available (*Fire Safety Risk Assessment – Means of Escape for Disabled People (Supplementary Guide)*).

This guide provides important information on the specific issue of planning escape for disabled people and provides the practical guidance missing from BS 5588 Part 12.

A fire risk assessment needs to demonstrate that, as far as is reasonable, all needs of 'relevant persons,' with respect to evacuation, have been considered, and this includes disabled people. It is accepted that with respect to people with no mobility or sensory limitations, buildings and procedures are based upon self-evacuation and evacuation times; travel distances, warning systems etc. are based upon this principle.

However, step two in the process of producing a fire risk assessment is to identify the people at risk and as a consequence their needs with respect to being aware of and evacuating from a safety risk scenario (step one relates to identifying the hazards).

The production of PEEPs is described within the guide and a matrix is provided to assist in drawing up the plans. The danger, particularly with respect to the evacuation of disabled people, is that it becomes a 'tick box' process without practical understanding of an individual's needs and the risks that exist.

The guide states that all staff should be given the opportunity to have a PEEP. There are a number of hidden disabilities such as heart, respiratory, or stress-induced conditions, which may affect a person's escape capability in an evacuation situation. Also, people who are excessively obese may potentially not be considered as disabled under the DDA or consider themselves as such, although any mobility restriction as a consequence may satisfy guidance with respect to the definition of a disability.

An evacuation plan is only of benefit if it is applied effectively and correctly. Information is key; both in that provided by a disabled person, and training and awareness of the Responsible Person and any helpers or persons involved with the implementation of a plan.

Communication and preparation of PEEPs for regular building users will be relatively easy to achieve. However, much of the information in the guide is based upon known and regular building users. Where the visitor is unknown, the system will be more difficult to formulate and a 'broad brush' approach will be necessary to consider all potential situations and disabled persons' needs. General evacuation plans for visitors, based upon a range of predicted scenarios, would need to be drawn up. In these instances, training of building staff in access awareness and the evacuation procedure contained within the plans is particularly critical.

Standard evacuation plans are described as being offered for people at entry / reception points. Staff awareness of these plans and clear indication to visitors of the plans' existence is paramount.

In uncontrolled visitor situations, the general evacuation plans should be broad ranging and flexible. As the guide suggests – 'Training for staff is vital … In order to do this they should receive disability etiquette training'.

The guide highlights that, when it comes to fire safety, certain expectations with respect to access, privacy, and dignity may need to be compromised. This is not an excuse to reduce standards, and the 'reasonableness' consideration with respect to how a disabled person is treated prior to or during an emergency must be a consideration.

The *Supplementary Guide for Means of escape for Disabled People* provides guidance for the Responsible Person to identify and address the risks to a disabled person and ensure their safe evacuation.

General considerations

It could be contended that, under the Equality Act, discrimination has not arisen with respect to evacuation until such times as an emergency occurs. However, it is important to appreciate that most disabled people are only too aware of their own safety and vulnerability in an emergency situation. As a consequence, any risks or failings with respect to fire and emergency safety strategies could be deemed to be a failing both under the Equality Act and the Fire Safety Order, even without an incident occurring.

The guide does acknowledge that when planning for means of escape it is 'planning for exceptional circumstances'. There are many instances where fire safety has been used as an argument for reduced levels of accessibility. Although in limited cases this may be potentially valid, in principle this would only be deemed acceptable when all 'reasonable' options have been exhausted and a 'real' safety risk cannot be ignored.

It must be appreciated that, in providing safe evacuation practices, there may be a need for the otherwise normally expected dignity and independence of disabled people to be compromised and the nature of the circumstances of any claim under the Equality Act in this respect are likely to take this into account. This does not preclude any evacuation plan being drawn up and applied sensitively.

All evacuation practices carry a risk. The guide recognises that unnecessary escapes should be avoided. People using stairs en masse and in haste increase the risk of falls etc. These risks are increased significantly for people with mobility and visual restrictions. It is therefore important that practices and, particularly, false alarms, are kept to a minimum.

This should, however, be without prejudice to the fact that evacuation processes for some disabled people may be complex and involve a number of people. As a result, careful and practised routines are necessary in order to ensure that safe and successful evacuation occurs.

Good communication systems to keep disabled people informed of the status of an emergency would be of benefit, but

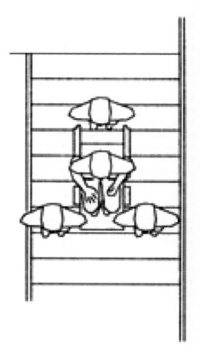

Figure 2.

only if the communication procedures in an emergency are robust. For example, a wheelchair user at a refuge with two-way communication may await confirmation of the need to evacuate prior to making a risky assisted descent down stairs. Early indication as to whether the emergency is a false alarm can avoid a stair descent, which can put the disabled person and potentially their assistant at risk of injury.

Good communication assists in ensuring that the right number of trained helpers reach the point where they are needed, and gives reassurance that the presence is known of a person in need of help and an assisted evacuation is imminent. Waiting helplessly whilst a fire alarm sounds is particularly stressful.

When it comes to evacuating a disabled person with a mobility difficulty via a staircase there are four methods advised:

1. Use of proprietary evacuation chairs.
2. Carry down in the person's wheelchair.
3. Carry down using an office chair.
4. Carry down using wheelies.

The evacuation strategy should not only be determined by individual needs but will be affected by the building design. Realistically, carry down strategies are only possible if the stairs are suitably wide and not too long (see Figure 2).

See also: Accessible environments, p.18; Building Regulations, p.76; Fire, means of escape, p.314; Personal Emergency Evacuation Plans (PEEPs), p.569.

Sources of further information

All organisations must ensure that means of escape to a place of ultimate safety, supported by sufficient numbers of competent persons, are in place to effect an evacuation without the assistance of the Fire and Rescue Services, as stated by the Regulatory Reform (Fire Safety) Order 2005.

In addition, disabled people must be treated as equally as non-disabled people, as stated by the Equality Act 2010.

PEEPs for Professionals is a two-day accredited course designed to give students expert knowledge of the law, its implications, the impact of barriers on the evacuation of disabled people, and how to implement effective means of escape for both disabled and vulnerable people.

Visit www.workplacelaw.net/training/course/id/78 for more information.

The *Fire Safety Risk Assessment Supplementary Guide – Means of escape for Disabled People* can be downloaded free of charge from www.communities.gov.uk

Display Screen Equipment

Andrew Richardson, URS Scott Wilson

Key points
- Use of DSE constitutes an adverse health condition.
- Workers using or operating DSE can suffer postural problems, visual problems, and fatigue and stress.
- Employers must identify users or operators – those whose normal work is to habitually use DSE. Laptops and homeworkers are included.
- Employers must carry out a risk assessment of these people's work, using trained personnel.
- Employers must analyse workstations and ensure they meet minimum standards.
- Employers must provide breaks and variety in the DSE users' or operators' work.
- Employers must provide and pay for eye and eyesight tests, and provide spectacles (where needed for screen-viewing distance), if employees request.
- Employers must provide training and information.

Legislation
- Health and Safety at Work etc. Act 1974.
- Health and Safety (Display Screen Equipment) Regulations 1992 (as amended by the Health and Safety (Miscellaneous Amendments) Regulations 2002).
- Workplace (Health, Safety and Welfare) Regulations 1992.
- Provision and Use of Work Equipment Regulations 1998.
- Management of Health and Safety at Work Regulations 1999.

DSE users and operators
The Regulations define what is deemed to be DSE, who is deemed to be a 'user' or 'operator' and what a workstation comprises. The criteria for determining who can be designated as a user or operator state that generally it will be appropriate to classify people as a user or operator if they:

- normally use DSE for continuous or near-continuous spells of an hour or more at a time;
- use DSE in this way more or less daily;
- have to transfer information quickly to or from the DSE; and
- need to apply high levels of attention and concentration; or
- are highly dependent on DSE or have little choice about using it; or
- need special training or skills to use DSE.

DSE risk assessment
The Regulations require that all employers carry out a suitable and sufficient risk assessment on workstations and it is suggested that a suitable way is to use an ergonomic checklist. Within the appendices of the Regulations is an example of a checklist that can be used. The checklist aids the employer in assessing the following.

Facts

- The symptoms most commonly leading to time off work in display screen equipment (DSE) users are headaches (absence reported by 7.9% of all respondents) followed by back pain (where the equivalent figure was 4.9%).
- Informal surveys indicate that one in three computer users may have the early symptoms of repetitive strain injuries (RSI).
- Contract Research reported that 55% of DSE users thought their laptops were heavy – carrying them for considerable distances had led to back, neck, and shoulder strains and injuries.
- Contract Research reported that 75% of DSE users have visual tiredness after using their laptop display for just 20 minutes.
- An HSE funded report (RR561) found high prevalence in DSE users of self-reported symptoms, such as headaches (52%), eye discomfort (58%), and neck pain (47%); other symptoms such as back (37%) and shoulder (39%) pain were also frequently reported.

Display screen
- Are the characters clear and readable?
- Is the text size adequate?
- Is the image stable?
- Is the screen size suitable?
- Can you adjust the screen brightness and contrast?
- Can the screen swivel and tilt?
- Is it free of glare and reflections?
- Are there window coverings in an adequate condition?

Keyboard
- Is it separate from the screen?
- Does it tilt?
- Is there a comfortable keying position?
- Does the user have a good keyboard technique?
- Are the keys easily readable?

Mouse / trackball
- Is it suitable for the task?
- Is it close to the user?
- Is there wrist / forearm support?
- Does it work smoothly?
- Can the software adjust the pointer speed and accuracy?

Software
- Is the software suitable for the task?

Furniture
- Is the work surface large enough?
- Can the user reach all of the equipment?
- Are the surfaces free from glare?
- Is the chair suitable and stable?
- Does the chair adjust in terms of back height and tilt, seat height, swivel mechanism and castors, etc.?
- Is the back supported, are the arms horizontal and eyes the same height as the top of the VDU?
- Are the user's feet flat on the floor?

Environment
- Is there room to change the position?
- Is the lighting suitable?
- Is the air comfortable?
- Is the heating level comfortable?
- Are the levels of noise comfortable?

The user / operator
- Have they any other problems?
- Have they experienced discomfort?
- Do they know of their entitlement to eye and eyesight testing?
- Do they take regular breaks?

Trained personnel, generally meeting the following criteria, must carry out the DSE risk assessments. They must be familiar with the main requirements of the DSE Regulations and should have the ability to:

- identify hazards and assess risks;
- draw upon additional sources of information on risk as appropriate;
- draw valid and reliable conclusions from assessments and identify steps to reduce risk;
- make a clear record of the assessment and communicate findings to those who need to take action and to the worker concerned; and
- recognise their own limitations as to the assessment so that further expertise can be called upon if necessary.

These assessments must be reviewed whenever:

- a major change occurs to the software, equipment or workstation;
- the workstation is relocated;
- the environment is changed; or
- there is a substantial change to the tasks or amount of time using DSE.

The purpose of the risk assessment is to reduce the risk of the workforce suffering postural problems, visual problems, and fatigue and stress.

Managing DSE work

Regulation 3 clarifies that the Regulations apply to all workstations, not just those used by users or operators. The Regulations require all employers to ensure their workstations meet the requirements set out in the schedule encompassed within the Regulations. This includes the use of laptop computers and homeworking.

The Regulations set out the criteria for ensuring that the daily work routine of users will not contribute towards any of the risks identified earlier and advises on the nature and timing of breaks. If you are a user or operator, Regulation 5 places a duty upon the employer to provide eyesight tests and special corrective appliances (normally spectacles) for DSE use. The Regulations require that all users or operators be provided with information on a number of relevant points including:

- the DSE Regulations themselves;
- risk assessment and means of reducing risk;
- breaks and activity changes;
- eye and eyesight tests; and
- to have initial training and training when the workstation is modified (Regulation 6).

Appendices within the Regulations give guidance on the use of laptop computers and the use of mouse, trackball or other pointing devices.

See also: Eye and eyesight tests, p.301; Furniture, p.340; Workplace health, safety and welfare, p.768; Work-related upper limb disorders, p.772.

Sources of further information

L26: Work with Display Screen Equipment: www.hse.gov.uk/pubns/priced/l26.pdf

HSG 90: The law on VDUs: An easy guide: www.hse.gov.uk/pubns/priced/hsg90.pdf

---

Dress codes

Jackie Thomas, Berwin Leighton Paisner LLP

Key points

- Employers seek to apply dress codes to their employees for many reasons. In doing so, however, it is important that employers consider any potentially discriminatory implications as dress codes have historically been challenged under both the Sex Discrimination Act 1975 (SDA) and Race Relations Act 1976 (RRA). These characteristics are now all protected under the Equality Act 2010.

- Furthermore, since the European Convention on Human Rights has been incorporated into UK law by way of the Human Rights Act 1998 (HRA), it may also be possible to challenge the application of a dress code on the basis that it infringes the employee's human rights.

- To avoid potential liability, employers should ensure that the policy applies in an equivalent way to both men and women and that any requirements imposed are reasonable when balancing the rights of the employee and the requirements of the employer's business.

- One factor that is relevant to this issue is whether the employer uses a dress code for health and safety reasons.

Legislation

- Human Rights Act 1998.
- Equality Act 2010.

The Equality Act came into force in October 2010. It is a consolidating Act, which brings together the old, separate discrimination legislation (for example, the Sex Discrimination Act 1975 and the Race Relations Act 1976) into one new Act. While the new Act does change the law in some respects it is unlikely to have an effect on existing case law on dress codes.

Why might a dress code be needed for health and safety grounds?

Dress codes are applied for a number of reasons. One of these reasons is health and safety. An example of this is the use of uniforms (often including both hats and gloves) for roles that involve the preparation of food. Further, employees in medical roles may be asked to modify

what they wear. An example is the NHS policy that was introduced in 2008, which requires all medical staff who come into contact with patients to either roll up their sleeves or wear short sleeves, and prevents them from wearing jewellery, watches and false nails. This policy is based on attempts to limit the spread of bugs and, particularly, superbugs, in UK hospitals.

Potential race discrimination claims

When policies are used for health and safety reasons, it is quite common for them to be challenged as amounting to race discrimination (previously either under the Race Relations Act or under the Employment Equality (Religion or Belief) Regulations 2003, and now under the Equality Act). This is because policies that require the wearing of particular clothes for both sexes are sometimes likely to conflict with religious requirements.

Facts

- 64% of those employers operating a dress code policy relax their dress codes rules at times, while just under a third (31%) do not.
- 66% of UK companies and public sector bodies found that while standards of dress are becoming less formal, the policing of what is and is not acceptable clothing for the office is being tightened up.
- A significant minority of employers (27%) say that while their dress code policy is observed, it still has to be policed.
- More than two-thirds (67%) of dress code policies now have the force of the employment contract behind them.
- Employees have to wear a uniform or overalls at less than half of organisations (46%).
- Two-thirds of companies support people in meeting the dress codes of their religion.
- Only around one in ten employers with a policy impose restrictions on religious dress or jewellery.
- Health and safety is a common reason for applying a dress code but it is not the most widely adopted reason by UK employers.

It is also possible for race discrimination claims to arise if a dress code has a disparate impact on a particular racial group. This is a claim of indirect discrimination, to the extent that the proportion of a particular racial group can comply is considerably smaller than employees outside of that group who can comply and the policy cannot be justified as a proportionate means of achieving a legitimate aim. In the case of dress codes on health and safety grounds, the employer is likely to establish a legitimate aim, but may find it more difficult to establish that the policy is a proportionate means of achieving that aim.

An example that resulted in a claim was a policy that required a Sikh to shave his beard for health and safety reasons. The employee claimed that this amounted to indirect discrimination (in that it was more difficult for Sikhs as a racial group to comply). However, the Tribunals held that the employer's actions were justifiable as the policy was in place for reasons of food hygiene. To the extent that such a requirement could not be justified it would be discriminatory.

A further example of this is a Muslim NHS radiographer who resigned because she said that the requirement to have bare arms discriminated against Muslims because it contradicted religious rules on dress. This case was not taken to an Employment Tribunal but it has reflected complaints raised against the policy by other Muslims. Had the case been litigated, the Trust would have needed to show that the policy was a proportionate means of achieving a legitimate aim (assuming that the employee was able to establish that the policy did disadvantage Muslim employees).

Finally, in the 2011 case of *Dhinsa v. Serco and Others*, an Employment Tribunal found that it was justifiable to prevent a Sikh prison officer from wearing a Kirpan (a Sikh ceremonial knife). Although it was found that the policy did place the claimant at a disadvantage on the grounds of his religious or beliefs, the employer was found to have a legitimate aim in securing the safety of staff, prisoners and visitors. The Tribunal went on to consider whether the ban on wearing the knife was a proportionate means of achieving this. To do so, it considered a number of factors, including statistics showing the number of assaults in prisons using knives or similar weapons, steps that the employer had taken to minimise the impact to the employee, and the fact that the Tribunal found no breach of the claimant's human rights. This is a good example of the balancing exercise that employers need to follow when relying on health and safety, in order to demonstrate that the policy is being applied in a proportionate manner.

> *See also*: Catering: food safety, p.117; Personal Protective Equipment, p.576.

Sources of further information

Equality and Human Rights Commission: www.equalityhumanrights.com

Driving at work

Kathryn Gilbertson, Greenwoods Solicitors LLP

Key points

- Employers need to manage the use of both the company car driver, and the person using his own vehicle for business, using risk assessments and a driving for work policy.
- Working time rules apply to drivers.
- Motorists can be prosecuted for driving while using a handheld mobile phone. Employers may also be prosecuted if they require their employees to use their phones when driving.
- All vehicles should be taxed, insured, kept in a roadworthy condition and have an up-to-date MOT where necessary.
- If a work vehicle is used as a workplace by more than one person it must be smoke-free at all times.

Legislation

- Health and Safety at Work etc. Act 1974.
- Road Traffic Act 1988.
- Management of Health and Safety at Work Regulations 1999.
- The Road Transport (Working Time) Regulations 2005.
- The Health Act 2006.
- Road Safety Act 2006.
- The Smoke-Free (Vehicle Operators and Penalty Notices) Regulations 2007.
- The Highway Code.

Road traffic law and the Highway Code

These lay down certain rules and restrictions (e.g. speed limits) and are normally enforced by the police and the courts. While the driver of the vehicle will primarily be held responsible for any offence, employers may also be liable, for instance in setting schedules that are so tight that the driver would consistently be breaking the speed limits if he attempted to meet them. The Magistrates Act 1980 may also be relevant to employers in England and Wales who aid, abet, counsel or procure an offence. Employers are responsible for ensuring their company vehicles are properly taxed and insured.

Health and safety legislation

Employers should manage at-work road journeys and other on-the-road work activities within their usual health and safety protocols. Occupational drivers or people employed to work on or by roads should be offered the same protection as those working within fixed workplaces. Employers also have a responsibility to ensure that others are not put at risk by the work activities of their employees.

Any breach of an employer's statutory or regulatory duties towards its employees giving rise to criminal liability may also be relied upon by a civil claimant as evidence of an employer's breach of duty in a negligence action and indeed in support of a claim for constructive dismissal.

There is a growing emphasis on safety and the responsibility of employers when any work-related vehicle accident is investigated. In light of this, it is important that employers implement a health and safety policy for driving at work, and keep abreast of the developments affecting workplace driving.

The main areas are:

- working time;
- mobile phones;
- regulatory compliance;
- risk assessment; and
- smoking.

Working time

The Road Transport (Working Time) Regulations 2005

The Regulations cover mobile workers who will, in the main, be drivers and accompanying crew involved in road transport activities in a vehicle that is required by EU laws to have a tachograph (Council Regulation 3821/85 on recording equipment in road transport). The Regulations include the following provisions:

- A mobile worker's working time shall not exceed an average 48-hour working week, typically calculated over a four-month reference period.
- A maximum of 60 hours may be worked in a single week (provided that the average working week does not exceed 48 hours).

- There is a ten-hour limit for night workers over a 24-hour period.
- Workers cannot work more than six consecutive hours without taking a break. If working between six and nine hours, a break of at least 30 minutes is required. If working over nine hours, breaks totalling 45 minutes are required. Each break may be made up of separate periods of not less than 15 minutes each.
- The Regulations affect self-employed drivers. Other drivers who fall outside the scope of the new Regulations, such as drivers of smaller vehicles or drivers exempt from the EU Drivers Hours Rules, are covered by the WTR; for example, the 48-hour average working week and the need for adequate rest. However, unlike the WTR, employees covered by the new Regulations cannot 'opt out'.
- Employers must monitor working time and should do what they can to ensure the limits are not breached. Records need to be kept for two years. Generally speaking, annual leave / sick leave cannot be used to reduce the average working time of a mobile worker. For each week of leave that is taken, 48 hours' working time must be added to their working time; for each day's leave, eight hours must be added to working time.

Case studies

Certain vehicle accidents in the workplace are RIDDOR reportable. Work-related driving accidents are the largest single cause of all reportable workplace accidents – accounting for, according to RoSPA, over 800 deaths in 2009. Failures in implementing proper health and safety procedures cost Britain's employers up to £6.5bn every year.

R v. Melvyn Spree (2004)

This case illustrates a combination of working time and dangerous driving that resulted in a manslaughter conviction for the company director.

Melvyn Spree, a road haulage director, was jailed for seven years after one of his lorry drivers fell asleep at the wheel and killed three motorists. It was held that Melvyn Spree, a director of Keymark Services, encouraged and enabled his drivers to work dangerously long hours, through fraudulent record-keeping and tachograph tampering. Melvyn Spree's fellow director, Lorraine March, was also jailed for conspiracy offences and Keymark Services was fined £50,000 for manslaughter.

Police investigations found that the driver, Stephen Law, was partway through an 18-hour shift when the accident occurred. The police also discovered systematic abuse of working hours' restrictions. Drivers were rewarded with a profit-share scheme. Melvyn Spree showed drivers how to jam tachographs and to keep false records of working times that demonstrated legal compliance.

R v. Knapman and Legg (2005)

In December 2005, Raymond Knapman, Partner at R&D Drivers, was prosecuted for manslaughter but acquitted after it could not be established that a fatal accident involving the deaths of two lorry drivers was caused by excessive hours or a heart attack by one of the drivers. However, Knapman was subsequently charged with eight counts of obtaining property by deception due to consistently requiring drivers to work excessive hours. Knapman pleaded guilty to the offences and to a breach of Section 3(2) of the HSWA for failing to ensure the health and safety of persons not in his employment and was sentenced to two-and-a-half-years' imprisonment in January 2006.

- If no employer exists, the agency, employment business or even the worker themselves should monitor working time.

For further guidance see *'Working time'* (p.758).

Mobile phones

Motorists can be prosecuted for driving while using a handheld mobile phone. Drivers committing this offence will be liable to pay a fixed penalty or a fine on conviction in court. The offence also attracts three penalty points.

The Regulations apply in all circumstances other than when the vehicle is parked, with the engine off. This means that the prohibition applies even if a vehicle has paused at traffic lights, stopped in a temporary traffic jam, or is in very slow-moving traffic.

The definition of 'handheld' means a mobile phone or other device that is held at some point during the course of making or receiving a call or fulfilling some other interactive communication function. An interactive communication function includes sending or receiving oral or written messages; facsimile documents; still or moving images; or accessing the internet.

Hands-free products, which do not require drivers to significantly alter their position in relation to the steering wheel in order to use them, have not fallen foul of the change in the law.

Employers' liability for mobile phone use whilst driving

The Regulations also created an offence of 'causing or permitting' another person to drive while using a handheld phone or other similar device. Employers may be prosecuted if they require their employees to use their phones when driving. The DfT has stated that employers cannot expect their employees to make or receive mobile phone calls while driving. This must be reflected in the company's health and safety policy and risk management policy. Employers will not be liable simply for supplying a telephone or for telephoning an employee who was driving. However, employers must send a clear message to employees that they are forbidden to use their handheld mobile phones while driving and their employer will not require them to make or receive calls when driving.

Employers should inform their staff that, when driving, handheld mobile phones should be switched off, or, if switched on, the calls should be left to go through to voicemail, and that a safe place to stop should be found to check messages and return calls. Company policy should specify that using a handheld phone or similar device while driving is a criminal offence and will be treated as a disciplinary matter.

Hands-free mobile phones

Hands-free kits are widely available and the use of these kits is still legal. However, employers should be aware that this does not mean that drivers will be exempt from prosecution altogether if they use hands-free kits. Dangerous and careless driving can still be committed as separate offences under the Road Traffic Act 1988.

Research shows that using a hands-free phone while driving distracts the driver and increases the risk of an accident. Therefore, many businesses have banned them outright. Employers who install hands-free kits should balance the commercial advantage of this with the potential risk of future liability, were an employee to cause an accident while speaking on the phone and driving.

General guidance

- Switch off the phone while driving and let it take messages.
- Alternatively, leave the phone switched on and let the calls go to voicemail.
- Alternatively, ask a passenger to deal with the call.
- Find a safe place to stop before turning off the engine and picking up the messages and returning calls.

Regulatory compliance (the vehicle risk assessment)

All vehicles should be taxed, insured, kept in a roadworthy condition and have an up-to-date MOT where necessary. Employers should check that privately-owned vehicles are not used for work purposes unless they comply with the above and are additionally insured for business use.

Risk assessment

Risk assessments should be carried out on:

- the driver;
- the vehicle; and
- the route.

The driver must be competent. Regular reviews of driving licences will identify those with repeat endorsements, which may suggest that they should no longer be allowed to drive on company business. Equally, receiving a report on someone's medical condition could mean that they are no longer fit to drive. Full insurance and vehicle safety checks / servicing should be carried out on a regular basis and documents retained. A corporate checklist could be issued covering those frequently required checks – condition of tyres, lights and bulbs, wiper blades, screen wash and jets, oil etc.

The route needs to be assessed to identify whether drivers:

- can reach the appointment without excessive speeds;
- need to incorporate an overnight stop; and
- may suffer tiredness as a result of a long drive and long working day.

Managing work patterns, namely the route and driver, effectively can reduce the amount of unnecessary journeys and reduce the potential for a culture of risk-taking through tiredness or unachievable targets.

Smoking

All enclosed public places and workplaces must be smoke-free. These include company cars and hire vehicles. If a work vehicle is used as a workplace by more than one person it must be smoke-free at all times. The legislation does not extend to private vehicles.

If a private vehicle is used for work and the employee doesn't ever use it with others, it is permissible to smoke in that vehicle. There is no guidance available at present with regard to vehicles that are used for primarily private journeys but sometimes used for business together with others.

Owners or managers of smoke-free premises will be guilty of an offence if they fail to prevent people from smoking. 'No smoking' signs must be displayed in all work vehicles. Managers should as a matter of good practice require that vehicles be smoke-free.

Conclusion

It is essential that there is a driving at work policy and risk assessment in place so that compliance with health and safety and working time legislation can be seen to be actively implemented and ongoing. This will not only assist in any HSE investigations but will also help to protect against any civil claims.

See also: Corporate manslaughter, p.191; Mobile phones at work, p.496; Risk assessments, p.624; Smoking, p.653; Vehicles at work, p.686.

INDG 382 Driving at work: managing work-related road safety:
www.hse.gov.uk/pubns/indg382.pdf

Managing Occupational Road Risk: The RoSPA Guide:
www.rospa.com/drivertraining/morr/

Workplace Law's *Driving at Work Policy and Management Guide v.5.0* helps
you cover yourself and your staff and ensure that your employees keep to the
highest standards of safe driving at work. This comprehensive new edition of
the policy and management guide updates several elements of the original
including the implications of recent legislation such as the Health Act 2006, the
Road Safety Act 2007 and the Corporate Manslaughter and Corporate Homicide
Act 2007. If your business hasn't already got a driving at work policy in place, or
your current policy is not up-to-date, this is an essential publication. The policy
highlights the issue of liability should prosecution occur following a driving at work
accident, and who might face prosecution as a result. For more information visit
www.workplacelaw.net.

Electricity and electrical equipment

Mahendra Mistry, Bureau Veritas UK Ltd

Key points

- Electricity is invisible, odourless and inaudible. Its presence can be detected by switching on a known load, using voltage-detecting instruments or when someone inadvertently touches it.

- Accidents involving electric shock are divided into two categories; direct contact and indirect contact. Direct contact is when someone touches a live conductor, i.e. a bare wire or terminal. Indirect contact is when someone touches a metal part of an appliance when there is an earth fault on the appliance, e.g. the body of a washing machine. Measures have to be taken to prevent persons receiving an electric shock from direct and indirect contact.

- Whilst contact with live electrical supplies does not always result in fatal injury, the chances of such an outcome are higher than with other types of accident. In any event, the consequences may still be quite severe, for example, deep tissue burns, muscle spasms from the shock, fire and/or explosions. It is also worth noting that an electric shock to someone working at height can result in a serious fall.

- The nature of portable appliances is such that they are often subject to mechanical abuse; for example a drill being lowered by dangling it on its cable; tools being accidentally dropped. This probably explains why about 25% of all reportable electrical accidents involve portable electrical equipment.

- To reduce the risk that damaged or faulty portable equipment might harm someone, the Electricity at Work Regulations 1989 require that portable electrical equipment (or any portable appliance) is tested at appropriate regular intervals. The intervals are not prescribed but are left to the judgement of competent persons. Guidance on suggested intervals can be obtained from the HSE and the Institute of Electrical Engineers (IEE).

Legislation

- Electricity at Work Regulations 1989.
- Provision and Use of Work Equipment Regulations 1998.
- Management of Health and Safety at Work Regulations 1999.

The Electricity at Work Regulations 1989 emphasise the duties and responsibilities of the 'duty-holder' to ensure that all electrical installations and equipment are selected, installed and maintained at all times in a safe condition to prevent danger. The duty-holder is anyone who looks after a premises, including an MD, director or manager, or carries out any electrical work.

It also states that only competent persons, who have the technical and practical knowledge of the electrical equipment, and have had sufficient training and experience in that class of work, should work on such electrical equipment and installations.

> **Facts**
> - Electricity kills and injures people. Around 1,000 electrical accidents at work are reported to the HSE each year and about 25 people die of their injuries.
> - The HSE states that 25% of all reportable electrical accidents involve portable appliances.
> - A few milliamps (one-thousandth of one amp) is sufficient to cause serious injury or death. The energy in a 60W lightbulb is sufficient to electrocute four people simultaneously.

New guidance document on safe electrical installations

The 17th edition of the Wiring Regulations, BS 7671: 2008, came into force on 1 July 2008 and applies to all new electrical installations. Compliance will enable companies to demonstrate conformity with the relevant parts of the Electricity at Work Regulations, should an incident occur.

As a result of the changes, residual current devices that prevent electric shocks must now be installed on socket outlets and tested annually. Consideration must also be given to emergency escape and warning systems, whilst facilities provided to ensure life support systems must have back-up power.

Newly-installed equipment must not emit dangerous levels of magnetic fields. In addition, the installation's documentation must be up-to-date, and inspection and testing must be undertaken periodically by a competent person.

A major step forward has been the reduction of disconnection times to reduce electric shock exposure and fire risk. Consideration has also been given to climate change by enabling the monitoring and control of energy use.

Causes of electrical faults

All electrical installation deteriorates over time. Factors affecting the rate of deterioration include the environment (dusty, wet, vibration, temperature, etc.), loading, utilisation of plant (24/7 operations), mechanical damage, general wear and tear and the level of maintenance.

The harm caused by faulty electrical equipment can vary from personal injury to property damage. Typical causes of such faults are:

- damaged or worn insulation;
- inadequate or sloppy systems of work;
- over-rated protection (fuses, circuit breakers);
- poor earthing of appliances;
- carelessness and complacency;
- overheated apparatus, e.g. poor ventilation;
- earth leakage;
- loose contacts and connectors;
- inadequate ratings of circuit components;
- unprotected connectors; and
- poor maintenance and testing.

How to avoid accidents involving electricity

The Electricity at Work Regulations 1989 state that 'work on or near to an electrical system shall be carried out in such a manner as not to give risk, so far as is reasonably practicable, to danger'. In addition, they require that:

'No person shall be engaged in any work activity on or so near any live conductor

---

Compliance update: important amendment to BS 7671: 2008

In July 2011, the first update to BS 7671: 2008 was issued to introduce significant changes to the Standard, with an effective date for full implementation of 1 January 2012. When this amendment comes into force it will impact all those responsible for electrical installation and safety in the facilities, estates, health and safety and electrical markets.

The most significant changes to the Standard see the introduction of four new specific Sections (444, 534, 710, 729). These deal with measures against electromagnetic disturbances, protection against overvoltage, medical locations and operating in gangways. Also of great importance – and of much wider relevance – is the addition of a new 'Electrical Installation Condition Report' to replace the previous 'Periodic Inspection Report'.

The new Condition Report will become the output from any periodic installation test and it will detail in clear language exactly what has been found and what faults, if any, need to be corrected. The coding of faults has also been changed by the new amendment, from four to three: C1 identifies a danger present, C2 for when there is a potentially dangerous condition, and C3 when there is a recommendation to improve the safety of the installation.

Applauding the new amendment, Managing Director of PHS Compliance, Paul Caddick commented:

"The new Condition Report is a very welcome introduction for improved safety. Previously it was extremely difficult for anyone but a qualified electrician to interpret the findings of test results and consequently a dangerous proportion were never acted upon. The style and format of this new report delivers the important information in an immediate and useful way. The duty-holder can quickly grasp what the test report is telling or advising and then take appropriate action swiftly to rectify faults."

(other than one suitably covered with insulating material so as to prevent danger) that danger may arise unless:

- it is unreasonable in all the circumstances for it to be dead; and
- it is reasonable in all the circumstances for him to be at work on or near it while it is live; and
- suitable precautions (including, where necessary, the provision of suitable protective equipment) are taken to prevent injury.'

This latter requirement makes it clear that working on live equipment is not something that is OK to do just because you are an electrician. It prohibits work on live equipment other than in exceptional circumstances, e.g. where other risks to life might exist if the supply were isolated. It should be remembered that testing is also live working and that many faults can be detected safely with the power switched off.

As previously mentioned, many electrical accidents involve the use of portable equipment. The position of the user, for example on a ladder, the tight grip on the equipment, damp conditions, etc. can all conspire to increase the risk of serious or fatal injury. It is therefore important to ensure that:

■ the equipment is fit for purpose, e.g. suitable protection from weather conditions;
■ low voltage equipment is used where possible, e.g. 110 volt transformers, battery powered tools etc.;
■ the proper protective devices are used, i.e. residual current devices for personal protection, fuses / circuit breakers for protection of equipment and wiring;
■ staff are adequately trained and competent, e.g. in simple inspection procedures;
■ a fault reporting system exists;
■ routine inspection and maintenance is undertaken by a competent person; and
■ PPE is used where appropriate.

Testing of fixed electrical installations

Fixed electrical installations, i.e. the wiring, distribution (fuse) boards, lighting, etc. also need to be tested periodically. This is because all installations deteriorate over time due to loading, overheating, environment, mechanical damage, wear and tear, age, and poor maintenance. Much of what makes up the installation is hidden, perhaps in roof voids, conduits, walls, etc. and thus any developing faults may be overlooked until it is too late. Many fires are caused by cables overheating in attics or between floors. It is therefore necessary for installations to be tested by a competent person at appropriate intervals, for example at least every five years for commercial premises and three years for industrial locations. Periodic

inspection and testing is an essential part of a maintenance programme and assists the duty-holder in meeting with parts of the requirements of maintenance as required by the Electricity at Work Regulations 1989.

Portable Appliance Testing (PAT)

Portable equipment is defined by the HSE as equipment that is 'not part of a fixed installation but may be connected to a fixed installation by means of a flexible cable and either a socket and plug or a spur box or similar means'. Included in the definition are the extension leads that are often used to supply electricity to portable equipment. Testing of portable electrical equipment should be carried out at appropriate regular intervals in order to ensure safety in use.

Since there is no statutory inspection period for such equipment, many employers have set inspection regimes requiring all portable electrical equipment to be tested annually. This may well be inappropriate (and unnecessarily costly), since a personal computer is unlikely to suffer the same rigorous use as an electrically-operated disc cutter or power drill. The frequency of PAT therefore should reflect the usage of the equipment and will depend on factors such as nature of the task, frequency of use and environmental conditions (e.g. indoor / outdoor use). It should also be noted that it is highly beneficial to train appliance users to carry out simple visual inspections.

Dealing with electric shocks

The critical factor when dealing with a person receiving an electric shock is to isolate the victim from the supply. This of course poses the risk that the person helping may suffer an electric shock in the process, e.g. by grabbing the victim whilst the electricity is still passing through him/her. It is therefore extremely important that where there is a significant risk of electric shock occurring, for example in switch-rooms, adequate first aid measures are in place. These would be in addition to general requirements for first aid provision in the workplace. First aiders should be trained as a minimum to:

- protect themselves and others;
- adequately assess the situation – making sure that it is safe to approach the casualty;
- isolate the casualty from the electrical supply, either by isolating the electrical supply or by pulling the casualty clear with a dry insulator, e.g. rubber gloves, rope, newspaper, wooden pole;
- stand on an insulating surface, e.g. rubber mat, dry wood;
- avoid touching the casualty's bare skin;
- give first aid treatment for electrocution, e.g. cardio-pulmonary resuscitation; and
- summon assistance.

See also: First aid, p.325; Health and safety at work, p.361; Personal Protective Equipment, p.576; Portable Appliance Testing, p.588.

Sources of further information

HSE – Electrical safety at work: www.hse.gov.uk/electricity/index.htm

Institution of Engineering and Technology (incorporating the Institute of Electrical Engineers): www.theiet.org

17th edition of the Wiring Regulations, BS 7671: 2008: www.hvnplus.co.uk/files/rough_guide_wiring_regs_jan_08.pdf

Emergency procedures and crisis management

Peter Power, Visor Consultants (UK) Ltd

Key points

- This chapter focuses on the important actions to take as soon as a likely emergency or crisis starts to unfold where prompt actions during the acute phase of any situation will considerably determine the amount of damage to reputation, business, and property as well as injuries to people that might otherwise follow. It is important to distinguish between Emergency Procedures (EP) and Crisis Management (CM):

  - EP refers to tactical level prepared drills that operate as 'knee-jerk' procedures to an event, such as rehearsed escape or other evacuation procedures applied in emergences, such as fires when an alarm is sounded. They do not require executive decision-making capabilities. Such reactions are driven more by compliance with health and safety rules than anything else. This includes reference to a range of legislation, such as the Management of Health and Safety at Work Regulations.

  - CM refers to inherently abnormal, unstable and complex situations that represent a threat to the strategic objectives, reputation or existence of an organisation. Crises, in this sense, present organisations with complex and difficult challenges that may have profound and far-reaching consequences. Unlike EP, CM does require executive decision-making capabilities. Scenarios include a dramatic loss of reputation, supply chain fracture or market collapse. Depending on the possible impact, it also includes loss of data, fires, floods and other likely catastrophes.

- The existing BSI Standard 25999, specifying Business Continuity (BC) Management, refers to 'incidents', which are defined as 'situations that might be, or could lead to, a business disruption, loss, emergency or a crisis'. Incidents can therefore lead to emergencies and/or crises and, within the workings of BC, incidents are more tactical than strategic. Crises, on the other hand, tend to be more strategic and cannot be solved by just using prepared EP drills, or BC plans to move to a recovery site.

- At the time of writing, the British Standards Institute (BSI) and UK Government Cabinet Office are about to issue new UK-wide guidance on CM. This is referred to as Publicly Available Specification 200 (PAS 200), soon to be made available from the BSI. When available, it will be guidance rather than a standard. As such, it is a general advice to help establish good practice.

Legislation

- Management of Health and Safety at Work Regulations 1999 and associated Regulations.
- Civil Contingencies Act 2004.
- Regulatory Reform (Fire Safety) Order 2005.

The Management of Health and Safety at Work Regulations 1999 require every employer to establish appropriate procedures to be followed in the event of serious and imminent danger to persons at work. This includes the duty of employers or controllers of premises to ensure that any necessary contacts are arranged with external services, particularly with regard to first aid, emergency medical care and rescue work.

Previously the Fire Precautions (Workplace) (Amendment) Regulations 1999 required an employer – or controller of premises – to carry out a fire risk assessment of their premises. This legislation was revoked in April 2006 with the introduction of the Regulatory Reform (Fire Safety) Order 2005 (RRO). The duty to carry out fire risk assessments continues to be expressed in the new legislation.

The Civil Contingencies Act 2004 is currently undergoing a review. In the Explanatory Memorandum to the Civil Contingencies Act 2004 (Amendment of List of Responders) Order 2011 it states:

'The Guidance that has been issued under Section 3 of the Act, Emergency Preparedness, will be updated later in 2011 once an ongoing review of the Act has been completed. We will issue interim guidance directly to those affected by the proposed changes highlighted in this Explanatory Memorandum'.

Crisis management

The capability to manage a crisis should not be seen as something that can simply be developed as and when needed. It requires preparation in advance. CM needs a systematic approach that creates structures and processes, trains people to work within them and is evaluated and developed in a continuous, purposeful and rigorous way.

In developing the capability there will be many opportunities for synergy with ordinary business management processes and BC arrangements. The development of a CM capability should be viewed as a mainstream activity that is relevant and proportionate to whatever the organisation does and where and how it does it. In addition, crisis managers should be aware that:

- each possible solution at the time may have severe medium- to long-term consequences of one form or another;
- they may have to choose the 'least bad' solution, which often means that every choice comes with a penalty of some sort;
- the development and maintenance of CM capability should be included in the organisation's governance and strategy review processes;
- crises place exceptional demands on them and their support teams, at a time when they may already be under pressure of time and intense scrutiny; and
- they should be aware of support staff working at extraordinary levels of activity and, possibly, dealing with distressing issues.

CM also requires an element of 'horizon scanning', which is a systematic examination of potential threats, opportunities and future developments, which may have the potential to create new risks or change the character of risks already identified.

Crises are often produced by risks that were not identified, or were not identified on the scale at which they presented, or were created by an unforeseen combination of interdependent risks. They develop in non-predictable ways and the response usually requires genuinely creative, as opposed to pre-prepared, solutions. Indeed, it is argued that pre-prepared solutions – of the sort designed to deal with more predictable and structured incidents – are unlikely to work in complex and ill-structured crises. They may, in fact, be counter-productive.

CM is a multi-level management process, which deals with situations or events that are outside the normal management capability of the organisation at the time of the event.

In very serious cases, when the build-up of damaging events rapidly passes a certain point, normally designated as the border of 'business as usual,' a state of chaos often erupts that will need an even more rapid response to counter. This response should focus on the speedy transition from routine to crisis management (CM) styles commensurate with the threat to the business and need to tackle the actual crisis. Equally important is the return to normality at the earliest opportunity, which is the second priority for CM.

In any crisis situation there has to be a simple and quickly formed structure to:

- confirm;
- control;
- contain; and
- communicate.

This structure will then give urgent directions to trigger the correct layer of emergency response; BC, or in some cases coordinate the whole response without BC (e.g. likely reputation rather than physical damage).

The place where, conceptually, BC overlaps the actual incident is the point where resources are normally most critical. Where external / market / media perception and pressures overlap the incident will often be the point where the reputation of the target organisation will be in the balance. In the centre of this should exist CM.

Very often organisations have not necessarily taken the wrong actions in terms of CM, but probably did the right ones too late, by which time the crisis itself sets the pace and you might end up following events, rather than getting in front and stopping the spread.

This is the stage where demands for urgent decisions will be at their highest, yet, concomitant with this, accurate information about what's going on will be very low. In a nutshell, it's the high risk / low delay in decision-making feature as opposed to the low risk / high delay in decision-making dilemma: Should a crisis team wait until they get 100% of the facts before making any decision? This will probably result in a good decision being taken too late, in which case it ends up being a bad decision.

Over the past few years there have been signs emerging that the gap between CM and BC is starting to narrow and combine with emergency procedures, especially when we face dramatic threats such as terrorism, flooding, violent protestors, immediate reputation damage caused by instant media portrayal and so on.

CM should be a process that can be rapidly triggered and should be activated (and often stood down if no further action is needed) whenever a likely crisis starts to become apparent so that:

- no time is lost identifying the risk and what to do; and
- you can start pulling the right levers to stimulate reactions at the other non-risk facing end, to help you contain and control events. For example, more physical resources, re-routing routes to your building and so on.

The first person or people to act as a crisis manager / CM team probably have to ask themselves five urgent questions to start with:

1. What do we know?
2. What don't we know?
3. What would I like to know?
4. How can I find out?
5. How long have I got?

From the answers to such simple questions a more accurate picture of events should emerge. However, a CM structure needs to be owned, rehearsed and updated as an extension of routine management so it can quickly be called up, but populated only according to anticipated requirements. It also fits comfortably with enterprise risk management to ensure that a uniform approach to risk identification, measurement and treatment is utilised across the whole organisation. By adopting this proactive approach to managing risk and crises, organisations can move from a 'silo' management approach to a deeper integration of its various businesses to (a) start the alarm ASAP if a crisis approaches, and (b) get a pan-company response going without delay.

The CM priorities are derived from the principle that, at the very least,

two activities need to be dealt with concurrently:

1. CM – any crisis needs to be recognised for what it is, and what it might cause. It might easily create other crises that could have a negative impact on reputation. The crisis or incident needs to be addressed and handled effectively, since damage to reputation represents one of the most serious impacts of any crisis.
2. Core business – the position of all core business activities needs to be established and a decision taken on what to do next – perhaps to trigger the BC plan to continue operations, albeit on a reduced scale.

These activities involve three key types of managerial decisions to separate the main groups of tasks to reduce overload risk, whilst a clear decision route is maximised. As far as possible this should reflect routine management structure, but modified since truly routine structures do not work for non-routine events:

- Strategic / executive direction.
- Controlling the incident.
- Operational tasks.

People not forming the core of any CM team should not feel obliged to communicate if they have nothing direct to contribute, or feel left out if their opinion is not sought at this stage.

CM is about going through decision-making processes in minutes rather than hours and avoiding mere discussions. It is also very much about action-centred leadership. It's important that any CM team:

- understands how media awareness is a key part of effective crisis management;

- appreciates the appetite and constraints of the media;
- accepts the ramifications of good or bad media handling;
- decides when, or if, to stand in front of a camera or microphone and who does it (normally the most senior person available and as close to the 'site' as possible);
- understands how to deal with invited and uninvited media intrusion; and
- understands how to choose the best, if not ideal, conditions to face the press.

Emergency procedures

Emergency evacuation (leave the premises) plan

Any fire risk assessment (*see below*) should enable the employer, or controller of premises where more than one business is based, to formulate an emergency evacuation plan in the event of fire, and most major incidents or accidents. It should, however, be recognised in today's climate that potential terrorist actions may be significant in the premises you occupy. Plans should, in these circumstances, be made for action in the event of a bomb threat (see '*Bomb Alerts*', p.56) or other terrorist activity. It may well be necessary to carry out an additional assessment with this in mind, since the evacuation procedures for these threats are often different to those for fire.

An emergency evacuation plan, regardless of the cause of the evacuation, would comprise the following four stages:

1. How staff are enabled to recognise an emergency.
2. Communication of the emergency to affected staff.
3. Preparation of staff and/or the building for the evacuation.
4. Actions of staff and others (e.g. visitors) in the emergency evacuation.

The above stages would lead to a plan that would cover, as a minimum, the following points:

- What action should be taken by persons discovering an emergency.
- Who is responsible for making contact with the emergency services and the method of doing so.
- Identification of type of warning alarms (bells / klaxons / sirens) to be used (this is important as confusion can arise over what different alarms mean).
- Location of call points, escape routes, duties and identities of persons with special responsibilities and assembly points, i.e. places of safety.
- Procedures for dealing with the emergency services on arrival (including providing information on risks to them, e.g. presence of asbestos, flammables, etc.); and the training that should be provided for all employees or occupiers.

Emergency 'invacuation' (stay inside) plan

Strictly speaking, no such term exists in the English language, but invacuation is progressively becoming obvious to describe in one word the necessity to actually do the opposite of evacuation and keep people inside a building, rather than get them to leave. This is especially true in cases of a bomb threat. In such a scenario it is quite likely that a knee-jerk evacuation will cause hundreds of people to spill on to the streets and therefore be exposed to greater danger from, for example, a vehicle bomb.

The UK Government's policy for communicating about terrorist threats is to issue a warning if it is the best way to protect any community or venue facing a specific and credible threat. Public safety is the absolute priority. Such advice will be issued immediately if the public needs to take specific action. In the event of an incident the message is 'go in, stay in, tune in'.

The 'go in, stay in, tune in' advice is recognised and used around the world. It was developed by the independent National Steering Committee on Warning and Informing the Public as being the best general advice to give people caught up in most emergencies. It follows that creating an invacuation plan makes great sense.

In 2004 the Government published a public information leaflet called 'Preparing for Emergencies, What You Need to Know' (see *Sources of further information*) which provided common sense advice for a range of different emergency situations, including some information on countering terrorism. With this in mind the key points of invacuation are:

- Before you use a public address system or get line mangers to pass the message to stay in, think about (a) what you are about to say and (b) what you will say. A calm voice in authority works better – especially if you can get hold of the most senior person in the building to make any announcement.
- This is not a time to use words such as 'We would like all staff to consider it a reasonable idea not to leave the building'. If it's necessary to keep everyone inside (and under the umbrella of Duty of Care that includes visitors, contractors and everyone else) it's also necessary to use clear language such as 'This is … speaking. The police have asked us to stay inside … for the time being we have shut all doors, windows … we must all now stay inside … we have plenty of fresh water, clean air here, etc."
- The point above about fresh water, or clean air, warm clothing, ease of communication via the telephone,

access to TV sets and so on is very important as a means to attract people to stay inside, rather than risk injury outside.

- If you have an enhanced protection area to minimise the damage of bomb blast, use it.
- Keep all communication lines available to you and be aware that sometimes people inside will take camera images on their mobile phones and they quickly get passed to TV news stations – who will ask for them.
- Make it clear you and/or your team are in command. Give frequent (e.g. every 30 minutes) updates, even if you have not had any more information. It's vital to reassure people who could easily panic.

Information and training

The plans that have been made will only be successful if properly implemented. It is therefore important that staff and others, e.g. occupiers, visitors, etc. are adequately trained or provided with information on what to do in an emergency. When a fire alarm sounds, staff often question whether it is a real emergency or just a drill. Staff should never be in doubt about such matters since delay in evacuation obviously increases the risk to safety.

Information should therefore be available to staff in written form and in a language that they understand. Regular evacuation drills should take place and the performance of these should not only be recorded but should also be reviewed. The frequency of these drills will be determined by the level of risk identified in the assessment.

Use of fire extinguishers

In the workplace, most buildings (and in particular, office blocks) have a number of fire extinguishers. It is not unusual though, to hear staff say "I've never been trained in how to use these," or "I've never had the chance to use one of these". It may be difficult to arrange for staff to actually practise using extinguishers on real fires and thus care needs to be exercised if fire action notices state that persons should tackle the fire with such equipment. Staff can, however, be trained on the choice of, and correct way to use, the proper extinguishers, for example matching the extinguisher to the type of fire. It must be stressed that they should not put themselves at risk in tackling a fire (see 'Fire extinguishers,' p.306).

Equipment testing

All equipment that could be involved in an emergency evacuation should be tested by competent persons at the required intervals (varying from weekly to annually). This will include:

- detection and alarm system;
- firefighting equipment;
- automatic door release units; and
- emergency lighting.

Records

Although frequency may vary with individual circumstances (i.e. based on the risk assessment), emergency evacuation drills should usually be carried out at least every six months for most buildings. This would usually be every three months if night working at the premises is involved.

The responsibility for ensuring that evacuation drills take place lies with management. The persons appointed to actually coordinate the evacuation, for example, fire wardens, fire marshals, etc., need to report the results of any drill to management in order to identify any shortfalls in the arrangements. Records of drills should be kept by management and should contain such information as:

- exact location of the drill;
- date and time;

- total number of participants;
- the evacuation time;
- any problems identified;
- actions required to be taken; and
- date of next drill.

In addition, records of equipment testing should be maintained.

Fire risk assessment

It is the duty of employers to conduct fire risk assessments of premises under their control.

These need not always be complicated exercises but should always involve the following key steps:

- Identification of any fire hazards – such as readily combustible materials, highly flammable substances and sources of heat.
- Identification of any persons who are especially at risk, and making special provisions for persons with disabilities.
- Identification of where fire hazards can be reduced or removed.
- Recording the findings.

The competent person undertaking the assessment should address the following issues:

- Legal requirements;
- Means of escape;
- Fire alarms / fire detection;
- Special risks, e.g. disabled persons, risks to neighbours, etc.;
- Firefighting equipment;
- Emergency lighting;
- Maintenance;
- Housekeeping; and
- Warning / emergency notices.

EP and CM

All emergency and crisis plans should focus, inter alia, on at least three key groupings of people – the vulnerable, victims (including survivors, family and friends) and responder personnel.

- Vulnerable people may be less able to help themselves in an emergency than self-reliant people. Those who are vulnerable will vary depending on the nature of the emergency, but plans should consider those with mobility difficulties (e.g. those with physical disabilities or pregnant women); those with mental health difficulties; and others who are dependent, such as children.

- Victims of any dramatic event, which includes not only those directly affected but also those who, as family and friends, suffer bereavement or the anxiety of not knowing what has happened. The impact of such stress / trauma will exist for years if not decades afterwards, so beware of anniversaries, delayed inquests, TV documentaries, etc. that will undoubtedly trigger a recurrence of memories relating to whatever drama initially took place.

- Responder personnel should also be considered. Plans sometimes place unrealistic expectations on management and personnel. Organisations should ensure their plans give due consideration to the welfare of their own personnel. For instance, the emergency services have health and safety procedures that determine shift patterns and check for levels of stress.

As obvious as it sounds, emergency plans should include procedures for determining whether an emergency has occurred, and when to activate the plan in response to an emergency. This should include identifying an appropriately trained person who will take the decision, in consultation with others, on when an emergency has occurred.

The maintenance of plans involves more than just their preparation. Once a plan has been prepared, it must be maintained systematically to ensure it remains up to date and fit for purpose at any time if an emergency occurs.

It may be that multiple organisations can develop a joint emergency plan where the partners agree that, for a successful combined response, they need a formal set of procedures governing them all. For example, in the event that evacuation is required, the police would need carefully pre-planned cooperation from various other organisations such as fire and ambulance services and the Local Authority, as well as involvement of others such as transport organisations.

UK Government advice

The Civil Contingencies Act 2004 identifies three pieces of legislation pre-dating this Act which were introduced separately in Britain and Northern Ireland under sector-specific legislation operated by the HSE and HSE Northern Ireland. These relate to major accident hazards at industrial establishments (Control of Major Accident Hazards Regulations (COMAH)), to hazardous pipelines (Pipelines Safety Regulations) and to radiation hazards (Radiation (Emergency Preparation and Public Information) Regulations (REPPIR)).

These sector-specific Regulations have established multi-agency emergency planning regimes in cooperation with the operators. To avoid duplication, the Civil Contingencies Act Regulations provide that the duty to maintain plans under the Act does not apply to emergencies that are dealt with by these pieces of legislation.

In the event of a major emergency, whether accidental or deliberate, the Government will start up a central News Coordination Centre (NCC) to try and ensure that messages and information to the public, stakeholders and the media are clear, consistent, timely and accurate.

The NCC works closely to support the central crisis centre, and liaises with communication advisors within the centre, the police Strategic Coordination Centre (SCC) at the scene, and the Government Liaison Team (GLT).

Since 1996 a national Media Emergency Forum (MEF) – a large working group consisting of senior media editors, government representatives, Local Authority emergency planners, emergency services, police and private industry, has regularly met on a voluntary basis to discuss communication issues arising from specific emergencies, and in identifying ways to improve communications in the future.

> *See also*: Bomb alerts, p.56; Business Continuity Management, p.92; Fire, means of escape, p.314; Personal Emergency Evacuation Plans (PEEPs), p.569.

Sources of further information

UK Government advice: preparing for emergencies: www.direct.gov.uk/en/HomeAndCommunity/InYourHome/Dealingwithemergencies/Preparingforemergencies/index.htm

British Standards Institute (PAS 200): www.bsigroup.com/en/Standards-and-Publications/How-we-can-help-you/BSS/-/-/BSS/PS/Our-services/Current-projects/PAS-200

Energy performance

Colin Malcolm, Workplace Law, and Gavin Miller, Building Compliance Associates Ltd

Key points

- Buildings account for approximately 40% of the UK's carbon emissions, and non-dwellings account for roughly half this total.
- Businesses therefore have a very important role in supporting the ambitious carbon reduction targets set by the Government, principally through the Climate Change Act 2008.
- There are a number of existing legislative controls that address energy in buildings, including taxation, trading, performance transparency and plant efficiency, whilst non-mandatory schemes such as sustainability certification labels are also commonplace.
- In addition, financial savings and stakeholder carbon reporting are, in the current climate, two of the most important energy performance drivers for business.
- The Carbon Reduction Commitment Energy Efficiency Scheme, which encapsulates many of the above issues, is addressed in '*The CRC Energy Efficiency Scheme*' (p.215) and is not discussed further here.

Legislation

- Building Regulations 2000 (as amended).
- Energy Performance of Buildings Directive.
- Energy Performance of Buildings (Certificates and Inspections) Regulations 2007 (as amended).
- Climate Change Act 2008.
- Energy Act 2008.

Energy Performance of Buildings Directive

The Energy Performance of Buildings Directive (2002/91/EC) was introduced to promote the improvement of energy performance of buildings, and is the main legislative instrument at European Union level to improve building energy performance. The Directive requires that member states must apply minimum energy performance requirements for new and existing buildings, energy performance certification and the inspection of boilers and air conditioning systems in buildings. In England and Wales, the Energy Performance of Buildings (Certificates and Inspections) Regulations 2007, as amended by SI 1900 (2009) and SI 1456 (2010) address these requirements and similar regulations are in force in Scotland and Northern Ireland.

The main requirements of the Regulations are as follows.

Energy Performance Certificates

The requirements for Energy Performance Certificates (EPC) were introduced to inform and educate potential buyers or tenants about a buildings energy performance. The assessment addresses both the building fabric and the main building services (heating, lighting, air conditioning) to produce an intrinsic asset rating that indicates how energy efficient the building has been designed.

Key requirements for EPCs include:

- Energy Performance Certificates are required when a commercial building over 50m² is built, sold or rented.
- EPCs must be provided free and made available by the owner to prospective buyers and tenants.
- EPCs are valid for ten years but must be renewed if modifications to the property are made.
- The EPC has two parts – a graphic rating and a recommendations report.
- Energy Performance Certificates must be produced by an accredited energy assessor.

Display Energy Certificates

Display Energy Certificates (DECs) are required for buildings that are occupied by a public authority or institution providing a public service to a large number of people and frequently visited by the public. Examples include libraries, hospitals, schools and government buildings.

The purpose of DECs is to raise public awareness of energy use and to inform visitors to public buildings about the energy use of a building. The DEC is based on the actual amount of metered energy used by the building over a 12-month period and is accompanied by an advisory report listing energy improvement recommendations.

Key requirements for DECs include:

- Display Energy Certificates are required for public buildings over 1,000m².
- The DEC is valid for one year and the advisory report is valid for seven years.
- They must be displayed in a prominent place and clearly visible to the public.

- The DEC and advisory report must be produced by an accredited energy assessor.

Air conditioning inspections

Inspections of air conditioning systems are required under the Regulations primarily to improve efficiency and reduce the electricity consumption, operating costs and carbon emissions.

The inspection includes a report that highlights opportunities for improvements to system operation. The inspection will be particularly relevant for old, oversized or poorly controlled systems and also systems using refrigerants subject to phase-out under separate regulations. Key requirements for air conditioning inspections include:

- For all systems first put into service on or after 1 January 2008, the first inspection must have taken place within five years of the date when it was first put into service.
- For other air conditioning systems, where the effective rated output is more than 250kW, the first inspection must happen by 4 January 2009.
- For other air conditioning systems, where the effective rated output is more than 12kW, the first inspection must happen by 4 January 2011.
- From 4 January 2011, if the person in control of the air conditioning system changes and the new person in control is not given an inspection report, the new person in control of the system must ensure the air conditioning system is inspected within three months of the day that person assumes control of the system.
- Inspections must be a maximum of five years apart.
- An energy inspection of an air conditioning system must be carried out by an accredited energy assessor.

Although a separate piece of legislation, there is some alignment with the Fluorinated Greenhouse Gases Regulations 2009 (FGG Regulations 2009), and updated requirements for companies and staff who work on F-Gas equipment.

Energy Performance of Buildings Directive Recast

On 19 May 2010, a recast of the Energy Performance of Buildings Directive was adopted by the European Parliament and the Council of the European Union in order to strengthen the energy performance requirements and to clarify and streamline some of the provisions from the 2002 Directive it replaces. The key changes are summarised below:

- By 31 December 2020, all new buildings must be nearly zero energy buildings. Where the building is to be owned and occupied by a public sector authority, then this requirement will apply after 31 December 2018.
- Where an existing building undergoes major renovations, the energy performance of one of the following must be upgraded – the entire building; the building unit that has been renovated, or the building elements.
- The 1,000m² threshold for major renovation has been deleted and this will take effect when the national regulations have been implemented.
- A harmonised calculation methodology for minimum energy performance requirements is set out in the Directive, and MS have to justify if the difference between current requirements and cost optimal requirements is more than 15%.
- When considering the economic feasibility of alternative systems, particular attention is to be paid to district heating or systems that would be based entirely or partially on renewable energy. The analysis of

alternative systems may be carried out for individual buildings, groups of buildings or for common typologies of buildings in the same area.
- Public authorities will be encouraged to take into account the leading role which they should play in the field of energy performance. Amongst other things, this may include implementing the recommendations in an EPC for buildings owned by them, within its validity period.
- Display Energy Certificates to be displayed in buildings larger than 500m² occupied by public authorities and frequently visited by the public, dropping to 250m² after five years.
- When a building is offered for sale or rent, the energy performance indicator of the EPC for the building is to be stated in the advertisement.
- The requirement that an EPC is displayed in commercial premises larger than 500m² frequently visited by the public and where an EPC has previously been issued.
- A more detailed and rigorous procedure for issuing energy performance certificates will be required
- Member states have to strengthen the quality of inspection of boilers, heat and AC systems.
- Member states will be required to introduce penalties for non-compliance. The penalties provided for must be effective, proportionate and dissuasive.

The transposition of the EPBD recast should be effective from July 2012.

Building Regulations

The Building Regulations set standards for design and construction that apply to most new buildings and many alterations to existing buildings. The legislative framework is principally made up of the Building Regulations 2000 and the Building (Approved Inspectors etc.) Regulations

2000. Both have been amended several times in recent years, the latest in March 2010 by the Building and Approved Inspectors (Amendments) Regulations 2010. The updated Regulations are the first since an agreement to only amend the Regulations every three years and no Part can be amended in consecutive reviews with the exception of Part F and Part L.

The latest Regulations made changes to several elements, including Part G (Sanitation, hot water and safety and water efficiency), Schedule 2A (Competent persons scheme) and Schedule 2B (non-notifiable work), which came into effect on 6 April 2010.

Changes to Part F (Ventilation), Part J (Heat producing appliances) and L (Conservation of fuel and power) came into effect on 1 October 2010. Practical guidance on ways to comply with the functional requirements in the Building Regulations is outlined in a series of Approved Documents. See *'Building Regulations'* (p.76) for more information.

Revised Approved Documents have been published for Part F, Part J and Part L (L2A new buildings other than dwellings and L2B existing buildings other than dwellings)

The next revision of Building Regulations Part L is anticipated in 2013, bringing in more elements of Government Policy as well as the 'recast' elements of the Energy Performance of Buildings Directive.

Climate Change Act 2008
The Climate Change Act 2008 is a comprehensive and far reaching piece of legislation that commits the UK to reducing overall carbon dioxide emissions by 34% by 2020 and 80% by 2050. Proposals are currently being considered to implement mandatory reporting of greenhouse gases

through powers under the Companies Act 2008. The Government either has to implement this measure, or explain to Parliament why it has not done so, by 6 April 2012. A decision in favour of mandatory reporting will impact on business and particularly so for those not currently collating, reporting or managing greenhouse gas emissions. A consultation to business and interested parties, managed by DEFRA, was undertaken outlining possible options. It closed on 5 July 2011 with a decision expected some time later in the year.

Energy Act 2008
The Department of Energy and Climate Change (DECC) has used powers in the Energy Act 2008 to introduce a system of feed-in tariffs (FIT) to incentivise small scale low carbon electricity generation.

From 1 April 2010 businesses installing eligible renewable or low carbon plant from 50kW up to 5MW can receive a guaranteed fixed payment for the electricity generated and exported to the grid, as well as receiving payments for electricity consumed on site. The tariffs have been introduced to increase renewable energy generation in the UK and to facilitate more sustainable energy production methods. Solar photovoltaic, hydroelectricity, Micro Combined Heat and Power and anaerobic digestion are examples of the technology available.

This is part of the Government's 'Green Deal' and currently consists of two key elements – the FIT described above, along with the Renewable Heat Initiative (RHI). The Renewable Heat Incentive (RHI) will, when implemented, support generation of heat from renewable sources at all scales and will provide long-term financial support to renewable heat installations to encourage the uptake of renewable heat.

See also: BREEAM, p.67;
Building Regulations, p.76; The
CRC Efficiency Scheme, p.215;
Environmental Management
Systems, p.286; Environmental risk
assessments, p.292.

Sources of further information

The Carbon Trust: www.carbontrust.co.uk

Communities and Local Government: www.communities.gov.uk

Department of Energy and Climate Change: www.decc.gov.uk

Planning portal: www.planningportal.gov.uk/england/professionals/en/

Approved Document L2A:
www.planningportal.gov.uk/england/professionals/en/1115314231806.html

Approved Document L1B:
www.planningportal.gov.uk/england/professionals/en/1115314231799.html

Power efficiency: www.powerefficiency.co.uk

Home Information Packs: www.homeinformationpacks.gov.uk/

Energy Saving Trust: www.energysavingtrust.org.uk/

F-Gas Regulations: www.workplacelaw.net/news/display/id/36191

GHG Reporting: www.defra.gov.uk/consult/2011/05/11/ghg-emissions/

FIT and RHI: http://bit.ly/r8mZW8

The Energy Performance of Buildings Directive

James Tiernan, Bureau Veritas UK Ltd

Key points

- The Energy Performance of Buildings Directive (EPBD) is a European Directive that came into force in January 2002, aimed at helping to achieve the CO_2 emissions reductions required under the Kyoto Protocol. It comprises five main themes – namely inspections, certification, training, procedures and information campaigns.

- Implementing the EPBD in England and Wales is the responsibility of the Department for Communities and Local Government (DCLG), supported by the Department for the Environment, Food and Rural Affairs (DEFRA) and the Department of Energy and Climate Change (DECC). In Scotland, the Building Standards Division is responsible for implementation. In Northern Ireland the Department of Finance and Personnel (DFPNI) supported by the Department for Social Development (DSDNI) is responsible.

- The EPBD has since been incorporated into UK law (the Building Regulations 2010 incorporate Regulations 3-6, 7, 9, and 10 of the EPBD), and sets in place various requirements aimed at improving the energy efficiency of buildings notably:
    - A valid Energy Performance Certificate (or EPC) is required whenever a building is constructed, sold or let.
    - A valid Display Energy Certificate (or DEC) is required for any public building over 1,000m².
    - Qualifying air conditioning systems must be inspected every five years.
    - Qualifying boilers should undergo regular inspection.

- Non-compliance can result in the award of financial penalties by Trading Standards.

- Under the Directive all new buildings will be required to be zero carbon by 2020.

- At the time of writing the implementation of the so-called EPBD Recast, which makes changes intended to further reduce energy consumption from buildings, was underway; however the current UK Government has delayed various pieces of regulation proposed by the previous administration that would have brought the EPBD into legislation, with further announcements expected later in 2011.

Legislation

- Energy Performance of Buildings Directive (EPBD) (2002).
- Building Regulations 2010.

Energy Performance Certificates (EPCs)

Regulation 7 of the EPBD requires member states to provide Energy Performance Certificates (EPCs). Under the Regulation, EPCs should be prepared for both domestic and commercial properties. A central register has been established to record all EPCs electronically. By December 2010, 5.7 million domestic EPCs and 210,000 non-domestic EPCs had been lodged.

The landlord or current owner is responsible for ensuring a building has a valid EPC whenever that building is constructed, sold or let (although changes to this are part of the EPBD Recast discussed later). The EPC is produced by an accredited energy assessor using an approved piece of software, and considers performance of the building envelope, location, orientation, building services and energy supplies.

The energy assessor calculates the energy performance and environmental performance of the buildings, giving each a current and potential score out of 100, and representing them graphically using the familiar bar graph format. A rating ranging from 'G' (worst performance) to 'A' (best performance) is awarded, depending on the numerical score. In addition to the ratings, the EPC must also contain recommendations that could be implemented to improve the energy and environmental performance, which are typically grouped as low, medium or high cost.

An EPC is valid for ten years once issued, but may need to be updated should the building be sold or let within that period and significant changes have been made that could affect the rating achieved. A valid EPC may not be required in the following situations (although further advice is recommended to ensure compliance):

- Lease renewals, extensions or surrenders;
- Compulsory purchase orders;

- Sales of shares in a company where buildings remain in company ownership;
- Temporary buildings with a planned time of use less than two years;
- Standalone buildings with a total useful floor area of less than 50m^2 that are not dwellings; and/or
- Industrial sites, workshops and non-residential agricultural buildings with low energy demand.

Trading Standards is responsible for enforcement of EPCs, and failure to produce one when required can result in a fine of at least £500, or of 12.5% of the rateable value of the building up to a total of £5,000. Where this cannot be calculated, a default penalty of £750 can be applied.

Display Energy Certificates (DECs)

Display Energy Certificates have also been implemented under Regulation 7 of the EPBD. DECs are required to be produced by all public authorities or institutions providing public services. To date, 72,000 DECs have been registered.

DECs are required to be displayed in a prominent location by the occupier in all buildings over 1,000m^2 that are 'occupied in whole or part by public authorities and by institutions providing public services to a large number of persons and therefore frequently visited by those persons'. Public authority buildings include, for example, Central Government departments or agencies, local Government, NHS trusts, schools, HE and FE, police, courts, prisons, MOD and statutory regulatory bodies, even if there is no public access. Institutions that provide public services include, for example, public museums and swimming pools, but not hotels or retail outlets. Where it is not clear if a building should display an EPC or not, further guidance should be sought.

A DEC can only be produced by an accredited energy assessor, but unlike an EPC it is based on actual historical energy consumption rather than calculated energy performance. A DEC must therefore be updated every 12 months to take the most recent energy consumption data into consideration.

The penalty for failing to display a DEC prominently and clearly visible to the public where required is £500, with a further fine of £1,000 for failing to possess a valid Advisory Report.

There are no DECs in Scotland. Instead, the Energy Performance (Asset Rating) is required to be supplied in public buildings.

Air conditioning inspections

Air conditioning systems are responsible for a large proportion of the total energy consumption of many commercial buildings – up to 30% in some cases. AC inspections are intended to cut the costs and CO_2 emissions of AC systems by identifying potential improvements to the operation of the system, checking they are operating at optimum efficiency, and identifying opportunities for replacing or upgrading existing AC systems with newer, more energy efficient, or more appropriately sized, systems.

Under the EPBD any AC system installed on or after 1 January 2008 must be inspected within five years of being commissioned. Any system with a rated output of over 250kW installed before this must have been inspected by 4 January 2009, while systems with a rated output of greater than 12kW must have been inspected by 4 January 2011.

In Scotland, existing systems rated more than 12kW have until 2013 to complete initial inspections. An approved survey approach has been prepared by the

Chartered Institute of Building Services Engineers (CIBSE) known as TM44. In all cases inspections are to be completed in accordance with the standard set out in TM44.

For the purposes of AC inspections, one or more AC units within a building (defined as 'a roofed construction having walls, for which energy is used to condition the indoor climate', and a reference to a building includes a reference to 'a part of a building which has been designed or altered to be used separately') controlled by a single person (i.e. who controls the technical functioning of the system not who can just change the temperature settings – typically the landlord in a multi-let office for example) are considered to comprise a single AC system.

Where a change in the person in control of the AC systems occurs (for example the landlord) and no AC inspection reports are handed over, the new responsible person has a period of three months in which to ensure all qualifying AC systems are inspected.

An AC inspection can only be conducted by a suitably accredited energy assessor, who undertakes a visual assessment of the whole system (for example, AC equipment, air handling systems and controls) and produces a report containing the following details:

- Details of the current system installed.
- Recommendations for improving efficiency of the existing system.
- Recommendations for replacing inefficient equipment or refrigerants.
- Details of any faults or maintenance requirements identified as well as recommendations regarding the existing maintenance regime.
- Recommendations for reducing the requirement for cooling.

It should be noted that the requirement to undertake AC inspections under the EPBD is a separate requirement from fluorinated gas inspections and checks (so called 'F-Gas'), that is stipulated under separate legislation. See 'Air conditioning and refrigeration systems' (p.32) for more information.

Currently the penalty for failing to produce a valid air conditioning report when requested by a trading standards inspector ranges from £300 to £5,000.

Currently there is no requirement to register TM44 reports centrally like EPCs and DECs; however, it is likely that some central electronic register will come into force in the near future.

Boiler inspections

While the EPBD stipulates that boilers above a certain size threshold should be subject to a period inspection and an accompanying one-off inspection of the whole heating system, there is a caveat that provision of advice regarding the 'replacement of boilers, other modifications to the heating system and on alternative solutions' can be provided instead.

The UK Government has, for the time being at least, opted for the second option and so has committed to targeting energy efficiency improvements through a programme of information and advice. Therefore there is currently no requirement for boiler inspections to be undertaken for energy efficiency purposes (notwithstanding other mandatory safety inspections).

Proposed EPBD Recast

Following a recent consultation various changes are, at the time of writing, in the process of being implemented (the so called 'Recast'). The stated intention of the EPBD Recast is to:

- extend the scope of the original Directive;
- strengthen certain provisions;
- clarify other aspects; and
- give the public sector a leading role in promoting energy efficiency.

Key proposals of the Recast:

- Requirement for DECs to be displayed in buildings larger than 250m² that are occupied by a public authority.
- Requirement for EPCs to be displayed in commercial buildings larger than 250m² that are frequently visited by the public and where an EPC has previously been produced on the sale, rent or construction of that building.
- Requirement for the energy performance of existing buildings of any size that undergoes major renovations to be upgraded in order to meet minimum energy performance requirements (so called 'consequential improvements'). Currently, there is a threshold of 1,000m².
- Minimum energy performance requirements to be set in respect of technical building systems, e.g. boilers, air conditioning units, etc.
- The establishment of common European principles for definition of low and zero carbon (LZC) buildings.
- Requirement to set targets for increase in LZC buildings with separate targets for:
  - new and refurbished dwellings;
  - new and refurbished commercial buildings; and
  - buildings occupied by public authorities.

Following the 2010 change in UK Government, there have been numerous changes to the planned legislation that was proposed by the previous Government, and at the time of writing the exact time and nature of changes remain unclear.

However, the Government must make provision for the EPBD recast under UK law by July 2012, and then bring the required changes into force by January 2013. Further clarifications are expected from the Department for Communities and Local Government in the latter half of 2011.

See also: Air conditioning and refrigeration systems, p.32; Corporate Social Responsibility, p.202; Environmental Management Systems, p.286; Waste management, p.707.

Sources of further information

Bureau Veritas UK and Ireland: www.bureauveritas.co.uk

Department for Communities and Local Government: www.communities.gov.uk/

Chartered Institute of Building Services Engineers: www.cibse.org

Environmental civil sanctions

Anna Willetts, Greenwoods Solicitors LLP

Key points

- Criminal law has not always allowed flexibility in enforcement of environmental legislation.
- In some instances, regulators have been forced to choose between issuing warning letters and cautions at one level, and taking criminal proceedings at another. This has often led to an imbalance in enforcement whereby criminal law has not been flexible enough to deal proportionally with offences, or ensure fines adequately cover the costs of non-compliance.
- However, further to a review about the effectiveness of such legislation, new civil sanctions regulations have been introduced to provide environmental regulators with a more flexible and balanced set of tools to deal with environmental offences.
- The Environment Agency in England and Wales has been using the new enforcement powers, called civil sanctions, since 4 January 2011.

Introduction

Civil sanctions can be used against a business committing certain environmental offences, as an alternative to prosecution and criminal penalties of fines and imprisonment. They allow the Environment Agency (and other regulators) to take action that is proportionate to the offence and the offender, and reflect the fact that most offences committed by businesses are unintentional.

The EA will still be able to use criminal punishments for serious offences, but the Government believes civil sanctions will make environmental law enforcement more flexible and effective for both regulators and businesses.

Legislation

The new civil sanctions system stems from the Regulatory Enforcement and Sanctions Act 2008, and provides the Regulator (the Environment Agency in this case) with an alternative set of powers to punish offenders via a civil route, rather than in the criminal courts.

A fairer approach to environmental enforcement?

Under the current criminal law regime, environmental offences may be harshly punished, for example under the Environmental Permitting Regulations 2010. Penalties under this legislation may involve a fine of up to £50,000 or imprisonment in the Magistrates' Court, or an unlimited fine and imprisonment if heard in the Crown Court. As environmental offences are 'strict liability' offences, even a minor breach of a Permit can result in a prosecution by the Environment Agency. Offences may be simply paperwork offences, and committed in ignorance, rather than purposely attempting to pollute or harm the environment, but currently may still result in a harsh penalty.

Alternative civil sanctions

In terms of environmental enforcement, in many ways civil sanctions do appear to offer a fairer and more practical solution to punishment of offences. Some environmental offences are of very low

Case studies

Scenario 1 – Hazardous waste

A company intentionally mixes hazardous and non-hazardous waste so it can be disposed of as non-hazardous, which would incur lower disposal costs and bring significant financial benefit. The site does not have a permit or exemption to allow mixing of this nature. An employee who has not received adequate training and is not wearing the appropriate protective equipment is injured when he comes into contact with the waste. The company has previous convictions for similar offences.

Likely enforcement response: Although a Variable Monetary Penalty is available, the nature of the offence, which resulted in injury, and the previous offending history, mean that this case is clearly deserving of prosecution.

Scenario 2 – Hazardous waste

During an inspection, a high street garage is unable to find hazardous waste consignment notes for the removal of waste oils. The company explains that they suspect the documents may have been thrown away accidentally by a junior member of staff during an office clear-out. There is no evidence of previous offending – on previous visits, the documents have been available.

Likely enforcement response: The company offers an Enforcement Undertaking which includes a commitment to, as far as possible, replace the missing notes, introduce new training and establish improved procedures to ensure there is no repeat. The company is clearly acknowledging its mistakes and making a commitment to change. The Enforcement Undertaking is accepted as it is considered it will deliver the necessary behaviour change and a more formal sanction is not required.

Scenario 3 – Producer responsibility

A company is registered under the Producer Responsibility (Packaging Waste) Regulations. When submitting its annual return it becomes clear that it has failed to purchase enough packaging recovery notes to meet its obligation. As a consequence, it commits an offence which is associated with a considerable financial saving from the cost it has avoided in not purchasing the notes. The company has produced the required number of notes in the past. However, a sanction is necessary to maintain confidence in the regime and to reassure companies who have met their obligations.

Likely enforcement response: Prosecution is considered inappropriate because of the company's previous good record but the size of the financial benefit made by the company means that a Variable Monetary Penalty is justified.

Source: EA.

magnitude, such as failing to complete paperwork on time, which does not, of itself, harm or damage the environment.

Civil sanctions may be a fairer approach to environmental crimes at the lower end of the scale, and would of course mean there is no criminal record. A small monetary punishment, such as a Fixed Penalty Notice of a few hundred pounds, for example, could be enough to remind the offender not to commit the offence again, especially if committed inadvertently.

Many of the 'penalties' that could be used under the civil sanctions may actually benefit the environment, far more than a criminal Court prosecution – which generally results in a fine – ever could.

Civil sanctions penalties
Some of the penalties that may be issued by the Regulator in lieu of criminal prosecutions include the following.

Compliance notice
A requirement to take specified steps within a stated period to secure that an offence does not continue or happen again.

Example: A waste management company allows litter to escape from a landfill site. A compliance notice could require them to install high litter fences and ensure that twice-daily litter picks are undertaken by site staff, within four weeks.

Restoration notice
A requirement to take specified steps within a stated period to secure that the position is, so far as possible, restored to what it would have been if no offence had been committed.

Example: An industrial unit allows pollution to escape into a river, creating a thick sheen of material on the surface of the water, killing 100 fish and reducing the biological oxygen demand (BOD) of the water. A restoration notice could require them to remove the polluting material, replace all the fish, and bubble oxygen through the river to increase BOD to previous levels.

Enforcement undertaking
These enable a person, who a regulator reasonably suspects of having committed an offence, to give an undertaking to a regulator to take one or more corrective actions set out in the undertaking.

Example: A recycling company is alleged to have caused dust and noise problems from its recycling facility. An enforcement undertaking could be offered by the company to the Regulator to install noise-reducing equipment and dust-suppression equipment which is maintained at frequent intervals, and to pay a sum of money to the local community environmental fund for environmental projects.

Stop notice
A requirement for a person to stop carrying on an activity described in the notice until it has taken steps to come back into compliance.

Example: A skip operator has been illegally using spare land as a place to store skips of waste overnight. A stop notice could be issued to 'stop' his skip business operating until he has removed all the skips from the land.

Civil Sanctions – lesser penalties?
It is often asked whether companies that fall foul of environmental legislation will suffer greater or lesser penalties under this regime. This very much depends on the offences. It is expected that for minor offences, such as omitting to use correct paperwork or forms, would result

in a lesser penalty under the civil regime than under the current criminal regime. A monetary penalty, of for example up to £1,000, would be a much lower penalty than the costs and time involved with defending a prosecution in the criminal Courts, as well as the potential for being found guilty and the associated criminal record.

It should be noted, however, that the planned monetary penalties have a maximum of £250,000, which would of course, for a single offence, be much higher than the £50,000 maximum under the Environmental Permitting Regulations for example. It will potentially be a balancing act as to whether greater or lesser penalties are imposed, and it should not be forgotten that it is at the discretion of the Regulator whether the civil penalties are used, on a case by case basis.

The main drive behind the civil sanctions is to retain criminal sanctions for the most serious cases, and to lead to reduced Court time being used up for the more minor offences. The point of these sanctions is to allow regulators to distinguish more effectively between those with a good general approach to compliance and those who tend to disregard the law, and hopefully punish the latter more harshly.

This should keep the minor and 'accidental' offenders out of the Courts, and save valuable Court time for the most serious and consistent offenders.

Operators should bear in mind that in using the civil sanctions regulators must apply a criminal standard of proof – i.e. they must be satisfied 'beyond reasonable doubt' that an offence has been committed before using them, except for enforcement undertakings and stop notices.

Recent use of the civil sanctions

The Environment Agency is the first enforcement body to use the new civil sanctions, and has recently accepted a financial offer from an engineering and information technology company headquartered in London – for packaging waste offences – as an alternative to criminal prosecution. This funding will be given to specified organisations to drive environmental improvements in local communities.

In addition to implementing a string of improvements to comply with Packaging Waste Regulations, the company has offered to fund 'environment improvements and community benefits' equivalent to the cost of the offences committed. The cost of the Environment Agency investigation and future monitoring are covered.

At the time of writing, 30 offers have been made by a range of organisations that have broken environmental regulations and proposed innovative ways of responding responsibly to the offence, including setting up a local community recycling awareness scheme, donations to a range of environmental charities, and provision of funding for a local school environmental project.

Conclusion

Oliver Letwin, the Cabinet Office Minister for government policy, has said that the sanctions will lead to sloppy regulation, and that powers to issue civil sanctions will be stripped from regulators and handed to the courts. He said regulators imposing sanctions directly without recourse to the courts was 'intolerable'. Sanctions would be based on flimsy evidence if the current approach continues, he believes. However, it is recognised that only regulators deemed by the Better Regulation Executive, part of the business department

(BIS), to be transparent and accountable, with proportionate and consistent enforcement, can issue sanctions.

At the moment, however, the civil sanctions remain in place, and it remains to be seen whether the Coalition Government will actually go as far as removing this legislation, which was enacted under a Labour Government.

> *See also:* Biological hazards, p.51; Contaminated land, p.182; Corporate Social Responsibility, p.202; Environmental risk assessments, p.292; Hazardous waste, p.700; Waste management, p.707.

Sources of further information

EA: www.environment-agency.gov.uk/business/regulation/116844.aspx

The *IEMA Foundation Certificate in Environmental Management* is one of the UK's most sought-after qualifications for organisations and individuals looking to get to grips with the basics of environmental management.

The course is designed to raise your competence if you are new to environmental management or if you have a wide range of premises and facilities management responsibilities.

This IEMA Foundation Certificate is a recognised qualification that will help you to:

- understand the importance of integrating environmental issues into the business process and the potential business benefits of doing so;
- have an appreciation of the driving forces leading the introduction of environmental management systems into an organisation and their potential benefits;
- outline the structure of environmental legislation;
- understand the potential impact of key pieces of legislation on business;
- appreciate the role of the key regulatory agencies;
- outline how an environmental review is conducted;
- identify significant environmental impacts of organisations;
- understand the purpose and nature of an environmental audit, and how an audit is carried; and
- demonstrate an understanding of the key elements of ISO 14001.

Workplace Law Environmental runs a programme of IEMA Foundation Certificate courses via classroom learning in central London. We also run a unique e-learning learning course that you can study when you want where you want. Our e-learning course is a popular and cost effective alternative for organisations with several members of staff looking to get certified. For more information visit http://iema.workplacelaw.net/iema/foundation-certificate.

Environmental Management Systems

Sally Goodman, Bureau Veritas UK Ltd

Key points

- 'Environmental responsibility' is a term commonly used in boardrooms throughout the UK. This highlights that effective management of environmental issues is an area of concern for many businesses, from large FTSE 100 companies to smaller organisations.
- This interest in environmental responsibility is generated for different reasons within each company. However, they all aim to achieve the same goal – to minimise and control potentially negative environmental impacts and risks.
- Many organisations in the UK are achieving this goal by the implementation of an Environmental Management System (EMS).

What is an Environmental Management System?

An EMS is a powerful tool for the identification and management of environmental risks. It also provides a mechanism for delivering performance improvements, effecting resource savings and promoting environmental best practice.

EMS Standards

There are two main EMS standards to which an organisation can receive external certification: the EU-based Eco-Management and Audit Scheme (EMAS) and the International Standard ISO 14001: 2004. Both Standards represent best practice in environmental management and have been widely accepted by UK businesses.

Up to the end of December 2009, at least 223,149 ISO 14001: 2004 certificates had been issued in 159 countries and economies. Annual growth is stabilised at almost the same level as in 2008 – 34,334 in 2009, compared to 34,242 in 2008, when the total was 188,815 in 155 countries and economies. The number

of certificates in the UK was 10,912, an increase of 15% since 2008.

ISO 14001 is applicable to any type and size of organisation, anywhere in the world. It can be applied to individual sites, or to entire organisations. Some organisations choose to implement ISO 14001 in a phased manner, site by site, or plant by plant, or subsidiary by subsidiary. Others design the EMS at corporate level and require each site or plant to interpret and implement the EMS accordingly. Which approach is chosen depends very much on resources and the existing company culture.

There is an alternative EMS standard aimed primarily at SMEs, companies that are more likely to have limited resources for environmental management. It is BS 8555: 2003: '*Environmental Management System. Guide to the phased implementation of an environmental management system including the use of environmental performance evaluation*', which is a phased approach to implementing an EMS.

A property management company with a number of regional offices around the UK wanted to implement an EMS and gain accredited certification to ISO 14001. Certification was increasingly becoming a condition of tendering, especially for public sector work, so there was a clear commercial imperative. The process adopted, with the help of Bureau Veritas Solutions, was (after carrying out a comprehensive initial environmental review) to develop a framework EMS.

A central EMS Manual, Register of Significant Environmental Aspects and Register of Environmental Legislation, was developed, along with environmental procedures that were then implemented at each of the regional offices. Objectives were developed by the Steering Committee chaired by the MD and each office developed its own targets within these objectives. Each regional office had an Environmental Coordinator (usually the Office Manager) who was responsible for facilitating the achievement of the targets, collecting KPI data and carrying out internal audits. A 'Roadshow' was taken round all the offices during the implementation phase to raise awareness and encourage buy-in, an e-learning awareness package for all staff was developed, and focused training was provide to the office managers and to the system auditors. Certification was achieved within 14 months. By setting clear targets, demonstrating senior management commitment and investing in training, the company showed significant savings in energy, waste and paper.

When all the phases (one to six) are successfully completed (with certification at each stage), then ISO 14001 is achieved and there is also the option of going on further to EMAS. Certification is obtained via the IEMA Acorn Scheme, which is officially recognised by the UK Government and offers accredited recognition for organisations evaluating and improving their environmental performance through the phased implementation of an EMS using BS 8555. Acorn focuses on environmental improvements that are linked to business competitiveness, and is flexible so that all types of organisation, whatever their size, can participate.

The six phases of BS 8555 are:

1. Commitment and establishing the baseline.
2. Identifying and ensuring compliance with legal and other requirements.
3. Developing objectives, targets and programmes.
4. Implementation and operation of the EMS.
5. Checking, auditing and review.
6. EMS acknowledgement.

EMAS is based on an EU Regulation and has been operative since 1995. The main difference between EMAS and ISO 14001 is that EMAS requires a formal initial environmental review and the additional production of a periodic public environmental 'statement' (in reality, a detailed report), which describes an organisation's environmental impacts and data, its environmental programme, the involvement of stakeholders, and progress in achieving improvement in performance.

The REMAS project (See *Sources of further information*) showed that the adoption of an accredited certified environmental management system improves site environmental management activities and that there is evidence that overall environmental management is better under EMAS than under other systems. A more recent study, in October 2009, reinforced this view.

In 2009 the EMAS Regulation was revised and modified for the second time. Regulation (EC) No 1221/2009 of the European Parliament and of the Council of 25 November 2009 on the voluntary participation by organisations in a Community eco-management and audit scheme (EMAS) was published on 22 December 2009 and entered into force on 11 January 2010. Changes from the previous version are as follows:

- Transitional registration procedures to facilitate introduction of EMAS III.
- Revised audit cycles to further improve applicability for small organisations.
- Single corporate registration to ease administrative and financial burdens on organisations with sites in more than one Member State.
- Cluster approach to provide specific assistance to clusters of organisations in the development and implementation phases of EMAS registration.
- Environmental core indicators to thoroughly document environmental

performance and to help organisations apply adequate indicators.

- Reference documents to provide guidance to organisations and to facilitate practical implementation of EMAS requirements.
- Single EMAS logo to communicate EMAS in one coherent and distinctive way.
- 'Global EMAS' to encourage global uptake of the scheme by making EMAS certification possible for organisations and sites located outside the EU.
- Information and promotion activities of EU Member States and European Commission to support EMAS III.

Currently (June 2011) 4,615 organisations and 8,011 sites are EMAS-registered, of which 58 organisations and 288 sites are in the UK.

Whichever Standard is used as the basis for an EMS, there are three underlying principles that form the backbone of the system:

1. Compliance with applicable environmental legislation.
2. Prevention of pollution.
3. Continual improvement.

EMAS, ISO 14001 and BS 8555 specify that an effective EMS should be based around the Plan – Do – Check – Act model.

Planning
Including development of an environmental policy, the identification and evaluation of environmental aspects and impacts, the development of objectives and targets, and the preparation of an 'environmental management programme' (action plan).

Implementation
Including defining roles and responsibilities, assessing competence and delivering training, communications,

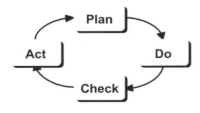

documentation and document control, operational control and emergency preparedness.

Checking and corrective action

Including monitoring of performance (including compliance with legislation), specifying corrective action, record keeping and internal auditing.

Review

Strategic level review ('management review').

The basic development process for an EMS is illustrated below.

Why implement an EMS?

The use of EMSs to control environmental risks is not a new concept. EMSs have been externally reviewed since 1992, with the number being externally certified within the UK growing all the time. There are many recognised and continued benefits from implementing an EMS:

- A more systematic approach to business management.

- Reduced risk of prosecution and improved relationships with regulatory authorities.
- The confidence to do business in what before may have been viewed as high-risk areas or processes.
- Financial benefits in terms of, for example, reduced waste and energy bills.
- Competitive advantage over organisations with less developed risk management procedures.
- Improved public profile and better relationship with stakeholders.
- Improved staff motivation, retention and attraction.
- Sets management requirements rather than absolute environmental requirements. This means the Standard can be applied to any organisation and the EMS can be tailored to reflect the individual requirements of that organisation.
- ISO 14001 is an internationally recognised certifiable Standard, which provides a level of assurance of the Environmental Management System through an external audit process.

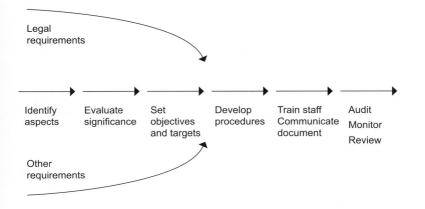

Legal requirements

Identify aspects → Evaluate significance → Set objectives and targets → Develop procedures → Train staff / Communicate / document → Audit / Monitor / Review

Other requirements

- The Standard allows the organisation to provide demonstrable performance and performance improvement over time. Ability to set long-term goals with clear actions to deliver and progress.
- Increased reliability of data on which decisions and actions are based.
- Allows for provision of clear governance, management accountability and reporting associated with company activities.
- Can contribute to staff commitment and motivation (and can be of benefit during recruitment).
- Is increasingly a requirement in tenders.

Types of EMS

An EMS can take many forms, from detailed and prescriptive procedures, to simple flowcharts. It can also be delivered and communicated in many different ways, from paper copies of procedures to electronic systems held on company intranets.

The type of EMS that an organisation chooses depends on the size and culture of the organisation and the existing communications process. One example that a large multi-site organisation may opt to use is an internet-based software package. Such packages can assist an organisation in rapidly establishing its EMS or to maintain and keep live its existing system. Alternatively, an SME may opt to install a simple flowchart-based system to provide clear instructions that all levels of staff can understand.

Integration

Although an EMS can be designed as a stand-alone system, it is now best practice to integrate the environmental management requirements with health and safety and quality management systems. It is also increasingly common to find EMSs being integrated with systems for improving corporate social responsibility

or corporate governance (see *'Corporate Social Responsibility'*, p.202).

Certification

Certification is a process by which an independent third party audits a management system against a recognised standard. In the UK, certification bodies are accredited (given a licence to operate) by the United Kingdom Accreditation Service (UKAS). UKAS is the sole national body recognised by government for the accreditation of testing and calibration laboratories and certification and inspection bodies.

There are many certification bodies operating in the UK, among which are the British Standards Institution (BSI), Bureau Veritas Certification, DNV, Lloyd's Register Quality Assurance (LRQA) and SGS.

Links with CSR

Corporate Social Responsibility (CSR), increasingly termed Corporate Responsibility (CR), is a collective term that brings together a company response to social (employee, health and safety and community), environmental and ethical issues and risks to deliver greater value to the company. Many companies recognise this with the publication of a CSR report.

CSR confers business benefits to all organisations and stakeholder groups in a number of areas. A coherent CSR strategy based on integrity, sound values and a long-term approach will offer clear business benefits. These cover a better alignment of corporate goals with those of society; maintaining the company's reputation; securing its continued licence to operate; and reducing its exposure to liabilities, risks and associated costs.

A robust CSR strategy should also support reporting against the requirements of the Business Review mandated by the

UK Government (under the Companies Act 1985, OFR (Repeal) Regulations). Of importance is how the CSR strategy identifies and covers all non-financial risks and opportunities that can affect business performance. The critical aspects to this process are to identify the material non-financial risks, ensure that processes are in place to manage these risks, and to explicitly link non-financial risk management to business performance.

As understanding and support for the concept of CSR grows, and consumers become increasingly aware of the issues, more and more companies are building socially responsible elements into

marketing and brand identity. Addressing these issues can improve performance whilst also acting as a differentiator to attract customers and corporate partners and create competitive advantage. An EMS is an essential part of a CSR strategy for any sector, although for those sectors with greater environmental impact, the EMS will have greater visibility within the CSR programme.

See also: Corporate Social Responsibility, p.202; International Standards, p.416.

Sources of further information

EMAS Helpdesk: http://ec.europa.eu/environment/emas/index_en.htm

EMAS Toolkit for small organisations: http://ec.europa.eu/environment/emas/toolkit/

ISO 14000 Information Centre: www.iso14000.com

International Organisation for Standardisation (ISO): www.iso.org/iso/home.htm

REMAS: http://remas.iema.net/

UKAS: www.ukas.com

Acorn Scheme for BS 8555: www.iema.net/ems/acorn_scheme

Certification bodies:

British Standards Institution (BSI): www.bsi-global.com

Bureau Veritas Certification: www.bureauveritas.co.uk

DNV: www.dnv.com

Lloyd's Register Quality Assurance (LRQA): www.lrqa.com

SGS: www.uk.sgs.com

Environmental risk assessments

Colin Malcolm, Workplace Law

Key points

- Environmental risk assessment is a process to identify and qualify the environmental risk associated with a particular activity, process, service or behaviour. The risk assessment will inform on the significance and probability of harm and therefore establish the options for effective management and operational control.
- These assessments sit at the heart of environmental management programmes and whilst some can be informal, the majority are documented, structured and auditable.
- Environmental risk can be a difficult concept to qualify without a formal assessment process, as in a typical organisation environmental issues range from tangible scenarios, such as hazardous chemical management, to far less quantifiable issues, such as statutory nuisance.

Legislation

- Environmental Protection Act 1990.
- Environmental Protection (Duty of Care) Regulations 1991.
- Environmental Permitting (England and Wales) Regulations 2007 (as amended).
- Environmental Damage (Prevention and Remediation) Regulations 2009.

Why manage risk?

All activities carried out by individuals or business interact with the environment, and it is important to be able to separate out the activities that can cause harm, hence requiring some form of control, from those that do not. Carrying out an environmental risk assessment empowers businesses to make informed decisions on compliance strategies, pollution scenarios, managing internal resources, defining operating parameters and engaging with stakeholders.

There is an increasing amount of environmental regulation that applies to business and this is one of the main drivers for managing risk. Some regulatory schemes, such as planning and the Environmental Permitting (England and Wales) Regulations 2007 (as amended) for example, explicitly require environmental risk assessments to be carried out and submitted to the regulatory bodies for consultation prior to the activity being authorised. However, for the majority of regulations, the onus is on the business to put in place effective risk assessment measures to ensure compliance is maintained.

The recently released Environmental Damage Regulations 2009 raise an interesting point here, as they can apply where there is the potential for environmental damage to occur; in other words before an incident actually happens. The Regulations also cover an incident once it has occurred, but the preventative aspect is particularly relevant as it directly links to environmental risk assessments, or rather the lack of effective risk management on a site. The Regulations, however, are designed to only apply to the most serious cases of environmental damage.

Environmental performance is becoming increasingly relevant in the mainstream business climate, particularly as significant liabilities may result from ineffectual risk management practices. This is principally relevant where the omission or error has subsequently caused the occurrence of a reportable environmental incident. Stakeholder interest in how a business manages its environmental responsibilities has reached unparalleled heights, and whilst many opportunities exist for businesses to prosper through strong environmental leadership, the opposite is also true for underperforming businesses.

For example, poor environmental performance can damage confidence in a broad spectrum of issues relevant to the majority of stakeholder groups, including:

- reputation and image;
- investor confidence;
- insurance premiums;
- regulatory scrutiny;
- tender submissions;
- local community relations;
- staff morale; and
- financial penalty.

The potential impacts from not effectively managing environmental risk can therefore extend significantly and can leave a long-term unwanted legacy for the business concerned.

When to manage risk

All businesses have an impact on the environment, so it is logical that all businesses need to carry out some form of environmental risk assessment. There is a clear need for distinction between activities that can cause significant pollution, such as an oil refinery, and those that cannot, such as a small office for example, but this should not influence the decision as to whether to carry out the assessment. What the above example illustrates is that different business activities have varying

hazards, risks and vastly different potential to cause harm; this issue is addressed below.

Businesses that have not carried out an environmental risk assessment of their activities are either ignorant of the issues or taking a calculated risk to ignore their responsibilities. Both these approaches are inadequate and are potentially exposing the business and the environment to unquantifiable risk.

Once a risk assessment has been completed, it will usually only need to be updated periodically. Typical issues that pre-empt updates include:

- introducing a new product, substance, process or activity to the business;
- a change to existing or new regulation;
- engaging contractors or new staff; and/or
- a change to local sensitive receptors or the means by which a polluting material can reach a sensitive receptor.

It is good practice for all businesses to have carried out an environmental risk assessment and to have a clear understanding, and management, of their environmental risks.

How to manage risk

The completed risk assessment process will give management reassurance that its activities and operations are being managed in a way that prevents or minimises risk to the environment, humans or other receptors. There are a number of different methodologies available to support the risk assessment process, ranging from complex dispersion modelling techniques (such as those used by air quality scientists) to a simple assessment of the connectivity between a hazard and a receptor. For all risk assessments, it is important to clearly define the scope, purpose and how and by whom the results will be used to inform environmental management practices.

The most simple and effective method for initially identifying risk scenarios is by using the source – pathway – receptor model. This model works by identifying potential sources of hazards (for example, such as acid storage), then the receptors that could be affected if they came into contact with the acid (such as water courses and aquatic organisms). The model then identifies the potential pathways between the source and the receptor (such as a drainage system or impermeable sloping surfaces). This assessment is typically carried out under normal, abnormal and emergency scenarios, which provide the widest possible information on the potential for risks and options for control measures.

Once the source – pathway – receptor model has been completed the next step is to prioritise risk, and this can be completed using a simple high, medium, low scale matrix to assess two key variables:

1. The severity if pollution occurred.
2. The likelihood of pollution occurring.

Using the above example, the storage of acid (the source) in a bunded area close to a surface water drain (the pathway) has the potential to cause pollution of the watercourse (the receptor) and if this occurred, severity will be high as acid can alter the pH of water and small variances can significantly impact aquatic organisms. In addition, to cause or knowingly permit polluting matter to enter a watercourse is prohibited under relevant water legislation.

However, the pathway is not currently viable due to the secondary containment

provided by the bund; hence the likelihood will not be high. This does not automatically mean, however, that likelihood will be low and the issue should be discounted from further risk investigation, as it is only the integrity of the secondary containment that will prevent an unplanned leakage from reaching the watercourse.

An abnormal event such as a forklift truck accidentally breaching bund integrity can quickly change the situation, hence the risk assessment should consider the widest possible range of scenarios. The whole risk assessment process should be documented and readily available for reference, audit or revision.

Conclusion

Environmental risk assessments should be integrated into normal business practices and act as one of the primary measures to inform management about the number, type and scale of environmental risk associated with their activities. The assessments should initially be very thorough, with revisions frequent enough to account for and respond to relevant changes that may impact on their accuracy. The results of the risk assessment should be disseminated to senior management, who must be aware of the risk and also to ensure adequate resources are made available to eliminate or control risks to an acceptable level.

See also: Corporate Social Responsibility, p.202; Environmental Management Systems, p.286; Risk assessments, p.624; Waste management, p.707.

Sources of further information

Workplace Law's *Introduction to Managing Environmental Compliance* course provides comprehensive coverage of what can be a complex and confusing issue for organisations that need to comply with a raft of UK and European legislation. The course provides an insight into relevant legislation governing environmental law and practice, with practical guidance on what are the key issues to be aware of to ensure legal compliance.

Improving energy efficiency, reducing carbon output, and minimising waste have become key themes in the UK regulatory regime. Understanding the framework of environmental management and its impact on your organisation is essential to ensure that you are fully up-to-date and compliant in this area.The course will examine all the key issues your organisation needs to address today in order to be compliant, as well as helping you to identify and manage those environmental risks to your business.

For more information visit http://environmental-management.workplacelaw.net/

Comment ...

Does the polluter pay?

Colin Malcolm, Workplace Law

All businesses interact with the environment. These interactions typically result in negative impacts at both a local and a global level, and the scale of impact can vary greatly according to many factors. This includes:

- the business sector;
- type of activity undertaken;
- the size or location of the business; and
- the effectiveness of internal environmental management controls.

So do businesses pay for the pollution and environmental impacts they cause? Most organisations are acutely aware of how many regulations currently exist to protect the environment, and that it is the fastest growing area of law in the UK.

Environmental law has evolved over time to address and regulate the many ways in which businesses can cause environmental impacts. Currently, the regulatory framework for environmental law is very broad, with many different types of law in place to protect the environment, such as:

- the prevention of pollution;
- minimising harm to humans from environmental pollution;
- conserving natural resources;
- informing, educating and encouraging behavioural changes;
- holding those responsible for environmental pollution accountable;
- attributing a financial cost to pollution; and

Colin Malcolm has an MA in Environmental Management, is a Full Member of the Institute of Environmental Management and Assessment (IEMA) and sits on the IEMA Full Membership assessment panel. He is experienced in supporting industrial and commercial sector clients in a breadth of environmental management and sustainability issues.

- recovering the costs associated with remediation of pollution incidents.

Complying with environmental law typically incurs a financial cost to business, such as obtaining and maintaining compliance with a permit, or disposing of waste at a landfill site.

It can therefore be argued that many businesses already pay for the pollution they cause. But do they?

The polluter pays principle is an overarching principle of UK and International environmental law, and is referenced in Article 174 of the European Union Treaty. It states that:

'Community policy on the environment shall aim at a high level of protection, taking into account the diversity of situations in the various regions of the Community.

'It shall be based on the precautionary principle and on the principles that preventive action should be taken, that environmental damage should as a priority be rectified at source, and that the polluter should pay.'

As a first step in understanding whether businesses are fully held to account for the impacts that result from their interactions with the environment, we'll look at the most topical of current environmental issues – climate change.

Climate change is one of the most relevant issues in current affairs and it needs to be seen in much broader terms than just the environmental impacts such as increasing global average temperatures, rising sea levels and more frequent extreme weather events. This is because these, and other projected impacts, have the potential to significantly disrupt our very way of living and, as such, climate change is one of the frontrunners in the development of international policy covering social and economic issues and environmental factors.

Since the time of the Industrial Revolution, the concentration of greenhouse gases (principally carbon dioxide, methane and nitrous oxide) in the atmosphere has risen significantly. The overwhelming majority of scientific evidence points towards human activities as the main contributor to this increase, hence also responsibility in part for the currently quantifiable effects of climate change – and for those effects projected to occur in the future.

"Wherever the revenue goes, environmental taxes and other types of economic instruments will play an increasingly significant role in the future direction of environmental policy."

The estimated financial costs of dealing with the actual and predicted effects of climate change are vast. For instance, the 2006 Stern Report estimated the annual costs of stabilising levels of carbon dioxide equivalent between 500 and 550 parts per million by 2050 to be approximately 1% of global GDP. To put this figure into context, 2010 world GDP was just under 63,000,000 million US dollars.

So who will pay for the work needed to deliver climate change mitigation and adaptation, now and in the future?

All businesses contribute to the release of greenhouse gases to the atmosphere. Even typically 'low' environmental risk sectors, such as office-based organisations, will contribute to climate change – directly, through activities such as the combustion of natural gas, transportation and the use of refrigerant gases in air conditioning systems; and indirectly, from the use of electricity, waste recovery and disposal activities.

Having established the link between business and environmental issues at both a global and a local level, a review of environmental law highlights some interesting examples of whether the existing framework of environmental regulation delivers 'the polluter pays principle'. One of the main ways in which environmental law can work towards the polluter pays principle is through the use of economic instruments. These are defined as 'the means by which decisions or actions of government affect the behaviour of producers and consumers by causing changes in the prices to be paid for these activities'.

For example, for the last ten years all businesses have paid a tax on the consumption of electricity, gas and certain other fuels. This tax, introduced by the Climate Change Levy Regulations 2001, was designed to encourage energy efficiency and reductions in greenhouse gases, and is paid directly through energy bills, with the tax being offset by reductions in National Insurance contributions.

The Finance Act 1996 introduced a tax on the disposal of waste to landfill – known as the landfill tax. Currently set at £56 per tonne the standard rate will continue to increase by £8 per tonne on 1 April each year until 2014 (£80/tonne) at which time a price floor will exist whereby the tax will not drop below £80/tonne until 2019/20.

There are several other examples where economic instruments are used as a delivery mechanism for environmental regulation. This includes the European Union Emissions Trading System, a cap and trade scheme targeting the energy-intensive industries throughout Europe; and the UK-wide Carbon Reduction Commitment Energy Efficiency Scheme, which although originally formulated as a cap and trade scheme, has now been redesigned as a carbon tax.

One of the main debates concerning the use of economic instruments as a mechanism to recover costs from businesses that cause pollution is what happens to the revenue generated. Tax 'hypothecation' refers to the earmarking or ring fencing of revenues for a specific purpose; and prominent examples of environmental taxation schemes such as

the climate change levy and landfill tax have not been subject to hypothecation.

In 2009 the Environmental Audit Committee asked the Treasury to confirm whether its definition of an environmental tax was one in which the revenues are explicitly hypothecated to environmental ends.

The Treasury confirmed that its definition of environmental tax is not linked to any hypothecation of revenue:

'Where the Treasury refers to environmental taxes, it means the climate change levy, aggregates levy and landfill tax – those taxes that were introduced primarily to have an environmental impact. Each of these taxes was introduced alongside a cut in National Insurance Contributions as part of the shift from 'goods' to 'bads'. Government's spending priorities are not, in general, determined by the way in which the money is raised. Hypothecating revenues to particular spending programmes imparts inflexibility in spending decisions and can lead to a misallocation of resources, with reduced value for money for taxpayers.'

Economic instruments such as environmental taxation are an extension of the polluter pays principle and are designed to encourage widespread changes in behaviour. Although not hypothecating environmental taxes may make sense from an economic standpoint, it does raise the potential for their purpose, in certain quarters, to be perceived as a 'stealth tax', leading to distrust towards their principal objective.

Wherever the revenue goes, environmental taxes and other types of economic instruments will play an increasingly significant role in the future direction of environmental policy. The majority of environmental law in the UK now has its origins from within the European Union and casting an eye over EU environmental policy emphasises this point:

'Greater use should be made of market-based instruments, including quota trading schemes, taxation measures and subsidies, to achieve environmental and other strategic objectives.' (EU Commission)

Another area where the polluter pays principle can be seen to be appearing in environmental law is through civil sanctions. Introduced through the Regulatory Enforcement and Sanctions Act 2008, civil sanctions provide a suite of additional sanctions to deal with environmental offences.

Civil sanctions do not replace any existing regulatory functions; rather they are designed to be complimentary to the criminal law, but provide much greater flexibility and proportionality in the available enforcement options, particularly for lower level breaches.

One of the principal planned outcomes of civil sanctions is to adequately compensate those who have been affected by environmental pollution, rather than simply attributing a financial penalty and this clearly links quite closely with the polluter pays principle. For example, restoration notices provide a framework whereby the polluted area can be restored to the position that would have persisted if no offence had been committed. So if a business pollutes a watercourse, part of its remedial action will be to restore the watercourse to its original state, and possibly even provide an alternative whilst the original is being cleaned up.

So how is the concept of the 'polluter pays principle' relevant to the role of Facilities

Management, and what should the sector be doing to address its increasing prominence in environmental law?

Legal compliance is of course the primary mechanism for enforcing the polluter pays principle on business through environmental law. Because of this the facilities management sector needs to ensure that a clear understanding of current and future environmental law is maintained, and that this is translated into a robust and auditable compliance strategy in all aspects of their work.

The Landfill Tax, Climate Change Levy and Carbon Reduction Commitment were introduced as economic instruments in environmental law which attribute a financial cost related to pollution, and the impact to business from each of these three laws can be influenced by the unique positioning and day-to-day responsibilities of the facilities manager. The most evident examples of this are in the areas of energy and waste minimisation programmes, which if properly implemented will improve efficiency, reduce environmental impacts and deliver cost savings.

Meeting the 'polluter pays principle' should also encourage the facilities management sector to align its service offering and business development strategy more closely to managing environmental issues.

So does the polluter pay?

In many respects, businesses already do, and have been doing for several years, as the examples of the climate change levy and the landfill tax demonstrate. Clearly the revenues received from these schemes – approximately £700m per year in the case of the climate change levy – are substantial, but they become less significant when compared with the estimated costs of climate change stabilisation according to the highly influential Stern Report. There is also the complex issue of environmental tax hypothecation to take into consideration, and whether there is enough transparency on where environmental tax revenues are ultimately allocated.

One of the key issues in this debate is the ability to accurately attribute a cost to the pollution caused by business, and then implement an equitable mechanism to recover payment. Existing schemes, such as the climate change levy, have been criticised by certain business sectors on this particular point; as even though a corresponding cut in employers' National Insurance costs was designed to ensure revenue neutrality, those organisations with a high energy usage and low employee scenario have typically been harder hit than those organisations with the reverse scenario.

The new civil sanctions regulations will be interesting to watch. Civil sanctions provide the framework for the polluter pays principle to be applied on an individual, case by case basis. This should result in a more transparent extension of the polluter pays principle, particularly when compared to some of the broader, all encompassing, approaches such as environmental taxation which have been criticised in some quarters as delivering less transparent outcomes.

One thing that is clear is the commitment of the European Union to a greater use of market-based instruments such as taxation and trading schemes in environmental policy. This will inevitably mean businesses will be subject to more environmental regulation under the banner of economic instruments. This, to a certain extent, may fit the definition of the 'polluter pays principle', as costs will be recovered from businesses that pollute the environment.

However, in the present climate, the ability of the environment to withstand pollution is being severely challenged, hence a true test of the polluter pays principle rests on two factors. Firstly, whether costs can be recovered proportionately, which taxes tend not to do. Secondly, will the revenue generated effectively address the wider costs to society of dealing with the consequences of environmental pollution?

Eye and eyesight tests

Andrew Richardson, URS Scott Wilson

Key points
- Use of DSE constitutes an adverse health condition.
- Users or operators of DSE can suffer visual fatigue or headaches.
- DSE does not cause permanent damage to eyesight, but pre-existing eye conditions may be accentuated.
- Employers must provide and pay for eye and eyesight tests if employees request.
- Tests must be carried out by a registered ophthalmic optician or medical practitioner.
- Employers must provide spectacles where needed for screen-viewing distance if employees request.

Legislation
- Health and Safety at Work etc. Act 1974.
- Opticians Act 1989 (as amended in 2005).
- The Sight Testing (Examination and Prescription) (No. 2) Regulations 1989.
- Health and Safety (Display Screen Equipment) Regulations 1992 (DSE Regulations) as amended by the Health and Safety (Miscellaneous Amendments) Regulations 2002.
- Management of Health and Safety at Work Regulations 1999.

Statutory requirements
Regulation 6 of the Management of Health and Safety at Work Regulations requires consideration to be given to carrying out health surveillance of employees where there is a disease or adverse health condition identified in the risk assessment. It has been identified that use of DSE by someone deemed to be a 'user' or 'operator' does constitute an adverse health condition.

The DSE Regulations require employers to inform users or operators about the risks to their health, and that they are entitled to an appropriate eye and eyesight test if they request it. The employer has a duty to provide tests to employees already designated as a user or operator and to employees who are being recruited or transferred to be users or operators. The employer must pay the reasonable costs of these tests. *Note: DSE users are not obliged to have the test.*

The Opticians Act 1989 and the Sight Testing (Examination and Prescription) (No. 2) Regulations 1989 set out the requirements for eye and eyesight tests.

Eye tests for DSE users
The purpose of the eye test is to improve the comfort and efficiency of the user by identifying and correcting any vision defects specific to DSE use. There is no evidence that DSE usage causes permanent damage to eyesight; what will occur is that pre-existing eye conditions will be accentuated, which can lead to temporary visual fatigue or headaches.

The test should include the test of vision and examination of the eye, and should

Facts

- 68% of DSE (Display Screen Equipment) users suffer eyestrain (19% 'frequently').
- According to TUC reports, 29% of respondents who reported problems with eyes, such as dryness or irritation of the eyes, said it was made worse by work.
- According to BBC reports, British adults spend up to 130,000 hours during their lifetime sitting in front of a computer or TV.
- Specsavers tested people's eyesight at roadshows around the country. Out of 546 drivers who agreed to participate in a voluntary sight check, 25 were unable to read a numberplate at a minimum distance; one person had to be taken to within two metres of the numberplate before he was able to read it; 53 drivers were borderline; and 51 drivers admitted that they had never taken an eye examination.
- Contract Research reported that 75% of DSE users have visual tiredness after using a laptop for just 20 minutes.
- 40% of employees are unaware that they can have a free eye examination from their employer if they regularly use a computer monitor in their job.

take into account the nature of the DSE work carried out. Only a registered ophthalmic optician or registered medical practitioner should carry out the test.

If an employee requests a test for the first time, it must be carried out as soon as is practicable. If an employee is to be transferred to a user or operator post, then the test must be carried out before commencement in that post. For people being recruited, once they are definitely going to be an employee, the test should be carried out before they commence any work that meets the user/operator criteria.

After the first test, the employer must be guided by the optician or doctor as to the frequency of subsequent tests.

If the tests show that the employee requires special corrective appliances (normally spectacles) specifically for distances when using DSE, then the employer must provide a basic special corrective appliance. Normal corrective

appliances are those spectacles prescribed for any other purpose, but the employer has no liability for these.

Eye tests for drivers

Professional drivers such as commercial vehicle and bus drivers must have a full eyesight test, as part of a full medical, before they can pass their driving test. They are then subject to re-test every five years after the age of 45, and yearly after age 65.

To comply with a 2006 EU Directive, the UK Government has until January 2013 to bring in legislation to extend this system to all car and motorcycle drivers. The DVLA carried out a public consultation on its proposals in spring 2011 but, at the time of writing the results have yet to be published.

Eyecare vouchers

It is worth considering how to manage the provision of eye tests and corrective spectacles for DSE users. Many employers just adopt a passive approach, and reimburse employees' claims.

An alternative is to set up an eyecare voucher scheme. This means providing employees with vouchers that they can spend on eye tests and corrective spectacles.

Such schemes simplify administration so can be cheaper than reimbursement, and demonstrate a caring attitude. Vouchers that are widely accepted allow employees the greatest flexibility and choice; for example they will probably be able to continue to use their existing optician. Research has shown that eyecare voucher schemes encourage more employees to take care of their eyesight, with the added benefit to the employer of reduced sickness downtime.

See also: Display Screen Equipment, p.244; Driving at work, p.250; Health surveillance, p.396.

Sources of further information

L26 Work with Display Screen Equipment: www.hse.gov.uk/pubns/priced/l26.pdf

Facilities management contracts

Marc Hanson, Berwin Leighton Paisner LLP

Key points

- Like all other forms of contract, a facilities management contract is essentially a legally binding and enforceable bargain between two parties.
- Each party contributes something to the bargain; the facilities management contractor – the provision of certain services; and the client – payment for those services.

Negotiation

For a contract to be legally binding, there must be an offer from one party, an unconditional acceptance of that offer by the other party, and consideration provided by each party for the promise made by the other party. A client's invitation to tender is not usually an 'offer' – it is usually no more than an offer to negotiate.

A facilities management contractor's tender to carry out the services will, usually, amount to the initial 'offer'.

When the client accepts the facilities management contractor's tender and each party gives consideration, then, provided both parties have an intention to be legally bound, a legally enforceable contract will come into place. Offers and acceptance can be made in writing, orally or by conduct.

It is of course unusual for facilities management contractors' tenders to be accepted without qualification by a client. There may be areas of extensive negotiation, e.g. in relation to scope of services and fees. Every time each party provides any revised proposals, then each revised proposal will take effect as a 'counter-offer'.

When eventually all outstanding points have been agreed, one party will invariably 'accept' the other party's final 'offer'.

When does the contract start?

The process of negotiating a facilities management contract can be protracted. In many cases, services may be provided to the client and payment may be made without any form of contract having been signed. Where relationships subsequently deteriorate, it can be extremely difficult to establish whether there was ever a binding contract in place and, if there was, on what terms it was made.

Whether a binding contract exists in such circumstances will depend on whether the parties managed to agree all the key terms of the contract and whether the terms of the alleged contract included all terms that would be essential for a contract to exist. It would be unlikely that there was a contract agreed if key elements of the contract were still outstanding; for example, if the exact scope of the services was undecided or if a price had not been agreed. However, a contract can still come into effect where certain points in the contract terms are still to be agreed, provided that the key elements have been finalised and agreed.

Scope of contracts

When drafting facilities management contracts, it is important to ensure that they cover the complete understanding and agreement between the parties. As such, it is necessary to include not only a list of the services to be provided by the supplier but also a mechanism for dealing with changes to the services and also any details as to what equipment or facilities are to be provided to the supplier by the client in relation to the services.

Payment

Careful thought needs to be given as to how payment to the facilities management contractor will be structured. Will it be on the basis of a lump-sum price, by prime cost or by reference to a schedule of rates? If the price is to adjust, then a mechanism needs to be set out allowing for this, detailing the circumstances in which adjustments will be made. Careful consideration also needs to be given to any mechanism to be included in the contract that would allow the contract price to be adjusted to reflect performance or non-performance by the supplier of the services.

Service levels

Service levels should be included in the contract against which the performance by the supplier can be assessed. Consideration needs to be given as to how poor performance is dealt with and whether the liability of the supplier under the contract is to be limited in any way.

Duration

The duration of a facilities management contract is of critical importance, and this should be clearly stated in the contract, together with the circumstances in which it can be extended or terminated by either party.

Facilities management contracts should also address other key areas such as compliance with statutory requirements, transfer of undertakings provisions, insurance requirements and provisions dealing with dispute resolution.

See also: Contractors, p.187.

Sources of further information

Workplace Law's *Facilities Management Contracts 2008* follows the success of two editions of the *Guide to Facilities Management Contracts*. Now fully updated, it provides an introduction to contract law as it relates to FM contracts, and a guide to common contractual law provisions. This extended publication looks at new areas of law including amendments to the Construction Act, amendments to CDM Regulations, revised TUPE Regulations, revised EU Procurement law, dispute resolution procedures, and European standards for facilities management. For more information visit www.workplacelaw.net.

Fire extinguishers

Andrew Richardson, URS Scott Wilson

Key points

- Requirements for fire extinguishers are covered under the Regulatory Reform (Fire Safety) Order 2005 and equivalent legislation in Scotland and Northern Ireland. The Order requires the Responsible Person to:
  - assess the fire risk in the workplace;
  - provide reasonable firefighting equipment;
  - check that people know what to do in the event of fire; and
  - ensure that fire safety equipment is maintained and monitored.

- Vehicles carrying dangerous goods are required to be fitted with fire extinguishers; there is no legal requirement for cars, but it is recommended.

Legislation

- Health and Safety at Work etc. Act 1974.
- Regulatory Reform (Fire Safety) Order 2005 and equivalent legislation in Scotland and Northern Ireland.
- United Nations European Agreement concerning the International Carriage of Dangerous Goods by Road, ECE/TRANS/202, Vol.I and II(ADR 2009).
- The Fire Safety (Employees' Capabilities) (England) Regulations 2010.

The Regulatory Reform (Fire Safety) Order (RRO) repealed the Fire Precautions Act 1971 and revoked the Fire Precautions (Workplace) Regulations 1997, as amended in 1999. It came into force on 1 October 2006. The main impact of the FSO was to move the emphasis of fire legislation towards prevention – any fire certificate issued under the Fire Precautions Act 1971 has ceased to have effect.

A small loophole in the FSO was closed on 6 April 2010 when the Fire Safety (Employees' Capabilities) (England) Regulations 2010 came into force. This re-established the duty of employers to take into consideration workers' health and safety capabilities when entrusting fire safety tasks to them.

Responsible Person

The FSO establishes the key role of the Responsible Person. Anyone in full or partial control of premises may be a Responsible Person. They might be, for instance, an employer for those parts of premises over which they have any control, a managing agent or owner for shared parts of premises, or shared fire safety equipment such as fire detection systems or sprinklers, or an occupier, such as self-employed people or voluntary organisations when they have any control.

It may be clearly evident who is the Responsible Person but, equally, several people may have some responsibility. In such cases, the Responsible Persons will need to work together to ensure that the arrangements they make are complete and consistent.

Risk assessments

A Responsible Person has a duty to assess the fire risk in the premises. A fire risk assessment carried out under the former Fire Precautions (Workplace) Regulations 1997 would be an excellent start, provided it has been kept up to date; relatively modest revisions would be required to take account of the wider scope of the RRO. The Responsible Person must carry out a fire risk assessment but could get another competent person to do it for them. Even so, the Responsible Person retains the duty to comply with the RRO.

The objective of the fire risk assessment is to devise arrangements to ensure that all persons on or near the premises can safely escape if there is a fire. Unlike previous legislation, the Order requires consideration of all persons who might be on the premises, whether or not they are employees – visitors, contractors, cleaners, members of the public and so on must be thought about. Furthermore, special arrangements may be needed for anyone who has a disability or who may need special help, e.g. persons with reduced mobility or hearing or vision problems.

There are five steps to a fire risk assessment:

1. Identify the potential fire hazards.
2. Identify people at risk.
3. Evaluate the risks, remove or reduce, and protect and decide if existing control measures are adequate.
4. Record findings and actions, plan, inform, instruct and train people.
5. Review and revise those assessments periodically.

The findings and actions from fire risk assessments are usually compiled into a fire action plan.

Classes of fire

In order to fight a fire, the following must first be determined:

- Is the firefighting equipment suitable for the fire risk?
- Is the equipment located correctly?
- Does the equipment have the correct signage?
- Have personnel been trained to use the equipment provided?

In order to determine what firefighting equipment is suitable, it is necessary to identify the classes of fire that may occur in the workplace. There are six classes into which all fires will fall:

- *Class A*: Fires involving solids (wood, paper, plastics, etc., usually organic in nature).
- *Class B*: Fires involving liquids or liquefiable solids (petrol, oil, paint, wax, etc.).
- *Class C*: Fires involving gases (LPG, natural gas, acetylene, etc.).
- *Class D*: Fires involving metals (sodium, magnesium, any metal powders, etc.).
- *Electrical fires*: Although not deemed as a class, electrical equipment fires form a separate category.
- *Class F*: Fires involving cooking fats / oils.

Once the class(es) of fire are identified, the appropriate firefighting equipment can be selected. This can include portable extinguishers, hose reels, sprinkler systems, hydrant systems or fixed firefighting systems. Selection should follow the guidance in BS 5306-8: 'Fire extinguishing installations and equipment on premises. Selection and installation of portable fire extinguishers – Code of practice.'

Type	Colour code	Suitable for
Water	Red	Class A fires only
Foam	Cream	Class B fires but can also be used on Class A
Dry powder	Blue	Class B fires but can also be used on Class A and electrical
Halon	Green	Do not use*
Carbon dioxide	Black	Electrical fires and small Class B fires
Special powder	Blue	Class D fires
Wet chemical	Canary yellow	Class F fires

Effectively banned after 31 December 2003 under the Montreal Protocol because of their ozone-depleting properties. If you have any of these in your workplace, you should arrange to have a replacement system installed as soon as possible.

Firefighting equipment

Portable extinguishers
Portable fire extinguishers come in a variety of types, as shown in the table below.

All are coloured red, but each may have a colour-coded panel to aid in identifying the type.

Fire extinguishers have limitations – they can be used only on the appropriate class of fire, they are of limited duration, and they have a limited range.

The fire extinguishers should be sited:

■ where they are conspicuously and readily visible;
■ on all escape routes;
■ close to specifically identified danger areas;
■ close to room exits, inside or outside dependent on risk;
■ at the same location on each floor;
■ grouped together to form a fire point;
■ no further than 30 metres from any person;

■ with their handle 1.1 metres from floor level; and
■ away from extreme heat or cold.

Fire extinguishers should be maintained in accordance with BS 5306-3: 'Fire extinguishing installations and equipment on premises. Code of practice for the inspection and maintenance of portable fire extinguishers.' This details monthly inspections, annual inspections, and maintenance and discharge test requirements. Those entrusted to carry out the maintenance must be competent, e.g. ISO 9001 certified or BAFE Approved.

Hose reels
Hose reels are primarily utilised on Class A fires. An adjustable nozzle allows the water to be in a jet or spray form: a jet can be used at the base of a fire whereas a spray allows a larger area of coverage and can be used for protection of personnel. Hose reels obviously have a continuous supply of water, which is provided in greater quantity than from an extinguisher, and hence they have a greater range.

Their limitations are:

- greater physical effort is required to operate them;
- they wedge open fire doors, allowing possible smoke spread;
- they should be used only on Class A fires;
- they are a trip hazard; and
- there is a tendency to remain fighting the fire for longer periods.

Hose reels should be provided for every 800 square metres of floor area or part thereof and should be in prominent and accessible locations at exits so that they can extend to all parts in all rooms. Hose reels should be inspected monthly and an annual test carried out to check their full functional capability.

Training

The Health and Safety at Work etc. Act 1974 requires employers to provide information, instruction, training and supervision to ensure the health, safety and welfare of employees. The Fire Safety (Employees' Capabilities) (England) Regulations 2010 (and equivalents in Scotland, Wales and Northern Ireland) state that employers need to take into account the capabilities of their employees before entrusting fire safety tasks, and that they should have adequate health and safety training and be capable enough at their jobs to avoid risk.

Firefighting is a high-risk activity. Where an employer's fire procedure instructs employees to tackle the fire, that employer has a legal obligation to ensure they are adequately trained. Most instructions to tackle fire usually end with 'if it is safe to do so' – unless you are trained, you will not be able to judge whether it is safe. The local fire brigade will be able to give advice on the training available and about local variations to emergency procedures, if any, in your area.

Fire extinguishers in vehicles

Vehicles carrying dangerous goods are required to be fitted with fire extinguishers under ADR 2009. Two dry powder extinguishers are usually required; the level of provision varies from 2kg to 12kg, depending on the vehicle weight.

There is no legal requirement to carry a fire extinguisher in a car, but it is recommended by RoSPA and other sources. The extinguisher should be dry powder or foam type, and manufactured to BS EN: 3 1996. Sizes vary but are typically in the range of 0.6-1.0kg.

See also: Fire safety in non-domestic premises, p.317; Firefighting, p.322; Risk assessments, p.624.

Sources of further information

The Government has produced a series of 14 guidebooks to assist the Responsible Person to carry out a fire risk assessment in different types of premises. The guides can be purchased through or downloaded from the Department for Communities and Local Government's website at www.communities.gov.uk/fire/firesafety/firesafetylaw/aboutguides/.

For SMEs with low risk environments such as offices, there is an Entry Level Guide, which can be downloaded from the same site: www.communities.gov.uk/publications/fire/regulatoryreformfire

---

Case review

Lessons from Penhallow

Alan Cox

The Penhallow fire has been described as the worst hotel fire for nearly 40 years in the UK, which tragically resulted in the loss of three lives. Should it have even happened in the first place?

In May 2011, O&C Holdsworth plc, the owners of the Penhallow Hotel, were fined £80,000 for breaching fire safety regulations and ordered to pay £62,000 in costs. This followed a fire at the hotel in Newquay in August 2007, which destroyed the building and killed three people. It was described by firefighters as "the worst British hotel fire in 40 years".

The investigation by Cornwall Fire and Rescue Service revealed a number of breaches of fire precautions, the most serious of which related to the fire risk assessment, which had not been carried out in accordance with the Regulatory Reform (Fire Safety) Order 2005 (known as the 'RRO').

The inquest into the fire reached an open verdict. However, evidence was presented to suggest that the fire may have been started deliberately. This was not made available publicly, and there do not appear to have been any witnesses to substantiate arson as the probable cause of the fire.

Conflicting evidence was given at the inquest and even the experts indicated that the only reason arson could be considered the probable cause was simply that they had ruled out other possible sources.

Alan Cox started his career with Warwick County Fire Service in 1963 and has had a long and distinguished career in operational and fire safety. He has produced a number of books and specialist technical videos. He is a qualified Fire Service Inspecting Officer, Member of the International Institute of Risk and Safety Managers (MIIRSM), Tech IOSH and Qualified Fire Investigator.

To some extent the question as to whether it was an accident or arson is academic. If the standard of fire precautions had been to an acceptable level, guests would still have been able to escape safely, unless an accelerant had been used to start the fire (no evidence was produced to indicate this).

In my opinion, incompetence played an important part in this tragic fire. I should state that I did not have any official role in this investigation and the information that I have obtained has been at second hand: speaking to some of the people involved; attending part of the inquest; and responses to my Freedom of Information (FoI) requests. Despite the magnitude of the fire, there has been no official fire investigation report.

There is also a statutory duty to provide a means of escape from fire. A breach of this duty could render the occupier liable for a fine under current law. Employers and controllers of premises remain under a duty to provide routes of escape and emergency exits, which are to be kept clear at all times.

There are also further offences implemented by the RRO, of which the Responsible Person should take account. This includes not taking adequate fire precautions, failure to do a proper risk assessment, or failure to provide adequate means of escape. Penalties could include both a fine and imprisonment.

Codes of practice – means of escape

In addition to the legislative requirements, there are also British Standards or Approved Codes of Practice (AcoP), which contain suggested ways of working towards many safety issues. If a Responsible Person does not comply with an AcoP, it will, in the case of a prosecution, be up to that person to demonstrate that the legal obligations were complied with in some other way.

Special considerations

Particular regard should be given to evacuating disabled people or those with restricted mobility in the event of a fire, and Responsible Persons should consider the potential types of disability and how these may affect any means of escape. The use of wheelchairs or crutches, for example, or the possibility of dealing with persons who have impaired senses, should feature in the fire risk assessment. (See 'Personal Emergency Evacuation Plans (PEEPs)', p.569.)

In addition, there should normally be alternative means of escape from all parts of the workplace, particularly in larger buildings or those housing many people. For this reason it is also important that any different escape routes are independent of one another. The RRO has provided a more appropriate means for the assessment of risks, in particular for those with special needs or where there are children, for example, as is the case with hospitals, children's homes or hostels, taking into account any inexperience, immaturity or lack of awareness of risks. Risk assessments can therefore be adjusted appropriately.

The basic considerations for any means of escape are:

- the likely time available for escape;
- the time needed for escape;
- the number of people to be evacuated;
- the distance required to travel in order to get out;
- the time of day it could be; and
- whether any assistance is required.

Building design

New buildings are often designed to keep fires in 'pockets' of the building and to contain fires in certain areas by using fire doors and other similar equipment. Consideration of the design of a building, whether old or new, is an essential factor when looking at means of escape. Architects are also legally required to consider, eliminate where possible, and then minimise health and safety risks in building design.

There have been some new British Standards published over recent months that have further developed recommendations and guidance on the design, management and use of buildings to achieve acceptable levels of fire safety for all people in and around buildings. BS 9999: 2008 provides a Code of Practice for fire safety in the design, management and use of buildings, and BS 8300: 2009 provides a Code of Practice on the design of buildings and their approaches to meet the needs of disabled people. These, among other relevant requirements, ought to be considered when buildings are designed, altered, extended or the use changed.

Fire doors

Fire doors are intended to keep out and contain fire. They should be fitted with effective self-closing devices, provided they are kept locked and labelled 'Fire Door – Keep Locked Shut'. Some doors may be linked to a fire warning system and activation of a fire alarm or smoke detector may trigger the door to shut or open as appropriate. These should be labelled 'Automatic Fire Door – Keep Clear'.

Emergency lighting

Emergency lighting is another important consideration. It ensures that people can get out safely if there is no natural daylight along escape routes, or if evacuation is at night. Emergency lighting should function notwithstanding the failure of normal lighting and also if there is a localised failure that could cause a hazard. In any event, emergency lighting should:

- show escape routes clearly;
- provide light so people can move along the route safely towards final exits; and
- enable fire alarm call points and firefighting equipment to be readily located.

Emergency exit signs

Emergency exit signs should be placed at exits to floors where stairs begin, at final exits from the premises, and should be illuminated if necessary. There are many precautions that can be taken in building design and used to help reduce the risk of injury from fire, when looking at the means of escape. Regional Fire Authorities' websites can provide useful advice in carrying out fire risk assessments. There are many simple measures that employers can take in providing adequate means of escape from fire as well as ensuring compliance with relevant legislation. Effective means of escape can be provided where an appropriate risk assessment has been carried out and acted upon. The five basic steps to always remember are:

1. Identify fire hazards.
2. Identify people at risk.
3. Evaluate the risks.
4. Record the findings.
5. Review and revise the findings.

See also: Accessible environments, p.18; Disability access and egress, p.238; Fire extinguishers, p.306; Fire safety in non-domestic premises, p.317; Firefighting, p.322; Personal Emergency Evacuation Plans (PEEPs), p.569.

Sources of further information

All organisations must ensure that means of escape to a place of ultimate safety, supported by sufficient numbers of competent persons, are in place to effect an evacuation without the assistance of the Fire and Rescue Services, as stated by the Regulatory Reform (Fire Safety) Order 2005. **PEEPs for Professionals** is a two-day accredited course designed to give students expert knowledge of the law, its implications, the impact of barriers on the evacuation of disabled people, and how to implement effective means of escape for both disabled and vulnerable people. Visit www.workplacelaw.net/training/course/id/78 for more information.

Fire safety in non-domestic premises

Kate Gardner, Workplace Law

Key points

- Historically, fire safety legislation developed in response to particularly tragic fires that resulted in huge casualties. As a result, fire safety provisions were scattered over many pieces of legislation.
- However, fire safety reforms have been in place since 1 October 2006 that have simplified, rationalised and consolidated existing fire safety legislation.
- Since the Regulatory Reform (Fire Safety) Order 2005 came into force, compliance has become easier, with a single fire safety regime that applies to all workplaces and other non-domestic premises.

Legislation

- Management of Health and Safety at Work Regulations 1999.
- Regulatory Reform (Fire Safety) Order 2005.

The Regulatory Reform (Fire Safety) Order 2005 (RRO) and the Management of Health and Safety at Work Regulations 1999 (MHSWR) require employers to carry out a fire risk assessment, and to provide and maintain such fire precautions as are necessary to safeguard those who use the workplace. They also require employees to be provided with relevant information, instruction and training about fire precautions.

Responsible Person

Under the RRO, responsibility for fire safety is that of the 'Responsible Person' for the building or premises. The 'Responsible Person' means, in relation to a workplace, the employer, if the workplace is to any extent under his control.

If the premises are not a workplace, the Responsible Person is the person who has control of the premises (e.g. as occupier)

in connection with the carrying on of his business. The Responsible Person could also be the owner of the premises, if the person in control of the premises does not have control of the business.

The Responsible Person should assess the risks of fire and take steps to remove or reduce those risks. The RRO also imposes obligations in respect of firefighting, fire detection, emergency routes and exits, and procedures to deal with serious and imminent danger.

Regulatory Reform (Fire Safety) Order 2005

The focus of the RRO is an approach based on risk assessment, where the Responsible Person decides how best to identify and address the risks in the premises.

The RRO applies to England and Wales only. As fire safety is a matter within the devolved competence of the Scottish Parliament, the RRO does not extend to Scotland.

The RRO applies to most non-domestic premises used or operated by employers, the self-employed and the voluntary sector. It makes no difference whether the business is operating for profit or not.

The RRO does not apply to domestic premises, offshore installations, ships, aircraft, locomotive or rolling stock, trailers/semi-trailers, mines / boreholes, fields, woods or agricultural land (provided the land is situated away from the undertaking's main buildings).

General fire precautions in relation to premises are covered by the RRO. This includes:

■ reducing the risk of fire and the risk of fire spread on the premises;
■ means of escape from premises;
■ ensuring that, at all material times, the means of escape can be safely and effectively used;
■ means for fighting fires;
■ means for detecting fire and giving warning in the case of fire; and
■ arrangements for action to be taken in the event of fire, including measures for the instruction and training of employees and to mitigate the effects of fire.

If, however, further hazards or potential risks are identified, these should equally be dealt with as part of the fire risk assessment.

Relevant Persons
People who are owed duties under the RRO (e.g. employees, tenants) are known as Relevant Persons. A Relevant Person is any person who may lawfully be on the premises or any person in the immediate vicinity of the premises who is at risk from fire.

General duties
The fire safety regime is based on an assessment of risk, the removal of hazards and the protection of persons from any hazards that remain. Its aim is to seek to prevent fires from starting and to mitigate the effect of fire.

In respect of employees, the Responsible Person owes a duty to take such general fire precautions as will ensure (so far as is reasonably practicable) their safety. In respect of Relevant Persons who are not employees (e.g. visitors, contractors etc.), the duty is to take such general fire precautions as may reasonably be required, in the circumstances, to ensure that the premises are safe.

Risk assessment and specific duties
To comply with the specific obligations imposed by the RRO, the Responsible Person must decide what general fire precautions he needs to take. That person must ensure that a suitable and sufficient assessment of risks to Relevant Persons is undertaken. This will be different for every different premise.

Hazards that are identified should be removed or reduced, so far as is reasonably practicable, and any residual risks avoided. For those premises where a dangerous substance is liable to be present, the risk assessment must also address ten specific considerations detailed in Part 1 of Schedule 1 of the Order.

Special consideration must also be given to people under the age of 18. This specifically includes considering their inexperience, lack of awareness of risks, immaturity and the extent of safety training provided to the young person.

The purpose of the risk assessment is to ensure that the risk of fire is either removed or reduced as far as possible. It may be the case that a risk of fire remains after initial measures have been taken and protection is necessary to safeguard Relevant Persons due to the features of the premises, the activity carried on there, any hazard present or any other relevant circumstances. In such cases, protection can be provided by appropriate fire precautions in the form of:

- means for detecting and warning of fire (Article 13);
- means for fighting fire (Article 13); and
- means of escape through emergency routes and exits (Article 14).

The premises (and any facilities, equipment and devices provided in connection with general fire precautions) must be subject to a suitable system of maintenance. They must be maintained in an efficient state, in efficient working order and in good repair (Article 17).

There must also be procedures (including safety drills) to be followed in the event of serious and imminent danger, with a sufficient number of competent persons nominated to implement those procedures (Article 15).

Most of the time, the arrangements for the effective planning, organisation, control, monitoring and review of these preventive and protective measures must be in writing. This applies where the Responsible Person has five or more employees, a licence required by any enactment is in force in relation to the premises or an alterations notice is in force (see below).

Enforcement

The main enforcing body is the local fire and rescue authority, although the HSE will enforce the RRO in respect of nuclear installations, construction sites and ships under construction or repair.

Enforcement notices

The Enforcing Authority (through the service of enforcement notices) will be able to require work to be carried out to ensure that people are safe (Article 30).

Prohibition notices

Where the use of premises involves or may involve serious risks to Relevant Persons, the Enforcing Authority may prohibit or restrict the use of the premises by serving a prohibition notice, which will usually stay in force until such time as the premises have been made appropriately safe (Article 31).

Alterations notices

Certain premises may pose a serious risk to Relevant Persons due to the premises' individual features, such as the purposes for which they are used and their particular hazards. A building may also develop to house new practices as a business changes. In either case, the Enforcing Authority may serve an alterations notice on the Responsible Person (Article 29) if it considers that the fire risk assessment currently in place has not been revised to take account of those developments.

An alterations notice will require the Enforcing Authority to be notified, in advance, of:

- changes to the premises;
- changes to services, fittings or equipment in or on the premises;
- an increase in the quantities of dangerous substances that are present in or on the premises; or
- changes to the use of the premises.

Offences

It is an offence to fail to comply with the requirements of the RRO, an enforcement, prohibition or alterations notice. It is also an offence to obstruct an inspector in the exercise of his powers. The level of penalty can be compared with existing fire safety law. For example, in respect of the main duties imposed by the RRO, upon summary conviction a fine not exceeding

£20,000 may be imposed. Upon conviction on indictment, the penalty is an unlimited fine, or imprisonment for a term not exceeding two years, or both.

Fire risk assessments

The purpose of a fire risk assessment is to identify the measures that need to be taken to comply with Part 2 of the RRO.

There are five steps that a Responsible Person will need to take:

1. *Identify potential fire hazards in the workplace.* This involves identifying potential sources of ignition such as smokers' materials, naked flames, electrical, gas or oil-fired appliances, machinery, faulty electrical equipment, static electricity, potential arson, etc. The Responsible Person should then identify sources of fuel (combustible materials) including flammable substances, wood, paper and card, plastics, rubber, foam and flammable gases (such as liquefied petroleum gas). The assessment must consider the workplace as a whole, including all work processes, outdoor locations and areas that are rarely used or visited.

2. *Decide who might be in danger (e.g. employees, visitors, disabled persons).* The assessment must identify who is at risk in the event of fire, how they will be warned and how they will escape. Locate where people may be working (whether at permanent or occasional workstations) and consider who else is at risk such as customers, visiting contractors or disabled people. These individuals may be unfamiliar with your fire precautions and will be at higher risk.

3. *Evaluate the risks arising from the hazards and decide whether you have done enough to reduce the risk or need to do more.* You should consider:

- the likelihood of fire occurring and whether it is possible to reduce the sources of ignition or minimise the potential fuel for a fire;
- the fire precautions you have in place and whether they are sufficient for the remaining risk and will ensure everyone is warned in case of fire (i.e. fire resistance and structural separation, fire detection and warning systems);
- the means by which people can make their escape safely or put the fire out if it is safe to do so (i.e. means of escape and means of fighting fire); and
- maintenance and testing of fire precautions to ensure they remain effective.

4. *Record the findings.* Where you employ five or more employees, you must record the significant findings of your assessment, together with details of any people you identify as being at particular risk.

5. *Keep the assessment under review and revise it when necessary.* Changes to the workplace that have an effect on either fire risks or precautions should trigger a review of your risk assessment. Examples that may lead to increased risks or new hazards include changes to work processes, furniture, plant, machinery, substances, building layout or the numbers or classes of people likely to be present in the building.

It is important to note that in addition to duties under the RRO, duties also exist in relation to conducting risk assessments in respect of an organisation's business and care should be taken that any findings implemented are compatible with the fire safety measures put in place.

See also: Disability access and egress, p.238; Fire extinguishers, p.306; Fire, means of escape, p.314; Firefighting, p.322; Personal Emergency Evacuation Plans (PEEPs), p.569; Risk assessments, p.624.

Sources of further information

The Regulatory Reform (Fire Safety) Order 2005:
www.workplacelaw.net/news/display/id/10738

Whether you are a small or large organisation, fire and its consequences pose a serious and significant risk to your business. The message from the Regulatory Reform (Fire Safety) Order 2005 and the fire authority is clear – the management of fire safety falls to the responsible person (e.g. an employer, a manager, a landlord or tenant, a charity or voluntary organisation). Under the RRO the Responsible Person must ensure that a 'suitable and sufficient' fire risk assessment for each of their premises has been carried out by a competent person. For low risk premises, many companies will be able to train an individual to take on the role of competent person from within their own organisation. Workplace Law's *Fire Risk Assessment course* is designed to meet the training needs of such persons. Visit http://fire-safety-training.workplacelaw.net for more information.

Firefighting

Kate Gardner, Workplace Law

> **Key points**
> - Since 1 April 2006, employers, or those who have control of non-domestic premises, have had a statutory duty to ensure that there are appropriate means of fighting fires. The employer or controller of the non-domestic premises is known as the 'Responsible Person'.
> - The Responsible Person also has to provide suitably trained people to operate non-automatic firefighting equipment (FFE).
> - It is up to the Responsible Person to identify the firefighting requirement for its premises as part of the fire risk assessment it is obliged to conduct and maintain in relation to those premises.

Legislation

- Provision and Use of Work Equipment Regulations 1998.
- Management of Health and Safety at Work Regulations 1999.
- Fire and Rescue Services Act 2004.
- Regulatory Reform (Fire Safety) Order 2005.
- Equality Act 2010.

Firefighting systems

FFE can be either active or passive, and is designed to extinguish fires by:

- starving the fire of fuel;
- smothering the fire to remove or reduce the concentration of oxygen; or
- cooling the fire to the extent that it cannot support combustion.

Active systems

Active systems are those such as sprinklers, gas floods and portable fire extinguishers, which become active only when a fire occurs or with human intervention.

Passive systems

Passive systems form part of the structure of the building and involve the division of the building into fire-resisting compartments.

Where compartments are breached to allow movement through the building or the provision of services, 'fire stops' can maintain the integrity of the fire resistance of the compartment.

Breaches in compartment walls for services etc., should be kept to a minimum and all openings should be protected with intumescent materials (i.e. pipe wraps, seals, etc.) or fire dampers.

All measures taken should comply with BS 9999: 2008 (Code of practice for fire safety in the design, management and use of buildings).

Firefighting equipment – statutory requirements

The Responsible Person has responsibility to determine what FFE is required and to provide and maintain appropriate FFE in those premises.

Non-automatic equipment must be placed so that it is easily accessible to those who need to use it, and approved signs must indicate its location.

As part of the risk assessment, the Responsible Person may have to consider the provision of additional FFE to assist with the evacuation of people from the premises and ensure there are competent people to use FFE. When assessing the requirements, the Responsible Person could consult his local Fire and Rescue Service for specific advice.

FFE also falls within the requirements of the Provision and Use of Work Equipment Regulations 1998 (PUWER), which prescribes the following:

■ Equipment should be suitable for the purpose for which it is provided, bearing in mind the risks posed by the activities and people in that work area. The equipment provided should take account of the activities and the number of people working in or using an area, as well as the environmental conditions (e.g. adverse weather conditions or the presence of corrosive chemicals).

■ Equipment should be capable of being used only for the purpose for which it is provided. It is essential that the correct type of firefighting equipment is provided (e.g. no water-based extinguishers where there is electrical equipment) and if equipment is used correctly it must not increase the risks to the user or others.

■ Equipment should be regularly inspected and maintained in an effective state. Inspection procedures should take account of the likelihood of equipment being damaged, moved or abused, and the need for more frequent maintenance than that specified by the manufacturer should be considered as part of a risk assessment.

■ People who may have to use the equipment should receive adequate information and instructions in its use. This is particularly important as, unless they are confident in the equipment's capabilities and limitations, people may fail to use it or use it incorrectly in a fire situation, putting themselves and others in danger.

Types of firefighting equipment

Within the workplace there will be different types of firefighting equipment. All workplaces must be equipped to fight fire.

In the event of a fire, staff should only attempt to fight the fire if they are suitably trained and it is safe to do so. If the fire can be extinguished, that risk to others will be removed.

Portable fire extinguishers

These are the most common types of firefighting equipment found in all workplaces. Each extinguisher must be correctly colour-coded. Currently within the UK there are two types of colour coding.

Prior to 1997, all extinguishers were uniquely colour-coded. Since January 1997, EC standards have dictated that all new extinguishers must now have red-coloured bodies, with a 5% band that is colour-coded according to the extinguisher's contents. Red/Black means red body, black banding. See *'Fire extinguishers'* (p.306).

Maintenance

Article 17 of the RRO requires the Responsible Person to ensure that a suitable 'system of maintenance' is in place, to maintain FFE in an efficient state, good working order and in good repair. As a minimum, maintenance of FFE should be undertaken in accordance with the manufacturers' instructions.

Training

Article 21 of the RRO, Regulation 13 of the Management of Health and Safety at Work Regulations 1999, and Regulation 9 of PUWER require employees (and others who may need it) to be provided with adequate training prior to being entrusted to use FFE.

Disability discrimination

Sections 6 and 21 of the Disability Discrimination Act 1995 (now superseded by the Equality Act 2010) require employers to make reasonable adjustments to the workplace to facilitate the use of work equipment, etc., by disabled people.

As part of their risk assessment, employers will have to consider the needs of disabled employees and others with regard to the type, size, location and siting of FFE and, where necessary, they will have to make 'reasonable adjustments' to ensure the safety of their employees and visitors.

Considerations may include re-siting, or reducing the size of, fire extinguishers so that wheelchair users can easily access them.

In addition, the following Code of Practice should be consulted: BS 8300 (Design of buildings and their approaches to meet the needs of disabled people).

See also: Fire extinguishers, p.306; Fire, means of escape, p.314; Fire safety-in non-domestic premises, p.317.

Sources of further information

The Regulatory Reform (Fire Safety) Order 2005 came into force in October 2006 and with it the requirement to have a strategy to evacuate all the occupants within a building. The fire service cannot be relied upon to assist with evacuation. As such, businesses are recognising the importance of having fully trained fire wardens as part of their emergency evacuation strategy. Having fire wardens has shown to be the quickest, most effective way to evacuate a building. It is also seen as a proactive approach – fire wardens can be used to identify fire hazards and dangers before a fire occurs. Workplace Law's *Fire Warden Training course* can include a practical fire extinguisher session if required. Visit www.workplacelaw.net/training/course/id/15 for more information.

First aid

Andrew Richardson, URS Scott Wilson

Key points

- The Health and Safety (First Aid) Regulations 1981 require employers to provide adequate and appropriate equipment, facilities and personnel so that first aid can be given to their employees if they are injured or become ill at work.
- When people at work are injured or fall ill, they must receive immediate attention.
- An ambulance must be called in serious cases.
- Employers need to assess what their first aid needs are.
- The minimum first aid provision on any worksite is:
  - a suitably stocked first aid box; and
  - an appointed person to take charge of first aid arrangements.
- First aid provision needs to be available at all times people are at work.

Legislation

- Health and Safety at Work etc. Act 1974.
- Health and Safety (First Aid) Regulations 1981.
- Offshore Installations and Pipeline Works (First Aid) Regulations 1989.
- Management of Health and Safety at Work Regulations 1999.

In October 2009 the HSE introduced updated training and approval arrangements for workplace first aiders. The second edition of the Approved Code of Practice and guidance for the Health and Safety (First Aid) Regulations 1981 (see *Sources of further information*) gives details of a new training regime, which gives employers more flexibility in deciding their first aid provision and recommends refresher training annually so that first aiders can maintain their skills.

Requirements and guidance

The aim of first aid is to preserve life and to reduce the effects of injury or illness suffered at work. The Regulations place a duty upon all employers to assess the first aid requirements within their workplace, to appoint competent personnel, and to provide equipment and facilities to enable first aid to be given to their employees if they are injured or become ill at work.

To assess the first aid requirements within each workplace, the employer should consider:

- the hazards and risks in the workplace;
- how many people are located in their premises;
- previous accident / incident history;
- the nature and distribution of the workforce;
- the remoteness of the workplace in relation to emergency medical services;
- the needs of travelling, remote and lone workers;
- employees working on shared or multi-occupancy sites; and
- annual leave and other absences of first aiders and appointed persons.

It should be noted that first aid provision for anyone other than employees, including the public, cannot be made under the Health and Safety at Work etc. Act 1974. The HSE strongly recommends employers to consider it, but employers should be aware that employer's liability insurance will not cover litigation as a result of first aid to non-employees; however, public liability insurance may.

The Regulations stipulate that first aid equipment must be provided. The minimum requirement is a first aid box, the contents of which will depend on the risks identified in the risk assessment. The HSE publishes guidance on suitable contents of first aid boxes (*see below*).

There is no standard list of items to put in a first aid box. It depends on what you assess the needs are. However, as a guide, and where there is no special risk in the workplace, a minimum stock of first aid items would be:

- a leaflet giving general guidance on first aid, e.g. the HSE's Basic advice on first aid at work;
- 20 individually wrapped sterile adhesive dressings (assorted sizes);
- two sterile eye pads;
- four individually wrapped triangular bandages (preferably sterile);
- six safety pins;
- six medium-sized (approximately 12cm x 12cm) individually wrapped sterile unmedicated wound dressings;
- two large (approximately 18cm x 18cm) sterile individually wrapped unmedicated wound dressings; and
- one pair of disposable gloves.

You should not keep tablets or medicines in the first aid box.

First aid rooms must be provided where the risk assessment has identified they are necessary, usually in high-risk working environments. The guidance details what is reasonably expected to be provided in such a room.

When the employer assesses the risk and identifies the need for personnel to give first aid, then sufficient numbers of people of the appropriate competency, in the appropriate locations, need to be arranged. The guidance offers suggestions as to the competency and numbers of personnel required, based on whether the workplace is low, medium or high risk, and how many people are employed at the site.

The competency of personnel falls into two categories – first aiders and appointed persons. The Regulations detail the criteria that each of these positions must meet, including qualifications and training – the HSE only specifies training for first aiders.

The HSE's new training regime (2009) identifies two levels of training:

1. First Aid at Work (FAW).
2. Emergency First Aid at Work (EFAW).

FAW qualified personnel are able to provide a higher level of first aid than EFAW qualified personnel. Organisations offering this training must be approved by the HSE.

The Regulations place a duty upon the employer to inform all its employees of the arrangements made in connection with first aid. The guidance suggests the setting-up of a procedure for informing staff that would include the details of the first aid provision and how employees are told of the location, equipment, facilities and personnel. It also suggests the provision of notices that are clear and easily understood to relay all or some of the information. Finally, it suggests that first aid information is included in induction training to ensure all new employees are aware.

See also: Accident investigations, p.29; Workplace deaths, p.218; Health and safety at work, p.361.

Sources of further information

L74 First Aid at Work: www.hse.gov.uk/pubns/priced/l74.pdf

L123 Health care and first aid on offshore installations and pipeline works: www.hse.gov.uk/pubns/priced/l123.pdf

INDG 214 First Aid at Work – Your Questions Answered: www.hse.gov.uk/pubns/indg214.pdf

INDG 347 Basic Advice on First Aid at Work: www.hse.gov.uk/pubns/indg347.pdf

The Health and Safety (First Aid) Regulations 1981 require employers to identify the requirements for first aid within their workplaces. This includes the provision, where appropriate, of fully qualified first aiders, i.e. those who have successfully undertaken an HSE-approved four-day First Aid at Work course.

Workplace Law's *First Aid at Work training course* meets the legislative criteria and successful candidates will be awarded an HSE-approved certificate that is valid for a period of three years. Visit www.workplacelaw.net/training/course/id/17 for more information.

Fit notes

Sophie Applewhite and Chloe Harrold, Loch Associates Employment Lawyers

Key points

- Officially known as a Statement of Fitness for Work, the fit note was introduced on 6 April 2010.
- This followed the recommendations of a major review carried out by Dame Carol Black, National Director for Health and Work, in response to growing concerns that the sick note system wasn't working.
- Employers had complained that the sick note system was too inefficient, as all it required GPs to do was state whether an employee on sick leave for longer than seven calendar days was 'unfit for work'.
- The fit note is designed to enable a gradual return to work for employees who have been on long-term sick leave, as well as cutting the cost of sickness absence to employers.
- It provides for two outcomes – a patient can be declared 'unfit for work', or 'may be fit for work'. GPs can then advise employers on ways that employees can be helped – by a reduction in hours, for example, changes to duties, or an adaptation to their working environment.

Legislation

- Social Security (Medical Evidence) and Statutory Sick Pay (Medical Evidence) (Amendment) Regulations 2010.

Introduction

The 'fit note' regime was introduced on 6 April 2010, despite concerns raised during consultation that stakeholders would not have sufficient time to familiarise themselves with the new rules.

The main changes to the pre-existing 'med 3' sick note regime include:

- The fit note allows a GP to select one of two options: 'You are not fit for work' or 'You may be fit for work taking account of the following advice'. It was hoped that this option would facilitate discussion between the employer and employee about suggested changes on return to work.

- The fit note lists four standard changes that could be made to an employee's work environment to help accommodate their return to work. The fit note also gives doctors the opportunity to detail alternative changes should this be appropriate.
- Within the first six months of absence, the maximum duration a fit note can be issued for is three months. After the first six months, a fit note can be issued for 'any clinically appropriate period'.
- If an employer is unable to facilitate a change or an adjustment, the detail given on the fit note is sufficient evidence that the employee has a condition that prevents him or her carrying out the role. It is not necessary for the doctor to issue a revised fit note.

The fit note is designed to enable a gradual return to work for employees who

---

Facts

■ It is estimated that the fit note will save £240m within ten years.
■ Six months on, 70% of GPs believed the fit note had made a positive difference in helping their patients back to work, with 48% stating the fit note helped them increase the occurrences of recommending a return to work.
■ The UK economy lost 190 million working days to absence in 2010, an average of 6.5 days per employee, an increase on 6.4 days in 2009.

Sources: Department for Work and Pensions (DWP), Confederation of British Industry and Pfizer Absence and Workplace Health Survey.

have been on long-term sick leave, as well as cutting the cost of sickness absence to employers. Importantly it does not alter the current self certification regime and employees are still entitled and required to self-certify for the first seven days of sickness absence.

Reaction to the regime
Concerns were initially expressed regarding the practicality of doctors providing a detailed assessment in respect of an employee's workplace when they have no knowledge of the workplace in question. Doctors are not specifically trained in occupational health and will only have the patient's explanation of their role and the workplace to rely on. This may not be comprehensive enough for the doctor to provide a proper assessment and, if so, employers may struggle to implement measures that are not detailed enough to apply to specific workplaces. Research published six months after the introduction of the fit note suggested that a staggering 89% of GPs had not received training in health and work within the past 12 months.

There have also been concerns that employees are being forced to return to work before they are ready, which may cause their health to deteriorate. This could arise from a misunderstanding of the employee's role and workplace on the part of the doctor. It is yet to be established if a doctor would be liable for the employee's deterioration in health in this situation.

Statistics published in July 2010 by the DWP showed that three-quarters of applicants for Employment and Support Allowance (ESA) made between October 2008 and November 2009 were found either fit for work or ceased their application before assessment was completed. Chris Grayling, Minister for Employment, stated that:

'*The vast majority of people who are applying for these benefits are being found fit for work. These are people who under the old system would have been abandoned on incapacity benefits*'.

Under the new fit note regime, it was hoped that applicants for ESA will now be given specific support and suggestions to assist with their return to work. However, DWP research carried out six months after the introduction of the fit note suggested that only 23% of GPs felt their knowledge of the benefits system was up to date.

A positive result?
Early indications were that the implementation of the fit note regime

would do little to convince doctors and employers that it will assist in reducing long-term absences. Doctors indicated that they would feel forced to take on an occupational health role and employers would be expected to facilitate suggestions from doctors that did not reflect an understanding of the practicalities of specific workplaces.

However, the DWP research suggests that in fact GPs consider fit notes to have made a positive difference. Almost half of the GPs surveyed said the fit note regime has made them more likely to suggest a return to work as an aid to recovery.

However, at a welfare reform conference in March 2011, Lord Freud criticised the effectiveness of the regime. The Welfare Reform Minister stated that many GPs are refusing to complete the fit notes and that there is an inherent contradiction in requiring GPs to complete fit notes. GPs are supposed to be advocating the health of their patient, but instead are required to become 'policemen' by completing the fit note. Lord Freud went on to suggest that many long-term sickness absences are linked to the workplace environment itself, making it very difficult for doctors to establish the real barrier to a recovery and return to work. Although there were hopes that the fit note would become electronic by the end of 2010, this has not transpired, and Lord Freud stated that although there were plans to establish an e-fit note, he could not elaborate on when this might happen.

Conclusion

Despite the encouraging results of the DWP six-month GP survey, figures concerning sickness absence show an increase for 2010. It may be that the impact of fit notes will be more long-term and that with some finessing of the regime, greater effectiveness will follow.

The DWP has invited members of the Chartered Institute of Personnel and Development to provide feedback on fit notes intended to survey the success of the scheme, from an employer's point of view. Responses are invited from organisations in the private, public and voluntary sectors that have had at least one fit note since last April. It will be interesting to see how the results of this compare to the survey of GPs carried out.

The fit note regime is still in its early stages but its operation so far appears to have gone slightly better than expected when first introduced. It remains to be seen whether fit notes will stand the test of time and further scrutiny.

See also: Alcohol and drugs, p.35; Health and safety at work, p.361; Health surveillance, p.396; Mental health, p.482; Occupational health, p.519; Stress, p.659; Workplace health, safety and welfare, p.768.

Sources of further information

The Department for Work and Pensions website has a page dedicated to the fit note which provides a 'fit note explained' guide and provides links for further information for healthcare professionals, employers and to employees and patients: www.dwp.gov.uk/fitnote

Forklift truck safety

David Ellison, The Fork Lift Truck Association (FLTA)

Key points

- There are many different types of forklift truck and it is really important that the right truck is used for the task to be performed.
- Forklift trucks can be very dangerous. They are involved in more serious incidents than any other form of workplace transport.
- Managers have a formal responsibility for planning their lifting operations, training the operators and others involved in these operations, maintaining and regularly inspecting the equipment to be used.

Legislation

- Lifting Operations and Lifting Equipment Regulations 1998 (LOLER 98).
- Provision and Use of Work Equipment Regulations 1998 (PUWER 98).

The term 'forklift truck' can be used to include a small hand pallet truck, which hardly lifts at all, through to container handling equipment, which may lift more than 50 tonnes many metres in order to stack loaded containers. There are many types of equipment involved, each with their own peculiar hazards. Some are fitted with highly specialist attachments and others carry their operator high up into dedicated racking systems. Some must operate on ultra-smooth surfaces and others carry loads over the roughest terrain. However, the same basic rules and regulations apply to all of them – and their managers.

General requirements

'If you are an employer you have a duty to ensure that work equipment provided for your employees comply with PUWER 98.' (PUWER 98 Regulation 3)

'Every employer shall ensure that work equipment is so constructed or adapted as to be suitable for the purpose for which it is used.' (PUWER 98 Regulation 4)

For forklift trucks, particular considerations include the following.

- *Ventilation.* This is important for the selection of motive power and operations in confined spaces. (PUWER 98 Regulation 4)
- *Carriage of employees.* Only suitable equipment may be used for carrying people. (PUWER 98 Regulation 25)
- *Falling Object Protection (FOPS).* Most forklift trucks will require FOPS, commonly called overhead load guards. They protect the operator from falling objects. (PUWER 98 Regulation 25)
- *Roll-Over Protective Structures (ROPS).* Forklift trucks fitted with a vertical mast (most standard counterbalance and reach trucks) do not need specific ROPS. Other types of truck, such as variable reach trucks, do need ROPS. (PUWER 98 Regulation 27)
- *Restraining systems.* Restraining systems, usually in the form of seat belts, are fitted to help keep the operator in the cab in the event of

an overturn, hence avoiding 'mouse-trapping' – where an operator is caught between the truck structure and the ground. Restraining systems are usually required to be fitted to counterbalance trucks, but not reach trucks. There is no legal requirement for sear belts to be worn. Instead, it is a matter for risk assessment and employer policy. However, the HSE recommends that seat belts, where fitted, should always be worn. (PUWER 98 Regulation 27 and HSE Information Sheet MISC 241)

■ *Fitting of lights, flashing beacons, reversing alarms and mirrors.* There is generally no absolute requirement for such items to be fitted, but all of these items may be significant aids to safety. (PUWER 98 Regulations 24 and 28)

There is a need for compliance with the Machinery Directive and for equipment to be CE-marked. (PUWER 98 Regulation 10)

Safe operating

'Every employer shall ensure that every lifting operation involving lifting equipment is properly planned by a competent person, appropriately supervised, and carried out in a safe manner.' (LOLER 98 Regulation 8)

'Every employer shall ensure that all persons who use work equipment have available to them adequate health and safety information and, where appropriate, written instructions pertaining to the use of the work equipment.' (PUWER 98 Regulation 8)

It is particularly important to ensure that handbooks are provided when used forklift trucks are purchased.

Good practice

The HSE publication 'Safety in Working with Lift Trucks' (HSG 6) has a short section covering the law but is not, in itself, a regulation-based document. It does, however, offer sound advice based on good practice. This publication is likely to be absorbed into L117 during 2012 (*see below*).

Medical standards for lift truck operators

HSG 6 provides advice to medical and occupational health professionals about the medical fitness of operators of forklift trucks.

De-rating

Where appropriate, the safe working load of the lifting equipment should be reduced to take into account the environment and mode in which it is being used (LOLER 98 Regulation 8). With forklift trucks, de-ration will be necessary when any attachment is fitted – even, for example, a temporary work platform for lifting people. A revised working plate should be fitted. De-ration can only be carried out by the manufacturer of the truck or a formal representative.

Working platforms

The use of working platforms on forklift trucks is covered by a number of regulations. These are brought together, with associated guidance, in the HSE Guidance Note PM28 (*see Sources of further information*).

Public roads

The use of forklift trucks on public roads, which may include forecourts, etc., is covered by a number of Regulations, which can be difficult to source. There are also issues concerning driving licences. FLTA Technical Bulletin 03, available from www.fork-truck.org/publications, brings together all of this information.

Maintenance

'Every employer shall ensure that work equipment is maintained in an efficient

state, in efficient working order and in good repair.'
(PUWER 98 Regulation 5)

It is recommended that a formal maintenance contract is taken out to cover this requirement.

Equipment may need to be checked frequently to ensure that safety-related features are functioning correctly. The frequency of maintenance will depend on the intensity of use (normally measured in operating hours), the working environment and type of operation. This should be discussed with the maintenance provider when agreeing the maintenance contract. If any of the main factors change, the frequency of maintenance may also need to be changed.

Pre-use checks
'You should ensure that employees have appropriate training and instructions so that they are able to ensure that the lifting equipment is safe to use.'
(LOLER 98 Regulation 8)

This requirement is usually formalised in daily or pre-shift checks. A number of systems are readily available to assist with this.

Records of maintenance and pre-use checks should be retained so that documents covering six to 12 months are available.

Inspection and thorough examination
Thorough examination is not part of maintenance and will not usually be included in a maintenance contract. Thorough examination is a legal requirement similar to an MOT test for a car or lorry. However, the frequency of examination and other factors are not straightforward.

'Every employer shall ensure that work equipment exposed to conditions causing deterioration, which is liable to result in dangerous situations, is inspected and thoroughly examined.'
(LOLER 98 Regulations 6 and 9)

In this context, use of the equipment causes deterioration. Most forklift trucks will require regular thorough examinations under LOLER. Items such as hand pallet trucks will require regular safety inspections under PUWER.

See 'Lifting equipment' (p.448) for the general requirements for thorough examinations.

The way in which LOLER 98 relates to forklift trucks is not always easy to work out. A sound, clear and comprehensive guide is available at www.thoroughexamination.org.

Training
The need to train operators of forklift trucks is obvious, but not always followed. Training for operators of hand pallet trucks is also needed as the improper use of such equipment causes many debilitating injuries. Managers and supervisors of operators also need training so that they can recognise and correct poor practice.

Perhaps less obvious is the need to provide training for other employees who have to enter areas where forklift trucks are operating. A high proportion of deaths and serious injuries involve other employees and visitors who are hit by forklift trucks or their loads.

'Every employer shall ensure that all persons who use work equipment have received adequate training for purposes of health and safety, including training in the methods which may be adopted when using the work equipment, any risks which

such use may entail and precautions to be taken.'
(PUWER 98 Regulation 9)

Advice on training for forklift truck operators is provided in the HSE Approved Code of Practice (L117). This includes the selection of people for training, the types of training required, the importance of training records and details of the accrediting bodies approved by HSE.

Refresher training

'There is no specific requirement to provide refresher training after set intervals, but even trained and experienced lift truck operators need to be re-assessed from time to time.'
(L117)

Specific instances where re-assessment leading to refresher training may be required include where operators may not have used trucks for some time, are occasional users, appear to have developed unsafe working practices, have had an accident or near miss, or where there is a change in their working practices or environment.

See also: Construction site health and safety, p.177; Lifting equipment, p.448; Vehicles at work, p.686; Work equipment, p.752.

Sources of further information

Safety in Working with Lift Trucks – HSG 6:
www.hse.gov.uk/pubns/books/hsg6.htm

Rider-operated Lift Trucks: Operator Training. Approved Code of Practice and Guidance – L117: www.hse.gov.uk/pubns/books/l117.htm (This publication is currently under review. It is likely to be combined with HSG 6 and re-published in 2012.)

Fitting and Use of Restraining Systems on Lift Trucks – HSE Information Sheet MISC 241: www.hse.gov.uk/pubns/misc241.pdf

Working Platforms (Non-Integrated) On Fork Lift Trucks – HSE Guidance Note PM 28: www.hse.gov.uk/workplacetransport/pm28.pdf

Thorough examination of forklift trucks: www.thoroughexamination.org

Fork Lift Truck Association: www.fork-truck.org.uk

Fuel storage

Sally Goodman, Bureau Veritas UK Ltd

Key points

- The Control of Pollution (Oil Storage) (England) Regulations 2001 Regulations apply to most industrial, commercial, and institutional sites storing more than 200 litres of oil, and private dwellings storing more than 3,500 litres.
- The Regulations set design standards for fuel tanks, to minimise risk of leakage and spillage.
- The Regulations do not apply:
    - if the oil is waste oil according to the Environmental Permitting (England and Wales) Regulations 2010;
    - if the container is situated within a building or wholly underground;
    - if the premises are used for refining oil, or for the onward distribution of oil to other places; and/or
    - on any farm if the oil is used for agricultural purposes.

Legislation

- Building Regulations 2000.
- Control of Pollution (Oil Storage) (England) Regulations 2001.
- Environmental Permitting (England and Wales) Regulations 2010.

Guidance

The Environment Agency produces a number of Pollution Prevention Guidance notes (PPGs), each of which is targeted at a particular industrial sector or activity, and gives advice on the law and good environmental practice. PPGs are valuable sources of advice for industry and the public. Those relevant to the Regulations considered here are:

- PPG2: Above Ground Oil Storage Tanks; and
- PPG8: Safe storage and disposal of used oils (see *Sources of further information*).

There is also extensive guidance on the Environment Agency website: www.environment-agency.gov.uk/business/topics/oil/default.aspx

Overview

Oil is toxic to plants and animals, can damage rivers and groundwaters, and destroy natural habitats and drinking water supplies. Spills caused by inadequate storage facilities and/or poor operating practices are common, and very expensive to remediate. Most incidents are caused by oil leaking from tanks or pipework, or tanks being over-filled. In order to address this problem, the Government imposed minimum design standards for fuel tanks via the Control of Pollution (Oil Storage) Regulations 2001, which are now mandatory for all fuel storage facilities.

Oil and fuels are the second most frequent type of pollutant of inland waters in England and Wales that are reported to the Environment Agency.

Reported oil pollution incidents have halved compared to ten years ago, since the introduction of the Oil Storage Regulations in England. There are now around 3,000 pollution incidents involving oil and fuels every year. Although some of

Case studies

In June 2010 a food manufacturer was fined £23,500 and ordered to pay £7,950 costs after admitting two charges that resulted in the pollution of the River Avon. Worcester Magistrates Court heard that Environment Agency officers found large pools of oil on the river near Evesham. Investigations confirmed that the company was the source of rapeseed oil entering the watercourse. The company reported a spill of about 5,000 litres of rapeseed oil at its premises and said a bund around the oil tank had not contained the leak. As a result, about 800 litres entered the River Avon. Environment Agency officers noted that the oil had escaped from a storage tank via a flexible pipe, fixed in place with a jubilee clip, which had become detached. The oil had leaked into the company's surface water drains and discharged into the surface sewer and ultimately into the River Avon. An Environment Agency spokesperson said:

"The company had poor knowledge of their own site drainage. They did not have a plan of what to do in the event of a spillage at the site and, unfortunately, this resulted in the pollution incident."

In 2010, a Somerset cellophane manufacturer was fined £30,000 and faced a £1m clean-up bill after gas oil spilled from a leaking storage tank. The tank corroded after being converted from storing fuel oil to gas oil, and the oil leaked into a containment bund. The bund, designed for the heavier fuel oil, was unsuitable for the gas oil and oil soaked into the ground.

In 2004 the River Loddon was polluted by thousands of litres of oil from a Wokingham district school. Swans and other wildlife were poisoned after the oil drained out of a defective storage tank at Maiden Erlegh School in Earley and ran to the river a mile-and-a-half away. Wokingham District Council, which runs the Silverdale Road comprehensive, was fined £18,000 over the leak, plus another £3,327.07 in costs and compensation. Investigators traced the oil back to Maiden Erlegh School, where there were two 2,500 litre storage tanks for heating oil. They believed a sight gauge, a plastic tube clipped to the tanks used to see the level of the oil, fell over and the oil had drained out of one because a switch was not properly closed off. A bund then failed to contain the leaking oil and it found its way into Thames Water's surface water system and got pumped into the river. The sight gauges were in fact redundant because they had been superseded by dial gauges and should have been replaced. It is believed that the tube flopped down because it softened in 30 degree-plus temperatures in the summer.

these affect land, the vast majority affect the water environment. On average an oil spill costs a typical business up to £30,000 in fines, clean up charges and production losses.

Failure to comply with the Regulations is a criminal offence and may lead to prosecution. It is important to bear in mind that in almost every case there would be even larger penalty costs relating to returning the polluted area back to pre-incident levels; this would include re-stocking rivers etc. The Environment Agency and other related bodies will also pass on their investigation costs.

Similar legislation exists in Scotland and Northern Ireland. The Water Environment (Oil Storage) (Scotland) Regulations 2006 apply in Scotland. These Regulations are different to the Oil Storage Regulations England and if you store oil in Scotland you should check the Scottish Environment Protection Agency (SEPA) website for details. The Control of Pollution (Oil Storage) Regulations (Northern Ireland) 2010 apply in Northern Ireland. They have a phased implementation and have different requirements to the Oil Storage Regulations England. If you store oil in Northern Ireland you should check the Northern Ireland Environment Agency (NIEA) for details. At the present time there are no equivalent regulations for Wales.

Storage facilities subject to the Regulations

The Regulations apply to all external, above-ground containers of more than 200 litres' capacity used for the storage of fuel oil and petrol. They do not apply to private dwellings with a capacity of less than 3,500 litres, petrol stations, refineries, farms, or tanks situated entirely within a building or wholly underground.

Waste mineral oils are also exempt under the Environmental Permitting Regulations 2010, under exemption S1 in Section 2 of Chapter 5 of Schedule 3 to the Environmental Permitting Regulations 2010 (EPR). Waste mineral oil storage is limited to three cubic metres, it must be in a secure container and you must provide secondary containment. Waste oil storage above three cubic metres must have an Environmental Permit from the Environment Agency.

Requirements of the Regulations

All storage tanks subject to the Regulations are required to have the following precautions / design features in place:

- The tank must be of adequate strength to store oil without leaking or bursting at full capacity, and positioned to minimise the risk of damage, e.g. by vehicles. The tank should also avoid significant risk locations (within ten metres of a watercourse or 50 metres of a well or borehole).
- The tank must be located within an impermeable secondary containment system (SCS) such as a bund or drip tray. The SCS must be able to hold at least 110% of the tank capacity. Where the SCS contains a number of tanks, it should be able to accommodate 110% of the largest tank's capacity, or 25% of the combined capacity, whichever is the greater volume.
- The bund should have no direct outlet discharging on to a yard or unmade ground, or connecting to any drain, sewer or watercourse.
- Any pipework that passes through the bund wall must be sealed into the bund with a material that ensures the bund remains leak-proof.
- Valves, filters, sight points or vent pipes must be located within the bund.
- External fill points must have at least a drip tray.
- All external pipes must be properly supported and positioned to minimise the likelihood of accidental damage from impact or collision.
- Underground fill pipes must have a number of control measures – including a continuous leak detection system, no mechanical joints and a sleeve to prevent damage. If not, they must be tested for leaks every ten years (every five years if they have mechanical joints).

- An automatic overfill prevention device must be present if the filling operation is controlled from a place where it is not reasonably practicable to see the tank and the vent pipe (if present). Mechanical joints must be readily accessible for inspection under a hatch or cover.
- Where mobile bowsers are used:
  - filling points, pumps and valves must be securely locked; and
  - controls must be in place for flexible pipes.
- A number of other detailed requirements are included in the Regulations, such as the positioning of sight gauges, fill points, vent pipes and other ancillary equipment.

Exemptions on existing tanks, installed before the Regulations came into force, were in place up to 1 September 2005, but these exemptions no longer apply. Companies are therefore encouraged to assess all existing plant, as well as new or recently installed facilities.

Underground tanks

Underground tanks are not subject to the Regulations; however, they are a high risk to the environment as they are difficult to inspect and leaks may not be immediately obvious. DEFRA has published a Code of Practice for the good design, operation and management of underground storage tanks, which would be considered good practice where implemented. In the event of a release from the tank, if a company were following the guidance it would be considered as a mitigating factor during any subsequent litigation.

Measures to control releases from an underground tank might include:

- groundwater-monitoring wells;
- leak detection devices;

- installation of new tanks or other maintenance;
- training; and
- implementation of emergency response procedures.

Careful consideration should be made when installing or removing such tanks. The Environment Agency provides guidance – *PPG27: Installation, Decommissioning and Removal of Underground Storage Tanks* – which should be consulted prior to any such work.

Section J of the Building Regulations

It is unlikely that a fire could be started by a fuel storage tank and its contents. However, it does need to be protected from a fire that may originate nearby.

For example, tanks should be sited:

- on a solid base, level and at least 42mm thick, that extends a minimum of 300mm around the footprint of the tank;
- 1,800mm away from non-fire rated eves of a building;
- 1,800mm away from openings (such as doors or windows);
- 1,800mm away from any appliance flue terminals;
- 1,800mm away from a non-fire rated building or structure (i.e. garden shed); and
- 760mm away from non-fire rated boundaries (i.e. wooden fence).

If it is not possible to meet these requirements, a fire rated barrier with at least 30 minutes protection should be provided, extending 300mm higher and wider than each applicable face of the tank.

See also: Building Regulations,
p.76; Environmental Management
Systems, p.286; Environmental risk
assessments, p.292.

Sources of further information

DEFRA: www.defra.gov.uk

DEFRA – Code of Practice for the good design, operation and management of
underground storage tanks:
www.defra.gov.uk/environment/quality/water/waterquality/ground/petrol.htm

Environment Agency: www.environment-agency.gov.uk

PPG2: Above Ground Oil Storage Tanks:
http://publications.environment-agency.gov.uk/pdf/PMHO0909BQSQ-E-E.pdf

PPG8: Safe storage and disposal of used oils: http://publications.environment-
agency.gov.uk/pdf/PMHO0304BHXB-b-e.pdf?lang=_e

PPG27: Installation, Decommissioning and Removal of Underground Storage
Tanks: www.environment-agency.gov.uk/static/documents/Business/ppg27.pdf

The Building Regulations: www.planningportal.gov.uk/buildingregulations/

Furniture

Phil Reynolds, FIRA

Key points

- Furniture is an unexpectedly complex commodity to specify because it encompasses many different requirements and crosses many different areas. There is more to getting it right than glancing through the pages of a glossy catalogue and basing choice on appearance and price.

- Furniture, and the office in general, is key to making a workforce feel comfortable, better motivated and less likely to absenteeism. Providing adequate working space can reduce the chance of work-related injuries from over reaching, repetitive strain etc.

- The Display Screen Regulations require employers to provide suitable seating and desks for all employees using computer equipment.

- Employers have a responsibility to provide a safe working environment with regard to structural stability, fire safety, cabling and, increasingly, the science of ergonomics. Personnel need to know how to adjust furniture to their physique and the tasks they perform. Company image and regulatory compliance apart, a workforce operating in a comfortable environment is likely to be better motivated and less prone to absenteeism than one that has to 'make do and mend'.

- In addition, the environmental impact of furniture is becoming increasingly important, and specifiers should bear this in mind when selecting their supplier.

- So how does the facilities manager or specifier decide which products to choose? The first, and most fundamental, requirement is a test certificate. Has the product been tested to the relevant standards – British, European or international – by an accredited test laboratory?

Legislation

- Health and Safety at Work etc. Act 1974.
- Furniture and Furnishings (Fire) (Safety) Regulations 1988, amended 1989 and 1993.
- Electricity at Work Regulations 1989.
- Workplace (Health, Safety and Welfare) Regulations 1992.
- Health and Safety (Display Screen Equipment) Regulations 1992 as amended by the Health and Safety (Miscellaneous Amendments) Regulations 2002.
- Provision and Use of Work Equipment Regulations 1998.
- Management of Health and Safety at Work Regulations 1999.
- Regulatory Reform (Fire Safety) Order 2005.
- Equality Act 2010.

Structural stability

Check that all tables, chairs and storage furniture have been tested to the appropriate standards. This should prove that the item is suitably strong, durable and stable in use to minimise the risk of any accident.

Fire safety

There is no legislation regarding flammability for upholstered office and

Case study

A council worker who suffered a slipped disc after using a broken chair at work received £10,000 in compensation in August 2009. Kay Fagg underwent a back operation after Southend-on-Sea Borough Council failed to replace her chair when it broke in early 2003. The 62-year-old now has long-term lower back pain and has been forced to retire.

She was working as a sheltered housing officer at the council when the chair's wheels stopped working, which meant it could no longer be moved easily. Despite complaining to her employers, she was never given a replacement.

It wasn't until Mrs Fagg was diagnosed with the slipped disc and returned to work after the operation that she was given correct seating.

Ann Vinden, Unison's Head of Local Government in the Eastern Region, said:

"Kay Fagg's employers should have replaced her chair as soon as they realised there was a problem. The solution was simple, but her injury was left to evolve, until it got so bad she had to have a serious operation and retire from a job that she loved. Employers must listen to staff, take health and safety checks seriously and resolve issues as soon as they start."

Kam Singh, from Thompson's Solicitors, added:

"It is unforgiveable that Mrs Fagg was forced to use a broken chair for almost nine months. Office workers who use computers are protected by strict health and safety rules, which mean they must be provided with equipment which can prevent them from suffering from injuries like this. Mrs Fagg's bosses should have listened to her concerns, carried out a risk assessment and replaced her chair."

contract furniture; this is different from the domestic market, where there are strict requirements designed to protect people in their homes. Instead, current legislation demands that the employer / landlord has the responsibility to decide the fire hazard rate of the building based on location and use and carry out suitable risk assessments for the area, which should cover the furniture as well. Once a rating has been set (usually 'low' for offices, rising to 'high' in institutions such as prisons and student accommodation), the upholstered furniture used must meet these requirements. The relevant British Standard is BS 7176: 2007 +A1: 2011 'Specification for resistance to ignition of upholstered furniture for non-domestic seating by testing composites'. In a new or refurbished building it is wise to consult with your local fire officer prior to specification.

Desk and storage items are not normally covered by fire safety requirements; however Clause 12 of BS 5852: 2006 'Methods of test for assessment of the ignitability of upholstered seating by smouldering and flaming ignition sources' can be used to assess the fire retardant properties of non-upholstered seating.

Cabling with desks

If a desk is provided with cabling (power or data) or provision for it to be fitted, the item should conform to specifications contained within BS 6396: 2009 'Electrical systems in office furniture and educational furniture'. This covers the basic requirements for electrical cabling and should ensure an appropriate degree of safety. It may also fall under the Electricity at Work Regulations if the system is directly wired into the building's electrical supply.

Environment and sustainability

The legality of all timber imported into the UK will soon be subject to EU Regulations designed to prevent illegal logging, to become effective in 2013; however, in the interim all wood and wood-based materials should be of legal origin and obtained from well managed and sustainable forests. The Forest Stewardship Council (FSC) and the Programme for the Endorsement of Forest Certification (PEFC) are the only global forest certification schemes operating within the UK that offer independent third-party certification of the 'chain of custody' process from forest gate to finished product.

All furniture should be purchased from manufacturers working towards sustainable production. Preference should be shown towards suppliers that manage their significant environmental impacts, preferably through the use of management systems such as ISO 14001: 2004.

Furniture manufacturers should have a credible 'end of life' policy, designing and manufacturing with a view for their products to be recycled, and using increasing amounts of recycled materials in their production of goods.

Furthermore, preference should also be given to those suppliers that are embracing the wider principles of sustainable development, including their social impacts and health and safety management. These would include manufacturers committed to sustainable objectives equivalent to those required for full membership of the UK Furniture Industry Sustainability Programme (FISP), which would demonstrate furniture manufacturers' intention of complying with all of the above.

The UK Government has also introduced Government Buying Standards for furniture, which are designed to improve the sustainable purchasing credentials of central government. These requirements are mandatory for all central government furniture purchasing.

Ergonomics

The science of ergonomics is fitting equipment and tasks to people, and not the other way around.

The whole installation – desks and chairs used in conjunction with computer equipment – should conform with the Health and Safety (Display Screen Equipment) Regulations 1992, so look for compliance with BS EN ISO 9241:1999 Part 5.

Incorrect posture can cause a multitude of long-term health problems, including back pain and repetitive strain injury. Personnel should be instructed in how to adjust their furniture correctly so that it meets their specific needs – it is a waste of money to buy expensive, adjustable furniture and then not train people how to operate the controls, but it often happens.

Some forward-looking employers are specifying sit–stand desks that enable the user to alternate between sitting and standing – an approach advocated by many ergonomists including Levent Caglar, Senior Ergonomist at the Furniture Industry Research Association (FIRA).

The marketplace is flooded with products claiming to be 'ergonomic', many of which fall far short of any decent ergonomic criteria. To give specifiers a reliable measure of true ergonomic performance, FIRA has instigated an Ergonomics Excellence Award. Items must meet, or in some instances exceed, the relevant British and European standards together with ergonomic criteria set by FIRA.

In addition, the Disability Discrimination Act 2005 imposes extra requirements on employers to provide suitable furniture for both disabled visitors and employees. This includes not only work desks and chairs, but also reception counters and visitor areas.

Current standards

Tables and desks: strength, stability and safety requirements

- BS EN 15372: 2008: 'Furniture. Strength, durability and safety. Requirements for non-domestic tables.'
- BS EN 527: 2011: Part 1 'Office furniture. Work tables and desks. Dimensions.'
- BS EN 527: 2002: 'Part 2 Office furniture. Work tables and desks. Mechanical.'
- BS EN 527: 2003: 'Part 3 Office furniture. Work tables and desks. Methods of test for the determination of the stability of the mechanical strength of the structure.'

Seating: strength, stability and safety requirements

- BS 5459: 2000 + A2; 2008. Part 2: 'Office pedestal seating for use by persons weighing up to 150kg and for

use up to 24 hours a day including type-approval tests for individual components.'
- BS EN 1335: 2000. Part 1: 'Office furniture. Office work chair. Dimensions.'
- BS EN 1335: 2009. Part 2: 'Office furniture. Office work chair. Safety requirements.'
- BS EN 1335: 2009. Part 3: 'Office furniture. Office work chair. Safety test methods.'
- BS EN 14703: 2007: 'Furniture. Links for non-domestic seating linked together in a row. Strength requirements and test methods.'
- BS EN 15373: 2007: 'Furniture, Strength, durability and safety. Requirements for non-domestic seating.'
- BS EN 13761: 2002: 'Office furniture. Visitors' chairs.'

Storage: strength, stability and safety requirements

- BS 4875: 2007. Part 7: 'Strength and stability of furniture. Methods for determination of strength and durability of storage furniture.'
- BS 4875: 1998. Part 8: 'Strength and stability of furniture. Methods for determination of stability of non-domestic storage furniture.'
- BS EN 14703: 2004. Part 2: 'Office furniture. Storage furniture. Safety requirements.'
- BS EN 14703: 2004. Part 3: 'Office furniture. Storage furniture. Test methods for the determination of stability and strength of the structure.'
- BS EN 14704: 2004: 'Office furniture. Tables and desks and storage furniture. Test methods for the determination of strength and durability of moving parts.'

Screens: strength, stability and safety requirements

- BS EN 1023: 1997. Part 1: 'Office furniture. Screens. Dimensions.'

- BS EN 1023: 2000. Part 2: 'Office furniture. Screens. Mechanical safety requirements.'
- BS EN 1023: 2000. Part 3: 'Office furniture. Screens. Test methods.'

Cable management
- BS 6396: 2009: 'Electrical systems in office furniture and educational furniture. Specification.'

Flammability
- BS 5852: 2006: 'Methods of test for assessment of the ignitability of upholstered seating by smouldering and flaming ignition sources.'
- BS 7176: 2007 + A1:2011: 'Specification for resistance to ignition of upholstered furniture for nondomestic seating by testing composites.'

Ergonomics
- BS EN ISO 9241: 1999. Part 5: 'Ergonomic requirements for office work with visual display terminals (VDTs). Workstation layout and postural requirements. Office accessories.'

Other items
- FIRA Standard PP045: 2003: 'Strength and durability of VDU platforms and support arms.'

> *See also*: Display Screen Equipment, p.244; Work related upper limb disorders (WRULDs), p.772.

Sources of further information

Furniture Industry Research Association (FIRA): www.fira.co.uk

Forest Stewardship Council (FSC): www.fsc-uk.org

Programme for the Endorsement of Forest Certification (PEFC): www.pefc.co.uk

Furniture Industry Sustainability Programme (FISP):
www.fira.co.uk/consultancy/environment/fisp

Government Buying Standards: www.defra.gov.uk/sustainable/government/advice/public/buying/products/furniture/index.htm

Gangmasters

Jonathan Exten-Wright, DLA Piper

Key points

- In lay terms, a gangmaster is a person who organises and oversees the work of casual manual labourers whose services are often supplied to businesses operating in sectors such as agriculture and food processing.
- Until 2004, people knew little of gangmasters and their work. However, the death that year of at least 21 Chinese cockle pickers at Morecambe Bay brought gangmasters into the public eye. The incident highlighted that, in some instances, gangmasters were exploiting labourers, who were often migrant workers. They were being required to work long hours in dangerous or unhygienic conditions and were being deprived of various employment rights, including being paid less than the national minimum wage.

Legislation

As a result of the Morecambe Bay tragedy, the Gangmasters (Licensing) Act 2004 came into force in April 2005. The Act, together with supplemental secondary legislation:

- introduced a licensing scheme for gangmasters;
- created an offence of operating without a licence;
- created an offence of engaging the services of an unlicensed gangmaster;
- created an offence of using false documentation; and
- established the Gangmasters Licensing Authority (GLA) to operate the licensing scheme, set licensing conditions, and maintained a register of licensed gangmasters. The GLA also enforces criminal offences under the Act.

Licensing scheme

The Act prohibits a person from acting as a gangmaster unless they are licensed to do so by the GLA. For the purposes of the Act, a gangmaster is an individual or business that:

- supplies labour to someone operating in one or more of the licensable sectors, which are agriculture, forestry, horticulture, shellfish gathering and food or drink processing and packaging;
- uses labour to provide a service to one of the licensable sectors (for example harvesting or gathering agricultural produce); or
- uses labour to gather shellfish.

'Using labour' includes where a worker is retained on an employment contract or a contract for services but also where someone makes arrangements with a worker that requires the worker to follow their instructions and that determines where, when or how the work is carried out.

Applicants for a licence and existing licence holders must comply with the GLA licensing standards in order to be granted and retain a licence. Compliance with the standards is assessed through inspections conducted by a GLA officer. All new applicants for a licence are inspected and existing licence holders can be inspected at random or on the basis of risk assessment.

The licensing standards require that:

- The licence-holder, the person who is the principal authority under the licence, and any person named on the licence, must at all times act in a fit and proper manner. As part of its assessment, the GLA considers the principal authority's competence and capability to hold a licence.
- The licence-holder must comply with PAYE, NI and VAT obligations, must pay the national minimum wage, and must provide workers with itemised pay slips.
- The licence-holder must prevent forced labour and mistreatment of workers. This includes not retaining identity papers, restricting a worker's ability to work elsewhere, or withholding wages.
- A licence-holder who provides accommodation for workers must ensure the property is of adequate quality and properly licenced (if required) and must allow time for the worker to find alternative accommodation when the provision of accommodation by the licence-holder ends.
- Workers must be able to exercise their rights not to be discriminated against and to, for example, rest breaks, maximum weekly working hours, and trade union membership.
- The licence-holder must comply with health and safety requirements including carrying out risk assessments, ensuring proper instruction and training, ensuring adequate and appropriate personal protective equipment, ensuring access to sanitary facilities, washing facilities, drinking water and facilities for rest/ consuming food and drink, and only using vehicles and drivers that are properly registered and licenced.
- A licence-holder must not charge workers a fee for work-finding services and must not make finding work

conditional on the worker using other services or hiring / purchasing other goods provided by the licence-holder.
- Written terms should be agreed between the licence-holder and any worker and between the licence-holder and any labour user. Written records should be kept by the licence-holder of workers and of labour users. Licence-holders should only use sub-contractors who also hold a current GLA licence.

Operating without a licence

It is a criminal offence to operate as a gangmaster without a GLA licence, and the maximum potential penalty is ten years in prison and a fine.

Recent information reveals that the GLA has made 12 successful prosecutions under the Act. A significant case involving the prosecution of a dairy farmer for using labour supplied by an unlicensed gangmaster is due before the Courts in autumn 2011. In terms of the power to revoke licences, GLA records show that approximately 110 licences have been revoked between March 2007 and August 2011.

Engaging the services of an unlicensed gangmaster

A business that operates in one of the licensable sectors, for example a food processing company, and is supplied with labour by a gangmaster (a labour user) must only use GLA licensed labour providers. It is a criminal offence for a labour user to use an unlicensed gangmaster, and the maximum potential penalty for this offence is six months in prison and a fine. It will be a defence, however, for a labour user to demonstrate that they took all reasonable steps to ascertain if the gangmaster had a licence and that the labour user did not know, and had no reasonable grounds to suspect, that the gangmaster did not hold a valid licence.

To safeguard against the risk of prosecution, before agreeing to use labour supplied by a gangmaster, a labour user should check the GLA public register, which lists all licence holders and applicants. The GLA also recommends that labour users should make an 'active check'. Under this system, the GLA retains a record that the labour user has made a check, informs them of any changes to the status of the gangmaster, and informs them if the gangmaster is inspected by the GLA. It will also be useful for labour users to refer to DEFRA's reasonable steps guidance (see *Sources of further information*), which sets out steps to take to verify the status of a gangmaster.

See also: Health and safety at work, p.361; Health and safety management, p.388; Outdoor workers, p.528; Workplace health, safety and welfare, p.768.

Sources of further information

Gangmasters Licensing Authority: http://gla.defra.gov.uk/

DEFRA – Guidance on the steps that a labour user can take to ensure a labour provider is licensed: http://gla.defra.gov.uk/embedded_object.asp?id=1013034

Gas safety

Simon Toseland, Workplace Law

Key points

- Persons in control of premises must ensure the safe condition of gas installations and appliances.
- The law requires that only Gas-Safe Registered engineers can work on gas appliances.
- Annual checks of gas appliances and systems are required where used in residential properties.
- Regular checks must be made of gas appliances and systems used in non-residential properties, although the law does not prescribe the inspection interval.

Legislation

- Health and Safety at Work etc. Act 1974.
- The Gas Safety (Installation and Use) Regulations 1998.
- Management of Health and Safety at Work Regulations 1999.
- Regulatory Reform (Fire Safety) Order 2005.

The Regulations apply to the installation, maintenance and use of gas appliances (both portable and fixed), and fittings and flues in domestic and commercial premises.

Regulation 3(3) of the Gas Safety (Installation and Use) Regulations 1998, as amended, states:

'No employer shall allow any of his employees to carry out any work in relation to a gas fitting or service pipework and no self-employed person shall carry out any such work, unless the employer or self-employed person, as the case may be, is a member of a class of persons approved for the time being by the Health and Safety Executive for the purposes of this paragraph.'

Compliance

Any person in control of premises, either landlord or managing agent, has a duty to make sure that gas appliances, fittings and flues are installed, maintained and repaired in accordance with the appropriate standards and legislation. By doing this, not only will responsible persons be complying with the law, they will, more importantly, ensure that such equipment can be safely operated.

Failure of the above duty may lead to prosecution of the person responsible for maintenance, with the attendant potential for civil action in the event of an injury or damage to property occurring.

Gas Safe Register

On 8 September 2008, the HSE awarded a ten-year contract to the Capita Group Plc to provide a new registration scheme for gas engineers from 1 April 2009; The Gas Safe Register.

The previous scheme, the Council of Registered Gas Installers (CORGI), had been in place for more than 17 years. During this time, the number of domestic

Case study

Thomas Dalton was given a sentence of 12 months' imprisonment at a York Crown Court hearing in Leeds in May 2009 after pleading guilty to two breaches of the Health and Safety at Work etc. Act 1974. The Court heard that Dalton, a self-employed plumber, was not a registered gas installer and was not competent to carry out work on gas fittings.

Dalton was served with a Prohibition Notice under the Health and Safety at Work etc. Act on 19 April 2005, which specifically prohibited him from carrying out any work on gas fittings. In spite of this he carried out gas fitting work at The Avenue, Harrogate on 6 May 2005, and as a result was sentenced to three months' imprisonment in December 2005. He admitted breaching the notice, as well as 13 further offences of failing to comply with gas safety regulations.

In January 2007 he deliberately ignored these warnings. He broke the prohibition notice again by installing a gas boiler at Cornwall Road, Harrogate in such a way that the flue and the pipe work put the householders' lives at risk, as carbon monoxide could have leaked into the property. After the hearing, HSE Inspector, Kate Dixon, said:

"Thomas Dalton has repeatedly put lives at risk by carrying out gas work without the proper accreditation or training, and whilst being prohibited from such work. The HSE wants to advise anyone who has had gas work carried out by this man since 2005, that they should arrange for a Gas Safe registered installer to check his work to make sure it's safe. The HSE will not hesitate to prosecute individuals who break the law in this way and put people's lives at risk."

gas-related fatalities fell significantly. However, a 2006 review of domestic gas safety involving industry stakeholders (including gas engineers and their representatives) and consumer groups identified no room for complacency and a strong case for change.

Practical advice for employers

The following points may be useful for employers requiring work to be carried out on gas appliances or systems:

■ Gas Safe Registered engineers are required to carry an ID card when undertaking work on gas appliances. The card carries photographic ID along with their registration number.

■ The ID card shows the type of gas work they can carry out, including whether or not they are allowed to work on commercial gas systems. This is an important point because being Gas Safe Registered does not necessarily mean that a person is competent to inspect, repair or maintain every gas appliance.

■ If there is some doubt about a person's competency, checks can be made by calling 0800 408 5500 or visiting the gas safe website, www.gassaferegister.co.uk.

■ If you have doubts about the safety of the work undertaken by a Gas Safe Registered engineer, then you can request that such work is inspected by a Gas Safe Register Inspector.

- For landlords, records of the safety check must be kept for two years and a copy given to each existing tenant within 28 days of the check being completed. A copy must also be issued to any new tenants before they move in.

Checks on gas appliances and systems used in non-residential properties need not necessarily be made annually. It is, however, required that the gas plant and equipment is inspected and maintained regularly. Where a new supply is installed to premises, it must have emergency controls to isolate the supply.

It is good practice to maintain a plan of the premises, showing the location of meters, valves, risers, gas appliances, isolation valves, etc. Uncontrolled gas escapes should be reported to the National Grid Gas Emergency Freephone Number: 0800 111 999.

- It is illegal to install instantaneous water heaters that are not room sealed or fitted with a safety device that automatically turns the gas supply off before a dangerous level of poisonous fumes builds up.

Record keeping
- For residential properties, records of safety checks should be held by the landlord or managing agent for at least two years.
- Tenants are entitled to see, upon request, records of safety checks that have been carried out.
- For commercial properties, records must be kept (for an undetermined period) to demonstrate that systems have been installed and regularly maintained to a suitable standard.

See also: COSHH: Control of Substances Hazardous to Health, p.207; Fuel storage, p.335.

Sources of further information

L56 Safety in the installation and use of gas systems and appliances: www.hse.gov.uk/pubns/priced/l56.pdf

HSE: www.hse.gov.uk/gas/index.htm

HSE gas safety line: 0800 300 363.

Gas Safe Register: www.gassaferegister.co.uk

Glass and glazing

Hayley Saunders, Shoosmiths

Key points

- There are a number of Regulations that impact on glazing and glass safety.
- The Workplace (Health, Safety and Welfare) Regulations 1992 address the issue of glazing in Regulation 14 and window design in Regulations 15 and 16.
- The Regulations deal with issues of safe operation and installation, use and cleaning.

Legislation

- Workplace (Health, Safety and Welfare) Regulations 1992 (and Approved Code of Practice).
- Building Regulations Part L (2010).

Since 1975, the Health and Safety at Work etc. Act 1974 (HSWA) has required employers, the self employed and certain people who have control over workplaces to ensure, so far as is reasonably practicable, the health and safety of anyone who may be affected by their work activities. So if glazing constitutes a risk, reasonably practicable measures need to be taken to deal with it.

The HSWA does not specifically mention glazing, but on 1 January 1993 the Workplace (Health, Safety and Welfare) Regulations 1992 came into force to implement the EC Workplace Directive. Regulation 14 includes requirements for glazing. The Regulations apply to a wide range of workplaces including factories, offices, shops, schools, hospitals, hotels and places of entertainment. They do not apply to domestic premises used for work, or to construction sites.

Glazing requirements

Regulation 14 of the Workplace (Health, Safety and Welfare) Regulations 1992 requires that, where necessary for reasons of health or safety, doors and low-level glazing in a workplace shall be glazed in a safety or robust material, or shall be protected against breakage. This does not mean that all glass in existing workplaces must be replaced with safety glass. A risk analysis must be carried out to see where there is danger – and that danger must be reduced so far as is reasonably practicable.

The Regulation requires that every window or other transparent or translucent surface in a wall, partition, door or gate should, where necessary for reasons of health or safety, be of a safety material or be protected against breakage of the transparent or translucent material; and be appropriately marked or incorporate features to make it apparent.

The HSE's revised Approved Code of Practice, paragraph 147 (see *Sources of further information*), says that particular attention should be paid to doors, and door side panels, where any part of the glazing is at shoulder height or below, and to windows where any part of the glazing is at waist height or below.

These cases are the same as those referred to in Approved Document N of the Building Regulations as 'critical locations'. Therefore, glazing that meets the requirements of Document N (or BS 6262: Part 4) for these critical locations should meet the requirements of Regulation 14.

Regulation 14 also says that glazing in critical locations must be marked, or incorporate appropriate design features, to make its presence apparent – the objective being to avoid breaking the noses of people who might otherwise walk into the glazing, not realising it was there.

The Regulation only expects action 'where necessary for reasons of health or safety'. So you need to assess every window or other transparent or translucent surface in a wall, partition, or door or gate to establish whether there is a risk of anyone being hurt if people or objects come into contact with it, or if it breaks.

Assessing the risks

The assessment needs to take into account all relevant factors such as the location of the glazing, the activities taking place nearby, the volume of traffic and pedestrians, and any previous experience of incidents.

Reducing the risks

Various options are given in the Approved Code of Practice (ACoP) to Regulation 14 on how to reduce the risk of accidents at critical locations.

What needs to be done will depend on the extent of the risk in individual circumstances. You may need to:

■ use glazing that is inherently robust, such as polycarbonate or glass blocks;
■ use glass that, if it breaks, breaks safely – e.g. glass that meets BS EN

12600 'European classification of safety glazing materials';
■ use thick ordinary glass that meets certain thickness criteria;
■ reorganise traffic routes (either for people or vehicles) to avoid the risk of glazing being broken;
■ mark the glazing to prevent people bumping into it; and/or
■ limit the area of glazing.

The Glass and Glazing Federation (GGF) has recently published a new Guidance Note on compliance with Regulation 14. The note provides helpful guidance on how to conduct a risk assessment and implement control measures.

Window design requirements

Regulation 15 focuses on window design. It requires that the operation of opening, closing or adjusting a window must not expose the operator to any risk. According to the Glass and Glazing Federation (GGF), this means that 'appropriate controls', such as window poles, must be provided where necessary; opening restrictors must be provided if there is a danger of falling out of the window; and the bottom edge of an opening window must be at least 800mm above floor level.

A further refinement is that opening windows must not project into an area where passers-by are likely to collide with them.

Regulation 16 requires that all windows and skylights shall be designed so they can be cleaned safely. Essentially, this means that, if they cannot be cleaned from floor level or other suitable surfaces, windows must be designed to be cleanable from the inside.

The ACoP to Regulation 16 makes a variety of recommendations about the safe use of ladders, cradles and safety harnesses. It also cross-refers to BS 8213: Part 1 2004 'Windows, doors and rooflights: Code of practice for safety in use and during cleaning of windows' (see Sources of further information) including door-height windows and roof windows.

Glazing options

So, what are your options when it comes to ensuring that the regulatory requirements are fulfilled in the most cost-effective way?

Choices include toughened or laminated glass, or the use of safety film.

Toughened glass is a glass that has been modified by thermal treatment to give strength, safety up to BS EN 12600 Classification 1.C1 Class A, and improved resistance to heat. Its light transmission is equal to ordinary glass.

Laminated glass consists of two or more sheets of ordinary or heat-treated glass, bonded together under heat and pressure by interlayers of transparent polymer. There are various types of laminated glass, including laminated safety glass, laminated security glass and bullet-resistant glass. Its most significant feature is that if the glass fractures on impact, fragments will remain bonded to the plastic interlayer. This minimises the risk of serious cuts from flying glass and maintains a protective barrier.

Film is a tough sheet of micro-thin high-clarity polyester that can be applied easily *in situ* to the interior or exterior of existing windows, glass doors and partitions. It is available in single-ply and multi-ply formats. The correct application of film on glass can upgrade the original glazing to meet the requirements of

Government Regulations on health and safety and British Standards.

Energy efficiency requirements under the Building Regulations

Amendments to Part L of the Building Regulations came into force on 1 April 2002, bringing new requirements for energy efficiency in buildings. The Building Regulations Part L were amended in October 2001, and new approved Documents L1 and L2 – giving approved guidance on how the new Part L can be complied with – were published at the same time. These Documents were updated and republished in 2006 and more recently in 2010.

The new Regulations will apply to new buildings, and to extensions and alterations to existing ones. Local Authority building control departments are a traditional source of help on how organisations can meet the Regulations.

Fenestration Self-Assessment Scheme (FENSA)

To assist the effective implementation of the requirements of Part L, a scheme for the self-certification of replacement glazing has been set up by the GGF. This scheme is known as the Fenestration Self-Assessment Scheme (FENSA).

Since 1 April 2002 all replacement glazing has come within the scope of the Building Regulations, so that anyone who installs replacement windows or doors has to comply with strict thermal performance standards. The Building Regulations have controlled glazing in new buildings for many years, but this represents only a very small percentage of the total building stock.

When the time comes to sell property – says the GGF – the purchaser's solicitors will ask for evidence that any replacement glazing installed after April 2002 complies with the new Building Regulations. There will be two ways to prove compliance:

1. A certificate showing that the work has been done by an installer who is registered under the FENSA Scheme.

2. A certificate from the Local Authority saying that the installation has approval under the Building Regulations.

See also: Building Regulations, p.76.

Sources of further information

Glass and Glazing Federation: www.ggf.org.uk

Building Regulations Approved Document L: www.planningportal.gov.uk/ buildingregulations/approveddocuments/partl/approved

HSE: Workplace health and safety: Glazing – Code of Practice: www.hse.gov.uk/pubns/indg212.htm

BS 8213-1:2004: Windows, doors and rooflights. Design for safety in use and during cleaning of windows, including door-height windows and roof windows. Code of practice:
http://shop.bsigroup.com/en/ProductDetail/?pid=000000000019976198

Head protection

Simon Toseland, Workplace Law

Key points

- Employers are required to provide head protection for employees who are at risk of sustaining injuries to the head either from things falling on them or by hitting their head against something.
- This general duty arises under the Personal Protective Equipment (PPE) Regulations 1992 in addition to the specific duty imposed for construction workers under the Construction (Head Protection) Regulations 1989.
- Many employees who work at height complain that a safety helmet will not help them if they fall on their head. Whilst it is undoubtedly true that safety helmets will afford little protection in the event of a fall, many such workers are (by nature of the work they are doing), at significant risk of hitting their heads, e.g. scaffolders, steeplejacks, slingers, etc.
- The fact should not be overlooked that the construction industry does not have the monopoly on head injuries. In recent times there have been some serious accidents involving the use of quad bikes in agriculture and forestry.
- With the exception of Sikhs wearing turbans, all employees must wear such head protection where a risk assessment identifies the need to do so.
- Employees need to understand where and when the wearing of head protection is a requirement and management need to ensure compliance. Employers can make rules governing the use of head protection and should bring these rules in writing to the attention of all those who may be affected.
- It is not a requirement under the above Regulations to provide visitors to site (e.g. potential house buyers, deliverers, etc.) with head protection. However, the general duties to others under the Health and Safety at Work etc. Act 1974 apply, and thus where a risk of head injuries might exist it will be necessary to do so.

Legislation

- Health and Safety at Work etc. Act 1974.
- Construction (Head Protection) Regulations 1989.
- Personal Protective Equipment Regulations 1992.
- Management of Health and Safety at Work Regulations 1999.

Selecting the right head protection

Because of the variety of head protection available, it is usually possible to find head protection equipment (e.g. helmets or bump caps), that suit the needs of both the task being undertaken and the wearer. Head protection should fit properly (adjustable), be as comfortable as possible for the wearer (ventilated, possibly cushioned) and compatible with the type of work to be done. For example, a mechanic working under a car in a pit may find that a bump cap is less of a nuisance in such a restricted space than a peaked safety helmet. It is very important that a helmet does not interfere with other PPE (e.g. ear defenders or eye protectors), and therefore an integrated approach may be necessary, such as the combinations used by forestry workers when operating chainsaws.

Hard hats are divided into three classes:

1. *Class A.* For general service. They provide good impact protection but limited voltage protection. They are used mainly in mining, building construction, shipbuilding, lumbering, and manufacturing.
2. *Class B.* For electrical work. They protect against falling objects and high-voltage shock and burns.
3. *Class C.* Designed for comfort, these light-weight helmets offer limited protection. They protect workers from bumping against fixed objects but do not protect against falling objects or electric shock.

Maintenance and replacement

As with all PPE, the employer is responsible for ensuring that it is adequately maintained. Users should be trained to the appropriate level in carrying out maintenance and inspection in order to identify for themselves any problems with the equipment. Self-employed persons have the same duties in regard to their own equipment. Hard hats should be replaced wherever there are visual signs of damage or defect, or following impact. However, most manufacturers recommend replacing hard hats every five years regardless of outward appearance. If work conditions include exposure to higher temperature extremes, sunlight, or chemicals, hard hats should be replaced after two years of use.

Always:

- wear the helmet the right way round;
- wear a chinstrap where necessary;
- wear the helmet so that the brim is level when the head is upright; and
- keep a supply of helmets for visitors.

Never:

- use your helmet for carrying materials;
- paint it or use solvents for cleaning;
- store it in direct sunlight;
- modify, cut or drill it; or
- share your head protection.

Risk assessment

If a significant risk of head injuries exists, then employers are required to adopt a hierarchy of control measures. As with risk control in general, the provision of PPE (thus head protection) is always to be regarded as a last resort. Measures should be taken to reduce the risks to as many people as possible; therefore it may be possible to:

- prevent access to areas where head injuries could be caused;
- provide protected routes;
- avoid the movement of loads overhead;
- fit some form of protection (e.g. guardrails, toe-boards) to prevent falling objects; and/or
- mark and cushion fixed hazards such as pipes and low access points.

See also: Construction site health and safety, p.177; Health and safety at work, p.361; Working at height, p.399; Ladders, p.425; Personal Protective Equipment, p.576; Slips, trips and falls, p.650.

Sources of further information

INDG 174 A short guide to the Personal Protective Equipment at Work Regulations 1992: www.hse.gov.uk/pubns/indg174.pdf

Case review

Landmark cases in health and safety

Workplace Law

Stress and mental health

Walker v. Northumberland County Council (1994)
John Walker was employed by Northumberland County Council as a social services officer. He suffered two nervous breakdowns, which he blamed on the pressure of his job and stressful working environment.

The Council had a duty of care towards Mr Walker to ensure that, due to the potentially distressing nature of his job, it provided adequate care for and gave attention to the state of his mental health; it did not do this and so did not notice when his mental health began to deteriorate.

It was argued that an employer's duty of care should extend to providing psychiatric care, should their employees feel that they need it due to work over load, distressing job nature or both.

The Council did not have the necessary policies in place to detect Mr Walker's deteriorating mental health and, when he returned from a long period of absence due to sickness, it did not have an adequate 'return to work' programme to aid him back into life in the workplace and to help the stress not to occur again.

Walker received £175,000 in compensation and Northumberland Council was found guilty of negligence in terms of allowing

 workplace law
health and safety

Workplace Law's Health and Safety team provide regular case reviews on historic, landmark and current case law, analysing judgments and giving critical commentary on the outcomes of each case. Visit www.workplacelaw.net for more information, where you can search for cases by subject, name, date or key word, by using our simple case review finder. Each case review details an overview of the facts, the outcome of the case, and expert analysis of the implications of the case, as well as a link to the official judgment. Here we look at a few important cases relating to mental health, the duty of care, and reasonable practicability.

Walker's mental health to deteriorate to such an extent without addressing the issue. This case did not set out any new principles of law per se; the emphasis of the case lies within bringing to the attention the seriousness of occupational stress and, therefore, the serious penalties that companies could incur if they allow it to go on in their workplace.

The case demonstrated that an employee could be successful in claiming damages due to stress against a company, and forced companies to recognise the seriousness of stress and the damage it can cause an employee.

Barber v. Somerset County Council (2004)

Leon Barber was employed in 1995 as the Head of the Maths Department in a Somerset secondary school. During his time at the school, his role changed and his hours were increased; due to this he began to find himself extremely stressed and becoming depressed.

Mr Barber informed his employer of the stress that he was feeling and raised concerns that he felt it may go further and cause depression-related illness. Despite the concerns being raised, however, his employer took no action to change the situation / job role or hours back to what he felt was manageable.

Eventually Mr Barber became ill with stress-related illness and depression and suffered a nervous breakdown.

He decided to sue the school for negligence to his condition, despite him trying to raise his concerns on a number of occasions.

Originally, Mr Barber won his case and was given £100,000 in compensation for personal injury, after the nervous breakdown that he suffered, which be believed was caused by the negligence of his employer (the secondary school).

Following this, however, the Court of Appeal overturned this decision; it said that some responsibility had to lie with the employee to make the employer completely aware of the gravity of the situation as employers cannot be expected to guess how serious a situation might be, if they are not told the full details of what is happening.

The case was then taken to the House of Lords in 2004, which awarded Mr Barber back £72,500 of his compensation. Its decision was that, although responsibility should lie with the employee to let the employer know if work-related stress was causing them to feel ill, it felt that Mr Barber had made his employer aware of his feelings towards the situation and they ignored it. The House of Lords was quick, however, to say that it still upheld the decision by the Court of Appeal that employees did have a responsibility to make the employer aware of the situation.

This case was important in putting some of the responsibility back with not only the employer, but also the employee. It was recognised that the employer did still have a large duty of care to the employee and had a legal duty to protect them from occupational stress should the situation be presented to them; however, employers cannot mind-read and should not be expected to guess that an employee is at risk of stress and stress related illnesses – they need some indication from the employee as to the situation.

HSE v. West Dorset Hospitals NHS Trust (2003)

In July 2003, workers at a West Dorset NHS Hospital lodged complaints against their working environment stating that they were being forced into working longer hours and that this was causing them to become stressed. A complaint was also lodged to the HSE against the NHS hospital from a former employee detailing that it failed to deal with claims of harassment and bullying whilst the person was employed by it.

When the HSE investigated, it found that the hospital had not undertaken any risk assessments for stress in the workplace; as a result the staff were left over-worked and feeling stressed. In addition to this, the managers were not being provided with adequate training surrounding health and safety in the workplace, which could have been a reason for the lack of compassion and risk assessments of stress-related tasks in the hospital.

Following this, the investigation found that not only was the hospital not assessing the risk of stress in the workplace, it also did not have any procedures / policies in place to manage stress should it occur.

Following the investigation by the HSE, it was decided that the NHS hospital was to be issued with a notice demanding that it significantly reduce the levels of stress faced by staff.

If it did not follow the proceedings of the notice and reduce stress levels by March of the following year, the Trust would face legal action and unlimited fines.

The HSE's actions were hugely important here as it moved stress into the criminal area of law, away from the civil disputes that it had once fallen in; by doing this it treated stress in the workplace as serious as any other health and safety hazard.

By doing this, it made companies aware of the serious implications of neglecting staff and so aiding stress in the workplace. The notice showed that there is a constructive way of dealing with occupational stress that can be effective in preventing and – where prevention is not possible – managing occupational stress.

The last survey done by the senior staff at the NHS hospital in Dorset showed significant improvements to the decrease in the amounts of stress that employees were facing and that they felt that stress in the workplace was being much better managed.

The notice proved to be the catalyst that the NHS and other organisations needed to better manage stress in their workplaces.

Duty of care

Donoghue v. Stevenson (1932)
In 1928, May Donoghue and an unnamed friend bought – along with other items – a ginger beer from a café in Paisley, Scotland. Donoghue drank some of the contents from the opaque bottle of ginger beer and her friend then poured the rest out into her own glass. Whilst the ginger beer was being poured, what appeared to be a decomposed snail came out of the bottle along with the contents.

As Donoghue had already drunk some of the ginger beer without realising, she visited her doctor and proscribed to later have gastroenteritis and be suffering from severe shock. In 1929, Donoghue took the manufacturer of the ginger beer in Paisley to court, to claim £500 of damages that she incurred as a result of drinking the ginger beer.

Before this case, a manufacturer had not been sued, as it was assumed that no duty of care was owed to Donoghue as there was no contract between her and the manufacturer to suggest so. There had been legal cases in the past, which were similar and where the claim had not been granted by the Inner House of the Court of Sessions, as there was no legal authority to allow otherwise.

Another issue with the case was that there was no obligation of care between the café owner and Donoghue, as Donoghue had not ordered or paid for the drink herself; there may have been an obligation of care between Donoghue's friend and the café owner, as she had ordered the drink and paid for it; however, she was unharmed by the contents, as she did not drink any of it before the snail was discovered.

Donoghue was allowed to appeal her case at the House of Lords by claiming – and being granted – poverty; thus not being able to pay the other side's legal costs if she should lose.

Donoghue's counsel argued that a manufacturer who produces something that is meant for human consumption should have a duty of care to the person who consumes it that it would not cause any damage to the said person – i.e. the manufacturer has a duty of care to make the product fit for human consumption. The opposing side argued that there was no authority for such a principle of law.

After being reviewed, it was ruled by the House of Lords that a company has a duty of care to those humans who will consume a product made by them, to make sure that the product is fit for human consumption and that it will not cause harm. This case was a landmark for future cases where the company seemed to neglect their line of duty to make their products safe for people to consume.

The case was then sent back to the Court of Sessions in Scotland for the ruling to be carried out.

The manufacturer died shortly before the ending of the case and Donoghue's claim was settled outside of the court for much less than the original £500.

The definition of 'reasonably practicable'

Edwards v. The National Coal Board (1949)

Mr Edwards was a miner employed by the National Coal Board, who was killed in an accident at work after the supporting structure of a mine roadway collapsed on top of him. The case was taken to the Court of Appeal.

The National Coal Board argued that it was too expensive and impractical to shore up every roadway in the mines. Many of the roadways that were often used were supported by timber frames to make them safe for the miners; however, the one that Mr Edwards was travelling through had no timber supports.

It was decided that it was not necessary to add timber support to every single roadway, but only those that appeared to need it. This was the first ruling to establish the need for a risk assessment to be carried out in order to establish the cost, time and trouble to mitigate a risk balanced against the risk and the severity of the damage that it might cause.

The courts also changed the language, which they used when setting out regulations to the mines; it was said that the roadways should be supported by timber where 'reasonably practical' rather than where 'possibly able', as they said that 'reasonably practical' was a much narrower term than the latter.

This established the concept of 'reasonable practicability' and it enabled the mines – and later other corporations – to effectively manage the risks involved; thus to reduce the number of accidents (fatal and otherwise) within the mines.

Health and safety at work

Kate Gardner, Workplace Law

Key points

- The Health and Safety at Work etc. Act 1974 (HSWA) imposes general duties on employers, the self-employed, controllers of premises, manufacturers and employees to ensure health, safety and welfare.
- Legislation is supported by Approved Codes of Practice (ACoPs) and guidance notes. Accepted industry standards are also important.
- Employers have a duty to employees in respect of their health and safety and may also be liable for negligent acts committed by fellow employees acting in the course of their employment.
- Occupiers of premises owe a duty of care to both lawful visitors and trespassers.

Legislation

- Occupiers' Liability Acts 1957 and 1984.
- Employers' Liability (Compulsory Insurance) Act 1969.
- Health and Safety at Work etc. Act 1974.
- Employers' Liability (Compulsory Insurance) Regulations 1998.
- Management of Health and Safety at Work Regulations 1999.

Overview

The HSWA imposes general duties on all employers and the self-employed to ensure the health and safety of those who may be affected by their business activities, and on employees to look after their own safety. It also allows outdated, prescriptive legislation to be replaced by objective-setting Regulations, supported by ACoPs and guidance notes.

The European Union instigates many changes in UK health and safety legislation by issuing EU directives that Member States are required to implement by passing their own legislation. The HSWA allows such requirements to be implemented in the UK as Regulations.

Case law has developed alongside legislation, imposing duties of care on employers and the self-employed to look after the health, safety and welfare of their employees and the health and safety of others affected by their business activities.

Breaches of health and safety legislation in the workplace can give rise to criminal investigation and liability.

In addition, whilst generally speaking the fact of a health and safety breach does not create a right to a civil claim, the reality is that the circumstances of the accident and related evidence are likely to be used in any civil personal injury claim.

Health and Safety at Work etc. Act 1974

Before the introduction of the HSWA, health and safety legislation had developed in a piecemeal fashion, providing specified industries and hazardous working activities with a set of prescriptive rules to follow. In 1974 this approach was replaced by the HSWA, which is now the cornerstone of modern health and safety legislation.

The HSWA imposes duties on everyone at work – employers, the self-employed, and employees. The principal duties it imposes are as follows.

Section 2: Duties of employers to employees

Employers must ensure the health, safety and welfare of their employees, subject only to the defence of 'so far as is reasonably practicable'.

To discharge its duty, an employer must provide (so far as reasonably practicable):

- safe plant and safe systems of work;
- arrangements for the safe use, handling, storage and transport of articles and substances;
- adequate information, instruction, training and supervision;
- safe places of work, including safe access and egress; and
- a safe working environment with adequate welfare facilities.

All employers must make and review a suitable and sufficient assessment of the risks of their activities to employees (Regulation 3 of the Management of Health and Safety at Work Regulations 1999).

An employer with five or more employees must prepare and regularly review a written health and safety policy statement, to set out how health and safety is managed in the organisation. This may include the organisational structure and detailed arrangements for health, safety and welfare. The employer must also bring it to every employee's attention.

Employers must also have in place such arrangements as are necessary to effectively plan, organise, control, monitor and review any preventive and protective measures.

Employers must appoint competent persons to assist with the measures necessary for ensuring health and safety (Regulation 7 of MHSWR) and must also consult with employee representatives (including trade unions) when making health and safety arrangements.

Section 3: Duties of employers to others

An employer must conduct his business so as to ensure that non-employees are not exposed to health and safety risks. Again, this duty is subject to the defence of 'so far as is reasonably practicable'. If the employer is a self-employed person, then he must also, so far as is reasonably practicable, conduct his business to ensure that he is not exposed to such risks. Non-employees include, but are not limited to, contractors, visitors and members of the public.

An employer must make and review a suitable and sufficient assessment of the risks of their activities to persons not in their employment and who may be affected by the business activities (Regulation 3 of MHSWR).

Section 4: Duties relating to premises

Any individuals or organisations with total or partial control of work premises must, so far as is reasonably practicable, ensure the health and safety of all non-employees who work there, to the extent of their control.

This means that landlords and managing agents may be responsible for the safety of those working in the common parts of buildings (e.g. cleaners, maintenance staff, etc.), whilst non-domestic tenants will be responsible for the health and safety of any person in the areas covered by their lease.

Section 6: Duty of manufacturers

Anyone who designs, manufactures, imports or supplies articles or substances for use at work must ensure, so far as is reasonably practicable, that those articles are safe for their intended use.

Section 7: Duties of employees

While at work, employees have a duty:

- not to endanger themselves or others through their acts or omissions; and
- to cooperate with their employer, e.g. by wearing protective equipment.

Section 8: Misuse of health and safety equipment

No person (whether an employee or not) shall misuse anything provided in the interests of health, safety or welfare.

Section 36: Individual liability

Where an offence is committed due to an act or default of some other person (not being an employee), that person shall be guilty of the offence and may be charged and convicted of it, whether or not the employer is also charged.

This means that an individual employee can be charged with a health and safety offence without the company being charged of that offence.

Section 37

Where an offence by the company is proved to have been committed with the consent, connivance or by the neglect on the part of any director, manager or company secretary (or similar person), then he or she will also be guilty of the offence and may be prosecuted personally.

Section 40

An employer who intends to rely on a defence of reasonable practicability in a health and safety prosecution is required to prove that they have done everything reasonably practicable (or everything practicable for some offences) to safeguard the health and safety of employees, non-employees or members of the public.

This means that the burden of proof is on the employer to prove that adequate prevention methods were in place.

Reasonably practicable

Many health and safety duties require the duty-holder to do everything possible to ensure the health and safety of others, subject only to a defence of 'so far as is reasonably practicable'. This phrase means doing less than absolutely everything physically possible (i.e. everything 'practicable') and involves a balance to be struck between the risk to health and safety (in terms of the likelihood of harm occurring and the potential consequences) and the inconvenience and cost in terms of time, money and other resources of overcoming that risk.

If the costs are disproportionate to the risks then the test is satisfied and the employer will be more positively able to assert the defence that it was not reasonably practicable to do more to protect against the risk.

Assistance on what is reasonably practicable comes from ACoPs and guidance documents issued by the HSE. Relevant British Standards industry guidance and common accepted industry practices should also be relevant considerations.

In practice, however, the defence of reasonable practicability is very difficult to satisfy, as it is often the case that when an incident is viewed with hindsight an additional measure with a limited resource implication may have been taken in relation to avoid that incident. In addition, the courts have also determined that the standard of care is the same, regardless of the size of the company and its available resources.

Employers should record their risk assessments and the decisions to implement or reject certain safety measures. Since safety measures must be proportionate to the risk they are averting, the first step is to identify and assess the risk, after which the available control measures should be identified and assessed. If the time and costs involved in the control measure are disproportionately high in comparison with the risk involved, then in theory the duty to do everything reasonably practicable will be satisfied, even though the measure is not implemented. In practice, considerable evidence of the reasons for this decision will be required if this defence is to be used in relation to a health and safety incident.

The difficulty for many employers in making this judgement is that the question of whether the correct balance is reached is one that only a court can definitely decide after looking at all the evidence in each case.

Regulations

Numerous Regulations have been made under the provisions of Section 15 of the HSWA that impose detailed obligations on employers and those controlling work activities. The most important of these are covered in other chapters of this Handbook.

Approved Codes of Practice

ACoPs have a 'quasi-legal' status. Although they do not provide definitive interpretation of legislation (only the courts can do that), compliance with the relevant ACoP does provide good evidence of compliance with the relevant statutory duty, and, crucially, can be evidence of

doing everything 'reasonably practicable'. Similarly, if an employer cannot show that he has followed an ACoP in relation to a health and safety incident, the employer must show that it has discharged its relevant health and safety obligations in some other way, or risk prosecution.

Prosecution

Health and safety prosecutions take place in the criminal courts, starting formally with the receipt of a summons to appear at the Magistrates' Court. This is usually issued in a court near to where the accident occurred. Generally, the case may be heard in the Magistrates' Court, where the maximum penalty that can be imposed is a £20,000 fine and/or (for a small number of charges) six months' imprisonment for each charge. Cases that are complex, or result from a more serious outcome, will be committed (referred) by the magistrates to the Crown Court, where the maximum penalty rises to an unlimited fine and/or two years' imprisonment (again for specific charges only).

Maximum penalties under the HSWA and the Regulations are set out in Table 1.

Mitigation and aggravation

Sometimes in health and safety investigations, the measures that an employer takes may amount to either a complete or partial defence to the charges. However, if a prosecution does follow, the same information may also amount to mitigation of the offence committed. Providing evidence of relevant mitigating factors can assist in a reduction in the fine imposed.

The courts have given guidance on the particular factors that amount to mitigating or aggravating features. This can increase or decrease the level of any fine, subject to the Court's discretion. These include the following.

Breach	Magistrates' Court	Crown Court
HSWA offences except those specified below	Maximum 12 months' imprisonment, a maximum £20,000 fine, or both	Maximum two years' imprisonment, an unlimited fine, or both
Associated Regulations[1]	Maximum 12 months' imprisonment, a maximum £20,000 fine, or both	Maximum two years' imprisonment, an unlimited fine, or both
Contravening requirements imposed specifically in relation to public inquiries or special investigations – S.33(1)(d)	A fine not exceeding £20,000	Magistrates' Court only
Obstructing an inspector – S.33(1)(h)	Maximum 51 weeks' imprisonment, or a fine not exceeding £20,000	Magistrates' Court only
Contravening any notice issued under Section 27(1) (general powers of HSE to obtain information) – S.33(1)(i)	A fine not exceeding £20,000	An unlimited fine

[1] Where no other specific penalty is specified.

Table 1. Maximum penalties available under HSWA and associated Regulations.

Mitigating factors
- A prompt admission of guilt.
- That the defendant fell only slightly short of meeting the test of reasonable practicability.
- A good safety record.
- Positive steps taken by the company to remedy deficiencies, including the cost involved in implementing these.

Aggravating factors
- Failure to heed previous warnings.
- Whether death or a serious injury resulted from the breach.
- Whether profit was put before safety.
- A deliberate breach.

Civil actions
Anyone who suffers injury or ill health as a result of work activities may be entitled to bring a personal injury claim against those responsible for compensation. To be successful, the injured party (the claimant) must prove that:

- the defendant owed him a duty of care;
- the duty was breached; and
- the injury was a foreseeable result of the breach.

The existence of a duty of care is generally easy to prove in an employer–employee relationship, since the employer has a duty to provide under civil law:

- safe plant and equipment;
- safe systems of work;
- safe workplaces with safe access and egress; and
- competent fellow workers.

These civil law duties are similar to the duties under Section 2 of the HSWA. Comprehensive risk assessments that are up to date will help to provide good evidence in defence of any claim that the employer has breached any of the above duties.

Occupiers' Liability Acts

Occupiers (those in control) of premises are under duties contained in the Occupiers' Liability Acts 1957 and 1984:

- The 1957 Act imposes a duty to take reasonable steps to ensure that lawful visitors to the premises are safe from dangers due to the state of premises.
- The 1984 Act imposes a duty to take reasonable steps to ensure that trespassers are not injured as a result of dangers arising from the state of premises. This is a slightly lower standard of care than that for lawful visitors.

A breach of these Acts is not a criminal offence and is only actionable in civil law in a claim for compensation.

Vicarious liability

An employer may be responsible for the negligent acts or omissions of employees committed in the course of their employment. A claimant can sue an employer on the basis of vicarious liability, provided he can show that the employee was negligent and this caused his injury. However, an employer will escape liability if it can show the employee was acting 'on a frolic of his own' outside the course of his employment.

Damages

Successful claimants in civil claims will usually receive compensation in the form of a one-off lump sum. This is assessed under a number of headings:

- Pain and suffering;
- Damage to clothing, property, etc.;
- Loss of earnings;
- Medical or nursing expenses;
- Other out-of-pocket expenses, such as additional travel or medication; and/or
- Inability to pursue personal / social / sports interests or activities.

Most personal injury claims are paid for out of insurance (employers' liability or public liability), subject to any exclusions or excesses under the policy. The Employers' Liability (Compulsory Insurance) Act 1969 and 1998 Regulations require employers to hold a £5m insurance cover for claims brought by their employees.

> *See also*: Health and safety at work, p.361; Health and safety enforcement, p.373; Health and safety management, p.388; Trade unions, p.674.

Sources of further information

HSE – Worker involvement: www.hse.gov.uk/involvement/index.htm

HSE – Consulting employees on health and safety: www.hse.gov.uk/pubns/indg232.pdf

workplace law
health and safety

IOSH accredited training from Workplace Law

IOSH Managing Safely Certificate

Choose from one of three study modes to suit your individual needs:

1. Classroom

Delivered in the classroom over a four-day block.

2. E-learning

Study when you want

3. Refresher

Revisit key topics from the Managing Safely course

NEW public course

IOSH Working Safely Certificate

IOSH has completely revamped the 'Working Safely' course with a view to meeting the needs of the modern workplace. No longer is this course 'death by PowerPoint', but rather a course that represents a new approach to health and safety training.

Booking line: 0871 777 8881
Buy online: www.workplacelaw.net/iosh

Health and safety consultation

Susan Cha, Kennedys

Key points

- There is a general duty on every employer, imposed by Section 2(6) of the Health and Safety at Work etc. Act 1974 (HSWA), to consult with safety representatives of trade unions.
- For the purposes of discharging that duty, the Safety Representatives and Safety Committees Regulations 1977 (the 1977 Regulations) set out details as to the appointment of safety representatives, together with their functions and the employer's duty to consult.
- Where employees are not in groups covered by trade union safety representatives, an employer has a duty to consult those employees under the Health and Safety (Consultation with Employees) Regulations 1996 (the 1996 Regulations).
- As a consequence, an employer might have to consult employees under both sets of Regulations where different employees have different representation in the workplace. A number of other Regulations, such as the Quarries Regulations 1999 and the CDM Regulations 2007, include specific requirements regarding consultation with employees.

Legislation

- Health and Safety at Work etc. Act 1974.
- Safety Representatives and Safety Committees Regulations 1977.
- Health and Safety (Consultation with Employees) Regulations 1996.
- Management of Health and Safety at Work Regulations 1999.
- Quarries Regulations 1999.
- Regulatory Reform (Fire Safety) Order 2005.
- CDM Regulations 2007.
- Health and Safety (Offences) Act 2008.

Safety representatives and committees

For the purposes of the appointment of safety representatives or committees and the duty to consult, a recognised trade union is one where the employer, or two associated employers, recognises the union for the purpose of collective bargaining.

The recognised trade union may appoint safety representatives from among the employees in all cases where an employer by whom it is recognised employs one or more employees.

A person ceases to be a safety representative for the purposes of these Regulations when:

- the trade union that appointed him notifies the employer in writing that his appointment has been terminated; or
- he ceases to be employed at the workplace; or
- he resigns.

An employer has a duty under Section 2(7) of HSWA to establish a safety committee with a function of keeping under review the measures taken to ensure the health and safety at work of its employees where requested in writing by at least two safety representatives. The committee should be

Case study

Walker v. N. Tees and Hartlepool NHS Trust (2008)

The EAT's judgment in this case gives some guidance in relation to how much time off should be permitted to employees for training. Mr Walker was a union safety representative who had had a request denied for additional time off to attend a 36-week course for one day a week. His employer argued that the course was not necessary and, if it was wrong on that, his allocated two half-days per week were sufficient for the discharge of his safety representative functions and for reasonable training. Although not deciding on the specific facts of the case – the EAT remitted it to the ET for rehearing – the EAT did reinforce the fact that a two-stage test should be applied in such circumstances, involving:

1. Is the training suggested reasonable in all the circumstances?
2. Has the employer permitted such time for it as is necessary?

The EAT also focused on the Health and Safety Commission's Code of Practice – 'Time Off for the Training of Safety Representatives 1978' – which, among other things, suggests:

- following basic training further training should be undertaken which is necessary to meet changes in circumstances or legislation; but also
- the number of safety representatives attending courses at any one time should be reasonable taking into account the occupational requirements of the employer.

Interestingly, the EAT asked the ET specifically to consider whether it was reasonable to expect Mr Walker to attend the course, using up his weekly allocation of time for 36 weeks, while delegating his safety representative function to a colleague; something that the EAT considered possible on the facts of the case.

established not later than three months after the request is made.

Functions of safety representatives
As well as having a function in relation to the employer's duty to consult in accordance with Section 2(6) of HSWA, safety representatives also have other functions set out in the 1977 Regulations. The most important of these are to investigate potential hazards and dangerous occurrences or the causes of any accidents in the workplace. Others

include investigating complaints by any employee they represent relating to that employee's health, safety or welfare at work, and to make representations to the employer arising out of their investigations or from complaints received or on general matters affecting health and safety at the workplace.

Additional functions are to carry out inspections of the workplace and to receive information from HSE inspectors and consult with those inspectors as the representative of the employees.

Employer's duty to consult safety representatives

In addition to the general duty to consult safety representatives for the purposes of effectively promoting and developing measures to ensure the health and safety at work of employees, the 1977 Regulations state that every employer shall consult safety representatives in 'good time' regarding a number of specified matters. These are as follows:

- The introduction of any measure at the workplace that may substantially affect the health and safety of the employees whom the safety representatives concerned represent.
- Arrangements for appointing or, as the case may be, nominating persons in accordance with the Management of Health and Safety at Work Regulations 1999 or the power to nominate safety representatives under the Regulatory Reform (Fire Safety) Order 2005.
- Any health and safety information that the employer is required to provide.
- The planning and organising of any health and safety training that the employer is required to provide.
- The health and safety consequences for the employees of new technologies into the workplace.

The 1977 Regulations also provide for inspection of the premises and of certain documents by safety representatives on reasonable notice.

Employer's duty to consult employees

The 1977 Regulations apply where there is a recognised trade union. The 1996 Regulations created a duty to consult those employees who do not have the benefit of safety representatives.

The matters on which the employer has to consult are effectively identical to

those on which he must consult safety representatives where he recognises a union, and which are set out above.

The persons to be consulted are:

- the employees directly; or
- in respect of any group of employees, one or more persons in that group who were elected, by the employees in that group at the time of the election, to represent the group for the purposes of such consultation (these are referred to as 'representatives of employee safety').

When the employer consults these representatives, he should inform the employees of the names of the representatives and the group of employees represented by them.

An employer shall not consult a person as a representative if:

- that person has notified the employer that he does not intend to represent the group of employees for the purposes of consultation;
- that person has ceased to be employed in the group of employees that he represents;
- the period of his election has expired and that person has not been re-elected; or
- the person has become incapacitated from carrying out his functions under the 1996 Regulations.

If an employer has been consulting representatives and then decides to consult the employees directly, he should inform the representatives and the employees of this.

The 1996 Regulations also impose a duty on the employer to provide employees and representatives with such information as is necessary to enable them to participate fully and effectively in the consultation. Where the employer consults representatives, he should also make available information contained in any records he is required to keep for the purposes of RIDDOR (so long as the records relate to the workplace or group of workers the representatives represent, and any information relating to specific individuals is not shared without their consent).

Functions of representatives of employee safety

These are very similar to those of appointed safety representatives. They are:

- to make representations to the employer on potential hazards and dangerous occurrences at the workplace that either affect, or could affect, the group of employees they represent;
- to make representations to the employer on general matters affecting the health and safety at work of the group of employees they represent and in particular in relation to those matters on which the employer has a duty to consult; and
- to represent the employees they represent in consultations at the workplace with inspectors of the HSE.

Payment for time off and training

Both the 1977 and the 1996 Regulations provide that the employer shall permit safety representatives or representatives of employee safety to take such time off with pay during working hours as is necessary for performing their functions. In reality, employers must adopt a reasonable approach to this issue. Some representatives will try to overstep the mark, and in those circumstances employers will need to be able to demonstrate that their response is reasoned and proportionate.

In relation to training, the 1977 Regulations provide that the employer will permit safety representatives paid time off to undergo training. The situation is more onerous for employers in relation to representatives under the 1996 Regulations. In those circumstances, the employer shall ensure that representatives are provided with such training in respect of their functions as is reasonable in all the circumstances and it will be the employer who has to bear the cost connected with that training, including travel and subsistence. Again, a good dose of common sense will be the most valuable commodity if there is disagreement over the type and cost of training.

Employer's liability for breaches

The penalties for breaches of health and safety offences have increased substantially following the implementation of the Health and Safety (Offences) Act 2008. Many offences are now punishable with a maximum fine of £20,000 or a custodial sentence or both.

A breach of the general duty under Section 2(6) of HSWA is a criminal offence punishable by a maximum fine of £20,000 in a Magistrates' Court or a custodial sentence not exceeding six months, or both. In the Crown Court the penalty is an unlimited fine or a custodial sentence not exceeding two years, or both. In the absence of other serious failures, it would be surprising were an organisation to be prosecuted for a single, relatively benign breach of its obligations in this area.

A breach of the Regulations is punishable in the same way. Again, as the Regulations are principally 'administrative' in nature, i.e. they do not directly involve risk, prosecution for their breach will be unlikely unless associated with other failings.

A breach of the general duty under Section 2(6) of the 1996 Regulations cannot result in civil liability for a breach. Civil liability can, however, follow a breach of the 1977 Regulations, and both sets of Regulations allow an employee to take action in an Employment Tribunal over an employer's failure to provide paid time off.

See also: Health and safety at work, p.361; Health and safety enforcement, p.373; Health and safety management, p.388; Trade unions, p.674.

Sources of further information

HSE – Worker involvement: www.hse.gov.uk/involvement/index.htm

HSE – Consulting employees on health and safety: www.hse.gov.uk/pubns/indg232.pdf

Health and safety enforcement

Kathryn Gilbertson, Greenwoods Solicitors LLP

Key points

- Inspectors have wide powers of investigation.
- Enforcement is usually secured through the service of prohibition or improvement notices.
- A prohibition notice can require work to stop immediately, and will continue to have effect during an appeal against the notice (unless the Tribunal hearing the Appeal directs otherwise).
- The effect of an improvement notice is suspended pending the outcome of any appeal against the notice.
- Criminal prosecutions can be brought following accidents for breaches of health and safety legislation as well as failure to comply with notices.
- Conviction usually results in a fine (which can be substantial) and an order to pay prosecution costs – neither of which is recoverable from your insurance company.
- Some offences can result in imprisonment (for individuals) for up to two years.
- Directors and managers can also be prosecuted and convicted in connection with offences committed by their company.
- The HSE intends to recover the costs of enforcement action such as improvement and prohibition notices through Fees Regulations.

Legislation

- Health and Safety at Work etc. Act 1974 (HSWA).
- Police and Criminal Evidence Act 1984 (PACE).
- Health and Safety (Enforcing Authority) Regulations 1998.
- Corporate Manslaughter and Corporate Homicide Act 2007.
- Health and Safety (Offences) Act 2008.

Overview

The HSE is the body responsible for promoting health and safety at work, including enforcement. However, for some types of workplace the relevant Local Authority is the Enforcing Authority.

The Health and Safety at Work etc. Act provides powers of investigation, which are in some respects even more stringent than those of the police, since no separate warrant is required to enter premises and persons can be compelled to answer relevant questions and to sign a declaration as to the truth of those answers.

Enforcing Authorities regularly serve improvement and prohibition notices to require employers to take remedial action. Contravention of such a notice is a serious criminal offence.

A further sanction for breaching health and safety legislation is criminal prosecution and conviction of the individual or organisation responsible. This could include a company director, secretary or manager being personally liable.

Although the principal penalty is a fine (which not uncommonly exceeds £100,000 for very serious offences), individuals can be sent to prison for up to two years if convicted of certain offences and an unlimited period if convicted of manslaughter.

Enforcing Authorities

The general rule is that the HSE is the Enforcing Authority. However, health and safety is enforced by the relevant Local Authority in certain workplaces, including:

■ offices;
■ retail outlets;
■ catering services;
■ exhibitions; and
■ sports grounds and theatres.

In this chapter, an 'Inspector' means an officer of the relevant Enforcing Authority, whether it is the HSE or Local Authority.

Powers of investigation

Section 20 of the HSWA gives inspectors wide powers to investigate suspected health and safety breaches. These include the power to:

■ enter and search premises without a warrant;
■ direct that the whole premises, or any part of them, be left undisturbed;
■ take measurements, photographs, recordings or samples;
■ take possession of items for testing or use as evidence;
■ interview any person; and
■ demand copies of documents as evidence.

Section 20 interviews

Inspectors can require interviewees to answer questions and sign a declaration that those answers are true. Although there is no right to remain silent, the evidence obtained cannot be used to prosecute interviewees or their spouse.

It can, however, be used against any other individual or company, including the interviewee's employer.

PACE interviews

Where an Inspector suspects that a particular individual or a company may have committed an offence, he should not question or continue to question that individual or company about the matter other than under caution. This will usually result in the individual suspect or a person with authority to speak on behalf of a company (as the case may be) being requested to attend a formal cautioned interview under PACE. Interviewees must be cautioned before the interview begins. The interview is generally tape-recorded and may also be video-recorded. The general form of caution is in the following terms: 'You do not have to say anything, but it may harm your defence if you do not mention, when questioned, something you later rely on in Court. Anything you do say may be given in evidence.'

Unlike interviews conducted under Section 20, answers to questions in PACE interviews may be used to prosecute the interviewee or the company that the interviewee represents.

There is a right in law to refuse a PACE interview or to remain silent but an adverse inference may be drawn by the Court at trial where an interviewee declines to answer or incorrectly answers when questioned and at trial gives evidence or different evidence on the subject of questions asked at interview.

Where a prospective interviewee would prefer not to physically attend a PACE interview, and if the Enforcing Authority is agreeable, then the interview may be conducted in writing. In these circumstances, the answers will have the same legal effect and consequences as if a face-to-face interview had taken place. See '*Interviews under caution (PACE)*' (p.421) for more information.

Voluntary statements

Inspectors often ask individuals with relevant information to provide a voluntary written signed statement (which may become admissible in Court under Section 9 of the Criminal Justice Act 1967).

Voluntary statements offer no protection to their makers and can be used against the individual concerned and others, subject in the case of the maker to the discretion of the trial judge or magistrates allowing an uncautioned statement being used against the maker. They are frequently compiled by an Inspector as part of his investigation with a view to the maker of the statement being called at trial as a witness for the prosecution. The maker of the statement is the person who decides what it contains and should not be pressurised into saying something that may not be true.

Informal enforcement action

Enforcing Authorities have a wide range of tools available to them when considering what type of enforcement action to take. In instances where there has been a minor breach of the law and the Inspector doesn't consider it serious enough to warrant formal action then the Authority may often provide the duty-holder with information and advice either in writing or face-to-face. This can include a warning that in the Inspector's opinion the duty-holder is in breach of the law.

Enforcement type notices

Inspectors can serve improvement and prohibition notices where service of such a notice(s) is justified. An improvement notice is a notice requiring a person to remedy a health and safety breach within a specified period. A prohibition notice is a notice prohibiting a person from carrying

on certain activities that, in the Inspector's opinion, involve or will involve a risk of serious personal injury until such time as corrective action is taken. A prohibition notice can and usually will be served to have immediate effect.

There is a right of appeal against these notices – the appeal is to an Employment Tribunal and must be made within 21 days of service of the notice.

One important difference between these types of notice is that on appeal the effect of an improvement notice is stayed until the appeal is determined, while a prohibition improvement takes effect even though an appeal is outstanding, unless the Employment Tribunal orders to the contrary. Failure to comply with an improvement or prohibition notice is a criminal offence, punishable by a fine of up to £20,000 and/or up to six months' imprisonment (if the matter is heard in a Magistrates' Court) or an unlimited fine and/or up to two years' imprisonment (if the matter is heard in the Crown Court).

Formal caution and conditional caution

A caution is a statement made by the Inspector and accepted by the duty-holder that there has been a breach of the law. It is an admission by the duty-holder that they have committed an offence for which there is a realistic prospect of conviction but the caution is offered as an alternative to court proceedings.

Criminal prosecution

There are two main types of health and safety offence – for breach of the Act or the Regulations. Most offences are triable either in the Magistrates' Court or the Crown Court. The main determining factor as to which court should hear the case is the seriousness of the offence.

Burden of proof

These are criminal proceedings and so the burden of proof falls on the prosecution to prove beyond all reasonable doubt that the defendant committed the offence charged. However, for breach of duties under HSWA or safety regulations where the duty is based on what is reasonably practicable (as many are), there is a reverse burden of proof on the defendant – who must show, on the balance of probabilities, that what it did was reasonably practicable in the circumstances.

Most health and safety offences are offences of strict liability, which means that the prosecutor is not required to prove that the defendant intended to commit the offences or that, through recklessness, they committed the offence.

Level of penalty

The Sentencing Guidelines Council issued definitive guidance on the level of fines to be imposed when health and safety offences cause death. The judge is expected to impose a fine of £100,000 for such cases. Seldom will the court reduce that level of penalty but it is subject to the defendant's genuine ability to pay. The current largest fine against a single company is £15m (against utility company, Transco).

A convicted defendant will generally also be ordered to pay the prosecution's costs, which in substantial cases are usually considerable in amount and on occasions even more than the fine itself. Although insurance policies will sometimes cover the defendant when prosecuted, they do not cover the payment of fines or prosecution costs. In some instances it is also possible for a court to order the disqualification of individuals from being directors of companies or taking part in the management of a company for up to 15 years.

Personal liability

Section 37 of HSWA provides that, where a company has committed a health and safety offence and the commission of the offence is proved to have occurred with the consent or connivance of, or is attributable to the neglect of, a director, company secretary or manager or a person purporting to act in such capacity (for example a shadow director), the individual as well as the company shall be guilty of the offence and liable to be proceeded against and punished accordingly.

The purpose of this offence is to increase the pressure on companies to comply with health and safety law, by putting at risk of prosecution senior officials.

Enforcement policies

The HSE and Local Authorities publish their own Enforcement Policies. Inspectors refer to these when deciding what health and safety enforcement action to take. The Enforcement Policy provides Inspectors with guidance about the principles to follow and the factors they must take into account when considering prosecutions.

One of the aims of the Enforcement Policy is to hold duty-holders, including directors and managers, to account where they have failed to carry out their responsibilities. The policy usually states that Inspectors should identify and prosecute individuals if a prosecution is warranted.

The HSE enforcement policy, known as HSE 41, was published in February 2009. The HSE has also published an operational circular regarding its policy on prosecuting individuals. The circular states that, where a body corporate has committed an offence, there is likely to be some personal failure by directors, managers or employees. In deciding whether to prosecute individuals, Enforcing

Authorities must consider whether there is sufficient evidence to provide 'a realistic prospect of conviction' and whether a prosecution is in the public interest.

In general, prosecuting individuals is stated to be warranted where there have been substantial failings by them, such as where they have shown wilful or reckless disregard for health and safety requirements, or there has been a deliberate act or omission that seriously imperilled their own health or safety or that of others.

In July 2004 the HSE issued a guidance document entitled 'Investigating Accidents and Incidents'. The guidance suggests that health and safety investigations form an essential part of compliance with obligations under the Management of Health and Safety at Work Regulations 1999. It also emphasises the importance of identifying 'root causes' of past failures (of which a written record should be kept) and suggests that there are almost always failings at managerial level. However, in practice it is likely that organisations will no doubt be reluctant to produce a document that can later be used as evidence against the organisation or its directors or managers in a health and safety prosecution.

Any business that has an accident or incident should consider taking legal advice as soon as possible. This advice should be sought to minimise the likelihood that any report produced can be used in evidence against the company involved.

HSE proposal for cost recovery

The HSE has published its proposal to extend the existing fees regulations to include proposed Fees Regulations, which would place a duty on the HSE to recover costs where duty-holders are found to be in material breach of health and safety law.

Costs will be recovered from the start of the intervention where the material breach was identified, up to the point where the HSE's intervention in supporting businesses in putting matters right has concluded.

The underlying policy of recovering costs for the HSE's interventions through the introduction of fees where there is a material breach of the law has been agreed by Government and is therefore not in question in this consultation.

Instead this consultation is intended to elicit views on how the HSE would recover the costs of the work it undertakes.

The consultation began on 22 July 2011 and ended on 14 October 2011.

> *See also*: Accidents investigations, p.29; Construction site health and safety, p.177; Corporate manslaughter, p.191; Workplace deaths, p.218; Health and safety at work, p.361; Health and safety inspections, p.381; Interviews under caution (PACE), p.421.

Sources of further information

HSE: Enforcement action: www.hse.gov.uk/enforce/index.htm

HSE Enforcement Policy: www.hse.gov.uk/pubns/hse41.pdf

Health and safety inductions

Rob Castledine, Workplace Law

Key points

- The first few days of anyone's new job is an important opportunity for an organisation to outline its safety culture – that's to say how health and safety is viewed and dealt with by the organisation.
- A safety induction should form part of an overall training plan for a new worker and may need to address a number of issues, including corporate induction ('Welcome to the company'), local (site) induction, fire evacuation, accident reporting and any other specific issues associated with their role.
- Inductions also need to consider full-time / part-time employees, temporary / agency workers, permanent onsite contractors and visiting workers.

Legislation

- Health and Safety at Work etc. Act 1974.
- Management of Health and Safety at Work Regulations 1999.

Introduction

An induction should be viewed as a process rather than a one off, first day experience. An effective health and safety induction is the first opportunity that an organisation has to explain to the new worker what its safety culture is all about. 'Safety culture' encompasses the company's attitudes, beliefs and approaches to health and safety in the workplace.

Key elements of an induction – the 'basics'

Site / building security
- Access into the site / building, signing in/out and links to fire procedures.
- Restricted areas.

Fire procedures
- What to do in the event of a fire or fire alarm activation in the workplace.
- Fire escape routes and the place of assembly.

- The organisation's policy on fighting fires and types of fire fighting equipment provided.
- Any special procedures regarding gas suppression systems (e.g. in IT server rooms).
- Arrangements for any out of hours / shift working.

Accidents and fire aid arrangements
- What to do in the event of an accident – reporting forms.
- Locations of first aid equipment, first aid rooms.
- Details of local first aid personnel (first aiders, appointed persons).
- Other emergency equipment – eye wash kits, burns kits.
- Dealing with others types of event – incidents and near misses.

Welfare facilities and other issues
- General tour of the site / building, showing toilets, washing facilities, canteen, etc.
- Smoking policy and any designated smoking areas (where relevant).
- Staff notice boards showing relevant health and safety information.
- Any safety / employee safety reps and their role in dealing with safety issues.

Methods of induction training

Induction training can take all sorts of forms and formats and it's really down to what suits the organisation and its employees best.

A 'traditional' induction would tend to be a formal training course where the new starter is taken through the usual safety induction and maybe a company DVD is shown.

More recently, online training methods have become more popular, whereby the new worker works through an e-learning type course covering the basics of health and safety at work. These systems can also help with record keeping and tracking who does what.

Whichever method is used, it's important to ensure the new worker has understood the key points, perhaps by way of a question and answer element or even by a short assessment / test.

Stages of induction

An effective induction should be viewed as a process comprising the following.

Week One, Day One

- The basics regarding fire / first aid (as described above).

Week One (by the end of the week)

- The company's health and safety policies and relevant procedures.
- General roles of employees and expectations that the company has with regards to health and safety.
- Basic information regarding any safe systems of work relevant to their job.
- Job specific information – dos and don'ts.
- Any training plans showing what else needs to be covered.

Month One (end of)

- Further, more detailed, information on the company's health and safety policies and relevant procedures.
- More specific training on the tasks being undertaken.
- Any gaps in their knowledge or understanding.
- Review of progress to date by their line manager / supervisor.

Other considerations

As with most types of safety training, the organisation should consider the needs of any new workers, as in some cases 'one size' might not fit all. Below are some specific examples of where induction training should be tailored to suit the needs of the employee:

- *Language barriers* – in some situations, workers may not speak or be able to understand English, so the organisation should consider whether information will need to be translated or interpretated for the new worker.
- *Learning difficulties* – the style of language and types of words used might need to be considered – how would we deal with someone who cannot read or write, or where a person's reading ability is significantly lower than everyone else?
- *Mobility / disability* – an effective induction should be able to cater for any new workers who may have mobility issues or any special considerations such as wheelchair users or blind / deaf people.
- *Temporary / agency workers* – depending on how long they might be with the organisation their induction could be limited to the Day One basics.
- *Returning employees* – depending on when they were last with the organisation, it could well be that their induction can be considerably shortened.

Record keeping

Of course, it goes without saying that induction training should be recorded to show what has been given, when it was given and that the recipient has understood it.

Conclusion

Every new worker must be given some form of induction training, and the extent to which the training is required is very much dependant on what kind of job / role they will be doing, the type of environment in which they will be working and the types of risks they might be exposed to.

See also: Construction site health and safety, p.177; Fire, means of escape, p.314; Health and safety at work, p.361; Health and safety management, p.377; Health and safety policies, p.393; Risk assessments, p.624; Training, p.678; Visitor safety, p.697; Workplace health, safety and welfare, p.768; Young persons, p.774.

Sources of further information

ING 345 'Health and safety training – what you need to know':
www.hse.gov.uk/pubns/indg345.pdf

Briefing – Health and safety induction training:
www.workplacelaw.net/news/display/id/11615

Health and safety inspections

Chris Platts, Rollits

Key points

- There are numerous bodies tasked with enforcing health and safety legislation. In summary, the HSE concerns itself with factories, agriculture, mines and quarries, docks and railways to name just a few.

- Local Authorities oversee, amongst others, shops, offices, schools and colleges, and hospitals. A Local Authority Inspector can generally only exercise his powers within the geographical limit of the Local Authority that employs him. HSE inspectors do not have the same geographical constraints.

- Each Enforcing Authority appoints suitably qualified individuals to be Health and Safety Inspectors. Each appointment must be in writing and must specify the powers that the Inspector can exercise. You have the right to ask the Inspector to produce evidence of his appointment.

- These, therefore, are the individuals who both exercise the statutory powers summarised here and who enforce the criminal liability which is envisaged by health and safety legislation.

- Never underestimate the situation when an Inspector visits. A Senior HSE Inspector once said to me that the second he walked into a workplace he 'put on his enforcement hat'. He could not have made a clearer statement of intent, and given the considerable array of powers which he enjoyed it is not surprising.

- It is the case that those organisations that have sound and compliant health and safety practices are likely to be much better positioned to deal with the Inspector in the exercise of his powers. Nevertheless, as the potential for fallout from the exercise of the Inspector's powers can be huge, whether in terms of reputation, or finances, in resources or time, even these organisations should have a good working knowledge of the powers at the disposal of an Inspector. This is particularly important if the visit has been prompted by an accident. This chapter aims to arm you with this knowledge.

Powers of a health and safety inspector

The powers available to an Inspector are set out in Section 20 of the Health and Safety at Work etc. Act 1974 (HSWA). As one Court has observed, the section is "obviously intended to contain wide powers". In summary, they are as follows.

Power to enter premises – HSWA 1974 Section 20(2)(a)

'At any reasonable time (or, in a situation which in his opinion is or may be dangerous, at any time) to enter any premises which he has reason to believe it is necessary for him to enter for the purpose mentioned in sub section (1) above.'

This is the Inspector's general power to come on to your premises in order to secure compliance with health and safety legislation. He must be exercising his general health and safety powers, and should be in a position to show a reason that would support his subjective belief that it is necessary to enter premises. In reality this is a wide ranging power that permits the Inspector to visit at any reasonable time or at any time where a dangerous situation exists. The Inspector does not need to give you notice of his visit and many visits are unannounced!

Power to examine and investigate – HSWA 1974 Section 20(2)(d)

'To make such examination and investigation as may in any circumstances be necessary for the purpose mentioned in subsection (1) above.'

This permits the Inspector to make such examination and investigation as he feels is appropriate. The exercise of this power is not necessarily limited to the premises in question. As the words 'examination' and 'investigation' are not defined, it is difficult to know what limits there are on this power and it should be regarded as giving the Inspector a wide discretion.

Power to ensure things are left undisturbed – HSWA 1974 Section 20(2)(e)

'As regards any premises which he has power to enter, to direct that those premises or any part of them, or anything therein, shall be left undisturbed (whether generally or in particular respects) for so long as is reasonably necessary for the purpose of any examination or investigation under paragraph (d) above.'

It is often the case after an accident that a 'no go area' is imposed by the Inspector or a stipulation that an item of equipment has to be left untouched. This Section is designed to allow the Inspector to consider the evidence and possibly to bring in another Inspector with specific expertise. The Inspector must serve a written notice and this must refer to this power and indicate the duration of the restriction. This period can be extended by serving a further notice.

Measurements, photographs and records – HSWA 1974 Section 20(2)(f)

'To take such measurements and photographs and make such recordings as he considers necessary for the purpose of any examination or investigation under paragraph (d) above.'

This is a self-explanatory and again wide-ranging power and it should be noted it is not necessarily related to or limited to the premises in question.

Articles and substances – HSWA 1974 Section 20(2)(g)

'To take samples of any articles or substances found in any premises which he has power to enter, and of the atmosphere in or in the vicinity of any such premises.'

As the word 'articles' is not defined, the likely inference is that this power permits an Inspector to consider equipment as well as any substance, but that he can only take away the sample and not the article or substance. The power is intended to cover atmospheric testing. Substance includes any solid or liquid or anything in the form of a gas or vapour. The Inspector must show a clear link to the premises that he has entered. No set procedures are specified and you should demand clear information as to what power is being used. No notice of such action is required.

Power to dismantle or subject to testing – HSWA 1974 Section 20(2)(h)

'In the case of any article or substance found in any premises which he has power to enter, being an article or substance which appears to him to have caused or be likely to cause danger to health or safety, to cause it to be dismantled or subjected to any process or test (but not so as to damage or destroy it unless this is in the circumstances necessary for the

purpose mentioned in sub-section (1) above).'

By virtue of this power, the Inspector can carry out relevant tests. The additional stipulation is that if such power is exercised on premises then the person in control of those premises has the right to require any such testing to be carried out in his presence. This does not mean that you can demand the Inspector waits for you to come back from another location! This is a very wide power.

Take possession of and detain dangerous articles or substances – HSWA 1974 Section 20(2)(i)

'In the case of any such article or substance as is mentioned in the preceding paragraph, to take possession of it and detain it for so long as is necessary for all or any of the following purposes, namely: (i) to examine it and do to it anything which he has power to do under that paragraph; (ii) to ensure that it is not tampered with before his examination of it is completed; (iii) to ensure that it is available for use as evidence in any proceedings for an offence under any of the relevant statutory provisions .'

This power allows the Inspector to remove and retain any article or substance so that appropriate examination can be carried out or with a view to ensuring it is not tampered with so as to retain its authenticity as evidence for use in any subsequent proceedings. The Inspector is required in these circumstances to leave with a responsible person a notice detailing the article or substance and the power he has exercised. He should also provide a similar sample marked as such.

Power to require answers – HSWA 1974 Section 20(2)(j)

'To require any person whom he has reasonable cause to believe to be able to give any information relevant to any examination or investigation under paragraph (d) above to answer (in the absence of persons other than a person nominated by him to be present and any persons whom the Inspector may allow to be present) such questions as the Inspector thinks fit to ask and to sign a declaration of the truth of his answers.'

This wide-ranging power exceeds any power which, for example, the police or Revenue and Customs officers have have. It should be distinguished from the situation where an individual gives a voluntary statement to the Inspector. Here the Inspector has the right to ask questions of individuals about whom there is a reasonably held belief that they can provide relevant information, and then to compel the individual to answer. That individual is entitled to be accompanied by someone else. So is the inspector. The individual is required to sign a declaration that the answers given are true. It is an offence to supply untrue information or to fail to answer at all. However, any answers given to the Inspector using this power are not then admissible in subsequent legal proceedings against the individual concerned. The Inspector can ask questions in writing as well as conduct a face-to-face meeting. It should be recognised that any answers can still be used in evidence against another individual or an employer.

Power to require production, inspection and take copies – HSWA 1974 Section 20(2)(k)

'To require the production of, inspect, and take copies of or of any entry in: (i) any books or documents which by virtue of any of the relevant statutory provisions are required to be kept; and (ii) any other books or documents which it is necessary for him to see for the purposes of any examination or investigation under paragraph (d) above.'

Using this power the Inspector can require production of documents. The power does not extend to removing originals but merely taking copies unless copying facilities are unavailable, whereupon the Inspector can take the originals and return them once inspected and copied. Any documents released and copied in this way should be marked as such by you. The Inspector can specify in quite general terms the type of document he is seeking. This can place an onerous burden on you to search and produce relevant copies of documents which meet the Inspector's criteria. Only documents that are protected by legal privilege are exempt from this requirement. In broad terms these are documents that are prepared during a solicitor / client relationship or with the predominant purpose of dealing with forthcoming litigation.

Power to require facilities – HSWA 1974 Section 20(2)(l)

'To require any person to afford him such facilities and assistance with respect to any matters or things within that person's control or in relation to which that person has responsibilities as are necessary to enable the Inspector to exercise any of the powers conferred on him by this section.'

This in essence is a general requirement to be helpful and unobstructive. This does not, however, mean that you should shy away from taking any relevant professional advice as to how to react to an Inspector's requests, or considering critically all requests made of you.

Any other power which is necessary – HSWA 1974 Section 20(2)(m)

'Any other power which is necessary for the purpose mentioned in sub section (1) above.'

This is a very important 'sweeping up power' as one Judge has accurately

categorised it. Its practical relevance, in particular in an age of computer information and data, is that the Inspector may well call for production of information that is kept electronically but nowhere else. This may extend to any type of electronic device, or memory device, where information is saved.

Power to deal with cause of imminent danger – HSWA 1974 Section 25(1)

'Where, in the case of any article or substance found by him in any premises which he has power to enter, an Inspector has reasonable cause to believe that, in the circumstances in which he finds it, the article or substance is a cause of imminent danger of serious personal injury, he may seize it and cause it to be rendered harmless (whether by destruction or otherwise).'

A power that permits the Inspector to deal with causes of imminent danger. There are additional procedural requirements that apply whereby the Inspector should, if practicable, take a sample of any offending substance and provide a portion of that sample to the person responsible for the premises.

Improvement notices – HSWA 1974 Section 21

If, after using any of his statutory powers and carrying out an investigation, an Inspector feels further action is required (implicitly believing informal action is inappropriate) he can serve an Improvement Notice where there is a suspected, possibly continuing, breach of health and safety legislation. The notice should specify what breaches are alleged to be occurring, and include reasons for the opinion and require the person to remedy the contravention within a set period of time. Where you object to the terms of the notice you have a right of appeal within 21 days. The terms of the notice are suspended until the appeal is disposed of.

Prohibition notices – HSWA 1974 Section 22

Here the Inspector can bring activities to a halt where he feels that they involve a risk of serious injury by serving a prohibition notice. Again, there are procedural requirements whereby the Inspector needs to specify his opinion and the matters that give rise to a risk of injury. Ultimately the notice directs that the relevant activities shall not be carried out unless the matters specified in the notice are remedied. A prohibition notice can take effect immediately or at the end of a period specified in the notice. Again, there is a right to appeal within 21 days of service of the notice. You should consider taking advice when served with either type of notice. There have been a number of cases that guide the approach that the Tribunal should take when deciding appeals. These matters may not have been taken into account by the inspector.

But beware...

HSE Inspectors do not have a general power of search, nor can they obtain a search warrant. The police, however, could obtain a warrant if they are involved in the investigation. It is important not to unwittingly release information and/or documents. In cases where documents are seized unlawfully it may be appropriate to make an immediate application for an injunction.

Help yourself

It can easily be seen that the Inspector has a considerable array of powers. It is an offence to obstruct an Inspector in the exercise of his powers and also to contravene an improvement or prohibition notice. An unlimited fine or two years' imprisonment can result from more serious offences. So clearly you should tread carefully. But you can help yourself. An awareness of an Inspector's powers will help you deal with an investigation visit. Forewarned is most definitely forearmed!

> *See also*: Accident investigations, p.29; Health and safety at work, p.361; Interviews under caution (PACE), p.421; Safety inspections and audits, p.628.

Sources of further information

HSE – What to expect when a health and safety inspector calls: www.hse.gov.uk/pubns/hsc14.htm

"Comment ... "

Counting the cost of safety failings

Gordon MacDonald, HSE

It's well recognised that good health and safety is good business. But some firms may seek to gain short-term competitive advantage by cutting corners and putting workers and the public at risk.

However, as early as April 2012, this may be set to change when, for the first time, those who break health and safety laws will pay their fair share of the costs to put things right – and not the public purse. This will provide a further incentive to operate within the law. And it will help level the playing field between those who comply and those who don't.

The HSE will recover all of the costs of an inspection should a material breach of the law be identified, and where there is a formal requirement to put things right. This will be added to the cost of any follow-up work which, for example, may include a return visit by an inspector to check that the business has rectified the problem.

Law-abiding businesses will be free from fees and will not have to pay a penny as a result of an HSE inspection. The cost recovery scheme will not apply to purely technical breaches either.

For example, failure to properly display the health and safety law poster in an otherwise well-run firm would normally be dealt with by verbal advice, and costs would not be recovered. However, inadequate guarding of machinery, which could result in significant injury to employees and requires the HSE to do

Gordon MacDonald joined the HSE in 1978 as a Factory Inspector, covering the Merseyside and Cheshire areas. Subsequent moves took him into Nuclear Directorate as Head of Strategy and Research, his first SCS post, and then to Field Operations Directorate as Regional Director for Yorkshire and the North East. He was appointed to the HSE's Senior Management Team as Head of Hazardous Installations Directorate in February 2009, where he leads some 500 staff in regulating the offshore and on shore oil and gas, chemicals, pipelines, explosives, mines and biological agents sectors.

more than give verbal advice, would attract cost recovery.

A resolution process to deal with any disputes over cost recovery will be established.

Cost recovery may sound like a radical idea to some, but it is actually a principle already operated by other regulators in Britain. And the principle of 'fee for intervention' has already been agreed by the Government.

The HSE recovers its costs in a range of industries under existing schemes and has considerable experience of making these systems work. For instance, the HSE already recovers its costs for intervention activity under an existing 'permissioning' regime, which applies to major hazards activities.

This includes top tier sites under the Control of Major Accident Hazards Regulations 1999 (as amended) (COMAH), offshore gas and oil installations, licensed nuclear installations and some pipelines activities.

But there's still an opportunity to shape the detail of how the proposals could work in the remaining sectors of the economy. Over the next three months, we're inviting employers, worker representatives and anyone else with an interest to submit their views to help make the introduction of 'fee for intervention' as successful as possible.

We're proposing to recover our costs from the start of the intervention, where the material breach was identified, up to the point where the HSE's intervention in supporting businesses in putting matters right has concluded. The HSE's fees are currently estimated at £133 per hour.

The underlying issue is one of fairness – those who the HSE needs to intervene with to put right their failings should pay the costs.

This is not only fair, it's good business, and is good for business too.

> "Law-abiding businesses will be free from fees and will not have to pay a penny as a result of an HSE inspection. It will help level the playing field between those who comply and those who don't."

To take part in the consultation visit:
www.hse.gov.uk/consult/condocs/cd235.htm

Health and safety management

Kate Gardner, Workplace Law

Key points

The Management of Health and Safety at Work Regulations 1999 require employers to manage health and safety by assessing risk. These risk assessments should include measures for:

- identifying the hazards associated with work activities and workplaces;
- assessing the risks from hazardous work activities and workplaces;
- implementing risk avoidance and risk control measures, using standard principles of prevention;
- providing effective systems to plan, organise, control, monitor and review preventive and protective measures in place;
- providing health surveillance where required by specific regulations or by risk assessment;
- appointing competent health and safety advisor(s);
- implementing emergency procedures where appropriate;
- consulting, informing and training employees;
- cooperating with other employers and their employees where appropriate; and
- ensuring the health and safety of young persons and new and expectant mothers.

Legislation

- Health and Safety at Work etc. Act 1974.
- Management of Health and Safety at Work Regulations 1999.

Overview

The Health and Safety at Work etc. Act 1974 (HSWA) imposes a framework of objective-setting legislation on employers, the self-employed and others controlling workplaces or work activities.

This means that employers must manage health and safety in the same way that they manage any other commercial activity (e.g. finance), and, increasingly, emphasis is placed on encouraging organisations, through legislation and campaigns, to integrate health and safety into the organisation's daily activities, rather than

it being 'bolted-on' to the company's operations.

The general duties imposed by the HSWA are supported by more detailed provisions in the Management of Health and Safety at Work Regulations 1999 (MHSWR) relating to the identification and control of risks in the workplace.

Risk assessments

The MHSWR require employers to make a suitable and sufficient assessment of the risks to the health and safety of their employees and others affected by their activities. As circumstances affecting work activities change, such as a change in a process, or purchase of new equipment, these risk assessments need regular review and revision.

Specific risk assessments

Some health and safety regulations (e.g. COSHH and the Regulatory Reform (Fire Safety) Order) require specific risk assessments to be completed in a certain format and to include specific information. Where specific risk assessments are required, a separate risk assessment under the MHSWR may not need to be completed.

Preventative measures

Schedule 1 of the MHSWR sets out the general principles of prevention that should be used when considering measures to prevent or control exposure to health and safety risks:

- Eliminate or avoid risks as far as possible.
- Evaluate risks that cannot be avoided.
- Combat risks at source.
- Adapt work to the individual by reviewing the design of workplaces, choice of work equipment and working methods.
- Adapt to technical progress (through regular reviews of risk assessments).
- Replace the dangerous with non-dangerous or less dangerous (e.g. equipment or substances).
- Develop a coherent overall prevention policy.
- Give priority to collective measures (which protect all those exposed to the risk) over individual protection.
- Give appropriate instructions to employees.

Health and safety systems

Models of health and safety management systems have existed for a number of years, including, for example, HSG 65 (the HSE guide on successful health and safety management), Approved Codes of Practice, British Standards and ISO standards.

HSG 65 recommends and Regulation 5 of the MHSWR requires every employer to make and give effect to appropriate arrangements for the effective planning, organisation, control, monitoring and review of preventive and protective measures.

Effective accident and incident investigation forms an essential part of this process. Investigation of near misses is equally important as it stands as a stark warning of where safety can be improved and provides an opportunity for processes to be reviewed and improved before an accident occurs.

Health surveillance

Employers must ensure that their employees are provided with such health surveillance as is appropriate, having regard to the risks to their health and safety identified by risk assessments. The level, frequency and procedure of the health surveillance should be determined by a competent person acting within the limits of their own training and experience. Health surveillance records should be kept, even after the employee has left the employer's service, and in some instances for over 40 years.

Appointing competent advisors

Every employer must appoint one or more competent persons to assist in relation to compliance with statutory health and safety requirements. A person is competent if he has sufficient training and experience, or knowledge and other qualities to enable him to be practical, reasonable, to know what to look for and how to recognise it (*Gibson v. Skibs A/S Marina and Orkla A/B and Smith Coggins Ltd* (1966)). There is no specific requirement for formal qualifications. However, the Institution of Occupational Safety and Health (IOSH) recommends a diploma-level qualification associated with full membership of the professional institutions, as does the International Institute of Risk and Safety Management (IIRSM).

It is often preferred by Enforcing Authorities for organisations to appoint a competent person from within the employer's organisation wherever possible, rather than relying solely on external consultants. Employees will know their jobs better than anyone else, and while fresh eyes to a situation can often help, the employee's involvement in that process is crucial.

However, if you are not confident of your ability to manage all health and safety in-house, or you need to know that what you are doing is necessary, sensible and proportionate for your business, you may require a consultant. The Occupational Safety and Health Consultants Register (OSHCR) will help you find a consultant with experience in your work. The register enables you to search different categories and provides you with a profile of each consultant, if required. As a duty-holder, it is your responsibility to ensure that if you use a consultant, they are competent and suitable, based upon your particular business needs. See 'The Occupational Safety and Health Consultants' Register' (p.521) for more information.

Emergency plan

Employers must establish procedures to deal with serious and imminent danger and appoint competent persons to implement those procedures insofar as they relate to evacuation. Employers must consider all potential dangers, e.g. fire, bomb threats, terrorism and public disorder.

Employers must also arrange contacts with the necessary emergency services (police, fire and ambulance services) as appropriate.

Information and training

Every employer should provide its employees and others working on its premises with comprehensible information on the risks to their health and safety (identified by risk assessments), together with details of the relevant preventive or protective measures.

Employees should be trained upon their induction and whenever working arrangements or conditions change (e.g. following the introduction of new machinery or a revised risk assessment). In some instances there should also be regular refresher training.

However, it is important to note that in dealing with work equipment safety, it is not sufficient to select preventative measures that rely solely upon the provision of such information, instruction, training and supervision. Instead employers must use (in this order of priority):

1. The provision of fixed guards.
2. The provision of other guards or protection devices.
3. The provision of jigs, holders, push-sticks or similar protection appliances.

Cooperation and coordination

Where two or more employers share work premises (e.g. contractors working in the client's premises) they must cooperate with each other to ensure the health and safety (including fire safety) of all persons working on the premises. This may include the appointment of and cooperation with a nominated person appointed to coordinate joint health and safety arrangements – see 'CDM' (p.137).

Employees' duties

Employees must:

- comply with any requirements or prohibitions imposed by their employer;
- use machinery, equipment, dangerous substances and safety devices

provided in accordance with any training they have received; and
■ inform their employer of serious and imminent danger or shortcomings in health or safety arrangements.

New or expectant mothers

Where the employer's activities could pose a risk to pregnant employees (or their babies), the employer must carry out a specific risk assessment, then take preventive or protective action to minimise that risk. If such action would not avoid the risk, then the employer must:

■ alter the employee's working conditions or hours of work if it is reasonable to do so to avoid the risk;
■ if this is not possible, offer her alternative work in accordance with Section 67 of the Employment Rights Act 1996; or
■ if this is not possible, suspend the employee from work for as long as necessary to avoid the risk.

The employer must go through the above process if the employee has a certificate from her doctor or midwife indicating that she should not work at night. An employer is not under any duty to alter a woman's working conditions until informed in writing that she is a new or expectant mother. See 'Pregnancy' (p.597).

Young persons

Before employing any young person, every employer must carry out a risk assessment that takes into account:

■ the inexperience, lack of awareness of risks and immaturity of young persons;
■ the layout of the workplace or workstation;
■ the nature, degree and duration of exposure to physical, biological and chemical agents;
■ the form, range and use of work equipment and the way in which it is handled;

■ the organisation of processes and activities; and
■ the extent of health and safety training provided.

In particular, no young person should do work:

■ beyond his physical or psychological capacity;
■ involving harmful exposure to toxic, carcinogenic or mutagenic substances, or to radiation;
■ involving a risk of accidents that young persons may not recognise or avoid, owing to their lack of attention, experience or training; or
■ where there is a risk to health from extreme cold or heat, noise or vibration.

Before employing a child (anyone under 16), the employer must provide a parent of the child with information on the health and safety risks identified by a risk assessment, details of the preventive and protective measures in place, and any information on shared workplaces and measures required for coordinating between two or more employers.

A young person who is over 16 may be employed to carry out work in a hazardous environment where the work is necessary for his training, provided that he is properly supervised and any risk is reduced to the lowest level reasonably practicable. See 'Young persons' (p.774).

Civil liability

Regulation 22 of the MHSWR was amended in 2002 to allow it to be used to bring some civil claims – in respect of the employer's failure to comply with the duties in the MHSWR – to safely assess and control the risks to young persons or to expectant mothers or women who are breastfeeding.

Other breaches of the MHSWR and offences committed contrary to the HSWA do not confer a right to bring a civil claim for damages.

However, where there is a civil claim for damages, the Court of Appeal ruled in *Peter Nixon v. Chance option Developments Limited*, that where there is a breach of the employer's statutory duty there can be no contributory negligence on the part of the employee.

See also: Health and safety at work, p.361; Health and safety enforcement, p.373; Health and safety inspections, p.381; Health and safety policies, p.393; The Occupational Safety and Health Consultants' Register, p.521; Pregnancy, p.597; Risk assessments, p.624; Young persons, p.774.

Sources of further information

HSG 245: Investigating accidents and incidents – A workbook for employers, unions, safety representatives and safety professionals: www.hse.gov.uk/pubns/priced/hsg245.pdf

The Workplace Law website has been one of the UK's leading legal information sites since its launch in 2002. As well as providing free news and forums, our Information Centre provides you with a 'one-stop' shop' where you will find all you need to know to manage your workplace and fulfil your legal obligations.

It covers everything from CDM Regulations to updates on the first case to be tried under the Corporate Manslaughter and Corporate Homicide Act, as well as detailed information in key areas such as asbestos and fire safety.

You'll find:

■ quick and easy access to all major legislation and official guidance, including clear explanations and advice from our experts'
■ case reviews and news analysis, which will keep you fully up to date with the latest legislation proposals and changes, case outcomes and examples of how the law is applied in practice;
■ briefings, which include in-depth analysis on major topics; and
■ WPL TV – an online TV channel including online seminars, documentaries and legal updates.

Content is added and updated regularly by our editorial team who utilise a wealth of in-house experts and legal consultants. Visit www.workplacelaw.net for more information.

Health and safety policies

Susan Cha, Kennedys

Key points

- Section 2(3) of the Health and Safety at Work etc. Act 1974 (HSWA) provides a duty on every employer to prepare and, as often as may be appropriate, revise a written statement of his general policy with respect to the health and safety at work of his employees and the organisation and arrangements for carrying out that policy, as well as to bring the statement to the attention of all his employees.

- Those employers with fewer than five employees are exempt by virtue of the Health and Safety Policy Statements (Exceptions) Regulations 1975.

- An employer is unlikely to fulfil all its duties by the use of a generic statement. The policy should relate to that particular business and the health and safety issues that are relevant to it.

Legislation

- Health and Safety at Work etc. Act 1974.
- Health and Safety Policy Statements (Exceptions) Regulations 1975.
- Management of Health and Safety at Work Regulations 1999.

Regard should be had to the Management of Health and Safety at Work Regulations 1999, which provide that:

- every employer should make and give effect to such arrangements as are appropriate, having regard to the nature of his activities and the size of his undertaking, for the effective planning, organisation, control, monitoring and review of the preventative and protective measures; and
- where the employer employs five or more employees, he should record the arrangements.

These Regulations also set out the requirement for every employer to undertake a risk assessment that is 'suitable and sufficient' of the risks to the health and safety of his employees; and the risks to the health and safety of

persons not in his employment arising from the conduct by him of his undertaking. Where there are five or more employees the employer is again required to record any significant findings of the assessment.

The undertaking of a risk assessment and the creation of arrangements as set out above, together with the putting in place of a health and safety policy, all of which should be recorded in writing, can be seen as supplementing one another.

The extent of the policy

The statement should involve a statement of the organisation's overall commitment to good standards of health and safety and usually include a reference to compliance with relevant legislation. In order to demonstrate that there is a commitment to health and safety at a high level, the statement should preferably be signed by the most senior member of the board, such as the CEO or someone in a similar position of seniority.

There are no set rules as to what should be included in a health and safety policy statement. However, it might typically include a commitment to achieving high standards of health and safety in respect of its employees; a similar commitment to others involved or affected by its activities, e.g. delivery partners, visitors, clients and members of the public; recognition of its legal obligations under HSWA and related legislation and a commitment to complying with these obligations; reference to specific obligations, e.g. to provide safe and healthy working conditions, equipment and systems of work; a commitment to consultation with its employees; and reference to the importance of health and safety as a management objective.

The policy should have particular regard to the type, extent and scope of the business. Drafting of the policy should be relatively straightforward in the case of a small business that is perhaps operating in a benign office environment as compared to a much larger enterprise that uses potentially dangerous machinery or industrial processes. There the policy would be expected to be far more wide-reaching and detailed to reflect the increased level of risk.

Where an organisation has in place one overarching document containing a statement of its general policy and commitments, it is advisable for the document to separate out those commitments that relate to health and safety.

In its publication, *Introduction to Health and Safety – Health and Safety in Small Businesses* INDG 259 (rev 1) (see *Sources of further information*), the HSE publishes a specimen health and safety policy, together with a risk assessment, which would be suitable for adoption by a small business. It should be noted that these do require real input regarding the risks and the arrangements of the particular business.

Review of the policy

It is important to note that it is not just a case of formulating a policy and never looking at it again – effect must also be given to it in the way an organisation manages its health and safety duties and responsibilities.

The health and safety policy should be, like other health and safety materials such as risk assessments and method statements, a living document. It should be regularly reviewed and this is made clear in Section 2(3) of HSWA. This is especially true where the nature or extent of operations change or new relevant legislation is introduced.

Employers' liability for breaches

- A breach of a general duty under Section 2 of HSWA can be punishable by up to a £20,000 fine in the Magistrates' Court and/or a custodial sentence not exceeding six months. In the Crown Court the penalty is an unlimited fine or a custodial sentence not exceeding two years, or both.
- A breach of the Management of Health and Safety at Work Regulations 1999 is a criminal offence punishable by the same sanctions.
- In reality, it is very rare in the absence of other transgressions for organisations to be prosecuted for a failure to have a health and safety policy in place. However, it can and does happen that such a failure is prosecuted among others or, prior to that, impacts adversely on the decision whether or not to prosecute in the first place.

See also: Health and safety at work, p.361; Health and safety management, p.388; Risk assessments, p.624.

Sources of further information

INDG 259: *Introduction to Health and Safety – Health and Safety in Small Businesses*: www.hse.gov.uk/pubns/indg259.pdf

Under UK law, every employer with five or more employees must have a written health and safety policy. Workplace Law's **Health and Safety Policy v.3.0** is a template for building a customised policy for your organisation. For more information visit www.workplacelaw.net.

A health and safety policy sets out your general approach, objectives and the arrangements you have put in place for managing health and safety in your business. It is a unique document that says who does what, when and how.

Your health and safety policy and procedures do not need to be complicated or time consuming; they inform staff and others about your commitment to health and safety, and simply describe how you will implement and monitor your health and safety controls.

Workplace Law can assist you in drafting documents that are easy to use and as simple as possible, using the following techniques:

- Using your own in-house font style and company logos to maintain consistency;
- Developing an electronic document retrieval process and incorporating links to health and safety 'good practice' information and guidance;
- Providing in-house training to your employees on the new policy / procedure; and
- Carrying out follow-up audits to establish the effectiveness of the new procedure and make any recommendations for improvement.

Visit www.workplacelaw.net for more information.

www.workplacelaw.net

395

Health surveillance

David Sinclair, Metis Law

Key points

- Employers who expose their employees to certain chemicals, physical agents, materials or ergonomic risks may be required to undertake systematic, regular and appropriate health surveillance on those employees.
- Health surveillance may be either specified in Regulations or covered by the umbrella provisions of health and safety legislation. Employers must offer night workers health assessments.
- Where health surveillance is required, it should be undertaken only by competent people, who in many cases must be medically qualified.
- Employers are required to provide adequate information to employees on health surveillance provisions, results and the records they keep. Records may have to be kept for up to 50 years.

Legislation

- Health and Safety at Work etc. Act 1974.
- Opticians Act 1989.
- Sight Testing (Examination and Prescription) (No. 2) Regulations 1989.
- Health and Safety (Display Screen Equipment) Regulations 1992.
- Manual Handling Operations Regulations 1992.
- Data Protection Act 1998.
- Working Time Regulations 1998.
- Management of Health and Safety at Work Regulations 1999.
- Control of Substances Hazardous to Health Regulations 2002.
- Control of Vibration at Work Regulations 2005.
- Noise at Work Regulations 2005.
- Control of Asbestos Regulations 2006.
- Equality Act 2010.

Specific and non-specific duties

Health and safety regulations can specify mandatory health surveillance – e.g. Regulation 22(2) of the Control of Asbestos Regulations 2006 and the Noise at Work Regulations 2005 – where employers expose their employees to certain biological hazards, chemicals or physical agents (e.g. asbestos, lead, noise, radiation, or vibration).

In such circumstances, the relevant regulations will specify the type, level and frequency of the surveillance to be undertaken, along with details on what records are to be kept by the employer and for how long. Regulation 7 of the Working Time Regulations 1998 imposes a duty on employers to offer night workers a free assessment of their health and capacity to carry out the work they are to be given, prior to them undertaking that work.

In circumstances where there is no specific duty on the employer to carry out health surveillance, the employer has general duties under Section 2 of the Health and Safety at Work etc. Act 1974 and Regulation 6 of the Management of Health and Safety at Work Regulations 1999 to carry out appropriate health surveillance. This general duty applies where the employer's risk assessments identify that:

- there is an identifiable disease or adverse health condition related to the work;
- there is a valid technique available to detect indications of the disease or condition;
- there is a reasonable likelihood that the disease or condition may occur under the particular conditions of the work; and
- health surveillance is likely to further the protection of the health and safety of the employees concerned.

Health surveillance can only be carried out in the above circumstances where the techniques used to undertake the surveillance pose a low risk to the employee.

Employers may need to carry out health surveillance in the following situations:

- Post-accident (or during long-term illness);
- On forklift truck and other machinery operators; and
- On drivers to test for colour blindness.

Employers should be extremely careful in undertaking pre-employment health surveillance, so that if, for example, they require candidates to complete pre-employment health questionnaires, they do not discriminate against disabled candidates in breach of the Equality Act 2010. Employers should seek expert assistance in deciding what surveillance is needed and who is competent to provide that surveillance.

Objectives
The objectives of health surveillance are to:

- protect the health of individual employees by detecting, as early as possible, adverse changes that might

be caused by exposure to hazardous substances;
- help to evaluate the measures taken to control exposure to health hazards; and
- collect, keep, update and use data and information for determining and evaluating hazards to health.

Procedures
There are a number of health surveillance procedures that employers can use, including:

- biological monitoring, i.e. taking samples of blood, urine, breath, etc. to detect the presence of hazardous substances;
- biological effect monitoring, i.e. assessing the early biological effects in exposed workers;
- clinical examinations by occupational doctors or nurses to measure physiological changes in the body of exposed people, e.g. reduced lung function; and
- medical enquiries (often accompanied by a medical examination) by a suitably qualified occupational health practitioner to detect symptoms in people.

Competent people acting within the limits of their training and experience should determine the appropriate level, frequency and procedure to be followed.

For most types of health surveillance the appropriate competent person will be a suitably qualified occupational medical practitioner, occupational health nurse or occupational hygienist.

Once health surveillance has been started, it must be maintained throughout the remainder of the employee's period of employment, unless the risks to which the employee is exposed and the associated health effects are rare and short-term.

Display screen equipment

Regulation 5 of the Health and Safety (Display Screen Equipment) Regulations 1992 (DSE Regulations) places a duty on employers to provide, when requested to do so, an eye or eyesight test to employees who are about to become (or who are already) display screen users.

Eye and eyesight tests are defined in Section 36(2) of the Opticians Act 1989 and the Sight Testing Examination and Prescription (No. 2) Regulations 1989, which specify what examinations the doctor or optician should perform as part of the test. Although the employer only needs to provide the eye or eyesight test when requested to do so, he is under a duty by Regulation 7(3) of the DSE Regulations to provide employees with adequate information about the risks to their health and their entitlement under Regulation 5.

Records

Where health surveillance is undertaken in compliance with particular Regulations, those Regulations will state what data is to be collected and the minimum period for which information is to be stored. Other health surveillance records should be kept:

- for the period specified in the Regulations; or

- for three years after the end of the last date of the individual's employment (the date after which the employee cannot normally bring a claim against the employer), whichever is the longer.

Employers will need to provide employees with access to their personal health records and copies of such records may have to be provided to the Enforcing Authorities. To comply with the employer's duty to provide information to employees (and others who might be affected), employers should provide the appropriate people with the general results of health surveillance, but keep confidential individuals' surveillance data.

Data gathered during health surveillance is regarded as 'sensitive data' within the meaning of Section 2 of the Data Protection Act 1998. As such, all health surveillance data must be processed in accordance with the requirements of that Act. Detailed advice should be sought as to these requirements.

See also: Display Screen Equipment, p.244; Occupational cancers, p.512; Occupational health, p.519.

Sources of further information

HSE: Understanding health surveillance at work: www.hse.gov.uk/pubns/indg304.pdf

Workplace Law's *Occupational Health 2008: Making the business case – Special Report* addresses the issues of health at work, discusses the influence of work on health and highlights the business case for occupational health services at work. The Special Report focuses on the advantages of occupational health services, and the benefits they can provide to a company, in terms of financial savings, increased employee morale, and improved corporate image. For more information visit www.workplacelaw.net.

Working at height

Jagdeep Tiwana, Bond Pearce LLP

Key points

- Although the number of accidents whilst working at height has continued to fall year on year, such accidents are still a major cause of workplace deaths and injuries in the UK. Falls from height are the most common cause of death from an accident at work.
- The risks from working at height and the procedures to minimise these risks are well-known. For this reason, specific legislation has been introduced.
- The Work at Height Regulations 2005 (WAH Regulations) governing work at height consolidate previous legislation and implement European Council Directive 2001/45/EC.
- Unlike previous legislation, which was only triggered by work above a height threshold of two metres, the WAH Regulations apply to work at height in any place where a person could be injured falling from it, even if it is at or below ground level.

Legislation

- The Health and Safety at Work etc. Act 1974.
- Lifting Operations and Lifting Equipment Regulations 1998.
- Management of Health and Safety at Work Regulations 1999.
- Work at Height Regulations 2005 (as amended).

The Work at Height (WAH) Regulations require employers to ensure that:

- all work at height is properly planned and organised;
- the work is carried out safely; and
- employees undertaking the work are trained and competent.

The WAH Regulations are broad, covering not only work at height but also falling objects, fragile roofs and equipment. Put simply, the extent of the risk and the wider ranging application of the WAH Regulations mean that employers must consider every aspect of their business that involves working at height,

whether it is something as complicated as undertaking building work or as straightforward as changing a light bulb. Working at height is defined as any work carried out at any place where a person can fall and injure themselves, which can include:

- working above ground level using a platform or scaffold;
- working on a roof where there is risk of falling through a fragile surface;
- working at ground level where there is risk of falling in a hole on the ground;
- working on the back of a lorry unloading goods;
- climbing fixed structures; and
- using a ladder.

The WAH Regulations apply to all work at height where there is a risk of fall liable to cause personal injury. Employers, the self-employed and any person who controls the work of third parties (such as contractors) have duties under this legislation. There are certain exemptions for shipping, off-shore installations and

Case study

The supplier of the office unit, Mobile Mini UK Ltd, was fined £80,000 and ordered to pay costs of £8,000 after being found guilty of breaching Section 2(1) of the Health and Safety at Work etc. Act 1974.

On 30 January 2008, employee David Boulton was unloading a temporary accommodation unit. He was standing on top of the unit in order to attach a sling from a crane when he fell on to the road, suffering fatal head injuries.

HSE Inspector, Tony Woodward, said:

"Our investigation showed the company's systems were fundamentally flawed. There were safety procedures that were not followed, and those systems that were in place were so cumbersome that employees found them difficult to follow.

"To make matters worse, the company wasn't even checking to see if the staff were following the procedures they had put in place, with workers being allowed to use sub-standard equipment. For example, a ladder used in this incident was one of the worst I have ever seen."

A builder was prosecuted by the HSE after two of his employees fell from the upstairs of a building when a temporary work platform collapsed above a stairwell.

At the time of the incident, the stairs had been removed and a temporary platform had been built over the gap to enable access to a small area above the stairwell. Two scaffold planks were rested on top of a piece of wood that had been screwed into a partition wall. They were also secured at the other end.

One employee was on the makeshift platform but the screws holding the planks in place gave way when a second worker joined him. Both men fell through the stairwell to the ground below, with one suffering a fracture.

Malcolm Shaun Foyle admitted breaching Regulation 4 of the Work at Height Regulations 2005, was fined £2,000 and ordered to pay costs of £1,000.

HSE Principal Inspector, Andrew Kingscott, said:

"Falls from height remain the largest cause of fatal and serious injuries in the construction industry. These two workers could have been much more seriously injured than they were."

docks. It does not apply to the provisions of paid instruction in relation to sports or certain recreational activities. The Regulations were amended in 2007 to cover the activities of climbing and caving instructors.

Facts

- Of the 42 fatal incidents in construction in 2009/10, 60% (25) were caused by falls from height, and two fatalities were caused by falls below two metres.
- Falls from height caused 1,019 major injuries in 2009/10.
- 29% (1,777) of all reportable incidents in construction in 2009/10 were caused by falls from height.
- 63% of all reported incidents are caused by 'low falls' (e.g. below two metres).

Duties under the WAH Regulations

Employers are under a duty to do all that is reasonably practical to prevent someone from falling. In fulfilling this duty, the employer must adopt a risk control hierarchy for managing work at height (including the selection of equipment).

The hierarchy for managing and selecting equipment for work at height has to be followed systematically.

1. The first step must always be to try to *avoid* work at height where possible, for example by carrying out the job on the ground or using long handled tools. Working at height may be designed out, for example by installing pipes and cables at ground level.
2. Take suitable and sufficient measures to *prevent*, so far as is reasonably practicable, any person from falling. Consider first working from an existing place of work and using existing means of access or egress that will provide protection such as a sound flat roof with permanent guard rails. If this is not possible, choose the most suitable work equipment for the task, such as scaffolds, cherry pickers or scissor lifts. Ladders are at the bottom of the list and they should only be used after the risk assessment considers this equipment suitable for the job.

3. If the risk of people or objects falling is not eliminated after step two then consider the provision of work equipment that will *minimise* the distance and the consequences of the fall. This can be achieved by using nets, airbags or fall-arrest systems.

This hierarchy means that employers must first consider if work at height can be avoided altogether. If not, consideration must be given to the use of equipment and other means that would minimise the causes and consequences of a fall, should one occur.

When planning to undertake work at height, regard must be given to:

- the relevant risk assessments;
- how the work can be carried out safely;
- adverse weather conditions; and
- steps to be taken in the event of an emergency.

The following factors should be taken into consideration when assessing the risks of working at height:

- The type of work activity, for example if it is light or heavy tasks.
- The equipment and material to be used – hand tools, dangerous substances, etc.

- The duration of the work; any job that will last more than 30 minutes should not be carried out from a ladder.
- The location where the work is to take place, including any overhead power lines or open excavations.
- The working environment, weather, lighting, floor condition etc.
- The condition and stability of existing work surfaces.
- The physical capabilities of workers.

In addition, only staff who are trained must undertake the work and, even then, when undertaking the work, they must be supervised by a competent person. When considering how the work should be done, thought needs to be given to the place where it is undertaken. For example, if at all possible, access to the elevated area should be from an existing place of work or access. If this is not possible, the most suitable equipment needs to be determined. This includes equipment to protect workers as well as provide access. To this end, the WAH Regulations emphasise the need to implement collective protection measures as opposed to personal protection measures. Collective measures are easier to use, and have the added benefit of protecting everyone and needing less maintenance and training compared to personal protective equipment, which requires a high level of training and maintenance, and only protects the user.

In relation to access equipment, the schedules to the WAH Regulations set out specific requirements that must be applied to equipment such as ladders and scaffolding. So, for example when using ladders, consideration must be given to:

- using the correct type of ladder;
- procedures to ensure that it is used safely, e.g. placed on and against a fixed unmovable surface;

- ensuring employees are trained how to use ladders safely;
- regular maintenance and inspection to ensure that they remain fit for the task;
- how to prevent unauthorised and improper use; and
- any ancillary risks that the use of the ladder may give rise to. In particular, they should take into account any risks to employees and third parties using the area in the vicinity of where the work is to take place.

See '*Ladders*' (p.425) for more information.

In relation to platforms, rigorous inspection regimes are required for platforms used for or to access construction work from which a person could fall more than two metres. Such platforms will generally be scaffolding. Inspection records should be kept at the construction site until the work is completed and thereafter for a further three months. See '*Scaffolding*' (p.638) for more information.

The WAH Regulations also regulate fragile surfaces (generally roofs) and require certain safeguard measures to be put in place such as suitable platforms, and covers and guardrails.

Although not rocket science, it is surprising how often these simple steps are disregarded.

Where work is carried out at height that could result in falling objects, it is necessary to ensure the area adjacent and below where the work is being carried out is clearly cordoned off to ensure that no unauthorised persons can enter this area.

WAIT

The HSE's latest initiative is WAIT – the Work at height Access equipment Information Toolkit. WAIT has been developed to help businesses understand the key issues when working at height and the factors to consider when selecting the most appropriate and safest type of access equipment.

On its website (see *Sources of further information*), users can enter details of the type of work that is being carried out, and results will display, showing the most appropriate types of safeguards in order to carry out the task. Parameters include:

- the height of the working platform;
- the duration of the work;
- the time between equipment movements;
- the available access to the job;
- the work activity itself; and
- if the equipment is required to be freestanding.

Depending on the options selected, the tool will provide the best solution for the job – using ladders, mobile trestle towers, platforms, fixed scaffold and so on. The aim is to demonstrate that any job will have more than one solution, and to show where the use of certain equipment is appropriate or inappropriate.

> *See also*: Accident investigations, p.29; Construction site health and safety, p.177; Ladders, p.425; Personal Protective Equipment, p.576; Scaffolding, p.638; Slips, trips and falls, p.650.

Sources of further information

INDG 401 (rev 1) – The Work at Height Regulations 2005 (as amended) A Brief Guide: www.hse.gov.uk/pubns/indg401.pdf

HSE: www.hse.gov.uk/falls

HSE WAIT: www.hse.gov.uk/falls/wait

HIV and AIDS

Robert Dillarstone and Lisa Jinks, Greenwoods Solicitors LLP

Key points
- Employers and employees need to understand what HIV and AIDS are.
- Employers need to be aware of their liability under various employment-related laws as well as health and safety legislation.
- Employers should implement an HIV and AIDS policy.

Legislation
- Health and Safety at Work etc. Act 1974.
- Management of Health and Safety at Work Regulations 1999.

What are HIV and AIDS and what are the real risks for the workplace?

It is important for employers and employees to understand what HIV and AIDS are as there are many common misconceptions. AIDS stands for Acquired Immune Deficiency Syndrome. It is caused by the Human Immunodeficiency Virus (HIV), which attacks the body's natural defence system and leaves it open to various infections and cancers.

HIV is mainly contained in blood. There is a minimal risk of it being contained in other bodily fluids such as urine, saliva and sweat unless these are contaminated with infected blood. HIV is not spread through normal social interaction such as sharing cutlery or toilets – it is transmitted through sexual intercourse or direct exposure to infected blood through accidental contamination.

The risk of infection at work is very low for the majority of workplaces. The types of occupation where the risk is higher include healthcare, custodial (e.g. prisons), education, emergency services, hair and beauty and plumbing.

There is no reason to treat workers with HIV differently from other workers. People who have the virus but have not developed AIDS will not usually be ill and their ability to work will normally be unaffected. There is often a time lag of many years before their ability to do their job will be affected. In many instances this time lag would be longer than people stay in their job on average in ordinary circumstances.

Those who develop AIDS will have severe illnesses inevitably affecting performance and should be treated in the same way as anyone with any life threatening illness.

As part of an employer's legal duty to protect the health and safety of persons at work, an employer is required to carry out a risk assessment to evaluate the risk of infection from HIV and take sufficient steps to minimise such risk. Employers are also required to give adequate information to their workforce on such risks and explain how these are being addressed in line with HSE guidance (subject to an employer's obligation not to disclose information about individual medical conditions).

First aiders

There are no reported cases of infection arising from first aid. However, first-aiders will need to be reassured of the low risks of being infected. The best way to do this is to provide up-to-date advice on HIV and review first aid training and procedures.

Facts

- There are approximately 86,500 adults in the UK living with HIV.
- It is not known what proportion of those will progress to AIDS and the incubation period between infection and onset of AIDS can be very long.
- During this time, the individual is unlikely to be ill and may not even be aware of the infection.
- Although there is no known cure for AIDS, HIV symptoms, such as swollen lymph glands, weight loss and minor infections, can be treated with anti-retroviral drugs and enable HIV positive people to lead healthy lives.

Policy document and implementation

Employers are advised to draw up a policy on HIV and AIDS so that, if a problem arises, this can be dealt with in accordance with the policy. The policy should be developed in consultation with employee representatives. Once finalised, managers should be provided with appropriate training and an employee awareness programme implemented. The policy will vary depending upon the type of organisation but could include:

- a brief description of HIV and AIDS and how HIV is transmitted;
- the organisation's position on HIV testing;
- an assurance of confidentiality;
- a guarantee that absenteeism or other AIDS-related work issues are to be treated like any other serious illness;
- assurances that colleagues are expected to work normally with such workers and that any refusal to work normally will be dealt with and if appropriate under the disciplinary procedure;
- identifying help available;
- first aid procedures; and
- provisions for overseas travel – the risks of infection through inadequate medical practices as well as sexual encounters.

It is also advisable to make express reference to HIV / AIDS in anti-discrimination and harassment policies.

See also: Health and safety at work, p.361; Occupational health, p.519.

Sources of further information

INDG 342 Blood-borne viruses in the workplace – guidance for employers and employees: www.hse.gov.uk/pubns/indg342.pdf

Protection against blood-borne infections in the workplace: HIV and hepatitis: Stationery Office (ISBN 13: 9 780113 219537): www.tsoshop.co.uk

National AIDS Trust: Advice for Employers: www.nat.org.uk/Media%20library/Files/Policy/2010/HIV-Work-employers.pdf

Homeworking

Dale Collins, Bond Pearce LLP

Key points

- The Health and Safety at Work etc. Act 1974 (HSWA) places specific duties of care on the employer, on the self-employed and on employees.
- Under HSWA, employers have a duty to protect the health, safety and welfare of their employees and other staff members working at an employer's workplace.
- This duty extends to all employees who work either at, or from, their home. As a general guide, therefore, employers should treat both the work area and the equipment used in an employee's own home as though they were in the main office.
- This approach should be reflected in the employer's employment policies and guidelines as well as in the Home Working Agreement made and signed between the employer and the employee before homeworking is approved.

Legislation

- Health and Safety at Work etc. Act 1974.
- Health and Safety (First Aid) Regulations 1981.
- Electricity at Work Regulations 1989.
- Reporting of Injuries, Diseases and Dangerous Occurrences Regulations 1995.
- Provision and Use of Work Equipment Regulations 1998.
- Management of Health and Safety at Work Regulations 1999.

Health and safety

The Provision and Use of Work Equipment Regulations 1998 cover the use of work equipment in the home. Generally, employers are only responsible for the equipment that they supply to their employees. For employees who work at or from their own homes, it is accepted current best practice that all equipment for use in the home office should be supplied by and remain the property of the employer, largely for reasons of data security and because of the responsibility for health and safety. In some situations it may be appropriate for employees to use their own equipment, but, if this occurs, the equipment should first be assessed from a health and safety viewpoint, as well as for its suitability to the work-related task(s) involved. Where computers and any other electrical equipment are used by the employee in the home, these are covered by the provisions of the Electricity at Work Regulations 1989.

Work furniture needs to be adjustable to provide correct and comfortable working heights. A good-quality, comfortable, adjustable chair is especially important. In order to reduce the risks of Repetitive Strain Injury (RSI), Work-Related Upper Limb Disorders (WRULDS), or other injuries due to poor work furniture, the employer is advised to supply all working furniture and any other occupational health equipment for use in the employee's home.

Lighting should be reviewed, especially with regard to glare, as home lighting arrangements are unlikely, on their own, to be fully adequate for office work.

Case study

While unfortunately some Local Authorities have not yet taken into account that homeworking reduces traffic and generally involves no alterations other than provision of electrical sockets and telephone lines, other Authorities adopt a sensible approach.

Oldham Borough Council, for example, recognises homeworking formally in its planning guidelines, and regards home offices as ancillary changes of use. The Oldham document is available to other planning authorities, which can use it as a blueprint for their own guidelines if they wish.

Babergh District Council in Suffolk has prepared its own leaflet, 'Working from Home – balancing the issues' because "we often get asked questions about homeworking and we are aware that there are people who don't really want to ask the question".

For many low key activities at home, planning permission may not be required. If required, proposals will be judged against current Local Plan policies and government planning guidance. Protection of the environment and maintenance of safe and peaceful residential areas will be of prime consideration. Specific regard will be made of the suitability of the premises, their surroundings and the intended use.

Businesses operating from home may well expand and careful consideration of the implications of such growth will be needed. It must not be presumed that because a particular low-key activity had previously not required planning permission, that growth of such a business, at that same location, will be acceptable. If necessary, the Council will try to aid such businesses by assisting them to find more suitable alternative property.

Source: Babergh District Council.

Risk assessments

Health and safety authorities and HSE inspectors have wide powers of inspection and enforcement and can visit employers. They also have the right to visit an employer's homeworkers in their home offices, in order to ensure that risks from working at home are being properly managed. In general, most HSE legislation includes provisions for workplaces in the home. Under the Management of Health and Safety at Work Regulations 1999, the employer is required to assess the risks and hazards that might be present in the workplace, be this in the office or at a member of staff's home office.

Employers with more than five workers have a legal requirement to carry out and record a conventional health and safety workplace risk assessment on a homeworker's home offices. This can be done either by the employer or by the homeworker himself (e.g. using a self-assessment checklist), if s/he is suitably trained. A hazard is literally anything that may cause harm. A risk is the likelihood that someone will be harmed by the hazard. The risk assessment involves the following five steps:

1. Identifying hazards that may cause harm, however small (such as keeping potentially harmful substances out of children's reach).
2. Deciding who might be harmed and how (e.g. the homeworker, members of the household, visitors).
3. Assessing the risks and taking appropriate action (e.g. deciding what steps must be taken to eliminate or reduce the identified risks).
4. Recording the findings – what steps have been taken to reduce or eliminate risks? Inform the homeworker, or anyone else affected by the work, of the findings.
5. Check the risks from time to time and take steps if needed, especially if there is a change in working procedures.

The HSE produces a free guidance booklet on safety for homeworkers. The Institution of Occupational Safety and Health (IOSH) has an excellent datasheet on its website, including a homework premises assessment form, stressing the importance of adequate training and of regular reassessment of the risks (see *Sources of further information*).

In September 2011, the Low Hazard Workplaces (Risk Assessment Exemption) Bill had its second reading in the House of Commons.

This Bill seeks to exempt employers from the requirement to produce a written risk assessment in respect of low hazard workplaces and the premises of those working from their own home with low hazard equipment.

The working environment

Employers should put in place a system for their homeworkers to report accidents or hazards, as there would be in a conventional workplace. Practical experience within the Telework Association suggests that the following areas also often need attention:

■ There should be a sufficient number of power sockets, avoiding overuse of extension leads, trailing cables and adaptors. Home offices may need rewiring for more sockets – have the homeworker's installation checked by an electrician.
■ The use of IT equipment usually requires an additional two power outlets, and one or two telecoms sockets. Safely stowing cabling is important.
■ Electrical equipment needs to be checked for safety (e.g. all cable grips in place, no burn marks on plugs or cracked sockets).
■ Shelves should be conveniently situated so that when heavy files are placed and replaced there is no risk of stress on the spine or overbalancing.
■ Office chairs and tables should all be of the appropriate height and adjustability for long periods of work.
■ If the homeworker wears reading glasses, the prescription should be correct for close work. Anyone working with computers should have their eyes tested, and the optician should be informed of the computer work.
■ Spotlights and anglepoise-type lamps are generally less tiring than fluorescents in small spaces. Light levels should be about 350 lux.
■ Computer screens should be positioned at right angles to windows. Blinds to prevent sunlight making screens hard to read should be installed where needed.
■ Temperatures should be as near as possible to 18.5°C. Small home offices can easily overheat

because IT equipment generates heat. Temperatures may become uncomfortably hot in summer unless adequate ventilation can be provided.

- Adequate ventilation is also important where equipment such as laser printers may give off ozone or other fumes.

- Psychologically, most homeworkers prefer to be situated so that they can see out of a window if possible, although, as noted above, it is important to avoid problems with glare and reflection on computer screens.

- Rest breaks are vital. There are now a number of software packages that can be set up to remind homeworkers to take frequent breaks and so interrupt their more concentrated work environment.

First aid and accidents in the workplace

Under the Health and Safety (First Aid) Regulations 1981, employers are required to ensure that they supply adequate First Aid provisions for their homeworking employees. The exact provisions will depend on the specific nature of the work task(s) being conducted and the hazards and risks involved.

Under the Reporting of Injuries, Diseases and Dangerous Occurrences Regulations 1995 (RIDDOR 1995), employers have a duty to find out about accidents, injuries, dangerous incidents or diseases arising from work-related activities. This might require putting in place procedures for homeworkers to be able to report such events and incidents to their employer.

Planning and Building Regulations

Planning permission

Setting up a home office constitutes a 'change of use' in strict planning terms. However, so far as planning departments are concerned, the average homeworker

is unlikely to require planning permission, particularly if he is not creating a nuisance to neighbours.

Surrey County Council provided the following advice to its own homeworking staff, who worked at home under its Surrey Workstyle Programme:

"Teleworking at or from home does not represent a significant change of use of a building likely to cause a nuisance or hazard to your neighbours. Unless you intend to make structural alterations to accommodate your working area, or extra noise, pollution etc. is generated because you are working at / from home, there is no requirement for planning permission."

Some other councils differ on whether home offices constitute a 'material' or 'ancillary' change of use (turning an outhouse into a garage and car repair workshop is rather more material than putting a computer into a spare bedroom).

Material changes of use to a property require permission; ancillary changes or temporary changes probably do not. Decisions on whether the change of use is 'material' are based on whether it will cause increased traffic, changes to the visible appearance of the property, nuisance such as noise or smells, or unsocial working hours.

Business rating

So far as the associated matter of rating is concerned, a number of factors will determine whether the space in your home used as an 'office' will be liable for business rates. These will include the extent and frequency of the business use of the room (or rooms) and any special modifications made to the property. Each case is considered individually, usually through a visit from your local Valuation Office – which you should contact for

Planning permission may not be required	Likely to require planning permission
Dual business and domestic use of a single internal room or outbuilding for an office, not requiring significant deliveries or callers	Exclusive business use of one or more rooms or employment of staff
Occasional use of domestic kitchen for seasonal produce	Commercial kitchen and catering
Child minding activity for a few children	Regular day-nursery or crèche
Occasional sale or servicing of cars belonging to the household	Regular sale or repair of vehicles for non-residents
Occasional meetings	Regular formal meetings
Using outbuilding or garage for low-key hobby activities	Using garage or outbuilding for manufacturing and other processes
Keeping and breeding a few small animals	Boarding of animals for fees or keeping significant numbers of animals
Overnight parking a single small trade van or taxi on a private driveway	Parking of heavy goods vehicles or coaches on site
Use of one or two bedrooms for bed and breakfast	Guest house or nursing home

Source: Babergh District Council.

further details. If your property needs to be assessed for business rates, your Valuation Office will work out a rateable value for the part used for non-domestic purposes, which may lead to a reduction in the council tax figure for the domestic part of the premises (see 'Business rates', p.102).

A Lands Tribunal appeal judgment, *Tully v. Jorgensen* (2003), found in favour of a full-time homeworker who used one room entirely as an office. The case concerned use by Mrs Tully of a back bedroom of her home for the purposes of her employment in the Inland Revenue, in order to cope with the effects of a serious spinal disability.

Although the room was used for work purposes between 20 and 36 hours a week for 44 weeks and was fully equipped with office equipment supplied by the Revenue, the appeal by Mrs Tully against the decision of the LT was allowed.

The impact of this judgment is that if one works from home, uses office equipment, has not made structural alterations and does not employ people from the premises, then business rates may not be levied.

A practical piece of advice for homeworkers is to ensure that their activities do not cause annoyance to neighbours – and hence visits from council officials are much less likely to occur.

Building a workspace
If the establishment of the home office involves any building work, such as

conversion of a loft space, there are strict Building Regulations that must be adhered to, mainly relating to means of escape in case of fire. Loft ladders and space-saver stairs are not favoured because they require familiarity of use for safe passage.

The homeworker may need to upgrade the floor between the loft and the rest of the house to give half an hour of fire resistance.

Another alternative to loft conversion, which has been successfully used by a number of homeworkers, is a personal office, which, depending on specification, can cost around £5,000 and may not need planning permission.

> *See also*: Building Regulations, p.76; Electricity and electrical equipment, p.256; Risk assessments, p.624; Work equipment, p.752; Work-related upper limb disorders, p.772.

Sources of further information

The Telework Association: www.telework.org.uk

The datasheet 'Teleworking' can be downloaded free from the Institute of Occupational Health and Safety's website: www.iosh.co.uk. The datasheet includes a homework premises assessment form and stresses the importance of adequate training and regular reassessment of the risks.

Insurance

Hugh Merritt, Hogan Lovells

Key points

- Employers are required by law to insure motor vehicles and their own liability to employees.
- There is a raft of other policies available to insure against the unexpected in the workplace – this chapter summarises some of the most common options.
- Any business involved in the marketing or sale of insurance or other intermediary services in relation to insurance may be required to obtain authorisation from the Financial Services Authority.

- Employers' Liability (Compulsory Insurance) Act 1969.
- Road Traffic Acts 1988 and 1991.
- Employers' Liability (Compulsory Insurance) Regulations 1998.
- Insurance Mediation Directive.
- Motor Vehicles (Compulsory Insurance) (Information Centre and Compensation Body) Regulations 2003.
- Employers' Liability (Compulsory Insurance) (Amendment) Regulations 2008.

Compulsory insurances

Employers' liability

With some exceptions, the Employers' Liability (Compulsory Insurance) Act 1969 requires employers to insure their liability to their employees for personal injury, disease or death sustained in the course of their employment in Great Britain. There is currently a penalty of up to £2,500 per day that the employer does not have insurance.

Employers are legally required to insure for at least £5m. The Association of British Insurers (ABI) advises that, in practice, most policies offer £10m minimum cover.

Employers are required to display a certificate of employers' liability insurance at each place of work, although measures were brought into force in October 2008, which meant that certificates no longer had to be displayed as hard copies, or kept for the previously-enforced minimum of 40 years. The Employers' Liability (Compulsory Insurance) (Amendment) Regulations 2008 now allow employers to display the certificate online, so long as it is readily available to all employees. The Regulations also allow employers to keep the certificate displayed in the workplace if this is easier.

Most organisations frame copies of their certificates and hang them in staff kitchens or other communal areas. There is a penalty of up to £1,000 for failure to display and provide a copy of a certificate to an inspector on request.

Employers' liability insurance must be issued by an authorised insurer; namely an issuer authorised by the Financial Services Authority (FSA). A register of authorised insurers can be found on the FSA website.

Employers do not need employers' liability insurance for employees working outside of Great Britain (including where on secondment abroad), but the local law of the relevant jurisdiction should be checked for equivalent regulations.

Since 28 February 2005, employers' liability insurance is no longer compulsory for companies with a single employee who owns 50% or more of the company. Family businesses that only employ close family members are also exempt from the requirement; however, this exemption does not apply to family businesses incorporated as limited companies.

Motor
Third-party motor insurance (as a bare minimum) is compulsory. Employers should also take steps to make sure that employees using private cars for work-related activities are insured to do so. The EU Motor Insurance Directive requires fleet policy-holders to register their vehicles with the Motor Insurance Database. For more information visit www.mib.org.uk.

Other workplace insurances
There are a number of types of insurance policies that businesses should consider to cover a variety of risks. This chapter summarises some of the most common ones.

Buildings and contents
If you own your business premises then you should consider taking out a suitable insurance policy that covers damage from a variety of causes. Standard policies typically cover risks including fire and lightning, explosion, riot, malicious damage, storms and floods. An 'all risks' policy provides cover against other risks, including accidental damage. Insurance policies do not cover wear and tear, electrical or mechanical breakdown or any gradual deterioration specified in the policy. Tenants should ask their landlords who is responsible for insuring

the premises. This will normally be the landlord.

The ABI advises that:

- business premises should be insured for the full rebuilding cost (including professional fees and the cost of site clearance) and not just for the market value;
- stock should be insured for its cost price without any addition for profit;
- plant and business equipment can also be insured, on either an 'as-new' or an indemnity basis, in which case wear-and-tear will be taken into account when settling claims; and
- contents are usually covered against theft, provided there has been forcible and violent entry to or exit from the premises. Damage to the building resulting from theft or attempted theft will also normally be covered.

Business interruption
Business interruption insurance should be enough to cover any shortfall in gross profit caused by damage to property (e.g. fire). It should also cover any increase in working costs, and restart costs.

Directors' and officers' liability
Directors' and officers' liability insurance would reimburse the individual director or company officer for personal liability. Alternatively, it can compensate the company itself where the company has reimbursed a director or officer for a personal liability.

Employees
- Employers can take out key-person insurance, which would compensate the company in the event of an injury or death preventing a key member of staff from working.
- Fidelity guarantee insurance would cover you for the costs of employee dishonesty (e.g. theft). This would not

normally be covered under buildings and contents insurance.

- Make sure that personal assault is covered in your contents insurance. This will provide compensation for you or your employees following injury during theft or attempted theft of money.
- Some employers provide income protection insurance, private medical insurance, life insurance, or personal accident and sickness insurance as an employee benefit.

Legal expenses

Legal expenses insurance would meet the cost of bringing or defending legal action. Most notably, legal expenses insurance is useful in insuring against Employment Tribunal claims, or court action taken by, for example, HM Revenue & Customs.

Professional indemnity

Professional indemnity insurance protects advisers who give professional advice – e.g. solicitors, accountants, architects and building surveyors.

Public liability

Public liability insurance would pay damages to members of the public for death, personal injury or damage to their property that occurs as a result of your business. It also covers associated legal fees, costs and expenses.

This type of insurance is unlikely to be necessary in the case of, for example, a web design business, but is necessary if the public has access to the business premises. Public liability insurance is compulsory for some businesses.

Product liability

Product liability insurance covers compensation awarded as a result of damage to property or personal injury caused by a product manufactured or (in some circumstances) supplied by a

business. Most businesses have cover of between £1m and £5m, although a common level of cover is around £2m.

Important considerations

- Insurance is subject to insurance premium tax, which cannot be set off against VAT. Life insurance, pensions and income protection insurances are exempt.
- It is important for employers to bear in mind that fines as a result of prosecution (e.g. by the HSE for a health and safety offence) cannot be insured against.

Selling insurance

Wide-ranging Regulations (under the EU Insurance Mediation Directive) governing the selling and administering of insurance policies by intermediaries came into force on 14 January 2005.

The Regulations apply to any business that purchases or arranges insurance for a third party or assists a third party in the preparation of an insurance claim. The Regulations only apply where the business provides such services by way of business, which in most cases means where remuneration is received for those services.

The following are examples of activities that may fall within the ambit of the Regulations:

- Contractors who obtain insurance on behalf of, or handle claims for, third parties, e.g. developers or funders;
- Employers who obtain project insurance on behalf of the other parties involved with the project;
- Insurance arrangements for joint venture companies or other group companies;
- Landlords who take out insurance in joint names with tenants; and

- Property management service companies that arrange insurance on behalf of clients.

Any businesses carrying out such activities, unless otherwise exempt, are required to obtain authorisation from the FSA or become an appointed representative or introducer-appointed representative of a firm authorised by the FSA. It is a criminal offence to carry out any of the regulated activities without obtaining authorisation or becoming an appointed representative or introducer-appointed representative, even if the business is contractually obliged to provide these services.

The FSA provides detailed guidance to businesses involved in the marketing or sales of insurance and other intermediary services in the form of the FSA Handbook, which is available online, though the subject matter is extremely complex, and professional advice is recommended. The FSA also provides advice online for consumers.

See also: Directors' responsibilities, p.227; Driving at work, p.250; Health and safety at work, p.361; Occupiers' liability, p.524.

Sources of further information

Association of British Insurers: www.abi.org.uk

Motor Insurers' Information Centre: www.miic.org.uk

Financial Services Authority: www.fsa.gov.uk

FSA Handbook: http://fsahandbook.info/FSA/html/handbook

International Standards

Kate Gardner, Workplace Law

Key points
- Every organisation would like to improve the way it operates, whether that means improving the quality of a product or service, advancing its environmental performance, improving its energy efficiency or ensuring that its health and safety provisions address changing legislation and protect the workforce.
- Management system standards can provide the necessary controls to address risks and monitor and measure performance in your business, as well as enhancing its image and reputation.

Quality management

In a competitive market environment it is essential that organisations provide products and services of the highest possible quality in order to meet growing customer demands. Improved quality cannot happen immediately but is a continual process that evolves over time and with experience.

ISO 9001 is by far the world's most established quality framework, currently used by around 1,064,000 organisations in 178 countries worldwide. It sets the standard not only for quality management systems, but management systems in general.

Part of a larger family of quality management standards, ISO 9001 is published by BSI in the UK as BS EN ISO 9001 'Quality Management Systems. Requirements.' This Standard can help all kinds of organisations to succeed through improved customer satisfaction, staff motivation and continual improvement.

ISO 9001 specifies requirements for a Quality Management System where an organisation:

- needs to demonstrate its ability to consistently provide products that meet the requirements of the customer and any applicable regulatory requirements; and
- aims to enhance customer satisfaction through the effective application of the management system. This includes processes for continual improvement of the system and the assurance of conformity and applicable regulatory requirements.

Many organisations using ISO 9001 also go on to seek independent certification by an assessment provider such as BSI. This can be used to make a statement to existing and potential new customers about the organisation's commitment to quality.

ISO 9001 is designed to be compatible with other management systems standards and specifications, on subjects such as occupational health and safety, environmental and energy management.

Environmental management

With more environmental controls and requirements being placed on organisations, it is essential that management can demonstrate to customers, owners and other stakeholders that they are implementing responsible environmental management practices.

ISO 14001 was one of the first Standards to address environmental management and is now widely used around the world. In the UK it is published by BSI as BS EN ISO 14001: 2004 'Environmental management systems. Requirements with guidance for use.'

The Standard is designed to address the delicate balance between maintaining profitability and reducing environmental impact. With the commitment of the entire organisation, ISO 14001 can enable achievement of both objectives. The business benefits of implementing these environmental management practices include cost savings, reduction in resource wastage, legislative compliance and a greener organisation profile.

ISO 14001 specifies the requirements for an environmental management system (EMS). Rather than stating specific environmental performance criteria itself, ISO 14001 provides a framework for an organisation to control the environmental impacts of its activities, products and services, and to continually improve its environmental performance. It applies to those environmental aspects that the organisation can control and over which it can be expected to have an influence.

Implementation of ISO 14001 should help an organisation implement, maintain and improve an environmental management system. As with ISO 9001, external certification to the Standard can be used to demonstrate conformance to customers and other stakeholders.

Energy management
Today more than ever, effective energy management is a crucial issue for the

success of any organisation. A BSI survey carried out in 2009 showed that of 800 public and private sector UK organisations, only half rated their energy management practices as good or very good. This is despite 78% saying that energy management is either important or very important to their senior management team. This demonstrates a willingness to engage with energy management issues but a shortfall in actual delivery.

For many, the answer is an Energy Management System – a framework for the systematic management of energy. EN 16001 is the management system standard that will potentially enable organisations to make energy cost savings and reduce their greenhouse gas emissions. Published as BS EN 16001 by BSI in the UK, the Standard demonstrates how to establish the systems and processes necessary to improve energy efficiency across all operations. EN 16001 will help in the implementation of planned actions cited in the EU Energy Services Directive (2006) and requires organisations to take into account relevant legal and legislative obligations, such as the CRC Energy Efficiency Scheme (p.215).

BS EN 16001 'Energy management systems – Requirements with guidance for use' specifies the requirements for an energy management system which requires the development of an energy policy, identification of an organisation's past, present and future energy consumption, as well as the development of an energy monitoring (metering) plan. Analysis of actual versus expected energy consumption allows businesses to put plans in place to help improve efficiency. Rather than prescribing exactly how operations should be run, BS EN 16001 provides the framework that will enable effective energy management. The framework goes beyond considering a range of technical solutions to areas of major energy consumption: it begins the process of behavioural change needed to embed energy efficiency considerations in everyday decision-making.

The management system format of Plan – Do – Check – Act ensures that BS EN 16001 can be used by any organisation, irrespective of size, structure or complexity, benefiting both multi-national corporations and small and medium-sized enterprises. The requirements of the Standard can also be aligned with those of other established and widely used management systems, such as quality and environmental management.

Organisations wishing to demonstrate compliance with the requirements of BS EN 16001, or to assure their customers that they have an appropriate energy management system in place, will be able to do so via independent certification.

Occupational health and safety

In 2009/10, 1.3 million people suffered from an illness they believed was caused or made worse by their current or past work, with 28.5 million work days lost as a result of work-related ill health and 5.1 million due to workplace injury. As a result of statistics such as these, recent years have seen greater emphasis being placed on the management of occupational health and safety in legislation, which all employers must embrace. Any organisation with five or more employees must carry out health and safety risk assessments and act upon their findings.

BS OHSAS 18001 'Occupational health and safety management systems. Requirements' is the internationally recognised assessment specification for an occupational health and safety management system. It is designed to enable an organisation to control its occupational health and safety risks and improve its performance. Like other management systems, BS OHSAS 18001 does not state specific occupational health and safety performance criteria.

It is intended to provide a framework that allows an organisation to consistently identify and control its health and safety risks, reduce the potential for accidents, aid legislative compliance and improve overall performance. OHSAS 18001 was developed by a selection of leading trade bodies, international standards and certification bodies to address a gap where no other third-party certifiable international standard exists.

Integrated management

ISO 9001, ISO 14001, EN 16001 and OHSAS 18001 are designed to be compatible. BSI now offers an integration management system requirements specification, PAS 99 (Publicly Available Specification), which enables the alignment of management system processes and procedures into one holistic structure. Integrated management is suitable for any organisation, regardless of size or sector, looking to integrate two or more of their management systems into one cohesive system with an aggregated set of documentation, policies, procedures and processes.

See also: BREEAM, p.67; The CRC Energy Efficiency Scheme, p.215; Energy performance, p.270; Environmental Management Systems, p.286; Health and safety management, p.388; Occupational health, p.519.

Sources of further information

British Standards Institute: www.bsigroup.com

OHSAS 18001 is an international standard enabling businesses to measure and demonstrate their commitment to occupational health and safety. Workplace Law's *Understanding OHSAS 18001 course* enables you to understand the benefits and limitations of the OHSAS management system, and practical guidance on how to achieve accreditation. When you are OHSAS 18001 certified your clients can be confident that you manage occupational health and safety, as well as confirming that you exercise due diligence and so reduce the risk of a firm's directors facing Corporate Manslaughter prosecution. For more information visit www.workplacelaw.net.

Interviews under caution (PACE)

Esther Woodhouse, Greenwoods Solicitors LLP

Key points

- If you remember nothing else in the panic of being asked to attend an interview under caution, ensure legal advice is obtained before and during the interview process. There are significant benefits to be gained by attending with a legal representative even if your mind set is 'I don't need one. I haven't done anything wrong'.
- The interview is usually tape-recorded or a contemporaneous record is made. Remember, whatever is said during an interview under caution may be used in evidence if your case is brought to trial. Be cautious about your responses.
- The 'caution' says 'You do not have to say anything. But, it may harm your defence if you do not mention when questioned something which you later rely upon in Court. Anything, you do say may be given in evidence'. It can be a minefield not knowing whether to answer questions or remain silent. Each set of circumstances will be very different and there is no right or wrong answer. However, answering questions is not always the best policy. You have the right to remain silent and exercising this right can protect your interests and may actually avoid advancing the prosecution case against you or the company.
- Some interviews may be conducted as a voluntary interview, for example at the HSE local office. In these circumstances, don't automatically assume you are required to attend. Understanding your legal rights is vitally important to you and your employer.

Legislation

- Police and Criminal Evidence Act 1984 (PACE 1984).
- PACE 1984 – Codes of Practice A-H.
- Criminal Justice and Public Order Act 1994, sections 34-38.
- Criminal Justice and Police Act 2001.
- Serious Organised Crime and Police Act 2005.
- Criminal Justice Act 2006.

Introduction

For many, visiting the police station is sadly a regular, even daily, occurrence and no doubt such individuals will become very familiar with the protocol and interview process over time, becoming immune to the many time-consuming processes, procedures, red tape and the fraught situations which can and do arise in such an environment. However, for many, being arrested for a criminal offence or simply being asked to attend a voluntary interview before the HSE on behalf of their company will not be within their comfort zone and will be a daunting prospect giving rise to extreme anxiety, concern and sleepless nights.

Police procedures can be protracted and the extent of police powers is vast and wide. Did you know that upon arrest the police have power to conduct a personal search of your person, they may search your home and they may seize your property? Are you aware that DNA and fingerprints will be taken with or without your consent?

Voluntary interviews should cause less anxiety, albeit still be a cause for concern, particularly if you have never attended one before. These types of interviews are still conducted formally but you are not under arrest and are free to leave at any time. Understanding your rights is vitally important.

Legal rights

The Police and Criminal Evidence Act 1984 (PACE 1984) provides a legislative framework by which the police deal with crime in England and Wales. It is aimed at balancing the powers of the police with the rights of the individuals it may concern. It deals with every aspect of a criminal investigation including powers of arrest, period of detention in custody, access to legal advice, taking of non-intimate/intimate samples, fingerprints, DNA and photographs, etc. In accordance with the Act, Codes of Practice have been developed that govern the exercise of those powers. The Codes are defined as follows:

- *Code A*: Stop and search.
- *Code B*: Searching of premises and seizure of property.
- *Code C*: Detention, treatment and questioning.
- *Code D*: Identification.
- *Code E*: Audio recording of interviews with suspects.
- *Code F*: Visual recording of interviews.
- *Code G*: Power of arrest.
- *Code H*: Detention, treatment and questioning of persons under S.41(8) Terrorism Act 2000.

Code C is the most relevant to interviews as its main purpose is to ensure that all persons who are suspected of committing an offence, and others who are in police custody, are dealt with fairly, properly and in accordance with the law.

Prior to the commencement of an interview under caution the following three basic rights should be provided:

1. The right to obtain the advice of a solicitor prior to and during the interview process.
2. The right to notify someone that you are being interviewed.
3. The right to consider the rule book called the Codes of Practice.

Objective of solicitor / legal representative in the interview

In understanding why it is important to you and/or your business to instruct a legal representative you may find it useful to understand their role. In the first instance they are not merely there to act as an observer and, unfortunately, are not there to answer questions on your behalf. Their primary objective is to protect and advance your legal rights.

The objective is a continuing role throughout the interview and wherever it may be necessary. In accordance with Section 58(1) of PACE 1984 and Code C:6D (notes for guidance) *'a suspect arrested and held in custody at a police station or other premises shall be entitled, if he so requests, to consult a solicitor privately at any time'*.

The service you receive from your legal representative not only deals with your private consultations and advice on preparing for the interview but will extend to the following important areas:

- They will determine the legality of your detention and/or the purpose of the interview.
- They will seek to obtain 'adequate' and 'sufficient' disclosure which will enable them to assess the evidence against you and be in a position to advise you appropriately on whether it

is in your interest to answer questions or remain silent.

- They will be experienced at intervening during the interview process (if necessary).
- They will seek clarification of any issues or questions which appear ambiguous.
- They will be experienced to challenge the police and/or investigating officers if they appear to be acting unfairly or asking inappropriate questions. Many investigators will often go a 'step too far' in obtaining evidence to support their case when in reality this is nothing more than a tactic to induce you to making a confession!
- They will take an accurate note of the disclosure, your instructions, the advice provided and your answers during interview. Although a tape may be provided of your evidence in due course it may be some months before such is provided so the notes of your evidence are important to the advancement and progression of your case.

Interview under caution

Interviews are dealt in accordance with PACE 1984 and Codes of Practice, in particular Code C which deals with the detention, treatment and questioning of suspects. If you are a suspect in custody Code C1:1 specifies that 'all persons should be dealt with expeditiously and released as soon as the need for detention no longer applies' (Code C 1.1).

Interview procedure guidance

- The solicitor/legal representative will obtain information from the investigating officer relating to the circumstances of the arrest / allegation to determine the legality of the proceedings.
- Disclosure is sought to determine the strength of the prosecution evidence.
- You will provide your initial instructions on the disclosure produced.

- Advice will then be provided on whether to answer questions or remain silent.
- If you decide to remain silent in the interview by answering 'no comment' this does not preclude you from producing a document known as a 'prepared statement'. This strategy can be used during or after the interview and/or before charge. Its aim is to set out in brief terms the nature of your defence, avoiding adverse inferences from a failure to mention facts. It protects your interests and avoids you being put at risk of substantial questioning and scrutiny by the investigator. Other deciding factors may include a weak prosecution case, limited disclosure or as a suspect, you would not be able to stand up to questioning for reasons of anxiety or illness.
- The interviewing officer will commence the tapes, caution you before questioning and seek to clarify whether you understand the significance.
- The interview will commence with questions being put to you which relate to the nature of the allegation you are suspected of committing.
- At the conclusion of the interview you will be given the opportunity to add anything else you feel is relevant to your case.
- The tapes will be turned off and sealed in your presence and you will be asked to sign various interview management forms completing the interview process.

The caution

Prior to answering questions in interview a suspect must be cautioned (PACE 1984, Code C: 10.4). The caution says:

'*You do not have to say anything. But, it may harm your defence if you do not mention when questioned something which you later rely upon in Court. Anything you do say may be given in evidence.'*

PACE 1984, Codes of Practice 10.1 specifies:

'Where there is a person whom there are grounds to suspect of an offence they must be cautioned before any questions about an offence, or further questions if the answers provide the ground for suspicion, are put to them if either the suspect's answers or silence, may be given in evidence to a court in a prosecution.'

If a police / investigating officer fails to caution a suspect prior to interview, any evidence obtained during the interview may become inadmissible and excluded from any subsequent trial.

What does the caution mean?

The first part of the caution allows you to exercise your right of silence. You do not have to say anything! Remember, anything you do say may be used in evidence.

If you choose to exercise your right of silence, inferences will only be drawn if you later seek to rely upon facts during proceedings which you could have reasonably provided to the investigator at the time of the interview. These must be facts which were in your knowledge / possession at the time of the interview and would not extend to facts which came

into your possession after the interview. The effect of a suspect's failure to mention facts when questioned and/or charged is related by the Criminal Justice and Public Order Act 1994 ss34-38.

Finally, anything you do say will have been tape-recorded or recorded contemporaneously so either a copy of the tape may be played in Court or the recorded note produced in evidence.

Conclusion

At the end of the interview process you will either be released from custody pending further investigation or immediately charged with an offence. This decision is taken by the Crown Prosecution Service, who are the lawyers for the police. If you have attended a voluntary interview then it is likely to take many weeks, even months, before a decision is made on whether a prosecution is likely to proceed.

See also: Accident investigations, p.29; Competence and the 'Responsible Person', p.158; Corporate manslaughter, p.191; Workplace deaths, p.218; Emergency procedures and crisis management, p.261; Health and safety inspections, p.381.

Sources of further information

Police and Criminal Evidence Act 1984: www.workplacelaw.net/news/display/id/10154

PACE Codes of Practice: www.homeoffice.gov.uk/police/powers/pace-codes/

Ladders

Jagdeep Tiwana, Bond Pearce LLP

Key points

- Ladders account for the greatest number of major injuries to employees due to falls, and the HSE considers that misuse of ladders in the workplace is in part due to the way they are used in the home.
- It has therefore issued a guide for employers on the safe use of ladders and stepladders.

Legislation

- Health and Safety at Work etc. Act 1974.
- Manual Handling Operations Regulations 1992.
- Reporting of Injuries, Diseases and Dangerous Occurrences Regulations 1995 (RIDDOR).
- Provision and Use of Work Equipment Regulations 1998.
- Management of Health and Safety at Work Regulations 1999.
- The Work at Height Regulations 2005 (as amended).

Main legal duties

The Health and Safety at Work etc. Act 1974 imposes a duty on all employers to ensure the health and safety of all employees and third parties. The duty includes taking measures to control risks to such persons when using ladders or stepladders.

The Work at Height Regulations 2005 (as amended) go further by stating that an employer must ensure that a ladder is used for work at height only where a risk assessment under Regulation 3 of the Management of Health and Safety at Work Regulations 1999 has been carried out. For the use of ladders to be permitted, the risk assessment must show that the use of more suitable work equipment is not justified because there is a low risk and the work will be of short duration, or there are existing features on site that the employer cannot alter.

The hierarchy of control

The HSE advises that the first step in the safe use of ladders or stepladders is to assess whether this is the most suitable access equipment, taking into account the 'hierarchy of control', as set out in the Work at Height Regulations 2005 (as amended). The hierarchy of control requires that:

- work at height is avoided wherever possible;
- where it cannot be avoided, work at height should be carried out using equipment or measures that prevent falls from height; and
- where is it not possible to prevent falls from height, work equipment or other measures should minimise the distance and consequences of a fall, should one occur.

Once it has been established that work at height is necessary, an assessment should be carried out as to whether a ladder or stepladder is the most suitable access equipment compared with other options. See 'Work at height' (p.399) for further details of what is required under the Work at Height Regulations.

Selection of ladders

Only those ladders and stepladders that have sufficient stability should be used (taking into account the worst case scenario and worst type of surface conditions to be encountered). The HSE recommends Class 1 Industrial or EN131 ladders or stepladders for use at work. Ladders and stepladders should also be stored and maintained in accordance with manufacturers' instructions.

Safe places for using ladders or stepladders

Ladders or stepladders should only be used on firm and level ground or clean, solid non-slippery surfaces, free from oil, moss or loose material, so the feet can grip and where:

- they can be secured;
- the restraint devices can be fully opened and any locking devices engaged; and
- they can be put up at the correct angle of 75˚.

Ladders should not be allowed to rest against weak upper surfaces (e.g. glazing or plastic gutters).

Ladders used for access to another level should be tied and stepladders should not be used for this purpose unless designed for such use.

Ladders or stepladders should only be used in areas where they will not be struck by vehicles (and should be protected with suitable barriers or cones if necessary).

They should also not be used in locations where they could be pushed over by other hazards such as doors or windows. If this is impractical, the HSE advises that a person should stand guard at the doorway, or inform workers not to open windows until they are told to do so.

Safety checks

It is imperative to establish that the ladder or stepladder is in a safe condition before it is used. A daily pre-use check should be carried out. A ladder or stepladder should only be used if it has no visible defects. Checks should be carried out in accordance with manufacturers' instructions and should include checks to ensure:

- the ladder or stepladder is suitable for work use. (Class 1 or EN 131 ladders or stepladders. Domestic Class 3 ladders or stepladders are not normally suitable for use at work);
- ladder stability devices and other accessories are in working order;
- ladder and stepladder feet are in good repair (not loose, missing, splitting, excessively worn, insecure, etc.) and clean (as these are essential for preventing the base of the ladder slipping). Ladder feet should also be checked when moving from soft / dirty ground (e.g. dug soil, loose sand / stone, a dirty workshop) to a smooth, solid surface (e.g. paving slabs), to ensure the foot material and not the dirt (e.g. soil or embedded stones) is making contact with the ground; and

■ the ladder or stepladder has been maintained and stored in accordance with the manufacturer's instructions.

All ladder or stepladder checks should be recorded. Ladders that are part of a scaffold system need to be inspected every seven days.

Use of ladders and stepladders

There are a number of factors that should be taken into account when using ladders or stepladders. These are set out below:

■ Ladders or stepladders should only be used for light (not strenuous) work and should be in one position for a maximum of 30 minutes.

■ Wherever possible, a handhold should be available. Where this cannot be maintained (either at all or for a brief period of time only), other measures will be required to prevent a fall or reduce the consequences, should a fall occur.

■ In relation to stepladders, where a handhold is not practicable, a risk assessment will be needed to justify whether use of the stepladder without a handhold is safe or not.

■ The user should be able to maintain three points (hands and feet) of contact at the working position.

■ The user should never overreach and should keep both feet on the same rung.

■ The user should never overload the ladder or stepladder, i.e. the person and whatever they are taking up should not exceed the ladder's maximum load.

■ A detailed manual handling assessment will be required where a worker is carrying more than 10kg up the ladder or steps.

■ The user should avoid working in a way that imposes a side loading – the steps should face the work activity wherever possible.

■ The user should avoid holding items when climbing, for example by using a belt.

Training

Employees should only use a ladder or stepladder if they are competent. Users should be trained and instructed in how to use ladders and stepladders safely. They should also know how to inspect a ladder before using it and should in particular be aware of the following:

■ The need to ensure that the ladder or stepladder is long enough.

■ In the case of ladders, the top three rungs should not be used, and ladders used for access should project at least one metre above the landing point and be tied. Alternatively, a safe and secure handhold should be available.

■ In the case of stepladders, the top two steps of a stepladder should not be used unless a suitable handrail is available on the stepladder and the top three steps of swing-back or double-sided stepladders should not be used where a step forms the very top of the stepladder.

■ The ladder or stepladder rungs or steps must be level. This can be judged by the naked eye. (Ladders can be levelled using specially designed devices but not by using bits of brick or whatever else is at hand.)

■ The weather should be suitable – ladders and stepladders should not be used in strong or gusting winds (the manufacturer's safe working practices should be followed).

■ Users should wear robust, sensible footwear (e.g. safety shoes / boots or trainers). Shoes should not have the soles hanging off, have long or dangling laces, or slippery contaminants on them.

■ Users should be fit – certain medical conditions or medication, alcohol or drug abuse could stop them from using ladders safely.

- Users should know how to tie a ladder or stepladder properly.

When using a ladder or stepladder, users should not:

- move it while standing on the rungs / steps;
- support it by the rungs or steps at the base;
- slide down the stiles;
- stand it on moveable objects, such as pallets, bricks, tower scaffolds, excavator buckets, vans or mobile elevating work platforms; or
- extend a ladder while standing on the rungs.

Electricity
Ladders or stepladders should never be used within six metres horizontally of any overhead power lines, unless the line owner has made them dead or protected them with temporary insulation. If this is a regular activity, the employer should ascertain whether the lines can be moved. Non-conductive ladders or steps should always be used for any necessary electrical work.

> *See also:* Construction site health and safety, p.177; Head protection, p.355; Health and safety at work, p.361; Working at height, p.399; Risk assessments, p.624; Scaffolding, p.638; Slips, trips and falls, p.650; Work equipment, p.752.

Sources of further information

INDG 402 – *Safe use of ladders and stepladders: An employer's guide:* www.hse.gov.uk/pubns/indg402.pdf

INDG 405 – *Top tips for ladder and stepladder safety:* www.hse.gov.uk/pubns/indg405.pdf

HSE: www.hse.gov.uk

Landlord and tenant: lease issues

Kevin Boa, Pinsent Masons Property Group

Key points

- Leases are documents that pass exclusive legal possession/control or exclusive occupation of land or premises for an agreed period in return for rent, usually paid monthly or quarterly.
- The landlord (who is not necessarily the freeholder) retains his title and receives the rent.
- The tenant pays the rent and enters into covenants (obligations) restricting and regulating his use of the premises, and other aspects as well as positive obligations, including repairs and decoration.
- A lease may be transferred to a new tenant subject to restrictions within the lease. Usually the landlord's consent is required but is not to be unreasonably withheld.

Legislation

- Law of Property Act 1925.
- Landlord and Tenant Act 1927.
- Leasehold Property (Repairs) Act 1938.
- Landlord and Tenant Act 1954.
- Landlord and Tenant Act 1985.
- Landlord and Tenant Act 1988.
- Landlord and Tenant (Covenants) Act 1995.
- Commonhold and Leasehold Reform Act 2002.
- Land Registration Act 2002.

Advantages of leases

The granting or taking of a lease is an alternative to holding a freehold interest in property. There may be the following advantages to both parties:

- The landlord retains an interest in the property.
- The landlord can enforce both positive and negative obligations against the tenant.
- The lease may have investment value to the landlord, with the payment of rent creating an income stream.
- Payment of rent at regular intervals may be a more manageable cost

to the tenant than raising finance to acquire the freehold.

- The tenant can have flexibility as to the duration of his interest in the property.
- Leases are an effective means of splitting up the disposal of a multi-occupied building.

Heads of terms

- Heads of terms may be agreed by the parties before a lease or agreement for lease is entered into.
- Heads of terms represent agreement in principle as to the basic matters to be contained within the lease and any other related documents. The more detailed the heads of terms, the less argument there will be in agreeing the terms of the lease.
- Heads of terms will generally not be legally binding between the parties, although, at a practical level, they may be difficult to renegotiate once specific principles have been agreed. It is therefore important for a landlord or tenant to involve surveyors and lawyers at an early stage.

Commercial Lease Code

Many of the larger commercial landlords in England and Wales have adopted the Code for Leasing Business Premises in England and Wales 2007 (Lease Code 2007 – see *Sources of further information*), which is a non-binding code of best practice for landlords of commercial premises. It also provides a useful guide for tenants in lease negotiations and model heads of terms. The Code has government support, but it is voluntary and many commercial landlords choose not to offer leases that are Code-compliant.

The surveyor

Commercial property surveyors will generally represent each party's interest, and each party will usually pay his own surveyor's fees on the grant of a lease.

The surveyors will advise their clients on and negotiate such matters as:

- the premises to be let;
- amount of the rent;
- length of the lease;
- options to break the lease early;
- whether the tenant has statutory security of tenure;
- rent reviews;
- rights over common parts;
- management of the building;
- parking;
- use of the premises;
- restrictions on assignment;
- who is responsible for repairs;
- insurance responsibilities; and
- tenant's guarantee.

These should all be covered in the heads of terms.

Agreement for lease

The parties may enter into an agreement for lease in advance of the lease itself being granted. An agreement for lease is simply a contract between the parties for the landlord to grant and the tenant to accept a lease at a future date or following certain agreed conditions being satisfied within a set period.

The form of the lease has to be agreed and attached to the agreement for lease. Once an agreement for lease has been exchanged, neither party may unilaterally withdraw from the transaction without being in breach of contract (see *'Buying and selling property'*, p.107).

Leases for three years or fewer taking immediate effect do not have to be in writing. However, oral leases are highly undesirable because of the difficulty of proving the terms of the lease.

An agreement for lease can have advantages to both parties:

- It creates certainty that the lease will be granted, on the given date or on satisfaction of any relevant conditions. This assists both parties in forward planning.
- The agreement for lease represents a tangible asset for the landlord from an investment perspective, and the tenant will be able to prepare for his future occupation of the property.
- A landlord may wish to be sure he has a tenant for his premises before acquiring them or carrying out works to them.

Conditional agreement

An agreement for lease may be made conditional on any matters that either party requires to be satisfied before being obliged to complete the lease. Examples include:

- obtaining planning permission;
- carrying out of works by one of the parties;
- obtaining consents needed under a superior lease; and
- obtaining a licence to carry out the tenant's business (e.g. sale of alcohol).

Where an agreement for lease is conditional, the terms of the conditions are of fundamental importance. It is important that surveyors' and legal advice is obtained.

Principal issues in the lease

Parties
The lease will be granted by the landlord (or lessor) to the tenant (or lessee). The landlord will want to be satisfied that the proposed tenant has the financial status to pay the rent and other sums due under the lease and to perform the various tenant's covenants.

Generally speaking, from an investment standpoint, the stronger the tenant the more valuable the lease will be to the landlord.

The landlord will usually want to see evidence of the tenant's financial status. This may include, for instance, providing copies of accounts and references. If the tenant is a new company without a proven track record, or is not perceived to be financially strong, the landlord may require additional security, such as third-party guarantees or a rent deposit.

Property to be leased
The lease will need to clearly define the extent of the premises. This is particularly the case where the letting is only part of a building. Relevant issues include the extent to which parts of the exterior, roof, foundations and other structural parts are to be included or excluded. This will also affect the tenant's repairing obligations, which will generally be by reference to the extent of the premises that are leased. Access and parking must also be agreed.

Term
The length of the term is a basic issue that is likely to affect the other terms of the lease. From the tenant's standpoint,

the length of term will be influenced by the tenant's need for the premises in question, and any plans that the tenant may have for the future. New lease accounting standards may have an impact on the length of the term and the tenant's obligation to pay stamp duty land tax (SDLT), which can be considerable in the case of longer leases (e.g. SDLT for commercial premises with a rent of £100,000 a year plus VAT will cost the tenant over £15,000 for a 20-year lease, under £4,000 for a five-year lease and zero for a one-year lease). Break rights (rights to terminate early) do not reduce the tax.

From the landlord's perspective, a longer lease, with regular upwards-only rent reviews, will be more valuable from an investment standpoint, creating greater certainty as to the landlord's future income stream and an enhanced capital value.

Security of tenure
Subject to specific exclusions, business leases granted for more than six months benefit from security of tenure under the Landlord and Tenant Act 1954. This entitles the tenant to the grant of a new lease on the same terms and conditions other than rent (subject to reasonable updating) at the end of the original agreed term, unless the landlord has specific grounds to resist this right. The landlord can, subject to restrictions, override the tenant's rights on various grounds, including showing an intention to demolish or reconstruct the property and requiring possession for its own business.

Subject to these and other exceptions and to the tenant complying with various procedural requirements, the courts have power to intervene if the parties cannot agree on a new lease.

As the procedures for invoking these rights are both complex and prescriptive, it is important that you seek legal advice on any proposed renewal at the relevant time.

Contracting out

It is possible for the parties to contract out of the security of tenure provisions in the lease. To do so, the landlord must send a notice to the tenant, in a prescribed form. The notice explains the effect of contracting out of the security of tenure provisions. Two weeks later the tenant must sign a declaration to the effect that he has understood the terms of the notice and has agreed to contract out of security of tenure. The two-week period can be waived if the tenant signs a statutory declaration before an independent solicitor.

Contracting out of security of tenure is an important issue in negotiating a new lease and obtaining legal advice is recommended.

Licences

It is sometimes suggested that the landlord can prevent the Landlord and Tenant Act 1954 from applying by granting an occupation licence instead of a lease, so that the tenant does not have exclusive possession of the property which he occupies.

Licences are to be treated with caution by landowners as such arrangements cannot be guaranteed to be legally effective and the landlord could inadvertently end up having a protected tenant with security of tenure. However, they are often used for short-term arrangements in business centres.

Break options

Either the landlord or the tenant may be granted a right to terminate the lease before the expiry of the agreed term. From a landlord's standpoint, break options granted to a tenant may reduce the investment value of the lease. A tenant may, nevertheless, want the flexibility to terminate the lease early, either at a fixed date or dates, or possibly through a rolling break on or after a given period during the term.

The landlord may expect some incentive to agree to this, in the form of a higher rent or a compensation payment if the tenant subsequently exercises the break right.

Many landlords seek to impose conditions on the exercise of tenants' break rights, e.g. performance of tenant's covenants. These are strictly construed and constitute a trap for the unwary tenant as they can render the break right illusory.

Superior leases

If the landlord does not own the freehold to the property, but instead occupies the premises under an existing lease, then it will be necessary for a tenant to take account of the terms of any superior leases. This will be important for the following reasons:

- He must ensure that the superior lease lasts longer than the new under-lease.
- The superior lease(s) may require that consents are obtained from the superior landlord(s), in order to permit the underlease to the under-tenant, or to permit other matters such as changing the use or carrying out alterations.
- The landlord may want to pass on obligations under the superior lease by requiring the undertenant to perform these. From an under-tenant's standpoint this may not be acceptable or reasonable. For example, repairing or service charge obligations in a superior lease may be geared to a much longer term than the sub-lease.

- Where the superior lease contains restrictions affecting the use or occupation of the premises, the under-tenant will generally be subject to these same restrictions. It is important, therefore, that an under-tenant knows what these are and that it is able to comply with them.
- An under-tenant will want to ensure that its intermediate landlord does not breach its obligations to its superior landlord(s), as this could give rise to a potential forfeiture action by the superior landlord, which in turn could prejudice the under-tenant's position.

Rent

Rent is important for both the landlord and the tenant. It affects the affordability of the premises to the tenant and the income stream received by the landlord.

It is common for rents of commercial properties to be calculated by reference to a rate per square foot. In addition to the rate itself, the basis of measurement will need to be agreed. Surveyors commonly measure premises on either a net or gross basis. Your surveyor or solicitor will be able to advise you further on these issues. Other issues to consider include the following:

- Is the tenant to receive an initial rent-free period? This is particularly common where the tenant needs to carry out fitting-out works.
- Is the rent exclusive or inclusive of other payments, such as insurance premiums or service charges? Exclusive rents are most common in commercial leases, but a rent that is inclusive of insurance premiums and service charges may be appropriate for both parties on a short-term letting or where the premises are a small part of a much larger building.
- Will value-added tax be payable on the rent?

Rent reviews

If the lease is for more than five years, it is common for the landlord to require the rent to be reviewed at periodic intervals. The most common review cycle in today's market is for five-yearly reviews, although other cycles as frequent as three-yearly are sometimes agreed. It should be noted, however, that cycles more frequent than every five years can increase the tenant's stamp duty land tax costs.

Rents can be reviewed by reference to a fixed percentage increase or to an index such as the Retail Price Index. The most common system, however, remains upward-only review to the open market rent at the relevant review date. On each review date the rent becomes the rent that a new tenant would pay for those premises.

'Upward only' means that the rent will be increased to the open market rent, but, if rental values stay the same or fall during the review period, the rent will remain the same and will not fall. This means that the landlord's income stream is protected against falls in rental value in the future. Although the Government is keen to promote flexibility in the letting market, including 'upward or downward' reviews, this has yet to receive general acceptance. However, for the time being the threat of legislation in this area has receded.

Tenant's covenants

The issues of most frequent concern are as detailed below.

Permitted use

The landlord will generally want to restrict the use of the premises to a given category of use or uses (such as retail or office), to allow him to retain some control. If the permitted use is too restrictive, then this could have adverse rent review consequences for the landlord. The tenant needs to check that the permitted use gives sufficient flexibility for its present purposes and for any changes that might be needed in the future or for an assignee.

Alterations

The degree of control over alterations required by the landlord is likely to be influenced by the nature of the building and the length of the term. A landlord will not want the tenant to have the ability to carry out alterations that may adversely affect the future letting of the premises, and may also require the tenant to reinstate any alterations at the end of the term. The tenant will, again, need to ensure there is sufficient flexibility for any changes that may be needed, either now or in the future.

A distinction may be drawn, for instance, between alterations affecting either the structure or the exterior (which are generally barred) and those, such as the erection of internal partitioning, that affect only the interior and are non-structural (which are generally permitted with the landlord's consent).

Repairs

The lease will govern the extent to which the parties are liable for repairs. An institutional landlord will frequently want the lease to be on a full repairing and insuring basis (known as an FRI lease). The tenant would be responsible for any repairs that arise during the term, and the landlord would be entitled to have the premises back at the end of the term – sometimes these would be in a better state of repair and condition than they were at the start. Conversely, a tenant may look to limit his repairing obligations, particularly if the lease is for a relatively short term and the tenant's interest is therefore limited.

Other issues to consider include the following:

- Should there be a schedule of condition attached to the lease, providing evidence of the premises' condition at the start of the lease, and for the tenant's obligations to be limited accordingly?
- With new or recently constructed premises, will the tenant have the benefit of warranties from the building contractor, services sub-contractor and professionals responsible for the design of the premises and supervision of the work?
- The tenant may seek to avoid liability for inherent defects, covering defects arising from deficiencies in the building's initial design or construction.

When considering repairing obligations, it is important to bear in mind not only the tenant's obligations to repair the premises themselves, but also its liability to contribute to the cost of repairs to other parts of the building through the service charge.

Assignment and subletting

Unless there are special circumstances, such as a very short term or very generous break rights, a tenant will generally want the ability to assign the remaining period of the lease to a third party, or possibly to underlet either the whole or part. When the lease is assigned, the assignee becomes the landlord's tenant. A landlord will wish to keep control over who can occupy the premises in the future and to know that any substituted tenant will be able to pay the rent. It achieves this by making conditional the tenant's right to assign or sublet the premises. The landlord will always require the right to approve any new tenant or undertenant, but most leases stipulate that such approval cannot be unreasonably withheld or delayed.

Insurance

The lease should state who is responsible for insuring the premises and who pays the premiums. The landlord will frequently want to retain control for insuring the building but to pass on the premium, or an appropriate proportion of it, to the tenant. This is particularly the case in a multi-occupied building or where the landlord has a portfolio of properties that are insured on a block policy. If there is a superior lease, this may also specify insurance requirements. A landlord will also want there to be loss-of-rent insurance so that the landlord's income stream is maintained if the premises are damaged or destroyed so as to be unfit for occupation. The tenant normally pays the cost of this.

In recent years there has been a debate with regard to uninsurable risks, in particular terrorism, or flooding in a flood plain area. Currently most leases do not permit suspension of rent or termination by the tenant in such circumstances. Unless leases are amended, there is considerable risk to the tenant.

Position following destruction or damage

The parties need to consider whether either the landlord or the tenant (or both) should have any rights to terminate the lease if the premises are destroyed by an insured risk. If the lease is not terminated, the tenant will want to ensure that rent due is suspended and that the landlord has an obligation to reinstate the premises as quickly as possible so that the tenant can resume occupation.

Service charge

The lease may require the tenant to pay a 'service charge'. This is particularly common in the case of a multi-occupied building, which may include common parts, or where there are services such as security, reception facilities or cleaning. The service charge may also cover repairs and maintenance to the building as a whole, so that each of the tenants contributes an appropriate proportion of these costs. Service charge provisions are also common where premises are part of a larger estate, such as a retail or industrial park. In those instances, where there are other services such as car parking facilities, additional security and landscaped areas, the service charge payable by each tenant can be a considerable sum.

Particular issues include:

- the services that the landlord is required to provide;
- the basis of calculation of the cost and the likely amount of each payment;
- the method of payment;
- the ability of the landlord to vary, add to or suspend services;
- the ability of the landlord to set up a fund for anticipated capital expenditure;
- the tenant's right to check the landlord's expenditure; and
- any cap on the level of the service charge.

With commercial property, there are no statutory restrictions limiting the level of service charge that can be imposed, or requiring the landlord to undertake services at the most competitive cost. Issues such as these need to be dealt with in the drafting of the lease. The landlord's ability to pass on particular costs needs to be examined carefully against the length of the lease and the condition of the building (and sometimes the estate) of which it forms part.

Assignment of an existing lease

The lease may be sold or transferred, possibly at a price according to its value, to a new tenant (known as the 'assignee'). It is likely that consents will be needed from the landlord, in relation both to the assignment itself and to any changes that the new tenant wishes to make, such as changing the use of the property, or making alterations. The landlord will usually want his fees paid by the tenant, who may try to pass the cost on to the assignee.

The terms of the assignment need to be agreed between the assignor and the assignee – this is similar to the arrangements outlined in *'Buying and selling property'* (p.107).

The tenant will usually have no right to renegotiate the terms of the lease itself at this stage to accommodate the assignee.

In leases granted before 1996 the tenant will remain liable to the landlord for performance of the covenants in the lease, even after the lease is assigned. Generally this is not the case for leases granted after 1995, but the tenant may still have to guarantee the immediate assignee's performance of the covenants in the lease. This is a complex area of law on which legal advice should be taken.

Land registration

Leases of seven or more years must be registered at the Land Registry, as must leases taking effect more than three months from the date of grant. Exceptionally, shorter leases must also be registered where there are express rights within them over other property (e.g. common parts or services) as otherwise these rights are not enforceable. From mid-2006 registered leases have been required to contain certain prescribed clauses otherwise the Land Registry will refuse to register them. The Land Registry also has requirements in relation to the

plan to the lease; this must be to scale and have a north point, and it must be signed by the landlord.

Key questions

Preliminary issues

- Is a new lease to be granted or will the tenant acquire an existing one? If it is an existing lease:
  - Is there a superior lease?
  - Are any consents required from the freeholder or from superior landlords?
  - What professional advice will be needed to assist in the negotiation of the heads of terms and the lease documentation?
- Generally, has the proposed landlord or assignor a proper title?

Is an agreement for lease necessary or desirable?

- When does the tenant want to occupy?
- Are there any conditions that need to be satisfied before the lease can be granted or assigned?
- Are any works required prior to the grant or assignment of the lease?
- If works are required, who will carry them out, and at whose cost?

What are the costs?

- Surveyor's fees.
- Legal costs.
- Search fees.
- Survey / valuation fees.
- Stamp duty land tax.

Rent and other payments

- What is the initial rent? Will there be a rent-free period?
- Are there to be rent reviews? If so, at what intervals and on what basis?
- Are there any other expenses such as service charges and insurance premiums?
- Is there a premium?
- Is VAT payable? Can the tenant recover this?

Terms of the lease

- Who will be the landlord and the tenant?
- Does the landlord require guarantors or other security?
- What will be the contractual term?
- Will the tenant have security of tenure?
- Will either party get break rights?
- What is the extent of the property to be let?
- Does the lease provide sufficient control for the landlord, and sufficient flexibility to the tenant, as to issues such as permitted use, alterations and alienation?

- What are the arrangements concerning insurance and following events of damage or destruction?
- What services will the landlord be responsible for carrying out?
- What will the tenant have to contribute by way of service charge?

See also: Business rates, p.102; Landlord and tenant: possession issues, p.438; Property disputes, p.600; Shared premises and common parts, p.648.

Sources of further information

Code for Leasing Business Premises in England and Wales 2007 (Lease Code 2007): www.leasingbusinesspremises.co.uk

The Workplace Law website has been one of the UK's leading legal information sites since its launch in 2002. As well as providing free news and forums, our Information Centre provides you with a 'one-stop shop' where you will find all you need to know to manage your workplace and fulfil your legal obligations.

It covers everything from CDM, waste management and redundancy regulations to updates on the Carbon Reduction Commitment, the latest Employment Tribunal cases and the first case to be tried under the Corporate Manslaughter and Corporate Homicide Act, as well as detailed information in key areas such as energy performance, equality and diversity, asbestos and fire safety.

You'll find:

- quick and easy access to all major legislation and official guidance, including clear explanations and advice from our experts;
- case reviews and news analysis, which will keep you fully up to date with the latest legislation proposals and changes, case outcomes and examples of how the law is applied in practice;
- briefings, which include in-depth analysis on major topics; and
- WPL TV – an online TV channel including online seminars, documentaries and legal updates.

Content is added and updated regularly by our editorial team who utilise a wealth of in-house experts and legal consultants. Visit www.workplacelaw.net for more information.

Landlord and tenant: possession issues

Michael Smith, Pinsent Masons Property Group

Key points

- The Landlord and Tenant Act 1954 (as amended by the Regulatory Reform (Business Tenancies) (England and Wales) Order 2003) gives certain tenants of business premises the right to remain in occupation of their premises after the term of their lease expires and to apply to court for a new tenancy of those premises.

- A lease with the protection of the 'security of tenure' provisions in the Act will continue beyond the contractual termination date of the lease until either the landlord or the tenant serves a notice under the Act to bring it to an end (or a surrender of the lease is agreed).

- Following the service of notice under the Act, either the landlord or the tenant is entitled to apply to the court for a new lease to be ordered. The tenant will be entitled to a new lease unless the landlord is able successfully to apply to terminate the lease or to oppose an application by the tenant for a new lease on one or more of the grounds set out in the Act (*see below*).

- It is still open to the landlord and the tenant to negotiate voluntarily the terms of a new lease of premises without either party having to serve a notice commencing the formal lease renewal process under the Act. Provided that the tenant remains in occupation of the premises for business purposes, the lease will still be 'protected' under the Act.

- The Act does not protect a tenant who does not continue to occupy the premises for business purposes. Failure by a tenant to observe its lease covenants could also entitle a landlord to oppose the renewal of the tenant's lease.

Legislation

- Landlord and Tenant Act 1954 (as amended by the Regulatory Reform (Business Tenancies) (England and Wales) Order 2003).

Landlord's notice (Section 25)

A landlord of business premises can take steps to terminate a business tenancy by serving on the tenant a notice under Section 25 of the Act. The landlord must give between six and 12 months' notice to the tenant and stipulate in the notice a termination date not earlier than the contractual termination date of the lease.

A landlord must use the appropriate form of notice or a 'form substantially to the same effect' and the landlord should take care to use the correct version of the form, depending on whether the landlord does or does not oppose the grant of a new tenancy to the tenant.

If the landlord does not oppose the grant of a new tenancy to the tenant, he must set out his proposals for the terms of the new tenancy, including the property to be demised, the new rent, and other terms in the schedule attached to the notice itself. If the landlord does oppose the tenant's right to a new lease, the ground(s) of opposition must be stated in the notice (*see below*).

Tenant's notice (Section 26)
It is also possible for a tenant to start the lease renewal process by serving on the landlord a request for a new tenancy under Section 26 of the Act.

The request must state the date on which the new tenancy is to begin, which must be between six and 12 months after the making of the request and not earlier than the contractual expiry date of the lease. If the landlord wishes to oppose any application by the tenant for a new lease of the premises, he must serve a counter-notice within two months, stating the ground(s) of opposition on which the landlord intends to rely.

Applications to Court
At any time after the service of the landlord's Section 25 notice, and up until the termination date stipulated in the notice itself (or where the tenant has served a Section 26 request, any time after the landlord has served a valid counter-notice and up to the date immediately before the date specified in the Section 26 request), the tenant is able to make an application to court for a new tenancy of the premises.

Alternatively, if the tenant has not made an application to court and the landlord does not oppose the grant of a new tenancy to the tenant, the landlord can make an application to court itself for the grant of a new tenancy to the tenant.

Where there has been no application by the tenant or the landlord to renew the lease and the landlord has served a Section 25 notice (or counter-notice to a Section 26 request) stating that he is opposed to the grant of a new tenancy, the landlord may apply to court for a 'termination order'. This is an order by the court that the lease be terminated.

The Act specifies that the deadline for making any of these applications is the date specified in the landlord's Section 25 notice or the tenant's Section 26 request. This 'statutory period' can, however, be extended by written agreement between the landlord and tenant before the current deadline expires. There is no limit in the Act on the length of the extensions to the statutory period that can be agreed between landlord and tenant or the number of extensions that can be agreed.

Grounds of opposition
The seven grounds of opposition, which entitle the landlord to oppose the grant of a new lease if he can prove one or more of them, are:

1. Tenant's failure to repair;
2. Tenant's persistent delay in paying rent;
3. Substantial breaches of other covenants by the tenant;
4. Landlord willing to provide suitable alternative accommodation;
5. Tenancy is of part only and the landlord wishes to let property as a whole;
6. Landlord intends to demolish or reconstruct the premises; and/or
7. Landlord intends to occupy premises for his own business.

Grounds (1), (2), (3) and (5) confer a discretion on the court whether or not to order a new tenancy even if the ground is made out.

Grounds (4), (6) and (7) are not discretionary. If the landlord proves the requirements of the ground, the court must refuse to order a new tenancy.

If the landlord successfully opposes the grant of a new lease on grounds (5), (6) or (7) (known as the 'no fault' grounds for possession, i.e. no fault on the part of the tenant), it is required to pay compensation to the tenant equivalent to one or two times the rateable value of the premises as at the date of service of the landlord's Section 25 notice or the tenant's Section 26 request. Two times the rateable value of the premises (or 'double compensation') applies when the tenant or its predecessors in the same business have been in occupation for at least 14 years prior to the termination of the current tenancy.

Excluding security of tenure

It is possible for the landlord and tenant of business premises to agree to exclude the security of tenure provisions of the Act in respect of a business lease by following the exclusion procedure described in the Act.

In summary, the landlord must serve a prescribed form of notice on the tenant no fewer than 14 days before the tenant enters into the tenancy, or becomes contractually bound to do so. The tenant (or a person authorised by the tenant) must then make a declaration in the prescribed form (or substantially in that form) before he enters into the tenancy or becomes contractually obliged to do so.

If the landlord's notice is given fewer than 14 days before the tenant enters the tenancy or becomes contractually obliged to do so, the tenant (or a person authorised on his behalf) must make a statutory declaration in a prescribed form, which must be sworn before a solicitor or commissioner for oaths.

Once the 'security of tenure' provisions of the Act have been excluded in this way, the procedures described above concerning the service of notices and the making of an application to court will not apply and the tenant must leave the premises when the contractual term of the lease expires (unless of course the

landlord has granted, or entered into an agreement to grant, a further lease of the premises to the tenant).

Forfeiture

The 'security of tenure' provided by the Act does not protect a tenant who commits breaches of the lease, and leases invariably contain a forfeiture clause that allows the landlord to terminate the lease in the event of certain acts of default by the tenant.

There are two ways in which a landlord can forfeit a lease; either by:

1. issuing and serving court proceedings on the tenant for possession of the premises; or
2. by 'peaceable re-entry' whereby the landlord re-enters the premises and takes back possession by changing the locks.

Where the breach involves a failure to pay rent, the landlord will ordinarily have a contractual right under the lease to forfeit (by peaceable re-entry) after a period of usually 14 or 21 days from when the rent fell due.

Where the breach involves something other than non-payment of rent, the landlord must serve a notice on the tenant under Section 146 of the Law of Property Act 1925, which notifies the tenant of the breach; requires it to be remedied within a reasonable time (if the breach is capable of remedy); and requires the tenant to pay compensation for the breach. If the tenant still fails to remedy the breach within a reasonable time, the landlord can at that stage proceed to forfeit the lease, either by peaceable re-entry or the service of proceedings at the landlord's election.

Following forfeiture, the tenant (or any sub-tenant) does have the right to apply to court to ask for 'relief from forfeiture', which is effectively an order that the lease be reinstated. This is a discretionary remedy of the court, which will normally be granted provided that the tenant's breaches are remedied and the landlord's costs are paid. The court does, however, have a very wide discretion to impose, as a condition of granting relief, whatever other conditions it considers appropriate in the circumstances.

Peaceable re-entry / waiver

Forfeiture (particularly by peaceable re-entry) can be perceived as an aggressive and sometimes risky step for a landlord to take. Before taking this step (or serving a Section 146 notice as a precursor to forfeiture), a landlord must ensure he has not 'waived' the right to forfeit a lease, for instance, by demanding or accepting rent after a breach has been committed by the landlord.

If a landlord does attempt to re-enter the premises once the right to forfeit has been waived, the re-entry will be unlawful and the landlord may be sued for damages by the tenant who has been unlawfully excluded from the premises. A landlord also has a duty of care in respect of any items belonging to the tenant that remain in the premises after a re-entry.

It is a criminal offence for a landlord to re-enter if he is aware that someone is in the premises who is resisting the landlord's re-entry. The landlord cannot peaceably re-enter residential premises and must obtain an order from the court to recover possession.

See also: Dilapidations, p.221; Landlord and tenant: lease issues, p.429; Occupiers liability, p.524.

Sources of further information

The Workplace Law website has been one of the UK's leading legal information sites since its launch in 2002. As well as providing free news and forums, our Information Centre provides you with a 'one-stop shop' where you will find all you need to know to manage your workplace and fulfil your legal obligations.

You'll find:

- quick and easy access to all major legislation and official guidance, including clear explanations and advice from our experts;
- case reviews and news analysis, which will keep you fully up to date with the latest legislation proposals and changes, case outcomes and examples of how the law is applied in practice;
- briefings, which include in-depth analysis on major topics; and
- WPL TV – an online TV channel including online seminars, documentaries and legal updates.

Content is added and updated regularly by our editorial team who utilise a wealth of in-house experts and legal consultants. Visit www.workplacelaw.net for more information.

Legionella

Giles Green, Tetra Consulting Ltd

Key points

- Legionnaires' disease is always serious and is fatal in approximately one case in eight. It is preventable.
- Health and safety legislation requires the risk of legionnaires' disease to be controlled and applies to employers, the self employed or anyone in control of premises (the duty-holder).
- Formal guidance specifies in detail what is suitable and sufficient to constitute all reasonably practicable precautions to control the risk.
- Successful prosecutions have resulted in substantial fines and costs. Manslaughter prosecutions brought before the Corporate Manslaughter Act were unsuccessful.

Legislation

- The Health and Safety at Work etc. Act 1974.
- The Notification of Cooling Towers and Evaporative Condensers Regulations 1992 ('Notification Regs').
- The Reporting of Injuries, Diseases and Dangerous Occurrences Regulations 1995 (RIDDOR).
- The Management of Health and Safety at Work Regulations 1999 (Management Regs).
- The Control of Substances Hazardous to Health Regulations 2002 (COSHH).

Guidance

- The official Approved Code of Practice (ACoP), approved by the Health and Safety Commission, is the Legionnaires' disease – The control of legionella bacteria in water systems approved code of practice with guidance from the Health and Safety Executive. This combined document is universally known as 'L8', although strictly this designation applies only to the ACoP. The ACoP has special legal status and demonstrating that its provisions have been followed can be used as a defence, but it follows

that demonstrating its provisions have not been followed can be used in a prosecution, obliging the duty-holder to demonstrate compliance with the law in some other way.
- There is also a British Standard, BS 8580: 2010 Water quality – Risk assessments for Legionella control – Code of practice (BS 8580).
- In addition, guidance on spa pools (hot tubs) can be obtained from the Health Protection Agency, endorsed by the Health and Safety Executive, entitled Management of Spa Pools Controlling the Risks of Infection (Spa Pool Guide).

The cause and route of infection

Legionella bacteria occur in most fresh water, usually at such low levels that they are harmless. Once inside a water system, however, if conditions favour growth, they can multiply to levels that pose risk of infection, but this level does not equate directly to risk as there are other factors, including the degree of exposure and the susceptibility of those exposed, so there is no sharp divide between a 'safe' and 'unsafe' level.

At temperatures below 20°C, legionella do not grow, they remain viable but dormant; from about 20°C to about 45°C they multiply, especially in the 32-42°C range, and undetectably low numbers can increase to significant levels in a few days. From about 45°C to about 50°C, multiplication ceases but the bacteria are not killed; at temperatures above about 50°C legionella start dying at a rate that increases with temperature. Almost 100% kill can be achieved within a few minutes at 60°C.

The mechanism of infection is generally accepted as by inhaling minute droplets of water contaminated with legionella in a spray or mist known as an aerosol, such as is generated by a shower, cooling tower, hot tub or any of many other installations. There have been cases attributed to aspiration, or 'gagging' whilst drinking, allowing contaminated water to enter the lungs, but these are uncommon and usually associated with people in care who were lying down to drink.

Strategy

COSHH and the Management Regs require hazards such as legionella to be considered in the context in which they constitute a danger, i.e. the risk they pose; the simple and familiar concept of risk assessment. If the duty-holder does not have the expertise to carry out a risk assessment, the Management Regs require that expertise be brought in by training or by appointing a competent individual, either directly or from outside the organisation.

If the findings of the risk assessment are that the risk is insignificant, no further action is required beyond making sure that it remains so, but in many cases the risk will be deemed to be significant, so a scheme of control is required. L8 contains model schemes of control for cold and hot water systems and cooling tower systems and very brief guidance for several other systems, whilst the Spa Pool Guide provides very detailed guidance for spa pools and the findings of the risk assessment should be used to decide whether these models are suitable, whether they should be supplemented or could be reduced. In any case, COSHH requires the scheme of control to prevent or control exposure to the hazard, ideally by eliminating the hazard (in this case legionella bacteria), rather than allowing the hazard to remain and placing barriers such in its way. Personal protective equipment is identified as a last resort and is to be used in conjunction with other control measures.

As the requirement refers to exposure to the hazard, there may be exceptional instances where the hazard is allowed to remain, for example where there is no mechanism for generating an aerosol and no likelihood that there would be.

Specific controls

At the time of writing, not one case of legionnaires' disease has been attributed to any water system that was designed, installed, maintained and operated correctly; in every case one or more faults, be they physical, operational or human, have been the underlying cause, whatever the actual mechanism. The specific controls required vary from system to system, but all are based on the strategy of avoiding the conditions under which legionella multiply from harmless source levels to harmful levels in water systems.

Cold and hot water systems are controlled by keeping the cold water cold and the hot water hot, maintaining a moderate water throughput to displace any water which would otherwise have time for legionella to multiply, and keeping the water and system reasonably clean.

Where water temperatures cannot be maintained outside the range that prevents multiplication, such as in showers or heated water for use by anyone especially susceptible to scalding, the other controls must be relied upon and may require additional emphasis.

Cooling tower systems usually operate within the temperature range in which legionella multiply and spa pools operate close to the optimum of 37°C, so the controls almost always applied are chemical. It is, in theory, possible to use controls other than chemicals in cooling tower systems and spa pools; however, it would then fall to the duty-holder to demonstrate beyond reasonable doubt that the alternative regime achieved and maintained control at least as well as a conventional chemical regime (or that it maintained control adequately and had some additional substantial advantage).

Other systems identified by risk assessment as constituting a significant risk require controls that match the risk. Guidance in L8 is rather lacking in detail on many of these and additional guidance has been published by the Water Management Society – see *Sources of further information*.

Turning theory into practice

Given unlimited time and other resources, it might be within the capability of many duty-holders to control legionella using in-house personnel, or even alone. However, the combination of knowledge of water systems, the ecology of legionella, the aetiology of legionnaires' disease and the controls (including in many cases chemical controls) mean that specialist help is usually required. This can create a further problem, as the duty-holder is responsible for ensuring that anyone providing specialised help is competent to do so, which is difficult for the duty-holder to determine without at least some of that specialised knowledge.

See also: Working at height, p.399; Lifting equipment, p.448.

Sources of further information

L8 Legionnaires' Disease: the control of legionella bacteria in water systems: www.hse.gov.uk/pubns/priced/l24.pdf

Water Management Society: www.wmsoc.org.uk

There is a quality management scheme specifically for legionella control service providers, the Legionella Control Association www.conduct.org.uk . This is recognised by the HSE and is mentioned in L8, but, like any quality management scheme, it does not guarantee quality and responsibility remains with the duty-holder. Workplace Law's *Legionella Policy and Management Guide v.4.0* has been published to help employers in England and Wales ensure that they comply with their duties under law, and to provide a clear record of the policy and procedure. Visit www.workplacelaw.net for more details.

Lift safety

Kathryn Gilbertson, Greenwoods Solicitors LLP

Key points

- Lifts must be examined at statutory intervals – at least every six months if the lift is used at any time to carry people, or every 12 months if it only carries loads.
- All passenger lifts should be fitted with safety features to assist trapped occupants in the event of a mechanical breakdown.
- Building managers have a responsibility to keep records relating to the operation, maintenance and repair of passenger lifts in premises under their control.
- The safety of lifting equipment in general (including machinery for lifting loads, as well as people) is covered under the Lifting Operations and Lifting Equipment Regulations 1998.
- The greatest risk relates to the maintenance – rather than use – of lifts.

Legislation

- Health and Safety at Work etc. Act 1974.
- Lifts Regulations 1997.
- Lifting Operations and Lifting Equipment Regulations 1998.
- Management of Health and Safety at Work Regulations 1999.

Lift Regulations 1997

While the Regulations mainly govern the manufacture of passenger lifts and lift components, they are of interest to building managers who are responsible for the safe use of lifts and lift maintenance. The Regulations were introduced to ensure that manufacturers designed enhanced safety features into lifts, including:

- a means of two-way communication in the event of a breakdown;
- adequate ventilation during a prolonged stoppage; and
- the operation of emergency lighting systems during a power failure.

The Regulations require building managers to keep documentation such as:

- an instruction manual containing the plans and diagrams necessary for use;
- a set of guidelines relating to maintenance, inspection, repair, periodic checks and rescue operations; and
- a log book in which repairs can be recorded.

Lifting Operations and Lifting Equipment Regulations 1998 (LOLER)

If you are a lift owner or someone responsible for the safe operation of a lift used at work, you are a 'duty-holder' under LOLER. You have a legal responsibility to ensure that the lift is thoroughly examined and safe to use.

A thorough examination is a systematic and detailed examination of the lift and all of its associated equipment by a competent person.

Competent person

It is unlikely you will have the necessary competence in-house to undertake the lift examination. Accreditation by the

---

Case studies

On 14 May 2010, Holmes Place and ThyssenKrupp Elevator UK, were both fined £233,000 over health and safety breaches in relation to a 32-year-old's death.

The lift at Broadgate Health Club in the City of London had dropped the day before the tragic accident and had allegedly been repaired. It dropped again on 12 March 2003, killing Polish-born Katarzyna Woja. Ms Woja was the last person leaving the lift; she became trapped in its doors and was crushed to death when it dropped.

It is thought there was either a hydraulic problem, or the lift's so-called brain – a programmable logic controller – froze or crashed. The Court was told that between 7 January 2002 and 11 March 2003, there were 41 separate call-outs in relation to the lift that killed Ms Woja – compared with an average of three per year for a well-maintained lift. There were also numerous occasions when the lift failed and gym staff simply reset it without calling ThyssenKrupp.

It was the prosecution's case that the cause of the lift's frequent problems should have been identified, or the lift taken out of service.

Two residents of a block of flats in Southampton were killed when they fell against the lift doors following an incident that took place in the early hours of the morning. Both men plunged about 30 metres and died.

The incident happened because the lift doors opened in the style of a cat flap due to inadequate fixings on the lower rail of the opening.

The HSE noted during its investigation that the maintenance regime needs to take particular account of the door retaining system. Further, that maintenance is carried out on a regular and frequent basis, and that lift landing entrances and lift doors are designed and constructed to withstand the anticipated use.

United Kingdom Accreditation Service to BS EN 45004 is an indication of competence. Most insurance companies can recommend accredited inspecting organisations.

Examination should take into account the condition and operation of:

- landing and car doors and their interlocks;
- worm and other gearing;
- main drive system components;
- governors;
- safety gear;
- suspension ropes;
- suspension chains;
- overload detection devices;
- electrical devices (including earthing, earth bonding, safety devices, selection of fuses, etc.);
- braking systems (including buffers and overspeed devices); and
- hydraulics.

Notification of defects

You should receive a written and signed report of the examination within 28 days but if there is a serious defect that needs to be addressed you should expect to receive it much sooner. Notification of a serious and significant defect will necessitate the lift being taken out of service until the fault has been addressed. The competent person is also legally required to send a copy of the report to the Enforcing Authority.

Risk assessment

It is the ongoing maintenance of lifts that poses the greatest hazard. Regulation 5 of LOLER covers the safety of people working on lifting equipment, including provisions to prevent falling. A risk assessment should be undertaken before maintenance work on lifts is carried out and the maintenance instructions provided by the manufacturers and installers should be given careful consideration.

Permit to work

The person responsible for the building should ensure that a permit-to-work system is in place to control access to the lift shaft and to dictate a safe system of work.

Emergency procedures

Action on what to do in the event of people being trapped in lifts should be clearly documented. Nominated competent employees must be trained and regularly practise any procedures.

Signage

Lifts should not be used in the event of a fire, unless they have been specifically designed to do so and signs should be displayed to this effect. Many lifts are only supported by two-way communication in the event of breakdown during normal office hours. In order to control the risk of people becoming trapped in lifts outside these times, signage to this effect should also be displayed.

See also: Working at height, p.399; Lifting equipment, p.448.

Sources of further information

INDG 339 Thorough examination and testing of lifts – guidance for lift owners: www.hse.gov.uk/pubns/indg339.pdf

L113 Safe use of lifting equipment: www.hse.gov.uk/pubns/priced/l113.pdf

BS 7255: 2001: Code of Practice for safe working on lifts: http://shop.bsigroup.com/en/ProductDetail/?pid=000000000030067175

Lift and Escalator Industry Association: www.leia.co.uk

Lifting equipment

Dave Fray, Bureau Veritas UK Ltd

Key points

- Lifting equipment comes under the Lifting Operations Lifting Equipment Regulations 1998 (LOLER).
- 'Lifting equipment' is defined as work equipment for lifting or lowering loads, and includes attachments used for anchoring, fixing or supporting it.
- A lifting operation is an operation concerned with the lifting or lowering of a load. A load in this context also includes people.
- Accessories for lifting are work equipment for attaching loads to machinery for lifting. Examples include single items such as a shackle, slings, eyebolts, swivels, clamps and lifting magnets. It could also be an assembly of items such as a lifting beam and multiple slings.

Legislation

- Health and Safety (Safety Signs and Signals) Regulations 1996.
- Lifting Operations and Lifting Equipment Regulations 1998.
- Provision and Use of Work Equipment Regulations 1998.
- Supply of Machinery (Safety) Regulations 2008.

Equipment marking

Items must be distinctly marked with the safe working load (SWL) / the working load limit (WLL). Where the SWL/WLL varies, dependent on the configuration, then it must be marked to indicate its SWL/WLL at each configuration. Where it may be possible to mistake similar items, they should be marked with individual unique identification numbers. Relevant machinery and lifting accessories manufactured after 1995 should exhibit a 'CE' marking.

Inspection criteria

- All employers must have lifting equipment examined when it is first put into service.
- Where the safety of the equipment depends upon the installation conditions, it must be examined after installation and before being put into use for the first time and after reassembly at a new site or in a new location.
- Lifting equipment must be examined during its lifetime when it is exposed to conditions that cause deterioration.
- Lifting equipment must be examined each time that exceptional circumstances that are liable to jeopardise the safety of the lifting equipment have occurred.

Inspection frequency

- Lifting equipment used for lifting persons – at least every six months.
- Lifting accessories – at least every six months.
- Other lifting equipment – at least every 12 months.
- In any case, in accordance with an examination scheme.
- Each time that exceptional circumstances that are liable to jeopardise the safety of the lifting equipment have occurred.
- The user should have enough knowledge to examine the equipment for obvious defects each time it is used.

Employers' duties

- Lifting equipment must be of adequate strength and stability for each load, particularly when stress may be induced at mounting or fixing points.
- A load, including anything attached to it and used in lifting, must be of adequate strength.
- All lifting operations must be properly planned by a competent person.
- Lifting operations must be appropriately supervised and safe.

Users' duties

The user must:

- ascertain the centre of gravity and arrange each lifting operation;
- identify the weight of loads to be lifted; and
- only use the correct rated equipment for lifting.

Competent person

The competent person carrying out the thorough examination must have such appropriate practical and theoretical knowledge and experience of the lifting equipment to be thoroughly examined as will enable them to detect weaknesses and to assess their importance in relation to the safety and continued use of the lifting equipment.

In order to undertake thorough examinations, the competent person must have sight of either:

- the Declaration of Conformity; or
- the original test certificate; or
- the last thorough examination report.

If paperwork is not available, it is the responsibility of the competent person to decide whether a test is required or not, or if a thorough examination will be sufficient.

Proof load testing

A proof load must not be applied without reference to the manufacturer or the relevant British Standard.

Reporting of defects

The competent person must:

- report all defects immediately to the owner of the equipment;
- advise the employer immediately if a severe defect is found. This should be in writing and contain relevant information about the faulty equipment and the repairs that are required to be effected; and
- send a copy of the defect report to the relevant Enforcing Authority within a 28-day timetable or as soon as is practicable. This is limited to cases where there would be an existing or imminent risk of serious personal injury arising from failure of the equipment should anyone attempt to use it.

Examination report

The report should contain:

- the name and address of the employer for whom the examination was made;
- the premises address where the examination was undertaken;
- identification marks or number of the equipment examined;
- the date of the last thorough examination;
- the SWL/WLL;
- whether it is a report after first installation, a report within six or 12 months, in accordance with a written scheme, or after the occurrence of exceptional circumstances;
- the date of the next thorough examination;

- any parts that are found to be defective or in need of attention, the required repairs and the date by which they should be repaired;
- details of test required or undertaken;
- the name of competent person making the report and address of their employers; and
- the date of the report.

Hired lifting equipment

Both the hirer and the hire company have responsibilities when lifting equipment is brought on to premises. The hire company must ensure that adequate evidence is transferred with the lifting equipment to show that the last thorough examination has been carried out. The Responsible Person hiring the equipment must ensure that he has selected suitable equipment and that evidence has been obtained that the last thorough examination has been carried out.

See also: Work equipment, p.752.

Sources of further information

INDG 290 Simple guide to the Lifting Operations and Lifting Equipment Regulations: www.hse.gov.uk/pubns/indg290.pdf

HSG 6 Safety in working with lift trucks: www.hse.gov.uk/pubns/books/hsg6.htm

GS6 Avoidance of danger from overhead electrical power lines: www.hse.gov.uk/pubns/books/gs6.htm

PM28 Working platforms on forklift trucks: www.hse.gov.uk/workplacetransport/pm28.pdf

Lighting

Liz Peck, Society of Light and Lighting

Key points

- In general, there is little in the legislation governing lighting in the workplace, and what the law does require is mostly qualitative (sufficient and suitable) rather than quantitative. The basic requirements are set out in BS EN 12464:1 (*Lighting of Indoor Workplaces*) and BS EN 12464:2 (*Lighting of Outdoor Workplaces*).

- The other major provisions are found in the Health and Safety at Work etc. Act 1974 and the Health and Safety (Display Screen Equipment) Regulations 1992, but other legislation such as the Disability Discrimination Act 2005 (and subsequently the Equality Act 2010) also has to be complied with.

- The Building Regulations, and in particular Part L on conservation of fuel and power, have provisions regarding the energy efficiency and carbon emissions of lighting in new and refurbished premises, but do not prescribe lighting standards. As part of the implementation of the Energy Performance of Buildings Directive and in fulfilment of Government policy to make buildings more efficient, a new edition of Part L came into force in October 2010.

Legislation

- Health and Safety at Work etc. Act 1974.
- Health and Safety (Display Screen Equipment) Regulations 1992.
- Regulatory Reform (Fire Safety) Order 2005.
- The Restriction of the Use of Certain Hazardous Substances (RoHS) in Electrical and Electronic Equipment Regulations 2005.
- Building Regulations (Northern Ireland) 2006 Part F.
- The Waste Electrical and Electronic Equipment Regulations 2006.
- Building Regulations, Part L – Conservation of fuel and power 2010.
- Building Standards (Scotland) 2010.
- The Equality Act 2010.

Health and Safety at Work etc. Act 1974

The primary requirement in the Act is that lighting should be 'sufficient and suitable'; only the Courts can define what that is,

and there are very few reported cases. The HSE publishes the guide *Lighting at work* (see *Sources of further information*), which indicates what it believes to be the minimum acceptable standards. However, these cover only very general work types, defined in terms of the visual difficulty of the task. These levels are for safety, not effective or efficient performance of the task.

Emergency lighting is now required by various provisions of the Act and it should be assumed that it is required in all workplaces and most other buildings used by the public. The detailed requirements are in British Standards (*see below*). The Regulatory Reform (Fire Safety) Order 2005 replaced many existing references to fire in legislation with one consolidated document (*see below*).

Health and Safety (Display Screen Equipment) Regulations 1992

These Regulations include provisions for lighting of workplaces where display screen equipment (DSE) is used. This covers far more than just offices. DSE is found in many environments, including factory equipment controls and on forklift trucks in warehouses. Again, the requirements of the Regulations are general, but are supplemented by an HSE guidance document, *Display screen equipment work. Guidance on regulations* – see *Sources of further information*. This covers far more than just lighting, but provides useful practical guidance on issues that the HSE expects to see addressed.

Equality Act 2010

Lighting is one of the areas to which 'reasonable adjustments' need to be made if otherwise disabled people are likely to be disadvantaged.

Regulatory Reform (Fire Safety) Order 2005

Over the years, legislation on fire safety had grown enormously and been included in many Acts and Regulations. The decision was therefore taken to bring it all together in one place. The Regulatory Reform (Fire Safety) Order 2005 does this. There is little change to the legislation other than the significant issue of transfer of responsibility from the fire service to the building owner. This means that all building owners and users (e.g. those who rent or lease buildings or occupy them under other agreements) have to carry out a risk assessment and implement appropriate measures. This work can be carried out in-house or sub-contracted to consultants.

Other legislative provisions

There are many small references to lighting in legislation covering specific industries, especially foodstuffs and catering, and heavy industry involving hazardous processes. These are too numerous to discuss in this chapter.

Building Regulations

Lighting is covered in the Building Regulations, Part L – Conservation of fuel and power, the updated version of which came into force in England and Wales on 1 October 2010. Supporting the Regulations are four Approved Documents:

1. *L1A*: Conservation of fuel and power in new dwellings.
2. *L1B*: Conservation of fuel and power in existing dwellings.
3. *L2A*: Conservation of fuel and power in new buildings other than dwellings.
4. *L2B*: Conservation of fuel and power in existing buildings other than dwellings.

Additionally, there are two Compliance Guides:

1. Domestic Building Services Compliance Guide.
2. Non-Domestic Building Services Compliance Guide.

The revised Building Regulations are also being used to provide the means for implementing the EU Energy Performance of Buildings Directive, which requires energy labelling of buildings (see '*Energy performance*', p.270).

Scotland and Northern Ireland have separate legislation dealing with building control.

The Building Regulations (Northern Ireland) 2006 Part F is similar to Part L (2006) for England and Wales. The Department of Finance and Personnel (DFP) has produced two technical booklets:

1. *F1: 2006* – Conservation of fuel and power in dwellings.
2. *F2: 2006* – Conservation of fuel and power in buildings other than dwellings.

Section Six of the Building Standards (Scotland) 2010 for both Domestic and Non-Domestic buildings covers the lighting requirements, and these differ from those detailed in Part L for England and Wales

Domestic premises in England and Wales put on the market must have an Energy Performance Certificate for the property, which identifies the potential for improved efficiency of the building services. Lighting is specifically listed and it is expected that the recommendations given by the certificate assessors will allow the owner to apply for improvement grants where appropriate.

Part L requires proper commissioning of new lighting installations and their controls, and this requirement has been strengthened in the new Approved Documents and Compliance Guides. The Society of Light and Lighting and the Chartered Institution of Building Services Engineers (of which the SLL is part) published *Commissioning Code L: Lighting* in 2003 (see *Sources of further information*) to give guidance on acceptable procedures.

Part M of the Building Regulations and the Disability Discrimination Act are also relevant, although neither says a great deal specifically about lighting. However, lighting should be suitable for the visually impaired and, where appropriate, the hearing impaired. This normally means careful control of glare, adequate luminance levels, good colour rendering and good modelling (e.g. to enable the deaf to lip-read). It is not recommended that lighting levels be increased above those in published guidance, except where a building is being specifically designed for the use of the partially sighted and the type of vision problem is known. It is normally appropriate instead to provide supplementary local lighting to meet the needs of the individual. This legislation should be read in conjunction with *BS: 8300 Design of buildings and their approaches to meet the needs of disabled people*, recently revised (see below).

WEEE and RoHS

The Waste Electrical and Electronic Equipment (WEEE) Regulations came into force on 2 January 2007. There are specific requirements for recycling lamps and luminaires, and restrictions on dumping lamps containing mercury or sodium in the Landfill Regulations. The lighting industry has developed a recycling management framework to deal with the specific issues raised by discharge lamps (including fluorescent and compact fluorescent) and luminaires. Further information can be obtained from Lumicom (www.lumicom.co.uk) on recycling of luminaires and Recolight (www.recolight. co.uk) for recycling of lamps.

It is worth noting that it has been suggested that current UK office specification practice may have to change as a result of the WEEE Directive. It will no longer be acceptable for almost unused luminaires installed as part of the landlord's scheme to be replaced when a tenant arrives. This may drive developers towards a shell and core approach, with no lighting installed until the tenant's needs are known. Developers wishing to promote themselves as having environmentally-conscious credentials will particularly have to take this into account.

The Restriction of the Use of Certain Hazardous Substances (RoHS) in Electrical and Electronic Equipment Regulations 2005 include mercury, which is widely used in discharge lamps and fluorescent tubes. Schedule 2 lists exemptions for lamps where the amount of mercury is less than defined quantities.

However, the Environment Agency has determined that fluorescent tubes are hazardous waste in England and Wales and as such should be recycled or disposed of at the limited number of landfill sites that can cater for mercury-bearing waste. The Scottish Environmental Protection Agency has similarly designated fluorescent tubes as special waste.

Practical procedures associated with domestic waste of fluorescent tubes and compact fluorescent lamps have yet to be implemented.

Environmental issues

In September 2007, the Department for Environment, Food and Rural Affairs (DEFRA) announced the Government's plan for the phased withdrawal of energy-inefficient light bulbs by 2012. On 13 April 2009, the Domestic Implementing Measures come into force with the timetable as shown in Table 1.

Some special interest charity organisations have expressed concern about health risks of the 'energy saving' alternatives.

As a form of fluorescent lamp, these contain a very small amount of mercury (typically around 3-5mg) and as such should not be disposed of with household waste but should be recycled. Many manufacturers have developed alternatives to the compact fluorescent 'energy saving' lamps using tungsten halogen and LED technology and there are now dimmable CFLs available.

Best practice guidance

The major relevant Standard is BS EN: 12464-1: 'Light and lighting. Lighting of workplaces. Part 1: Indoor workplaces' (2010). However, the general reference document, which incorporates the relevant information from BS EN: 12464-1, is the SLL's *Code for Lighting 2009*, available as a CD-ROM. This provides extensive guidance on the lighting of all types of building and associated spaces. An updated print publication is due to be released in late 2011.

Equivalent information appears in separate ISO / CIE standards for the lighting of indoor and outdoor workplaces. The *Code*

Clear lamps			
Stage	Date	Phasing out	Replacements
1	1.9.2009	All clear lamps >9501m (~80w GLS)	Energy Class C
2	1.9.2010	All clear lamps >7251m (~65w GLS)	Energy Class C
3	1.9.2011	All clear lamps >4501m (~45w GLS)	Energy Class C
4	1.9.2012	All clear lamps >601m (~12w GLS)	Energy Class C
5	1.9.2013	Increased quality requirements	Energy Class C
Review	2014		
6	1.9.2016	All clear lamps >601m	Energy Class B
Non-clear lamps			
Stage	Date	Phasing out	Replacements
1	1.9.2009	All non-clear lamps	Energy Class A

Table 1. Timetable for phased withdrawal of energy-inefficient lightbulbs.

for Lighting and other Society of Light and Lighting recommendations have already been updated to take account of its recommendations.

The European Commission has asked CEN (the European Standards Organisation) to draft a number of Standards covering aspects of the Energy Performance of Buildings Directive; these are currently in draft. Two are directly relevant to lighting and others will have secondary impacts. Details can be found from the BSI (www.bsi-global.com).

Neither British Standards nor the recommendations of professional bodies such as the SLL have any legal status except as best current practice, but mandated European standards currently have to be followed for projects coming within the scope of the Public Procurement Directives, unless a specific reason is given why they are not appropriate.

The SLL also publishes a series of Lighting Guides covering specific building types. *Lighting Guide 7: Office lighting* includes recommendations on ensuring the lighting is suitable for offices containing display screen equipment. The essential change from previous versions of Society recommendations on this topic is relaxation of the luminaire luminance limits when modern screens and software are in use, and a recommendation for ensuring that there is adequate light on the walls and ceiling. The former Luminaire Category Rating system has now been withdrawn as it is no longer relevant.

The current guidance on emergency lighting is in *BS: 5266 Emergency lighting*. Part 1 of this Standard implements EN 50172 and is the Code of Practice; Part 7 is the UK implementation of EN 1838. Other parts of the Standard cover specific issues such as wayfinding. Further parts

are in preparation. Especially important is regular planned maintenance and testing of emergency lighting, and this is provided for in the Standards.

Note: Emergency lighting should include luminaires outside the final exit to the building to assist adaptation to outside night-time lighting levels. This is particularly important for the elderly and the visually impaired.

The SLL has published *Lighting Guide 12: Emergency lighting design guide*, which covers all aspects of emergency lighting from design to maintenance.

Exterior lighting of areas such as public car parks and access roads, including those in shopping centres, is covered by BS: 5489-1 'Code of practice for the design of road lighting. Lighting of roads and public amenity areas' (2003). However, this document is difficult to apply in many cases as it is intended for use by road lighting engineers. Guidance on practical aspects of car park lighting is in preparation to supplement the Standard.

BS: 8300 'Design of buildings and their approaches to meet the needs of disabled people' has few recommendations directly related to lighting, but many of its provisions have indirect relevance, e.g. requirements for appropriate visual contrast between doors and door furniture.

Frequently asked questions

1. *Is it mandatory to provide daylight in offices?* The answer is yes, if it is reasonably practicable – unlike the situation in countries such as Germany where it is mandatory. However, the interpretation of 'reasonably practicable' is open to wide variation, and in practice the provision has no force.

2. *Do employers have to provide the recommended illuminance if the*

employee requests a lower level?
There is no authoritative answer to this question as it has not been tested in Court, but it would be unwise to let the luminance in the space drop below 200 lux since this is the minimum in the European Standard for continuously occupied spaces and is also the level recommended for office tasks in the HSE's guide, *Lighting at work*.

3. *Must I provide emergency lighting in schools?* Because most users of schools will be familiar with the layout and use in hours of darkness by school pupils is very limited, it is not generally necessary to provide emergency lighting. However, if the building will be used out of hours or by individuals who are not familiar with the building, e.g. evening classes, parents' evenings, etc., escape lighting should be provided in those parts of the building so used. However, the recommendations in the Schedule of the SLL's Code for Lighting are for the task area, and lower levels of lighting are suggested for surrounding areas. In offices, the task area will normally be only the area of the desk.

See also: Accessible environments, p.18; Building Regulations, p.76; Display Screen Equipment, p.244; Waste Electrical and Electronic Equipment: the WEEE and RoHS Regulations, p.737.

Sources of further information

HSG 38 Lighting at work: www.hse.gov.uk/pubns/books/hsg38.htm

L26 Display screen equipment work: www.hse.gov.uk/pubns/priced/l26.pdf

Society of Light and Lighting and the Chartered Institution of Building Services Engineers: *Commissioning Code L: Lighting* (2003). ISBN: 1 903287 32 4.

SLL's *Code for Lighting 2009*: Available as a CD-ROM (ISBN: 978 1 906846 07-7).

Society of Light and Lighting: www.sll.org.uk

SLL – *Lighting Guide 7: office lighting*. ISBN: 1 903287 52 9.

SLL – *Lighting Guide 12: Emergency lighting design guide*. ISBN: 1 903287 51 0.

British Standards Institute: www.bsi-global.com

Lightning conductors

Helen Nicholson and Jonathan Riley, Pinsent Masons Property Group

Key points

- A lightning protector system protects buildings by providing a low-resistance alternative between the highest point of the building and the earth.
- Various consents may be needed for the installation of a lightning conductor. Planning consent will probably not be required, but listed building consent, conservation area consent or scheduled monument consent may be required.

Legislation

- Town and Country Planning Act 1990.

Purpose of lightning conductors

In certain atmospheric conditions a build-up of static electricity results in a discharge between the sky and the earth and the 'strike' may hit the earth at its highest point. A lightning conductor channels the strike and directs it towards buildings or structures along the path of least resistance to earth.

A lightning protection system can reduce the amount of damage that may be caused to buildings by providing a low-resistance alternative between the highest point of the building and the earth. At its most basic, it usually takes the form of a metal rod installed at the pinnacle of the building and connects via a rod to the earth.

Standards and specifications

Benjamin Franklin invented lightning rods in 1747 when he realised that attaching a conductor to a rod could divert the 'strike' of lightning harmlessly to earth. His concept remains the basis of lightning-rod designs today.

Evidently, the lightning protection system has been improved and modified since Franklin's time. The relevant specifications used to be found in BS 6651: 1999 'Code of practice for the Protection of Structures against Lightning,' which was replaced in 2008 by a suite of European Standards designated BSEN 62305. Compiled in four parts, BSEN 62305 is more complex than its predecessor and addresses in far more detail the risk management aspect and consequences of lightning, the various levels of protection to be considered, the design requirements for compliance, as well as the zonal protection of electronic and electrical systems within the structure. One of the most critical differences between the old and new Standards is to make the protections of electronic equipment an integral part of the Standard and to emphasise the use of coordinated surge protection devices.

Part Three of the new Standard corresponds to the main body of former Standard BS 6651 and clearly advises strict adherence to the provision of a conventional lightning protection system – to the total exclusion of any other device or system that may enhance protection.

The principal components of a conventional structural lightning protection system, in accordance with BSEN 62305, are as follows:

- *Air termination network.* This network is the point of connection with a lightning strike. It typically consists of a meshed conductor made of copper or aluminium tapes covering the roof of the structure, designed to intercept the lightning.
- *Down conductors.* These carry the current from the lightning strike safely to the earth termination network. Rods are attached to the external façade of the structure.
- *Earth termination network.* This is the means of dissipating the current to the general mass of earth. They can take the form of earth electrodes in the form of rods and plates or earth rod clamps, all of which serve the purpose of establishing a low-resistance contact with the earth.

However, there are other lightning protection systems available, outside the scope of BS EN 62305, including the 'early streamer emission' device, which is a modern alternative to the rod. The device emits a stream of charged particles upwards, which helps to attract the 'strike'. The lightning is then safely passed to the earth through the rest of the system. However, not everyone in the industry supports the use of this system.

The effect of a lightning conductor had been improved by using an isotope named americium-241 (a sealed source of ionising radiation) coating to the spike. However, because all radioactive material is subject to control by the Environment Agency, its use on lightning conductors has not been permitted since February 2000. Any remaining sealed conductors should be removed with expert advice and guidance. Great care must be taken in particular in handling the sealed conductors and ensuring appropriate disposal.

The British Standards Institution (BSI) provides advice on other British Standards relating to electrical installations and good earthing practice. There are also various international standards that provide useful guidance. The European International Electrotechnical Commission (IEC) prepares standards that serve as a basis for national standardisation. The UK is a member of the IEC through the BSI and the IEC issued BSEN 62305 in 2006 replacing the IEC 61024: 'Protection of structures against lightning' and IEC: 61312: 'Protection against lightning electromagnetic impulse' standards.

Planning consents

Planning permission is not normally required to erect a conductor on most buildings since it is not a 'development' for the purposes of the Town and Country Planning Act 1990. However, if a lightning protection system is sufficiently complex and has a degree of permanence as well as affecting the external appearance of a building, it may require planning permission.

If the building concerned is listed, a scheduled ancient monument or within a conservation area, care needs to be taken as it will be subject to special controls.

Listed building consent from the appropriate local planning authority is required. Any works that affect the character of a building of special architectural or historic interest must be authorised. It would depend on each individual case whether the erection of a lightning conductor would have this effect as location and materials are relevant factors. While compliance with BSEN 62305 ensures that technical issues are adequately addressed, it does not necessarily ensure that aesthetic criteria are met. Therefore placing the conductor behind a buttress, or otherwise out of sight, could avoid affecting the character of the building.

If the building is an ancient monument, it could be damaged by the attachment of a conductor or a lightning protection scheme. Any addition to a scheduled monument needs prior consent and this includes any machinery attached to a monument (if it cannot be detached without being dismantled). In all the above cases reference should be made to English Heritage or Cadw (as appropriate) or the local planning authority before erecting a conductor on an ancient monument or a listed building, since unauthorised works to either are a criminal offence.

If the building is in a conservation area, there may also be controls affecting the erection of a lightning conductor, and a preliminary check should be made with the conservation officer of the relevant planning authority before proceeding with any work.

See also: Planning procedures, p.585.

Sources of further information

British Standards Institution: www.bsi-global.com/index.xalter

Environment Agency: www.environment-agency.gov.uk

Local Exhaust Ventilation (LEV) systems

Maria Anderson, Workplace Law

Key points

- People who supply, own and use LEV have legal responsibilities under health and safety legislation. This includes the employer of the people being protected by the LEV, the supplier of the equipment and service providers. The principal piece of legislation is the Control of Substances Hazardous to Health Regulations 2002 (as amended), known as the COSHH Regulations.
- Regulation 7 of COSHH imposes a duty on employers to ensure that the exposure of his employees to substances hazardous to health is either prevented or, where this is not reasonably practicable, adequately controlled. Considerations include either changing the processes and/or control exposure at source, including adequate ventilation systems such as LEVs.
- The COSHH Regulations apply to a variety of workplaces such as factories, workshops, construction industry, etc. Wherever chemicals or substances (including dusts) classified as dangerous are used in their activities, LEV is an engineering control system designed to reduce exposures to airborne contaminants such as dust, mist, fume, vapour, or gas in the workplace.

Legislation

- Health and Safety at Work etc. Act 1974.
- The Management of Health and Safety at Work Regulations 1999.
- The Control of Substances Hazardous to Health Regulations 2002 (as amended).
- Construction (Design and Management) Regulations 2007.
- Supply of Machinery (Safety) Regulations 2008.

Introduction

Dust, fibres, vapours and gases are very common in many workplaces. From farms to factories, the consequences of inhaling these dangerous substances can be very damaging to health, and some of these illnesses do not manifest themselves for many years, making it very difficult to control. Occupational asthma and other lung diseases are the main problems when employees are exposed to airborne contaminants. A COSHH risk assessment must be carried out where the principles of good practice are considered to ensure that reasonable practicable measures are in place to prevent and control exposure.

What is an LEV?

An LEV is an engineering controls system to reduce exposure to airborne contaminants such as dust, mist, fume, vapour or gas in a workplace. Most systems, but not all, have the following components:

- *Hood*: Where the contaminant enters the LEV.
- *Ducting*: This conducts air and the contaminant from the hood to the discharge point.
- *Air cleaner*: Filters help to clean the contaminated air; however not all systems require the air to be cleaned.

Principles of good practice for the control of exposure to substances hazardous to health (Schedule 2A to the COSHH Regulations):

1. Design and operate processes and activities to minimise emission, release and spread of substances hazardous to health.
2. Take into account all relevant routes of exposure – inhalation, skin absorption and ingestion – when developing control measures.
3. Control exposure by measures that are proportionate to the health risk.
4. Choose the most effective and reliable control option which minimises the escape and spread of substances hazardous to health.
5. Where adequate control of exposure cannot be achieved by other means, provide in combination with other control measures, suitable personal protective equipment.
6. Check and review regularly all elements of control measures for their continuing effectiveness.
7. Inform and train all employees on the hazards and risks from the substances with which they work and the use of control measures developed to minimise the risks.
8. Ensure that the introduction of control measures does not increase the overall risk to health and safety.

LEV is an important option for controlling exposure and should be considered during the design stage of a new or refurbished structure where a workplace is likely to use dusts etc. within it. LEV isan effective way to minimise the escape and the spread of substances hazardous to healthand is a preferred option to using personal protective equipment, which has its obvious limitations.

- *Air mover*: The extraction system is usually a fan that helps the air to move from the hood to the discharge point.
- *Discharge*: This releases the extracted air to a safe place; if clean at this point can be directly discharged to the surrounding atmosphere.

Issues to consider

An LEV should be specifically designed for the process. Sometimes an LEV is not the best option as the contaminant is too extensive or too large. Remember, in all cases other options such as elimination or substitution should be considered first.

During the design process it is important that the following issues are considered:

- The key properties of airborne contaminants.
- How gases, vapours, dusts and mists arise.
- How contaminant clouds move with the surrounding air.
- The processes in the workplace, which may be sources of airborne contaminants.
- The needs of the operators working near those sources.
- How much control will be required.

Properties of airborne contaminant

Type of airborne contaminant

- *Dust*: Solid particles such as grain dust, wood dust, silica flour, etc.
- *Fume*: Vaporised solid that has condensed, e.g. rubber fume, solder fume, welding fume.

■ *Mist*: Liquid particles small enough to be airborne, e.g. electroplating, paint sprays, steam.

■ *Fibres*: A solid particle with a length that is several times the diameter, e.g. asbestos or glass fibres.

■ *Vapours*: The gaseous phase of a liquid or solid at room temperature, e.g. styrene, petrol, acetone or mercury.

■ *Gas*: A gas at room temperature, e.g. chlorine, carbon monoxide, etc.

Particle size of contaminants

■ *Inhalable particles*: Small enough to be breathed in by the respiratory system, size of 0.01 µm, up to 100 µm in diameter.

■ *Respirable particles*: Smaller particles that can penetrate deeply into the lungs, size of maximum 10 µm.

■ *Non-inhalable particles*: Particles above 100 µm are too large to be breathed in, they normally settle on the floor and surfaces near the work activity.

The airborne cloud is not always visible; for example, respirable dust is not visible to the human eye, and mist and fume clouds are more visible than the equivalent concentration dust.

Movement of particles in air

Particles move with the air in which they are suspended, for example particles bigger than 100 µm travel when they are ejected but settle quickly on surrounding surfaces; particles of less than 100 µm may travel around the process but do not travel far; smaller particles float and remain suspended in the air, maybe for several minutes, and move with air currents, which puts in danger of exposure people that may not be directly involved in the process.

An LEV needs to remove both inhalable and large particles, to prevent contamination and build-up of particles that can be lifted by air movements entering the worksite from windows or vents.

Other properties to be considered are processes that generate substances that are:

■ flammable or combustible substances, with the potential to produce an explosive atmosphere when airborne;

■ corrosive or abrasive that may attack the components of the LEV;

■ sticky dust or mist that will with time partially or completely block the LEV system; and/or

■ materials that may condensate in the LEV system, blocking partially or completely the pipe work or the filters.

Processes and sources considerations

Effective application of an LEV requires good understanding of the process and sources. Process means the way airborne contaminants are generated, for example cutting, shaping, sanding or grinding. The source of the contaminant is generated by a process.

Types of sources are:

■ buoyant, e.g. hot fume.
■ injected into moving air, e.g. by a spray gun.
■ dispersed into workplace air, e.g. draughts.
■ directional from equipment, e.g. cutting disk.

Source strength will determine the area that the contaminant will reach, and the concentration of the cloud formed. The LEV must consider the source strength to ensure full efficiency.

One process can create several sources at different stages; for example a grinding process can create directional cloud produced directly from the grinding disc, secondly from the containment layer or protective guards; also the dust deposited on clothing and surrounding areas must be considered.

Design

Hood design
Effective LEV systems contain or capture all the contaminant cloud within the LEV hood, and conduct it all away from the breathing zone of the operator. The degree of enclosure will depend on the type and source of contaminant. Every hood should have its own airflow indicator such as a manometer to ensure continual effectiveness of the control measure.

Very large, very small or moving sources of contaminants are very difficult to control with LEVs, so in these cases it is advisable that before introducing an LEV system, the source of contaminant should be eliminated or reduced so far as is reasonably practicable. Only after these considerations have been evaluated and/or implemented should an LEV system be introduced in the process.

Choosing the right hood type is essential. It should consider:

- How much the hood constrains the contaminant cloud.
- How well the LEV induced airflow carries the contaminated cloud into the system.
- How little of the contaminant cloud enters the process operator's breathing zone.

Hoods have a wide range of shapes, sizes, and designs, and the main aim is to ensure that most of the contaminants are caught by the LEV system.

Hoods can be classified in the following three categories:

1. *Enclosing hoods*: Always more effective than capturing or receiving hoods. A full enclosure is where the process is completely enclosed, e.g. a glove box. A room enclosure will enclose the process and the operator, e.g. abrasive-blasting rooms or booths.
2. *Receiving hoods*: The process usually takes place outside the hood. The hood receives the contaminant cloud, which has a speed and direction given by the process, e.g. a canopy hood over a hot process in the kitchen.
3. *Capturing hoods*: This is the most common type of LEV hood and is sometimes called a captor or capture hood. Like the receiving hood the process and source are outside the hood, but in this type of hood the LEV system must generate enough air flow to draw in the contaminant air.

Hood design considerations
- Maximise the enclosure of the process and source.
- Ensure the hood is as close as possible to the process and source.
- Position the hood to take advantage of the speed and direction of the airflow from the source.
- Match the hood size to the process and contaminant cloud size.
- Separate the contaminant cloud from the worker's breathing zone as much as possible.
- Use ergonomic principles when designing the application of an LEV hood.
- Try out the LEV selected.
- Use observation, information and good control practice, to assess exposure control effectiveness, e.g. smoke or dust lamp, take measurements such as air sampling where necessary.

The rest of the LEV system should conduct the contaminated air away for cleaning or discharge; parts of the system that must be designed are ductwork, fans, air cleaners and discharge points.

Ductwork connects the components of a ventilation system and conveys the contaminated air from the LEV hood to the discharge point. It consists of some or all of the following:

- Ducting from the hood;
- Dampers to adjust or balance the flow in different branches of the LEV system;
- Bends, junctions and changes in the duct diameter;
- Markings including test points and hazard warnings for the duct contents, e.g. hot, acid, gas, etc.; and/or
- A connection to the air cleaner and air mover.

Ducts can be either circular or rectangular in cross sections. Circular are preferable because they are lighter, withstand greater pressure differences and produce less noise.

When deciding on the best material the type of contaminant must be considered. The material chosen should give the best resistance consistent with cost and practicability, and have sufficient strength and supporting structures to withstand likely wear and tear.

Fans and other air movers

The fan is the most common air mover; it draws air and contaminant from the hood through the ductwork to discharge point. Categories of fan include:

- *Propeller*: Often used for general ventilation, they are light and inexpensive to buy and run and will not produce much pressure.
- *Axial*: Are not suitable for dusts, they are compact but do not develop high pressures and cannot overcome the resistance to flow that many industrial applications require.
- *Centrifugal*: Are the most commonly used fans for LEV systems, they generate large differences in pressure and can produce airflows against considerable resistance.
- *Turbo exhauster*: Can generate the high suction pressure needed to power low volume high velocity systems; they are not conventional fans.
- *Compressed air driven air mover*: Are appropriate where electrically powered fans are unsuitable, e.g. where access is difficult or where there are flammable dusts or gases. They are expensive to run and produce high levels of noise.

Air cleaners for particles consist of fabric filters, cyclones, electrostatic precipitators and/or scrubbers.

- Fabric filters are suitable for dry dusts; the fabric may carry electrostatic forces which help to attract and retain dust. Particles are removed by impaction (large particles meet the surface of the filter), impingement (medium sized particles meet the fibres within the filter) or diffusion (where small particles are attracted towards the fibre).
- Cyclones consist of a circular chamber where the air throws particles out to the wall by centrifugal action.
- Electrostatic precipitators are suitable for fine dusts; they give dust and fume particles an electrical charge and attract them onto collecting surfaces with an opposite charge.
- Scrubbers; the equipment wets the particles, washing them out of a contaminant cloud. The equipment will wet the particles causing them to settle out in water providing a suitable disposal system.

Air cleaners for gases and vapours use destructive methods:

- *Thermal oxidation or incineration*: Gases are destroyed before discharge by burning or thermal oxidation. Heat recovery may offset fuel costs.

- *Packed tower scrubbers for substances that mix with water*: A tower is filled with packing to provide a large surface area, water or a reagent solution flows in at the top of the tower and contaminant air enters at the bottom, the fluid absorbs the contaminant and clean air is discharged at the top. To avoid bacteria growth such as legionella or smells the system must be cleaned regularly.
- *Recovery methods such as absorption*: Contaminated air passes through filters that remove gases and vapours, activated carbon filters are the most common, air is filtered before it is passed through the filter and then clean air is discharged to the atmosphere.
- *Discharge to the atmosphere*: Extracted air must not re-enter the building unless it has been proved that it is clean and free of contaminant.

Other issues to consider when installing an LEV system include:

- Noise generated by the LEV system, for example from the fan, ducts, air travelling and vibrating hoods.
- Thermal comfort; some LEVs create draughts that may cause discomfort to operators.
- Lighting that would depend on the type of work.
- Access to the system from the operator to carry out the routine activities, and for maintenance operators to carry out testing, cleaning, inspection or maintenance and repair.

Examination and testing

Every LEV system requires statutory 'thorough examination and testing' by a competent person. The examination and testing report must have a prioritised list of any remedial actions for the employer.

Routine checks keep the LEV system running properly. The frequency of routine checks would depend on the type of contaminant and process. A trained employee is able to make routine checks and report any defects to supervisors; any faults must be put right immediately to ensure effectiveness of the control.

The thorough examination serves as an audit to ensure continual effectiveness of the system, and the objective of testing is to find defects and have them remedied to regain control.

The COSHH Regulations require thorough maintenance examination and test of control measures at intervals so that control remains effective at all times. The maximum time between tests of LEV systems is set down in the COSHH Regulations and for most systems this is 14 months, but in practice this is normally taken to mean annually. If wear and tear of the LEV is liable to decrease effectiveness, tests may be required more frequently.

Examination can be carried out by an external contractor or internal competent employee. Thorough examination and testing of LEV involves three stages:

- *Stage 1*: A thorough visual examination to verify the LEV is in efficient working order, in good repair and in a clean condition.
- *Stage 2*: Measuring and examining the technical performance to check conformity with commissioning data.
- *Stage 3*: Assessment to check the control of worker exposure is adequate.

As standard the examiner should attach a test label to each hood when tested, e.g. a red label with fail or a green label with date of next test. The examiner should also produce a report where it is clearly stated the data collected during the test, and any maintenance required with a clear action plan. The employer should plan and schedule such repair and retest to assure control.

See also: Confined spaces, p.173; Health and safety at work, p.361; Machine guards and safety devices, p.470; Noise at work, p.505.

Sources of further information

HSE – Local exhaust ventilation: www.hse.gov.uk/lev/index.htm

HSG258 Controlling airborne contaminants at work: www.hse.gov.uk/pubns/books/hsg258.htm

Loneworking

Kathryn Gilbertson, Greenwoods Solicitors LLP

Key points

- Loneworkers are those who work by themselves without close or direct supervision. Risk assessment is, therefore, essential.
- There are a number of legal provisions that specify systems of working that require more than one person. These include:
  - Electricity at Work Regulations 1989.
  - Work in Compressed Air Regulations 1996.
  - Diving at Work Regulations 1997.
  - Confined Spaces Regulations 1997.
  - Control of COSHH Regulations 2002.
- Further guidance is offered to employers by BS 8484, a Code of Practice that recommends loneworker devices.
- There are other provisions that require work to be done 'under the immediate supervision of a competent person' or similar wording, which would suggest that the work, although carried out by one person, must be done in the presence of another.

Legislation

- Health and Safety at Work etc. Act 1974.
- Management of Health and Safety at Work Regulations 1999.

Who is at risk?

- People who work separately from others in factories, warehouses, shopping centres, etc.
- People working on their own in petrol stations, shops, small workshops, homeworkers, security guards, etc.
- Mobile workers working away from their fixed base, e.g. engineers, sales representatives, breakdown mechanics, social workers, estate agents; the list is not exhaustive.

Remember that loneworking is not a formal categorisation of work – anyone who stays late at the office to finish off a report, or who pops in at the weekend to prepare for the coming week is working alone.

Risk assessment

As an employer, you need to be fully aware of all loneworking that is going on in your organisation, whether it is undertaken by people who are employed by you directly (such as your sales force) or by people who work on your premises (such as your cleaners). A risk assessment should be carried out for loneworking as with other areas of risk in the workplace. A risk assessment for loneworking needs to take particular account of the specific hazards associated with the work task and of the people who are carrying it out. Every loneworking situation will be different, but some common issues to consider are:

- *Access to and egress from the place of work*. Can the loneworker get to and from the workplace safely? Is the work being carried out in a confined space?
- *Nature of the work*. What sort of work is being undertaken? Are loneworkers

Case study

A Borders country estate was fined £3,000 following a loneworking incident that resulted in a gamekeeper's death. The victim was a temporary gamekeeper who sustained serious injuries when he overturned his quad bike on a slope. He was eventually found 200 yards from the scene of the accident and it would seem had been trying to get to a nearby farmhouse to raise the alarm.

He had not been issued with a phone (although the normal gamekeeper had) and had no means of communication through which he could summon help. Nobody had noticed he was missing for 52 hours. The Prosecution arose because the gamekeeper's death had not been immediate and if he had means of communication he would have had the opportunity to summon assistance.

Mental Health Matters Ltd, a North East based charity, was fined £30,000 and ordered to pay £20,000 costs after one of its employees was killed by a service user. The victim was a support worker who was stabbed to death whilst visiting a patient at his home. The attack happened on her final day of her probation period.

The prosecution told the court that the patient's mental health was known to be deteriorating and that the company failed to respond to a number of warning signs. It also failed to provide the victim with the level of protection needed for this work.

The HSE commented "if Mental Health Matters had carried out a risk assessment, it would have resulted in the visiting arrangements being reviewed".

dealing with the public, where they might face aggressive or violent behaviour? Do they have to carry heavy items, or work in outdoor weather conditions?

■ *Location of work*. Where does the work take place? Where work is carried out by mobile workers or off site, the employer will have little control over first aid provision and emergency procedures. Does work take place at height?

■ *Time of work*. When does the work take place? We are all naturally tired first thing in the morning and last thing at night. Are there any increased risks related to the time of day, such as pub closing time or rush hour? According to recent research tired drivers are the cause of one in ten accidents.

■ *Use of work equipment*. What, if any, work equipment do they need to use? Use of electrical equipment or machinery will increase the risk. Check that they have been trained how to use it.

■ *People*. Who are the people who are working alone? You will need to consider their age, maturity, experience, health and fitness, and general state of mind. Where young people or new and expectant mothers are concerned, the risks will be increased.

An evaluation of the risks should highlight the control measures that are required to ensure work is carried out in a suitably safe manner. Some common control measures for lone workers are the following:

- *Redesign of the task to eliminate the need for loneworking.* This can be done, for example, by changing shift patterns to implement a buddy system where two people work together at all times.
- *Provision of information, instruction and training.* This might include training in the safe use of work equipment, or how to handle aggressive behaviour when dealing with the public.
- *Establishment of communication and supervision procedures.* To ensure that a manager is able to contact the worker at regular intervals; to make sure that arrangements in the case of an emergency have been put in place; and to check that a loneworker has arrived back safely once work has been completed.

- *Use of loneworker devices (LWD).* BS 8484 provides guidance on the best practises to be adopted when using electronic devices to transmit the location, identity and voice to a monitoring centre and request assistance or offer additional personal security.
- *Provision of mobile first aid facilities.* To ensure that loneworkers can deal with minor injuries themselves.
- *Health surveillance of loneworkers.* At regular intervals, to ensure that workers are fit and healthy to carry out the tasks required of them.

> *See also*: Health surveillance, p.396; Homeworking, p.406; Night working, p.502; Risk assessments, p.624.

Sources of further information

INDG 73 Working Alone – health and safety guidance on the risks of working alone: www.hse.gov.uk/pubns/indg73.pdf

An increasing amount of organisations are employing loneworkers, in all areas of industry and business. Flexible Working Regulations have enabled employees to enjoy the benefits of working from home; today's 24–7 culture means that more companies are open around the clock; and greater automation in industry has meant a shift from the traditional 9–5 working day. Loneworking can be of great benefit to a business – but it also has its problems. Health and safety issues that affect traditional employees still apply – in some cases more so. Employment issues such as data protection and absence management are also paramount, and must be considered when employing people who work alone.

Workplace Law's *Loneworking 2008: Special Report* helps you get to grips with these issues, to determine whether loneworking is a viable option for your business. Using practical case studies, checklists and assessments, the report discusses the pros and cons of loneworking, and what an employer must do to overcome the dangers, implement healthy and safe working policies, and ensures their loneworkers are being cared for just as well as those in the traditional workplace. For more information visit www.workplacelaw.net.

Machine guards and safety devices

Kathryn Gilbertson, Greenwoods Solicitors LLP

Key points

- Operation of dangerous work equipment requires the use of machine guards and other safety devices, as specified under the Provision and Use of Work Equipment Regulations 1998.
- Failure to ensure that machine guards are used properly is harder for employers to defend under health and safety legislation because the duty to comply is more stringent and does not allow them to take into account mitigating factors such as time, cost and inconvenience.

Legislation

- Lifting Operations and Lifting Equipment Regulations 1998 (LOLER).
- Provision and Use of Work Equipment Regulations 1998 (PUWER).
- Health and Safety (Miscellaneous Amendments) Regulations 2002.

Statutory requirements

Guards are fitted on machinery as a control measure to prevent the risk of accident or injury caused by contact with moving parts. Machine guarding is covered by the Provision and Use of Work Equipment Regulations 1998 (PUWER). Regulation 11 states that measures must be taken that:

- 'prevent access to any dangerous part of machinery or to any rotating stock-bar'; or
- 'stop the movement of any dangerous part of machinery or rotating stock-bar before any part of a person enters a danger zone'.

It is worth noting that this requirement under health and safety law is to do what is 'practicable' – not 'reasonably practicable' – to comply. What this means is that, unlike many of the employer's duties under health and safety law, there can be no argument about the time, cost or inconvenience it takes to make sure guards are used. The only justification can be whether there is no technical solution to protect workers from the dangerous machinery in question – an unlikely argument to win in the event of an accident.

Types of guard and safety device

There are a number of types of guard and control device to protect from dangerous machine parts, including the following:

- *Fixed guards.* These are always in position, and difficult to tamper with, but can restrict access and make cleaning difficult.
- *Adjustable guards.* These can be adjusted by the user, or automatically by the machine as work is passed through them. They allow better access, but increase accidental risk of contact with dangerous parts.
- *Interlocking guards.* These ensure that equipment can only be operated when a moveable part connects to the power source, so that they default to safe. A good example here is the door of a photocopier machine, which disconnects the power when opened.
- *Trip devices.* These detect the presence of an operator within a

> **Facts**
> - In the food and drinks industry, machinery and plant causes:
>   - over 30% of fatal injuries;
>   - over 10% of major injuries (e.g. requiring hospitalisation);
>   - over 7% of all injuries (i.e. major injuries and over-three-day absence injuries); and
>   - almost 500 injuries per year reportable to the HSE.
> - When questioned, over 90% say that all or most of their machines requiring guarding are fitted with emergency stop controls. Around 75% have a system for preventive maintenance of guarded machinery. Over 60% say they always check that new machinery complies with safety regulations.
> - Analysis of injuries investigated by HSE in the food and drink industries over a four-year period highlighted the main types of machinery involved:
>   - Conveyors – 30%.
>   - Forklift trucks – 12%.
>   - Bandsaws – 5%.
> - When considering conveyors:
>   - 90% of conveyor injuries occur on flat belt conveyors;
>   - 90% of the injuries involve well-known hazards such as in-running transmission parts and trapping points between moving and fixed parts; and
>   - 90% of accidents occur during normal forseeable operations – production activities, clearing blockages, etc.

danger zone to shut off power, using trip switches, pressure pads or laser sensors.

- *Other measures.* These include two-handed control devices and hold-to-run controls, which default to safe (for example by isolating the equipment from its power source) if the operator releases them.

Risk control measures

One of the most common areas of risk with the use of machine guards is human intervention by the operator. This can be as a result of human error, in that guards have not been properly reinstated following cleaning or maintenance; or it can be as a result of overconfidence or negligence, where an operator thinks he can work faster or better without the guard in place. For these reasons, it is imperative that employers monitor work with dangerous

machine parts closely and at regular intervals, carry out regular inspections to ensure the safe operation of guards and safety devices, and ensure that operators are provided with all the information, instruction, training and supervision that is necessary.

Recent and proposed changes

While PUWER remains substantially in force, some amendments were introduced under the Health and Safety (Miscellaneous Amendments) Regulations 2002, which affect the employer's approach to work with dangerous machine parts. The HSE had been concerned under the original Regulations that there was too much leeway for employers to simply provide training and instruction to workers as a control measure when using dangerous machinery, instead of starting with the safest option of fitting

Case study

On 6 July 2011, ThyssenKrupp Tallent Ltd was fined £16,000 and ordered to pay £5,972 in costs following an accident on a welding machine. The victim was using the machine to weld nuts on to car parts when her left hand middle finger became trapped between the electrode and another part of the machine. She fractured her finger and suffered a severe electrical burn.

The HSE identified that the machine had no jig fitted to hold the work piece in place, which meant that the left hand was very close to unguarded moving parts and that the company had failed to follow its own procedures.

The court also heard that the company had been prosecuted in 2009 following a previous machinery guarding incident at its Cannock site. Further, that it was a well-known risk in this industry and as such employers must ensure that they have effective guarding in place at all times.

On 4 July 2011, Ineos Enterprises Ltd was fined £12,000 and ordered to pay £6,607 towards costs for ignoring basic safety guidelines after a maintenance worker severely injured his right hand. The victim lost his ring finger and suffered damage to his middle and little finger after his gloved hand was pulled into machinery.

The court was told that Ineos failed to follow health and safety guidance that advises against wearing of gloves when using metalworking lathes. It had introduced a policy in May 2010 making the wearing of protective gloves mandatory. Several employees were reprimanded for not wearing gloves. The victim was reminded to wear his gloves by his line manager on the morning of the accident.

His gloves got caught in the rotating mechanism and dragged his hand into the machine. The HSE investigation highlighted deficiencies with assessing risk from its glove policy and lack of guarding.

On 10 February, Studleigh-Royd Ltd was fined £14,000 following two accidents in quick succession at Cranswick Convenience Foods. One victim had his left arm amputated after it became trapped in the rotating knives of a poorly guarded meat tenderiser.

Workers were in the habit of overriding a magnetic sensor interlock. Bypassing the interlock made it quicker and easier to feed meat into the machine but it exposed them to dangerous moving parts.

The HSE noted that an adequate knowledge and understanding of the law covering the work equipment, combined with a proper risk assessment, would have identified the need for improved guarding.

fixed guards. Under the 2002 amending Regulations, 'information, instruction, training and supervision' are now seen as an additional requirement, but not as one of the principal control measures to manage risk when working with dangerous machine parts.

The UK is required by the European Commission to report on the introduction of national legislation brought in to comply with European Directives.

Next steps

Employers should:

- select work equipment that is suitable for its intended use in respect of health and safety;
- specify clearly the health, safety and hygienic design requirements for the supplier to meet (including noise levels); and
- check that the equipment supplied meets your specification and the supplier has met their legal duties.

> *See also*: Accident investigations, p.29; Construction site health and safety, p.177; Workplace health, safety and welfare, p.768.

Sources of further information

INDG 291 Simple Guide to the Provision and Use of Work Equipment Regulations: www.hse.gov.uk/pubns/indg291.pdf

INDG 271 Buying new machinery
www.hse.gov.uk/pubns/indg271.pdf

INDG 270 Supplying new machinery
www.hse.gov.uk/pubns/indg270.htm

Manual handling

Maria Anderson and Simon Toseland, Workplace Law

Key points

- Manual handling has been defined as 'any handling task involving the human body as the power source'.
- Manual handling thus includes lifting, lowering, pushing, pulling, carrying and holding. The most common injuries that occur as a result of manual handling are strains, sprains and slipped discs, i.e. musculoskeletal disorders (MSDs).
- It should not be overlooked, though, that other injuries such as cuts, burns and bruises may also occur when handling items or equipment that may be hot, sharp or awkward. The HSE suggests that in 2009/10 some ten million working days were lost through MSDs.
- One of the most frequently asked questions is "What weight am I legally permitted to lift?" The simple answer is that there are only guidelines, and no limits. For example, a sack of potatoes weighing 25kg may be lifted safely by an individual who is properly trained and physically fit.
- By contrast, however, we have all heard of people who injured their back by lifting light weights such as a file in the office.
- It is therefore very important that employers look at their manual handling activities and try to ensure that employees are not put at unreasonable risk.
- Such prudent measures as manual handling training (which must be appropriate to the type of work) should be used. It is especially important to record the training and to provide refresher training at appropriate intervals.

Legislation

- Health and Safety at Work etc. Act 1974.
- Manual Handling Operations Regulations 1992.

Under the key legislation, both employers and employees have duties.

The employer's duties are to:

- avoid, so far as is reasonably practicable, the need for its employees to undertake any manual handling operations at work that involve a risk of their being injured; this can be either by elimination of the activity or by mechanisation;
- where it is not reasonably practicable to avoid the need for manual handling, which involves the risk of injury, assess the risk of all such manual handling operations;
- take appropriate steps to reduce the risk of injury to the lowest level reasonably practicable;
- provide information, instruction and training as necessary to reduce the risks to its employees;
- take appropriate steps to provide employees with general indications of the weight of each load and, so far as is reasonably practicable, precise information on the weight of each load and the heaviest side of any load; and
- ensure that any risk assessment shall be reviewed if:
  - there is reason to suspect it is no longer valid; or
  - there has been a significant change in the manual handling operations to which it relates.

If the review shows that changes are necessary, the employer should make them. The employer's obligations under the Regulations are thus ongoing.

The employee's duties are to:

■ take reasonable care of his own health and safety and those affected by his activities;

■ cooperate with his employer in health and safety issues;

■ make use of appropriate equipment in accordance with the training provided; and

■ follow appropriate systems of work laid down by his employer with regard to manual handling.

The risks presented by manual handling activities are affected by a variety of factors, not least the physical condition of the individual concerned. Hence in fulfilling the above duty, employees could reasonably be expected to advise their employers of any significant medical condition that might put them at greater risk, for example previous or current MSDs, injuries as a result of sporting activities, and so on.

Manual Handling Operations Regulations

The Manual Handling Operations Regulations 1992 (as amended) (MHO Regulations) came into force in January 1993 under the Health and Safety at Work etc. Act 1974, to implement a European Directive on the manual handling of loads, with some small changes made in 2002.

The MHO Regulations aimed to reduce the risk of injury from manual handling by imposing duties on employers and employees. This Regulation works closely with the Management of Health and Safety at Work Regulations 1999 (MHSAW), which require employers to carry out a suitable and sufficient risk assessment of all work activities, including manual handling.

Regulation 4 of the MHO Regulations sets out the duties to employers, and establishes a clear hierarchy of control measures, which the employer must follow.

Avoid the risks

Avoid, as far as is reasonably practicable, the need for an employee to carry out manual handling operations at work that can cause an injury.

This may be done by not doing the task at all, or redesigning it to move the load in a different way; for example, by automating or mechanising the process with lifts or conveyor belts.

To be deemed reasonably practicable, the cost of the measures would be proportionate to the benefits obtained by the changes.

Assess the risks

Employers are required to make a suitable and sufficient assessment of any hazardous manual handling operations that cannot be avoided. This assessment must consider all the manual handling operations and relevant individual factors.

Important issues to consider in MHO assessments:

- A generic risk assessment may be sufficient. However, more complicated operations may need more detailed assessments.
- Consider the wide range of operations, for example, in construction or maintenance.
- Consider peripatetic operations (taking place in different locations, e.g. delivering goods).

- Emergency services will need to apply everyday judgements to carry out risk assessments on the spot, using their professional knowledge (dynamic risk assessment). It will be essential to provide training to enable staff to carry out such assessments; however, this does not mean that a written risk assessment is not necessary.
- Moving and handling people is a more complex task that requires consideration of medical conditions and human rights.
- Always seek employees' contributions.

The assessment must be carried by a competent person, which may be a member of staff, an external expert familiar with the operations to be assessed, or a team of assessors. The people involved in the assessment must have the following skills:

- Clear knowledge of the Regulations' requirements.
- Familiarity with the operation to be assessed.
- The ability to identify the hazards related to the operation, including the ones that are not that obvious.
- Use of additional resources if necessary.
- Ability to draw valid and reliable conclusions.
- Identify realistic steps to reduce the risks.
- Make clear records of the assessment.
- Ability to clearly communicate the findings.
- Recognise their own limitations and seek help if deemed necessary.

Reduce the risks

So far as is reasonably practicable, reduce the risk of injury from manual handling operations. Where possible, mechanical assistance should be provided, for example, a sack trolley or hoist. Where this is not reasonably practicable, changes to the task, the load and the working environment should be explored, introducing more sympathetic systems of work.

TILE

An ergonomic approach must be used when designing the MHO, to make the job fit the person, rather than the other way around. For example, a bench height may be modified according to the height of the individual, or the load weight reduced in accordance to each individual's capability.

The level of detail of the assessment and control measures implemented will depend on the complexity of the task. A more thorough approach will look at all the manual handling hazards, taking into account the nature of the *task*, the physical capabilities of *individuals* involved in the task, the size and shape of the *load*, and the *environment* in which the lifting operation is being performed. Think TILE.

Task

The following aspects must be considered regarding the MHO:

- Does the MHO involve manipulating loads away from the body?
- Are there bodily movements, such as twisting, stooping or reaching upwards?
- Does the task demand excessive lifting or lowering distances?
- Does the load need to be carried for long distances?
- Does the load require excessive pushing or pulling?
- Does it entail frequent, repetitive or prolonged physical effort?
- Is there sufficient rest or recovery time?
- Is team handling necessary? Possibly because one person cannot carry out the task on their own (ensure that individuals have similar build, clear communication, enough space to

manoeuvre, one person takes charge and give instructions).

If any of these issues take place, the employer will have to try to reduce the risk, by using lifting aids or changing the layout or systems of work.

Individual capability

The MHO Regulations do not set a specific limit for the weight that an employee can be expected to handle.

This is because even small and light objects can cause ill health or injuries; for example, when repetitive tasks are involved, physical capabilities will need to be carefully considered, in particular if:

- the job requires unusual strength, height or agility;
- the operation will create a hazard to a pregnant woman or a person with a history of back problems or other ill health issues;
- the individual requires special information or training;
- the clothing, footwear or personal effects worn at the time of the operation are unsuitable for the task. For example, movement or posture may be affected by personal protective equipment or gloves may affect the level of grip of the load; and/or
- his/her knowledge and training are insufficient or inadequate, including knowledge on how to apply a good manual handling technique, knowledge about the load (weight, size) and risks and possible injuries caused by unsafe MHO and the mechanical aids available.

Load

In particular, they should consider if any of the characteristics of the load can be changed; for example, buying smaller bags to reduce the weight, or getting the product in bulk bags, to reduce the frequency of handling.

The main aspects to consider are:

- The weight of the load – is it too heavy?
- Is the load too bulky or unwieldy?
- Is it difficult to grasp?
- Is it unstable, or are the contents likely to shift?
- Does it contain sharp edges, or is too cold or hot?

Environment

Some areas may be less risky than others; an open warehouse may be a relatively safe area, but the likelihood of injury can increase if the area is poorly lit or there are obstacles in the way. Consider:

- space constraints that may prevent a good posture;
- uneven, slippery or unstable floors; for example, contaminated surfaces with oil, water or ice;
- variation in level of floors or surfaces; for example, steps or slopes;
- extreme temperatures or humidity levels;
- ventilation problems or strong winds;
- poor lighting conditions; and/or
- obstacles present on the route.

The assessment should be recorded and kept up to date, and should be reviewed if new information comes to light or if there has been a change in the manual handling operations.

The assessment should also be reviewed if a reportable injury occurs or when individual employees suffer an illness, injury or the onset of disability, which may make them more vulnerable to risk. See *Table 1* (*opposite*) for a quick guide to TILE.

Regulation 5 considers the duties of employees, who must take reasonable care of their own health and safety and of those affected by their activities; they must cooperate with their employers and use the equipment provided according to training and instructions given; and always follow the safe systems of work implemented by the employer to ensure safe manual handling operations.

Training programmes

Effective information and training complement a safe system of work and form an important part of the plan to reduce the risk of injuries caused by MHO. There is no specific guidance in what a training course on manual handling must include, but employers should make sure that their employees understand clearly how manual handling operations have been designed to ensure their safety.

A course should include:

- how injuries can occur;
- how to recognise hazardous operations;
- how to carry out safe manual handling, including good handling technique and the use of mechanical aids;
- how to deal with unfamiliar operations;
- how to use personal protective equipment if provided, such as gloves;
- the importance of keeping the workplace tidy; and
- knowing your own personal capabilities and limitations.

Employers should ensure they keep records of training to show who has been trained and when, and monitor the efficacy of the training by evaluating sick absenteeism and near-miss reports. There is no single correct way of handling a load, as a safe MHO depends on many different factors. However, the HSE has published guidance on good handling technique, and some important points need to be considered:

- Think before handling / lifting.
- Keep the load close to the waist.
- Adopt a stable position.

Factors affecting the risk of injury from manual handling	Questions to ask in the assessment
What task is being undertaken?	**Does the task involve:** ■ Holding or manipulating loads at distance from trunk? ■ Unsatisfactory bodily movement or posture? ■ Twisting or stooping? ■ Reaching upwards? ■ Excessive movement of loads, especially: ■ excessive lifting or lowering distances? ■ excessive carrying distances? ■ excessive pushing or pulling of loads? ■ Risk of sudden movement of loads? ■ Frequent or prolonged physical effort? ■ Insufficient rest or recovery periods? ■ A rate of work imposed by a process?
How well suited is the individual to the task?	**Does the job:** ■ Require unusual strength or height? ■ Create a risk to those who might be pregnant? ■ Create a risk to those who are known to have a health problem e.g. returned from relevant sickness absence? ■ Require special information or training for its safe performance?
What are the properties of the load itself?	**Is the load:** ■ Heavy, bulky or unwieldy? ■ Difficult to grasp, unstable, or with contents likely to shift? ■ Sharp, hot or in some other way potentially damaging?
Does the environment increase the risk?	**Are there workplace limitations such as:** ■ Space constraints preventing good posture? ■ Uneven, slippery or unstable floors? ■ Variations in level of floors or work surfaces? ■ Poor lighting conditions? ■ Weather conditions?
Other factors affecting the risk	■ Is movement or posture of the individual hindered by clothing or personal protective equipment?

Table 1. TILE.

- Ensure a good hold on the load.
- Moderate flexion (slight bending) of the back, hips and knees at the start of the lift.
- Do not flex the back any further while lifting.
- Avoid twisting the back or leaning sideways, especially while the back is bent.
- Keep the head up when handling.
- Move smoothly.
- Do not lift or handle more than can be easily managed.
- Put down then adjust.

Assessment tools

The manual handling assessment chart (MAC) is a tool developed by the HSE to help health and safety inspectors, employers and safety representatives assess the most common risk factors in lifting, carrying and team handling operations. The use of a MAC doesn't equate to a full risk assessment.

- Spend some time observing the task to ensure that what you see is representative of normal working procedures. Consult employees and safety representatives during the assessment process. Where several people do the same task, make sure you have insight into the demands of the job from the workers' perspective.
- Select the appropriate type of assessment (e.g. lifting, carrying or team handling).

- Ensure you read the assessment guide before you make your assessment.
- Follow the appropriate assessment guide and flow chart to determine the level of risk for each risk factor.
- Enter the colour band and corresponding numerical score on the score sheet. The bands determine which elements require attention.
- Add up the total score. This helps prioritise those tasks that need most urgent attention and checks the effectiveness of those improvements.
- Enter the remaining task information asked for on the score sheet.

- The levels of risk are classified as:

  - G – Green: Low level of risk.
  - A – Amber: Medium level of risk.
  - R – Red: High level of risk.
  - P – Purple: Very high level of risk.

The purpose of the assessment is to identify and reduce the level of risk of the task.

See also: Construction site health and safety, p.177; Health surveillance, p.396; Lifting equipment, p.448; Occupational health, p.519; Risk assessments, p.624; Training, p.678.

HSE – Better Backs Campaign:
www.hse.gov.uk/msd/campaigns/whybetterbacks.htm

INDG 143 *Getting to grips with manual handling: a short guide:*
www.hse.gov.uk/pubns/indg143.pdf

HSE's MAC tool: www.hse.gov.uk/msd/mac/index.htm

The Manual Handling Operations Regulations 1992 (as amended 2002) require employers to train staff involved in manual handling activities. With large numbers or high turnover of staff, it can sometimes prove more cost effective to have the training resource in your own company. Workplace Law's *Manual Handling: Train the trainer course* is designed to enable staff to acquire sufficient knowledge and training competence to be able to prepare and deliver in-company training to those staff at risk from manual handling activities, and to thus meet the above legislative requirements. It also incorporates the recent HSE guidance on 'labelling loads'. Visit www.workplacelaw.net for more information.

Mental health

Elizabeth Stevens, Steeles Law

Key points

- Employers are subject to a variety of legal obligations in respect of their employees' health and wellbeing. These obligations arise from health and safety legislation, the breach of which is a criminal offence, and also from the law of negligence, contract and discrimination. Injury to an employee's mental health is treated by the law in the same way as injury to physical health.
- The HSE defines 'workplace stress' as 'the adverse reaction people have to excessive pressure or other types of demand placed on them'. According to the HSE, stress is not an illness but a 'state'; it is only if stress becomes too excessive and prolonged that mental and physical illness may develop.
- Employers are not under a duty to eliminate all stress in the workplace, but once an employee has raised the issue of stress, an employer is under a duty to take steps to minimise the risk to the individual. A failure to do so could render the employer liable for a future personal injury claim and/or constructive unfair dismissal claim.
- Employers are also under certain duties in respect of those individuals whose mental health (whether or not impacted by work) amounts to a disability under the provisions of the Equality Act 2010.

Legislation

- Health and Safety at Work etc. Act 1974.
- Protection from Harassment Act 1997.
- Working Time Regulations 1998.
- Management of Health and Safety at Work Regulations 1999.
- Equality Act 2010.

Main cases

- *Sutherland (Chairman of the Governors of St Thomas Beckett RC High School) v. Hatton and others* (2002).
- *Barber v. Somerset County Council* (2004).
- *Essa v. Laing Ltd* (2004).
- *Nottinghamshire County Council v. Meikle* (2004).
- *Hartman v. South Essex Mental Health and Community Care NHS Trust and other cases* (2005).
- *Green v. DB Group Services Limited* (2006).

- *Hone v. Six Continents Retail Ltd* (2006).
- *Majrowski v. Guy's and St Thomas' NHS Trust* (2006).
- *Sayers v. Cambridgeshire County Council* (2006).
- *Intel Corporation (UK) Ltd v. Daw* (2007).
- *McAdie v. Royal Bank of Scotland plc* (2007).
- *Sunderland City Council v. Conn* (2007).
- *Dickens v. O2 PLC* (2008).
- *Cheltenham Borough Council v. Laird* (2009).
- *Veakins v. Kier Islington* (2009).
- *Thaine v. London School of Economics* (2010).

Health and safety legislation

The Health and Safety at Work etc. Act 1974 (HSWA 1974) places a duty on employers to ensure the health,

Case study

Green v. DB Group Services (UK) Ltd (2006)

Ms G worked as a company secretary for a commercial bank. During the recruitment process she disclosed to her employer that she had previously been treated for depression. After commencing employment, G alleged that she was being harassed and bullied by a group of four colleagues. She complained to her manager and to the HR department, but no effective steps were taken to deal with her complaints. G subsequently experienced problems with another colleague, who she claimed was conducting a sustained campaign against her, which was designed to undermine and humiliate her. She again complained to her manager and the HR department, but no formal action was taken. Following a holiday, she found that she was unable to walk through the doors of the bank's offices and she was hospitalised with a major depressive disorder. She eventually returned to work for a short period but, following a relapse, she took another period of sick leave and was eventually dismissed.

Her personal injury claim succeeded and she was awarded over £850,000 in damages. The High Court held that G's employer was vicariously liable for the harassment and bullying by G's fellow employees. The bank had failed to take adequate steps to protect her from what was regarded by the Court as a sustained campaign of bullying. The Court considered that a reasonable and responsible employer would have intervened as soon as they became aware of the problem and taken steps to stop the bullying. The defendant knew that the claimant had suffered depression in the past and was therefore more vulnerable than the population at large. The Court was also satisfied that the behaviour of the bank's employees amounted to harassment under the Protection from Harassment Act 1997, for which the defendant was also liable, although a separate award of damages for this was not made.

safety and welfare of their employees as far as is reasonably practicable. This includes taking steps to minimise the risk of stress-related illness or injury to employees. Under the Management of Health and Safety at Work Regulations 1999 employers are obliged to carry out an assessment of the risks to employees' health, including a suitable and sufficient risk assessment for stress. If, after completing an assessment, an employer believes there is a potential risk to employees' health, they should take all reasonable steps to limit this risk and to monitor the situation.

An employer's breach of the statutory duties imposed by the HSWA 1974 is a criminal offence, enforceable by the HSE but not directly actionable by individual employees. However, a breach of health and safety regulations may give rise to a civil liability for damages, where an employee can show that the employer's breach caused illness or injury to the employee. In addition, a failure by an employer to have due regard for the health and safety of its employees may amount to a fundamental breach of contract, entitling an employee to resign and bring a claim for constructive unfair dismissal.

HSE Management Standards

To assist employers in fulfilling their duties in respect of carrying out risk assessments, and to measure performance in managing work-related stress, the HSE has devised its 'Management Standards'. These are a set of best practice statements of management competencies, to provide a framework for dealing with workplace stress and to help employers meet their legal obligations. It is widely recognised that managers, through their management style and their role as 'gatekeepers' of working conditions, have a significant impact on levels of employee stress, and therefore play a crucial role in identifying and resolving any problems, and preventing unacceptable levels of stress occurring.

The Management Standards are voluntary, but may be used as evidence by the HSE of an employer's failure to comply with their duty to manage stress under the HSWA.

The HSE, in conjunction with the CIPD and Investors in People, has also developed a 'Stress Management Competency Indicator Tool,' to allow managers to assess whether they currently have the behaviours identified as effective for preventing and reducing stress at work. The aim is to help managers reflect on their behaviour and management style. Further tools designed to assist managers and those who train and support them in that role are also being developed. The HSE website has further details.

Working Time Regulations

The Working Time Regulations 1998 implement the European Working Time Directive (93/104/EC), which was adopted by the EC as a health and safety measure and is consistently interpreted as such by the Courts and Employment Tribunals. Long working hours are a recognised contributory factor to the incidence of work-related stress, and an employer's breach of its obligations under the Regulations is actionable by the HSE and in some cases (in relation to rest and leave entitlements) directly enforceable by the individual worker in the Employment Tribunal.

Employers have a duty to take all reasonable steps to ensure that the limits contained in the Working Time Regulations are complied with. This includes a maximum 48 hour average working week (which the employee can voluntarily contract out of) and limits on night working. The Regulations also provide entitlements to daily and weekly rest breaks and paid annual leave, which the employer is under a duty to ensure workers can take but is not obliged to force workers to take.

A failure by an employer to take reasonable steps to comply with the maximum 48 hour average working week is a criminal offence, punishable by a potentially unlimited fine. The HSE is responsible for monitoring compliance and expects employers to maintain records going back at least two years to show that the 48 hour maximum has been complied with in respect of all employees who have not opted out. Employers should bear in mind that the maximum includes time the worker spends working for other employers.

The ability of workers to opt-out of the maximum 48 hour average working week has been the subject of lengthy debate at a European level, with proposals put forward for amendments to be made to the Working Time Directive, which would result in the opt-out being removed. These proposals have, for now, been rejected, and it remains to be seen whether further proposals will be put forward to amend the Directive. (See 'Working time,' p.758.)

Common law duties

There is a duty implied into all contracts of employment that the employer will take reasonable care for the health and safety of its employees. If an employer breaches this duty and the employee has suffered psychiatric injury as a result, an employee may be entitled to bring a negligence claim against the employer (also known as a claim for personal injury). To do this, the employee must be able to demonstrate that his psychiatric injury was a *reasonably foreseeable* consequence of the employer's breach of duty. Much of the case law has dealt with the issue of whether an individual's psychiatric injury was reasonably foreseeable, as a consequence of bullying suffered by that individual and/or pressures of work.

Reasonable forseeability

Guidance on the issue of reasonable forseeability was given by the Court of Appeal in its landmark judgment in four conjoined stress cases – *Sutherland (Chairman of the Governors of St Thomas Beckett RC High School) v. Hatton and others* (2002). The main points of the guidance are as follows:

- The key question is whether this kind of harm (psychiatric injury) to this particular employee was reasonably foreseeable.
- Forseeability depends upon what the employer knows, or ought reasonably to know, about the individual employee. An employer is usually entitled to assume that an employee can withstand the usual pressures of

the job unless he knows of a particular problem or vulnerability.

- No occupation should be regarded as intrinsically dangerous to mental health.
- An employer is generally entitled to take what he is told by his employee at face value, unless he has a good reason to think to the contrary.
- To trigger a duty on the employer to take steps, the indications of impending harm to health arising from stress at work must be plain enough for any reasonable employer to realise that he should do something about it.
- The employer will only breach the duty of care if he has failed to take the steps that are reasonable in the circumstances, bearing in mind the magnitude of the risk of harm occurring, the gravity of the harm that may occur, the costs and practicability of preventing it, and the justifications for running the risk.
- The size, resources and scope of the employer's operation, and the need to treat other employees fairly, can all be taken into account when deciding what is reasonable.
- An employer can only reasonably be expected to take steps that are likely to do some good.
- An employer who offers a confidential advice service, with referral to appropriate counselling or treatment services, is unlikely to be found in breach of duty.
- If the only reasonable and effective step would have been to dismiss or demote the employee, the employer will not be in breach of duty in allowing a willing employee to continue in the job.

On appeal from the Court of Appeal's decision in *Sutherland v. Hatton* the House of Lords overturned one of the four cases, *Barber v. Somerset County Council* (2004) and in doing so emphasised that the guidelines set out by the Court of Appeal,

whilst 'useful practical guidance,' did not have statutory force.

The key point for employers in order to avoid the risk of a potential claim is to take such action as is reasonable to avoid exacerbating an employee's ill health, as soon as they become aware of any underlying vulnerability an individual may have to stress in the workplace. Possible remedial actions might include allowing the employee to take a sabbatical, redistributing work, extra training and counselling. What is reasonable for the employer to do will depend on factors such as the size of the employer, the resources available, and the impact on other employees.

Sutherland v. Hatton remains the leading case in the area of workplace stress, but subsequent decisions have taken a slightly different stance in relation to some of the guidance set out in that judgment.

Provision of counselling services

In the case of *Intel Corporation (UK) Ltd v. Daw* (2007) the Court of Appeal held that an employee's email to her manager stating that she was 'stressed out' and 'demoralised' and including two references to previous episodes of post-natal depression, was crucial to the issue of reasonable foreseeability. In the circumstances, urgent action should have been taken to reduce the employee's workload. The Court expressly rejected Intel Corporation's submission (following the guidelines in *Sutherland v. Hatton*) that its provision of a counselling and medical assistance service was sufficient to discharge its duty of care. The provision of such services and whether the employer has fulfilled its duty will depend on the facts of the case, and it is for the judge to decide in any particular case which parts of the *Sutherland v. Hatton* guidance are relevant.

This was confirmed by the Court of Appeal in *Dickens v. O2 PLC* (2008), in which the Court held that reference to the employer's counselling service was insufficient in the circumstances of the case. Having explained to her employer the difficulties she was experiencing and the severe effect on her health, the Court was satisfied that management intervention was required and the employee should have been sent home and referred to the employer's occupational health department.

Relevance of the Working Time Regulations

The claimant in *Hone v. Six Continents Retail Ltd* (2006), a pub landlord, claimed that he consistently worked around 90 hours per week, despite not opting out of the maximum 48 hour working week under the Working Time Regulations. He successfully used his employer's breach of the Regulations as part of his argument that his psychiatric injury had been reasonably foreseeable.

However, the High Court in *Sayers v. Cambridgeshire County Council* (2006) made it clear that the fact that an employee is working in excess of the 48 hour per week limit will not in itself render any resulting injury reasonably foreseeable.

Intrinsically stressful jobs

One of the conjoined cases in *Hartman v. South Essex Mental Health and Community Care NHS Trust and other cases* (2005) considered by the Court of Appeal was *Melville v. Home Office*. The employee in this case was a prison officer whose duties included the recovery of bodies of prisoners who had committed suicide. After helping to cut down a body and attempting revival in May 1998, Mr Melville developed a stress-related illness and eventually retired in 1999 on ill health grounds.

Before the Court of Appeal, the Home Office argued that since it knew of no particular vulnerability of Mr Melville it was entitled to assume that he was up to the normal pressures of the job. The Home Office was not successful in this argument and the Court of Appeal held that it was foreseeable that such an injury may have occurred to employees exposed to traumatic incidents. Home Office documents noted that persons whose duties involved dealing with suicides might sustain injuries to their health, and procedures were put in place in relation to post-incident care, which had not been properly implemented.

Discriminatory harassment

Discrimination legislation (since 1 October 2010, the Equality Act 2010) outlaws harassment on the grounds of sex, race, disability, sexual orientation, gender reassignment, religion or belief and age (the 'protected characteristics'). The harassment does not necessarily need to be directed towards the individual who brings a complaint, and the harassment can be based on an individual's perceived characteristic as well as their association with someone who has one of the protected characteristics. The legislation provides that an employer will be liable for workplace harassment based on one of the protected characteristics, carried out by his employees, unless he has taken reasonable steps to prevent this from occurring. Reasonable steps might include, for example, implementing an equal opportunities and anti-bullying policy, and training managers in dealing with complaints of bullying and harassment.

The Equality Act 2010 introduced a harmonised definition of harassment and extends the protection for employees from harassment carried out by third parties. Employers can potentially be liable for the harassment of employees by a third party, if it is aware that such harassment has occurred on two or more occasions and it has not taken reasonably practicable steps to prevent it. *Note: the Government is intending to consult over the removal of the 'unworkable' third party harassment provisions.*

In the case of *Essa v. Laing Ltd* (2004) the Court of Appeal confirmed that personal injury, which includes psychiatric injury, arising from acts of discrimination does not need to be reasonably foreseeable in order for employees to recover damages. The employee only needs to prove that the discrimination caused the injury to occur. In appropriate cases, an Employment Tribunal will discount the award of compensation paid to a claimant where their psychiatric ill health has been caused by a combination of factors, not all of which are the employer's responsibility (see for example, *Thaine v. London School of Economics* (2010)).

Protection from harassment

The House of Lords confirmed in the case of *Majrowski v. Guy's and St Thomas' NHS Trust* (2006) that employers can be held liable for workplace bullying under the Protection from Harassment Act 1997 (PHA). For this to apply, claimants only need to show they have suffered anxiety or distress as a result of the harassment, rather than a recognisable psychiatric injury in order to bring a negligence claim. It is necessary for claimants to establish a 'course of conduct,' in contrast to a one-off incident of harassment, which can be sufficient to bring a claim under discrimination legislation.

In a subsequent case, *Green v. DB Group Services Limited* (2006), the High Court upheld another claim for workplace bullying under the PHA 1997, but made no separate award of damages, since it had

taken into account the anxiety caused by the harassment in assessing the amount of compensation awarded in respect of the employer's negligence (a figure of over £850,000 – see *Case study*).

More recently, in *Sunderland City Council v. Conn* (2007), the Court of Appeal overturned a County Court decision that a manager's conduct towards an employee amounted to harassment under the PHA 1997. There were only two alleged incidents, the first of which did not 'cross the boundary from the regrettable to the unacceptable' and was not sufficiently serious to be regarded as criminal. There had therefore been no 'course of conduct' necessary to establish a claim under the Act.

In *Veakins v. Kier Islington Ltd* (2009), however, the Court of Appeal concluded that the primary focus should be on whether the conduct is oppressive and unacceptable, as opposed to merely unattractive, unreasonable or regrettable. The County Court in this case had erroneously focused too heavily on whether the conduct would sustain criminal liability. In its judgment, the Court of Appeal observed that since the case of *Sutherland v. Hatton*, it has become harder for employees to succeed in a negligence action based on stress at work and therefore claims under the 1997 Act are more prevalent. However, the Court noted that whilst there is nothing in the language of the Act that excludes workplace harassment, it should not be thought 'from this unusually one-sided case' (*Veakins*) that stress at work will often give rise to liability for harassment. In the great majority of cases, according to the Court of Appeal, the remedy for high-handed or discriminatory misconduct by or on behalf of an employer will be more fittingly in the Employment Tribunal.

Unfair dismissal

Ill health is a potentially fair reason for dismissing an employee, as it relates to their capability to do a job.

In the case of *McAdie v. Royal Bank of Scotland plc* (2007) the employee's stress-related illness was attributed to the conduct of the employer. However, the Court of Appeal agreed with the EAT that the employer could still fairly dismiss the employee for ill health capability in these circumstances. Medical evidence demonstrated that the employee had no prospect of recovery from the illness and she had expressly stated that she would never return to work. There was no real alternative to dismissal.

However, the Court of Appeal accepted that the cause of the employee's incapacity was a relevant factor to take into account and approved of the Employment Appeal Tribunal's suggestion that employers should 'go the extra mile' in finding alternative employment for an employee who is incapacitated by the employer's own conduct, or they should put up with a longer period of absence than they would do in normal circumstances.

Disability discrimination

If an employer is considering dismissing an employee who is suffering from a mental illness they must consider whether the illness may constitute a disability under the Equality Act 2010 (previously the Disability Discrimination Act 1995). The Equality Act protects those with physical or mental impairments from discrimination, provided that impairment satisfies the test for a disability under the Act. Stress itself is not an illness, but a stress-related condition could be an impairment within the meaning of the Act. Employers are under a duty to make reasonable adjustments where any arrangements made by the employer place a disabled person at a substantial disadvantage compared to non-disabled employees. Where an employee is suffering from a mental impairment, reasonable adjustments might include a phased return to work after sickness absence, a reallocation of duties, reduced working hours, mentoring or counselling.

Disabled employees are not generally entitled to additional sick pay, unless the employer has caused sickness absence to be prolonged as a result of its failure to make reasonable adjustments. This was the case in *Nottinghamshire County Council v. Meikle* (2004), which was not a case dealing with an employee's mental health but could be applied equally to those with stress-related illness whose employer does not take the necessary steps to enable the individual to return to work. If an employer dismisses an employee for a reason related to their disability they may be guilty of disability discrimination unless they are able to show that the dismissal was justified. A medical report should be obtained in order to establish the employee's likely prognosis and the effectiveness of any potential adjustments before any decision is taken to terminate an individual's employment.

Pre-employment checks

In view of the potential liabilities faced by employers in relation to employees suffering from mental health problems, it might appear prudent to carry out pre-employment checks to establish whether an individual has any pre-disposition to such impairments. This should be done with caution, since a refusal to employ individuals with previous mental health issues is likely to result in successful claims for disability discrimination.

Even greater caution is required following the implementation of the main employment provisions of the Equality Act

2010 on 1 October 2010. Section 60 of the Act restricts the ability for employers to ask questions about an individual's health before an offer of employment is made. Any such questions are permitted for very limited purposes only, including to establish whether any adjustments are required to the recruitment process itself, and in order to ascertain whether the individual can carry out tasks 'intrinsic to the work concerned'. The extent to which health-related questions are permitted has not yet been subject to judicial scrutiny, and case law is awaited to clarify the exemptions.

Questions that do not fall within the permitted exemptions may lead to a presumption of discrimination if a disabled candidate is refused the job. Section 60 does permit health-related enquiries to be made after a job offer has been made, and the offer can be made conditional on satisfactory responses to those enquiries. However, employers should remember that they will be subject to a duty to make reasonable adjustments in respect of any prospective disabled employee, and a decision taken to withdraw a job offer on these grounds runs a high risk of being challenged as discriminatory by the individual.

This provision in the Act relating to pre-employment medical questions is intended to assist disabled people to overcome the well-documented prejudice of some employers towards recruiting those with disabilities, particularly those with a history of mental health issues. In any event, the use of such questionnaires has not always been effective in practice, as demonstrated by the case of *Cheltenham Borough Council v. Laird* (2009) in which the former managing director of the Council had been granted ill health retirement following a depressive illness. The Council subsequently brought proceedings for negligent and fraudulent

misrepresentation against Mrs Laird, on the grounds that she had not disclosed her previous episodes of stress-related depression on her pre-employment medical questionnaire. The High Court rejected the Council's claims, on the basis that the answers provided by Mrs Laird had not been false or misleading. The questionnaire was poorly drafted and the wording did not expressly require Mrs Laird to disclose information about her previous history of stress and depression.

NICE guidance

In November 2009, the National Institute for Health and Clinical Excellence (NICE) issued new guidance for employers on promoting mental wellbeing at work. The guidance was developed in recognition of the importance of work in promoting mental wellbeing and is intended to assist employers to meet their legal duties to protect the health of employees. It also recognises that the current financial climate has the potential to increase mental health problems in employees because of worries about job insecurity and unemployment.

The guidance attempts to reduce the number of working days lost due to work-related mental health conditions, including stress, depression and anxiety. It makes a number of recommendations designed to improve the management of mental health in the workplace, including the prevention and early identification of problems. The recommendations, aimed at organisations of all sizes, include the following:

■ Ensure systems are in place for assessing and monitoring the mental wellbeing of employees so that areas for improvement can be identified and risks caused by work and working conditions addressed. This could include employee attitude surveys and information about absence rates, staff turnover and investment in training and development, and providing feedback and open communication.

- If reasonably practical, provide employees with opportunities for flexible working according to their needs and aspirations in both their personal and working lives.
- Strengthen the role of line managers in promoting the mental wellbeing of employees through supportive leadership style and management practices.

The NICE guidance also states that the HSE Management Standards (*see above*) may provide a valuable tool in implementing the guidance.

Conclusion

To protect themselves against HSE enforcement action and potential claims by employees, employers should consider the following practical steps:

- Organise and conduct suitable risk assessments of potential stressors.
- Make counselling facilities available to employees.
- Show a receptive and flexible response to complaints.
- Follow the HSE Management Standards and NICE Guidance.

- Be cautious in asking questions about health before making an offer of employment and remember the duty to make reasonable adjustments to prospective employees.
- Provide a written health and safety policy to employees, which includes a section on how to deal with stress.
- Put a bullying and harassment policy in place and ensure employees are aware of their obligations under the policy.
- Provide training to managers in dealing appropriately with employees suffering from stress and mental health issues, including the requirement to consider reasonable adjustments.
- Ensure sickness absence procedures take into account potential disabilities arising from mental impairments.
- Obtain an up-to-date and detailed medical report before considering dismissing an employee with a mental impairment.

See also: Occupational health, p.519; Stress, p.659; Working time, p.758.

Sources of further information

HSE (work-related stress): www.hse.gov.uk/stress

MIND: www.mind.org.uk

Mental health foundation: www.mentalhealth.org.uk

The Shaw Trust: www.tacklementalhealth.org.uk

NICE: www.nice.org.uk

Mobile phone masts

Michael Smith, Pinsent Masons Property Group

> **Key points**
>
> - There are many opportunities for property owners to obtain revenue from the siting of mobile phone masts (and other telecommunications equipment) on buildings, and joint ventures between building owners and mobile network operators are now quite common.
> - There are also other opportunities for building owners to work with telecommunications companies to increase the services that are offered both to tenants of their buildings and to visitors, often sharing the resulting income with the operator.
> - You should ensure that the documentation is suitable for your needs. Operators will proffer a standard document, but you should take professional advice to ensure that the document represents the terms that have been agreed and does not prejudice your property interests.

Legislation
- Landlord and Tenant Act 1954.
- Telecommunications Act 1984.
- Communications Act 2003.

Mobile phone masts
The first opportunity relates to GSM mobile network operators. Almost all commercial property owners will have received requests from time to time from the four GSM mobile network operators; namely O2, Orange, T-Mobile and Vodafone, to permit installations of mobile network apparatus forming part of their networks on rooftops or other such structures.

The apparatus typically consists of a base transceiver station (BTS), normally housed in a small cabin or equipment room, together with an array of antennae used for receiving and transmitting signals both from subscribers and between cells forming the operator's network.

The arrangements in each case will also include a power supply for the BTS, and an arrangement for the connection of the BTS to the fixed-wire network. This is usually achieved by way of a fibre-optic connection from the BTS direct to a fixed-wire network (usually BT), or by use of a microwave dish which sends the signal to the nearest convenient point where it can be connected to the fixed network.

Third-generation networks
A second opportunity available for property owners is to work with the five third-generation (3G) licensees (the four GSM licensees plus Hutchison 3G) in connection with the construction of their 3G networks. The technology involves cells of a much smaller size than is the case with GSM traffic. There is now an increasing need for many more base stations as the operators have been keen to recover the substantial sums that they paid to the Government for the spectrum licences in 2000. As a result of encouragement from OFCOM and the European Commission, and also for cost saving purposes, certain mobile network operators and third-generation licensees are entering into site- and network-sharing agreements with each other. Accordingly, requests to install electronic

---

Case study

The Bridgewater Canal Company Ltd v. GEO Networks Ltd (2010)

In this case the Court held (on an appeal from the decision of an arbitrator) that the consideration payable to a landowner by the operator of a communications network under the Electronic Communications Code (the Code) in exchange for the right to carry out work on land included payment not only for the right to undertake the works on the land itself but also for the right for the works to remain in place on the land once the initial works had been carried out.

The facts of the case were that GEO Networks Limited (GEO) had the benefit of an existing duct housing a fibre optic cable for which it paid rent, which ran beneath the Bridgewater Canal, owned by the Bridgewater Canal Company (the 'company'). GEO laid an additional fibre optic cable through the existing duct, which did not require any invasive works to the canal itself. The company did not object to the laying of the additional cable per se, but claimed that it was entitled to payment of a sum under the Code to reflect the value of the right to leave the new cable in place in addition to the right to install the new cable in the first place. The arbitrator had held that the company was not entitled to demand such a payment. The Court disagreed.

The Court held that the consideration under the relevant paragraph of the Code had to take into account the right to retain the works on the land after the works had been completed, as well as the right to complete the physical works in the first instance. This required the payment of an additional sum to the rent that GEO already paid to the company. This additional amount had to be fair and reasonable, so as not to amount to a 'ransom', but it would take into account everything that GEO acquired by laying the additional cable.

---

communications apparatus on rooftops or other structures may be made in joint names on occasion.

Tower companies

A third opportunity for property owners is to deal with a set of new intermediary players, known as 'tower companies,' which, instead of building one base station tower for one operator, build larger towers and rent the antenna space out to multiple operators. The property owner can either enter into a partnership agreement with a tower company in respect of the rights to site towers on its buildings or other

property, or appoint the tower company as a managing agent.

Other opportunities

Other opportunities for joint ventures between property owners and telecoms companies include the following:

- *Installation of in-building systems.* These will permit users within a limited area, for example, shopping centres, airports, holiday camps etc., to receive services without interference. The necessary infrastructure is expensive to install (the equipment for a regional shopping centre might cost in the region of £500,000) and therefore tends to be shared between operators. The landowner benefits from receiving a share of the revenue and making the property more attractive to visitors, tenants and other users of the property.

■ *Installation of high-speed connections.* These allow organisations to have permanent connections to the internet. Larger organisations can afford their own connections, but it is also possible for smaller organisations to obtain such facilities by joining forces. Landlords of office buildings are in an ideal position to provide such services to their tenants, in partnership with so-called building local exchange carriers (BLECs). Once again, the landlord can expect to share the revenues while the BLEC bears the installation costs. Such a system can be expected to make a building more attractive to potential tenants.

Typical contents of a telecommunications agreement

The contents of telecoms agreements vary according to the circumstances, but typically will include provisions regulating:

■ the type of arrangement to be created (lease, licence and so on) and how long it is to last for;
■ the exact location of the site;
■ the amount of the rent, whether it is subject to review and, if so, on what basis;
■ any rights of way (or other rights – e.g. power or fibre) that the operator will need;
■ the obligations to be undertaken by each party;
■ whose responsibility it is to obtain planning permission if this is required;
■ the type and amount of equipment to be installed;

■ whether sub-letting or sharing is to be permitted and, if so, on what terms; and
■ if the arrangement is by lease, whether it should be excluded from the security-of-tenure provisions of the Landlord and Tenant Act 1954.

Operators have standard agreements that they proffer, but they may not be suitable for every transaction and will need to be checked carefully with the help of professional advice.

Electronic Communications Code

Telecoms operators enjoy various rights under the Electronic Communications Code ('the Code'), which is part of the Telecommunications Act 1984 (as amended by the Communications Act 2003). These rights can make it difficult to remove masts and other telecommunications equipment from property at the end of an agreement.

Furthermore, under the Code, operators may be able to install equipment on land even without the owner / operator's consent and the Court has given some guidance on the principles that apply to the payment of compensation / consideration to landowners in such circumstances; most recently in the case of *The Bridgewater Canal Co Ltd v. Geo Networks Limited* (2010) (see *case study*). Professional advice will be needed on the implications of the rights under the Code in connection with the particular building as well as the terms of the agreement itself.

Health concerns

There have been some well-publicised health and safety issues in the recent past in relation to mobile phones network equipment. Research is continuing, but for the present the perceived risks arising from the equipment will need to be managed and apportioned between the parties within the agreement.

See also: Planning procedures, p.585.

Sources of further information

The Workplace Law website has been one of the UK's leading legal information sites since its launch in 2002. As well as providing free news and forums, our Information Centre provides you with a 'one-stop shop' where you will find all you need to know to manage your workplace and fulfil your legal obligations.

It covers everything from CDM, waste management and redundancy regulations to updates on the Carbon Reduction Commitment, the latest Employment Tribunal cases and the first case to be tried under the Corporate Manslaughter and Corporate Homicide Act, as well as detailed information in key areas such as energy performance, equality and diversity, asbestos and fire safety.

You'll find:

■ quick and easy access to all major legislation and official guidance, including clear explanations and advice from our experts;
■ case reviews and news analysis, which will keep you fully up to date with the latest legislation proposals and changes, case outcomes and examples of how the law is applied in practice;
■ briefings, which include in-depth analysis on major topics; and
■ WPL TV – an online TV channel including online seminars, documentaries and legal updates.

Content is added and updated regularly by our editorial team who utilise a wealth of in-house experts and legal consultants. Visit www.workplacelaw.net for more information.

Mobile phones at work

Lisa Gettins, Sarah Lee and Heyma Vij, BPE Solicitors

Key points

- Use of mobile phones at work raises two important issues – the health and safety ramifications of driving whilst conducting a telephone conversation, and the general disruption that mobile phones cause in a working environment.
- The law has important implications for both employers and employees. Employers should have a detailed and practical policy that explains the law in clear and unambiguous language.
- Employers are strongly advised against purely paying 'lip service' to a mobile phone policy, given that employers will be liable if they encourage, pressure or require staff to use a handheld phone whilst driving.
- As such, it is not enough to simply issue the policy. Employers need to ensure it is being implemented.

Legislation

The Road Vehicles (Construction and Use) (Amendment) (No. 4) Regulations 2003 (the Regulations) came into force on 1 December 2003 and render it unlawful for individuals to drive a motor vehicle on the road if using a handheld mobile telephone, including any device that performs an interactive communication function by transmitting and receiving data, other than a two-way radio.

Mobile phones and driving

Practical implications

There has been some debate over what exactly is meant by 'handheld'. The Regulations define a mobile telephone or other device to be handheld if it 'is, or must be, held at some point during the course of making or receiving a call or performing any other interactive communication function'. Therefore, logically, not only would this include making or receiving calls but it would also include receiving or responding to text messages, other messages including video imagery or photographs, or accessing the internet.

In terms of what constitutes holding a phone, perhaps not surprisingly, this would include positioning the phone between an individual's shoulder and ear! The law in this area does not allow for what some employers and employees consider to be 'clever little loopholes'.

The Regulations apply to vehicles, whether they be public service vehicles such as buses or taxis, or private hire vehicles such as airport cars and coaches, together with all other vehicles that are used on the road including cars, vans, goods vehicles, heavy goods vehicles and motorcycles.

Therefore, in terms of producing a mobile phone policy, employers should make it clear to employees that they are not required to make or receive telephone calls whilst driving, even if they have a hands-free kit. In any case, the policy should make it absolutely clear that employees should never use handheld mobile phones whilst driving and that they will never be required to do so.

Exemptions

Again, there is little in respect of loopholes and the Regulations apply even when vehicles are stopped at traffic lights or level crossings, including traffic jams and the like. The only real exemption to using a handheld phone is for genuine emergency calls to 999 and this is only when it is not practical or it is unsafe to pull over and stop the vehicle.

In order to operate a phone whilst driving and not be in contravention of the Regulations the individual is required to use hands-free equipment. This involves using a bona fide cradle, which can also be placed on the handlebars of a motorcycle. The key point is that the person in question does not hold or attempt to cradle the phone themselves.

Health and safety

Employers need to be aware that the Regulations state clearly that no person shall 'cause or permit' any other person to drive a motor vehicle on the road whilst that person is using a handheld mobile phone or device. However, obligations under the Regulations are not all that needs to be considered.

In light of the clear health and safety obligations that employers are under to manage risks relevant to their employees, the advice has to be that staff should be told, by way of a contractual document, where relevant, or a non-contractual policy, that they are strictly prohibited from using handheld mobile phones whilst driving and that to do so is likely to result in disciplinary action. In any event, prudent advice would be that staff are recommended not to use a mobile phone even if there is a hands-free kit in the vehicle – the reason being that even though the Regulations may not be contravened, health and safety laws may well be. Telephone conversations between employer and employee can often be very detailed in nature, requiring thought and analysis on behalf of the employee, which invariably is likely to reduce the amount of concentration to which that employee is turning to the road.

To reiterate, employers should go further to protect themselves by being aware of their duty to provide a safe working environment under the Health and Safety at Work etc. Act 1974, and being aware that both employer and employee can be prosecuted for careless or dangerous driving if causation shows that the person in control of the vehicle was distracted and not showing due care and attention.

Mobile phones in the office

Leaving the issues of mobile phones and driving to one side, the other important point to consider is the effect, and perhaps more importantly, the disruption, that can be caused by mobile phones in the office.

Ring tones can be loud and irritating. Further, if employees are allowed to keep mobile phones on in the office they may embark on conversations with friends and family that may not be appropriate. Employees who have casual conversations with their friends about what they may be doing at the weekend could arguably, in some circumstances, cause offence to other members of staff.

Several employers have had to deal with grievances raised by other members of staff following inappropriate conversations that have taken place in the office, often sparked by supposedly 'light hearted' conversation with friends.

Thinking logically, a mobile telephone connects people to pretty much everything in their personal life from parents, husbands, wives and also dealing with issues such as paying gas bills and ordering new ring tones!

Practically, it is advisable for employers to have a policy that mobile phones should be switched off or on silent during office hours, albeit this policy should be exercised sensibly if an employee informs you that they are expecting an urgent call. Any policy should be used consistently throughout the company.

Conclusion

In summary, it would be sensible to promote a policy within the company that phone calls, dealing with text messages, emails and the like should only be dealt with when employees are parked away from the road in a safe place. The policy should further promote that mobile phones be switched off or turned to voicemail whilst driving so that messages can be picked up safely at the end of any particular journey.

Employees can be prosecuted for driving while using a handheld mobile phone. Employees committing this offence will be liable to pay a fixed penalty or a fine on conviction in court. Since 27 February 2007 three penalty points are applicable for using a handheld mobile phone whilst driving, together with a £60 fine.

> *See also*: Driving at work, p.250; Vehicles at work, p.686.

Sources of further information

Workplace Law's fully-revised *Driving at Work Policy and Management Guide v.5.0* helps you cover yourself and your staff and ensure that your employees keep to the highest standards of safe driving at work. The work has already been done for you – simply download and customise our comprehensive user-friendly MS Word formatted documentation and insert your company's details where highlighted, saving you time and money on tedious research and drafting.

As with all policy documentation from Workplace Law, this policy has been fully reviewed by our qualified lawyers to provide expert guidance that you can trust. Visit www.workplacelaw.net for more information.

Music licensing

Christine Geissmar, PPL, and Barney Hooper, PRS for Music

Key points

- Just about every workplace in the UK where music is played can find itself covered by wide-ranging copyright laws.
- Most businesses are aware of the benefits of music, but many are unaware of the legal requirements relating to its use in the workplace.

Legislation

- Copyright, Designs and Patents Act 1998.

PPL and PRS for Music

Copyright protects music in different ways and under the Copyright, Designs and Patents Act 1988, two separate licences are usually required whenever you play recorded music in public – one from PPL and one from PRS for Music.

PPL provides music licensing solutions for businesses that play recorded music in public. Established in 1934, PPL carries out this role on behalf of tens of thousands of performers and record companies, without retaining a profit. PPL's licensing of the public performance of recorded music enables businesses to lawfully use millions of recordings for very little effort.

PRS for Music carries out a very similar role, but collects for the songwriters, composers and music publishers, in respect of the rights in the musical compositions embodied in recorded music. Also a not-for-profit company, the society has 80,000 members and has been operating since 1914.

PPL and PRS for Music are two separate companies. Whilst carrying out similar functions, the two licensing organisations operate independently, represent different

rights holders and have separate tariffs, terms and conditions.

Why does a workplace need a music licence?

Like any other aspect of business, music has to be paid for. Under the Copyright, Designs and Patents Act 1988, if recorded music is 'played in public' (i.e. played in any context other than a domestic one) every play of every recording requires the permission of the record company that controls the copyright in the recording and the permission of the creator of the music in that recording.

If PPL and PRS for Music did not exist, businesses that wanted to use music would be required to contact potentially thousands of record companies, writers, composers and music publishers to individually obtain their permission, before being able to play music lawfully.

'Playing music in public' has a wide legal meaning. It is not limited to playing music at business premises where members of the general public have access (such as pubs, shops and gyms) but also such premises such as offices, factories and warehouses where music is being played to staff.

It is the legal responsibility of the proprietor to ensure that all their business premises

> **Facts**
> - Music can increase profits – research shows that more than a third of customers would be willing to pay 5% more for products and services from businesses that play music.
> - Music means more customers – 81% of customers that visited hair salons said that they would like to hear music. 70% would pay more to go to a restaurant that plays music.
> - Music can improve customer service – over a quarter of callers who were left holding in silence for one minute thought they had been holding for over five minutes on the phone, compared to 0% of callers who listened to on-hold music.
> - Music works for your staff – 66% of employees believed that music made them feel more motivated at work, with over a quarter stating they would be less likely to take sick time if music was played at work. 77% said they were more productive when good music was played, and this rises to an overwhelming 83% among warehouse workers.
> - In 2010 the size of the UK music industry was estimated at £3.8bn.

are appropriately licensed for playing recorded music in public. The proprietor is liable for any acts of infringement within the premises; therefore it is their duty to ensure that the premises are accurately licensed and that staff do not use recorded music within unlicensed premises.

Penalties for infringing copyright laws

Where a business or organisation requires a licence but does not obtain one, it is infringing copyright by playing recorded music in public. PPL and PRS for Music each regularly visit premises around the country to assess licensing requirements and ensure their information is up to date and complete.

Businesses requiring a music licence should contact both PPL and PRS for Music to ensure they have what they need. Both licensing organisations can help explain the requirements and ensure companies get the licences that are right for their business.

If a business continues to use music without having the correct licences then they could be taken to court; however this is always a last resort and both PPL and PRS for Music will work with a business to ensure they are licensed correctly. If a case did go to court, then a business could be ordered to stop playing music until it has paid for the relevant music licences, and they may also have to pay legal costs.

Exemptions

A PPL licence is not required where a business or organisation does not play recorded music, or only does so in a domestic context (which would include closed family events such as weddings or birthday parties).

Where only live music is played, a PPL licence is not needed (as the recording is not being used) but a PRS for Music licence will be needed to cover the use of the compositions being performed live.

The process of obtaining a licence

The process of obtaining a licence from both PPL and PRS for Music is quick and easy.

To apply for (or find out more about) a PPL licence, you can either ring the PPL New Business Team on 0207 534 1070 or visit www.ppluk/en/Music-Users and click on the 'apply now' button. Licence fees must be paid in full before a licence is issued.

PRS for Music can be reached on 0800 0684828 or by visiting its website: www.prsformusic.com.

The cost of a licence

The cost of both a PRS for Music and PPL licence depends on the kind of business, what the size of the business premises is (which is measured in different ways depending on the business type), and how music is being played.

When introducing and revising tariffs, PPL seeks to charge a fair and reasonable fee. PPL and PRS for Music regularly consult with trade bodies and users in the affected business sectors as part of the tariff review process.

If, within the first year of having a PPL licence, a company's circumstances change, they can contact the PPL New Business Team to discuss. If a company has had a PPL Licence for longer than a year, they should contact the PPL switchboard on 0207 534 1070.

For more information on the benefits a licence can bring to a business, please visit www.musicworksforyou.com.

See also: Noise at work, p.505.

Sources of further information

PPL: www.ppluk.com

PRS for music: www.prsformusic.com

www.musicworksforyou.com

Night working

Pinsent Masons Employment Group

Key points

- Night workers attract special protection limiting their shifts and requiring the completion of regular health assessment; therefore it is important for employers to ascertain whether they employ workers who would be classified as night workers. If so, they should check:
  - how much working time night workers normally work;
  - if night workers work more than eight hours per day on average, whether the amount of hours can be reduced and if any exceptions apply;
  - how to conduct a health assessment and how often health checks should be carried out;
  - that proper records of night workers are maintained, including details of health assessments; and
  - that night workers are not involved in work that is particularly hazardous.

Legislation

- Working Time Regulations 1998.
- Management of Health and Safety at Work Regulations 1999.

Working Time Regulations

The Working Time Regulations provide basic rights for workers in terms of maximum hours of work, rest periods and holidays. Night workers are afforded special protection by the Regulations. Depending on when they work, workers can be labelled 'night workers'.

Night workers

A 'night worker' is any worker whose daily working time includes at least three hours of night time:

- on the majority of days he works; or
- sufficiently often that he may be said to work such hours as a 'normal course', i.e. on a regular basis.

Employers and workers can agree a proportion of annual working time that must be worked during night time in order to qualify as a night worker in a collective or workforce agreement. However, the agreement cannot exclude workers who would otherwise count under the 'normal course' test.

Night time

In the absence of any contrary agreement, night time is defined as the period between 11 p.m. and six a.m. Another definition of night time hours can be determined in a relevant agreement, provided it lasts at least seven hours and includes hours between midnight and five a.m.

Night work limits

An employer must take all reasonable steps to ensure that the normal hours of a night worker do not exceed an average of eight hours for each 24 hours over a 17-week reference period (which can be extended in certain circumstances).

The average eight-hour limit applies to the 'normal hours of work' performed by the worker, not the hours actually worked.

> **Case study**
>
> *R v. Attorney General for Northern Ireland ex parte Burns* (1999)
>
> This case discussed the definition of 'night worker' and, in particular, the meaning of 'normal course'.
>
> The claimant was asked to change to a shift system, which would involve some night work, which she agreed to. She started on the new shift system but when the night work starting causing her some health problems she insisted on moving to a day shift and subsequently was dismissed.
>
> The UK had not implemented the Working Time Directive at the time, despite the fact that it should have done. The claimant brought judicial review proceedings against the Government for its failure to implement the Directive in time, since its failure left her unable to claim compensation against her employers.
>
> As part of the proceedings, the Northern Ireland High Court had to determine whether the claimant was a 'night worker' for the purposes of the Directive. The claimant worked the night shift one week in every three and in that one week of night shifts she worked three hours during night time. The Government argued that the claimant did not work nights in the 'normal course of her employment'.
>
> The court held that the words 'as a normal course' of employment meant no more than as 'a regular feature' of employment. Since the Directive had contemplated that a worker may work only three hours in the night time to be classified as a night worker, the definition of night worker should not be limited to only those workers who work in the night time exclusively or even predominantly.
>
> 'Normal course' therefore means 'on a regular basis', and is a wide definition. However, occasional or ad hoc work at night does not make a person a night worker.

Overtime was originally excluded from the calculation of normal hours but it has been included since 2003 where it is obligatory and guaranteed or regularly worked.

Therefore, normal hours includes contractual hours plus any compulsory or regular overtime, but not ad hoc overtime or a reduction in hours because of holidays, sickness, maternity or other reasons.

The average is calculated over a 17-week rolling reference period but a collective or workforce agreement may be used to stipulate fixed successive periods.

The other provisions in the Regulations relating to rest breaks and holidays apply equally to night workers.

Special hazards

Where a night worker's work involves special hazards or heavy physical or mental strain, there is an absolute limit of eight hours on any of the worker's working days. No average is allowed. Work involves a 'special hazard' if either:

- it is identified as such between an employer and workers in a collective agreement or workforce agreement; or
- it poses a significant risk as identified by a risk assessment that an employer has conducted under the Management of Health and Safety at Work Regulations 1999.

Health assessment

All employers must offer night workers a free health assessment before they begin working nights, and thereafter on a regular basis. Workers do not have to undergo a health assessment, but they must be offered one.

All employers should maintain up-to-date records of health assessments. A health assessment can comprise two parts – a medical questionnaire and a medical examination. It should take into account the type of work that the worker will do and any restrictions on the worker's working time under the Working Time Regulations. Employers are advised to take medical advice on the contents of a medical questionnaire.

Recent evidence has emerged linking night working and cancer, especially breast cancer amongst female night workers. A number of possible theories for why this might be have been put forward, including artificial light that night workers are regularly exposed to, and the fact that their body clock is out of sync with the environment. Employers should be alert to this and ensure that their workers know that drugs are available to help shift workers adjust to the time changes and to promote sleep.

New and expectant mothers

New and expectant mothers have certain special rights in relation to night work. See *'Pregnancy'* (p.597).

See also: Loneworking, p.467; Occupational cancers, p.512; Occupational health, p.519; Pregnancy, p.597; Working time, p.758.

Sources of further information

Guidance on night working is available from the DirectGov site: www.direct.gov. uk/en/Employment/Employees/WorkingHoursAndTimeOff/DG_10028519

Noise at work

Andrew Richardson, URS Scott Wilson

Key points

- Employers have a legal duty to safeguard their employees' hearing.
- Employers must assess the risks of hearing loss and implement risk control measures.
- There are three action levels of daily personal noise exposure.
- Employers must take at least the stated measures to reduce noise exposure, from providing ear protection or installing soundproof enclosures, to using quieter processes or equipment.
- Control measures may include health surveillance, where noise exposure levels are significant.
- Failure to control noise can lead to enforcement action by the local authority to prevent a 'statutory nuisance'.

Legislation

- Health and Safety at Work etc. Act 1974.
- Environmental Protection Act 1990.
- The Statutory Nuisance (Appeals) Regulations 1995.
- Provision and Use of Work Equipment Regulations 1998.
- Management of Health and Safety at Work Regulations 1999.
- The Control of Noise at Work Regulations 2005.

The 2005 Regulations replaced the 1989 Regulations and introduced new action requirements on employers; e.g. action to protect workers must now be taken at levels of noise five decibels lower than in the 1989 Regulations, and hearing checks are now required for workers regularly exposed to noise above 85 decibels. The Regulations only came into force in the music and entertainment sectors on 6 April 2008 and for seagoing ships on 6 April 2011.

Statutory requirements

Hearing loss can be greatly reduced if machinery manufactured is quieter, if employers introduce policies and risk control measures to reduce exposure to noise, and if employees utilise risk control measures.

Under the Health and Safety at Work etc. Act 1974 an employer has a legal duty to safeguard its employees. Under the Management of Health and Safety at Work Regulations 1999 there is a duty to carry out risk assessments, implement risk control measures and, where necessary, carry out health surveillance.

The Control of Noise at Work Regulations 2005 require the employer to carry out noise assessments using the following five basic steps:

1. Identify where there is likely to be a noise hazard.
2. Identify all workers likely to be exposed to the hazard.
3. Evaluate the risks arising from the hazard and establish the noise exposure.
4. Record the findings.
5. Review the assessments and revise as necessary.

Facts

- Over one million employees in Great Britain are exposed to levels of noise that put their hearing at risk.
- The HSE suggests that at least 25% of those people who report noise induced hearing loss also report having tinnitus. A further 10% reported tinnitus in the absence of hearing loss.
- One-third of Europe's workers are exposed to high levels of noise for more than a quarter of their working time, and almost 40 million workers (equivalent to the entire population of Spain) have to raise their voices above normal conversational levels in order to be heard for at least half of their working hours.
- The IES report, 'The Costs and Benefits of the Noise at Work Regulations' (1996) showed that 25.7% of firms provide audiometry in establishments with noise levels over 85dB(A).
- Estimates from the Self-reported Work-related Illness (SWI) surveys in 2005/06 estimated that 68,000 people ever employed in Great Britain were suffering from work-related hearing problems in this year.
- Hearing loss caused by exposure to noise at work continues to be a significant occupational disease. Recent research suggests 170,000 people in the UK suffer deafness, tinnitus or other ear conditions as a result of exposure to excessive noise at work.

Employers have an overriding duty to eliminate exposure to noise of their employees and others affected – ideally by combating the noise at source. Where elimination is not reasonably practicable, the employer's duty is to reduce the risk of damage to their employees' hearing from exposure to the noise to the lowest level reasonably practicable.

Action levels

The Regulations identify action levels at which various actions need to be taken by the employer. These levels include reference to daily personal noise exposure, which is defined as the personal exposure to noise at work (over an eight-hour day), taking account of the average levels of noise in working areas and the time spent in them but not including the wearing of any ear defenders or protectors. The 2005 Regulations have reduced the exposure action levels of the 1989 Regulations by 5dB(A). This is a significant change, since a drop of 3dB(A) represents a halving of the sound pressure level ('loudness') of the noise. The current action levels are as follows:

- *First action level.* A daily personal noise exposure of 80dB(A) and a peak value of 112 pascals.
- *Second action level.* A daily personal noise exposure of 85dB(A) and 140 pascals.
- *Limit value.* A peak sound pressure of 87dB(A) and 200 pascals. The limit value will take into account the reduction afforded by hearing protection.

If an employee is exposed to the first action level or above, but below the second action level, then the employer must provide suitable and sufficient personal ear protection. However, it is not compulsory for workers to use it.

If an employee is exposed to the second action level or above or to the peak action level or above, then the exposure to the noise must be reduced so far as is reasonably practicable (excluding the provision of ear protection). An example would be the provision of a soundproof enclosure. If it is not reasonably practicable to reduce the noise, then an ear protection zone must be demarcated and identified; ear protection must be provided to all workers likely to be exposed, the employer must ensure it is being worn, and the employer must provide information on how to obtain that protection.

With the exception of ear defenders issued because of exposure between the first and second action levels, the employer has to ensure that all other equipment and ear protection is utilised and maintained in a suitable manner. This will mean regular checks on its use and condition, and health surveillance.

Employees have a duty to comply with and use the measures the employer introduces and to report any defects or difficulties in complying with the Regulations. Information, instruction and training have to be provided for all employees likely to be exposed to the first action level or above.

Noise Abatement Notices

Under Section 79 of the Environmental Protection Act 1990 a Local Authority has powers to issue an Abatement Notice where noise is considered to be causing a 'statutory nuisance'. Noise that might constitute a statutory nuisance includes:

- noise emitted from premises so as to be prejudicial to health or a nuisance; and/or
- noise that is prejudicial to health or a nuisance and is emitted from or

caused by a vehicle, machinery or equipment in a street or, in Scotland, a road.

Equipment can include musical instruments and noise includes vibration from machinery.

The Noise Abatement Notice will specify what actions are required to stop the nuisance; these can range from simply 'turning the volume down' to making modifications to a building or installation. There is a statutory appeal process, which is set out in the Statutory Nuisance (Appeals) Regulations 1995.

In general it is not necessary to prove actual prejudice to the health of an individual to establish statutory nuisance. It is sufficient to prove that there is material interference with the personal comfort of residents in the sense that it materially affects their well-being.

Section 79 also states that it is the duty of every local authority, where a complaint of statutory nuisance is made, to take such steps as are reasonably practicable to investigate the complaint. Often this will involve the environmental health department of the Local Authority visiting the domestic property or properties in question to monitor the situation and take noise readings. There is no set decibel level at which noise becomes a statutory nuisance for the purposes of the Act.

If the noise does not stop or the conditions of the Notice are not complied with, then the recipient of the notice will have committed a criminal offence and may be prosecuted. The maximum fine for circumstances where the offence takes place on domestic premises is £5,000, or £20,000 if the offence occurs on industrial, trade or business premises.

Where a business is charged with breach of an Abatement Notice a defence is available if the company in question can demonstrate that they used the best practicable means to avoid the nuisance. This includes among other things:

- Having regard to local conditions and circumstances;
- The current state of technical knowledge;
- The financial implications;
- The design, installation, maintenance, manner and periods of operation of plant and machinery; and
- The design, construction and maintenance of any buildings and structures.

Even where a business has been operating out of premises that pre-date any domestic housing developments in the locality, and complies with any relevant legislation, it may still find itself subject to a Noise Abatement Notice should a complaint be made by a resident.

> *See also*: Construction site health and safety, p.177; Health surveillance, p.396; Vibration, p.689; Workplace health, safety and welfare, p.768.

Sources of further information

L108 Controlling Noise at Work: www.hse.gov.uk/pubns/priced/l108.pdf. This is the Approved Code of Practice for the 2005 Regulations and provides authoritative comprehensive guidance.

INDG 362 Noise at Work – Guidance for employers on the Control of Noise at Work Regulations 2005: www.hse.gov.uk/pubns/indg362.pdf.

DEFRA – *Bothered by Noise?*: http://archive.defra.gov.uk/environment/quality/noise/neighbour/documents/bothered-by-noise-060701.pdf

Nuisance

Kevin Boa, Pinsent Masons Property Group

Key points
- A nuisance may arise by reason of unjustified acts or omissions in the occupation or use of land or property.
- Some nuisances can be the subject of private civil law claims, whereas others constitute criminal offences or are enforced by Local Authorities.

Legislation
- The Environmental Protection Act 1990.
- Crime and Disorder Act 1998.

Private nuisance

Private nuisance is an unlawful interference with a person's use or enjoyment of land, or some right over, or in connection with it. Private nuisance is historically concerned with the regulation of land use between neighbours. You may encounter a nuisance either when you are adversely affected by a situation or when another party claims that a nuisance has been committed by you.

Private nuisance problems may lead to litigation or be settled by some form of mediation. Specialist legal advice is usually necessary. Evidence for or against the nuisance is always crucial, so formal records and other evidence should be kept. Certain defences are allowable to specific nuisances.

Generally only a person or organisation who has an interest in the affected land may take legal action. There must be provable damage, whether direct damage to land, interference or encroachment. Normally the damage caused must have been reasonably foreseeable by the person who caused the damage. Liability may lie with more than one party at the same time.

A number of specialist defences can be raised by a defendant, including prescription, which means that the nuisance has been actionable for a period of 20 years and the claimant has been aware of this during that time. Certain types of 'damage' may not be claimed. For example, personal injury claims cannot be made in private nuisance.

Examples of private nuisance
- Structural damage as a result of pile-driving.
- Resurfacing of a driveway so that water flows on to a neighbour's land causing damage.
- Allowing a building to become so dilapidated or infested that the building, or part of it, falls on to neighbouring land or the claimant's property is affected by vermin or damp.
- Encroachments, e.g. tree branches overgrowing neighbouring property or tree roots growing into neighbouring land. Tree problems are very often the cause of private nuisance actions. They are also often encountered in boundary disputes.
- Noise nuisance, e.g. from persistent loud music. This is a very commonly encountered nuisance and one where the Local Authority can be involved (see 'Statutory nuisance' below).
- Loss of amenity caused by smoke, fumes, vibration, smells or dust. With

regard to this type of private nuisance and noise nuisance, the locality must be considered as well as the conduct. While conduct may be a nuisance in one area, it may be tolerated in another. Compare (a) a residential estate with (b) an industrial estate.

Adopted nuisance

A nuisance need not have been originally created by a defendant if, with knowledge, he has adopted or acquiesced in it – e.g. a new occupier does nothing to prevent damage caused by encroaching tree roots.

Rylands v. Fletcher

The rule in *Rylands v. Fletcher* arises from an 1865 court case. This established a tort (civil wrong) whereby a defendant would be strictly liable for all foreseeable consequences where damage was caused by the escape of something dangerous which had accumulated on the land for some 'non-natural' purpose. The original case concerned a leak of water from a reservoir. The following principles apply:

- The absence of wilful default or negligence by the defendant is irrelevant.
- The substance that 'escapes' must as a result be likely to do mischief.
- The substance must have accumulated on the land and escaped from the land.
- Damages can only be claimed in respect of damage to the land or to objects on the land.
- The damage must have been foreseeable by the defendant as a result of the escape.
- Certain defences to strict liability are recognised – e.g. Act of God or acts by trespassers or strangers that result in the damage.

Examples of the application of Rylands v. Fletcher

- An electric current discharging into the ground.
- Explosives.

However, the escape of water from a sewer serving a block of flats was recently held to be non-actionable on the basis that the sewer was an ordinary incident of domestic activity.

Public nuisance

Every public nuisance is a criminal offence. An individual claimant who wishes to pursue a civil claim in public nuisance must prove that the nuisance affects a widespread class of people as opposed to an individual alone. The class of people affected can be broad, e.g. all the staff of a building or a group of neighbours.

A civil claimant must prove special damage, which goes beyond damage suffered by others affected by the same circumstances. For that reason private claims in public nuisance areas are rare. Any sort of damage can be the subject of compensation. Many public nuisance cases are concerned with the highway:

- Highway nuisance such as obstructions on the highway.
- Dangerous premises abutting the highway.
- Impeding rights of access to and from property adjoining the highway.

Anti-social behaviour orders

An Anti-Social Behaviour Order (ASBO) is a civil order under the Crime and Disorder Act 1998 and prohibits a person from acting in an 'anti-social' manner.

An ASBO is made by an application by a Local Authority, registered social landlord or police to the magistrates' court and the ASBO, if awarded, will last for a minimum of two years. Criminal penalties (including imprisonment) may result from a breach of the terms of an ASBO.

Statutory nuisance

Certain types of nuisance are governed by statute. The Environmental Protection

Act 1990 (EPA) covers many statutory nuisances and provides a procedure for enforcement. Almost all statutory nuisances are enforced by Local Authorities. They do not depend on a complainant having occupation rights (although usually he/she will).

The EPA covers a wide range of nuisances such as condition of premises, smoke emissions, emissions of fumes, gases, dust, steam and smells, problems due to the keeping of animals, and noise problems, both from premises and from vehicles and equipment. If a nuisance falls under the provisions of a statute, enforcement by a Local Authority is likely to be more efficient and certainly less costly than pursuing a claim in private nuisance. The Local Authority will require good evidence of the nuisance. It will usually inspect and, in cases of noise, will employ decibel meters.

Enforcement is by way of abatement notices. You may encounter these either by asking the Local Authority to take action to serve one or by having one served on you. There are specific procedures for service of notices, appealing against notices and offences in connection with them. Specialist advice will be needed. Failure to comply with an abatement notice (which is not successfully appealed) will constitute a criminal offence.

Examples of statutory nuisance

- Emission of smoke, fumes or gases to residential premises.
- Premises that are so dilapidated or neglected as to be prejudicial to health or a nuisance.
- Loud music being played several nights each week after 11 p.m.
- Nuisance from animals.

See also: Boundaries and party walls, p.63; Dilapidations, p.221; Landlord and tenant: possession issues, p.438; Property disputes, p.600; Trespassers and squatters, p.681.

Sources of further information

The Environment Agency: www.environment-agency.gov.uk

The Workplace Law website has been one of the UK's leading legal information sites since its launch in 2002. As well as providing free news and forums, our Information Centre provides you with a 'one-stop shop' where you will find all you need to know to manage your workplace and fulfil your legal obligations.

Content is added and updated regularly by our editorial team who utilise a wealth of in-house experts and legal consultants. Visit www.workplacelaw.net for more information.

Occupational cancers

Chloe Harrold and Pam Loch, Loch Associates Employment Lawyers

Key points

- A recent decision by the Danish Government has brought the topic of occupational cancers to the forefront and may result in similar claims around Europe.
- Employers are required to protect the long-term health of their employees as far as is reasonably practicable.
- Employers must conduct regular risk assessments.
- Prevention must take priority over protection.
- The top ten industry sectors contributing to occupational cancer and linked to deaths are construction, personal and household services (this sector includes repair trades, laundries and dry cleaning, domestic services, hairdressing and beauty), shift work, land transport, metal workers, painters and decorators in the construction industry, printing and publishing, wholesale and retail trades, mining, and manufacture of transport equipment.

Legislation

- Health and Safety at Work etc. Act 1974.
- Management of Health and Safety at Work Regulations 1999.
- Control of Substances Hazardous to Health (COSHH) Regulations 2002.
- Control of Asbestos Regulations 2006.

Occupational cancer is the term given to cancer caused by exposure to carcinogens in the workplace. At present, occupational exposure is the primary form of exposure to more than half of the chemicals, groups of chemicals, mixtures, and specific exposures in the human environment.

Some industries and occupations are recognised as presenting a higher risk of occupational cancer. There are synergistic effects between some occupational carcinogens and lifestyle factors; for example, occupational exposure to asbestos dramatically increases the likelihood of tobacco smokers developing lung cancer. The WHO 'Global Burden of Disease' study carried out in 2002 indicated that around 20-30% of the male and 5-20% of the female working-age population (aged between 15 and 64 years) may have been exposed during their working lives to lung carcinogens including asbestos, arsenic, beryllium, cadmium, chromium, diesel exhaust, nickel and silica. Worldwide, these occupational exposures account for about 10.3% of cancer of the lung, trachea, and bronchus.

The International Agency for Research on Cancer (IARC) classifies more than 60 exposures as cancer-causing and a further 55 as probably cancer-causing. Some of these exposures take place in occupational settings.

The cancers caused by exposure to carcinogens in the workplace can often take 30-50 years to develop. Studies into the incidents of occupational cancer therefore continue to reveal increasing

Case studies

The widow of a cabinet maker was awarded £375,000 in compensation following her husband's death in 2005 from nasal cancer caused by his work. Coinciding with this judgment, the HSE has announced it will be re-examining the occupational risks involved in working with wood dust following its last investigations carried out over ten years ago.

Samsung has been ordered to pay compensation to the families of two employees who died of leukaemia which the South Korean courts accepted was caused by their job. The two employees worked cleaning wafers on a production line and their families argued that the constant exposure to chemicals and ionising radiation caused their illnesses.

Samsung is refuting that there is scientific evidence which supports this finding and in 2010 the HSE told workers at a Scottish semiconductor plant, Greenock, that they were not at risk of cancer. However, there is some evidence that suggests that those working at semiconductor factories in the UK, and elsewhere, experience higher incidences of cancer. The HSE will continue to monitor health and safety in the semiconductor manufacturing industry but has no plans to carry out further research at Greenock at this stage.

The General Assembly of the State of Pennsylvania, USA, passed legislation in June 2011 that designates cancer as an occupational disease for firefighters. The Firemen's Association of the State of Pennsylvania welcomed the new law, which will allow both career and volunteer firefighters to claim Worker's Compensation if they develop cancer as a result of their work.

Cancer is known to develop as a direct result of exposure to carcinogens present in the emergency situations firefighters respond to. Research carried out by the University of Cincinnati and published in 2006 indicated that rates of testicular cancer were 100% higher in firefighters.

On the first Friday of July each year the British Lung Foundation (BLF) culminates its campaign to raise awareness of the risks of asbestos exposure. Action Mesothelioma Day started in 2006 when the BLF delivered a charter filled with 14,000 signatures to 10 Downing Street. This Charter calls for improved care and treatment for patients, better protection for employees and more funding for research.

numbers of those diagnosed with cancer despite many of the most dangerous carcinogenic materials being banned and regulations introduced to minimise other risks. The main list of cancer-causing substances is produced by The International Agency for Research on Cancer (IARC). This list contains all hazards evaluated to date, according to the type of hazard posed and to the type of exposure. There are several groupings. The TUC believes that all substances in

Group 1 and 2A should be removed from the workplace or, if that is not possible, exposure should be fully controlled. Caution should also be used to prevent exposure to substances in Group 2B.

- *Group 1*: The agent is carcinogenic to humans.
- *Group 2A*: The agent is probably carcinogenic to humans.
- *Group 2B*: The agent is possibly carcinogenic to humans.
- *Group 3*: The agent is not classifiable as to its carcinogenicity to humans.
- *Group 4*: The agent is probably not carcinogenic to humans.

The list can be found at: http://monographs.iarc.fr/ENG/Classification/index.php

Asbestos-related cancers

The most well-known incidence among the general public of occupational cancer is probably that caused by asbestos, which contributes the most to both attributable deaths and registrations (larynx (3), lung (1,937), mesothelioma (1,937), stomach (32)).

Asbestos is the generic name for a wide range of naturally occurring minerals that crystallise to form long thin fibres and fibre bundles. Asbestos was a material that experienced heavy commercial use, often as building / insulation material, particularly following the Second World War, until it was discovered in the 1970s to be a highly toxic and deadly material. Much legislation was passed in the 1970s and 1980s to restrict and regulate the use of asbestos and there were a number of highly publicised personal injury cases. Since these developments, the cases of cancer caused by asbestos dropped, as did public awareness of occupational cancer in general. However, asbestos remains the biggest industrial killer of all time, with studies suggesting that the diseases related to asbestos poisoning will eventually kill ten million people worldwide.

Mesothelioma is a form of cancer affecting the lining of the lung, and sometimes the abdomen and/or the heart. Several UK mesothelioma studies suggest that between 96% and 98% of male mesothelioma cases are due to occupational or paraoccupational exposure (e.g. exposure from living near an asbestos factory or handling clothes contaminated due to occupational exposure). The cancer caused by asbestos exposure often takes several decades to develop. This is clear from records that show there were 2,249 deaths in 2008 attributed to mesothelioma, an increase from 153 in 1968. According to Cancer Research (2010), the annual number of cases of mesothelioma in the UK is expected to peak in 2015.

The only known cause of this cancer is asbestos exposure, which, in the majority of cases, is likely to have taken place at work. As the link between asbestos exposure and cancer has been firmly established, cases and awards of compensation are common where negligence on the part of the employer can be established. Asbestos is considered so harmful that the HSE assumes there is no low level 'acceptable' threshold for exposure to the substance.

Pleural thickening and pleural plaques are conditions affecting the lungs, which are also consequences of asbestos exposure. Currently it isn't possible in the UK to pursue compensation claims for these conditions, following the landmark House of Lords ruling of *Rothwell v. Chemical Insulation & Co. and Ors.* (2006), which stated that sufferers should not be awarded compensation as the physical changes that result do not lead to any obvious symptoms.

However, in October 2011, Supreme Court judges ruled that the Damages (Asbestos-related Conditions) Act – legislation passed in Scotland in 2009, which offers those that have pleural plaques the opportunity to claim compensation – was enacted within the Scottish Parliament's scope of power, and should no longer be subject to legal contention. Insurers in Scotland may now face claims stalled whilst the Act was debated, and sufferers of pleural placques may be able to claim compensation.

Other occupational cancers

There are a large number of cancers that may arguably be attributable to the workplace environment; however, the Department for Work and Pensions Industrial Injuries and Disablement Benefit (IIDB) scheme sets out specific forms of cancer that are currently eligible for compensation. These are:

- Leukaemia (other than chronic lymphatic leukaemia) or cancer of the bone, female breast, testis or thyroid due to exposure to electromagnetic radiation or ionising particles (disease number A1);
- Acute non-lymphatic leukaemia due to exposure to benzene (C7);
- Skin cancer due to exposure to arsenic, arsenic compounds, tar, pitch, bitumen, mineral oil (including paraffin) or soot (C21);
- Sinonasal cancer due to exposure to nickel compounds (C22a) or due to exposure to wood, leather and fibre board dust (D6);
- Lung cancer due to exposure to nickel compounds (C22b) or due to work as a tin miner, exposure to bis(chloromethyl) ether, or to zinc, calcium or strontium chromates (D10) or due to silica exposure (D11);
- Bladder cancer due to exposure to various compounds during chemical manufacturing or processing, including 1-naphthylamine,

2-naphthylamine, benzidine, auramine, magenta, 4-aminobiphenyl, MbOCA, orthotoluidine, 4-chloro-2-methylaniline, and coal tar pitch volatiles produced in aluminium smelting (C23);
- Angiosarcoma of the liver due to exposure to vinyl chloride monomer (C24);
- Mesothelioma (D3); and
- Asbestos-related lung cancer (lung cancer with asbestosis (D8) or lung cancer and evidence of at least five years' asbestos exposure before 1975 in certain jobs (D8A)).

There are also certain industries that have a higher risk associated with contracting a work-related cancer, the majority of which can be found among blue-collar workers. Any manufacturers using asbestos will be putting their employees at risk, and with no worldwide ban on the use of asbestos this is a very real and present danger.

Those exposed to wood or leather dust are at increased risk of contracting nasal cancer, and the rubber industry shows a high proportion of bladder cancer. The occurrence of occupational cancers appear across the board and a 2006 US study found that staff employed at IBM factories had high rates of cancers linked to chemical and electromagnetic field exposure. In the UK, asbestos-related illnesses are high amongst former shipyard and other heavy industry workers.

Recent developments

Receiving widespread publicity in March 2009, the Danish National Board of Industrial Injuries took the decision to award compensation to women who had developed breast cancer after working long periods on shift work. In total, 38 women, all of whom had worked night shift patterns for more than 20 years, received compensation from the Danish Board. Of those, seven nurses received between

> **Facts**
>
> - In the UK, someone is diagnosed with cancer every two minutes. More than one in three people in the UK will be diagnosed with cancer in their lifetime. One in four will die from cancer. Over 300,000 people in the UK are diagnosed with cancer every year.
> - The total number of cancer registrations in Great Britain in 2003 attributable to occupational causes was 13,338.
> - Scientists estimate that around 5.4% of cancer deaths in the UK are attributable to occupational exposure to carcinogens.
> - 2007 figures show that the cancers contributing the largest number of estimated deaths in men are lung cancer (2,850), other and unspecified cancers (480), pleural and respiratory cancers (430), bladder cancer (310), leukaemia (230), and mesentery and peritoneal (180). In women, the only cancer contributing more than 100 estimated deaths is lung cancer (660).
> - In 2007, scientists from Stirling University said occupational cancer deaths could be around 24,000 a year – four times greater than the official estimate by the HSE of 6,000.
> - A survey by Hazards Magazine and the TUC states that a worldwide epidemic of occupational cancer is claiming at least one life every 52 seconds.
> - A report by the HSE in 2010, 'The Burden of Occupational Cancer in Great Britain,' has quantified for the first time the impact of occupation on the burden of cancer in Britain for all cancer sites and the carcinogens highlight the fact that many carcinogenic exposures in the workplace affect multiple 'cancer sites'.
> - Around half of all occupational cancer deaths in the UK are attributable to asbestos exposure.
>
> *Sources: HSE, Cancer Research UK.*

30,000 and one million Danish kroner (£3,700 to £123,000).

The Danish Board's move was prompted by the International Agency for Research on Cancer (IARC) publication in 2007, which found that women who work through the night on a regular basis could be more likely to develop cancer.

One major study by IARC reported a 36% increase for women who had worked night shifts for over 30 years, compared to those who never worked nights. However, this research is not conclusive and many organisations, including Breakthrough

Breast Cancer, have called for further research to more firmly establish and explain this link. The Danish case may well prompt an increase in claims against employers, both in the UK and elsewhere in Europe, particularly in relation to night-time working.

Although the scientific evidence remains a matter of debate, more detailed findings were published by the IARC in 2010 and a further study by the HSE is expected to be reported in 2011. If these findings are consistent with the IARC's 2007 report, there is likely to be a surge of claims from employees who realise that their illnesses may well have been caused by their employment. Any employee who was diagnosed within the last three years will be able to bring a personal injury claim against their employer, former or otherwise.

Employers' legal responsibilities

The prevention of workplace cancer has a much lower profile in the workplace than preventing injuries from risks such as falls from height or electrocution. This is despite the fact that only 220 to 250 workers die each year as a result of an immediate injury as opposed to the 15,000 to 18,000 that die from cancer.

The prevention of occupational cancer is specific because it relies heavily on legislation, since the population at risk can be relatively easily identified. There is a hierarchy of preventive measures and its prevention is very important because among certain groups of workers occupational risk factors may determine the majority of cancer cases. Furthermore, occupational exposures are avoidable hazards to which individuals are involuntarily exposed.

The Health and Safety at Work etc. Act 1974 requires employers to ensure the health of their employees as far as is reasonably practicable, and the protection of long-term health is included within this remit. The Management of Health and Safety at Work Regulations 1999 require employers to conduct risk assessments, and this includes risks of exposure to cancer-causing substances or conditions.

COSHH requires employers not merely to control hazardous conditions but

to prevent them. It is not enough to minimise the effect of a cancer-causing substance; it must, where possible, be removed altogether. Where a hazardous substance cannot be removed entirely then employers must ensure that they take all reasonable measures to minimise the risks and at all times must comply with the maximum exposure limits set out for carcinogens (among other substances).

Where an employer is not able to entirely remove a hazardous substance it should ensure it has a robust and regular risk assessment process in place. Health and safety policies warning employees of the risks and how they should be carrying out their role to minimise any risk of exposure should be carefully drafted and distributed to all staff.

If an employer has taken all reasonable measures to minimise risk, and has complied with its legal obligations in relation to exposure limits, the risk of successful action against it if an employee does contract a work-related cancer will be greatly minimised.

Conclusion

If employees are working in an environment where some exposure to hazardous material cannot be avoided, an employer must take all reasonable measures to minimise the dangers. Employees should be fully notified of all potential risks, and regular risk assessments and reviews should take place. These measures are not only important from the point of view of protecting employers from legal action but more importantly are paramount to aiding the prevention of devastating and potentially deadly occupational cancers.

See also: Asbestos, p.39; Construction site health and safety, p.177; COSHH: Control of Substances Hazardous to Health, p.207; Health surveillance, p.396; Night working, p.502; Occupational health, p.519.

Sources of further information

Hazards Magazine (a useful work cancer prevention kit consisting of various guides and resources): www.hazards.org/cancer/preventionkit/index.htm

HSE: www.hse.gov.uk/research/rrpdf/rr800.pdf

Cancer Research:
http://info.cancerresearchuk.org/cancerstats/causes/lifestyle/occupation/

World Health Organisation: www.who.int/occupational_health/publications

International Agency for Research on Cancer:
http://monographs.iarc.fr/ENG/Classification

Cancer Backup: www.cancerbackup.org.uk

Occupational Safety and Health Consultants Register

Simon Toseland, Workplace Law

Key points

- The Occupational Safety and Health Consultants Register (OSHCR) was introduced in response to the Government-commissioned report on the UK health and safety system, *Common Sense, Common Safety*, published in October 2010.
- This is the first time HSE and Local Authority Inspectors will be able to refer businesses to a central register of general health and safety consultants.
- Registration is voluntary for individuals who provide commercial advice on general health and safety management issues.
- Businesses will be able to search for a consultant by county, industry and topic.
- OSHCR Ltd has been established as a not-for-profit company. The intention is that once it is up-and-running, the HSE will no longer need to be directly involved.

Legislation

- Organisations are required to have access to competent health and safety advice and support under Regulation 7 of the Management of Health and Safety at Work Regulations 1999.

Introduction

The health and safety consultants register is there to assist and support an organisation's existing in-house health and safety representative. The scheme is voluntary. Health and safety consultants with the highest qualifications and experience are eligible to apply to the OSHCR. Organisations using a consultant from the register can have confidence that such persons are highly qualified and experienced, and have been recognised as such by the relevant professional bodies which, through their codes of conduct, will require them to only give advice that is sensible and proportionate.

Competence

Launched on 31 January 2011, the Occupational Safety and Health Consultants Register (OSHCR) is an online voluntary scheme facilitated by the HSE. The aim of the register is to provide a database of health and safety professionals who have been deemed to be competent to offer general health and safety advice within the UK.

Therefore, any employer who has a need for external health and safety support has the reassurance that they are dealing with someone who has adequate experience and qualifications and who is regularly maintaining their technical skills. In addition, the individual has made a commitment to provide sensible and proportionate advice.

This last point is particularly relevant to the HSE. One of the primary reasons that the register was established is because

Occupational Safety and Health Consultants Register

of the amount of inappropriate advice (by consultants) that has led businesses to spend a disproportionate amount of time and costs dealing with trivial risks and hazards, thus contributing to the perceived 'nanny state' culture that is often highly publicised. Furthermore, there have also been a small number of cases where the HSE has prosecuted consultants who have given advice that has put people at risk.

It has been one of the HSE's key objectives to reduce the burden of complying with health and safety law and encouraging a 'sensible' approach to risk management, long before Lord Young came along.

In fact, the OSHCR has been talked about for a number of years, but it was Lord Young's report, *Common Sense, Common Safety*, which gave the register momentum.

Costs
In order to achieve a place on the register, an individual must have a status recognised by one of the professional bodies participating in the scheme. In addition, the cost of registration is an annual fee of £60, or £30 if registered before 30 April 2011. For the health and safety consultant it enables them to demonstrate their professional standing and to be contacted by potential new clients.

Conclusion
Businesses choosing to use a consultant from the scheme must be mindful that the consultant they choose may not necessarily be competent to deal with any specialist requirements – asbestos, fire and occupational health are just a few that spring to mind.

It is, therefore, trusted that as a 'competent person' they will reliably inform the client they may need to look elsewhere.

Finally, there are many consultants and professional bodies who feel excluded. The APS, for example, has been running its own professional register for years, but its registered members are currently not eligible to apply.

From 30 September 2011, the way the HSE provides information services will change to a web based service, replacing the current HSE Infoline service. The Consultants register is viewed by some as an alternative with a reduced cost to the HSE.

See also: Competence and the 'Responsible Person', p.158; Health and safety inspections, p.381; Health and safety management, p.388.

Sources of further information

The OSHCR online register: www.hse.gov.uk/oshcr/index.htm

Common Sense – Common Safety:
www.hse.gov.uk/aboutus/commonsense/index.htm

Lord Young's report: http://www.number10.gov.uk/wp-content/uploads/402906_
CommonSense_acc.pdf

Occupational health

Greta Thornbory, Occupational Health Consultant

Key points

- 'Work is generally good for your health and wellbeing.' The authors of this statement, Waddell and Burton, added several provisos in that there are various physical and psychological aspects of work that are hazardous and can pose a risk to health and work should do the worker no harm.
- Conversely, employers want to employ people who will give them good service, who have the knowledge, skills and understanding to take on the roles and tasks required of them. Occupational health (OH) services are designed to support and help employers meet these requirements.
- This chapter will cover:
  - What is Occupational health (OH)?
  - How OH can help employers to fulfil their legal requirements.
  - The financial implications of health and safety at work whilst ensuring business viability.

Legislation

There is a great deal of legislation that employers are required to consider and comply with regarding the health, safety and welfare of employees; not only health and safety legislation but also the legislation that comes under employment law, all of which affects the health of the employee. All the health and safety legislation is under review by the present Government at the time of writing.

The Health and Safety at Work etc. Act 1974 (HSWA) is an overarching piece of legislation in that it sets out the duty of the employer to take care of the health, safety and welfare of their employees, and of others who may be affected by his work undertaking – so far as is reasonably practicable. It is from this main Act that most secondary health and safety legislation is derived, and singularly the most important is the Management of Health and Safety at Work Regulations, which charges employers with the duty to undertake a risk assessment in relation to the health and safety of employees.

What is OH?

In 1950 the Joint ILO (International Labour Organisation) / WHO (World Health Organisation) issued the first definition of OH, which was updated in 1995 to these three objectives:

1. The maintenance and promotion of workers' health and working capacity.
2. The improvement of working environment and work to become conducive to health and safety.
3. The development of work organisation and working cultures in a direction that supports health and safety at work and in doing so promotes a positive social climate and smooth operation and may enhance the productivity of the undertaking.

Defining 'health'

The most accepted definition is from the World Health Organisation, which defines health as 'a state of complete physical, mental and social wellbeing and not merely the absence of disease or infirmity'.

> **Case study**
>
> At an Employment Tribunal, when Dundee City Council was found in breach of the Management of Health and Safety Regulations, its Personnel Manager admitted that he did not understand the meaning of OH and the Tribunal itself struggled to define it during the course of the hearing. It would probably have been better if they had asked for an OH expert from one of the OH bodies to give an explanation and to demonstrate the business case. According to reports, the HSE has said that the appropriate use of OH expertise and resources is necessary to comply with statutory duties and will help with reducing work-related sickness absence.

Why occupational health?

OH has been and is promoted on all levels; the international perspective is supported by the WHO/ILO. In turn OH has been, and is, supported at a national level by all UK governments to a greater or lesser extent, although to date there is no legal requirement for employers or employees to have access to OH it is strongly recommended in much of the guidance issued from government departments.

OH also figures clearly in the Government strategies for health at the beginning of the 21st Century. Various government departments, together with the HSE, have produced a plethora of strategies and plans over the years. Today much of this is based on the work of Dame Carole Black from the recommendations in her report published in 2008. The report and all the up-to-date strategies can be found at the website: www.dwp.gov.uk/health-work-and-well-being/. Health, work and wellbeing is a cross government initiative that promotes the positive links between health and work. Companies can even download a tool that enables them to assess and help improve the health and wellbeing of their employees.

Key to the health, work and wellbeing initiatives is the management of absence from work by supporting employees with ill health, chronic conditions by enabling return to work and rehabilitation. This is where occupational health professionals can help and support both employers and employees as they have the specialist knowledge and skills to do this. The Government is in the process of piloting a number of projects with regard to return to work and rehabilitation following the introduction of the new 'fit note' which GPs are required to indicate what special considerations should be given to an employee returning to work after sickness absence. The mantra today being 'not work can't you do but what can you do'.

Small employers who are not in a position to employ dedicated occupational health professionals can now use a free Occupational Health advice line. However, to companies who have access to occupational health professionals there remains no change to their remit except to embrace the concept of public health and the setting up of a new Council for Work and Health. This council is chaired by Diana Kloss and its members are from all the branches of the professions that make up occupational health and health and safety. The Council is in its early days of work. For more details and up-to-date information on occupational health and health at work, visit the Health, Work and Wellbeing website.

The legal aspect

One legal requirement for employers is Employer Liability Compulsory Insurance (ELCI) and this is often quoted when challenging employers about their health, safety and wellbeing provision for employees. However, there are many costs not covered by the insurance. The issue here is that it is the cost of the insurance that is the problem, not that it is a compulsory legal requirement.

Financial aspects

If employers want 'maximum output for minimum outlay' then they need to appreciate the financial benefits of considering the health and wellbeing of employees, particularly the occupational health or the ill health that is caused or made worse by work.

OH professionals advise organisations on health assessment, health surveillance and monitoring, managing absence and general health and lifestyle issues. Every employer pays a premium for employers' liability compulsory insurance. This is to cover injuries and ill health experienced by employees whilst at work. It does not cover the whole scenario. For every £1 of insured costs of an accident or ill health

there will be another £10 of uninsured costs. The HSE describes ELCI as the tip of the iceberg. As the founder of easyJet said after being cleared of the death of five people in a tanker accident and a subsequent 11-year lawsuit: "If you think safety is expensive, try an accident".

OH professionals

OH professionals are mainly doctors and nurses who have undertaken specific training in the field of OH, usually to first degree or higher degree level. The Faculty of Occupational Medicine has developed an accreditation scheme for Occupational Health Services and employers. It is still in its early stages of development and over 200 organisations are going through accreditation at time of writing. This service can be accessed via www.seqohs.org.

See also: Health surveillance, p.396; HIV and AIDS, p.404; Mental health, p.482; Pregnancy, p.597; Smoking, p.653; Stress, p.659.

Sources of further information

World Health Organisation (WHO): www.who.int/en/

Working for Health: www.dwp.gov.uk/health-work-and-well-being/

Workplace Law's *Occupational Health 2008: Making the business case – Special Report* addresses the issues of health at work, discusses the influence of work on health and highlights the business case for occupational health services at work. For more information visit www.workplacelaw.net.

Occupiers' liability

Melissa Thompson, Pinsent Masons Property Group

> **Key points**
> - Occupiers of any premises owe duties to take reasonable precautions to protect lawful visitors and those who trespass on to the premises unlawfully.
> - The duties arise under the Occupiers' Liability Acts and under health and safety legislation.
> - Stricter protection will be owed to disabled people and children.
> - Occupiers should have stringent risk assessment procedures in place to ensure that all risks are quantified and minimised, irrespective of who is entitled to come on to the premises.

Legislation
- Occupiers' Liability Act 1957.
- Defective Premises Act 1972.
- Health and Safety at Work etc. Act 1974.
- Occupiers' Liability Act 1984.
- Protection of Freedoms Bill.

Who is an occupier?
An 'occupier' is somebody with sufficient control over the use of the premises. Individuals and companies can be occupiers and obvious examples include owners, tenants, employers and licensees.

The control need be neither exclusive nor entire. A landlord and his tenant can both be occupiers of premises where the landlord retains control of the common parts. The Defective Premises Act 1972 may also increase a landlord's obligations where he is obliged to maintain the premises.

Duty to visitors
'Visitors' includes all lawful visitors such as invitees, licensees and those with a contractual right to enter. The duty to visitors is governed by the Occupiers' Liability Act 1957. Occupiers must take such care as is reasonable in all the circumstances to ensure that any visitor is safe when using the premises for the purposes for which he or she is invited or permitted to be there. The law recognises that the physical or mental abilities of a disabled person or a child are decreased. Extra care will be required in assessing the risks and deciding the controls that are needed to ensure these groups are not harmed. Occupiers should, as a minimum, be able to provide (on demand) suitable risk assessments, disability access audits, maintenance records and access statements.

Reducing any danger
In the event of an incident and a subsequent claim, the Court will decide whether the occupier took sufficient steps to keep the visitor reasonably safe. The Court is likely to consider:

- the purpose of the visit;
- the occupier's knowledge of any dangers and whether the occupier gave any warning about those hazards;
- the conduct of the visitor, and whether the visitor exceeded the limit of his or her permission or right;
- the physical state of the premises, including measures such as barriers, signage and lighting; and

Case study

The case of *Piccolo v. Larkstock* (2007) highlights the need for landlords to carefully consider whether any occupier liability problems apply in relation to their tenanted properties, and if so to address them to avoid being caught in a negligence claim. In the case, Mr Piccolo was badly injured when he slipped on some wet flower petals on the floor outside a florist's shop in Marylebone Station. He sued the occupier tenant, Larkstock, along with the landlord operator of the station, Chiltern, for negligence and breach of duties under the Occupiers Liability Act 1957. Under the Act, an occupier of premises owes a duty to take such care as is reasonable in all circumstances to ensure that a visitor is reasonably safe in using the premises. The court found the tenant, Larkstock, liable on the basis that the presence of flowers on the station concourse created a foreseeable risk of someone slipping. Whilst Larkstock argued that it had in place an adequate cleaning system, the court did not agree, branding it a reactive rather than proactive system, with spillages often not being dealt with promptly enough. Whatever action Larkstock was taking it was deemed not to be safe enough to prevent accidents occurring.

Mr Piccolo was not so successful in his claim against the landlord. He had attempted to argue that the landlord:

- failed to take sufficient steps to ensure its tenant, Larkstock, had a safe and effective cleaning system;
- should have threatened forfeiture of Larkstock's lease in addition to merely sending letters reprimanding them for not efficiently cleaning up spillages; and
- should have instructed its own cleaners to deal with spillages.

Notwithstanding some reservations as to the landlord's actions, the court did not agree and found that the landlord was not in breach of its duties to Mr Piccolo. The tenant had taken responsibility for cleaning the area in question and the landlord had, in the Court's view, taken adequate steps to try to ensure the tenant operated a safe and efficient cleaning system. Therefore the Court considered that the landlord had taken reasonable steps to deal with the issue. However, the Court did acknowledge that the landlord could have taken more stringent steps to reprimand the tenant, in particular, by threatening forfeiture / termination of the lease if the tenant continued to fail in its duties as occupier of the premises. It would also be advisable for landlords to consider whether any specific tailoring of the general repair / cleaning covenant in a lease might be necessary at the lease drafting stage to address occupier liability issues.

Where there are common parts and communal concourses (as in the *Piccolo* case) a landlord should consider tightening the cleaning covenant by specifying, for example in the case of a florist, that the immediate concourse area must be kept free of spillages and wet petals at all times. This is likely to stand the landlord in even better stead should the tenant subsequently fail to comply with those obligations.

whether the accident was facilitated by an independent third party, such as a competent contractor.

A danger to a visitor may lead to liability unless the occupier can demonstrate it took reasonable care to negate the danger but the occupier will only be found liable if the injury suffered was likely to result from the danger. A Court will not look at matters with the benefit of hindsight. The duty to visitors is therefore not absolute and is subject to limits of reasonableness.

It is essential that occupiers can provide evidence of adequate risk assessments, appropriate inspections and full maintenance records. Whilst the Court is less likely to find that an occupier should have warned of obvious risks, it should be remembered that children are less aware.

Organised events

Organisers of events will owe a duty to visitors who attend their event. However, they may also owe a duty to others who help to organise the event or otherwise participate in the event such as exhibitors or bands. Organisers should consider the insurance arrangements of all concerned as otherwise there may be unforeseen consequences. Those participating in events should enquire into the organiser's insurance cover and require any insurance to cover any specific hazards.

Contributory negligence

A visitor who exceeds the level of their permission or fails to take reasonable care may have any compensation reduced if the behaviour is a factor in an incident, or the claim could fail completely if they were found to have behaved extremely carelessly for their own safety.

Duty to trespassers

A 'trespasser' is somebody who goes on to premises without any invitation or right.

The Occupiers' Liability Act 1984 imposes a duty on an occupier where the occupier:

- is aware of a danger or has reasonable grounds to believe a danger exists;
- has reasonable grounds to believe that the trespasser was in or will come into the vicinity of the danger; and
- is reasonably expected, in all of the circumstances, to protect the uninvited entrant against the risk.

Hence, there will be no liability where an occupier is unaware of a danger and has no reasonable grounds for believing that a danger exists. Nor will there be liability where it cannot be anticipated that a trespasser would be in the vicinity of a danger at a particular time of night.

The level of protection expected will depend upon the seriousness of the danger and the type of trespasser. For example, a trespasser will require more protection from a high voltage electrical installation than a pile of stones. Measures needed to protect an innocently roaming child will be more readily expected than those needed to protect a thief (although a thief can still expect protection in certain circumstances).

Particular care should be taken when children might trespass on to premises. Inadequate fencing, roof access, or equipment can easily entice children who are unlikely to heed warnings.

The risk assessment process is therefore important. For more information see *'Trespassers and squatters'* (p.681).

Employers' duties under the Health and Safety at Work etc. Act 1974

Employers are under additional duties, many of which stem from this Act. An employer should implement safe working practices and protect employees from certain risks. There is also a duty for an employer in control of work premises to ensure that the health and safety of anybody entering or using the premises is protected.

A breach of this Act does not form the basis for a claim for damages, but can lead to a criminal prosecution by an Enforcing Authority.

Occupiers' liability and protests

A recent amendment to the Protection of Freedoms Bill, to allow an individual or group the right to protest in a 'quasi-public' space, and which is progressing through Parliament, will potentially increase the duties owed to protestors by occupiers of such spaces.

A 'quasi-public' space refers to privately owned land that is open to the public but is subject to restrictions that the owner may impose (such as opening times), for example a shopping centre.

If the amendment becomes law, the protestors will be entitled to enter the premises to protest, which will mean that they may be classified as 'visitors' and be owed a duty of care. As such, owners of these 'quasi public' spaces will be required to keep both protestors and members of the public safe during a protest.

Conclusion

Occupiers should take all reasonable precautions to ensure the safety of all people entering their premises, or they could face the unwelcome prospects of compensation claims and prosecution. Taking all possible precautions to ensure the safety of all visitors is a matter of common sense. Trespassers might seem to be another matter, but this is not the case. You should consider not only those people who are invited on to your premises, but also consider the circumstances in which others might gain access. Reasonable measures to protect invitees and non-invitees and to avoid trespassers gaining access should be implemented.

See also: Disability access and egress, p.238; Insurance, p.412; Nuisance, p.509; Property disputes, p.600; Risk assessments, p.624; Trespassers and squatters, p.681.

Sources of further information

The Workplace Law website has been one of the UK's leading legal information sites since its launch in 2002. As well as providing free news and forums, our Information Centre provides you with a 'one-stop shop' where you will find all you need to know to manage your workplace and fulfil your legal obligations.

It covers everything from CDM, waste management and redundancy regulations to updates on the Carbon Reduction Commitment, the latest Employment Tribunal cases and the first case to be tried under the Corporate Manslaughter and Corporate Homicide Act, as well as detailed information in key areas such as energy performance, equality and diversity, asbestos and fire safety. Content is added and updated regularly by our editorial team who utilise a wealth of in-house experts and legal consultants. Visit www.workplacelaw.net.

Outdoor workers

Kathryn Gilbertson, Greenwoods Solicitors LLP

Key points

- Employees whose work requires them to be outside for long periods of time could be exposed to excessive amounts of sun.
- Too much sun is harmful to the skin. A tan is a sign that the skin has been damaged. The damage is caused by ultraviolet (UV) rays in sunlight.

Legislation

- Health and Safety at Work etc. Act 1974.
- Control of Artificial Optical Radiation at Work Regulations 2010.

Employers are responsible for the health and safety of their staff under the Health and Safety at Work etc. Act 1974. Employees who contract skin cancer could potentially launch a claim in negligence against their employers years after their job has ended, in a similar way to asbestos-related illnesses.

Cancer Research Campaign's latest mortality statistics show that there were over 2,000 deaths from skin cancer in the UK in 2008 and that the number of malignant melanoma cases is rising. Up to 90% of these deaths are preventable and many can be dealt with if diagnosed in time.

In 2005 the European Parliament ruled that members' governments must decide themselves whether businesses would be required to ensure their workers are protected from exposure to the sun.

The proposed EU Directive required employers to carry out daily risk assessments for the strength of the sun and could have made firms liable for any skin cancers suffered by their workers.

However, in September 2005, MEPs voted against the EU-wide standards for the protection of workers against sunlight. Voting on the Optical Radiation Directive, which was aimed at protecting workers from damage to their eyes at the workplace, MEPs rejected a proposal that the measures included in the Directive should cover natural sources of radiation (including sunlight) as well as artificial radiation (e.g. from lasers). Therefore, the Control of Artificial Optical Radiation at Work Regulations 2010 relate to health and safety requirements regarding the exposure of workers to the risks from artificial optical radiation only.

The HSE offers the following guidance for employers on the subject of workers and sun protection in its leaflet, 'Sun Protection: advice for employers of outdoor workers':

- Include sun protection advice in routine health and safety training. Inform workers that a tan is not healthy – it is a sign that skin has already been damaged by the sun.
- Encourage workers to keep covered up during the summer months – especially at lunchtime when the sun is at its hottest. They can cover up with a long-sleeved shirt, and a hat with a brim or flap that protects the ears and neck.

- Encourage workers to use a sunscreen of at least SPF 15 on any part of the body they can't cover up and to apply it as directed on the product.
- Encourage workers to take their breaks in the shade, if possible, rather then staying out in the sun.
- Consider scheduling work to minimise exposure.
- Site water points and rest areas in the shade.
- Encourage workers to drink plenty of water to avoid dehydration.
- Keep workers informed about the dangers of sun exposure.

- Encourage workers to check their skin regularly for unusual spots or moles that change size, shape or colour and to seek medical advice promptly if they find anything that causes them concern.

See also: Occupational cancers, p.512; Personal Protective Equipment, p.576; Radiation, p.602; Temperature and ventilation, p.666.

Sources of further information

INDG 337 – Sun Protection: advice for employers of outdoor workers: http://hse.gov.uk/pubns/indg337.pdf

INDG 147 – Keep your top on – health risks from working in the sun. www.hse.gov.uk/pubns/indg147.pdf

Cancer research: Sunsmart: http://info.cancerresearchuk.org/healthyliving/sunsmart/

Outsourcing

Louise Smail, Ortalan

Key points

Before outsourcing:

- Have clear objectives and understand the implications of outsourcing.
- Consider issues, aside from cost, that the outsourcing will bring.
- Is there a commitment from the organisation to manage the relationship?
- How will key people be affected by the outsourcing?
- What impact will it have on the organisation?
- Have a clear understanding about what services are to be provided.

Supplier selection:

- Guarantee to meet specific service levels in the contract.
- Has a proven track record in the service being outsourced.
- Issues around conflict of interest with other clients.
- Has intellectual property been considered?
- Ongoing training of staff and staff development.

Contract:

- Supplier's performance – KPIs.
- Flexibility for introducing contract variations.
- Flexibility to accommodate new services and projects.
- The means of resolving day-to-day problems.

Legislation

UK law does not specifically regulate outsourcing arrangements, but the following should be considered.

Public sector

Public sector outsourcing may be subject to UK Regulations that implement EC public procurement directives. Where this is the case, the awarding authority may be required to use the *Official Journal* of the EU to advertise the intention to outsource and ensure that all bidders are treated equally. The EU public procurement rules are likely to impact on the timing of the pre-contract procedure and influence the award criteria adopted.

Even if the outsourcing by public organisations is outside the public procurement legislation, the awarding authority should still generally seek to comply with the spirit of the legislation (*OJ C179/2, 1 August 2006*). The UK private finance initiative (PFI) legislation applies to certain public sector outsourcing arrangements. Notice should also be taken of other laws and guidance such as:

- detailed guidance published by the Office of Government Commerce: www.hm-treasury.gov.uk/d/managingrisks_deliverypartners.pdf
- Human Rights Act 1998.
- Local Government Acts 1999 – 2003.
- Freedom of Information Act 2000.

Financial services

The main piece of legislation regulating financial services is the Financial Services and Markets Act 2000 (FSMA). The statutory regulator is the Financial Services Authority (FSA) under the FSMA and issues rules and guidance. An FSA-regulated firm cannot delegate or contract out of its regulatory obligations when outsourcing, and must give advance notice to the FSA of any proposal to enter into an outsourcing arrangement and of any changes to such arrangements. There are specific FSA rules on outsourcing. There are additional requirements where firms outsource portfolio management for retail clients to a supplier in a non-EEA state.

There are no additional regulations related to IT, telecommunications, or business processes. It is important that any prospective supplier or customer should ensure that any proposed outsourcing is not subject to additional regulatory requirements in other sectors.

Benefits of outsourcing

Outsourcing has been an issue for many businesses for some time, both in the private and public sector. The current economic crisis has seen many businesses who previously may not have considered this looking closely at it as an option, and it may now form a new model for businesses during the foreseeable future. Outsourcing and partnering arrangements have the potential to deliver value well beyond cost savings, by opening access to talent and capabilities, whilst maximising business model flexibility. However, this is not without its challenges. Many companies are held back by cost benefit justification and their own lack of experience. Often when the projects involving outsourcing fail, the organisation's first inclination is to blame the service providers, with the

service providers thinking that the main cause of failure is poor collaboration with customers.

Many of the disadvantages can be avoided if organisations research the service provider and do not regard outsourcing simply as a money saving scheme, as this is not always the case. As a consequence, organisations should be certain that they have a valid reason for outsourcing and that they intend to liaise regularly with the service provider to avoid losing all control of the process.

The current economic downturn is a large concern for a lot of organsiations who are looking to protect their profit margins by reducing cost. Some consider restructuring their business or divesting loss-making assets. Many companies are considering or have already decided to outsource certain of their business processes, which should allow them to reduce costs, refocus core activities and to help transform their business.

When an organisation is thinking about outsourcing, they need to carry out a due diligence exercise, looking carefully at what it is they are going to outsource. They need to look closely at the internal costs and make sure that all the tax, legal and commercial issues have been fully explored. Then an organisation can go into the process with a full understanding of all the issues that they need to address. It may also be possible that at the end of this exercise outsourcing is no longer an attractive option and isn't taken up.

Processes that can be outsourced
- Most IT functions can be outsourced, including:
    - network management to project work;
    - website development; and
    - data warehousing.

- This can provide benefits by providing the latest technology and software upgrades without having expensive investments.
- *HR and business processes* – this can include activities such as recruitment, payroll and secretarial services and can provide access to specialist skills that you only pay for when they are used.
- *Finance* – auditing is usually outsourced anyway, but this can be the entire accounting function, including bookkeeping, tax management and invoicing.
- *Sales and marketing* –an agency can be used to handle marketing communications.
- *Health and safety* – there are consultants who specialise in health and safety compliance.
- Fire and security.
- Legal advice.
- Logistics.
- Installations and service.

Non-business-critical tasks can also be outsourced, such as cleaning, catering and facilities management.

Types of outsourcing
- *Direct outsourcing* – contract between outsourcing organisation and supplier.
- *Multi-sourcing* – outsourcing organisation contracts with many different suppliers – important to indirect outsourcing wherethe outsourcing organisation appoints a supplier who then immediately sub-contracts to a different supplier, possibly outside the UK.
- *Joint venture or partnership* – the outsourcing organisation and the supplier set up a joint venture company, partnership or contractual joint venture. These maybe as an off-shore entity.
- *Captive entity* – the outsourcing organisation outsources its processes to a wholly owned subsidiary.

- *Build operate transfer* – the outsourcing supplier contracts with a third party to build and operate a facility. This facility is then transferred back to the outsourcing organisation.

Choosing your outsourcing partner
It is important to investigate not just the references offered by the potential partner but ensure that they are not so popular and in such demand that they are overstretched and won't be able to supply the right staff. When interviewing and selecting your partner, make sure that the key staff who are suggested are actually the ones that will be part of your arrangement with them.

Legal issues and contract matters
Contracts will necessarily focus on ensuring the key performance indicators, to ensure that the service is provided in the way that the organisation expects. There are other issue that also need to be considered – maintenance agreements, software licenses, any assets – and that can include employees who are needed to provide the service and details of their ownership.

How long is the outsourcing contract for? What break points are included? It is important to have a point in the contract so that if it isn't working the contract can be broken without penalties.

Where the outsourced service is critical to the operation of the business, any losses that may be suffered if the supplier fails may be significant. Any liabilities that are imposed in the contract should be proportionate to the value of the contract. It is important to identify any areas where the outsourcing organisation's liability should not be subject to any limit. This could be in relation to indemnity in relation to intellectual property rights or TUPE issues, which are often unlimited because they represent the organisation's protection for unquantifiable third party liabilities, which the outsourcing organisation is able to prevent or control.

Privacy, confidentiality and intellectual property

It is important that there is a non-disclosure agreement and privacy clause as part of the contract. Employees of the organisation would be subject to this and so should the outsourcing organisation. It is also important to make sure that the organisation outsourcing the work has a clear understanding and contractual arrangement to ensure that its intellectual property is protected. Transfer of employees to the new provider is an important consideration. It is also important to consider obligations to the employees and also the liabilities for these employees at the end of the contract. The contract also needs to look at how assets that are used by the outsourcing organsiations are managed, and also what will happen to these at the end of the contract.

Consideration needs to be given to how any new intellectual property rights are to be dealt with, and who owns them. Some of this will depend on negotiation, how much is paid for the services, and which of the services are a bespoke solution. The organisation may have to consider granting a license.

- Identify confidential information and specify the type of security that is expected.
- List applicable privacy laws and regulations.
- Require the outsourcer to limit access to authorised personnel.
- Specify that the outsourcer shall be liable for complying with applicable laws and regulations.
- Exercise access and control over the information; impose restrictions on how information may be used,

transferred, or shared; and ensure the right to audit the outsourcer's security procedures.

Business continuity

Organisations whose business continuity plan has relied upon their own resources will now need to make sure that the outsourcer is also part of this process. They should be part of any arrangements and fully informed about their part in it and also take part in any exercises.

Managing the contract performance

Any service description in the contract should be legally enforceable. There needs to be a detailed description of the services to be provided and all the obligations on the outsourcing organisation should be clearly identified and what the organisation expectsof the outsourcing organisation (key performance indicators).

Terminating an outsourcing arrangement

Any outsourcing arrangement needs to include a mechanism for management of the relationship and procedures to be followed when problems arise. If these remedies are not successful then there should be an exit plan that considers the provision of the services for the duration of the notice period and any period including cooperation with the new outsourcer. There also needs to be a system for the return or transfer back of assets and software and the licences of intellectual property and the provision of information and know-how to the organisation or new outsourcer. Consideration will also have to be made about the treatment of employees and any obligations under TUPE and other relevant regulations.

Whatever organisations decide to do, they need to remember that although the work is outsourced still need to be managed as it remains with them.

Ten things to avoid

1. Selecting the wrong vendor.
2. Outsourcer's people do not understand the organisation – make sure that you get the staff that were promised, or equivalent experience and competence.
3. Stuck with only one vendor and a contract that doesn't work – ensure that the question is asked before outsourcing – what happens if this doesn't work?
4. Creeping specification – starting with a neat and easily understood set of services and then widening it to include many other issues and therefore making the contract limits and performance difficult to understand.
5. Security breaches and confidentiality arrangement.
6. Failure of the business continuity – disaster recovery system.
7. Employees not engaged with the process.
8. Difference governance models – make sure these are aligned before you start.
9. Increased costs – ensure that increases in costs, both for employees and services, are covered in the contract.
10. Misaligned reporting systems.

> *See also*: Business Continuity Management, p.92; Competence and the 'Responsible Person', p.158; Contractors, p.187; Facilities management contracts, p.304; Health and safety at work, p.361; Health and safety management, p.388; Waste management, p.707.

Sources of further information

National Outsourcing Organisation www.noa.co.uk

Complying with health and safety law – and looking after the health, safety and welfare of everyone connected with your organisation – can be time-consuming and complicated, diverting precious resources and attention away from your main focus – your core business. Workplace Law can help, because managing the health and safety of an organisation, like yours, is *our* core business.

The Workplace Law Health and Safety Support Contract provides you with total support, 365 days of the year, regardless of how many people you employ. Since every organisation is unique, so is our support contract – tailored to meet the specific needs of your business, your people, and the sector you operate in.

The level of support can be designed to either assist your health and safety team, or act as your competent person, as required under the Management of Health and Safety at Work Regulations 1999. We offer sensible, pragmatic advice concentrating on practical action to control significant risks – not over-responding to trivial issues. Visit www.workplacelaw.net for more information.

Overseas workers

Jan Burgess, CMS Cameron McKenna

Key points

- All employers who send employees to work overseas or who are responsible for hiring labour must have management systems and support services in place to assess and minimise the risks associated with working abroad.
- Risk assessments must consider both UK law and local law. These should take into account the environment where the work is to be performed, the worksite itself, travel to and from the worksite, and medical risks associated with the area.
- A failure to carry out appropriate risk assessments may result in prosecution under UK law if an employee is exposed to unnecessary risk, due to the fact that the initial breach (failure to risk assess) would have occurred in the UK, and not overseas. Breaches under the local law may also result in prosecution in the local country. There may also be a civil claim for compensation for an injury or a fatality.
- Employers' liability insurance should be extended to cover overseas accidents.

Legislation

- Health and Safety at Work etc. Act 1974.
- Management of Health and Safety at Work Regulations 1999.
- Corporate Manslaughter and Corporate Manslaughter Act 2007.
- Health and Safety (Offences) Act 2008.

Employers should also follow the guidance in the relevant Approved Code of Practice issued by the HSE.

Criminal liability

If employees are sent to work overseas, or if an employer has responsibility for contract staff working on their overseas worksite, it is important to know that if an employee or contractor is injured, killed or is put in a position that could have resulted in injury or death, that employer could be prosecuted in the UK, as well as in the country where the incident happened. Whilst the Health and Safety at Work etc. Act 1974 (HSWA) and associated legislation only applies to Great Britain, an employer could still be prosecuted for an overseas accident / incident if that employer failed to properly assess the risks associated with the work and did not take reasonably practicable steps to minimise the risk. The employer would have to have broken the law in Great Britain – for example, at the planning stage, when it decided to send an employee to work overseas.

All employers sending staff to, or employing staff in, overseas countries must be aware of their responsibilities and liabilities, along with potential penalties, under health and safety law in that local country. It is very common for overseas countries to deal with health and safety issues in an entirely different manner from the way they are dealt with in the UK and – in particular – the liabilities of directors and individual managers may be

Case studies

McDermid v. Nash Dredging and Reclamation Co Ltd (1987)

The claimant was asked to work on a tug in Sweden owned by the defendant's parent company. The claimant was seriously injured and sued his employer, claiming that it had breached its duty of care by delegating its responsibility to the parent company. The House of Lords held that the duty could not be delegated.

Square D Ltd v. Cooke (1992)

The claimant was sent to work in Saudi Arabia on premises occupied by another company. He had an accident whilst working there. He sued his employer. The Court of Appeal held that the UK employer was not responsible for the injuries whilst working in Saudi Arabia on premises occupied by another company. It would be too much to ask the employer to assume responsibility for daily events in relation to premises occupied by a third party abroad. However, the UK employer would be expected to consider matters such as the place where the work was to be done, the type of work, and the employee's suitability to do the work. The level of control that the UK employer would be expected to exercise would depend upon what was reasonable under the circumstances.

Palfrey v. Ark Offshore Ltd (2001)

The claimant was sent to work for a contractor in West Africa by his UK employer. He did not receive the necessary vaccinations for the country in question, contracted malaria and died. In between trips he had been to a UK travel clinic for a yellow fever vaccination but did not ask for any anti-malarial tablets. His widow sued his employer and the clinic. The court held that the employer was required to ascertain publicly available information in respect of health hazards that the employee would face, draw these to the employee's attention and give advice on appropriate medical steps to be taken. However the employer did not do that and the widow won the case.

Hopps v. Mott MacDonald Ltd (2009)

The claimant claimed damages against his employer and the Ministry of Defence for personal injuries suffered in a roadside bombing in Basra, Iraq. The claimant's employer was a consultancy, which provided contract personnel to perform civil engineering services. The claimant was a consultant engineer who volunteered to go to Iraq. The employer's personnel were located within a secure military base, with army protection both there and during site visits. Whilst travelling in a Land Rover (which was escorted by another Land Rover containing soldiers) the claimant was injured when the vehicle was struck by material from a bomb on the roadside. The claimant sought damages on the basis that there had

Case studies – *continued*

been a failure to take reasonable care for his safety; he contended that a risk assessment should have been performed to assess the suitability of the transport arrangements and the provision of security. He also argued that the vehicle should have been an armoured vehicle, because if so he would have suffered either no injury or significantly less injury.

His claim was ultimately dismissed, and it was held that although his employer did not carry out a written risk assessment, it did keep security under review. It was also clear from the evidence that the Army's advice at the relevant time would not have been that the level of risk would require use of armoured vehicles, and that, in any case, it could not be concluded that an armoured vehicle would have prevented or minimised the injury.

far more prominent (since in many foreign jurisdictions, there is no possibility for prosecution of a corporate entity).

Steps to reduce the risk of prosecution might include:

- *Maintaining records* – you should ensure that all plans and decisions to send workers abroad are documented and records retained to ensure evidence is available to show the basis of the decision.
- *Carrying out risk assessments* – you should ensure that risk assessments are performed in relation to any individual selected for work abroad. Ensure that all systems for assessing those risks are regularly reviewed and updated.

Civil liability

It is often the case that if an employer is based in the UK and there is an accident overseas, it can be sued in a UK court, even though the accident happened in a foreign country. The employer may also be sued in the courts of the country where the accident happened. In order to hear the case in the UK, the court must have jurisdiction. In most cases, jurisdiction will be accepted if the employer's registered

office or place of business is in the UK. If the employer is based in England, the English courts will hear the case. Likewise, if the employer is based in Scotland, the Scottish courts will hear the case.

Assuming the court is prepared to hear the case, it will also need to apply the relevant law, which may be set out in the employee's contract of employment. There are special rules for determining which law governs the contract in the absence of an express choice of law. In those circumstances, whilst the court is prepared to hear the claim, it may be required to apply the law of the jurisdiction where the accident occurred to establish liability and compensation payable.

Prosecution of individuals

In terms of the primary legislation – the HSWA, Section 37 – individual directors, officers, and company secretaries may be prosecuted for 'conniving, consenting or neglecting' in the offence committed by the company of exposing those in its employ to risk of injury. If the decision-making process took place in the UK, and individual directors or managers were neglectful in the decisions they took in sending workers abroad, who were subsequently injured, it is possible that the company and individuals could be prosecuted under UK legislation.

The Health and Safety (Offences) Act 2008 is of particular note in relation to penalties for health and safety breaches. The Act came into force on 16 January 2009 and applies to any offences committed after this date. Those who are prosecuted for health and safety incidents after this date can expect to see much more significant penalties being handed out by the courts. The Act extends to England and Wales, Scotland and, to a limited degree, Northern Ireland. Although it increases penalties for health and safety breaches, it does not impose any new obligations, and the duties of employers will remain as before under the HSWA and its associated regulations.

The new Act makes three main changes to the penalties for health and safety offences previously imposed. It raises the maximum financial penalties available to the courts; it makes imprisonment an option for a wider range of health and safety offences; and it makes certain offences currently only triable in the lower courts, triable in either the lower or the higher courts. This means that many more cases will now be open to an unlimited fine and a term of imprisonment.

The Corporate Manslaughter and Corporate Homicide Act 2007 makes specific provision for fatalities in the workplace. However, there is no provision for prosecution of any individuals (who could nonetheless still be prosecuted for common law manslaughter). The fatality must take place on UK territory (or an offshore platform in the North Sea, or a UK registered ship or aircraft) and therefore, should a worker be killed overseas, this particular piece of legislation would not operate. That said, prosecution under an equivalent law of the foreign territory in question may be possible.

It should of course be borne in mind (as noted above) that an employer or individual may only be prosecuted for an overseas accident / incident if the initial breach of the legislation occurred in Great Britain. That might occur where, for example, an employer failed to carry out an appropriate risk assessment at the planning stage when it decided to send an employee to work overseas.

Insurance

In the UK all employers are legally obliged to carry employers' liability insurance. However, this requirement does not extend to employees working abroad. There have been tragic cases of employees involved in serious accidents overseas, only to discover that their employers' insurance did not cover overseas incidents. It is therefore vital that employers ensure that their policy is endorsed to cover overseas risks. Employers should also check what is required by way of compulsory insurance in the country where they are sending the employee to work.

Checklists

In order to show that proper risk assessments were conducted, the following checklist may prove useful:

- Do you have a documented audit trail in relation to a particular decision showing managerial involvement?
- Did you ask to be briefed in local arrangements and particular circumstances that could adversely affect safety?
- Did you conduct a site inspection?
- Are there weekly reports submitted from the local area of operation detailing incidents, injuries, near misses, etc?

- Were you shown, or at least briefed, on the relevant contractual provisions relating to the scope of work with particular regard to safety management requirements?
- Do you keep training records for all team members? Are these reviewed and is training updated?
- Have you been briefed on the environmental, political and medical risks associated with the locale? This would include taking advice from the Foreign and Commonwealth Office, and taking medical advice on matters such as vaccination requirements etc.
- Do you keep written work instructions and/or operate a permit to work system? Are you satisfied that this is adequate for the work and that the system has been implemented and followed?
- Do you prescribe minimum language requirements for your teams to overcome any linguistic problems posed by having a mixed expatriate / local workforce? Are you satisfied that this has been complied with and is regularly monitored?
- Have you performed a safety audit in relation to local legal requirements? What steps were taken to remedy any deficiencies identified by that audit?
- If relevant, do you insist that team members perform toolbox talks? Do you keep written records of those toolbox talks?
- Do you spot-check individual team members returning to the UK to debrief them on health and safety matters overseas?
- Do you have a whistleblowing scheme in place? If you do, are employees encouraged to use it if they believe that there are health and safety issues that need to be addressed in the locale?
- Have you adequate insurance in place to cover the risks of overseas work?

- Are you fully aware of your health and safety liabilities and duties in the overseas country?

Travel risk assessment factors

The role of the country's regional manager or locally appointed agent will be important in dealing with the following questions:

- What is the country of destination?
- Where will the worker be staying? What type of accommodation will he/she be occupying?
- What occupational work activities will the worker be performing?
- What medication will the worker require?
- How much travel experience does the worker have?
- What access to medical services will the worker have in the country of destination?
- How will the worker be travelling to the country of destination? What type of travel will he/she be using in the country of destination?
- How long will the worker be staying in the country of destination? Will his/her family accompany him?
- What is the worker's medical and psychiatric history?
- What is the worker's lifestyle / behaviour? Will he/she require coaching and assistance on this prior to departure (e.g. alcohol consumption in certain countries)?
- Is the worker covered by insurance?
- What arrangements are in place for emergencies and for emergency evacuation?
- Is necessary security in place?
- How will contact be maintained at all times?

See also: Alcohol and drugs, p.35; Health and safety at work, p.361; Insurance, p.412.

Sources of further information

Foreign and Commonwealth Office: www.fco.gov.uk

Travel advice and information, including the latest travel updates and news:
www.fco.gov.uk/en/travel-and-living-abroad/

Overseas Security Information for Business (OSIB) – this is a free service jointly
run by UK Trade and Investment and the Foreign and Commonwealth Office.
OSIB provides authoritative, accessible and topical country specific information on
the key issues related to political, economic and business security environments:
www.ukti.gov.uk/osib

Packaging

Sophie Wilkinson, Shoosmiths

Key points

- Businesses that manufacture, use and handle packaging have legal obligations in relation to recycling and recovery of packaging waste if they have an annual turnover of more than £2m and handle more than 50 tonnes of packaging per year.
- These obligations apply to any 'producer' of packaging or packaging materials.
- Packaging is defined as a product made from any material (such as paper, glass, plastic, aluminium and wood) that is used for containment, protection, handling, delivery and presentation of goods, from raw materials to processed goods.
- There are specific targets that producers need to meet when recycling packaging made from paper, glass, metal, plastic and wood.
- Packaging must meet certain requirements before it can be placed on the market.
- Failure to comply with applicable legislation is a criminal offence and penalties can be large.

Legislation

- Directive 94/62/EC as amended by Directive 2004/12/EC on packaging and packaging waste.
- Packaging (Essential Requirements) Regulations 2003, as amended.
- Producer Responsibility Obligations (Packaging Waste) Regulations 2007, as amended.

Legal obligations on the producers of packaging

In 1994, the European Union adopted Directive 94/62/EC on packaging and packaging waste ('the Directive'). The Directive introduced a regime for the recycling and recovery of packaging waste and required packaging to comply with requirements regarding its design and composition. This included restrictions on the use of certain hazardous substances in the manufacture of packaging.

The main objectives of the 1994 Directive are to:

- increase the recycling and recovery of packaging;
- minimise the amount of packaging used;
- ensure packaging meets certain minimum standards; and
- reduce the amount of packaging sent to landfills.

The Directive has been implemented into UK law by the Producer Responsibility Obligations (Packaging Waste) Regulations 2007 ('the 2007 Regulations') and the Packaging (Essential Requirements) Regulations 2003 ('the 2003 Regulations').

The key feature of the 2007 Regulations is the focus on sharing the responsibility of meeting the recovery and recycling targets from production to retail. They apply to

Packaging

Facts

- At the end of its life, about 55% of packaging ends up as commercial and industrial waste and about 45% ends up as household waste.
- In 2007, packaging accounted for an estimated 4.7m tonnes, around one-fifth of the household waste stream, and around 5% of all waste sent to landfill.
- On average, 16% of the money spent on a product pays for the packaging.
- 3% of a household's annual energy use is taken up by packaging.
- Over 60% of the total plastic waste in Western Europe is comprised of packaging, which is typically disposed of within one year of sale.
- 24 million tonnes of aluminium is produced annually, 51,000 tonnes of which ends up as packaging in the UK.
- 70% less energy is required to recycle paper compared with making it from raw materials.
- In 2008, the UK recycled 61% of its packaging waste, a massive increase from the 28% achieved in 1997. This means over 6.6 million tonnes of packaging waste were diverted from landfill, and over 8.9 million tonnes of CO_2 equivalent emissions were avoided.

any 'producer' of packaging or packaging materials that has an annual turnover of more than £2m and handles more than 50 tonnes of packaging per year. Groups of companies need to consider the total amount of packaging and the turnover of all subsidiary companies to see if they are affected.

Packaging producers, for the purposes of the 2007 Regulations, are businesses who do some or all of the following:

- Manufacture raw materials for packaging;
- Convert the raw materials into packaging;
- Fill the packaging;
- Sell the packaging to the final user or consumer;
- Lease or hire out packaging;
- Act as a 'licensor' or pub operating business; and/or
- Import packaging (including packaged goods) into the UK.

The 2007 Regulations therefore cover a broad range of companies, carrying out a range of activities.

Packaging is defined as a product made from any material (such as paper, glass, plastic, aluminium and wood) that is used for the containment, protection, handling, delivery and presentation of goods, from raw materials to processed goods.

Producers can comply with their obligations in one of two ways:

1. By recycling and recovering the relevant quantities of packaging themselves and providing this in the form of Packaging Recovery Notes (PRNs) or Packaging Waste Export Recovery Notes (PERNs) obtained from accredited packaging waste reprocessors and exporters.
2. By joining a registered compliance scheme, which will comply with the producer's obligations on their behalf. Figures for 2005 suggest that 90% of producers choose to comply with their packaging obligations by joining a compliance scheme.

The targets for recycling packaging waste in 2011 and 2012 have been increased.

The 2003 Regulations require those responsible for packing and filling products into packaging to ensure that packaging meets certain requirements before it is placed on the market. These include:

- Packaging must be manufactured so as to permit reuse or recovery in accordance with specific requirements.
- Noxious or hazardous substances in packaging must be minimised in emissions, ash and leachate when the packaging is incinerated or landfilled.
- Packaging must not contain more than the maximum permitted levels of cadmium, mercury, lead and hexavalent chromium.
- Packaging volume and weight must be the minimum amount to maintain necessary levels of safety, hygiene and acceptance for the packed product and the consumer.

Failure to comply with the requirements of the 2007 Regulations is a criminal offence, punishable on summary conviction (in the Magistrates' Courts) by a fine not exceeding £5,000 or on conviction in the Crown Court to an unlimited fine. Fines are payable per offence (and there are a number of offences under the 2007 Regulations) and per year of non-compliance, so fines can be large.

The 2003 Regulations are enforced by trading standards officers, and failure to comply is punishable in the same way as the 2007 Regulations.

Although the packaging waste regime has been in place since 1997, there are still a number of businesses that are not aware they are covered by these requirements. The costs of compliance can be significant depending on the quantities and types of packaging handled. In July 2009, a major drinks manufacturer was fined £261,000

for eight counts of failing to register with the Environment Agency and eight of failing to recover and recycle packaging waste. Whilst the company had not deliberately flouted the rules, the Court ruled that they had been reckless in failing to observe them. From January 2011 the Environment Agency has, however, been able to use civil sanctions for certain packaging offences that occurred after 6 April 2010 in England and 15 July 2010 in Wales. The Environment Agency is now actively using its powers to grant civil sanctions and is no longer automatically seeking to prosecute businesses.

There are specific requirements where substances come into contact with food. The European Framework Regulation (EC) N0. 10/2011 on Materials and Articles Intended to come into Contact with Foodstuffs lays down the general safety requirements for all substances that come into contact with food and is implemented by the Materials and Articles in Contact with Food (England) Regulations 2010. These Regulations specify standards that must be met by materials coming into contact with food.

Currently there is particular concern in relation to bisphenol A (BPA). BPA is widely used in the production of plastics and can be used in a variety of products, including food contact materials. Directive 2011/8/EU bans the manufacturing of baby bottles containing BPA from 1 March 2011 and sales and imports of baby bottles containing BPA have been banned since 1 June 2011. The US and Canada have already taken action to limit BPA exposure. The EU Executive Commission said it would ban the manufacturing by 1 March 2011 and ban the marketing and market placement of polycarbonate baby bottles containing Bisphenol A by 1 June 2011. See 'Packaging and Bisphenol A' (p.545) for more details.

See also: Packaging and Bisphenol A, p.545; Recycling, p.605; Hazardous waste, p.700; Waste management, p.707.

Sources of further information

BIS – Guidance Notes on the Packaging (Essential Requirements) Regulations 2003: www.bis.gov.uk/assets/biscore/business-sectors/docs/p/11-524-packaging-regulations-government-guidance

DEFRA – Making the Most of Packaging: A Strategy for a Low-Carbon Economy: http://archive.defra.gov.uk/environment/waste/producer/packaging/documents/excec-summary-pack-strategy.pdf

Packaging and Bisphenol A

Claire Morrissey, Woodfines LLP

Key points

- Bisphenol A (BPA) based plastics have been in use for more than 50 years, and global production of BPA was estimated to be more than 2.2 million tonnes in 2009.
- It is an endocrine disruptor, i.e. it can mimic the body's own hormones, and therefore has the potential to cause negative health effects.
- BPA can leak from plastics into foodstuffs, the air and water. This means that most people have ingested BPA at some point in their lives.
- In 2008, the governments of several countries issued reports concerning the safety of the chemical on the basis that it could contribute to a number of health problems. Most specifically, concerns were raised about the exposure of babies to BPA via use of plastic baby bottles.
- On 25 November 2010, the European Union Executive Commission said it would ban manufacturing by 1 March 2011 and ban the marketing and market placement of polycarbonate baby bottles containing the organic compound bisphenol A by 1 June 2011.

Legislation

- Directive 2006/121/EC amending Directive 67/548/EEC – known as REACH (Registration, Evaluation Authorisation and Regulation of Chemicals) – this regulates companies manufacturing or importing into the EU chemicals in quantities at or above one tonne per year. It provides a chemical regulatory framework but does not include BPA. It does regulate many of the other chemicals used in the manufacture of plastics.
- Directive 2002/72/EC – known as the plastics directive, this deals with materials and articles made of plastic that are intended to come into contact with foodstuffs.
- Regulation 1935/2004/EC – requires that materials and articles containing BPA, such as some can coatings, do not make food harmful. The Regulations also make sure that they do not change the nature, substance or quality of the food.

- The Plastic Materials and Articles in Contact with Food (England) Regulations 2009 – these permit the use of BPA in the manufacture of plastic materials and articles intended to come into contact with food, provided that no more than 0.6 mg/kg migrates into the food.

Legal obligations on producers

Under the Plastic Materials and Articles in Contact with Food (England) Regulations 2009, it is a criminal offence for any person or company to fail to ensure that no more than 0.6 mg/kg of BPA migrates into food where the chemical is used in a product which comes into contact with food. The maximum penalty for such a breach, on summary conviction, is a term of imprisonment not exceeding three months or a fine not exceeding level five on the standard scale (£5,000).

However, companies using the chemical must ensure that they comply with the

---

Facts

- 50% of Europe's food is packaged in plastic.
- Bisphenol A (BPA) is an organic compound used in polycarbonate plastics, which are in turn used in baby and water bottles, medical and dental devices, household electronics and as coatings on the inside of food and beverage cans.

---

legislative provisions of the individual countries where they intend to manufacture and/or sell their products.

In the USA, concerns were raised about the use of BPA in September 2008. In March 2009 the six largest companies that sell babies bottles decided to stop using the chemical, and in May 2009 the first states in the USA started to limit and ban its use in children's products. In January 2010 the Food and Drug Administration (FDA) expressed some concern about the potential effects of BPA on foetuses, infants and young children. Efforts were made to establish what could be done to reduce exposure to BPA, but it was not banned. It is still used in infant formula and foods as well as a lining for food cans.

In Canada, the use of BPA in plastic baby bottles was banned in August 2010, following over two years of debate about the risks posed by the use of BPA, and fierce resistance by the chemicals industry. It has been added to the List of Toxic Substances and will be banned from all products.

In the EU in June 2008 an updated risk assessment report on the chemical concluded that it was safe and set a Tolerable Daily Intake level of 0.05 mg/kg bodyweight. Doubts were later raised about this opinion and individual member states took different stances. For example, in Belgium the chemical is banned in food contact plastics, whilst Denmark imposed a temporary ban in March 2010; in Germany and France many manufacturers withdrew products containing the chemical from the market. On 25 November 2010, the European Union Executive Commission said it would ban the manufacturing by 1 March 2011 and ban the marketing and market placement of polycarbonate baby bottles containing the organic compound Bisphenol A by 1 June 2011.

At present there is not enough evidence available for the European Union or the UK to determine the true risk of BPA to humans. However, on the basis that it is agreed that the chemical can interrupt the hormonal function of humans, there is enough evidence to cast doubt on its safety.

See also: Catering: food safety, p.117; Packaging, p.541.

Sources of further information

General information and resources about BPA: www.bisphenol-a.org

---

Parking

Kelvin Reynolds, British Parking Association

Key points

- It's important to remember that people don't park cars for the sake of it. Parking is a means to an end, not an end in itself. The car park is the gateway to the retail centre, the hospital, the university, the commercial office, and so on, and it's where your customers and visitors gain a first impression of the kind of organisation you are. It is also their last impression when they depart. If their experience of your parking operation – be they customers, clients or employees – is not a good one, their opinion of you will reflect that.
- There are, broadly, two different types of parking– public parking, which takes place on the public road and in car parks run by the local authorities, and public or private parking, which takes place on private land.
- Parking on the public road network is regulated by law, whereas parking on private land is unregulated but can be subject to a range of apparently unrelated laws in many situations.

Legislation

- Road Traffic Regulation Act 1984 (all UK)
- Traffic Management Act 2004 (in England and Wales)
- The Road Traffic Act 1991 (in Scotland)
- The Traffic Management (Northern Ireland) Order 2005
- Consumer Protection Regulations 2009.

Parking in public places

Parking places on the public road network and in local authority car parks is provided by local authorities using powers they have in the Road Traffic Regulation Act 1984; this same legislation gives local authorities throughout the United Kingdom the power to impose controls and restrictions on where parking can take place. Until 1994 all enforcement was undertaken by police and traffic wardens services. In 1994, parking enforcement was 'decriminalised' and local authorities were able to adopt enforcement powers in England, Scotland and Wales. Similar powers were introduced

in Northern Ireland in 2005. The vast majority of towns and cities in the UK have adopted these powers so that, nowadays, in general, parking on public roads or car parks is generally operated and enforced by local authorities.

Civil enforcement

On 31 March 2008, important changes were introduced to civil parking enforcement when the parking provisions of the Traffic Management Act 2004 became law, in England and Wales and formally established Civil Enforcement Areas (CEAs), which saw the first changes to on-street parking enforcement since 1994. The new arrangements were designed to provide motorists with a fairer and more consistent service across the country, helping raise industry standards in parking. Decriminalised parking enforcement is now called Civil Parking Enforcement (CPE), and parking attendants are known officially as civil enforcement officers (CEOs), although many are still seen as parking

Case study

From 2012, Nottingham City Council will be the first to impose a £250-a-year per space parking levy on employers with more than ten eligible parking spaces. Other Councils thought to be considering the levy include Bristol, York, Devon, Hampshire, Leeds, Bournemouth, South Somerset and Wiltshire. The Transport Act 2000 gives provision to local authorities to introduce workplace levy licensing schemes, meaning that they have the power to impose licenses and charges on workplace parking places, dependent on the number of spaces available.

The Forum of Private Business says it is "deeply opposed" to the scheme and believes it will provoke fury among business owners, who could be forced to pay tens of thousands of pounds each year for providing their employees with somewhere to park their cars.

Forum spokesman, Chris Gorman, said:

"When the Nottingham scheme was given the go-ahead last year, we said at the time that it would only be a matter of time before it spread to other towns and cities. Sadly, it appears those fears will soon be realised. In our view it's simply a stealth tax which will have a disproportionate impact on small businesses. It's the equivalent of charging homeowners to park on their own driveways and will increase parking problems in town centres and cities.

"Businesses already contribute enormous amounts to public services through existing taxes such as business rates. Whatever its supposed justifications, the danger is that these schemes could open the floodgates to a raft of new taxes and charges being levied on to companies to pay for things which were previously paid for through general taxation. While councils' finances are under pressure, this is a very short-sighted idea as companies are likely to avoid areas with a scheme in operation, meaning jobs, investment and therefore tax revenue will end up elsewhere."

It should be noted that employers are legally able to pass on the charges to employees if they so wish. If spaces are allocated to particular employees this is of course easier to do. Where employees share spaces then apportionment could be more complex, but at £250 a year this could be said to be around £1 per day per space in a typical 'nine to five, Monday to Friday' set-up, and perhaps this would make it easier to recover costs from employees who are provided with the parking spaces. The purpose of the charge is to discourage single car occupancy, encourage modal shift to other forms of transport, and any surplus revenues generated must be reinvested in public transport or similar schemes. In Nottingham the funds are being used to underpin the development of the local tram system.

attendants in our towns and cities. The original 'decriminalised' enforcement rules still operate in Scotland and Northern Ireland, although their governments are considering adopting elements of CPE in due course.

The introduction of CPE aimed to make councils more transparent about their parking management policies. Under the Regulations they must provide suitable training for everyone involved in parking enforcement, undertake regular reviews of policies through consultation with stakeholders, and communicate these policies effectively to the public. Annual parking reports are encouraged, which enables councils to explain why they manage parking, what their objectives are, and how successful they have been in meeting them. They must make clear that enforcement is based on achieving compliance with local parking regulations and not about the number of tickets that might be issued.

The Regulations allow independent adjudicators to refer cases back to local authorities where a parking contravention has taken place and can, in some mitigating circumstances, ask the local authority to consider cancelling the penalty charge. The Penalty Charges under CPE are 'proportional' to the seriousness of the contravention, with a higher and lower level penalty charge. For example, parking on a double yellow line is seen as being more serious than overstaying at a parking meter bay, and charged accordingly. Research and evidence from annual reports indicate that the benefits expected from the Traffic Management Act are being achieved as compliance is improving, and the total number of penalty charges issued by local authorities is reducing. Differential penalties also encourage enforcement to take place where it is most effective in maintaining traffic flow and improving road safety.

Adjudication

Adjudication of civil parking enforcement is undertaken by four agencies in the UK. The Traffic Penalty Tribunal covers England and Wales outside London. The Parking and Traffic Adjudication Service covers London. There are separate adjudication services for Scotland and Northern Ireland. Each adjudication service produces an annual report, which can be found on their websites. It provides details and statistics about the nature of enforcement and adjudication that takes place.

Managing parking outside Civil Enforcement Areas

Outside the towns and cities in England and Wales, mainly in rural areas where Civil Enforcement Areas have not been established, and in some parts of Scotland, parking is still enforced using the 1984 Road Traffic Regulation Act where local authorities make regulations to control and manage parking but it is enforced by police and traffic warden services. In these cases motorists do not have access to the statutory adjudication services associated with CPE but must contest their parking tickets through the central ticket office of the issuing police force and/or the Magistrates Courts.

Parking on private land

When parking on private land is required to be controlled and regulated there are two primary areas of law that are involved:

- Where parking is allowed and vehicles are invited on to the land this is covered by the law of contract.
- Where vehicles are not permitted on land then usually a trespass occurs in England and Wales.

Remember that private land includes not only office and factory premises, for example, but also many retail centres, supermarkets, railway stations, business parks, hospitals, universities, leisure centres, housing estates, residential property including individual homeowners land – the list goes on.

Landowners and their contractors or agents are permitted to undertake enforcement action, which commonly involves wheelclamping, vehicle removal or the issue of so-called 'private parking tickets'.

In managing parking schemes on private land, operators are generally acting within parameters of two separate contracts – firstly a contract with a landlord (assuming that the landowner is not directly undertaking the management of the parking scheme and has employed someone else to do this) to manage and operate the parking scheme on their behalf and, secondly, a contract with a motorist to provide facilities to park in a car park. These two contracts set out the authority for an operator to act in any particular way, so it is important that each contract is clear and legal.

It is well understood that a contract is a legally binding agreement between two or more parties. Contracts do not need to be written down; they can be made by the spoken word, in writing or by a combination of both. Where public parking takes place on private land the parking operator is normally the 'offeror' of the contract and the driver is the 'offeree'. The 'offeror' is providing parking and related services. The offer is accepted by the driver of the vehicle – the 'offeree' – when they enter on to the land.

So, for example, in a public parking place provided on private land, there would usually be a notice or information sign at the entrance, which indicates that it is private land but that public parking is permitted, subject to certain terms and

conditions, such as a time limit, or for a fee, or a combination, or other such reasonable terms and conditions. When a driver leaves a vehicle in the parking place they are deemed to have accepted the offer. They may be in breach of contract if they do not comply with all the terms and conditions.

Raising standards in public parking on private land

The BPA fully supports the raising of standards in control and enforcement of parking on private land, and is calling for better regulation of the whole private parking sector.

In its Master Plan for Parking the BPA states:

'We want to see an Independent Appeal Service for the entire private parking sector. We want to see legislation to provide for universal keeper liability for parking charges in all circumstances. We want to see the introduction of a standardised Penalty Charge Notice (PCN) and other statutory notices used in Civil Parking Enforcement.'

This can be achieved by properly defined 'parking tickets', fair and legitimate enforcement, access to independent adjudication for everyone and the introduction of universal keeper liability. We want the principles to be similar to those established by existing road traffic law throughout the UK and thus provide motorists with similar rights and responsibilities, whether they park on private land or a public road.

For years, headline stories in the media have often referred to 'cowboy clampers' and the like, unfairly bringing the majority of the parking industry into disrepute. The British Parking Association (BPA) works with the media to get them to understand that management and control of parking on private land is as essential as it is on the public highway – the main difference, of course, being that the latter is regulated whereas the former is not. The public, of course, do not understand the difference – and why should they? As far as they are concerned, they park a car in a car park and expect to be treated fairly and reasonably in every case.

Private landowners and car park operators are expected to emulate the standards and behaviours, and provide the services, of the regulated world, and yet must not mimic these directly for fear of being accused of 'passing off'; this is an extremely difficult balance to keep. It is one that the BPA strives to achieve for its members. It must be accepted that landowners and operators have a right to control and manage parking on private land. Equally, we believe they are entitled to a clear and unambiguous legislative framework to enable this, rather than to rely on a complex set of seemingly unrelated laws managed by a number of government departments.

We also want to ensure that Membership of an Accredited Trade Association is compulsory for all parking operators to undertake enforcement on private land. The BPA does as much as it can within the present law through its Approved Operator Scheme (AOS) and its Code of Practice, and we will continue to encourage our members to improve their service to their clients and to their customers, ensuring that all parking enforcement is fair, reasonable and legitimate.

Approved Operator Scheme (AOS)
Companies who wish to become a member of the BPA and who undertake any kind of parking management on private land must join our AOS, otherwise membership of the BPA is not possible.

Our AOS is intended for those companies and businesses that operate parking enforcement services on private land and unregulated public car parks. Operators may conduct a range of services in this sector, including, but not limited to, vehicle immobilisation and/or removal, ticketing, or services such as back office functions, data management and debt recovery. Landowners and businesses, especially those that provide public parking, are encouraged to employ only companies that are members of the BPA's AOS to manage their parking. This ensures high standards, plus fair and reasonable enforcement, and should mitigate the number of complaints of poor parking management.

In 2006 the BPA introduced its Code of Practice for vehicle immobilisation; in 2007 this was followed by a Code of Practice for the use of private parking tickets issued directly to the vehicle or the driver or through the post as a result of camera enforcement. Both of these codes have been superseded in 2009 by the Code of Practice for the Approved Operator Scheme (see below).

In 2007 the BPA also entered into a partnership with the DVLA to limit access to its records only to those companies that were members of an Accredited Trade Association. We did this to ensure that vehicle keepers' personal data was better safeguarded and only released to responsible car parking companies where it was absolutely necessary to aid and support proper enforcement action. The DVLA will not release data to a parking company that is not a member of an ATA. In practice this means the BPA, through its Approved Operator Scheme, as it is the DVLA's first ATA and currently the only one in the parking sector.

AOS members are required to comply with the Code of Practice and are subject to a stringent compliance audit upon joining and at least annually. In addition, we operate a complaints and sanctions scheme and members that do not comply with the terms of the Code will have sanctions applied and are liable to have their membership suspended or terminated. There are around 160 members in the Scheme and a list of current members can be found on the BPA website.

Where, for example, a business, local authority, retailer, university or a hospital contracts out parking enforcement on private land, that third party will be required to become a member of the AOS in order to gain access to DVLA vehicle keeper data.

The Approved Operator Scheme Code of Practice was revised and updated in May 2011. The revised Code continues to provide guidance on the management and enforcement of privately controlled parking areas and related activities, as well as incorporating recent changes to consumer protection law, making it a more robust document with a stronger framework for operators to follow and consumers to refer to. Increasingly it is becoming the industry standard and we hope that it will prove to be a blueprint for any government legislation that may be introduced in future. Similarly, the BPA has successfully piloted an appeals system for both the motorist and the operator, which it is hoped will provide a working model for an independent body to implement.

In addition to this, the BPA formed the Approved Operator Scheme Board in 2009 to continue the improvement of fairness and standards within the parking industry. Board members, who include consumer and business representatives and government observers, are firmly of the

opinion that the most effective way to bring about a step-change in the quality and fairness of parking enforcement on private land is to work together with the parking industry to raise standards, holding to account those operators whose standards fall below acceptable levels. The Board's objective is to hold the industry to account in improving standards of management for private off-street car parks. As such, the key areas of compliance, communication and complaints are high on the agenda, as well as an independent arbitration service. The Board would welcome applications for membership from landowners and/or their representative associations.

The full Code of Practice is available to download from the BPA's website. The following overview provides a quick summary of the key points:

- Act rationally; behave responsibly; charge reasonably.
- Provide staff in uniform who have been properly trained.
- Display clear, easily understood, signage in car parks warning of clamping or ticketing and the driver's contractual responsibilities.
- Charge reasonable rates that reflect the actual cost of the operation.
- Ensure that operatives are easily identifiable, and are licensed where necessary.
- Offer a dispute resolutions service.
- Store removed vehicles securely.
- Offer a 'grace' period, when pay and display tickets have expired.
- Provide a number of different payment methods for enforcement actions.

Remember the controls are subject to the underlying laws of contract. Proper enforcement or parking on private land is dependent on clear signage that is visible from all points of the car park.

Methods of parking enforcement on private land

Typically, parking enforcement on private land is managed using conventional systems such as the issue of parking tickets or the application of wheelclamps or other immobilization devices by staff employed either by the landowner or a contractor. Whilst these functions look similar to those employed by public authorities, they are unregulated and motorists do not enjoy the same rights. See 'Wheelclamping' (p.744) for more information. Other methods employed are ANPR and 'self-ticketing' and these are described below.

Automatic Number Plate Recognition (ANPR)

Any business that operates off-street parking as a private landlord / landowner will have two main concerns:

1. Protecting the property from unwanted trespass.
2. Ensuring that in protecting the property, legitimate users of the facilities are able to park unhindered in the private car park.

There are a number of methods by which this can be enforced, with the newest of these being Automatic Number Plate Recognition (ANPR) technology. As with all methods of enforcement of unregulated private land, charges enforced by ANPR are subject to the Law of Contract. They are also self-regulated through the BPA's Code of Practice for parking enforcement on private land and unregulated private car parks and through voluntary membership to the BPA's Approved Operator Scheme.

The principle is as follows:

- CCTV-style cameras are placed at the entrance and exit to a car park.
- Timed photographs are taken of the vehicle itself entering and leaving the car park, and also close-ups of the vehicle's number plate (its Vehicle Registration Mark or VRM).
- The duration of the stay of the vehicle is calculated from the times registered on the two sets of photographs.
- If a vehicle has exceeded the duration of stay permitted by the purchase of a ticket or on car park signage (e.g. 'maximum two hour stay for customers only'), then the driver of the vehicle may be required to pay an additional charge (which must also be stated in the car park's signage).
- The main difference with ANPR and other kinds of enforcement is that if a driver does contravene any of the terms and conditions of the contract, they should also be aware that they may not receive a notification attached to the vehicle at the car park but some time later in the post. Drivers should also be made aware that DVLA records will be checked to identify the Registered Keeper in the event of a breach of contract.
- Using the vehicle's registration mark or VRM, the operator can access the DVLA's Vehicle and Keepers' registers and send a notice to the registered keeper of the vehicle, requesting payment of the additional charges. (The Vehicle Registration Acts allow anyone with a 'reasonable cause' to obtain details of a vehicle keeper. It is commonly accepted that breach of contract for parking on private land is a 'reasonable cause'.)

The BPA's Code of Practice contains recommendations for the size, placement and information required for private car park signage, including stating the fact that the car park is monitored by ANPR technology and that the DVLA will be contacted to obtain keeper details in the event of a parking contravention occurring.

As with all new technology, there are issues associated with its use. For example, repeat users of a car park inside a 24-hour period sometimes find that their first entry is paired with their last exit, resulting in an 'overstay'. Operators are becoming aware of this and should now be checking all ANPR transactions to ensure that this does not occur.

Some 'drive in / drive out' motorists that have activated the system receive a notice from the ANPR operator even though they have not parked or taken a ticket. Reputable operators tend not to uphold these (unless advised differently by the landowner / landlord), but operators should also now be factoring in a sensible 'grace period' to allow a driver time either to find a parking space (and to leave if there is not one) or make a decision whether the tariff is appropriate for their use or not. This 'grace period' is, however, at the discretion of the landlord / landowner and will also vary in duration, dependent on the size / layout / circumstances of the car park.

The BPA has published a 'Parking Practice Note of Grace Periods and Observation Periods' (to confirm that a breach of contract or contravention has occurred) and it is important to understand the difference. See *Sources of further information*.

'Self-ticketing'

In some parking environments there are circumstances where conventional methods of enforcement are not appropriate. A very small parking area may not warrant a full parking contract, and the landowner may not wish to use vehicle immobilisation. In such a case, the use of self-ticketing is becoming popular. A landowner (of say a corner shop, or small rural pub for instance) will purchase a parking kit from a responsible operator who is registered with the AOS, which will enable them to manage the parking bays

on their site. Whilst adhering to the terms and conditions of the issuer of the pack (who in turn will adhere to the terms and conditions of the AOS), the landowner will issue the parking ticket and return a copy to the operator, who will access the DVLA's Vehicle Keeper Details database to pursue the outstanding ticket if necessary. The landowner cannot and should not have any access to these details.

It is important to remember that the company that produces and sells the parking kit and provides the notice processing service is responsible for ensuring that it and all of its agents and customers comply with the terms and conditions and obligations set out in the AOS Code of Practice for parking enforcement on private land and unregulated public car parks. Failure to do so will result in sanctions being applied to them and they are liable to have their membership of the AOS suspended or terminated as a result of the actions of third party 'self-ticketers'.

Safer Parking Scheme – Park Mark®

Car parks used by employers, staff and customers can all benefit from being a member of the Safer Parking Scheme (SPS). The scheme is run by the BPA for the Association of Chief Police Officers, and is aimed at reducing crime and the fear of crime in parking areas.

The Safer Parking Scheme is instantly recognisable by the Park Mark® logo and has been designed to create a benchmark standard for parking across the UK, establishing safer parking for drivers and vehicles. When using a Park Mark® car park, people can be confident in the knowledge that it is providing police approved safer parking; this means quality management, appropriate lighting, effective surveillance and a clean environment, all of which lead to safer parking. National statistics show that around 22% of vehicle crimes occur in car parks. Many parking facilities with the award have experienced a dramatic reduction in crime, and London, for example, has seen a reduction of crime in car parks across the capital by nearly 40% since 2001. Where facilities do not experience vehicle-related crime, they have been able to create an environment where motorists feel safer.

The scheme has approaching 5,000 member car parks in the UK, and this is increasing weekly. So good is the Safer Parking Scheme that it is now being adopted in Vancouver and other parts of Canada too, and the BPA is working closely with Canadian Direct Insurance and Vancouver Police Department to extend the scheme to other parts of Canada.

Lifecare plans (LCP) and asset management

In general, a car park owner owes a duty of care under statute and common law to any person who enters on to their premises, including employees, children and trespassers. Failure to comply with this duty can render the owner liable criminally (under health and safety law) and/or face civil actions (under the common law relating to duty of care). In the event of a court determining liability of a car park owner, the inclusion of a LCP may be relevant in establishing whether the owner has complied with its duty of care, if any. The recommendation by the Institute of Civil Engineers and endorsement by the HSE, that owners should have a LCP, is likely to hold significant weight in establishing the standard or care owed in the event that a claim is brought.

There are no exceptions, and failure to comply with the legal obligations may give rise to prosecution. The law in this area is complex and advice should be taken on best practice. If a car park structure is left unchecked, or has reduced levels of safety, problems can easily occur. This can be avoided by adopting a LCP, which is a strategic and managed approach to the inspection, maintenance and management of parking structures. Don't forget also that a 'car park structure' doesn't need to be a whole building; a simple retaining wall alongside a surface car park needs to be properly maintained too!

It is also a requirement to confirm that a lifecare plan exists, where appropriate, when considering entering a parking facility into the Safer Parking Scheme.

Components of a lifecare plan:

- Description of car park with age, photographs, as-built information, etc.
- Record of previous investigations, repairs, accidental damage, winter maintenance, etc.
- Records of daily surveillance and routine inspections.
- Condition survey and material testing.
- Structural appraisal (by a Chartered Structural Engineer).
- Priced repair / maintenance options.
- Recommendations for future action.
- Records of works undertaken and costs involved.

The BPA publishes a number of parking practice notes associated with lifecare plans and asset management, which give further guidance to owners and operators of parking structures.

We want to see greater emphasis on the need to ensure that parking structures are properly inspected and maintained. Owners and operators are encouraged to have financial mechanisms in place to fund routine structural assessments, lifecare planning and essential maintenance.

In its Master Plan for Parking 2011/12 the BPA says:

'We want to see a greater emphasis on the need to ensure that parking structures are properly inspected and maintained. Owners and operators should be encouraged to have a financial mechanism in place to fund routine structural assessments, lifecare planning and essential maintenance.'

The BPA is concerned that Britain's many ageing car parks are properly serviced and maintained. Funds should be made available by owners and operators to provide regular safety and structural inspections, which will identify defects and prompt repairs to minimise the risk of structural failure.

Nuisance vehicles – abandoned vehicles and trailers

Public perception of an area with abandoned and other nuisance vehicles is of deprivation and neglect, and is regarded by the police as 'signal crime'. From the perspective of running car parking services, the principal points to be considered are that abandoned and nuisance vehicles are usually untaxed, unregistered and not roadworthy.

There is little doubt that many of the problems with the fast removal of abandoned and other nuisance vehicles revolve around the very complicated set of legislation that governs them. No fewer than five Acts of Parliament or Regulations directly affect how they are to be classified, removed and restored to the owner, and when they can be destroyed. This does not include the human rights requirements or the European End of Life Vehicle Directive.

The main piece of legislation on vehicle abandonment is the Refuse Disposal (Amenity) Act 1978. This provides a clear

duty on the local authority for removing vehicles that they believe to be abandoned in their area. The local authority must delegate the removal of the vehicle to what are referred to as 'authorised officers'. These officers should be well trained and experts in the field; they should understand the legislation and be able to offer themselves as expert witnesses in cases of vehicle abandonment in courts. In legal terms, it is their decision alone as to whether a vehicle is abandoned or not. The local authority duty extends to all vehicles believed to be abandoned in the open air which, by definition, means a car park facility with at least one side being open. There are different routes for dealing with the vehicle if it has been abandoned on a road or private land. The BPA has published a 'Parking Practice Note on Abandoned and Nuisance Vehicles', and Keep Britain Tidy (formerly known as ENCAMS) also offers advice on this issue, which can be found on its website – see *Sources of further information* for details of both.

Parking for motorists with disabilities

The Equality Act 2010 is the primary legislation that encourages everyone to have due regard and make reasonable adjustments to meet the needs of motorists and passengers with a disability, including any changes to their car parks to ensure that there is no disadvantage to disabled people when using these services and facilities.

The Act covers the widest spectrum, including those with auditory and visual impairments, as well as those with specific mobility difficulties. Remember, not everyone who is disabled is a wheelchair user – employers should positively give consideration and take action to meet the varying needs of disabled people wherever change may be necessary, through consultation, education and constant review at every level.

When assessing the area needed for staff and customer parking, it should be noted that the current UK norm for parking spaces is 2.4 metres wide by 4.8 metres long. The space for manoeuvring (roadways) between bays is six metres. These dimensions are neither minimum nor written in tablets of stone, and may be revised to suit your particular needs, but remember that good access and wider bays aids efficient use of the parking area.

Government guidelines (*Inclusive Mobility* published by DfT) recommend that 6% of parking should be allocated to disabled people, unless otherwise covered by local planning regulations. The guidance also recommends how to identify these spaces, with special markings and signage. It is recommended that parking spaces for disabled people are 3.6 metres in width, where the difference (1.2 metres) is yellow hatched to enable sufficient access for wheelchair users. These spaces should carry the wheelchair logo on the surface of the bay and display the appropriate sign at a driver's eye level. Advice on how to ensure that you cater for the needs of people with disabilities can be obtained from organisations such as Disabled Motoring UK and RADAR, with whom the BPA works closely.

The BPA, DMUK and BCSC undertook major research in partnership with DfT in 2009, which indicated that the 6% one size fits all approach leads to oversupply in some situations and undersupply in others. It is expected that Inclusive Mobility will be superseded in 2012 with more flexibility in the guidance. The BPA recommends the following allocation.

The Blue Badge Scheme

Since its introduction in 1971, the Disabled Persons Badges for Motor Vehicles Scheme (commonly called the Blue Badge Scheme) has played an important part in traffic management for those with mobility impairment.

It is important to remember that the Blue Badge scheme offers statutory concessions on the public road network throughout the UK. There are no specific concessions off-street anywhere in the UK. Some local authorities extend these concessions to their public parking places and many private operators of public parking places provide facilities for people with disabilities, and the Blue Badge is used to confirm eligibility to those facilities. However, these facilities are provided by landowners to meet their obligations under the Equality Act 2010 and related legislation and not specifically the 'Blue Badge Scheme' rules.

Size of car park (no. of spaces)	Designated bay provision
1-50	Two + 3% total car park
51-200	Three + 3% of total car park
201-500	Four + 3% of total car park
501-1,000	Five + 3% of total car park
1,000+	Six + 3% of total car park

The Blue Badge scheme is designed to provide convenient access for disabled motorists to reserved, accessible parking, rather than to provide free parking. Many parking providers, such as private landowners, do charge – particularly where there is a barrier system for entry to the car park. Others do not – but this is purely discretionary and is up to the individual landowner.

If people with disabilities are to be charged for their parking then it is important to ensure that methods of payment are accessible to them and that the charging regime is fair. For example, remember that many people with mobility difficulties will have problems using conventional payment machines also – they may not be able to remove a ticket from a barrier ticket dispenser or make payment at an automatic pay point. Also remember that people with mobility difficulties will take longer to do things and it may be that they need additional time in the parking facility and thus charges should be set accordingly so as not to discriminate. For example, it may be that Blue Badge holders have a longer time limit where parking is free or perhaps have a reduced hourly rate where charges are made.

In Scotland the law has been changed recently to require all local councils to speak with landowners who provide public parking with a view to providing enforcement of parking spaces provided for disabled people. There is no obligation for the landowner to accept, but the conversation must take place. This means that, for example, if a local supermarket or hospital provides parking for people with disabilities it would be possible for those spaces to be managed and enforced by the local authority (or an agent) using statutory powers.

The aim of the Blue Badge Scheme is to ensure that these people are more mobile and have access to adequate and fair parking provision. However, the growing abuse and fraudulent use of the badges in recent times has threatened the integrity of the system. Accessibility is a key concern for everyone but, at the same time, misuse of the scheme cannot be left unchallenged. With this in mind, in February 2011 the Government announced a major programme of reforms for the scheme, the most comprehensive changes in 40 years. The reform programme was developed in consultation with disabled people, local authorities and other stakeholders, including the BPA, and aims to both crack down on drivers who abuse the scheme and make the scheme more sustainable for the future.

The reforms include a new badge design, a data-sharing system for local authorities, and improvements to the assessment of eligibility. The Badge's new design, which incorporates advanced printing and security features, ensure it will be harder to copy, forge or tamper with, and a more secure method of distribution and supply will guarantee that Badges are safely delivered to their intended recipient.

An improved method of production and distribution for the Badges will create a data-sharing system that will enable local councils to improve enforcement too. This new system will prevent potential misuse, for example, where multiple applications are made, or where names of the deceased are used. Also local Civil Enforcement Officers undertaking parking enforcement will be able to check whether a Badge on display in a vehicle is genuine or not, regardless of where it is issued in England, Scotland and Wales; Northern Ireland has yet to decide if it will join this service.

Current legislation in England, Wales and Scotland does not recognise on-street investigators; therefore their powers are limited, especially with regard to inspection and seizure of Blue Badges. In Scotland, Section 73 of the Transport (Scotland) Act 2001 empowers the police, police traffic wardens and council parking attendants to inspect Blue Badges. Only a uniformed police officer can seize a Blue Badge. Some local authorities in England deploy on-street investigators who do not necessarily involve the police when conducting investigations and reporting individuals for prosecution. This is based on the fact that they are in possession of sufficient evidence to prove the case and that the prosecution is in the public interest.

Investigations into Blue Badge fraud must be done at all times in accordance with a number of Acts and Regulations including:

- the Regulation of Investigatory Powers Act 2000 (RIPA) (with associated relevance to the Human Rights Act);
- the Regulation of Investigatory Powers (Scotland) Act 2000 (RIPSA) (with associated relevance to the Human Rights Act);
- the Data Protection Act; and
- the Freedom of Information Act.

The Blue Badge scheme is invaluable to people with mobility difficulties, and managing it correctly is vital to ensure that only the right people benefit from it. The scheme needs to be pro-actively managed, have consistent standards, and the ability to share information and best practice.

Hospital parking

NHS Trusts are increasingly under pressure to ensure that health funding is used to provide health care services. Parking at some hospitals is often very busy, with competing demands that cannot all be met. Staff, patients and their visitors have different needs and it is necessary to manage the parking to ensure that it is used effectively and efficiently, and priority is given to those most in need of it.

Pressure on spaces is the main reason behind many hospitals introducing parking charges. Hospitals in town centres and near other facilities have to ensure that their parking is protected against use by others, such as commuters, shoppers and local residents, and effective parking management is important in these circumstances.

The last ten years have seen a rise in the number of hospitals introducing some form of controlled parking, and with this there has been an increase in parking operators being brought in to manage and run facilities. Many hospitals and operators take a joint venture approach that allows them to make a projection of what the scheme will cost and the potential for any surplus. Where NHS Trusts do make a surplus income from parking operations, it is usually diverted to enhance patient care facilities or to support the parking charges for people with disabilities or with life-threatening illnesses who make frequent and regular visits to the hospital.

It doesn't always have to be a case of charging for parking. One option is to offer free parking for up to two hours, similar to the schemes run by supermarkets and retail parks. This can work well with smaller hospitals, and tickets can be made available if someone needs to stay longer. Problems with this system can arise if parking in the areas around the hospital are Pay and Display – people will then take advantage of the free parking at the hospital. For larger and city centre hospitals, the most common option is pay and display. With this system, hospitals have the flexibility to issue permits to staff or provide a reduced rate for parking. They can also offer permits to patients who are terminally ill or those visiting patients in the hospital for a long-term stay for the time they need.

The governments in Scotland and Wales have effectively abolished charging for hospital car parking spaces, but the subject is still very much under debate. In Wales, free parking is available at almost every hospital. In England, the Government has given the decision to local health chiefs to make on policy, and charges are common. This is similar to Northern Ireland, but a review of this policy is currently being undertaken.

The Hospital Parking Charter

In 2010, the BPA published a Charter for Hospital Parking, designed to help and encourage NHS Trusts to engage parking management systems that are fair for all. The BPA worked with the NHS Confederation and the Healthcare Facilities Consortium, which both consulted with their members to help inform the contents of the Charter.

Bringing together the interests of hospital car park users (staff, visitors and patients), Government, local authorities and commercial organisations, the Charter has been circulated to all NHS Trusts across the UK. Its aim is to strike the right balance between fairness for patients and visitors and hospital staff, as well as the Trust itself.

The Charter sets out guidelines to help Trusts and car park operators deliver effective and efficient parking for users – many of whom have particular needs. Hospital parking is subject to supply and demand like any other commodity or service, so charging is important, but Trusts should offer concessionary parking where appropriate.

Charges should not be introduced to generate income but, rather, to ensure that key staff, bona-fide patients and visitors are able to park at the hospital. Without income to support the car park maintenance – security, upkeep of facilities and staff – funds that should be dedicated to healthcare provisions would have to be used instead.

It is ultimately the patients and visitors who benefit from managed hospital car parks. It is absolutely critical that hospitals – particularly those within urban areas – set car parking charges at a rate that will deter commuters and shoppers who may abuse the system, as was the case before charges were introduced.

Since the Charter was created the BPA has established a special interest group for those working in the healthcare sector. This group is currently updating the Charter and a new version will be published in late 2011.

The parking profession

The British Parking Association takes development of the parking profession seriously. In 2007 the UK Skills Board for Parking was established to provide direction for the UK parking skills agenda. At the end of 2009 this board was combined with the board of the Institute of Parking Professionals to provide a more streamlined approach to:

- providing strategic direction for the delivery of the UK Parking Sector Skills Strategy;
- aligning the parking sector with the government skills agenda;
- collating research;
- developing partnerships with organisations that can support employers to carry the skills agenda forward;

- developing a qualifications and career pathway for all employees in the sector;
- supporting the Institute of Parking Professionals; and
- signpost opportunities (including funding) to employers so that they can upskill their workforce.

The board brings together expertise from a range of parking professionals and skills specialists to ensure that it achieves excellence in parking for all.

See also: Disability access and egress, p.238; Driving at work, p.250; Vehicles at work, p.686; Wheelclamping, p.744.

Sources of further information

The British Parking Association's website contains useful information related to different areas of parking: www.britishparking.co.uk

The BPA, on behalf of the Parking Forum, produces a number of Parking Forum Position Papers, which describe various aspects of parking and the associated benefits and challenges. These are available as free PDF documents from the Parking Forum website: www.parkingforum.co.uk

Additionally, BPA Parking Practice Notes, which provide guidance on a range of parking matters, are available to purchase. For more information contact Abdul Traore at Abdul.t@britishparking.co.uk

The BPA is committed to represent all that is best in parking control, management, enforcement and the maintenance of parking structures. Contact Alison Tooze at alison.t@britishparking.co.uk or 07974 576087 for more information about how we can help you in delivering excellent parking facilities at your premises.

Association of Chief Police Officers: www.acpo.police.uk

Inclusive Mobility: www.dft.gov.uk/transportforyou/access/tipws/inclusivemobility

Mobilise – Promoting mobility for disabled people: www.mobilise.info/

To find out where the nearest Park Mark® accredited car park is visit www.parkmark.co.uk. If you would like to find out more about the Scheme, contact the Safer Parking Scheme team on 01444 447318.

The Institute of Parking Professionals: www.theipp.co.uk

Keep Britain Tidy:
www.keepbritaintidy.org/KeyIssues/AbandonedVehicles/Default.aspx

Sources of further information

Secured by Design: www.securedbydesign.com

BPA – Code of Practice for Parking Enforcement on Private Land and
Unregulated Car Parks:
www.britishparking.co.uk/write/BPA_CodeofPractice_v8.pdf

Parking Practice Note of Grace Periods and Observation Periods:
www.britishparking.co.uk/write/Documents/Library/ppns/PPN%20023%20
Jun08%20-%20observation%20and%20grace%20periods.pdf

Parking Practice Note on Abandoned and Nuisance Vehicles:
www.britishparking.co.uk/write/Documents/Library/ppns/PPN%20019%20
January%2007-%20Nuisance%20Vehicles.pdf

RADAR: www.radar.org.uk/radarwebsite/

DfT – Blue Badge Reform Strategy for England:
www.dft.gov.uk/transportforyou/access/bluebadge/reform/

Hospital Parking Charter: If you would like a copy of the Charter or your
organisation would like to sign up to it, contact Dave Smith on 01444 447316 or
email dave.s@britishparking.co.uk

Comment ...

Please release me

Kelvin Reynolds, British Parking Association

In 2009, the then Labour Government introduced the Policing, Crime and Security Bill to tackle the problem of a lack of regulation in the clamping industry.

Some of the main issues of contention of those campaigning to prevent rogue practices include:

- the fact that clamping companies do not currently require a licence;
- a lack of industry-wide standards for signage; and
- penalties and the lack of any genuine appeals process.

In August 2010, the Conservative / Liberal Democrat Coalition Government demonstrated that it was prepared to go further than its predecessor by announcing that all clamping on private land would be banned in England and Wales, a measure that has been in place in Scotland since 1991.

The clamping ban would only apply to private land and will not change existing lawful authority, such as traffic enforcement by local authorities and police on highways.

While welcoming the changes, the British Parking Association has some concerns. We are fed up in the same way that everybody else is with the excesses of clamping and the horror stories that you hear in the media, day in, day out. But for the Government to simply come out with a statement that 'we will ban clamping' really

Kelvin Reynolds is Director of Operations and Technical Services at the BPA.

Kelvin joined the BPA in March 2004 where he took responsibility for managing the Safer Parking Scheme as well as the Association's development of technical services to its expanding membership. Kelvin has directed the BPA's involvement with the Government's development of Civil Parking Enforcement and the review of the Blue Badge Scheme for disabled people. He is also leading the development of the BPA's Approved Operator Scheme and its Codes of Practice and the BPA relationship with the DVLA.

doesn't suggest any understanding of the problem.

Landowners and property owners need to be able to manage their land and so we have lobbied long and hard and the Government has moved and is now in a position where it is saying we're going to ban clamping without 'lawful authority,' and that means there are a number of exemptions.

Managing parking on private land, whether by being there by invitation or whether you're trespassing, is subject to law of contract and they're saying that entering into a contract where you accept that you will be clamped is not lawful authority. They've quite clearly set that out in the Bill. So we're trying to understand just what that means.

What we're trying to say is that it is probably going to be more confusing after this Bill comes into being than it is now, and we don't think that's right.

Following months of consultation, the Freedom Bill was introduced in February 2011. Banning clamping on private land is just one provision within this Bill that covers a myriad of things, from the licensing of individuals to data protection to the registration of offenders to criminal record checks. There's a whole raft of things in there.

Typically a Bill has three stages in Parliament, both in the House of Commons and in the Lords, so I would think we're looking at 2012 before this appears on the statute books as an Act of Parliament. What shape it will be in when it gets there will be subject to all the debate that takes place during that time.

Government figures released prior to the publication of the Bill have shown that around 500,000 clampings take place every year on private land, with an average release fee of £112, amounting to total release fees of £54.9m.

Home Office Minister, Lynne Featherstone, introduced the intended ban by saying:

"The Government is committed to ending the menace of rogue private sector wheel-clampers once and for all… I think, on the whole, everyone will heave a sigh of relief that something that never really worked is now not going to have to."

One of the main causes for concern since the Coalition announced its intentions is that landowners and facilities managers will not have the tools to manage their land effectively if clamping is banned outright. The Government has suggested, however, that ticketing is a viable alternative.

> "We think the Government's plans are a charter for the selfish motorist, because effectively people will be able to park where they like, when they like, almost with impunity, and nobody will be able to do anything about it. That's not fair."

However, we've been able to demonstrate that ticketing isn't the panacea that the Government thinks it is, in that you have driver liability and that's become well known these days. Proving who the driver is can be quite difficult sometimes and so we've said that for ticketing to be a viable alternative, assuming that the ban comes in, you need to have some form of owner liability. That now exists on the face of the Bill, so we've managed to achieve that.

There are exemptions for barrier controlled car parks, which are also there because, we believe, we've been able to point out to Government that the simple ban on immobilisation of vehicles is too crude.

The British Parking Association, along with many others, is not satisfied with the proposals for parking regulations contained within the Bill as they are in their current format. We think the Government's plans are a charter for the selfish motorist, because effectively people will be able to park where they like, when they like, almost with impunity, and nobody will be able to do anything about it. That's not fair.

Let's get everyone involved in this profession to be properly registered, properly regulated and then we really will identify the rogues and the villains and we can outlaw those. This is going to make life difficult for lots of people and add lots of confusion to everyone.

We'd like the Government to give our approved operator scheme some legal backing, require everyone who is involved in parking management to be a member of an accredited trade association of one form or another, and then we'll be properly regulating the industry.

Until we do that there are opportunities for people to operate outside the codes of practice we've established.

We are very keen as an industry to make certain that parking management on private land is undertaken fairly, reasonably and responsibly. We absolutely abhor the excesses of clamping and removals and we want to work with Government to get that practice outlawed and to rid the country of it.

But in doing so, let's not throw the baby out with the bath water and prevent reasonable landowners all over the country from being able to manage the land that they need and use for pleasure, leisure, business and commerce.

Until the Freedom Bill receives Royal Assent, the current law regarding clamping will continue to apply.

Permits-to-work

Kathryn Gilbertson, Greenwoods Solicitors LLP

Key points

- A permit is a formal written system used to control certain types of work that are potentially hazardous.
- A permit is issued by an authorised person (AP) who understands the risks and the control measures required to be put in place.
- It is issued to a competent person (CP) in charge of the work stating that it is safe to work in the plant / area specified.
- The permit will detail precautions that must be observed by persons carrying out that work.
- The permit will have a specified time limit.
- Only the person who issued the permit should amend or cancel it.
- The competent person is responsible for the safe conduct of the work and must not break the conditions specified on the permit.
- The competent person must fully understand the requirements of the permit and inform everyone what they may or may not do.
- The competent person must return the permit to the authorised person for cancellation when the work is completed or when he has withdrawn from site. A permit is not a replacement for risk assessments and method statements.
- Permits should be retained for at least three years by the authorised person.

Legislation

- Health and Safety at Work etc. Act 1974.
- Electricity at Work Regulations 1989.
- Construction (Health, Safety and Welfare) Regulations 1996.
- Confined Spaces Regulations 1997.
- Management of Health and Safety at Work Regulations 1999.
- Control of Substances Hazardous to Health Regulations 2002.

Where permits apply

These are used to control high-risk activities and areas where specific hazards could be present. As such, permits are usually issued on the following occasions:

- Electrical work, including high-voltage electrical work;
- Lift works;
- Asbestos;

- Roof works;
- Work on scaffold towers and Mobile Elevated Working Platforms (MEWPs);
- Confined spaces;
- Demolition works;
- Excavation works;
- Pressure systems;
- Hot works; and
- Work that can be carried out only by removing normal control measures (e.g. live working on a supply to a critical piece of equipment).

Contents of permits

There are four stages in the process of permits-to-work, namely issue, receipt, withdrawal and cancellation. Thus, the following process should take place:

1. *Issue*. Detail safety precautions that have been taken. Authorised person gives declaration that it is safe to

Case study

On 18 July 2011, London Concrete Ltd was fined £16,000 and ordered to pay £9,397 costs following an accident at Gerrards Cross. An employee's arm was torn off by an auger as he was carrying out repairs.

The victim was attempting to repair the industrial 'cock screw' auger. He had incorrectly isolated the electrical power and had failed to complete the permits-for-work as required by the company's policy. The repair failed and on the following day both the victim and his supervisor failed to check that the power had again been isolated. The supervisor accidentally activated the machine and it tore off the victim's arm above the elbow.

The HSE investigation found that the victim had not been trained on the equipment that he was using. Further, the plant manager did not supervise the work correctly, which meant that permits-to-work were frequently not completed. They felt that the risk of such an accident occurring was completely foreseeable. Permits-to-work are designed to prevent just the sort of misunderstanding that existed between the manager and the fitter.

proceed with the job within specified limits.

2. *Receipt.* Declaration by competent person in charge of the work that he fully understands the requirements, and has informed everyone what they may or may not do.

3. *Withdrawal.* Declaration by the CP that the work has been done or discontinued and the staff under his control have withdrawn. A declaration that the work area is in a safe condition.

4. *Cancellation.* The AP signs to cancel the permit to work. AP makes the plant / area operations.

Why should I issue a permit to work?

A higher degree of safety is achieved through using a permit to work rather than verbal instructions. Verbal instructions can be misheard or misinterpreted. Thus, a permit to work system provides an additional level of safety for the authorising person / company.

Roof works
- Adequate means of access, either temporary or permanent (the use of crawling boards, roof ladders etc.).
- Testing of the roof's fragility.
- Edge protection.
- Preventing the hauling of materials or objects.
- Personal protective equipment.

Confined spaces
- Atmospheric monitoring.
- Emergency rescue procedures.
- Isolation of fluid or energy sources.
- Personal Protective Equipment, e.g. breathing apparatus.
- Additional supervision and monitoring.

Hot works
- Ensuring that sprinklers (if installed) are isolated and re-activated after the works.
- Housekeeping – removal of combustible materials from the area, e.g. paper, cardboard, flammable liquids, etc.

- Use of protective non-combustible curtains to protect property and person.
- Provision of suitable fire extinguishers.
- Fire trained person to re-visit area 30 to 60 minutes after the hot works have finished ensuring no smouldering embers or hot surfaces remain.
- Ensure hot work equipment is maintained and inspected.

General maintenance

Be conscious of the human factors:

- Poorly skilled workforce.
- Unconscious and conscious incompetence.

The following issues may contribute towards a major accident or hazard:

- Failure of safety critical equipment due to lack of maintenance;
- Human error during maintenance;
- Static or spark discharge during maintenance in an intrinsically safe zone;
- Incompetence of maintenance staff; and/or
- Poor communication between maintenance and production staff.

What do I need to do?

You should review your existing system and clarify how it works, the types of jobs permits are used for, and check that it is in operation. In particular, the responsibility and training of those involved in authorising and overseeing work undertaken under a permit should be reviewed. Only designated individuals should authorise work permits and should clarify who is responsible for specifying the precautions, making sure that these also cover contractors. The precautions need to specify requirements while the work is in progress, as well as requirements needed to cover the work itself, such as welding fumes, vapour from cleaning solvents, etc. It is usual to have plans and diagrams along with detailed work and method statements attached to permits-to-work.

A first aid kit and, if appropriate, emergency breathing apparatus, should be kept close at hand to the area in which the work is being carried out. For confined spaces, this could include a rescue harness and lifting equipment to retrieve the worker if an incident occurs.

See also: Confined spaces, p.173; Construction site health and safety, p.177; COSHH: Control of Substances Hazardous to Health, p.207; First aid, p.325; Risk assessments, p.624; Workplace health, safety and welfare, p.768.

Sources of further information

HSG 65 Successful health and safety management:
www.hse.gov.uk/pubns/priced/hsg65.pdf

HSE – Permit to work systems:
www.hse.gov.uk/comah/sragtech/techmeaspermit.htm

INDG 258 Safe work in confined spaces: www.hse.gov.uk/pubns/indg258.pdf

Personal Emergency Evacuation Plans (PEEPs)

Elspeth Grant, TripleAConsult

Key points

- It is essential that building managers reduce the risk of fire and its effects to a minimum for everyone, including disabled people, using environments under their control.
- Few non-disabled people would consider it sensible to stay in a burning building with no evacuation plan or escape route and it is just as inappropriate to expect disabled people to wait in refuges for 25 minutes until the Fire and Rescue Services arrive. It is therefore a requirement for the workplace manager or Responsible Person to ensure that there is effective escape for people with disabilities.

Legislation

The Regulatory Reform (Fire Safety) Order (RRO) replaced the requirement for fire certificates and in so doing moved the fire safety emphasis on to prevention by using a risk-based assessment method.

The legislation requires employers to:

- undertake a fire risk assessment;
- identify the significant findings of the risk assessment;
- provide and maintain fire precautions; and
- provide information, instruction and training to employees about fire precautions.

The RRO is intended to protect employees but must take account of other people present. Employers must take account of any duty of care they and their employees may have to other occupants of the building. Morally and legally, we need to consider the capabilities of all individuals using buildings under our management.

Penalties

When considering the options, building managers would do well to bear in mind the penalties for non-compliance with the legislation. These vary from case to case but in general non-compliance results in the following:

Equality Act 2010

A civil court case or Employment Tribunal, with the potential for fines and damage to brands through bad publicity; it is also very much driven by what is reasonable. The Equality and Human Rights Commission monitors disability issues and may take up individual cases if appropriate.

Regulatory Reform (Fire Safety) Order

Criminal prosecution with the potential for fines, imprisonment of the Responsible Person, and litigation.

Corporate Manslaughter and Corporate Homicide Act 2007

Criminal prosecution of either organisations or individuals resulting in fines or imprisonment.

Guidance

Current guidance (BS 9999: 2008 – see *Sources of further information*) means that 'stay put' fire evacuation plans for disabled people, involving waiting for assistance for the Fire and Rescue Services, are no longer defensible. Steps must be taken to provide evidence of due diligence should there be a death or injury from the effects of fire.

The new Standard, BS 9999: 2008, now provides a whole section on disabled evacuation and building managers, Responsible People and fire competent people must take particular notice of the statement in BS 9999: 2008 Section 46.1:

'Providing an accessible means of escape solution should be an integral part of the fire safety management process. Fire safety management should take into account the full range of people who might use the premises, paying particular attention to the needs of disabled people.'

'It is important to note that it is the responsibility of the premises management to ensure that all people can make a safe evacuation. The evacuation plan should not rely on the assistance of the fire and rescue service. This is an important factor that needs to be taken into account in building design.'

This approach is supported in Guidance from the Department of Communities and Local Government (CLG) on Means of Escape for Disabled People, which states very clearly that:

'Under current fire safety legislation it is the responsibility of the person(s) having responsibility for the building to provide a fire safety risk assessment that includes an emergency evacuation plan that includes an emergency evacuation plan for all people likely to be in the premises, including disabled people, and how that plan will be implemented. Such an evacuation plan should not rely upon the intervention of the Fire and Rescue Service to make it work.'

This is re-enforced in the CLG Guidance on *Fire Safety Risk Assessment – Sleeping Accommodation* Section 3.4.3:

'Once a fire has started, been detected and a warning given, everyone in your premises should be able to escape to a place of total safety unaided and without the help of the fire and rescue service'.

Preparing a PEEP

Effective means of escape has to be a combination of:

- physical fire protection measures for safe means of escape; and
- a clear, unambiguous, management strategy for the safe evacuation of disabled people.
- Building managers should be aware of:
- how many disabled people are in the building;
- the nature of their disabilities; and
- refuges and their location.

Design principles

The principles to be considered when designing escape strategies remain constant for all buildings and include:

- planning and protecting escape routes leading to safety, both horizontally and vertically;
- construction and surface finishes to be of adequate fire resistance;
- the segregation of high-risk areas; and
- the provision of means of giving warning of fire and, where appropriate, detecting outbreaks of fire.

The Equality Act 2010 makes it clear that it is illegal to discriminate against a disabled person in any shape or form. This not only clearly applies to access issues but to emergency egress and it is therefore important to implement these design principles in a non-discriminatory manner.

Structural attributes
Generic structural attributes should be implemented as standard during any design or refurbishment activity to assist all disabled people during an evacuation.

Emergency alarms
All fire alarms should have visual beacons to assist those with hearing impairments.

Egress stairwell
Stairwells should have clearly contrasting nosings, wall and handrails with a Light Reflectance Value (LRV) of at least 30 pts (preferably yellow walls and blue handrails) to assist those with sight impairments during an evacuation. Emergency lighting will further assist disabled people and the audible fire alarm should be reduced to ensure that noise levels are kept to a minimum to assist those with dementia, autism or mental health problems.

Clear print signage
All escape signage will be in clear print and utilise colour to assist those with learning difficulties, dementia, dyslexia, autism or mental health problems.

Resting / refuge areas (RAs)
These should meet BS 9999: 2008. RAs must be clearly identified and have two-way communication. RAs are not

intended as places where disabled people will stay during a total evacuation of the building but are to enable disabled people to rest during their descent to a place of total safety. Their existence must be supported by detailed evacuation plans, assisted escape devices and Personal Emergency Evacuation Plans (PEEPs) for anyone regularly visiting a building.

Emergency planning principles
A multi-faceted approach to an evacuation strategy should be taken through structural attributes; anticipatory evacuation plans for uncontrolled visitors; management processes to include the use of individual PEEPs; training and communication.

The requirements for evacuation strategies are different for different disabilities (not forgetting that some disabled people have multiple disabilities).

Wheelchair users:

■ Many have excellent upper body strength.
■ Only 5-7% can never leave their wheelchair.
■ Clearly defined refuge areas.
■ Evacuation chairs.
■ Lifesliders or stairclimbers.
■ Buddies.

Ambulant:

■ Buddies.
■ Clearly defined refuge areas.

Impaired vision:

- Clarity of environment.
- Familiarisation.
- Buddies.
- Signage – Braille.
- Lighting – Strobes, LED.
- Handrails with tactile bumps.

Impaired hearing:

- Auxiliary aids such as sound alertors.
- Sleep alertors.
- Fire alarms with flashing beacons.
- Buddies.

Learning difficulties:

- Clear, simple English.
- Use photographs or illustrations.
- Translated into Makaton if appropriate.
- Clearly visible.
- Buddies.

Epilepsy / aspergers:

- Episodes triggered by flashing lights:
  - Cognitive difficulties following an episode.
- Aspergers Syndrome:
  - Can find certain sounds and/or visual stimulation to be fixating or distressing.
  - Alarms can lead to overload or lock-in.
  - 'Overload' can be easily reached, which can trigger epilepsy.
- Vulnerable individuals can wander.
- Carer may need to comfort a distressed person before evacuation.

An underlying principle is that disabled people will evacuate the building using the same mechanisms as those definedin the standard evacuation procedures (i.e. for non-disabled people). This standard approach ensures simplicity and clear mechanisms to evacuation, thus reducing opportunities for confusion in high stress situations.

Up until recently, protocols suggested that disabled / vulnerable people should wait until non-disabled people had evacuated prior to starting their own evacuation. This is no longer deemed appropriate and disabled people are encouraged to commence evacuating as soon as an alarm sounds.

For reference and further information, refer to BS 9999: 2008 Section 46 and the Communities and Local Government Guidance Document on Means of Escape for Disabled People, both of which detail a number of different approaches.

As with any other plan or process-driven activity, always start with a high level strategy by considering the approach to:

- identification of evacuation barriers across the property portfolio (an access assessment in reverse);
- access of the environment by disabled people and an assessment of risk for this group of people in the event of fire;
- assignment of roles and responsibilities – who does what and when;
- integration with existing emergency planning;
- provision of evacuation devices and arrangement with suppliers;
- implementation of Personal Emergency Evacuation Plans (PEEPs); and
- monitoring and review of PEEPs processes.

Planning for everyone

Although PEEPs are seen as an evacuation plan for disabled people, building managers should consider applying the PEEPs process to anyone who could have a problem evacuating a building. This may include those who are pregnant; have a temporary impairment; injuries (such as a broken leg); are permanently disablement; have hearing, sight, ambulatory or cognitive impairment; and others such as children.

It is fair to say that once the process is agreed, the day-to-day issues become ones of data collation and management and it is strongly advised that an electronic web-enabled database should be utilised for this purpose.

The impaired person should always be actively involved in the identification and physical exploration of the escape routes available before confirmation that these are suitable. Any hazards that will prevent or restrict escape must be identified as well as all the areas that a disabled or vulnerable person would be reasonably expected to use.

Additional support mechanisms and management processes will be identified through the PEEP process. This process will identify specific mechanisms and/or the provision of equipment that will assist disabled people, where they are unable to proceed without assistance (via the non-disabled route) to a place of total safety.

In a controlled environment (i.e. office building or residential care home) a participatory approach should be taken to evacuation strategies and PEEPs, whereas in an uncontrolled environment (i.e. a shopping mall or hotel) then an anticipatory approach is required.

For scheduled visitors / contractors, enquiries should be made by the person they are visiting / working for. For unscheduled visitors, generic fire and emergency procedures will apply.

All PEEPs should integrate with existing processes and procedures. Existing staff and tenants should undertake a 'catch-up'

exercise with all staff and tenants. New staff, tenants, visitors and contractors should, preferably prior to taking up a post, be assigned to accommodation, or as soon as possible after arriving.

Communication and consultation

The different groups of people who should be considered and are likely to be present in a building are staff; contractors; visitors; residents; students; and/or customers – individuals and groups (hiring rooms, public events, etc.).

The non-disabled population will follow the escape routes or make their way out by the way they came. The general plan will need to be expanded upon to include disabled and vulnerable people.

Staff

The Human Resources or personnel department are responsible for ensuring that staff are provided with suitable escape plans. The induction process should be the start of this. Information should also be provided within the staff handbook. A system is required to ensure regular updating of plans.

Contractors

Where there are contractors working in the building the department that they are working for has initial responsibility for their safety. They should ensure that, where necessary, steps are taken to ensure that they are provided with a suitable escape plan, chosen from the standard set of plans for the buildings.

Residents

Where the accommodation is provided in a hotel, part of the booking-in procedure should include the offer of a suitable escape plan. Additional accessible information is required in each room adjacent to the evacuation procedures for all residents. In hostel accommodation or student dwellings etc., suitable Personal Emergency Evacuation Plans (PEEPs) should be provided by the Accommodation Manager based on the standard set of plans for the building.

Students / pupils
When a child or student is enrolled, their escape plan should be developed as a part of the admissions process. Care should be taken that all disabled students are provided with a plan if they need one.

Visitors
A system of standard plans should be set up and offered to visitors at the time they book into the building at reception.

If there is a booking procedure, and previous knowledge of the visitor can be gained, the opportunity to pre-plan the escape should be taken. Booking forms, meeting planners and tickets should all include information about the managed escape process.Part of the booking procedures for groups should include provision of standard plans. Where there are a large number of disabled people, it may be acceptable that the party organiser plays a role in the provision of suitable escape. Provision of suitable information should be a compulsory part of the booking process, and booking administration should facilitate this.

Implementation of a PEEPs process
To implement a PEEPs process within a controlled (i.e. office environment) environment, the following process should be used.

Level 1 (Identify)
Collate PEEPs information as part of standard HR set of forms. Any additional information received from other agencies will be collated with the returned information.

Level 2 (Analyse)
The appropriate manager must consult with the disabled person and assess whether specific evacuation assistance is required. The manager will take into consideration information of the physical environment gathered as part of the Access Assessment. If required, a PEEP for the individual must be prepared and further discussion undertaken with the disabled person to ensure that the plan is thoroughly understood. If auxiliary aids are to be purchased, then these should be ordered and delivered where possible prior to occupancy of the building by the disabled person.

Level 3 (Implement)
A PEEP must be provided to each individual in a suitable format. After a defined period, the manager must check that the PEEP is appropriate and fully understood by the individual. Checks must also be made that any buddy system is appropriate. Corrections as a result of the review must be reflected in the PEEP. Fire Wallets and Fire and Rescue Services information must be updated to reflect the PEEPs.

Level 4 (Review)
There must be a process for reviewing PEEPs on an ongoing basis with a minimum of a six-monthly review. All disabled people with progressive illnesses will have a specific review period set. All changes to PEEPs must be reflected in the Fire Wallet and Fire and Rescue Services data.

Training and communication
Applicable staff must be trained in emergency response procedures and be aware of how to assist disabled people. In certain instances, it will be necessary to train friends or buddies to assist a disabled person during an evacuation. A full consultation exercise must take

place with the disabled person, family and other agencies as appropriate during this process. It is essential that a communication strategy is implemented to ensure that disabled people fully understand and are able to execute the defined strategy for disabled people in the event of fire.

Conclusion

Addressing the issues of fire safety for disabled people is therefore very important for both legal and moral reasons.

It is essential that the high level Fire Strategy considers the needs of all occupants of the building and that any processes such as PEEPs are embedded in working practices which are both practical and cost-effective. Disabled and vulnerable people must be involved in setting up the evacuation strategies that must be reviewed on a regular basis, particularly after any refurbishment work or change in occupancy.

It is possible to devise an evacuation strategy which is both practical and cost-effective, which reduces risk to a minimum and demonstrates that the organisation has undertaken due diligence if there is an investigation following a fire.

> *See also*: Accessible environments, p.18; Disability access and egress, p.238; Fire, means of escape, p.314.

Sources of further information

United Kingdom Disabled People's Council: www.bcodp.org.uk

The Fire Safety Risk Assessment Supplementary Guide – Means of escape for Disabled People can be downloaded free of charge from www.communities.gov.uk

All organisations must ensure that means of escape to a place of ultimate safety, supported by sufficient numbers of competent persons, are in place to effect an evacuation without the assistance of the Fire and Rescue Services, as stated by the Regulatory Reform (Fire Safety) Order 2005.

In addition, disabled people must be treated as equally as non-disabled people, as stated by the Equality Act 2010.

PEEPs for Professionals is a two-day accredited course designed to give students expert knowledge of the law, its implications, the impact of barriers on the evacuation of disabled people, and how to implement effective means of escape for both disabled and vulnerable people.

Visit www.workplacelaw.net/training/course/id/78 for more information.

Personal Protective Equipment

Andrew Richardson, URS Scott Wilson

Key points

- Personal protective equipment (PPE) is worn or held by persons at work to protect them from risks to their health and safety.
- The Personal Protective Equipment at Work Regulations 1992 (PPEW) apply in most instances; there are six other sets of Regulations that include their own particular PPE requirements.
- PPE is at the bottom of the hierarchy of risk control measures; PPE should therefore be used only as a last resort.

Legislation

- Health and Safety at Work etc. Act 1974.
- Personal Protective Equipment at Work Regulations 1992.
- Management of Health and Safety at Work Regulations 1999.
- Personal Protective Equipment Regulations 2002.
- Work at Height Regulations 2005.

Personal Protective Equipment Regulations 2002

The PPE Regulations deal with the suitability of PPE brought to market. The Regulations place a duty on Responsible Persons who put PPE on the market to ensure that the PPE satisfies the basic health and safety requirements that are applicable to that type or class of PPE, and that the appropriate conformity assessment procedure is carried out. All new PPE should have the CE mark, identifying that the equipment satisfies certain safety requirements and has been tested and certified by an independent organisation.

Personal Protective Equipment at Work Regulations 1992

The PPEW Regulations exist to ensure that certain fundamental duties covering the provision and use of PPE are applied whenever PPE is required.

The Regulations define PPE as 'all equipment (including clothing affording protection against the weather) which is intended to be worn or held by a person at work which protects him against one or more risks to his health and safety'.

Protective clothing includes aprons, clothing for adverse weather conditions, gloves, safety footwear, safety helmets, high-visibility waistcoats, etc. Protective equipment includes eye protectors, life jackets, respirators, underwater breathing apparatus and safety harnesses. The Regulations do not apply to PPE provided under the following Regulations, which have their own specific requirements:

- Construction (Head Protection) Regulations.
- Ionising Radiations Regulations.
- Control of Substances Hazardous to Health Regulations.
- Control of Lead at Work Regulations.
- Control of Noise at Work Regulations.
- Control of Asbestos Regulations.

However, if a task requires an employer to provide PPE under PPEW and one of the above specific requirements, the PPE items must be compatible. The Management of Health and Safety at

Work Regulations 1999 require employers to carry out a suitable and sufficient risk assessment to enable the most appropriate means of reducing the risks to acceptable levels. When determining the most suitable risk control measures, there is a risk control hierarchy. PPE is the final category and should not be used unless the risks to health and safety cannot be adequately controlled in any other way.

The Regulations place the following responsibilities on the employer, who must:

- assess the risk;
- select suitable PPE;
- provide the PPE;
- maintain and replace it as necessary;
- provide information, instruction and training on the PPE provided; and
- provide a system to allow employees to report defects or loss of PPE.

Employers are not allowed to charge for PPE. The Regulations require the employee to take reasonable care of the PPE provided, and, under the Health and Safety at Work etc. Act, the employee has a duty to use the PPE. Types of PPE are discussed below.

Head protection

Head protection is used, for example, where there are low level fixed objects such as pipes, machines or scaffolding, a risk of falling objects, and general construction. Head protection must fit the user and be adjustable to make sure it does not fall off. It should be kept clean and undamaged without deep scratches or fractures. After impact it should be replaced immediately and never be worn back to front or customised with stickers, marks or holes. Bump caps should be used for less hazardous activities. They do not protect from falling objects but are useful to prevent bumps against fixed objects, and stopping hair being tangled in moving parts of machinery.

Eye and face protection

This can reduce harm from liquid or chemical splash, from flying objects like chippings and debris while using power-driven tools, dust, gas or liquid mists from machinery or high pressure cleaning activities, and radiant heat or sparks while welding or using ovens and furnaces. Some special PPE can protect as well from intense light or optical radiation from welding and lasers.

Safety spectacles with lenses in a metal and plastic frame are practical as they can incorporate corrective lenses or fit over prescription glasses. Goggles are made of plastic and have an elastic headband; they protect the eyes from all angles, some have ventilation, and are not suitable against gas or fine dusts. Faceshields have one large lens mounted on a head harness or a helmet; they protect the face but do not completely enclose the eyes.

Hand and arm protection

This protects against cuts and abrasions from handling sharp objects such as metal sheets or knives and against cold / hot environments or materials. PPE can be used to keep hands warm in cold weather when operating machines, as vibration white finger occurs more often and more severely when the hands are cold. Some hand or arm protection can cause allergies or sensitisation, e.g. latex gloves, and in this case it may be better to choose another material such as nitrile.

Protective clothing

This is used mainly while working with chemicals or sprays and where machinery and knives can cause punctures and cuts. Their correct use and selection can prevent electric shocks and they can resist the build up of static electricity in potentially explosive atmospheres. They can protect the user from cold, hot or wet environments.

Foot protection

Foot protection prevents feet and toes from been crushed by falling objects. They can also protect from sharp objects or nails present on the ground and from slips, trips and falls. The use of an anti-slip sole can reduce the likelihood of slipping on certain floors. Safety footwear can be provided with thermal insulation or made with materials that will resist chemicals or molten metal splash.

Drowning protection

Life jackets and buoyancy aids should be worn where there is a foreseeable risk of drowning when working near water. They should be of the correct size and suitable for the water and weather conditions.

High visibility clothing

Most high visibility clothing has a fluorescent yellow or orange background with bands of shiny materials designed to make the wearer easy to see under any light conditions. It is normally used where there is a risk of being hit by a moving vehicle or moving object.

Fall arrest equipment

When carrying out work at height, the Work at Height Regulations 2005 apply in addition to the PPEW Regulations. PPE for this class of work include fall arrest equipment such as safety harnesses and energy-absorbing lanyards.

Hearing protection

A full assessment must be carried out to ensure the correct selection of hearing protection. Earplugs fit into the ear canal to form a seal; they can be permanent or indefinite use, reusable only a few times or disposable to be used only once. Earmuffs are normally provided where there are major exposures to noise. They normally have hard plastic cups that will fit over and surround the ears and are sealed to the head by cushions.

Respiratory protection

This can be a tight-fitting facepiece that can be in the form of a full face mask, a half face mask or a filtering facepiece. If harmful dusts are present in the workplace, nuisance dust masks are not suitable as they are not classified as PPE. A suitable dust respirator should be provided and will normally have two head straps and will comply with Regulations. Respirators rely on filtering contaminants from the workplace air, and include facepieces and respirators and power-assisted respirators.

See also: Construction site health and safety, p.177; Head protection, p.355; Working at height, p.399; Work equipment, p.752.

Sources of further information

L25 Personal Protective Equipment at Work: www.hse.gov.uk/pubns/priced/l25.pdf

INDG 174 A Short Guide to the Personal Protective Equipment at Work Regulations: www.hse.gov.uk/pubns/indg174.pdf

INDG 367 Inspecting fall arrest equipment made from webbing or rope: www.hse.gov.uk/pubns/indg367.pdf

Pest control

David Cross, Rentokil

Key points

- Put simply, pest management or control is the destruction or prevention of unwanted pests.
- Many differing techniques are used, but the basic principles are common to most situations. These involve environmental management to exclude pests from sites, restrict access to food, water and harbourage, and, as a last resort, physical or chemical control.

Legislation

- Protection of Animals Act 1911.
- Public Health Act 1936 and 1961.
- Prevention of Damage by Pests Act 1949.
- Health and Safety at Work etc. Act 1974.
- Wildlife and Countryside Act 1981.
- Food and Environmental Protection Act 1985.
- Control of Pesticide Regulations 1986.
- Food Safety Act 1990.
- Wild Mammals Protection Act 1996.
- Biocidal Products Directive (98/8/EC).
- Control of Substances Hazardous to Health (COSHH) Regulations 2002.
- Food Hygiene (England) Regulations 2006 (there are similar regulations in Wales, Scotland and Northern Ireland).

What is a pest?

A pest is any animal that is found in the wrong place at the wrong time and whose presence could result in damage, contamination and/or transmission of disease. The most common pest species include those shown in the table below.

Common pest species	
Commensal Rodents Birds	Brown rat, black rat, house mouse. Feral pigeon, woodpigeon, Collard dove, jackdaw, jay, magpie, carrion / hooded crow, rook, lesser black-backed gull, Canada goose, Monk and Ring Necked parakeet. (Variations to the list of birds which may be taken under general licence are found in England, Scotland, Wales and Northern Ireland.)
Textile pests	Varied carpet beetle, fur beetle, common clothes moth, case-bearing clothes moth, and many more.
Stored product Insects	Grain weevil and beetle, flour beetle, book lice, warehouse moth, Mediterranean mill moth.
Public health Insects Vertebrate pests	Flies, fleas, cockroaches, bed bugs, wasps, ants. Rabbit, grey squirrel, mole, feral cat, fox and mink.

> **Facts**
> - Results from the national rodent survey (2008/09), published in 2010, suggest that the number of professional rat control treatments reported by UK local authorities fell slightly in 2008/09 compared to 2007/08, to around 465,000.
> - After the Second World War, bed bug populations appeared to decline to a point where infestation by them was rare. In the past ten years, however, there has been a significant resurgence in bed bug incidences to the point where they have, once again, become a significant pest of homes, hotels, hostels and care facilities.

Why do pests need to be controlled?

Financial reasons

The financial impact of pest activity can take many forms, from the contamination of food, rendering it unfit for retail sale, the direct consumption of stored food by insects, rodents, birds or other pests, damage to buildings by rodents gnawing through electric cables, water pipes or structural timbers, or by birds blocking ventilation pipes and guttering with nesting materials.

Damage to parks, gardens and recreational facilities such as sports fields and golf courses by the burrowing behaviour of rabbits and moles can also be significant, as can the effects on trade and reputation when pests become established in a commercial premises.

Health and disease

Most pests have the ability to carry or transmit disease, be responsible for allergic reactions, bite or sting.

Nuisance

In some cases, neither health nor finance are put at risk, but the pest becomes a nuisance because of a phobia, excessive noise caused by the persistent call of birds, or rendering a building unsightly with droppings.

Types of service available

- Contracted pest control services monitoring entire sites on a regular basis against specific pests, to ensure that the site remains pest-free.
- Employment of a pest control company to eradicate a localised pest problem such as wasps' nests, rodent or bird infestations.
- Training courses to help create 'awareness' amongst workplace staff of the problems that pests can cause, to recognise the signs and potential for infestation.
- Preventive proofing to buildings to deter pests from entering or roosting on sites.
- Supply of insect control devices such as fly killers.
- Supply of insect-monitoring devices, such as moth pheromone pots and flea traps, which are used to establish the presence and scale of an infestation.
- Fumigation of containers, commodities and entire buildings to eradicate deep-seated infestations.
- Use of controlled atmospheres to control insect pests in delicate items by reducing oxygen levels or replacing free air with Nitrogen.
- Localised heat treatments for the control of insects such as bed bugs in bedroom furniture.

Case study

In April 2010 Kentucky Fried Chicken (KFC) admitted breaching hygiene rules at a branch in Leicester Square. KFC admitted having inadequate pest control at the branch, failing to provide adequate handwashing facilities, and failing to keep the restaurant clean and in good order. The company was fined £11,000 for the breaches. Environmental health inspectors from City of Westminster Council found cockroaches, mice and flies during an inspection of the branch in 2008.

KFC spokeswoman, Nina Arnott, commented: "These charges date back to August 2008 and, as soon as we were made aware of the results of the inspection, we took immediate action to bring the restaurant back up to our strict hygiene standards."

She added that a new manager had been installed and the restaurant had performed well in recent inspections by environmental health officers.

Categories of legislation

Legislation covering pest control can be split into three categories.

Species protection legislation

Protection of Animals Act 1911

This Act provides general protection for domestic and captive animals and makes it an offence to do or omit to do anything likely to cause unnecessary suffering. This would include not providing food and water to animals confined in a cage or live capture traps.

Amendments to this Act include the prohibition of the use of poisons on any land or building except for the purpose of destroying insects, rats, mice and other small ground vermin.

Wildlife and Countryside Act 1981

This Act provides protection for certain animals and the environment. From a pest control point of view, the Act covers most of the legislation affecting birds, and lists birds that may be taken by authorised persons (see Table.) An authorised person

is the owner or occupier of the land on which the birds are to be controlled, or any person authorised by the owner or occupier.

Birds from this list may be controlled using live capture traps or by shooting (excluding weapons with a muzzle diameter greater than 1.25 inches) under a general licence which is updated and issued by the Department for Environment, Food and Rural Affairs (DEFRA). Certain species of bird from this list may also be controlled using stupefying baits under a specific, one-off licence, which is applied for and issued under the discretion of DEFRA's Wildlife Administration Unit.

Birds can only be controlled in order to:

- prevent spread of disease;
- protect public health and safety; and
- protect crops, livestock, forestry, fisheries and water.

Wild Mammals Protection Act 1996

This Act is designed to protect wild mammals against abuse and cruel treatment. It makes it an offence to mutilate, kick, beat, impale, stab, burn, stone, crush, drown, drag or asphyxiate any wild mammal with intent to inflict unnecessary suffering.

Environmental health legislation

Prevention of Damage by Pests Act 1949

This Act requires, as far as is practicable, that districts are kept free from rats and mice. It requires Local Authorities to carry out periodic inspections, to destroy rats and mice on their land, and to enforce the same duties on to owners and occupiers.

Public Health Act 1936 and 1961

This Act gives Local Authorities powers in relation to the removal of materials or rubbish from an area / premises, disconnection, sealing and removal of drains water pipes and sewers where they have been disconnected, taking necessary action for destroying or removing vermin and taking steps to deal with the nuisance or damage caused in built-up areas by pigeons.

Food Safety Act 1990

This Act makes it an offence to sell food for human consumption that fails to comply with food safety requirements. This would be the case if the food was contaminated with droppings or pest bodies.

The defence under the Food Safety Act is due diligence, which means that everything reasonably practicable must have been done to avoid contamination.

Food Hygiene (England) Regulations 2006 (similar Regulations exist in Wales, Scotland and Northern Ireland)

New EU food hygiene Regulations have consolidated 17 EU measures in the food hygiene area into just two; the Food Hygiene (England) Regulations 2006 provide the framework for these measures to be enforced. This legislation

is structured so that it can be applied flexibly in all food businesses, regardless of their type and size. It is now required that microbiological hazards along with other foreign body contamination should be considered as part of a Hazard Analysis Critical Control Point (HACCP) system. The control and prevention of pests should be a prerequisite of any HACCP system because of the risk of them carrying and spreading microbiological hazards.

The Food Safety (General Food Hygiene) Regulations 1995 and the Food Safety (Temperature Control) Regulations 1995 have both been superseded by these Regulations.

Legislation on pesticide use

The following legislation covers the duties placed on pesticide users and their employers.

Food and Environment Protection Act 1985

Part III of this Act is of direct concern to pest control as it provides for the making of Regulations concerned with the control of pesticides with a view to protecting the health of human beings, creatures and plants, safeguarding the environment and securing safe, efficient and humane methods of controlling pests.

Control of Pesticide Regulations 1986

These Regulations were introduced under the Food and Environmental Protection Act 1985 and stipulate that only approved pesticides may be advertised, supplied, stored or used in the UK and that only those with provisional or full approval may be sold.

Health and Safety at Work etc. Act 1974

This Act provides a comprehensive system of law covering the health and safety of people at work and members of the public who may be affected by activity at work.

Biocidal Products Directive (98/8/EC)

The Directive is intended to harmonise arrangements for the authorisation of pesticides used in public health pest control, wood preservatives, industrial preservatives, disinfectants and certain germicidal chemicals.

The Directive was implemented in March 2000, but full harmonisation will take several years to achieve.

Control of Substances Hazardous to Health (COSHH) Regulations 2002

These Regulations were originally introduced under the Health and Safety at Work etc. Act 1974 and make it necessary to assess the risk to health arising from work involving a hazardous substance and, if a risk is identified, to determine what precautions are required either to eliminate or to reduce the risk. This may involve the use of personal protective equipment for people working with the substance, or the exclusion of personnel from an area while a substance is being applied.

All pesticides should be considered as being potentially hazardous to health and assessed accordingly using information from the Manufacturers Material Safety Data Sheet and Manufacturers Label Instructions.

Recent and forthcoming developments

There are some major developments that will affect, or will be likely to affect, the pest control industry:

- A new General Licence (L-35) was introduced in England in January 2010 that allows owners and managers of food premises (and any person they authorise to act on their behalf) to catch (for the purpose of preserving public health or public safety and release unharmed) Robins, Blackbirds, House Sparrows and Starlings that have become trapped in the building.

- Herring Gulls and Greater Black-backed Gulls have been removed from the general licence for the preservation of human health and safety due to their numbers in natural habitats having declined in recent years. Control of these species now requires a specific licence acquired from Natural England, unless it is done by the removal or oiling of eggs (Herring Gulls only). Monk and Ring Necked Parakeets have now been added to the General Licence, and can be controlled with permission from an authorised person (refer to the information on the Wildlife and Countryside Act 1981 above).

- Sulfuryl Fluoride is available for use as a fumigant gas in empty flourmills, food and feed manufacturing units and stores in the UK and for the control of wood-destroying pests such as common furniture beetle and powder post beetle.

- The use of phosphine gas for the control of rabbits, rats and moles in their burrows is being tightly controlled under a new product stewardship scheme RAMPS (UK). Phosphine-releasing products may only be purchased by people who hold a qualification in the safe use, storage and disposal of the product. The new qualification is the Level 2 Award in The Application of Aluminium Phosphide for Vertebrates Control and is available from City and Guilds following training by a RAMPS approved trainer. The requirement to hold this qualification comes into force fully in 2015 but people are encouraged to begin training and assessment prior to this.

- Regulations for the disposal of hazardous waste came into force on 16 July 2005. These Regulations change the way in which waste can be categorised and disposed of. This is relevant to the pest control industry in terms of the disposal of pesticides, dead pests and their droppings, particularly where the latter have accumulated in significant quantities. For further details consult www.environment-agency.gov.uk.

- The WEEE Regulations came fully into force during July 2007. Additional costs will be associated with the disposal / recycling of fluorescent tubes and waste electrical goods. This is most relevant to the pest control industry in terms of the disposal of spent UV tubes from electronic fly control units. See 'Waste Electrical and Electronic Equipment: the WEEE and RoHS Regulations' (p.737) for more details.

- The Work at Height Regulations 2005 apply to all work at heights where there is a risk of a fall liable to cause personal injury. The Regulations place duties on employers, the self-employed and any person controlling the work of others. This is most significant to the pest control industry when applying anti-roosting and perching devices to building ledges to deter nuisance birds. The Regulations came into force on 6 April 2005; guidance notes are available from www.hse.gov.uk. See 'Working at height' (p.399) for more details.

- Changes are being made to the way that exposure to a variety of chemicals is measured. Previously, two principal measures of exposure were the Occupational Exposure Standard (OES) and the Maximum Exposure Limit (MEL), both of which have been replaced by a new measure, the Workplace Exposure Limit (WEL). Ultimately all documentation, including Material Safety Data Sheets (MSDSs) will be changed to show WELs rather than OESs and MELs. The HSE does not consider that this change merits the destruction and replacement of MSDSs for pesticide use and the change will be gradual.

- Pest control providers are being encouraged to demonstrate continuous professional development to show that individuals are kept up to date with developments and changing practices within the pest control industry. Membership of BASIS PROMPT (the Professional Register of Managers and Pest Technicians) or similar schemes will be a mandatory part of membership of the British Pest Control Association from the end of 2012.

> *See also*: Catering: health and safety issues, p.123; Working at height, p.399; Hazardous waste, p.700; Waste Electrical and Electronic Equipment: the WEEE and RoHS Regulations, p.737.

Sources of further information

Environment Agency: www.environment-agency.gov.uk

The British Pest Control Association: www.bpca.org.uk

National Pest Technicians Association: www.npta.org.uk

RAMPS (UK): www.ramps-uk.org

Planning procedures

Helen Nicholson and Jonathan Riley, Pinsent Masons Property Group

Key points

- Planning permission (subject to a number of exceptions) is required for material changes of use of land and operational development.
- An application for planning permission will be made to the Local Planning Authority (LPA).
- Failure to secure planning permission, or comply with planning conditions, may entitle the LPA to take enforcement action.
- There is a right of appeal to the Planning Inspectorate.

Legislation

- Town and Country Planning Act 1990 (as amended).
- Town and Country Planning (General Permitted Development) Order 1995.

When is permission required?

Planning permission is required for development, which means:

- operational development (which includes building, engineering, mining or other operations); or
- material changes in the use of any buildings or other land.

Operational development

Building operations include the demolition of buildings, the rebuilding of buildings, structural alterations of, or additions to, buildings and any other operations normally undertaken by a person carrying on business as a builder.

A building includes any structure or erection, and any part of the building, structure or erection, but does not include plant or machinery comprised in a building.

Not all structures are considered by the law to be buildings. The three primary factors are size, degree of permanence and physical attachment.

Material changes of use

In order to assess whether planning permission is required for a change of use, it is necessary to look at the primary use of a piece of land or building and the extent of any ancillary uses. Ancillary uses do not require planning permission. For example, the primary use of a building may be for office purposes, but there may be ancillary storage uses within that building that would not require planning permission. In addition, a slight or trivial change of use will not require planning permission. For example, a small amount of storage use in an office building would not require planning permission. In 2009, the Government introduced a new and streamlines application process for making non-material amendments to existing planning permissions without the need to apply for planning permission.

Demolition

Although demolition constitutes development, currently planning permission is only required for demolition of dwelling houses and adjoining buildings. Buildings in a conservation area, listed buildings and scheduled ancient monuments are subject to separate controls under other legislation and demolition of such buildings will require consent.

Exceptions

Legislation provides that certain operations or material changes of use are exempted from the need for planning permission. For example, internal or external improvements, alterations or maintenance work, none of which materially affect the external appearance of buildings, may not require planning permission. Some of these concessions are removed in respect of certain sensitive areas, such as conservation areas or national parks. Also, the Government has introduced a requirement for planning permission for mezzanine floors over a certain size threshold in retail premises.

The Town and Country Planning (General Permitted Development) Order 1995 grants an automatic planning permission for certain operational development and material changes of use, including certain industrial and warehouse developments, subject to a series of complicated conditions and restrictions. Government directions and planning conditions can withdraw these rights.

Securing a grant of planning permission

If a particular proposal requires planning permission, a formal planning application needs to be made to the relevant LPA. If the proposal is acceptable in planning terms, planning permission should be granted either conditionally or unconditionally. Usually the permission will automatically run with the land. A person who buys a piece of land which has the benefit of a planning permission can implement that planning permission and build in accordance with any approved plans. It may be necessary to secure the copyright in any approved plans.

Only two types of planning permission can be obtained:

1. Detailed planning permission.
2. Outline planning permission, which establishes the principle of development. Any such permission would include a condition requiring reserved matters (e.g. layout, scale, appearance, access and landscaping) to be submitted to the LPA within three years of the date of the outline permission.

Environmental impact assessment

In respect of certain types of development or activity, an environmental impact assessment will need to be submitted. The assessment is required to ensure that the effect of development on the environment, in particular of certain specified public and private projects, is taken into account as part of the decision-making process and before permission is granted. Following recent European case law, the need for environmental assessment must be considered both on the initial outline planning application and any subsequent reserved matters application.

Keeping permissions alive

All permissions are subject to strict time limits. If development is not begun within those time limits, the permission will lapse. In the case of a detailed permission, the usual time limit condition requires development to be started within three years from the date of the detailed permission. In the case of an outline planning permission, the usual time limit condition requires development to start no later than three years from the date of the outline planning permission or, if later, two years from the final approval of reserved matters. In October 2009, the Government introduced a mechanism to enable planning permissions that are due to lapse to be kept alive by an application procedure to extend the time within which the development must have been commenced.

Breach of planning control

A breach of planning control can take one of three forms:

- The carrying out of operational development without the benefit of planning permission.
- The carrying out of a material change of use of any building or land without the benefit of planning permission.
- The breach of a condition attached to a permission.

LPAs have wide powers of enforcement (including criminal sanctions), should development be carried out without any necessary planning permission or failing to comply with planning conditions.

Listed buildings and conservation areas

Buildings may be 'listed' as being of special historic or architectural interest. Any works that affect the character of a listed building require a listed building consent. Non-compliance with the listed building legislation can lead to enforcement action and criminal sanctions. LPAs must determine, after consultation, whether any part of their area should be designated as a conservation area. Such designation has the following consequences:

- Building design has to be of a high quality;
- Restrictions are placed on demolition; and
- Trees are protected.

Planning appeals

If the LPA refuses to grant planning permission, listed building consent or the modification of a planning condition, the applicant has the right to appeal to the Planning Inspectorate. An appeal can be pursued through written representations, a hearing or a public inquiry.

The appeal is determined by a planning inspector appointed by the Secretary of State for the Department of Communities and Local Government. He will be independent from the Local Planning Authority. The process may take many months or even years. In respect of certain more complex sites, the decision is made by the Secretary of State following receipt of a report by an appointed planning inspector.

Note that an LPA's decision to grant planning permission is open to challenge by means of an application for judicial review within three months calculated from the date of the Decision Notice. The judicial review procedure is commonly used where a third party seeks to quash the decision of the LPA. Only certain types of person have the right to make such an application, and only if there are grounds for a challenge.

> *See also*: Building Regulations, p.76; Property disputes, p.600.

Sources of further information

The Workplace Law website has been one of the UK's leading legal information sites since its launch in 2002. As well as providing free news and forums, our Information Centre provides you with a 'one-stop shop' where you will find all you need to know to manage your workplace and fulfil your legal obligations. Visit www.workplacelaw.net for more information.

Portable Appliance Testing (PAT)

Paul Caddick, PHS Compliance

Key points

- Electrical safety forms part of the obligation of any duty-holder (an employer or director, possibly a premises / facilities / health and safety manager) to provide a duty of care to those using or visiting the site they are responsible for. As the risk posed by electricity can be considerable – both by electrocution and fire – this obligation must be considered as a priority and diligence carefully evidenced. There is legislation in place that must be complied with, supported by British Standards and Approved Codes of Practice (ACoPs) guiding appropriate practice. Ultimately the duty-holder must ensure that electrical equipment used at work must be maintained in a safe condition, so far as is reasonably practicable.
- Safety of portable appliances may be achieved by firstly ensuring appropriate and safe usage, paying particular attention to the environment, the type of equipment, the method and frequency of its use. Regular visual inspections should be conducted and an appropriate regime for electrical testing implemented, to be carried out by a 'competent person' with suitable and regularly calibrated test equipment. All such inspections and tests should be documented.

Legislation

- Health and Safety at Work etc. Act 1974.
- Electricity at Work Regulations 1989.

The foundation for legislation concerning electrical safety is the Health and Safety at Work etc. Act 1974 (Section 2(2)(a)), which states that 'electrical equipment used at work must be maintained in a safe condition, so far as is reasonably practicable'.

This is supported by the Electricity at Work Regulations 1989 (Regulation 4), which states the requirement that 'Regular inspection and testing is necessary for all electrical installations and all equipment connected to the installation where this is necessary to prevent danger'.

Naturally, we have to ask what 'all reasonable and practicable steps' might

be, and to interpret this the electrical industry takes its guidance from the British Standard BS 7671: 2008, 17th Edition, which makes it clear that portable appliance testing must be conducted.

Why test and what to test

The fundamental fact is that employers should be keeping everyone on site safe, as far as is reasonably practicable. This is just as true in a school as on a building site, in an insurance office or at a petrol station forecourt. The level of risk at different sites may vary, so the recommended frequency of test for portable appliances may vary, but all organisations – whether private or public – are included in the requirement to prove diligence.

Electricity presents a hazard in the workplace and, as with all hazards, management of it should be dealt with

Facts

- Electricity can kill. Each year about 1,000 accidents at work involving electric shock or burns are reported to the HSE. Around 30 of these are fatal.
- Those using electricity may not be the only ones at risk: poor electrical installations and faulty electrical appliances can lead to fires that may also cause death or injury to others.
- The hazards:
    - Contact with live parts causing shock and burns (normal mains voltage, 230 volts AC, can kill);
    - Faults that could cause fires; and
    - Fire or explosion where electricity could be the source of ignition in a potentially flammable or explosive atmosphere.
- According to Norwich Union, electricity is the second largest cause of fires in commercial and industrial premises in the UK.
- PAT relates to all appliances that can be attached to the electrical supply via a plug and socket arrangement.

by first carrying out a risk assessment in order to identify what needs to be done:

- Identify the hazards;
- Decide who might be harmed and how;
- Evaluate the risks arising from the hazards and decide whether existing precautions are adequate or more should be taken; and
- If you have more than five employees, record any significant findings.

The HSE states that the risk of injury from electricity is strongly linked to where and how it is used. It makes it clear that some items of equipment can involve greater risk than others, and names extension leads as items particularly liable to damage.

PAT includes not only truly portable equipment such as an office fan or a drill; the scope is wider. It includes everything that is attached to the electrical wiring system via a 13A plug. A drinks machine, for example, is classed as a 'portable appliance', as is a freezer cabinet, a cooker extraction unit, a kettle or a

photocopier. The HSE defines a portable appliance as equipment that has a lead (cable) and plug and that is 'normally moved around' or 'can easily be moved' or equipment that 'could be moved'. The groupings used by the IET (the Institute of Engineering and Technology) include 'stationary', 'IT', 'moveable', 'portable' and 'handheld' – all are dealt with by PAT.

This breadth means that PAT usually concerns large numbers of individual tests (on many sites it can run to thousands) from Christmas tree lights to air conditioning units; all have a potential to introduce danger. In an office, for example, it's typical for each and every worker to have approximately seven portable appliances around them; a computer for instance, can mean three or four tests, and many people mistakenly count this as just one. Given the large number of items that can be associated to every worker it's perhaps not surprising that many organisations exclude so-called 'personal' items. However, these appliances pose exactly the same potential risk in terms of electrocution or fire as any other, so if they're connected on your site it's certainly questionable to exclude them.

Faulty, deteriorating or damaged

A common mistake is the assumption that brand new equipment is automatically safe and unnecessary to test. Electrical manufacturing is not without error and even when an item has just come out of the box it can still be dangerous. Product recalls are one source of evidence for this and statistics relating to fire are another.

Not only may equipment be faulty due to manufacturing defect, it is liable to damage. Cuts or abrasion can occur to the cable covering during movement of equipment or surrounding items. The plug may become damaged, the casing cracked, or pins bent. Where the cable enters the plug the sheath should be securely gripped and if this is loosened or damaged the individual wires may be subject to pulling forces that could cause detachment from internal pins. This could, for example, mean that the item was no longer earthed and potentially its casing could become live (for example, the metal casing of a washing machine).

The location of portable appliances can also cause damage, for example if placed too close to a heat source or in damp or wet conditions. Common examples include extension leads placed on top of radiators or cables on the floor (perhaps under the equipment or surrounding items) being submersed in spilt liquid. Heat, dust, vibration and damp or wet environments may cause damage to appliances.

Deterioration occurs to all items over time and wear may impact the safe operation of electrical appliances. The rate of wear can be directly linked to the frequency of movement, the type and frequency of usage. Flexes may be coiled or bent and hence become worn, and connections

can become loosened by movement such as regular plug insertion / removal from sockets.

In addition to damage and deterioration, users can influence or interfere with the safety of appliances. The way in which the portable appliance is used and maintained should, of course, be safe and appropriate. Care should be taken when moving appliances and consideration made for where they are situated. Fuse replacement, when necessary, should be with a correctly rated match, not a piece of wire or other metal conductor. This commonly-found practice can easily result in a potentially dangerous situation, allowing overheating to occur within the appliance and a fire to result.

Users have the potential to negatively impact the safety of portable appliances, but also may play a key role in safe operation. Regular visual inspection may form an integral part of any safety regime and many of the potential faults and risks identified can be found by looking critically at the appliance. Staff should therefore be encouraged to undertake a full visual inspection regularly, in some cases every time an appliance is used.

If users consider something to be wrong or potentially unsafe – for example if a flex appears damaged, wiring loose, or the equipment fails to operate normally – then they should call for support and stop using the item immediately. The matter should be reported to a supervisor or manager and the equipment should be labelled as faulty and taken out of use – the plug should be taken off to prevent further use and, if possible, the item removed and securely stored until repaired. This 'fault reporting system' should be documented, for example in the health and safety section of the employee handbook.

Testing regime

Electrical equipment and installations should be maintained to prevent danger. The HSE strongly recommends that this includes an appropriate system of visual inspection and, where necessary, testing. Records of the results of inspection and testing should be kept. Plug-top stickers are commonly used by engineers to mark individual items, allowing easy identification to evidence the satisfactory test, including test date and next recommended test date. The sticker will also provide traceability of the engineer's work, detailing who has conducted the test.

What must be proven is compliance with BS 7671 (17th Edition) and this should include a system of recorded inspections and tests, undertaken by a 'competent person'. This is defined as 'a person who possesses sufficient technical knowledge, relevant practical skills and experience for the nature of the electrical work undertaken and is able at all times to prevent danger and, where appropriate, injury to him/herself and others'.

In practice this means using a professional affiliated to an approved body such as the NICEIC (National Inspection Council for Electrical Installation Contracting – the electrical contracting industry's independent voluntary body for electrical installation matters throughout the UK) and/or an individual qualified to C&G 2377.

If you wish to retain a 'competent person' on staff then engineers can become qualified and C&G accredited at various technical colleges or specific training companies. They will need to be equipped with a regularly calibrated tester that can measure and report on earth continuity, insulation resistance and touch current. The cost of in-house expertise and equipment means that most organisations outsource their testing requirement to a suitable contractor. There is a large and unregulated market for such work and so testing buyers should take care to check credentials, including:

- Test engineers are qualified to C&G 2377;
- The supplier adequately monitors and can evidence the compliance of test engineers on site;
- Minor repairs will be included in the per-test price (plug top or fuse replacement, lead re-termination);
- Reporting is comprehensive, validated and swift / simple to access (for example, web based reporting); and
- Public liability insurance is in place to underwrite their work (recommended £10m level).

As already outlined, users of a site have their part to play in ensuring electrical safety. Induction training for all employees, for example, should include an instruction to carry out regular and full visual inspection of electrical appliances used. When plugging in portable appliances, users should be instructed to use outlets (sockets) that are close by, so that in an emergency disconnection is rapid. Overloading sockets should be avoided and extension leads and multi-socket adapters used with due care for safe loading and safe situation. This could be considered common sense but for the employer it's important to build a culture and systems that create robust safeguards. One of the most confusing areas for PAT concerns the frequency of test, as there is no statutory inspection period laid down. Frequency of test relates to the potential level of risk and therefore should reflect the environment, the specific situation that the equipment is being used in, and the type of equipment in use. The recommended frequency ranges from three- to 48-month intervals, depending upon classified environment and usage.

Equipment	Frequency
Construction sites 110V equipment	All equipment: every three months
Industrial including commercial kitchens	S, IT and M: every 12 months P and H: every six months
Equipment used by the public	S and IT: every 12 months M, P and H Class 1: every six months M, P and H Class 2: every 12 months
Schools	All Class 1 equipment: every 12 months All Class 2 equipment: every 48 months
Hotels	Class 1 equipment: S and IT: every 48 months M and P: every 24 months H: every 12 months
Offices and shops	Class 1 equipment: S and IT: every 48 months M and P: every 24 months H: every 12 months
Key: S = stationary equipment, e.g. vending machine IT = IT equipment, e.g. computer M = moveable equipment, e.g. extension lead P = portable equipment, e.g. fan H = handheld equipment, e.g. drill	

Table 1. Portable appliance test (PAT) frequencies.

Employers should regularly review their PAT programme and ensure that inspections and tests are being conducted at appropriate intervals – a suitable guideline to follow is that provided by the IET (see Table 1). For further information, refer to the IEE Code of Practice for In-service Inspection and Testing of Electrical Equipment, 3rd Edition (see Sources of further information). Note: The recommended frequency for PAT testing depends on the environment and the type of equipment. Recommendations for combined inspection and testing frequencies are summarised here. Full details are available in the IEE Code of Practice for In-service Inspection and Testing of Electrical Equipment, 3rd Edition.

Assuring insurance
In addition to the requirement for an employer to comply with legislation, there is the consideration of insurance. A fairly typical insurance policy condition requires the policy-holder to 'take all reasonable precautions to prevent or diminish loss, destruction or damage occurrence or cease any activity which may give rise to liability'. This very much echoes current legislation and, with an emphasis on

prevention, it's crucial for policy-holders to take a proactive approach to electrical safety. In the event of an accident or incident resulting in a claim, insurers would 'consider all material facts pertaining to the loss, including any statutory obligation of the business'.

Insurers will seek to examine electrical inspection records to ensure a regular and systematic maintenance regime is in place, in order to satisfy themselves that electrical equipment used has been maintained in a safe condition (in line with legislative requirements).

Conclusion

In summary, an appropriate regime for portable appliance inspection and testing should rightly be considered a 'must-have' part of the safety schedule for legislative compliance, and forms part of the duty of care required by any duty-holder. The approach to fulfilling this duty of care should begin with assessment of the potential risk to identify and evaluate hazards and establish who may be harmed, in order to establish adequate precautions. This risk assessment should be reviewed periodically and revised as necessary.

Safety of portable appliances may be achieved by firstly ensuring appropriate and safe usage, conducting regular and documented visual inspections, and then, where necessary, a 'competent person' should undertake testing using a suitably calibrated electrical tester to measure and record earth continuity, insulation resistance and touch current.

PAT forms part of a duty-holder's compliance with safety legislation concerning electricity. It should be noted that fixed wire installations should also be subject to inspection and testing in order to meet the relevant legislation, in line with the appropriate British Standard. As the HSE states:

'Installations which conform to the standards laid down in BS 7671: 2008 are regarded by the HSE as likely to achieve conformity with the relevant parts of the Electricity at Work Regulations 1989.'

See also: Display Screen Equipment, p.244; Portable Appliance Testing, p.588; Risk assessments, p.624; Work equipment, p.752.

Sources of further information

Institute of Engineering and Technology: www.theiet.org

National Inspection Council for Electrical Installation Contracting: www.niceic.com

Electrical Trade Association: www.eca.co.uk

INDG 236 Maintaining portable electric equipment in offices and other low-risk environments: www.hse.gov.uk/pubns/indg236.pdf

IEE: *Code of Practice for In-service Inspection and Testing of Electrical Equipment*, 3rd Edition (2008). ISBN: 978 0 86341 833 4.

Power lines

Anna Cartledge, Pinsent Masons Property Group

Key points

- Power companies can negotiate rights to install power lines and equipment with landowners.
- Power companies have compulsory purchase powers whereby the company can make an Order in respect of the land and rights needed for a power line, which will then need to be confirmed by the Secretary of State.
- The power companies can also apply to the Secretary of State for a necessary wayleave in respect of an existing or new power line.
- In either case, compensation will be payable to the landowner.
- Power companies have the right to fell or lop trees if they present a danger to power lines.

Legislation

- Electricity Act 1989.
- Town and Country Planning Act 1990 (as amended).
- Utilities Act 2000.

Overview

This chapter concentrates on who can install power lines, what agreements they enter into and how they can compulsorily acquire rights where landowners refuse to grant them voluntarily.

Who can install power lines?

The generation, supply and transmission of electricity are activities regulated by Section 6 of the Electricity Act 1989 as substituted by Section 30 of the Utilities Act 2000. No person (which includes companies) may carry out these activities without a licence issued by the Secretary of State, which may have conditions attached to them. It is only such licence-holders that are permitted to install electricity lines on or above land.

However, before installing any overhead power lines, licence-holders must also obtain the Secretary of State's consent

pursuant to Section 37 of the Electricity Act 1989 and this consent may again be subject to any conditions the Secretary of State considers appropriate. The most common licence-holders are the regional electricity companies, which took over the supply of electricity from the area boards following the changes introduced by the Electricity Act 1989. These include, for example, EDF energy, Powergen UK, Npower and other such well-known companies.

Agreements relating to power lines

Before a licence-holder can install a power line it also needs the consent of all landowners whose land will be subject to the installation of the power line. The most common forms of agreement giving effect to the licence-holder's right to install a power line are as follows.

Easements

Easements are rights in land that can be granted permanently or for a set number of years. An easement is a legal right attached to the land. Therefore it binds any future owner of the land upon which the power line has been installed and will benefit any successor to the licence-holder who has installed the power line. Where an easement is granted it will give the licence-holder the right to install the line and to retain it permanently and will generally include such ancillary rights as are reasonably necessary for the exercise or enjoyment of the rights granted (e.g. for maintenance, etc.).

Wayleaves

Wayleaves are rights over property granted by a landowner for payment. They are very common, especially where licence-holders require rights over residential properties. They are of a less permanent nature than easements and are personal to the parties to the agreement. It is important to note, however, that wayleaves can only be terminated in one of three ways; namely when the period of the wayleave expires; or if the landowner gives notice under the termination provisions of the wayleave; or if the ownership of the land changes and the wayleave ceases to be binding on the new landowner. Even in these circumstances, notice must be properly served on the licence-holder and the right to keep the power lines on the land can be obtained compulsorily (see below).

Following the proper service of a notice, if the licence-holder does not make an application to compulsorily acquire the right or does not negotiate a new wayleave, it must remove the power lines within three months of the service of the notice.

Compulsory powers

Where a licence-holder requires power lines and the consent of the landowner cannot be obtained, the Electricity Act 1989 provides a mechanism for the compulsory acquisition, on payment of compensation, of the necessary rights. The Electricity Act 1989 provides two routes for the acquisition of such rights.

Compulsory purchase

Schedule 3 to the Electricity Act 1989 permits the Secretary of State to authorise the compulsory acquisition of land or rights over land by a licence-holder where that land is required for any purpose connected with the activities it is authorised to carry out. This compulsory acquisition of rights extends to both the creation of new rights over land and the acquisition of any existing rights over land (as well as the compulsory purchase of the land itself). The acquisition of rights would include, for example, an easement for an underground power line. Before making an order for compulsory purchase, any interested party is afforded the opportunity to make objections. This can result in a public inquiry being held to determine whether the order will be beneficial.

Necessary wayleaves

Paragraph Six of Schedule 4 to the Electricity Act 1989 allows a licence-holder to apply to the Secretary of State for a necessary wayleave where:

- the licence-holder has given the landowner 21 days' notice to grant the wayleave and the landowner has failed to do so;
- the licence-holder has given the landowner 21 days' notice to grant the wayleave and the landowner has made it subject to conditions to which the licence-holder objects; or
- the landowner has given notice to the licence-holder requiring the licence holder to remove an existing power line in place under a wayleave.

Before making any order in relation to a necessary wayleave, the Secretary of State must give both parties the opportunity of stating their case.

Compensation

Whether land (including rights) is acquired compulsorily or by agreement, the licence holder must pay compensation to the landowner for the rights it has acquired in the land. The compensation, based on the provisions of the Land Compensation Act 1961, should take into account:

- the value of land taken, applying normal market rules;
- injurious effect on land such as dangers associated with electromagnetic fields;
- the visual impact of the power lines;
- loss of development value;
- loss of use of the land and the effect on quiet enjoyment of it; and
- the landowner's legal and surveyor's costs.

The rule-of-thumb guide is that the landowner should be in no worse a position than before the acquisition of the land or right. The compensation may be payable as a lump sum or as a periodic payment.

Other rights for licence-holders

The Electricity Act 1989 provides other rights for licence-holders, including the right to fell or lop trees causing or likely to cause an unreasonable source of danger due to their proximity to the power lines and the right to enter property to carry out such felling or lopping. Licence-holders also have the right to disconnect other services to property if required to allow maintenance works, together with the right to break up streets (private or public)

in order to repair and maintain electricity cables and power lines.

Practical issues

Unless the grounds for objection are very strong, it is sensible to deal with the licence-holder's request by way of agreement, although the compulsory powers will always be in the background should negotiations break down.

Aside from the property issues, appropriate indemnities should be sought from licence-holders to cover eventualities such as the death or injury caused to any person on the land and any potential nuisance caused by the existence of the power lines. Landowners should also ensure that the licence-holder makes good any damage caused to the land in the process of installing or maintaining the power lines.

Finally, landowners should be aware of Schedule 6 (as amended by the Utilities Act 2000) to the Electricity Act 1989, which contains the Public Electricity Supply Code. This provides that any person, intentionally or by culpable negligence, damaging or allowing any power line to be damaged, is liable on summary conviction to a fine.

See also: Electricity and electrical equipment, p.256; Mobile phone masts, p.492.

Sources of further information

The Workplace Law website has been one of the UK's leading legal information sites since its launch in 2002. As well as providing free news and forums, our Information Centre provides you with a 'one-stop shop' where you will find all you need to know to manage your workplace and fulfil your legal obligations. Visit www.workplacelaw.net for more information.

Pregnancy

Mandy Laurie, Dundas & Wilson

Key points

- Health and safety legislation protects new and expectant mothers from certain risks in the workplace.
- If reasonable actions cannot be taken to avoid identified risks, the employer must alter the new or expectant mother's work, or hours of work, if it is reasonable to do so and it would avoid identified risks.

Legislation

- Workplace (Health, Safety and Welfare Regulations) 1992.
- Working Time Regulations 1998.
- Maternity and Parental Leave etc. Regulations 1999.
- Management of Health and Safety at Work Regulations 1999.
- Work and Families Act 2006.
- Agency Workers Regulations 2010.
- Equality Act 2010.

Health and safety legislation

Risk assessments

The Management of Health and Safety at Work Regulations 1999 require employers to carry out a risk assessment for new or expectant mothers and their babies.

A new or expectant mother is an employee who is pregnant, who has given birth within the previous six months (which includes a stillborn birth after 24 weeks of pregnancy) or who is breastfeeding.

Provided the new or expectant mother has notified the employer (in writing) of these circumstances, the employer must:

- provide the new or expectant mother with information on any identified risks to her or her baby's health; and
- take reasonable actions to avoid identified risks.

The kind of risks that may be identified for new or expectant mothers can include:

- heavy lifting;
- long hours;
- standing or sitting for prolonged periods;
- chemical handling; and/or
- working in awkward spaces.

If reasonable actions cannot be taken to avoid identified risks, the employer must alter the new or expectant mother's work, or hours of work, if it is reasonable to do so and it would avoid identified risks. If it is not reasonable to alter a new or expectant mother's work, or hours of work, or, if in doing so, it would not reduce identified risks, the employer should offer the employee suitable alternative work. If this is not possible, the employee should be suspended on full pay for as long as is necessary to avoid risks. See Figure 1 *(overleaf)* for a risk assessment flowchart.

Night work

New or expectant mothers may continue to work nights, unless they have been issued with a medical certificate from a doctor or midwife providing an exemption from night work. In these circumstances, the employer should offer the new or expectant mother suitable alternative work before considering suspending (on full pay).

O'Neill v. Buckinghamshire County Council (2008)

Most employers are now familiar with the fact that a failure to carry out a risk assessment for a pregnant woman can amount to sex discrimination, or discrimination on the grounds of pregnancy and maternity. In this case the claimant argued that there was a general obligation on employers to carry out a risk assessment if an employee notifies them that they are pregnant. However, the EAT held that there was no general requirement to conduct a risk assessment, but rather three conditions must be met to establish whether an employer has a duty to conduct a risk assessment for a pregnant worker:

1. The employee notifies the employer in writing that she is pregnant.
2. The work is of a kind that could involve a risk of harm or danger to the health and safety of the expectant mother or her baby.
3. The risk arises from either processes, working conditions or physical, chemical or biological agents in the workplace.

In this case, the employee was unsuccessful in arguing that a disciplinary procedure should have been dropped once it was known that she was pregnant, and this should have been included in the risk assessment. The EAT did confirm that where the duty is triggered a failure to carry out the risk assessment will still result in an automatic claim for sex discrimination. The sensible course of action therefore suggests that employers should still conduct a risk assessment when there is any hint that conditions two or three above are met.

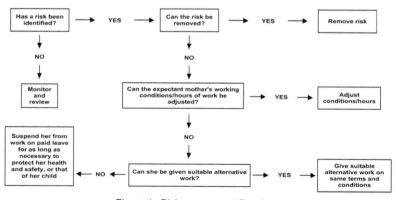

Figure 1. Risk assessment flowchart.

Rest facilities

The Workplace (Health, Safety and Welfare Regulations) 1992 require the employer to provide suitable rest facilities for new and expectant mothers. Any rest facilities must be equipped with an adequate number of tables and seating with backs for the number of persons at work likely to use them at any one time.

Agency workers

The Agency Workers Regulations 2010 came into force on 1 October 2011.

Amongst other things, the Regulations extend an employer's duty to carry out risk assessments (and implement the findings) to agency workers who have completed the requisite 12-week qualifying period.

See also: Health surveillance, p.396; Risk assessments, p.624.

Sources of further information

Electronic template maternity, paternity and adoption policies and management guides are downloadable from www.workplacelaw.net.

INDG 373 A guide for new and expectant mothers who work: www.hse.gov.uk/pubns/indg373.pdf

Workplace Law's *Guide to Flexible Working 2008* provides information on the formal legislative right to request flexible working, but also considers flexible working patterns in a wider sense. It covers the reasons why some employers are looking to introduce flexible working into their workplace, explores different flexible working patterns, including their benefits and disadvantages, and provides detail on the formal legislative right to request. For further details visit www.workplacelaw.net.

Property disputes

Michael Brandman, Blake Lapthorn Tarlo Lyons

Key points

- Keep copies of documents and records of telephone calls and other conversations.
- Act quickly. Some remedies are available only if you act without delay.
- Take professional advice at an early stage.
- Consider mediation as a means of resolving disputes. It is often quicker and cheaper than litigation. The Courts encourage mediation, and successful litigants have in some cases been refused costs against their opponents in circumstances where mediation has been offered and rejected. However, the Court of Appeal recently ruled that litigants cannot be forced into mediation against their will.

Legislation

- Landlord and Tenant Act 1927.
- Leasehold Property (Repairs) Act 1938.

Strategy

Always verify your rights before tackling the problem. Start by assembling and checking the relevant documents. They will often tell you what your position is.

In a landlord and tenant dispute, consider:

- the lease;
- rent review documents;
- licences for alterations;
- licences for change of use; and
- guarantees.

In a neighbour dispute, consider:

- title deeds; and
- planning documents.
- In disputes with the previous owner, consider:
- the contract;
- pre-contract enquiries and replies;
- the transfer or conveyance; and
- relevant correspondence.

Gather your evidence

You may need to be able to prove your case in court or in some other kind of dispute resolution process. Good-quality evidence will help you. It is important to:

- keep a written record of relevant events. Keep a log setting out times, dates and people involved as events unfold. This is essential where the events complained of are changing. Where there are, for example, building works, or where conduct of individuals is the problem, make a careful note of things that happen from day to day;
- take photographs before and after any alterations are made; and
- keep an accurate written record of telephone calls and other conversations.

Act quickly

Property disputes rarely just go away. The property does not move and the value of property makes it likely that disputes will continue until they are resolved by some sort of process.

Delay in reacting to circumstances may result in some emergency remedies being unavailable. Injunctions or restraining orders preventing further building works or demolition works, for example, can be obtained only where prompt action is taken. Delay may result in a change in the balance of power. It is almost always easier to prevent a building from being constructed, or from being demolished, than it is to obtain an order for the removal or reconstruction of such a building.

Take professional advice

The value of the property as an asset usually warrants good professional involvement. Surveyors are necessary in many disputes (e.g. repair and boundaries). A planning expert will help to win arguments about the right to construct or alter a building, even before an application for planning permission is made. Expert evidence will often be necessary if a dispute goes to court. Early involvement of expert professionals ensures that case preparation is carried out along the right lines from the beginning. Solicitors should be asked to provide an early assessment of the strengths and weaknesses of any case as soon as a problem arises. Weaknesses can then be addressed and strengths shown to the opponent in the best possible light. The aim is to find an early resolution.

Resolving disputes

The best strategy for dealing with the particular dispute needs to be identified as quickly as possible. The documents, evidence and expert advice will assist in this respect. If an injunction or restraining order is necessary to prevent a problem from becoming entrenched or

to avoid a major change in the position on the ground, then court action will be necessary. Lawyers will assist in the preparation of witness statements and expert evidence. Subject to Claimant undertaking being given, restraining orders can be obtained on the same day where the matter is extremely urgent, or more usually within three to seven days from the commencement of action.

Many disputes are settled by mediation, rather than through the Courts. This allows the parties to remain in control of the procedure and to create a solution that may be more flexible than possibilities available through the Courts. Mediation is a form of facilitated negotiation and is voluntary – no one can be forced into a solution against their will. It is almost always quicker and cheaper than litigation.

Statutory time limits

In some cases (e.g. applications for renewal of business leases and applications for judicial review of planning decisions) there are strict and very short time limits within which to bring proceedings. In such cases, obtaining prompt legal advice can be crucial. Mediation must not be allowed to delay the issue of proceedings when time is critical.

See also: Boundaries and party walls, p.63; Buying and selling property, p.107; Dilapidations, p.221; Landlord and tenant: lease issues, p.429; Landlord and tenant: possession issues, p.438.

Sources of further information

Centre for Effective Dispute Resolution: www.cedr.co.uk

Radiation

Stephen Day, Bureau Veritas UK Ltd

Key points

- There are two types of radiation – ionising and non-ionising.
- Specific regulations require employers to protect employees from the adverse effects of ionising radiation, either from irradiation or from radioactive contamination.
- Employers also have a general duty to protect employees from non-ionising radiation (e.g. over-exposure to the sun).

Legislation

- Radioactive Substances Act 1993. This Act has now been repealed in England and Wales but is still in force in Scotland and Northern Ireland. In England and Wales the Environmental Permitting Regulations 2010 now apply. Section 23 applies specifically to radioactive materials.
- Ionising Radiations Regulations 1999.
- Management of Health and Safety at Work Regulations 1999.
- Radiation (Emergency Preparedness and Public Information) Regulations 2001.
- The Transport of Dangerous Goods and Use of Transportable Pressure Equipment Regulations 2009.
- The Control of Artificial Optical Radiation at Work Regulations 2010.
- Environmental Permitting Regulations 2010.

Types of radiation

Radiation may be classified as either ionising or non-ionising:

- Non-ionising electromagnetic radiation (e.g. ultraviolet and radio waves) does not change the structure of atoms.
- Ionising electromagnetic radiation has enough energy to ionise or electrically charge atoms. Ionising radiation has sufficient energy to cause changes

within the DNA molecule and can therefore be a cause of cancer.

Non-ionising radiation

Non-ionising radiation (NIR) is the term used to describe the part of the electromagnetic spectrum covering two main regions, namely optical radiation (ultraviolet (UV), visible and infrared) and electromagnetic fields (EMFs) (power frequencies, microwaves and radio frequencies).

Optical radiation

Optical radiation is another term for light, covering ultraviolet (UV) radiation, visible light, and infrared radiation. The greatest risks to health are probably posed by:

- *UV radiation from the sun*. Exposure of the eyes to UV radiation can damage the cornea and produce pain and symptoms similar to that of sand in the eye. The effects on the skin range from redness, burning and accelerated ageing through to various types of skin cancer. Protective measures include minimising time of exposure, use of sun screen lotions, wearing clothes, including a hat, which cover the skin, and use of sunglasses.
- *The misuse of powerful lasers*. High-power lasers can cause serious damage to the eye (including

blindness) as well as producing skin burns. The main method of protection is to use engineering controls that prevent intra beam viewing.

The Control of Artificial Optical Radiation at Work Regulations came into force on 27 April 2010 and aim to protect workers from the risks to health from hazardous sources of artificial optical radiation (AOR).

Workers in Great Britain are generally well protected from AOR and the majority of businesses with hazardous sources know how to manage the risks effectively. Further work is only expected to be undertaken by those businesses that use hazardous sources of AOR and where the associated AOR risks have not already been reduced to as low as is reasonably practicable. The HSE has produced guidance to help duty-holders decide whether they are already protecting their workers or whether they need to do more under the new regulations – see *Sources of further information*.

Electromagnetic fields

Electromagnetic fields (EMFs) arise whenever electrical energy is used. For example, EMFs arise in our home from electrical appliances in the kitchen, from work processes such as radiofrequency heating and drying, and in the world at large from radio, TV and telecoms broadcasting masts and security detection devices. It has been known for a long time that exposure of people to high levels of EMFs can give rise to acute effects. The effects that can occur depend on the frequency of the radiation. At low frequencies the effects will be on the central nervous system of the body, whilst at high frequencies, heating effects can occur, leading to a rise in body temperature. In reality, these effects are extremely rare and will not occur in most day-to-day work situations. Employers

should conduct a risk assessment, as required by the Management of Health and Safety at Work Regulations 1999. The effects from these electromagnetic frequencies can be controlled by compliance with guidelines published by the Health Protection Agency (HPA).

Ionising radiation

Ionising radiation occurs as either electromagnetic rays (such as X-rays and gamma-rays) or particles (such as alpha and beta particles). It occurs naturally (e.g. from the radioactive decay of natural radioactive substances such as radon gas and its decay products) but can also be produced artificially. People can be exposed externally, to radiation from a radioactive material or a generator such as an X-ray set, or internally, by inhaling or ingesting radioactive substances. Wounds that become contaminated by radioactive material can also cause radioactive exposure.

The Ionising Radiations Regulations 1999 apply to a large range of workplaces where radioactive substances and electrical equipment emitting ionising radiation (X-rays) are used. They require employers to keep exposure to ionising radiation as low as reasonably practicable and to not exceed annual dose limits. Employers should bear in mind that two distinct types of hazard may exist – irradiation, i.e. the emission of penetrating radiation (gamma- or X-rays), and contamination, i.e. the presence of radioactive powders, liquids or gases that could be inhaled, ingested or absorbed into the body. The following hierarchy of control measures is recommended:

1. Engineering controls, including shielding and ventilation.
2. Procedural controls, e.g. restricted access and safe systems of work.

3. Personal protective equipment – this should be used as a last resort.

The Regulations contain specific requirements that may apply, depending on the nature of the work. They include the following:

- Employers must notify Enforcing Authorities before starting work.
- Employers must appoint a suitably qualified Radiation Protection Adviser and a local Radiation Protection Supervisor.
- Employers must set up either controlled or supervised areas where radiation hazards exist.
- Employers must assess whether any of their employees will fall into the category of classified workers, in which case arrangements must be made with an Approved Dosimetry Service and for annual medical examinations.
- Employers must produce a set of local rules that describe the safe operating procedure for working with the radiation source(s).

The Radioactive Substances Act / Environmental Permitting Regulations include the following requirements for users of radioactive materials:

- *Registration of holdings.* The user must be in possession of a Registration that specifies the type and amount of radioactive material under their control.
- *Authorisation for the accumulation and disposal of waste.* The user must be granted an authorisation by the Environment Agency that specifies the disposal route, the physical form and the limit / frequency of disposals.
- *Review of exemption orders for disposal of waste.* The consultation period has passed and new exemption orders are expected to be published towards the end of 2011.

See also: Health surveillance, p.396; Occupational cancers, p.512; Outdoor workers, p.528; Personal Protective Equipment, p.576; Risk assessments, p.624.

Sources of further information

HSE: www.hse.gov.uk/radiation.

INDG 337: Sun protection: advice for employers of outdoor workers: www.hse.gov.uk/pubns/indg337.pdf

HSE – Guidance for Employers on the Control of Artificial Optical Radiation at Work Regulations (AOR) 2010: www.hse.gov.uk/radiation/nonionising/employers-aor.pdf

Health Protection Agency Centre for Radiation Chemical and Environmental Hazards: www.hpa.org.uk/radiation

Recycling

Mike Lachowicz, Bureau Veritas UK Ltd

Key points

- Article 1(a) of the Waste Framework Directive defines waste as 'any substance or object … which the holder discards or intends or is required to discard'. The key point encapsulated in this definition is that 'waste' is relative: substances or objects that are waste to one holder would be considered a resource to another. This is an important concept that should be reflected in the waste management strategy of any organisation – in order to reduce waste, and improve business efficiency.

- Although the UK Government recognises the importance of a robust waste management strategy, some businesses display a more lax attitude towards their responsibilities, as few businesses are aware of the true cost of waste. Most businesses estimate waste costs at 1% of turnover but the real potential savings are about 4.5% – or £22bn a year. Studies conducted by Envirowise have consistently shown that the true cost of waste is ten times that of the cost of disposal to the business (e.g. EN505). Good waste management, including avoidance, reduction, re-use and recycling, has the potential to improve the bottom line performance of all businesses.

- In terms of potential benefits from more environmentally aware waste management, the most common are as follows:
  - Reduced waste disposal costs;
  - Reduced costs of resources used to store and manage waste;
  - Reduced procurement costs;
  - Compliance with waste legislation;
  - Compliance with non-legally binding Standards and Codes of Practice; and
  - Satisfaction of stakeholder requirements.

- The recent change in government has resulted in a series of new policies and initiatives. The Government Review of Waste Policy in England 2011, announced in June 2011, sets out its intention to continue its commitment to the waste hierarchy by increasing re-use and recycling; reduction of total waste and waste sent to landfill; use of End of waste Quality Protocols; to encourage resource efficiency as well as partnerships such as 'voluntary responsibility deals', a comprehensive Waste Prevention Programme and rewarding 'good behaviour' by businesses, the civil society and individuals.

Legislation

- Environmental Protection Act (Part II) 1990.
- Environmental Protection (Duty of Care) Regulations 1991 as amended by Waste (Miscellaneous Provisions) (Wales) Regulations 2011.
- Controlled Waste Regulations 1992, plus subsequent amendments (1993) as replaced by the Waste Management (England and Wales) Regulations 2006 and Waste (Scotland) Regulations 2005 and to be replaced by the Controlled Waste (England and Wales) Regulations 2011.

> **Facts**
>
> ■ In 2009, the UK deposited approximately 138 million tonnes of waste either on to or into the land.
> ■ Around 11% of total waste is attributable to municipal disposal, and around 17% is commercial and industrial waste.
> ■ Up to 60% of rubbish that is thrown away could be recycled.
> ■ It has been estimated that UK businesses lose up to 4.5% of annual turnover every year through avoidable waste.
>
> *Sources: DEFRA and EA.*

■ Environment Act 1995.

■ Landfill Tax Regulations 1996 (as amended 2011).

■ The Producer Responsibility Obligations (Packaging Waste) (England and Wales) Regulations 1997 (as amended 2000, 2007, 2008, 2010).

■ The Landfill (England and Wales) Regulations 2002 as amended by the Landfill (England and Wales) Regulations 2004 and the Landfill (England and Wales) (Amendment) Regulations 2005.

■ End-of-Life Vehicles (Storage and Treatment) (Scotland) Regulations 2003.

■ Environmental Protection (Duty of Care) (England) (Amendment) Regulations 2003.

■ Landfill (Scotland) Amendment Regulations 2003.

■ Packaging (Essential Requirements) Regulations 2003 (as amended 2006 and 2010).

■ The Special Waste Amendment (Scotland) Regulations 2004.

■ Hazardous Waste (England and Wales) Regulations 2005 (amended 2009).

■ Hazardous Waste Regulations (Northern Ireland) 2005 (as amended 2009).

■ List of Wastes (England) Regulations 2005 (as amended) and Explanatory Memorandum to the List of Wastes

(England) Regulations 2005 No. 895 (Wales).

■ Waste Electrical and Electronic Equipment Regulations 2006 and Waste Electrical and Electronic Equipment (Amendment) Regulations 2010.

■ Environmental Permitting (England and Wales) Regulations 2007 (as amended 2010).

■ The Waste Management Licensing Amendment (Waste Electrical and Electronic Equipment) (Scotland) Regulations 2007.

■ The Site Waste Management Plans Regulations 2008 (England) and the Site Waste Management Plans Regulations (Northern Ireland) 2011.

■ The Waste Batteries and Accumulators Regulations 2009, Waste Batteries and Accumulators (Charges) Regulations (Northern Ireland) 2009, Waste Batteries and Accumulators (Treatment and Disposal) Regulations (Northern Ireland) 2009 and Waste Batteries (Scotland) Regulations 2009.

■ The End-of-Life Vehicles Regulations 2003 and (Amendment) Regulations 2010.

■ End-of-Life Vehicles (Producer Responsibility) Regulations 2005 and (Amendment) Regulations 2010.

■ The Waste (England and Wales) Regulations 2011.

Business waste

The generation of waste arising from business activities and its management should be governed by the 'waste hierarchy,' which provides guidelines for increasingly environmentally responsible waste management. It represents a chain of priority for waste management, citing the prevention of waste as the most favourable option. If not possible, waste prevention is succeeded by minimisation, re-use, recycling, energy recovery, then disposal, which is considered the least favourable.

There are certain instances in which the order of the waste hierarchy is debated; for instance, there is some disagreement as to whether waste paper is best recycled or sent to an energy-from-waste incinerator, from a purely environmental perspective. However, the hierarchy generally provides a good guideline.

Although waste recycling is prominent, and heavily publicised and promoted in businesses, there is as yet no overriding legislation that enforces recycling in the UK. There are two key pieces of legislation that are potentially relevant, the first being the Producer Responsibility Obligations

(Packaging Waste) Amendment 2008, which impose on producers the obligation to recover and recycle packaging waste (and which implement Directive 94/62/EC), and secondly, the Waste Electrical and Electronic Equipment (WEEE) Regulations, which place certain obligations surrounding re-use of EEE upon suppliers. Other key EU legislation aimed at increasing recycling and transposed into UK law includes the End of Life Vehicles Regulations and the Waste Batteries and Accumulators Regulations. In addition, all businesses are now obliged to segregate commercial waste (though in practice this is often done by their waste contractor). Specific requirements to recycle are, furthermore, embedded in European Union (EU) legislation relating to certain industries and activities. Relevant legislation is discussed below.

Demonstrating compliance with current regulations will protect your business from prosecution. The overriding Directive that focuses on recycling and waste management throughout the EU is the Framework Directive on Waste (Council Directive 75/442/EEC as amended by Council Directive 91/156/EEC and adapted by Council Directive 96/350/EC). The contents of the Framework Directive are implemented in the UK through the Environmental Protection Act 1990, amended by the Environment Act 1995 and also by various principal regulations.

End of Life Vehicles (Producer Responsibility) Regulations 2005 (amended 2010)

Car manufacturers now have an obligation to take back old / scrap cars from consumers and ensure that more of the waste from scrap cars and light vans is recycled rather than landfilled. The Regulations establish responsibilities for vehicle manufacturers and professional importers to:

Avoidance and Minimisation

Reuse

Recycling

Recovery

Disposal

Maximum Conservation of Resources

- put in place collection networks to take back their own brands of vehicles, when those vehicles reach the end of their lives;
- since 2006, ensure that value re-used or recovered is at least 85% of the weight of their end of life vehicles (95% from 2015); and
- since 2007, provide 'free take-back' to last owners, who present their end of life vehicles for scrapping at collection networks.

The Regulations apply to passenger cars and light vans.

End of Life Vehicles Regulations 2003 (amended 2010)

The End of Life Vehicle Regulations 2003 apply to vehicles, end of life vehicles, their components and their materials. The Regulations place the main responsibilities for recycling and collection on dismantlers and scrap metal recyclers, outline materials and components that manufacturers cannot use in vehicle manufacture, as well as the requirement on manufacturers to provide material coding to facilitate identification of materials and components suitable for re-use and recovery.

The Packaging (Essential Requirements) Regulations 2003 (amended 2004, 2006, and 2009)

This Regulation implements Articles 9 and 11 of Directive 94/62/EC of the European Parliament and the Council on packaging and packaging waste. Companies must ensure that any packaging placed on the market by the organisation meets the minimum national standards necessary for safety, hygiene and consumer 'acceptance' purposes and that it is re-usable, biodegradable or recoverable by recycling, energy recovery or composting.

The Producer Responsibility Obligations (Packaging Waste) (England and Wales) Regulations 1997 (amended 2008 and 2010)

The Producer Responsibility Regulations 1997 implement the 1994 European Directive on packaging and packaging waste. The UK was required to recover a minimum 50% by weight of packaging material and to recycle a minimum of 25% by weight of the total waste flow by 2001. These Regulations differ from most other environmental legislation to date, as they require collaboration between a wide range of players throughout the packaging chain. The obligations are broad and affect not only producers of packaging and packaging materials, but also those who use the packaging around their products and those who sell packaging to the final user.

The Regulations, last updated in 2010, are designed to encourage more sustainable use of packaging by promoting recovery / recycling / minimisation – thus reducing the amount of packaging waste going to landfill and minimising resource wastage. A company is obligated if its turnover in the previous financial year was more than £2m per year and it handles (in aggregate) more than 50 tonnes of packaging material and/or packaging in a year. It is then required to ensure that a proportion of the packaging is diverted from the waste stream and recovered for re-use, recycling or incinerated with energy recovery. The exact nature of a company's obligation depends on the amount of packaging the organisation supplies, the materials it comprises and its role in the packaging chain. Recycling and recovery targets are set for raw material manufacturers, converters, packers or sellers.

Waste Electrical and Electronic Equipment (England and Wales) Regulations 2006 (amended 2007 and 2010) and the Waste Management Licensing Amendment (Waste Electrical and Electronic Equipment) (Scotland) Regulations 2007

These Regulations transpose the main provisions of Council Directive 2002/96/EC on waste electrical and electronic equipment, as amended by Council Directive 2003/108/EC. The Waste Electrical and Electronic Equipment (WEEE) Directive has recently been by far the most widely publicised legislation as WEEE is currently the fastest-growing waste stream in the UK. Volumes for business WEEE in London alone are estimated to be around 100,000 tonnes per annum. Producers who put electrical and electronic equipment (EEE) on the market in the UK must set up systems for the treatment of WEEE and are responsible for financing the costs of the collection, treatment, recovery and environmentally-sound disposal of the EEE items. In respect of any WEEE cited under these Regulations, the operator of a scheme must ensure that systems are set up to provide for the treatment of WEEE, using the best available treatment, recovery and recycling techniques. The WEEE Directive and Regulations will affect every business that uses electrical equipment in the workplace. The Directive places responsibility on business users along with producers for ensuring WEEE is correctly treated and reprocessed, encouraging the re-use of equipment over recycling. In practice, all retailers and distributors have to be linked to a WEEE recycling scheme, but most charge their customers for this service.

Restriction of the Use of Certain Hazardous Substances (RoHS) in Electrical and Electronic Equipment Regulations 2006 (amended in 2008)

Complementary to the WEEE Directive is the Restriction on the Use of Certain Hazardous Substances in Electrical and Electronic Equipment Directive (RoHS). This Directive aims to harmonise legislation controlling the use of hazardous substances in EEE across the EU and also seeks to reduce the environmental impact of WEEE by restricting the use of certain hazardous substances during manufacture. The RoHS Directive covers all products mentioned in the WEEE Directive except medical, monitoring and control equipment. The Directive requires the substitution of lead, cadmium, hexavalent chromium, polybrominated biphenyls (PBBs) and polybrominated diphenylethers (PDBEs) since 1 July 2006. Manufacturers need to understand the requirements of the RoHS Directive to ensure that their products, and their components, comply.

The Landfill (England and Wales) Regulations 2002 (amended 2005)

These Regulations set out a pollution control regime for landfills for the purpose of implementing Council Directive 99/31/EC on the landfill of waste ('the Landfill Directive') in England and Wales. The obligation to meet the requirements of the Regulations is mainly on landfill operators, but there are provisions that waste producers need to be aware of, such as:

- granting of permits for landfill;
- prohibition of certain wastes; and
- duty of care / waste transfer notes.

The Landfill Directive is helping to bring about a change in the way we dispose of waste in this country. It aims to reduce the pollution potential from landfilled waste that can impact on surface water, groundwater, soil, air, and also contribute to climate change. In England and Wales the Directive is applied under the Landfill (England and Wales) Regulations 2002 and was fully implemented by July 2009. The Directive sets demanding targets to reduce the amount of biodegradable municipal landfilled waste.

Since 30 October 2007, new rules have applied for non-hazardous waste. Liquid wastes are banned from landfill and all waste must be treated before it can be landfilled, factors that will force businesses to review their waste management strategies. 'Treatment' of non-hazardous waste is defined as either a physical, thermal, chemical or biological process, including sorting. To comply with these Regulations, businesses could either collect waste as individual waste streams or send to separate waste streams to be recycled, or make a request with their waste management contractor to sort the waste for recycling.

See also: Packaging, p.541; Waste management, p.605; Waste Electrical and Electronic Equipment: the WEEE and RoHS Regulations, p.737.

Sources of further information

Much of the Government's environmental information resources, including those related to waste recycling, are currently being re-organised into fewer, 'one stop shop' agencies.

A considerable amount of information on waste management can be found on the website for the Department for Environment Food and Rural Affairs (DEFRA), which offers information on environmental topics at www.defra.gov.uk/environment/waste/

The Envirowise programme has been merged with that of WRAP and can supply information and assistance via its helpline on 0800 585 794, or www.envirowise.wrap.org.uk/

The Waste and Resources Action Programme (WRAP) works in partnership to encourage and enable businesses and consumers to be more efficient in their use of materials and recycling at www.wrap.org.uk or contact the helpline on 0808 100 2040.

A considerable amount of information on waste management can be found on the EA's website: www.environment-agency.gov.uk/business/topics/waste/default.aspx

NetRegs is a partnership between the UK environmental regulators – the Environment Agency in England and Wales, SEPA in Scotland and the Environment and Heritage Service in Northern Ireland – which provides free environmental guidance for small and medium-sized businesses throughout the UK. Help and advice is available by business type, environmental legislation and environmental topic and has useful links to other websites: www.netregs.gov.uk/netregs

EA National Customer Contact Centre: 08708 506 506; SEPA: 01786 457710; NIEA: netregs@doeni.gov.uk

National Industrial Symbiosis Programme (NISP) works directly with businesses of all sizes and sectors with the aim of improving cross-industry resource efficiency via commercial trading of materials, energy and water and sharing assets, logistics and expertise. It engages traditionally separate industries and other organisations in a collective approach to competitive advantage involving physical exchange of materials, energy, water and/or by-products together with the shared use of assets, logistics and expertise. NISP is delivered at regional level across the UK: www.nisp.org.uk
Telephone: 0845 094 9501.

Registration, Evaluation and Authorisation of Chemicals (REACH)

Andy Gillies, Gillies Associates

Key points

- June 2012 marks the fifth anniversary of the coming into force of the REACH Regulation, which heralded a fundamental change in chemical safety regulation impacting across the EU and beyond. REACH stands for the Registration, Evaluation, Authorisation and Restriction of Chemicals. Regulation EC 1907/2006 (REACH).

- REACH and its supporting Regulations (e.g. EC 340/2008 – Fees and Charges, EC 440/2008 – Test Methods, plus a series of Regulations amending Annexes to REACH) are EU Regulations, which means that the legal instruments have 'direct effect' and apply directly to duty-holders in each member state. REACH replaces a complex framework of European Directives with a single system.

- The main REACH Regulation took effect on 1 June 2007. The aims of REACH are 'to ensure a high level of protection of human health and the environment, including the promotion of alternative methods for assessment of hazards of substances, as well as the free circulation of substances on the internal market while enhancing competitiveness and innovation' (*Article 1(1)*). REACH will furthermore give greater responsibility to industry to manage the risks from chemicals and to provide safety information that will be passed up and down the supply chain.

- REACH puts a responsibility on those who place chemicals on the market (mainly EU-based manufacturers and importers to the EU) to understand and manage the risks associated with their use. This is a fundamental change to the previous system of chemicals regulation under the Existing Substances and Notification of New Substances regime, where the regulatory authorities in each member state were responsible for carrying out the risk assessment based on information provided by the manufacturers.

- A key concept of REACH is 'No data, no market'. If substances are not registered with the European Chemicals Agency (EChA) then manufacturers or importers will not be legally allowed to place them on the EU market.

Legislation

Main European legislation

- REACH legal text in Regulation EC no. 1907/2006 on the Registration, Evaluation, Authorisation and Restriction of Chemicals.
- Regulation (EC) 340/2008 on Fees and charges payable to the European chemicals Agency.
- Regulation (EC) 440/2008 (plus amendments) laying out test methods under REACH.
- Regulation (EC) 1272/2008 on Classification, Labelling and Packaging of substances and mixtures (has a number of overlaps with REACH).

UK implementing legislation

- The REACH Enforcement Regulations 2008.

- The REACH (Appointment of Competent Authorities) Regulations 2007.

Key elements

- *Registration.* Manufacturers or importers will need to register any substance supplied into the EU market above one tonne per company per year.
- *Evaluation.* EChA will carry out evaluations of Registration Dossiers to ensure testing proposals are appropriate. National Competent Authorities will carry out annual in-depth evaluations of substances flagged as being of potential high risk.
- *Authorisation.* Substances of very high concern – category 1 or 2 CMRs (carcinogens, mutagens, and toxic for reproduction), PBTs (persistent, bioaccumulative, and toxic), vPvBs (very persistent and very bioaccumulative), and substances of equivalent concern such as endocrine disrupters – will require authorisation.
- *Restriction.* Any substance that poses a particular threat requiring Community-wide action may be subject to restrictions, ranging from partial restrictions on specific uses to total bans.

Registration

All substances, unless specifically exempted in the REACH Regulation, that are manufactured or imported above one tonne per company per year must be registered. 4,300 distinct substances were registered by nearly 25,000 companies by the first registration deadline at the end of 2010 (*see below*). Most of these registrations covered high volume chemicals manufactured or imported in quantities less than 1,000 tonnes per year. EChA expect to receive around 13,300 new dossiers covering 3,500 substances by the 2013 deadline.

Industry is required to submit registration dossiers for each substance to the European Chemicals Agency (EChA). There are two main components to the registration dossier – a technical dossier and a chemical safety report (CSR). The technical dossier contains relevant data on the physico-chemical properties and hazards of the substance as defined in Annexes VI to X of REACH depending on tonnage bands. Manufacturers or importers of substances manufactured or imported in quantities above ten tonnes per company per year must, as part of their registration dossier, carry out a chemical safety assessment and demonstrate adequate control of risks to humans and the environment throughout the substance life cycle from manufacture to final disposal. This is documented in the CSR. All companies that have pre-registered a substance will be invited to join a Substance Information Exchange Forum (SIEF) for that substance. The main purpose of SIEFs is to facilitate data sharing and joint submission of information to EChA.

Any new substance manufactured or imported into the EU must be registered with EChA before being placed on the market. Registration for existing substances (called 'phase-in substances'), which were pre-registered between June and December 2008, will be phased in over an 11-year period:

- Substances supplied above 1,000 tonnes a year per manufacturer or importer, or substances of very high concern, had to be registered by November 2010;
- Substances supplied between 100 and 1,000 tonnes a year must be registered by May 2013; and
- Substances supplied between one and 100 tonnes a year must be registered by May 2018.

Industry will also be required to prepare risk assessments and provide information on suitable control measures for using the substance safely to downstream users. This will be done through the provision of exposure scenarios for each identified use attached to extended Safety Data Sheets.

The first registration deadline for phase-in substances was 30 November 2010, and a major effort was made by manufacturers and importers to complete registration dossiers on time. Dossier submissions have to be made on the REACH-IT system using IUCLID 5 software. A number of issues arose that will need to be addressed during the second registration period:

- The actual number of registrations was less than predicted, which caused some anxiety for downstream users concerned about possible disruption in their supply chains.
- Nearly 90% of total registrations were joint submissions, but problems were encountered with SIEF activities (e.g. difficulties in identifying lead registrant, data sharing and business confidentiality issues).
- The majority of dossiers screened did not contain sufficient information to demonstrate the substances met the required conditions for definition as 'intermediate' (and hence benefit from reduced information requirements in the registration dossier). The guidance on intermediates and 'strictly controlled conditions' was updated by EChA in December 2010 and remains subject to debate between industry bodies and EChA.
- Classification and labelling notifications (linked to the CLP Regulations) for small quantity substances used in research give rise to confidentiality questions.

Registration is just the start of a long journey for chemical companies as they grapple with evolving guidance, respond to dossier evaluations, and get used to greater public access and scrutiny of data.

Evaluation

Two evaluation processes will be conducted under the Regulations. These are a Dossier evaluation and a Substance evaluation. Under the Dossier evaluation the European Chemicals Agency will scrutinise all animal testing proposals submitted with a registration dossier to make sure they meet the requirements in Annex IX and X and that no unnecessary testing is proposed. In addition, 5% of all registration dossiers at each tonnage level will be subject to a full compliance audit by the Agency to ensure the data submitted by industry is of suitable quality.

Under the substance evaluation, member states and the European Commission will agree on an annual list of priority substances to be assessed in-depth for possible regulatory action across the EU because of concerns about their properties. Member state competent authorities (e.g. the HSE in the UK) will carry out substance evaluations that may lead to restrictions on manufacture, supply or use. It may also lead to addition of the substance on to the priority list for authorisation or a proposal to change classification and labelling.

Authorisation

Industry will be required to obtain Community-wide authorisations for the use of substances of very high concern (SVHC). These are substances that may have very serious and often irreversible effects on human health and the environment. This will apply to substances identified as carcinogenic, mutagenic or toxic to reproduction (CMRs, category one and two); persistent, bioaccumulative and toxic substances (PBTs); substances that are very persistent and very bioaccumulative (vPvBs); and substances demonstrated to be of equivalent concern, such as endocrine disruptors. EChA and Member State CAs draw up a 'Candidate

List' of SVHC which are then prioritised for listing in Annex XIV of REACH (the 'Authorisation List').

Fifty-three substances had been identified on the Candidate List by the middle of 2011, including familiar chemicals such as acrylamide, aluminosilicate refractory ceramic fibres, dibutyl phthalate (DBP), hexabromocyclododecane (HBCDD), lead chromate and trichloroethylene. Fifteen substances from the Candidate List have been recommended by EChA for inclusion in Annex XIV to date. These include musk xylene (used as a fabric softener and in detergents), MDA (used in panels), HBCDD (flame retardant used in furniture and textiles), and three phthalates (DEHP, BBP, and DBP).

Authorisation will only be granted if the risks of the substance are adequately controlled and considered safe. Substances listed in Annex XIV may only be used by industry after submission of an application for authorisation of continued use. If adequate control is not possible, authorisation could still be granted on socio-economic grounds that its continued use outweighs the risks to human health and the environment. There will be a drive for progressive substitution by safer substances or technologies over time.

Restriction

Substances, either on their own or in a preparation (mixture) or article, which have been restricted may not be manufactured, placed on the market, or used except in strict compliance with the conditions of the restriction. Restrictions will be applied where action at European level is needed.

Implications for downstream users

Downstream Users (DUs) are defined in REACH as anyone using chemicals, although distributors and consumers are not classed as DUs (however distributors must ensure that safety information is passed down the supply chain with the substances they sell). REACH in general places no substantial legal duties on downstream users of chemicals. However, there are commercial, supply chain and operational risks for downstream users and it is important for DUs to familiarise themselves fully with REACH. EChA has published specific guidance for DUs, which gives detailed information on what needs to be done. Downstream users should provide details of their uses of the chemicals to their suppliers so that these uses are included in the registration dossier. In certain cases downstream users may need to produce their own Chemical Safety Report (CSR).

Downstream users have a right to request that the suppliers' chemical safety assessment covers their specific use. In addition, downstream users have a right to join a Substance Information Exchange Forum during the registration process. Duties under REACH also extend to those who produce or import articles. Articles are defined as 'an object which during production is given a special shape, surface or design which determines its function to a greater degree than does its chemical composition'. The key duties relate to the provision of information where the article contains over 0.1% (w/w) of substances deemed of very high concern (SVHC).

A DU also has notification obligations if he uses a substance outside the conditions of the supplier exposure scenario, or determines that the classification and labelling of his substance is different from that received from the supplier. They must communicate any new information on hazardous properties or risk management measures to the supplier. From the first half of 2011 DUs have started to receive extSDS from suppliers containing

Exposure Scenarios describing conditions of safe use. A legal duty applies to DUs to ensure that they are using the appropriate risk management measures (control measures) stated in the extended SDS. It is important that DUs check that the substances they use are registered (does it have a REACH registration number?) for their intended use, and that they are using the recommended controls.

It is also prudent for DUs to keep informed about developments regarding intended registrations for 2013, and proposed additions to the Candidate List. In time, some SVHC will be taken out of use altogether, and others will be subject to authorisation or restriction, so companies need to identify any supply chain issues in advance and make contingency plans.

CLP Regulation and REACH

The Classification, Labelling and Packaging (CLP) Regulation (EC 1272/2008) covers classification and labelling (C&L) requirements for chemical substances and mixtures and implements the UN Globally Harmonised System (GHS) of C&L. There is a significant overlap with REACH in terms of the requirements for classification and provision of Safety Data Sheets. A transitional period of five years has been set to implement all the provisions of CLP; CLP requirements applied to substances from 1 December 2010 and will be required for mixtures from 1 June 2015. All substances, regardless of quantity used, had to be notified to EChA and listed on the C&L Inventory by 3 January 2011.

Competent Authority

The HSE is appointed as the UK Competent Authority (UK CA) for REACH, working in conjunction with the Environment Agency and other bodies. Its responsibilities are far reaching and include:

- providing advice to manufacturers, importers, downstream users and other interested parties on their respective responsibilities and obligations under REACH;
- enforcing compliance with registration;
- conducting substance evaluation of prioritised substances and preparing draft decisions;
- proposing harmonised Classification and Labelling for CMRs and respiratory sensitisers;
- identifying substances of very high concern for authorisation; and
- proposing restrictions.

Enforcement

Member states are responsible for the enforcement of REACH in their jurisdiction. In the UK this is achieved through the REACH Enforcement Regulations 2008. Broadly speaking, the HSE will enforce duties on registration and supply chain up to the point of retail sale, with Local Authority Trading Standards enforcing for retail sale. Enforcing use-related duties will be done by a number of regulatory bodies, including HSE, EA, SEPA, Local Authorities and DECC.

The Government has proposed a best practice approach, which encourages duty-holders to comply. However, enforcement activity is more likely in situations where:

- manufacture, import, sale, supply or use of substances is conducted without the appropriate registration;
- a hazardous substance is used outside the terms of an authorisation or contrary to a restriction;
- there has been a failure to provide required information up and down the supply chain;
- there has been a failure to comply with other duties regarding information;

- there has been a failure to comply with the duty to apply recommendations such as in safety assessments; and
- there has been a failure to comply with the duties to cooperate and supply information.

Recent developments

The UK CA ran a pre-registration enforcement campaign and identified a significant number of companies that had pre-registered substances not listed on EINECS and that may not qualify for phase-in status. During 2010/11 the UK participated in an EU-wide enforcement initiative named REACH-EN-FORCE 1, focused on the 'no data, no market' principle, checking that valid registrations and adequate safety data sheets were in place. A follow-on programme, REACH-EN-FORCE 2, focused on obligations of downstream users and formulators started in April 2011. Market intelligence-led enforcement campaigns on specific substances have been run, for example looking at uses of ammonium dichromate and MDI (methylene diphenyl diisocyanate).

See also: Hazardous waste, p.700.

Sources of further information

DEFRA (UK responsible department): www.defra.gov.uk/environment/quality/chemicals/reach/index.htm

HSE (UK Competent Authority): www.hse.gov.uk/reach/index.htm

The HSE has provided a help desk that can be contacted at 0845 408 9575 or via email at ukreachca@hse.gsi.gov.uk

European Chemicals Agency: http://ec.europa.eu/echa/home_en.html

CEFIC REACH guidance: www.cefic.org/Industry-support/Implementing-reach/

DG Environment: http://ec.europa.eu/environment/chemicals/reach/reach_intro.htm

CIA 'REACH Ready': www.reachready.co.uk/

BOHS REACH guidance: www.bohs.org/groups/reach/

An unofficial consolidated version of the REACH Regulations, including all amendments up to early-2010, has been produced by the European Commission and is available at http://eur-lex.europa.eu/LexUriServ/LexUriServ.do?uri=CONSL EG:2006R1907:20090627:EN:PDF

Reporting of Injuries, Diseases and Dangerous Occurrences (RIDDOR)

Kathryn Gilbertson, Greenwoods Solicitors LLP

Key points

- From 12 September 2011, statutory reporting to HSE of work-related injuries and incidents under RIDDOR (the Reporting of Injuries, Diseases and Dangerous Occurrences Regulations 1995) will move to a predominantly online system.
- Revised online forms will make the reporting process quick and easy.
- Businesses will no longer report incidents by post or fax.
- The Incident Contact Centre will still take reports of all fatal and major incidents by telephone
- The HSE's Infoline telephone service will end on 30 September 2011.
- The seven online RIDDOR reporting forms will be:
  1. F2508 Report of an injury.
  2. F2508 Report of a Dangerous Occurrence.
  3. F2508A Report of a Case of Disease.
  4. OIR9B Report of an Injury Offshore.
  5. OIR9B Report of a Dangerous Occurrence Offshore.
  6. F2508G1 Report of a Flammable Gas Incident.
  7. F2508G2 Report of a Dangerous Gas Fitting.

Legislation

- Reporting of Injuries, Diseases and Dangerous Occurrences Regulations 1995.

What needs to be reported

You need to report:

- deaths;
- major injuries;
- accidents resulting in over three days' lost time;
- injuries sustained by a visitor or third party who is taken from your premises to hospital;
- diseases;
- dangerous occurrences; and
- gas incidents.

When it needs to be reported

If any of the above instances occur, the Responsible Person shall notify the relevant Enforcing Authority by the quickest practical means, usually a telephone call, and within ten days send a report to the relevant Enforcing Authority.

Death or major injury

If there is an accident connected with work:

- and your employee, or a self-employed person working on your premises is killed or suffers a major injury (including as a result of physical violence); or
- a member of the public is killed or taken to hospital;

you must notify the Enforcing Authority without delay. You should telephone the Incident Contact Centre and complete the appropriate form on the website.

Reportable major injuries are:

- fracture other than to fingers, thumbs or toes;
- amputation;

- dislocation of the shoulder, hip, knee or spine;
- loss of sight (temporary or permanent);
- chemical or hot metal burn to the eye or any penetrating injury to the eye;
- injury resulting from an electric shock or electrical burn leading to unconsciousness or requiring resuscitation or admittance to hospital for more than 24 hours;
- any other injury leading to hypothermia, heat-induced illness or unconsciousness; or requiring resuscitation; or requiring admittance to hospital for more than 24 hours;
- unconsciousness caused by asphyxia or exposure to harmful substance or biological agent;
- acute illness requiring medical treatment, or loss of consciousness arising from absorption of any substance by inhalation, ingestion or through the skin; and
- acute illness requiring medical treatment where there is reason to believe that this resulted from exposure to a biological agent or its toxins or infected material.

Over-three-day injury
- If there is an accident connected with work (including an act of physical violence) and your employee, or a self-employed person working on your premises, suffers an over-three-day injury you must report it to the Enforcing Authority within ten days.
- An over-three-day injury is one that is not 'major' but results in the injured person being away from work or unable to do the full range of their normal duties for more than three days.

Diseases
If a doctor notifies you that your employee suffers from a reportable work-related disease then you must report it to the Enforcing Authority. Reportable diseases include:

- certain poisonings;
- some skin diseases such as occupational dermatitis, skin cancer, chrome ulcer, oil folliculitis / acne;
- lung diseases including occupational asthma, farmer's lung, pneumoconiosis, asbestosis and mesothelioma;
- infections such as leptospirosis; hepatitis; tuberculosis; anthrax; legionnellosis and tetanus;
- other conditions such as occupational cancer; certain musculoskeletal disorders; decompression illness and Hand–Arm vibration syndrome;
- Bursitis of the knee or elbow (Beat knee or Beat elbow);
- traumatic inflammation of the tendons of the hand or forearm or of the associated tendon sheaths, e.g. Tenosynovitis or Tennis elbow;
- carpal tunnel syndrome; and
- Hand–Arm vibration syndrome.

The above list is not exhaustive and, if in doubt, employers should check on the RIDDOR website to see if a condition is reportable. It should be noted that a disease only needs to be reported if its occurrence has been caused by a work-related activity.

Dangerous occurrences
Certain types of incident are designated as dangerous occurrences. This is more than a 'near miss', and the Regulations clearly define the type of dangerous occurrences that should be reported. These include:

- Collapse, overturning or failure of load-bearing parts of lifts and lifting equipment;
- Explosion, collapse or bursting of any closed vessel or associated pipework;
- Failure of any freight container in any of its load-bearing parts;
- Plant or equipment coming into contact with overhead power lines;
- Electrical short circuit or overload causing fire or explosion;

- Any unintentional explosion, misfire, failure of demolition to cause the intended collapse, projection of material beyond a site boundary, injury caused by an explosion;
- Accidental release of a biological agent likely to cause severe human illness;
- Failure of industrial radiography or irradiation equipment to de-energise or return to its safe position after the intended exposure period;
- Malfunction of breathing apparatus while in use or during testing immediately before use;
- Failure or endangering of diving equipment, the trapping of a diver, an explosion near a diver, or an uncontrolled ascent;
- Collapse or partial collapse of a scaffold over five metres high, or erected near water where there could be a risk of drowning after a fall;
- Unintended collision of a train with any vehicle;
- Dangerous occurrence at a well (other than a water well);
- Dangerous occurrence at a pipeline;
- Failure of any load-bearing fairground equipment, or derailment or unintended collision of cars or trains;
- A road tanker carrying a dangerous substance overturns, suffers serious damage, catches fire or the substance is released; and
- A dangerous substance being conveyed by road is involved in a fire or released.

The following dangerous occurrences are reportable except in relation to offshore workplaces:

- Unintended collapse of any building or structure under construction, alteration or demolition where over five tonnes of material falls; a wall or floor in a place of work; any false-work.
- Explosion or fire causing suspension of normal work for over 24 hours.

- Sudden, uncontrolled release in a building of 100kg or more of flammable liquid; 10kg of flammable liquid above its boiling point; 10kg or more of flammable gas; or of 500kg of these substances if the release is in the open air.
- Accidental release of any substance that may damage health.

Note: Additional categories of dangerous occurrences apply to mines, quarries, relevant transport systems (railways etc.) and offshore workplaces. These categories can be found at schedule two of RIDDOR.

If any injury occurs and is reportable under one of the other categories in Regulation 3 then a dangerous occurrence should not be reported separately. If, however, the injury is not reportable under Regulation 3 then the dangerous occurrence must be reported.

Gas incidents
If you are a distributor, filler, importer or supplier of flammable gas and you learn, either directly or indirectly, that someone has died or suffered a 'major injury' in connection with the gas you distributed, filled, imported or supplied, then this must be reported immediately. If you are an installer of gas appliances registered with the Gas Safe Register (formerly CORGI), you must provide details (using form F2508G2 – *Report of a Dangerous Gas Fitting*) of any gas appliances or fittings that you consider to be dangerous, to such an extent that people could die or suffer a 'major injury', because the design, construction, installation, modification or servicing could result in:

- an accidental leakage of gas;
- inadequate combustion of gas; or
- inadequate removal of products of the combustion of gas.

For more, see '*Gas safety*' (p.348).

How to report incidents

All accidents, diseases and dangerous occurrences may be reported to the Incident Contact Centre (ICC). These reports will be processed by the ICC and forwarded to the relevant Enforcing Authority. Employers will receive a copy acknowledgement and incident number for each report. In the case of internet reports, they will receive a PDF version of the completed report. Certain incidents (e.g. fatality, major injuries) require the report to be made by the quickest practicable means (normally by telephone). Also bear in mind that fatalities will be treated as a scene of crime. If it is necessary to report an incident out of hours, you may need to contact the local Enforcing Authority, who will, if required, involve the police.

Record keeping

Records may be kept in any form, i.e. paper or electronically, and must be kept either where the work to which they relate is carried out or at the usual place of business of a Responsible Person. For accidents, records should be kept for a minimum of three years, including:

- Date and method of reporting;
- Date, time and place of event;
- Personal details of those involved; and
- A brief description of the nature of the event or disease.

All workplace accidents should be recorded in an accident book (B1510). This applies to organisations employing ten or more employees and is in addition to any requirement to report to the HSE or other Enforcing Authority.

Proposed changes

Following a three-month consultation with stakeholders in spring 2011, the HSE is seeking a change to the RIDDOR Regulations to extend the reporting threshold. The effect of this would be that employers would no longer have to report 'over-three-day injuries' from mid-2012.

The HSE board agreed to align the accident reporting time with that used in the MED3 Fit Note issued by GPs, so employers would only report 'over-seven-day' accidents in future. The HSE recommended the change at its August board meeting, accepting the results of a public consultation, which found a two-thirds majority in favour of extending the absence threshold. The changes also include extending the period during which duty-holders must notify a RIDDOR-reportable accident from ten to 15 days after the accident. The board will undertake a review in three years' time, to check for any negative impact on injury rates and RIDDOR reporting. The amendments to RIDDOR will need to be laid before parliament by next February, so that the new arrangements can come into force after April 2012.

See also: Accident investigations, p.29; Gas safety, p.348; Health and safety at work, p.361; Occupational cancers, p.512.

Sources of further information

L73 *A guide to the Reporting of Injuries, Diseases and Dangerous Occurrences Regulations 1995*: www.hse.gov.uk/pubns/priced/l73.pdf

Comment ...

Reviewing health and safety

Simon Toseland, Workplace Law Health and Safety

This year the HSE has made great strides in trying to change people's attitudes and perceptions about health and safety. A number of initiatives, which include 'Making the pledge' – businesses signing up to the HSE's policy to provide only sensible risk management in the workplace – the launch of the 'Health and safety made simple' website and, more recently, the 'Red tape challenge', are helping to dilute the opinion that good risk management needs to be a burden to business.

The HSE's budget will be cut by 35% by 2014/15, and this is bound to have a significant impact. We have already seen that the number of workplace fatalities has risen in 2011. There are of course a number of reasons as to why this might be the case, some suggesting that there is direct correlation with the drop in the number of HSE visits and prosecutions. I anticipate that the deterrent will come from the incurrence of higher fines and the HSE's new powers to recover the costs of its regulation where duty-holders are found to be in material breach of health and safety law. There are already indications that a CDM Coordinator will be given a greater role to play in undertaking site audits, to help fill the gaps left by the reduction of the number of inspectors.

I have had a number of stand-out moments this year. I was thrilled when one of our key clients – an international construction company – achieved accreditation with OHSAS 18001. Having set the agenda and orchestrated the

Simon Toseland is Workplace Law's Head of Health and Safety. Simon has over ten years' experience of delivering health and safety consultancy and training, is a Chartered Member of the Institution of Occupational Safety and Health, a Registered Member of the Association for Project Safety, and a Graduate Member of the Institute of Fire Engineers. In January 2011 Simon became approved on the Occupational Safety and Health Consultants Register.

programme for completion we actually hit our target after just 16 months, two months earlier than anticipated.

The launch of the NEBOSH Construction Certificate was also a key milestone for me. Having spent several months planning the developing of the course it was great to finally be able to deliver it in person. The wonderful feedback from the delegates on the launch course made the hard work all the more satisfying. I now look forward to preparing the NEBOSH International Certificate, which we will provide in an e-learning format.

One of the disappointments for me this year was the Government's review of health and safety, which was delivered in the form of Lord Young's Report. Although there were a number of positives in the report, for me it raised too many concerns. Lord Young attempted to bracket some sectors as 'low risk' environments and called for adventure licensing activities to be scrapped. I feel that this over-simplifies health and safety, and gives rise to a potential 'lowering of standards'.

I also think an opportunity has been lost to improve education, management, health and enforcement levels. Arguably the report also promotes an upsurge in HSE activity, which is already significantly stretched. Even now I still get asked for my views on the report, some 12 months after its release.

If given the opportunity to change anything, I would certainly like to see more detailed guidance on competency requirements for health and safety professionals. Much of the scaremongering and the overreaction to trivial risks has been caused by the poor advice given by so-called health and safety consultants. If more guidance was available on assessing competency for specialist areas, i.e. fire safety, I believe this would be a big improvement to the industry.

For 2012 I look forward to the continued growth of Workplace Law and my role continuing to change along with the requirements of our partners.

With the anticipation of an impending economic collapse firmly in the back of our minds, and as the message of sensible risk management begins to hit home, clients want a solutions-based service that is simple and represents good value. For me, being able to innovate will create our future. It doesn't have to be expensive or difficult; indeed it should be simple, true, empathetic and creative.

> "I wish I had a pound for every time someone asked me...
>
> ... Does CDM apply to this?"

S. Toseland.

@SToselandWPL

Risk assessments

Kathryn Gilbertson, Greenwoods Solicitors LLP

Key points

- Risk assessments help to protect our staff and our businesses. The process of risk assessment can become over complicated. However, this need not be the case since all that is really involved is looking at what can go wrong and trying to prevent it, e.g. ensuring spillages are cleaned up promptly so that people do not slip. For most risks, assessment is quite straightforward.
- When producing a suitable and sufficient risk assessment it is essential to involve the right people – those who understand the risk assessment process as well as those who are involved in the work or area being assessed. The end result should be the controls necessary (e.g. safe systems, precautions) to lower risks to as low a level as is reasonably practicable.
- The law does not require you to eliminate all risk, but you are required to protect people as far as is 'reasonably practicable'.

Legislation

- Health and Safety at Work etc. Act 1974.
- Management of Health and Safety at Work Regulations 1999.

The Management of Health and Safety at Work Regulations 1999 (MHSWR) introduced the need for employers to make a suitable and sufficient assessment of health and safety risks to employees and other persons affected by work activities. These, therefore, are the Regulations that require employers to conduct risk assessments in order to identify and manage their workplace risks.

Employers' duties under MHSWR include the requirement to carry out specific risk assessments when employing young persons (16- and 17-year-olds), and where employing women of childbearing age, to take into account any risks to such persons arising from their work. In addition to the duties under MHSWR, many Regulations (COSHH, Manual Handling Operations, Fire, Noise, and DSE Regulations) impose specific duties to conduct, as necessary, additional risk assessments. Professor Ragmar Lofstedt is currently conducting a review into health and safety legislation at the request of the Employment Minister, Chris Grayling. It is likely that some duties for specific risk assessments as detailed in separate regulations could be merged or removed. The report on Prof. Lofstedt's review is due to be published in the autumn of 2011.

Risk assessments, as with other duties in health and safety legislation, should be carried out by competent persons.

Why do we need risk assessments?

The main reason for conducting risk assessments is to ensure that we have adequately considered the things that can go wrong in the workplace and their likely effects. By so doing we can implement measures that will either reduce the likelihood of such events occurring, or, if the worst should happen, limit the severity of injuries that occur.

Case study

BUPA Care Homes (BNH) Ltd was fined £150,000 in January 2011 after a pensioner died at a nursing home in Birmingham. Brigid O'Callaghan (74) died after being strangled by a lap belt when she was left strapped in a wheelchair overnight. Staff at Amberley Court Nursing Home did not properly check on Mrs O'Callaghan, leaving her in a wheelchair in her room rather than helping her to bed. She was discovered dead the next morning by a member of staff, having slipped from the wheelchair to the floor, with the lap belt strap around her neck.

The home had failed to carry out a proper risk assessment and care plan, did not communicate her needs to staff, failed to ensure she could call for help and did not monitor whether night time checks were carried out. BUPA pleaded guilty to two breaches of Section 3(1) of the Health and Safety at Work etc. Act 1974. The first charge focused on the issues most closely connected to Mrs O'Callaghan's death and the second on the potential hazards for the other residents. The company was fined £150,000 in total and ordered to pay £150,000 in costs.

Adequate risk assessments are therefore fundamental to ensuring the effective management of health and safety risks at work. They should take into account:

- people;
- premises;
- plant; and
- procedures.

Since risk assessments are an absolute requirement under health and safety legislation, failure to conduct them is an offence that can be easily prosecuted. Such a failure often comes to light as a result of inspections or investigations by the relevant Enforcement Authorities. The competence of the persons conducting risk assessments may be called into question.

Approach to risk assessment

The HSE suggests that risk assessments should follow five simple steps, notably:

- *Step 1*: Identify the hazards.
- *Step 2*: Decide who might be harmed and how.
- *Step 3*: Evaluate the risks and decide on precautions.
- *Step 4*: Record your findings and implement them.
- *Step 5*: Review your assessment and update if necessary.

Principles of prevention

The best way to avoid risks is to remove the hazard completely. Unfortunately, that is not often an option but nonetheless the law requires that, as an initial step, that's what we should try to do. In other words we should apply a hierarchy of risk controls to a situation in order to arrive at the reasonably practicable measures we need. MHSWR suggests the following:

- *Avoid the risk completely* – change the design or the process.
- *Substitute* – use less hazardous materials, e.g. different chemicals.
- *Minimise* – limit exposure to individuals, perhaps by job rotation.
- *General control measures* – guarding, barriers or warning systems.
- *PPE* – the last resort because it protects only the individual.

Two general principle risk control measures should be applied:

1. Controls should give priority to protecting collective groups rather than individuals.
2. In general terms, the more that human behaviour is involved in the control measure, the more likely it is to go wrong at some point. This explains why PPE is often a weak control measure, since it depends on correct usage by the individual.

A hazard is anything that may cause harm, such as chemicals, electricity, working from ladders, an open drawer, etc., whilst the risk is the chance, high or low, that somebody could be harmed by these and other hazards, together with an indication of how serious the harm could be.

Categories of risk assessment

Generic and site or location-specific assessments

In many workplaces the risks that exist may be considered to be low or trivial, and few if any additional control measures would be necessary to protect people. This is often the case with repetitive work tasks. It may well be that one office environment is pretty much the same as another and therefore a generic risk assessment may well suffice for most of the activities. Caution should be exercised, however, in adopting this approach because incorrect assumptions can be made. It may therefore be necessary to conduct a site- or location-specific risk assessment in order to take into account differing hazards such as:

- differing work conditions;
- differences in work location, e.g. difficult access; or
- time constraints, e.g. pressure of work at different times of the year.

Generic risk assessments at times may result in forgetting that other risks exist. They should not be used as a basis for assuming that common sense will always be applied by employees. As with all risk control measures, it is necessary to maintain an appropriate level of monitoring in the workplace to ensure their continued effectiveness.

Qualitative versus quantitative assessments

Risk assessments may be qualitative in nature, e.g. a written report describing identified hazards and the recommended means of controlling related risks, or quantitative, where some form of numerical rating is applied to identify the level of risk. Some objective quantitative elements exist in risk assessments that involve measurement such as noise, hazardous substances, etc. but risk assessment is not usually an exact science. The choice of qualitative versus quantitative is often a matter of personal or organisational preference.

For a quantitative assessment, a matrix is often used, e.g. low / medium / high or 1–5 scales for likelihood and severity of an accident occurring. The drawback with a numeric approach is that often people are persuaded that the numbers are scientific and thus the risk assessment process becomes driven by the need to achieve a certain score rather than to effectively identify and control risks. The key to successful risk assessment therefore lies largely in the competence of those involved. Whatever choice is made regarding type or method of risk assessment, the results should always be consistent as well as being simple to understand and action.

Conducting risk assessments
The law doesn't define the term 'risk assessment' and neither does it suggest

any given template for conducting them. Whilst that may not seem much help to those conducting them, there is much guidance available regarding the type of control measures that are necessary in order to achieve the usually required standard of 'reasonably practicable'. Therefore, when conducting risk assessments, it is wise for the assessor to be aware of the information that is available for the type(s) of risks involved. Sources of such information include:

- Regulations, e.g. Work at Height Regulations 2005.
- Approved Codes of Practice (ACoP), which provide practical interpretation of the legislation for employers.
- Good practice guidance notes from the HSE, special interest groups and trade associations.
- Company's own health and safety policy and arrangements document (sometimes more exacting than the law itself).
- The people doing the job are invaluable; you need to know how things are actually done rather than just how they should be done.
- External consultants, e.g. asbestos specialists.

With regard to the cost of the control measures deemed necessary, a common definition of the term reasonably practicable is 'the cost in time, trouble, effort and money versus the level of risk that exists'. The cost bears no relation to the size of the business; it is purely dependent on the level of risk that exists. If an accident occurs, it may well be necessary for the employer and, by

extension, the assessor, to show that it was not reasonably practicable to have done more to prevent it. To do this, a minimum requirement would be to consult relevant information sources (particularly ACoP), and expertise and thus employ good practice.

Don't just file them away

Employers with five or more employees have a legal duty to record risk assessments in writing. That, however, should not be the end of the matter since the exercise has little value if the people affected are not aware of the key findings and control measures. These have to be communicated, whether by memos, training, team briefs, etc. The method of communication must reflect the needs of the audience and the seriousness of the risk that exists. When conducting a risk assessment, it is also important to set a date for review to check whether or not it is still adequate. Review would also be necessary as a result of changes in working practices, new plant, changes in legislation and especially as a result of an accident.

Finally, it is necessary to get out into the workplace and make sure that not only are the risk control measures in place, but they are also working.

> *See also*: Construction site health and safety, p.177; Health and safety at work, p.361; Working at height, p.399; Training, p.678.

Sources of further information

INDG 163 Five steps to risk assessment: www.hse.gov.uk/pubns/indg163.pdf

Safety inspections and audits

Rob Castledine, Workplace Law

Key points

- Monitoring health and safety standards can be categorised into two distinct types of activity – 'reactive' by way of incidents, visits and complaints, and 'active' by way of inspections, audits, surveys and other checks.

- Most organisations put a lot of effort into recording, investigating and responding to accidents and incidents at work. Whilst this is important, it's obviously better to be preventative through adopting a robust system of safety inspections and audits, rather than waiting for things to go wrong and then responding by way of an investigation.

- Organisations should adopt a range of different methods for monitoring safety standards and these should address a number of key areas including premises, plant / equipment, people and procedures – known as the 'Four Ps'.

- Inspection regimes should also ensure statutory requirements for inspections are being addressed and that inspections take into account the relevant risks that the workplace / work activity poses.

Legislation

- Health and Safety at Work etc. Act 1974.
- Safety Representatives and Safety Committees Regulations 1977.
- Management of Health and Safety at Work Regulations 1999.

Introduction

Maintaining good standards of health and safety in the workplace requires an employer to adopt a number of different methods and processes. Of course, it goes without saying that it's far better to be preventative in relation to workplace injuries, rather than simply responding to and dealing with accidents and incidents where someone has been injured or something has been damaged. By that time it's already too late.

Monitoring of safety standards in the workplace requires careful planning to ensure all areas and likely issues are addressed and that the inspection process

(whatever it is) gives some 'value' to the over safety culture. Far too often, a standard generic tick list is drawn up without any real thought over what the inspection is trying to achieve.

In relation to workplace inspections, health and safety legislation is often not clear enough and very rarely specifies the frequency and type of inspections that are required. It is therefore left to HSE Approved Codes of Practice, HSE Guidance Notes or industry standards to provide some further clarity on what might be required. Bear in mind also the particular circumstances or level of risks posed by a particular workplace or activity, or even the people who might be at risk could also determine how often and to what extent inspections might be required.

Monitoring

'Active monitoring' is the term the HSE uses to describe the range of methods that an organisation can use to check that good standards of health and safety are being maintained. In its publication, HSG 65 'Successful Health and Safety Management', the HSE explains that:

'.... monitoring reveals how effectively the health and safety management system is functioning and ... looks at both hardware (premises, plant and substances) and software (people, procedures and systems).'

OHSAS 18001 'Occupational Health and Safety Management Systems', published by the British Standards Institution (BSI) also emphasises the importance of checking, by way of monitoring the effectiveness of workplace controls using both inspections and audits.

Inspection vs audit

There can be confusion between the terms 'inspection' and 'audit' and whilst the terms might mean different things in different organisations, it's important to clarify the terms

In HSG 65, the HSE defines 'auditing' as 'the structured process of collecting independent information on the efficiency, effectiveness and reliability of the total safety management system and the drawing up plans for corrective action,' whereas 'inspections' are more concerned with checking day-to-day safety measures (i.e. a physical check on machine guarding, housekeeping standards, welfare facilities, condition of ladders and steps, etc.).

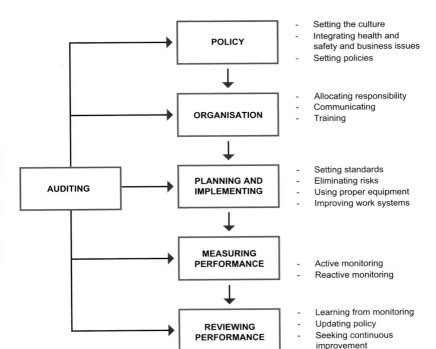

By inference, auditing is more focused on checking that safety systems are being effective (i.e. the safety policy and procedures), leaving inspections to monitor day-to-day standards.

What needs inspecting?

Adopting the HSE's suggestion that both hardware and software aspects should be checked, a logical approach is one based on the four Ps – premises, plant, people and procedures.

Premises

As many workplace injuries are caused by poor housekeeping, it makes good sense to have a robust programme for checking that the premises and workplace are maintained to acceptable standards. Also, as the Workplace (Health, Safety and Welfare) Regulations 1992 and the CDM Regulations 2010 lay down minimum standards for workplaces and construction sites respectively, any workplace inspection should be addressing the following issues.

- General housekeeping / tidiness
- Condition of stairs
- Signage / notice boards
- External yard areas
- Car parking
- Pedestrian / vehicle routes
- General lighting / ventilation
- Office areas / storage facilities
- First aid facilities
- Refuse areas
- External fire routes
- Specific hazardous areas –chemical stores, roof areas, plant rooms
- Welfare facilities – toilets, washrooms, canteen areas
- Fire routes / equipment
- Site security
- Loading / unloading bays
- Plant rooms
- Areas where asbestos has been identified

Plant and equipment

Any equipment / machinery that is used should be checked to ensure that it is safe to use. In terms of equipment, an organisation needs to consider fixed / portable equipment, personal protective equipment (such as respiratory protection and fall arrest equipment), vehicles, equipment provided in plant rooms, emergency / life-saving equipment and equipment that is used off site such as portable access scaffolds or hand tools.

Equipment inspections tend to fall into three typical categories.

Operator / user checks

Before being used to check that the equipment safe – e.g. ladders, stepladders, forklift trucks, access equipment and power presses.

Periodic maintenance

In the same way that a car gets serviced, certain types of work equipment will need 'servicing' to ensure that they continue to operate correctly. Maintenance inspections should be linked to a planned preventative maintenance programme (PPM). HSE guidance also details requirements for specific items – e.g. electrical equipment should be 'regularly inspected' and a portable appliance testing regime (PAT) will normally fulfil this requirement.

Statutory inspections

Normally undertaken by an Insurance Engineer, this is akin to a car's MOT. Examples include passenger lifts (six-monthly), forklift trucks (12-monthly), cranes (12-monthly), local extraction ventilation, e.g. welding extraction (14-monthly), eye bolts (12-monthly), and pressure equipment (between 12 and 24 months, depending on the type of equipment).

People

If people work safely then they won't be injured. However, of course, there could be many reasons why someone does not work safely. Inspection regimes that address a worker's safety practices (behavioural safety) can add real value to the overall safety effort. However, behavioural inspections or safety observations may not be directly applicable to all types of workplaces or workers and may only be suited to those organisations that are fairly advanced in terms of their health and safety processes.

Safety observations not only address the day-to-day safety issues (such as the wearing of safety equipment, working in a clean and tidy workplace and using the correct tool for the right job etc.) but also consider a person's attitude to what they might be doing (am I likely to be injured and if so, how badly?) or even whether there might be situations where they could be distracted.

Inspecting 'people' does not specifically have to address how they are working. There are a range of other 'people' issues that could be inspected or checked. Examples include pre-employment or medical surveillance, eye tests for display screen equipment users, hearing tests for people working in noisy areas or lung function tests for workers exposed to specific hazardous substances (such as iso-cyanate based paints or weld fumes).

Procedures

Inspecting procedures really covers checking that work is being done safely, so there could be a close link between safe systems of work and whether people are working safely (behavioural safely). These types of inspections will need to be more focused on actual work tasks or activities rather than the environment in which a

person is working. It's also good to check that relevant forms and paperwork are being correctly completed in the workplace (e.g. visitors' book, contractor inductions, maintenance logs, permits, etc.).

Who should be undertaking workplace inspections?

There's no hard and fast rule that covers who should undertake the inspection, but it goes without saying that anyone who is asked to undertake an inspection should know what they are doing. This includes having some knowledge of what they are looking for, understanding any forms or checklists that are used, and knowing what to do after the inspection, in terms of addressing any deficiencies.

There are three occasions when Trade Union Appointed Safety Representatives are legally entitled to undertake safety inspections:

1. If the workplace has not been inspected in the previous three months, which means that more frequent inspections can only be allowed with agreement from the employer.
2. If there has been substantial change in the conditions of work (e.g. after the installation of new equipment, new processes, building alterations, etc.).
3. Where there has been a notifiable accident or dangerous occurrence (e.g. reported under the Reporting of Injuries, Diseases and Dangerous Occurrences Regulations 1995 – a RIDDOR event).

Interestingly, non-union appointed safety representatives have no legal entitlement to inspect the workplace. However, in many workplaces it makes perfect sense to get Safety / Employee Representatives (Trade Union or otherwise) directly involved in safety inspections.

A robust safety inspection regime will ensure that people at different levels in the organisation get involved – senior manager safety tours, line manager / supervisor inspections and checks by individual operatives.

Frequency of workplace inspections
The following factors should be considered when deciding upon an inspection frequency:

- The extent of the risks involved – is it an office or high risk chemical plant?
- What are the implications if safety standards are not maintained – minor or major incident?
- How many people may be affected – few to many, employees or third parties?
- Recommendations from suppliers and service engineers.
- Statutory requirements.
- The extent and frequency of any other maintenance or inspection regimes – inspection by third parties, insurance engineers.
- The historical evidence of poor standards of workplace safety, e.g. equipment being tampered with or misused, or the previous housekeeping conditions in a particular area.

For most workplaces, quarterly or six-monthly safety inspections will probably be adequate, but bear in mind the above factors, which might lead an organisation to adopt a more frequent inspection schedule. Everyone is different and a blanket approach simply does not work.

Getting better value out of your risk assessments
It's also worth considering how an organisation can utilise any risk assessments that have been undertaken, as often these can be useful checklists in their own right.

A 'suitable and sufficient' risk assessment will have identified the hazards, who might be harmed and the controls required to minimise the risks. In the same way, a workplace inspection should check what hazards are present, who they pose a risk to, and how they should be controlled. So a well structured risk assessment can be easily turned into a well designed inspection checklist.

Inspection checklists and forms
Often, organisations fail to properly think through their workplace inspection forms, which can lead to confusion and misinterpretation.

Common mistakes include:

- Using a generic template where only some of the questions are relevant to the place being inspected – inspections should be risk based and should address the hazards / risk in that particular area.
- Being too brief or too detailed.
- Not giving any aide memoires or headings, so people are left to decide what to check for themselves.
- Not having space for observations or actions required.
- Having questions that are vague or misleading.

It can also be useful to have the questions structured in such a way that all positive answers (i.e. things are all 'OK') are clearly identifiable from any negative answers (i.e. things that need addressing). Consider also using Yes/No or ✓/x, as most people are naturally drawn to assuming that any 'No' or 'x' comments means that there is an issue that needs addressing.

Consider the two simple and abbreviated checklists overleaf, which will help to explain some of these points further.

Are floor coverings likely to cause a trip hazard?	No	
Are there any lights that are broken / defective?	No	
Is there sufficient ventilation?	Yes	
Are there any signs that cannot clearly be seen?	No	
Are there any fire routes that are obstructed?	No	
Fire extinguishers provided and accessible?	Yes	

Table 1. An example of an unstructured and complicated checklist.

Are floors free from trip hazards?	Yes	
Are all lights working and in good order?	Yes	
Is there sufficient ventilation?	Yes	
Are signs posted up and easy to read?	Yes	
Are fire routes clear and unobstructed?	Yes	
Are fire extinguishers provided and accessible?	Yes	

Table 2. An example of an easy to use and straightforward checklist.

It would be hard to interpret the overall findings of this inspection, as there's a mixture of Yes and No comments and as people are naturally drawn to the 'No's' it appears as if there are four issues (No's) that need addressing, whereas in fact, things are all OK and no corrective action is required. By rewriting the questions, it is far easier to see that the workplace is OK and that no action is required.

Record keeping
Records should be kept for as long as they need to be to show that the safety systems are working. Bear in mind that if anyone is likely to claim compensation as a result of a work-related injury or condition it could be up to three years from the date of the injury (or even three years from the date of diagnosis in relation to long-term exposure or health related matters – Limitation Act 1980). So in some cases your organisation might need to show that it was inspecting things several years ago and not just now! It's also important to be able to demonstrate that, where issues have been identified as requiring some corrective action, something has been done. It is therefore good practice to include a sign off part on the inspection form for a person to be able to acknowledge that the matter has now been put right (date, signature and brief description of action taken).

Conclusion
An organisation needs to ensure that there is the right balance between reporting and investigating incidents at work and being proactive in going out and checking things. Adopting a robust system of workplace / equipment inspections and safety audits should help to identify potential situations and conditions that could lead to accidents.

See also: Accident investigations, p.29; Construction site health and safety, p.177; Health and safety inspections, p.381; Health and safety management, p.388.

Sources of further information

HSG 65 – Successful Health and Safety Management:
www.hse.gov.uk/pubns/books/hsg65.htm

OHSAS 18001 Occupational Health and Safety Management Systems:

HSE – A Guide to Measuring Health and Safety Performance:
www.hse.gov.uk/opsunit/perfmeas.pdf

Safety signage

Kate Gardner, Workplace Law

Key points

- The purpose of safety signs is to warn and/or instruct people of the nature of certain risks and the measures to be taken to protect against them.
- Employers have a duty to display safety signage where risk remains after all other appropriate steps to avoid or control risks have been taken.
- Employers must provide and maintain any such safety sign.
- Employees must receive suitable and sufficient instruction as to the meaning of safety signs and what they need to do to comply with them.

Legislation

- Health and Safety (Safety Signs and Signals) Regulations 1996.
- Management of Health and Safety at Work Regulations 1999.

Statutory requirements

Employers are required to make a suitable and sufficient assessment of the risks to the health and safety of employees that they are exposed to whilst at work. In addition, employers have the same duty to non-employees involved in the workplace. This duty also extends to the self-employed in relation to their own wellbeing. These assessments must be reviewed if any element affecting the work changes or if the employer has any other reason to suspect that the assessment is no longer valid.

If, after assessing a certain risk, the employer believes there is some residual risk which cannot be avoided or reduced by other means, safety signage to warn of the hazard should be provided.

Signs have the purpose of either warning, instructing, or both, as to the nature of the risk and how the risk should be protected against. This is a decision to be taken by the employer and should only be taken

where there is no significant risk of harm remaining, having considered the nature and magnitude of the risks arising from the work concerned.

A safety sign is information or instruction about health and safety at work and may comprise a sign board, a safety colour, an illuminated sign, an acoustic signal, a verbal communication or hand signal. Where the risk relates to traffic control on an employer's premises, traffic signs of the same colour and style used on public roads should be put up where appropriate. Traffic routes are recommended to take the form of continuous white or yellow lines. There are separate regulations that relate to the supply of substances, equipment and products and to the transport of dangerous goods.

The 1996 Regulations on safety signs implement European law, which is an attempt to standardise safety signage across Europe, so that wherever in the EC the signs are placed, they have broadly the same meaning.

Employers have a further legal duty to provide information, instruction and training to employees on the safety signs used in the workplace. This should involve the

Meaning	Example	Sign type
Toxic material		Warning sign
Dangerous location		
No pedestrian access		Mandatory safety sign
Safety helmets must be worn		Mandatory safety sign
Emergency exit		Fire exit and equipment sign
Explosive material		Dangerous substance sign

measures to be taken in connection with safety signs, for example wearing safety footwear beyond a certain point.

Employers are also required to label pipes that contain dangerous substances where, for example, there is a discharge or sample point and therefore a risk of leakage. These are similar to those seen on receptacles containing dangerous substances.

Safety signs that the employer uses in the place of work must be large enough to be seen and understood and of the types specified below, depending upon the type of risk.

See also: Construction site health and safety, p.177; Fire extinguishers, p.306; Health and safety at work, p.361.

Safety signs and signals: www.hse.gov.uk/pubns/priced/l64.pdf

How well do you know your safety signs? Could you decipher the meaning of every safety sign without the aid of the accompanying written warning? Images alone can be unclear and confusing – creating a potential hazard for those whose first language is not English, who have a visual impairment or learning difficulties.

Workplace Law has created a quick quiz for members to take part in. We've displayed 20 regularly used safety images in an interactive quiz, which you can find at www.workplacelaw.net/news/display/id/15609. Take part to find out how much you really know!

Scaffolding

Alex Foulkes, Scafftag

Key points

Scaffolding must be inspected by a competent person:

- before it is put into use;
- at seven-day intervals until it is dismantled;
- after bad or excessively wet weather or high winds or another event likely to have affected its strength or stability; and/or
- after any substantial additions or other alterations.

Legislation

- Health and Safety at Work etc. Act 1974.
- Construction (Head Protection) Regulations 1989.
- Supply of Machinery (Safety) Regulations 1992 (as amended by SI 1994 No 2063).
- Health and Safety (Safety Signs and Signals) Regulations 1996.
- Lifting Operations and Lifting Equipment Regulations 1998.
- Provision and Use of Work Equipment Regulations 1998.
- Management of Health and Safety at Work Regulations 1999.
- Management of Health and Safety at Work Regulations 1999.
- Work at Height Regulations 2005.
- Construction (Design and Management) Regulations 2007.

Introduction

This chapter is intended to outline the procedures and checks that need to be put in place when erecting, using and dismantling scaffolding, to ensure the safety of those using the equipment, and those nearby.

Work at height

The Work at Height Regulations, consolidating previous UK and European legislation on work at height, came into force in April 2005. They apply to all work at height, both internal and external, where there is a risk of a fall liable to cause personal injury. The Regulations bring together all previously existing work at height regulations and state the minimum health and safety requirements for the use of equipment for work at height. There are now no height limits. These Regulations apply to all work at height, including below ground level and low height, where there is a risk of a fall liable to cause personal injury.

In response to the Regulations, the BSI (British Standards Institution) published the British Standard, BS 8437: 2005.

TG: 20

The TG:20 issued by the NASC is a technical guidance document on the use of the European Standard BS EN12811-1 that advises on all tube and fit scaffold design in the UK.

TG20:08 is based on the previous UK British Standard BS 5973: 1993 for access and working scaffolds and special scaffold structures in steel, which was withdrawn by British Standards because it was not compatible with the new European Standard BS EN 12811-1.

Fundamentally, the principles of BS 5973 remain unchanged and there are very few changes in the day-to-day work of a scaffolder. Although the scope of TG20:08 is generally wider than in previous guidance, it is the justification of the design that has altered rather than the scaffold structure itself.

While dealing with many common applications, TG20:08 defines a range of scaffolds, referred to as 'basic scaffolds', for which no further design is required to establish the capability of the scaffold.

Management duties

The Work at Height Regulations place duties on employers, the self-employed, and any person who controls the work of others (for example, facilities managers or building owners who may contract others to work at height). Such persons have a duty to ensure that work at height is properly planned, appropriately supervised with consideration to the complexity of the work, and carried out in a manner that is safe. This should include ensuring that:

- all work at height is properly planned and organised;
- those involved in work at height are competent;
- the risks from work at height are assessed and appropriate work equipment is selected and used;
- the risks from fragile surfaces are properly controlled;
- equipment for work at height is properly inspected and maintained; and
- erecting, altering and dismantling of complex designed scaffolding (e.g. suspended scaffolds, shoring, temporary roofs, etc.) should be done under the direct supervision of a competent person.

The person in charge of the work in hand must ensure that the fall of materials or objects from height is prevented, or where it is not reasonably practicable to prevent falling material, that steps are taken to prevent any person below being hit. To prevent use by unauthorised persons of incomplete scaffolds, relevant warning signs identifying the areas where access is not permitted should be displayed at the access points to these areas. Entrance to these areas by unauthorised persons should be controlled. Controls could be work permit systems, and/or barriers and warning signs.

The term 'competent person' referred to in many pieces of health and safety legislation is not legally defined, but is generally accepted as someone who has the necessary experience, training, qualifications and/or skills to do whatever task is required, competently.

For instance, the CITB (Construction Industry Training Board) defines a competent person as 'A person who has practical and theoretical knowledge together with actual experience of what they are to examine, so as to enable them to detect errors, defects, faults or weaknesses which it is the purpose of the examination or inspection to discover; and to assess the importance of any such discovery.'

Selection of work equipment

When selecting equipment for work at height, the duty holder is required to prioritise collective measures over personal measures. This means that collective protection offered by scaffolding, guardrails etc. should be given preference to personal fall protection systems such as work positioning, rope access and fall arrest. However, it does not prohibit the use of the latter if it is deemed the most appropriate work equipment, given the nature of the work to be carried out.

With regards to scaffolding, the BS EN 12811-1: 2003, NASC TG 20:08, Work at Height Regulations 2005 and Construction (Design and Management) Regulations 2007 state that:

- all components must be undamaged and serviceable;
- all tubes and fittings must comply with BS 1139 part 1. Tubes must not be bent, split, or distorted or corroded. Fittings must be regularly serviced and maintained; and
- all scaffold boards should comply with BS 2482 or to the National Association of Scaffolding Contractors Technical Guidance note TG5:91. Boards must not be excessively split, warped or knotted.

Other components

1. *Gin wheels.* To be secured on load bearing couplers S.W.L. Clearly marked 'LOAD NOT TO EXCEED 50 KGs'.
2. *Guard Rails (WAHR).* Minimum height of 950mm on working platforms, secured on the inside of the standards. An intermediate guard rail shall be positioned to limit any gap to 470mm.
3. *Toeboards.* Minimum height to 150mm fixed on all working platforms.
4. *Boarding.* To be close boarded and end butted throughout. Overhang of the boards of any thickness should not exceed four times their thickness and should not be less than 50mm e.g. details on 38mm board min. 50mm – max. 150mm.
5. *Transoms.* Maximum spacing 1.2m
6. *Sheeting / debris netting.* To be fixed only to structures designed for their use.
7. *Standards.* Centres dependent upon duty use (see Table *opposite*).
8. *Lift heights.* Centres not to exceed 2m, but base lift may be up to 2.7m max. for passage of pedestrians.

9. *Ties.*
   - Types of ties – box, lip. through, reveal and masonry anchor.
   - Class of ties:
     - Light duty 3.5kN in tension.
     - Standard 6.1kN in tension.
     - Heavy duty 12.2kN in tension.
   - Layout, on alternate standards on alternate levels max. 4m apart.
   - Masonry anchor ties should have 5% tested for pull out.
10. *Ledger joints.* Not more than 1/3 into a bay and be staggered.
11. *Standard joints.* Joints in standards should be staggered.
12. *Standard / ledgers.* Fixed with right angle couplers.
13. *Bracing (TG20:08):*
    - Ledger, façade and plan. Ledger bracing should be fixed at alternate pairs of standards in all lifts. Note: TG20:08 does take account of part ledger braced structures where ledger braces can be removed from a maximum of two levels at any one time.
    - Façade bracing, set at between 35° and 55° and shall be fitted at least every six bays. It should extend across two ledger braced bays (preferred), be in a continuous line, or fitted over a single bay.
    - Plan bracing, if the façade bracing is fitted over a single bay only and the scaffold is greater than 8m in height, then plan bracing must be fitted every four lifts and every 12 bays.
14. *Baseplates.* Generally placed below standards and 150mm x 150mm.
15. *Access hierarchy:*
    - Staircases.
    - Ladder access bays with single lift ladders.
    - Ladder access bays with multiple lift ladders.
    - Internal ladder access with a protected ladder trap.
    - External ladder access using a safety gate.

Duty	Class	BS 12811-1 Loading	Standard centres
Very light duty	Class 1	0.75 kN/m2	2.7m
Light duty	Class 2	1.50 kN/m2	2.4m
General purpose	Class 3	2.00 kN/m2	2.1m
Heavy duty	Class 4	3.00 kN/m2	1.8m
	Class 5	4.5 kN/m2	Designed
	Class 6	6.0 kN/m2	Designed

Risk assessment

A risk assessment must be completed prior to the start of any work; this will identify the appropriate equipment for the task, as well as giving guidance on equipment inspection requirements. Contractors should always be required to provide a method statement with a risk assessment, before commencing work.

When considering the nature of the work to be carried out, take into account:

- the working conditions, slopes, poor ground conditions, and obstructions;
- distance to be negotiated for access and egress;
- distance and consequences of a fall;
- duration and frequency of use;
- evacuation and rescue; and
- additional risks posed by the use, installation or removal of the work equipment.

Good practice

General access scaffolding

Scaffolding must be inspected by a competent person:

- before it is put into use;
- at seven day intervals until it is dismantled;

- after bad or excessively wet weather or high winds or another event likely to have effected its strength or stability; and/or
- after any substantial additions or other alterations.

A competent person must carry out general access scaffolding inspections. Written proof of the competence of persons used to inspect scaffolding should always be obtained by the person responsible for the scaffolding and inspection work. A written report must be prepared by the competent person. The report should be written out at the time of the inspection but if this is not possible it must be provided to the responsible person within 24 hours. A copy of the report should be kept on site and a further copy be retained for a period of three months from the completion of the work with the person on whose behalf the inspection was carried out.

The Blue Book from Scafftag provides a ready-made record management system with carbon copy reports sheets to enable shared records amongst all relevant parties.

Handover certificates should refer to relevant drawings. The scaffold inspection report should note any defects and corrective actions taken, even when those actions are taken promptly, as this assists with the identification of any recurring problems.

If a scaffold fails inspection this must be reported by the person carrying out the inspection, to the person responsible for the scaffolding, as soon as possible. The use of a Scafftag® visual tagging system attached to the scaffold would prohibit further use until the necessary remedial action and re-inspection has been completed.

The necessary remedial action must be carried out by the scaffolding company and a re-inspection carried out by the competent person before the scaffolding can be re-used. Where scaffolding is erected in an area accessible to employees or the public the following should apply:

- The minimum amount of equipment and materials should be stored in the vicinity of the scaffold.
- People should be prevented from walking under or near the scaffold by means of barriers and signs.
- All means of access to the scaffold from ground level, such as ladders, should be removed when scaffolding is left unattended.
- Consideration should be given to whether the area local to the scaffolding should be a designated hard hat area.

In addition to scaffolding structure inspections, scaffolding tie tests should also be carried out, prior to commencing work, by a suitably competent person other than the installer of the original fixings. In addition, a minimum of three anchors must be tested and at least five percent of the overall quantity (one in 20).

If, when using load test equipment, any of the ties fail to meet the safety margin a full inspection is required and the rate of proofing should be doubled to one in ten. Each anchor should also be tested to at least one-and-a-half times the required tensile load with no significant movement of the fixing. The use of a Structural Tie Test Tag attached to the scaffold tie will record the specific tie test date / test load applied / required tensile load and a unique reference number for ongoing monitoring. The use of a Structural Tie Warning Tag attached to the scaffold tie is a prohibition-only tag for those ties not being tested.

Mobile towers

Formal instruction and training must be provided for all those who erect tower scaffolds and these persons must be deemed competent. Training for erection and use must be provided by a competent person and should be recorded in the employee's training record. Towers should rest on firm level ground with the wheels or feet properly supported. Safe access to and from the work platform must be provided. Tower scaffolds must be inspected by a competent person and a record of the inspection be made and kept for three months after dismantling of the scaffold. Inspections are required:

- before first use;
- after substantial alterations;
- after any event likely to have affected its stability; and/or
- if the tower remains erected in the same place for more than seven days.

If a tower fails inspection this must be reported by the person carrying out the inspection to the person responsible for the tower, as soon as possible. The use of a Towertag® status tagging system attached to the tower would prohibit further use until the necessary remedial action and re-inspection has been completed. Where a tower is erected in an area accessible to employees or the public:

- People should be prevented from walking under or near the tower by means of barriers and signs.
- The minimum amount of equipment and materials should be stored on the scaffold tower.
- All means of access to the tower from ground level, such as ladders, should be removed when scaffolding is left unattended.
- Consideration should be given to whether the area local to the tower should be a designated hard hat area.

Other access equipment

The following requirements apply to all working platforms such as scaffolding, MEWPs (Mobile Elevating Work Platforms), cradles, trestles, podiums, steps and stairways. Such equipment should be:

- located on a stable and suitably strong surface;
- of suitable strength and rigidity for the purpose;
- if appropriate, prevented from slipping by attachment to another structure or surface (e.g. by an anti-slip device);
- capable of being erected and dismantled safely, without the risk of components becoming displaced and endangering others;
- stable while being altered or modified; and
- if wheeled structures, such as a tower scaffold, prevented from moving during work at height (e.g. wheel locks).

See also: CDM, p.137; Construction site health and safety, p.177; Working at height, p.399; Ladders, p.425; Lifting equipment, p.448.

Sources of further information

www.scafftag.co.uk

www.hse.gov.uk

www.nasc.org.uk

NASC Technical Guidance TG20: www.nasc.org.uk/tg20_guidance_update

Security

Royal Institution of Chartered Surveyors (RICS)

Key points

- This chapter deals with the threat to the physical security of buildings and their occupants, rather than other aspects of security such as financial crime and data theft.
- Security planning is an ongoing management activity that requires regular review to ensure that it remains relevant to the threat, building roles and organisation operations. The facilities manager will recognise the changes in building management that may initiate the need for a review in security. They will be in a position to advise the organisation that its operation may be vulnerable and where aspects of security may need to be reviewed.
- Security plans should be included within the business operational planning and management process, and be continuously monitored and reviewed as appropriate.
- Security plans should be reviewed periodically in response to changes in the threat, market position, and operational approaches, all of which may change the organisation's risk profile and vulnerability. The extent of the review will be influenced by the extent and depth of change. However, the entire security profile should be examined periodically to ensure that standards are maintained and systems are fit for role. Periods between reviews can be based upon the importance of the site and the operations it supports and prioritised accordingly.
- Security plans should be reviewed at least on an annual basis.

Ownership of security plans

Ownership of the security plans should be defined as part of this process and should sit at the most appropriate level of the host organisation, depending upon the potential risk to business. This could mean that a main board director has the key responsibility, but wherever the responsibility lies it is essential to identify that individual and thus define ownership as well as operational and reporting responsibilities.

Development of a security strategy

Any security strategy will be driven by the business needs of an organisation and will therefore be an essential aspect of the business cycle, particularly insofar as budget and risk are concerned.

It is important, therefore, to demonstrate that the strategy is proportional to the organisation's needs and operation, as an over-emphasis on security will be costly and will influence the perceptions and attitudes of customers, clients and employees, and could adversely impact upon the effectiveness of an organisation in delivering its core business product. In order to ensure that there is no ambiguity or uncertainty, the strategy should clearly articulate the organisation's culture and approach to security. The culture will reflect organisation practices and allow all parties to understand that security requires good communication across the organisation, with an opportunity for engagement and consultation.

The security strategy should be driven by the senior management team, with support throughout the management structure. The security strategy should be agreed and signed off by the most appropriate level in the host organisation, ensuring that all risks are understood and the risk of non-compliance is clearly demonstrated. The service delivery can then be managed against measurable outputs geared toward the business plan of an organisation.

Sector requirements

Regardless of the technologies and processes that an organisation may adopt, it is important to consider how the security strategy complements its needs and sector requirements in terms of regulatory expectations.

It is clear that those engaged in the leisure industry, for example, will need to adopt a low profile but effective security strategy, which protects their assets from a number of diverse threats. At the same time they will not wish to adversely affect the experience of their visitors who, although they will expect to be secure, will not normally expect to be affected by overt and robust physical security measures.

Those in the financial services will be concerned about the potential impact of the loss of data and information. They will also be influenced by the expectations of the FSA, which requires a risk-based and demonstrable approach to security. Equally, they may consider ISO 27001 accreditation as an important factor in terms of their management systems. As a consequence their strategy will reflect these requirements.

Organisations that are engaged with government contracts will be required to adopt a strategy that reflect the standards required to protect government assets.

Defining risks

Although it is advisable for a risk-based approach to be considered when developing the strategy, this should be done in the context of these predetermined standards that may be defined and imposed upon the organisation by a third party. Risks can be categorised in various ways but they are most commonly defined as those threats that, should they materialise into events, have the potential to cause loss in some form, to the organisation.

Typically they can be considered under the headings of:

- general acquisitive crime;
- terrorism;
- low level criminality and criminal damage;
- anti-social behaviour; and
- protest and disorder.

Within many organisations, natural threats such as flooding may also be considered. These risks can further be considered in relation to their impact upon the organisation's people and physical assets, its ability to maintain continuity of business, and its ability to continue to meet statutory and contractual liabilities.

Principles of a security strategy

Any security strategy should ultimately reflect the organisation's needs and be cost effective, while being sufficiently robust to counter the threats. A considered process of planning and consultation will ensure that security is proportional and that unnecessary and restrictive measures are not imposed upon the organisation, facilities, building construction and subsequent operation. Equally, it will ensure that security reflects the organisation's profile and culture while remaining compliant with regulatory requirements.

The ability to integrate the full range of measures, though not always possible, promotes a strategy of defence in depth. The individual measures serve to provide a specific form of protection to defeat a threat; however, where they overlap they are more likely to defeat a wider range of threats from the more determined attack.

This approach serves to support the following principles:

- *Deterrence* – the effect of security on an adversary's decision making process.
- *Detection* – the ability to identify and monitor an attack.
- *Delay* – time taken for an adversary to counter security measures to access an asset of value and ability to respond.
- *Detention* – the ability to respond to an adversary and detain them before escape.

The principles highlighted above are not exclusive and others could be equally valid, but they reflect the main issues in applying security in the built environment and across organisations.

Proportionality

It is advisable to ensure that security reflects not only the potential threats that an organisation may face, but also their attitudes to risk, while taking into consideration the issues of image, reputation and customer / client expectations. This will allow security measures, be they physical or procedural, to be applied in a manner that is proportional to the organisation and the protection of their assets.

Failure to address security in the context of the level of protection it offers can result in unnecessary expenditure and the possibility of creating an oppressive environment.

Integration and coordination

The integration of security measures will ensure that the diverse threats are mitigated by an interlocking series of measures that provide defence in depth and a layered approach. It is advisable to coordinate security procedures, ensuring that there is an effective response to the security alarms, incidents and compromises in order to reduce the probability of a successful attack and increase probability of the detection, and in doing so reduce the loss or compromise of assets. It is worth considering engagement with other organisations in the area, police authorities and security agencies to ensure that security is effectively coordinated and integrated with local initiatives.

Security issues should integrate with other management activities, including the organisation's operational, strategic and investment planning to ensure that it is considered at the appropriate level and included as part of the wider management activities.

See also: Bomb alerts, p.56; Insurance, p.412; Night working, p.502; Occupiers' liability, p.524.

Sources of further information

This chapter is taken from *Security 1st edition*, which follows on from the White Book, *The strategic role of facilities management in business performance,* published by RICS, and is one of a series that seeks to assist the chartered facilities management surveyor in delivering individual or collective services to clients. With acknowledgements to Gary Hellawell and Mark Whyte, Carillion (TPS) and David Parkinson, and the RICS Facilities Management Professional Group Board.

www.rics.org

BS 8220-1:2000: Guide for security of buildings against crime. Dwellings.

BS 8220-2:1995: Guide for security of buildings against crime. Offices and shops.

BS 8220-3:2004: Guide for Security of buildings against crime. Storage, industrial and distribution premises.

BS 7958:2009: Closed-circuit television (CCTV). Management and operation. Code of practice.

Shared premises and common parts

Kevin Boa, Pinsent Masons Property Group

Key points

- Tenants in occupation of shared premises are usually subject to a service charge imposed by the landlord for the provision of services. The services include repair and maintenance for which the landlord is responsible and insurance for the building or estate as a whole. The landlord will seek to recoup the cost of all services by the service charge.
- Tenants should carefully negotiate the provisions requiring them to pay for maintenance and repair so as to avoid incurring costs that may benefit the landlord's premises long after the expiry of their lease.

Service charge

The main issue for a tenant in shared premises is the payment of a service charge – what benefits it provides and obligations it imposes. The parties' interests will of course be contradictory. The landlord will want to try to recover all costs incurred in connection with the building and will want a free rein to decide what work is needed and who to charge it to. The tenant ideally will want to limit liability to basic services provided and for these to be reasonable amounts known in advance for budgeting purposes. A happy medium is found only by careful wording of the relevant provisions in the lease and a pragmatic working relationship between the landlord (or its managing agents) and the tenants.

Repair and maintenance

The lease will contain a clause requiring the landlord to maintain and repair the main structure and exterior of the building and the common parts. The tenant will usually be responsible for internal repair and may be required to decorate the interior, usually every five years, leaving it in the same condition as when taking on

the lease. Anything more onerous than this is an unnecessary burden on the tenant.

The tenant should ensure that the precise areas of the building contained in the common parts, for which the landlord has responsibility, are clearly identified in the lease to avoid later dispute and are sufficient for their use of the building. Such items as windows and doors could be either an integral part of the structure of the building or an internal fixture. The tenant should not wait for any dispute to arise before trying to decide who is responsible. Any major repair works carried out will be reflected in an increased service charge. Tenants of new buildings should be vigilant that they are not asked to contribute towards the cost of the original construction of the building.

Repair or improvement?

A tenant does not have any input as to whether repairs are carried out on a cheaper or more expensive basis unless there is a specific reference in the lease for the landlord to obtain approved estimates. The landlord therefore could seek to add some expensive long-term structural repairs to the service charge, which could be seen as an improvement rather than a necessary repair. Tenants under a short-term lease should not be responsible for extensive long-term repairs and should ensure that the wording of the lease makes it clear that they are not required to contribute.

Insurance

It is usually the landlord's responsibility to provide adequate insurance cover for the building. The cost of this is recouped via the service charge. With shared premises this will invariably be by way of a block policy. The tenant should ensure that the lease contains a clause providing that:

■ in the event of the building (or the means of access) being damaged or destroyed, the landlord is obliged to use the insurance monies to reinstate the building and make up the shortfall;
■ the rent be suspended until the building is reinstated; and
■ the tenant has an option to determine the lease if the building is not reinstated within a specified period.

Calculation of service charge

The lease will specify the proportion of the costs of the service charge. This may simply be by reference to a fair proportion, by reference to the relative floor size occupied by each tenant, or by some other method.

Tenants should ensure they are aware of exactly what services are included in the service charge (which should be scheduled in the lease), as, for example, a tenant on the ground floor does not want to pay a proportion of maintenance towards the lift (although in practice the landlord is unlikely to accept such a restriction).

The landlord can be required to certify, by way of a certificate issued by an accountant or surveyor, that the service charge is properly calculated in accordance with the lease.

Service Charge Code

In 2007 the Royal Institution of Chartered Surveyors published a non-binding code of practice on service charges. The Code promotes best practice and encourages landlords and property managers to adopt the Code. The Code cannot override the lease terms and in practice it has had little effect on property occupiers, but this could change if the property industry decides to adhere to the Code.

See also: Boundaries and party walls, p.63; Dilapidations, p.221; Landlord and tenant: lease issues, p.429; Property disputes, p.600.

Sources of further information

The Workplace Law website has been one of the UK's leading legal information sites since its launch in 2002. As well as providing free news and forums, our Information Centre provides you with a 'one-stop shop' where you will find all you need to know to manage your workplace and fulfil your legal obligations. Content is added and updated regularly by our editorial team who utilise a wealth of in-house experts and legal consultants. Visit www.workplacelaw.net for more information.

Slips, trips and falls

Jagdeep Tiwana, Bond Pearce LLP

Key points

- Slips, trips and falls are the most common cause of injuries at work; 95% of major slips result in broken bones.
- The HSE's campaign to reduce the number of slips, trips and falls in the workplace by 10% by 2010 concluded last year, having seen a reduction of serious injuries from nearly 11,000 to just over 10,000.
- However, this is still a significant number of injuries, many of which can be avoided by taking some simple steps.

Legislation

- Health and Safety at Work etc. Act 1974.
- Workplace (Health, Safety and Welfare) Regulations 1992.
- Reporting of Injuries, Diseases and Dangerous Occurrences Regulations 1995 (RIDDOR).
- Management of Health and Safety at Work Regulations 1999.

Legal duties

The Health and Safety at Work etc. Act 1974 imposes a duty on all employers to take steps to ensure the health and safety of their employees and third parties (such as customers or workmen). This duty includes taking steps to control risks to such persons from slips, trips and falls. Employees are also under a duty to behave in a responsible manner to ensure their own safety and that of others around them. Additionally, they must make use of any safety equipment provided by their employer.

The Workplace (Health, Safety and Welfare) Regulations 1992 impose the specific requirement that floors must be suitable and in good condition. They must also be free from obstructions and people must be able to move around safely.

The Management of Health and Safety at Work Regulations 1999 impose a duty on employers to carry out risk assessments, including those relating to hazards involving slips, trips and falls. Once these have been highlighted, employers must put into place measures to prevent the risk of accidents arising from slips, trips and falls.

Risk assessment

The HSE recommends five steps in risk assessment of slips, trips and falls:

1. Look for slip and trip hazards (e.g. uneven floors, trailing cables, slippery surfaces – when wet or otherwise).
2. Identify who may be harmed and how (pay particular attention to older or disabled people).
3. Consider the risks and whether current safety measures adequately deal with these.
4. Record findings (if five or more employees).
5. Review the risk assessment regularly (particularly where there has been an accident involving a slip, trip or fall or where there have been significant changes in the workplace).

The HSE has issued specific mapping tools to assist managers and safety representatives to prevent accidents involving slips, trips and falls. These can be found on the HSE website, together with other useful information (see *Sources of further information*).

Good system

The HSE has highlighted the following as the requirements of a good system to prevent slips, trips and falls.

Planning

An employer should identify key areas of risk and work with employees to identify areas on the site giving rise to risk of slips, trips or falls. Employers should also take care when selecting floor coverings and equipment to prevent or reduce slip and trip hazards. This can be done by choosing anti-slip flooring and fitting splash guards.

Training

Staff should be trained in how to avoid accidents involving slips, trips and falls, including cleaning up spillages and not placing trip hazards in the workplace. Staff should also be trained on how to use safety signage to warn of slippery floors etc. and to wear suitable footwear.

Organisation

Work activities should be organised in a way that minimises the risks of slips, trips and falls, and specific staff members should be given responsibility for ensuring that the workplace is kept safe. This can be done by ensuring the workplace is always well-lit, free from obstructions and tripping hazards (e.g. trailing cables) and

that all spillages are cleared up quickly to prevent slipping.

Control

Records should be kept to ensure good cleaning and maintenance operations are being used. Checks should be carried out regularly to ensure safe working practices are being used.

Monitor and review

Employers should regularly review accident records and identify any areas where current safety arrangements are deficient. Steps should be taken to remedy any deficiencies highlighted. Employees should be encouraged to report any safety issues.

Common causes of accidents

The HSE has identified the main factors that it considers contribute to slips, trips and falls:

- *Flooring*. Wet floors pose a well-known slipping hazard and ill-fitted or damaged floor coverings can lead to tripping. Floors should be regularly inspected and damaged flooring repaired. Spillages should be wiped up as soon as possible and safety signage used to warn of the hazards of wet floors.
- *Contamination*. Contamination from oil, grease or even rainwater can make floors very slippery and it is therefore important that floors are cleaned thoroughly and rainwater is mopped up as soon as possible and safety signage used.

- *Obstacles*. Simple measures such as keeping areas clear of obstructions and work areas tidy can reduce the number of accidents.
- *Cleaning*. All workplaces will need to undergo cleaning but this can create slip and trip hazards, so cleaning is best done when the least number of people will be exposed to the risks of slipping or tripping (e.g. after the premises in question close to the public). When this is not possible, access to wet areas should be stopped and cleaning carried out in sections, using signs and/or cones. It is important to note that these will warn of a hazard, but will not prevent people from entering the area. Appropriate amounts of cleaning products should be used to remove grease and oil from floors.
- *Human factors*. Instilling a positive attitude in staff, and training them to deal with any hazards as soon as they arise, will have a positive effect on workplace safety.
- *Environment*. Environmental factors such as lighting, weather and condensation can have an impact on the risk of people suffering a slip, trip or fall. For example, excessive glare from sunlight on a shiny floor may prevent people from seeing a tripping hazard, or badly lit stairs will present a hazard. This can be reduced by using high visibility non-slip step edges.

- *Footwear*. The HSE recognises that footwear can play an important part in preventing slips, trips and falls, while unsuitable footwear may contribute to an accident. Employers can reduce risks by providing non-slip footwear for staff or by implementing a footwear policy requiring employees to wear flat shoes that have a grip. This will not, of course, reduce the risk of slips or trips by customers or other third parties, and additional steps may still be required. By undertaking a risk assessment and implementing good systems, the majority of accidents involving slips, trips and falls can be avoided.

See also: Accident investigations, p.29; Catering: Health and safety issues, p.123; Construction site health and safety, p.177; Workplace deaths, p.218; First aid, p.325; Health and safety at work, p.399; Working at height, p.361; Reporting of Injuries, Diseases and Dangerous Occurrences (RIDDOR), p.618; Risk assessments, p.624.

Sources of further information

'Shattered lives' – HSE slips, trips and falls: www.hse.gov.uk/slips/

HSE Slips and trips e-learning package: www.hse.gov.uk/slips/step/index.htm

Smoking

Sally Cummings, Kennedys

Key points

- Smoking in prescribed places is against the law throughout the British Isles. Since 2004, Regulations have been phased in for each country within the UK, with the ban slowly taking effect at different dates across the different countries. The Republic of Ireland was the first to introduce the ban in 2004, whilst England was the last to follow suit, and has now been 'smoke-free' since July 2007.

- For all of the UK, it is therefore now against the law to smoke in virtually all 'enclosed' and 'substantially enclosed' public places and workplaces, meaning that previously designated indoor 'smoking rooms' have now been outlawed. The ban also applies to public transport and work vehicles used by more than one person.

- Employers and managers of smoke-free premises and vehicles have legal responsibilities to prevent people from smoking, namely to:
  - take reasonable steps to ensure staff, customers, members and visitors are aware that the premises and vehicles are legally required to be smoke-free;
  - display 'no smoking' signs in smoke-free premises; and
  - ensure that no one smokes in smoke-free premises or vehicles.

Legislation

Previously, smoking was purported to be governed by a combination of the Health and Safety at Work etc. Act 1974, the Management of Health and Safety at Work Regulations 1999 and the Workplace (Health, Safety and Welfare) Regulations 1992, which required employers to take steps to assess and minimise the risks to their employees and visitors from environmental tobacco smoke in the workplace.

These requirements, although still in existence, have been largely overtaken by the 'smoke-free' legislation, which, insofar as England is concerned, consists of the following Regulations:

- The Smoke-free (Premises and Enforcement) Regulations 2006 set out definitions of 'enclosed' and 'substantially enclosed' places and the bodies responsible for enforcing smoke-free legislation.

- The Smoke-free (Exemptions and Vehicles) Regulations 2007 set out the exemptions to smoke-free legislation and vehicles required to be smoke-free.

- The Smoke-free (Penalties and Discounted Amounts) Regulations 2007 set out the levels of penalties for offences under smoke-free legislation.

- The Smoke-free (Vehicle Operators and Penalty Notices) Regulations 2007 set out the responsibility on vehicle operators to prevent smoking in smoke-free vehicles and the form for fixed penalty notices.

- The Smoke-free (Signs) Regulations 2007 set out the requirements for no-smoking signs required under smoke-free legislation.

- The Tobacco Products (Manufacture, Presentation and Sale) (Safety) Regulations 2002 (as amended) set out the packaging and labelling requirements, including health warnings on cigarette packets.

The Health Act 2009, which contains provisions to ban the display of tobacco products and regulate tobacco vending machines, received Royal Assent on 12 November 2009. The Tobacco Advertising and Promotion (Display) (England) Regulations 2010 ban the display of tobacco products at the point of sale and come into force for large retailers (over 280 square metres of floor space) on 1 October 2011 and for smaller retailers from 1 October 2013. The ban on tobacco vending machines will come into force on 1 October 2011.

Similar legislation exists for Scotland, Ireland, Wales and Northern Ireland, and further information can be accessed via the websites listed in 'Sources of further information'.

The Scottish position is set out briefly below, but the focus of this chapter is on the application of the smoke-free legislation as it applies in England and Wales.

Scotland

On 30 June 2005, the Scottish Parliament passed the Smoking, Health and Social Care (Scotland) Act 2005, which introduced a complete ban on tobacco smoking in enclosed public places in Scotland from 26 March 2006.

The Act makes it an offence for those in charge of 'no-smoking premises' to knowingly permit others to smoke there. Although there are defences of taking reasonable precautions to prevent smoking and having no lawful means to prevent smoking, those convicted may be fined up to £2,500. It is also an offence to fail to display 'no-smoking' signs in such premises. The Act provides that culpable managers may also be prosecuted, as well as their employer companies. Those convicted of smoking in 'no-smoking' premises may be fined up to £1,000.

The Prohibition of Smoking in Certain Premises (Scotland) Regulations 2006 add further provisions relating to the provision and display of no-smoking signage, giving effect to Schedule 1 of the Act, which lists the types of premises that are prescribed to be no-smoking premises, defining key expressions such as 'premises' and 'wholly enclosed,' and setting out the levels of the relevant fixed penalties and other administrative matters.

The Scottish Parliament passed the Tobacco and Primary Medical Services (Scotland) Act 2010 on 27 January 2010, ending the display of tobacco in shops and banning cigarette vending machines in Scotland. This Act also contains provisions to ban proxy purchasing, introduce a register for tobacco retailers, and making it an offence for under 18s to purchase tobacco.

England and Wales
In England and Wales, smoking restrictions were slowly phased in from 2006, with an initial ban on smoking in NHS and Government buildings. This was followed by a ban on smoking in enclosed public spaces in 2007, and in licensed premises by 2008. The Health Act 2006 devolved regulation-making powers to the Welsh Assembly, and the bans took effect in April (Wales) and July (England) 2007. The requirements of both the English and Welsh smoke-free legislation are very similar.

As indicated above, employers and managers in charge of premises and vehicles to which the legislation applies should:

- take reasonable steps to ensure staff, customers, members and visitors are aware that the premises and vehicles are legally required to be smoke-free;
- display 'no-smoking' signs in smoke-free premises; and
- ensure that no one smokes in smoke-free premises or vehicles.

Premises
Premises:

- that are open to the public;
- that are used as a place of work by more than one person; or
- where members of the public might attend to receive or provide goods or services,

are to be smoke-free in areas that are enclosed or substantially enclosed.

Premises are 'enclosed' if they have a ceiling or roof and, except for doors,

windows and passageways, are wholly enclosed either permanently or temporarily. Premises are 'substantially enclosed' if they have a ceiling or roof but there is an opening in the walls or an aggregate area of openings in the walls that is less than half the area of the walls.

The ban therefore includes offices, factories, shops, pubs, bars, restaurants, private members clubs and workplace smoking rooms. A 'roof' also includes any fixed or movable structure, such as canvas awnings. Tents, marquees or similar are also classified as enclosed premises if they fall within the definition.

Vehicles
'Enclosed vehicles' are to be smoke-free at all times if they are used 'by members of the public or a section of the public (whether or not for reward or hire)'; or 'in the course of paid or voluntary work by more than one person, even if those people use the vehicle at different times, or only intermittently'.

For example, a delivery van used by more than one driver, or which has a driver and passenger, must be smoke-free at all times. It should be noted that the Regulations do not extend to vehicles used for non-business purposes.

Signage
All smoke-free premises must display a no-smoking sign in a prominent position at each entrance that:

- is the equivalent of A5 in area;
- displays the international no-smoking symbol in colour, a minimum of 70mm in diameter; and
- carries the words, 'No smoking. It is against the law to smoke in these premises,' in characters that can be easily read.

In addition, any person with management responsibilities for a smoke-free vehicle has legal duties to display a no-smoking sign in each enclosed compartment that can accommodate people.

Penalties

Smoking in a smoke-free premises or vehicle can attract a fixed penalty notice of £50 or a fine up to £200. Failure to display no-smoking signs in smoke-free premises and vehicles can attract a fixed penalty notice of £200 or a fine up to £1,000.

Failing to prevent smoking in a smoke-free premises or vehicle can lead to a fine up to £2,500.

The smoke-free legislation is enforced by a number of bodies, but primarily by district councils. There is no provision in the legislation for smoke-free offences to result in a review of a pub's licence. Councils are encouraged to approach enforcement by supporting businesses and the public to comply with the smoke-free law in the first instance, by providing advice, support and information. Formal enforcement action should only be considered where the seriousness of the case warrants it.

Exemptions

There are some limited exemptions, including the following:

- Private dwellings (with particular exceptions such as a communal stairwell);
- Designated bedrooms in hotels, guest houses, inns, hostels and members clubs (if they meet conditions set out in the Regulations); and
- Designated bedrooms / rooms used only for smoking in care homes, hospices and prisons (if they meet conditions set out in the Regulations).

Compliance

The first prosecution took place in England less than three weeks after the introduction of the ban in July 2007, resulting in fines of nearly £2,000, plus costs of approximately the same, imposed upon a licensee of a bar in Lancashire after he admitted failing to prevent smoking. In March 2009, the licensee challenged the sentence, seeking leave from the High Court to judicially review the decision on the grounds that the ban violated his human rights, and could bankrupt his business. The Court dismissed his application, refusing him such permission, on the basis that he did not have an arguable case.

In an earlier case in 2008, the High Court similarly ruled that Nottinghamshire Healthcare Trust had not breached long-term psychiatric patients' human rights by prohibiting them from smoking at a high security mental hospital.

However, statistics confirm that there is an ongoing high level of compliance with the legislation, and this trend has continued over the three years since the legislation came into effect, with the overall compliance rate for the three years since implementation standing at 98.3%.

Between January and June 2010, authorities in England carried out over 70,000 premises and vehicle checks for compliance with the ban and signage requirements. 98.4% of those inspected were found to be compliant in terms of the ban itself, whilst 97% appeared to comply with the obligations regarding signage.

During that same period of inspection, 371 written warnings were issued by the authorities upon premises / vehicle owners for failing to prevent smoking – equating to 0.5% of all premises / vehicles inspected. Enforcement against individuals smoking in smoke-free areas was higher, however, with 400 fixed penalty notices served, and with 28 Court hearings relating to contravention of the legislation being held in the first six months of the year alone.

Statistics released by the Department of Health in 2010, following a study conducted by the University of Bath, confirmed that the number of heart attack admissions to hospital fell by 2.4% in the first 12 months of the ban. The study suggested that the fall in emergency admissions may be linked to the general decrease in exposure of the public to second-hand smoke since the ban was introduced. Whilst the study only monitored the change in trends between 2000 and 2008, it is thought that it demonstrates a clear association between the smoking ban and a reduction in the rate of hospital admissions for smoking-related afflictions, thus lending further support to the Government's decision not to interfere with what appears to be an effective means of regulation.

International comparisons

In Dublin, there were a reported 2,000 job losses following the implementation of a ban on smoking in bars in March 2004, and the Irish Revenue Commissioner reported a 16% decrease in tobacco sales. Subsequent research suggests that airborne particulate matter, a feature of some pollution, decreased by up to 80% in pubs, and health and lung function of bar workers improved in the 12 months after the ban came into effect. The introduction of a ban on smoking in bars in New York allegedly saw a reduction in pollution levels following the ban but a short-term decrease of $28.5m in wage and salary payments.

In Norway, rather than implementing a complete ban, non-smoking areas and requirements for better ventilation are part of a 15-year staged process for the introduction of a ban. Norway banned smoking in bars, following the initiative in Ireland, and bar owners face the prospect of losing their licences if smokers are not persuaded to stop lighting up.

Other countries that have introduced bans or restrictions include Italy, Spain, Belgium, Sweden, the Netherlands, South Africa, Russia, Nepal and Tanzania. The State of California banned smoking in most indoor workplaces as long ago as 1995, with a temporary exemption for bars and casinos ending in 1998.

Introducing a smoking policy

All employers should by now have taken action to comply with the statutory smoking ban. In terms of the introduction of new or revised smoking policies to supplement or reinforce such action, an employer should consider both the form of the policy and the manner of its introduction. Whether a policy is reasonable or not will be a question of fact. However, as well as considering the implications of smoking bans, a prudent employer should take account of:

- the practicalities of the workplace;
- the nature of the business, including whether clients will be regularly visiting the building and whether employees are visiting clients at their homes;
- workplace opinion;
- assistance to smokers in adapting to the new policy;
- consultation with individuals and/or their representatives;
- ensuring that employees are fully aware of the possible sanctions for breach of the policy, including cross-references to the disciplinary procedure; and
- regular reviews of the policy for ongoing effectiveness.

Once a policy is in place it must be consistently enforced. Employees seeking to comply or trying to give up smoking should be offered support.

Employers might find it helpful to approach the smoking ban in a similar way as many of them do to the wearing of personal protective equipment. In other words, employers should:

- identify all those areas in which smoking will constitute an offence;
- alter the existing staff handbook or relevant procedures such that smoking in those areas will attract disciplinary sanctions;
- having publicised the alterations, make good on the promise to use them when there are reasonable grounds for doing so; and
- in the case of visitors, make them aware of the company's policy, and

eject them from the premises if the policy is breached. They should also take into account any increase in fire risk resulting from smokers gathering outside or even smoking surreptitiously. If practical, safe areas outside could be provided, distant from combustible materials and with suitably designed bins or buckets for smokers' waste.

See also: Driving at work, p.250; Occupational health, p.521; Vehicles at work, p.686.

Sources of further information

Helpful guidance, including downloadable signage, can be accessed at the following websites:

England – www.smokefreeengland.co.uk

Wales – www.smokingbanwales.co.uk

Scotland – www.clearingtheairscotland.com

Northern Ireland – www.spacetobreathe.org.uk

Stress

Heidi Thompson, Workplace Law

Key points

- All employers owe a legal duty of care to their employees. Injury to mental health is treated in the same way as injury to physical health.
- Sixteen general propositions for bringing any civil claim for compensation for stress were provided by the Court of Appeal and approved as general guidance by the House of Lords. These are listed below (see 'Criteria for civil cases').
- A successful claim must show that, on the balance of probabilities, an employer had knowledge or deemed knowledge of the foreseeability of harm to a particular employee, so that the lack of his taking reasonable steps to, as far as is reasonably practicable, alleviate the risk of or prevent that harm occurring constituted a breach of duty of care to the employee, and that this caused the injury or loss.
- The HSE has urged employers to carry out risk assessments and implement measures to eliminate or control workplace stress or risk criminal prosecution. The HSE's Management Standards on stress (a web-based toolkit to help businesses comply with their duties) were published on 3 November 2004. Employers will need to take on board this HSE guidance in order to provide best practice in health and safety.

Legislation

- Health and Safety at Work etc. Act 1974.
- Management of Health and Safety at Work Regulations 1999.
- The Equality Act 2010.

Main cases

- *Stokes v. Guest Keen and Nettlefold (Bolts and Nuts) Limited* (1968).
- *Walker v. Northumberland County Council* (1995).
- *Katfunde v. Abbey National and Dr Daniel* (1998).
- *Sutherland v. Hatton* (2002).
- *Barber v. Somerset County Council* (2004).
- *Hartman v. South Essex Mental Health & Community Care NHS Trust* (2005).
- *Mark Hone v. Six Continents Retail Limited* (2005).
- *Edward Harding v. The Pub Estate Co. Limited* (2005).
- *London Borough of Islington v. University College London Hospital NHS Trust* (2005).
- *Dickens v. O2* (2009).

Legal aspects of stress claims

All employers owe a legal duty of care to their employees. Injury to mental health is treated in the same way as injury to physical health.

Criteria for civil cases

A successful civil claim must show that, on the balance of probabilities, an employer had knowledge or deemed knowledge of the foreseeability of harm to a particular employee, so that the lack of his taking reasonable steps to, as far as is reasonably practicable, alleviate the risk of or prevent that harm occurring, constituted a breach of duty of care to the employee, and that this caused the injury or loss.

Facts

- In 2009/10 an estimated 435,000 individuals in Britain, who worked in the last year, believed that they were experiencing work-related stress at a level that was making them ill.
- Around 16.7% of all working individuals thought their job was very or extremely stressful.
- The annual incidence of work-related mental health problems in Britain in 2008 was approximately 5,126 new cases per year. However, this almost certainly underestimates the true incidence of these conditions in the British workforce.
- An estimated 230,000 people, who worked in the last 12 months, first became aware of work-related stress, depression or anxiety in 2008/09, giving an annual incidence rate of 760 cases per 100,000.
- In 2009/10, an estimated 9.8 million working days were lost through work-related stress. Every person experiencing work-related stress was off work for an estimated average of 22.6 days, which equates to 0.42 days per worker.
- The incidence rate of self-reported work-related stress, depression or anxiety has been broadly level over the years 2001/02 to 2008/09, with the exception of 2001/02 where the incidence rate was higher than the current level.

Source: HSE.

A stress injury is not as immediately visible as, for instance, a broken leg, and the 16 propositions put forward by Lady Justice Hale in the Court of Appeal judgment of *Sutherland v. Hatton* (and related cases) are still regarded as the best useful practical guidance as to whether or not a stress claim may be successful. (These propositions are listed in full at the end of this section.)

Nonetheless, every case does still depend on its own facts and in the later House of Lords case of *Barber v. Somerset County Council* Lord Walker preferred as a statement of law the statement of Swanwick J in *Stokes v. Guest Keen and Nettlefold (Bolts and Nuts) Ltd* that "the overall test is still the conduct of the reasonable and prudent employer, taking positive thought for the safety of his workers in the light of what he knows or ought to know".

State of knowledge

Knowledge of the employee and the risks they are facing is key in both leading House of Lords cases on workplace stress. Lord Walker went furthest in the *Barber* case and stated that, where there was developing knowledge, a reasonable employer had a duty to keep reasonably abreast of it and not be too slow to apply it. Where the employer has greater than average knowledge of the risks, he may be obliged to take more than the average or standard precautions.

Knowledge is critical in the area of what is or is not 'reasonably foreseeable' in a civil claim. This was reinforced by the Court of Appeal case of *Mark Hone v. Six*

Continents Retail Limited. In this case it was brought to the employer's attention that long hours were being worked and the employee was tired. It was held that it did not matter that the employer did not accept the level of the recorded hours as accurate as the fact the employee had been recording those hours was sufficient to indicate that he needed help and contributed to the "sufficiently plain indications of impending harm to health".

This may be contrasted with another Court of Appeal case of *Edward Harding v. The Pub·Estate Company Ltd* where, as manager, the claimant's hours were within his own control, no reduction in hours was requested, nor additional staff; complaints concentrated on working conditions. In this case, the Court of Appeal overturned judgment at first instance because they found no sufficient message was ever passed to the employers of a risk to the employee's health. This was reinforced through the recent case of *Dickens v. O2*, where the claimant was entitled to damages for psychiatric injury caused by work-related stress. The injury was reasonably foreseeable from the point at which the claimant described her severe symptoms to the employer and said she did not know how long she could continue before taking sick leave. Thereafter the employer breached its duty of care by failing to send her home and refer her immediately to its occupational health service.

On deciding if a psychological injury was reasonably foreseeable after the event, it has been held that this is "to a large extent a matter of impression" (*London Borough of Islington v. University College London Hospital NHS Trust*). Therefore, all factors that would go to make up such an impression should be monitored, such as working hours, increased workload, time off sick etc. as well as specific indications of stress from employees. This is therefore a matter to be considered on the individual facts of each case.

Part-time workers

However, time actually spent at work may well be a crucial factor in some cases. It was held in the Court of Appeal case of *Hartman v. South Essex Mental Health and Community Care NHS Trust* that it would only be in exceptional circumstances that someone working for two or three days a week with limited hours would make good a claim for injury caused by stress at work.

Confidential advice / health service

There are a number of precautionary measures outlined at the end of this chapter for employers to protect themselves against workplace stress claims and, in particular, the provision of a confidential advice service was thought likely to provide a good defence in the *Sutherland* case.

It was confirmed in *Hartman* that the mere fact that an employer offered an occupational health service should not lead to the conclusion that the employer had foreseen risk of psychiatric injury due to work-related stress to any individual or class of employee. An employer could not be expected to know confidential medical information disclosed by the claimant to occupational health. However, there may be circumstances where an occupational health department's duty of care to an employee requires it to seek his or her consent to disclose information that the employer needs to know, if proper steps are to be taken for the welfare of the employee.

Stress in other civil claims

Stress now raises its head more often in claims involving bullying and harassment, disability, discrimination and constructive dismissal. Failure to recognise and address stress issues in the context of these types of claim could result in significant liability for an employer. For instance, where an employee may establish that he falls within the definition of a disabled person under the Equality Act 2010 and an employer fails to make reasonable adjustments to the workplace for this disability, compensation would also be payable for the psychiatric or physical injuries occurring from stress suffered as a result of this.

Sixteen propositions for stress claims

A summary of the 16 propositions stated in *Sutherland v. Hatton* is provided below.

General

1. The ordinary principles of employers' liability apply.
2. There are no occupations that should be regarded as intrinsically dangerous to mental health.

Reasonable foreseeability

3. The threshold question to be answered in any workplace stress case was stated as: 'whether this kind of harm to this particular employee was reasonably foreseeable'. This has two components: (a) an injury (as distinct from occupational stress) that (b) is attributable to stress at work (as distinct from other factors).
4. Foreseeability depends upon what the employer knows (or ought reasonably to know) about the individual employees.
5. Factors likely to be relevant in answering the threshold question include:
   ■ the nature and extent of the work done; and
   ■ signs from the employee of impending harm to health.
6. The employer is generally entitled to take what he is told by his employee at face value, unless he has good reason to think to the contrary.

7. To trigger a duty to take steps, the indications of impending harm to health arising from stress at work must be plain enough for any reasonable employer to realise that he should do something about it.

Duty of employers

8. The employer is in breach of duty only if he has failed to take steps that are reasonable in the circumstances.
9. The size and scope of the employer's operation, its resources and the demands it faces are relevant in deciding what is reasonable; these include the interests of other employees and the need to treat them fairly (e.g. in any redistribution of duties).
10. An employer can only be expected to take steps that are reasonable in the circumstances.

Guidelines for employers

11. An employer who offers a confidential advice service, with referral to appropriate counselling or treatment services, is unlikely to be found in breach of duty.
12. If the only reasonable and effective step would have been to dismiss or demote the employee, the employer will not have been in breach of duty in allowing a willing employee to continue in the job.

(However, in light of the lead judgment of Lord Walker in *Barber* that there is a requirement for 'drastic action' if an employee's health is in danger, it may be said that in the absence of alternative work, where an employee was at risk, ultimately the employer's duty of care would not preclude dismissing or demoting the employee at risk.)

13. In all cases, it is necessary to identify the steps that the employer both could and should have taken before finding him in breach of his duty of care.

14. The claimant must show that the breach of duty has caused or materially contributed to the harm suffered. It is not enough to show that occupational stress alone has caused the harm; it must be attributable to a breach of an employer's duty.

Apportionment

15. Where the harm suffered has more than one cause, the employer should pay only for that proportion of the harm suffered that is attributable to his wrongdoing, unless the harm is truly indivisible. It is for the defendant to raise the question of apportionment.
16. The assessment of damages will take account of pre-existing disorders or vulnerability and of the chance that the claimant would have succumbed to a stress-related disorder in any event.

It is not the case that one or other of the tests is more important; all 16 have to be looked at in respect of each individual case.

Criteria for criminal liability

There is no specific statute or other regulation controlling stress levels permitted in the workplace; therefore broad principles of health and safety at work will be applied as set out in the Health and Safety at Work etc. Act 1974 (HSWA) and the Management of Health and Safety at Work Regulations 1999 (MHSWR).

Where no action is taken by an employer on stress, he may be deemed to have fallen short of his duty to take all reasonably practicable measures to ensure the health, safety and welfare of employees and others sharing the workplace and to create safe and healthy working systems (HSWA).

Additionally, there is the requirement to undertake risk assessments of stress and put in place appropriate preventive and protective measures to keep the employees safe from harm (MHSWR).

Any breach of an employer's statutory or regulatory duties under health and safety legislation towards his employees giving rise to criminal liability may also be relied upon by a civil claimant as evidence of the employer's breach of duty in a negligence action and, indeed, in support of a claim for constructive dismissal.

Risk assessments

West Dorset Hospitals NHS Trust was the first organisation to have an improvement notice issued against it with the requirement that it assessed and reduced the stress levels of its doctors or other employees or face court action and a potentially unlimited fine. More recently, Liverpool Hope University has been served with an improvement notice on similar grounds. The HSE has urged employers to carry out risk assessments and implement measures to eliminate or control workplace stress or risk criminal prosecution.

It is therefore important that risk assessments for stress are undertaken, regularly reviewed and recommended actions implemented. Unlike civil litigation, any criminal prosecution carries with it the threat of an unlimited fine and/or imprisonment.

A general risk assessment of potential 'stressors' at work should be sufficient for most businesses but should additionally take into account any discrete categories of employees, such as night workers, the young, and expectant mothers. But if an employer becomes aware of an employee at specific risk or who has raised any concerns, an individual risk assessment should be carried out for them, recommendations implemented and regularly reviewed.

Health and safety policy

It is further recommended that an employer's health and safety policy sets out guidance on how stress should be dealt with and a clear complaints-handling procedure. In this way a company can show that it has followed its own procedures in dealing with any complaints and implementing any actions.

Conclusion

In order to protect themselves against enforcement action as well as employee claims, employers are advised to organise risk assessments of potential stressors, to make facilities such as counselling and grievance procedures available to employees, and to show a receptive and flexible response to complaints. In addition, compliance with the HSE Management Standards / Guidance will assist in showing that an employer has met the reasonable standard of duty of care required.

Combating stress: an employer's checklist

- No employer has an absolute duty to prevent all stress, which can be as a result of interests outside work. However, once an employee has raised the issue of stress, an employer is under a duty to investigate properly and protect the employee as far as is reasonably practicable.
- Health monitoring – both through a confidential advice line and/or regular company medicals.
- Counselling – an employer who offers a confidential advice service, with referral to appropriate counselling or treatment services, is unlikely to be found in breach of duty. This is of course relative to the problem and the service provided but is a good indication that a proactive approach by an employer can protect him from

stress claims and enforcement action.
- Regular medicals – these are a useful tool in alerting employers of any risks. However, medical confidentiality has to be observed and express consent given by employees for their clinical information to be shared with employers.
- Dismissal – in the absence of alternative work, the employee deemed at risk should be dismissed or demoted.
- Written health and safety policy – clear guidance in a company's health and safety policy on how stress should be dealt with shows that the company is complying with the health and safety regulations to provide a safe working environment for employees and enables staff to follow a set procedure. It would also stand as a defence where an employee fails to disclose that he is suffering from stress because of ignorance of a company's procedures.
- Equally, a bullying and harassment code should be in force and there should be a clear complaints-handling procedure.
- Risk assessments should cover all workplace risks and should therefore include stress. HSE guidance on risk assessment can be found at www.hse.gov.uk/pubns/indg163.pdf.
- Risk assessments should be regularly reviewed and recommended actions implemented.
- Working time – employers can combat stress by monitoring and recording employees' working time with action being taken if the benchmark set out in the Working Time Regulations 1998 is breached.
- Implementation of HSE Management Standards / Guidance will assist in showing that an employer has met the reasonable standard of duty of care.

See also: Health and safety at work, p.361; Mental health, p.482; Occupational health, p.521; Working time, p.758.

Sources of further information

As part of the general duty to keep abreast of developing knowledge and practice, employers should be aware of the HSE's stress page at www.hse.gov.uk/stress/index.htm. This includes example stress policies and the HSE's Management Standards for workplace stress.

Additional HSE guidance includes an action pack (*Real Solutions, Real People* (ISBN: 0 7176 2767 5 priced at £25.)). The pack includes a guide for employers and employees alike and an introduction to the Management Standards. Other HSE guidance in the form of free leaflets include the following: *Tackling stress: the management standards approach – a short guide*; *Making the Stress Management Standards work: How to apply the Standards in your workplace*; and *Working together to reduce stress at work: A guide for employees*. These are available at the publications section of the HSE stress web page.

The Management Standards look at six key areas (or 'risk factors') that can be causes of work-related stress: 'demands,' 'control,' 'support,' 'relationships,' 'role,' and 'change'. The standard for each area contains simple statements about good management practice that can be applied by employers.

HSE guidelines such as these are voluntary and as such are not legally binding. They do, however, have evidential value. They assist the court in the interpretation of legislation and what the reasonable standard of duty of care owed may be. Therefore, compliance with these Management Standards / Guidance will assist in showing the court that an employer has met the reasonable standard of duty of care required.

Temperature and ventilation

Bob Towse, Heating and Ventilating Contractors' Association (HVCA)

Key points

- Requirements for the provision of ventilation to buildings are set out in the Workplace (Health, Safety and Welfare) Regulations 1992.
- These Regulations came into force on 1 January 1993 for new buildings and on 1 January 1996 for existing buildings.
- The Regulations apply to a wide range of workplaces – factories, shops and offices, schools, hospitals, hotels and places of entertainment, but not those workplaces involving construction work on construction sites, those in or on a ship, or below ground at a mine.
- The Regulations replaced several pieces of older law, including parts of the Factories Act 1961 and the Offices, Shops and Railways Premises Act 1963.

Legislation

- Workplace (Health, Safety and Welfare) Regulations 1992.

General requirements

The Regulations require the workplace and the equipment, devices and systems to be maintained (including cleaned as appropriate) in an efficient state, in efficient working order and in good repair.

The Regulations do not go into detail, but the Approved Code of Practice (ACoP) – Workplace Health, Safety and Welfare – and HSE information sheets give further guidance on how to comply – see *Sources of further information*.

Temperature in indoor workplaces

Regulation 7 requires that 'during working hours, the temperature in all workplaces inside buildings shall be reasonable'.

The ACoP suggests that, in the typical workplace, the temperature should be at least 16°C unless much of the work involves severe physical effort, in which case the temperature should be at least 13°C.

These temperatures would be considered by most building occupants to be below comfort levels. However, the ACoP defines a reasonable temperature as one that should secure the thermal comfort of people at work, allowing for clothing, activity level, radiant heat, air movement and humidity.

For air-conditioned buildings in the UK, the CIBSE Guide A recommends a dry resultant temperature of between 21°C and 23°C during winter and between 22°C and 24°C in summer for continuous sedentary occupancy.

It is recognised that room temperatures in buildings without artificial cooling will exceed the summer values for some of the time but should not exceed 25°C for more than 5% of the annual occupied period (typically 125 hours).

There is currently no maximum temperature for a workplace, a bone of contention for many indoor workers during summer time, when temperatures rise in buildings to conditions that are uncomfortable.

Productivity has been shown to decrease at temperatures of 25°C and above and the risks to workers' health increase as conditions deteriorate from those accepted as comfortable. Calls from organisations for an upper temperature limit have been made in the past. In July 2009, following pressure from USDAW (Union of Shop, Distributive and Allied Workers), the Government asked the HSE to review the Regulations on workplace temperature.

John Hannett, USDAW General Secretary, placed hope in a maximum workplace temperature being in force by the end of 2009, so that, by summer 2010, workers would be able to work comfortably and in the knowledge that they are protected in law.

In 2010 the voices of teaching union, NASUWT, were added to the call for a maximum workplace temperature, when findings from a survey revealed that 94% of members felt they had worked in excessively high temperatures during the summer and 83% felt they had worked in excessively cold temperatures in winter. In 63% of reported cases, the temperature issue was not resolved appropriately.

A response from Government was promised by the end of summer 2010; however, to date nothing further has been recommended.

Ventilation

Regulation 6 requires that 'effective and suitable provision shall be made to ensure that every enclosed workplace is ventilated by a sufficient quantity of fresh or purified air'.

It should be noted that this Regulation covers general workplace ventilation, not local exhaust ventilation for controlling specific hazardous materials or substances hazardous to health.

The ACoP states that workplaces should be sufficiently well-ventilated so that stale air, and air that is hot or humid because of the processes or equipment in the workplace, is replaced at a reasonable rate. In many cases, natural ventilation through windows or other openings may be sufficient, but mechanical ventilation or air conditioning may also be required to meet certain circumstances.

The ACoP also states that mechanical ventilation systems should always operate in a way that draws in some fresh air. One hundred per cent recirculation (no fresh air) is considered unhealthy.

All spaces that rely on mechanical means of ventilation should be supplied with outdoor air at a rate sufficient to dilute internally generated pollutants.

Ventilation should also remove and dilute warm, humid air and provide air movement that gives a sense of freshness without causing a draught. If the workplace contains process or heating equipment or other sources of dust, fumes or vapours, more fresh air will be needed to provide adequate ventilation.

The HSE guide, *General Ventilation in the Workplace* (HSG 202), confirms that a fresh air supply rate of eight litres per second per person should provide a clean and hygienic workplace in open-plan offices, shops and some factories. Higher fresh air supply rates of up to 32 litres per second per person are recommended for heavily contaminated buildings.

The Chartered Institution of Building Services Engineers (CIBSE) Guide A: *Environmental Design,* recommends the following outdoor air supply rates for sedentary occupants:

- Eight litres per second per person.

Supply air quality

Air introduced from outside a building should be free from any impurities likely to be offensive or cause ill health. Outdoor air is generally considered acceptable, provided that the air intake is not sited so that excessively contaminated air (such as might be found near flues, extract outlets or car parks) is drawn into the building.

The level of air pollution in some locations may mean that outdoor air is not suitable to introduce into a building unless it has first undergone adequate particle filtration.

Air that is recirculated should be adequately filtered before being redistributed within the building.

Maintenance

The ACoP requires that 'any device or system used to provide fresh air to a building or space should be maintained in an efficient state so as to ensure that the air produced or delivered is both suitable and sufficient for use within the workplace'.

The term 'efficient state' relates to good working order and not productivity or economy. In particular, the plant should be kept clean and free from any substance or organism that may contaminate the air passing through it.

The ACoP refers to the need to 'regularly and properly clean, test and maintain mechanical ventilation and air-conditioning systems to ensure that they are kept clean and free from anything that may contaminate the air'.

Depending on use, compliance with this duty is likely to require a suitable system of maintenance, inspection, adjustment, lubrication and cleaning, as well as the keeping of accurate records.

CIBSE and the Heating and Ventilating Contractors' Association (HVCA) each publish a number of guides intended to assist in this purpose.

See also: Air conditioning and refrigeration systems, p.32; Building Regulations, p.76.

Sources of further information

Chartered Institute of Building Services Engineers (CIBSE): www.cibse.org

Heating and Ventilating Contractors' Association: www.hvca.org.uk

INDG 244 Workplace health, safety and welfare – A short guide for managers: www.hse.gov.uk/pubns/indg244.pdf

GEIS 1 Heat stress in the workplace – What you need to know as an employer (GEIS1): www.hse.gov.uk/pubns/geis1.pdf

HSG 202 General Ventilation in the Workplace: www.ucu.org.uk/media/pdf/f/g/HSG202_-_Ventilation.pdf

Toilets

Hayley O'Donovan, PHS Washrooms

Key points

- First impressions count in any organisation, and the provision of clean and well maintained toilet and hand washing facilities is a basic requirement that visitors to any premises should receive.
- There are a number of regulations and standards that organisations must meet in this area, including workplace, health and safety and environmental standards.
- Companies that care about their reputation and their customers will always exceed the minimum maintenance and cleanliness practicalities needed to meet these requirements.

Legislation

Some of the main laws that impact on washroom provision and maintenance are as follows:

- Water Industry Act 1991 as principally amended by the Water Industry Act 1999 and the Water Act 2003.
- Workplace (Health, Safety and Welfare) Regulations 1992.
- Water Supply Regulations 1999 (updated 2010).
- Control of Substances Hazardous to Health (COSHH) Regulations 2002.
- Equality Act 2010 (which replaced most of the Disability Discrimination Act 1995).
- Waste (England and Wales) Regulations 2011 and the Waste Information (Scotland) Regulations 2010 (which replaces the Environmental Protection Act (Duty of Care) Regulations 1990 (amended 1991).

Workplace (Health, Safety and Welfare) Regulations 1992

One of the key regulations relating to washroom facilities is the provision of facilities specified by the Workplace (Health, Safety and Welfare) Regulations 1992. Within this legislation, the key regulations relating to washroom facilities are as follows.

Regulation 20 – sanitary conveniences

This advises that 'in the case of water closets used by women, suitable means should be provided for the disposal of sanitary dressings' at readily accessible places and that they and the rooms containing them are kept in a clean and orderly condition.

Regulation 21 – 'suitable, sufficient and accessible' facilities

Water and soap dispensers

The Regulation states that 'washing facilities ... must include soap or other suitable means of cleaning'. In addition, facilities must provide 'a supply of clean hot and cold, or warm, water (which shall be running water so far as is practicable)'.

Hand dryers and paper dispensers

The Regulation states that 'washing facilities ... must include towels or other suitable means of drying'.

Toilets

Toilet roll holders

The Approved Code of Practice for Regulation 21 advises that in the case of water closets, 'toilet paper in a holder or dispenser ... should be provided'.

Other legislation

There are a number of other provisions that must be applied. The most high profile of these is making reasonable adjustments in accordance with the Equality Act 2010, which replaced most of the Disability Discrimination Act 1995. The Act has very specific requirements for disabled WCs. Disabled WC facilities have to meet the needs of wheelchair users, the blind, visually impaired and ambient disabled as well as the many disabilities including sensory, physical and learning disabilities. The Act requires suitable facilities for managing personal hygiene, including accessible wash basins, soap dispensers, toilet paper and paper towels. These fittings and fixtures need to be designed and installed so they are readily accessible to a person in a wheelchair and when seated on the toilet.

In terms of waste disposal, under the Waste (England and Wales) Regulations 2011 there is a duty of care to ensure that all waste is handled and disposed of in a secure and correct manner. This is a legal responsibility, and breaching the duty of care is a criminal offence. Waste should only be handled by individuals or businesses that are authorised to deal with it by the Environment Agency. A record should be kept of all waste received or transferred through a system of Waste Transfer Notes (a document that must be completed and accompany any transfer of waste between different holders and held for two years). The local Environmental Health Officers can issue 'Statutory Notices' requesting the production of all documentation relating to the removal of waste. Failure to comply

with this legislation can lead to a fine of up to £5,000.

In addition, under the Water Industry Act 1991, as principally amended by the Water Industry Act 1999 and the Water Act 2003, no items should be flushed that could cause a blockage within the sewer or drain, which is a criminal offence under Section 111.

Complying with Standards is one thing, but there are good reasons to see compliance merely as the minimum standard people expect. That is why many organisations, especially those with a vested interest in promoting hygiene, such as restaurants and bars, go way beyond what is demanded of them by law or common practice.

Poorly maintained washrooms can cast doubt on a company's professionalism. People – including customers, employees, suppliers and other visitors – expect a company with a good reputation to have well cared for facilities. This is the area that legislation sometimes fails to reach and that can only be achieved by looking at the design, specification, cleaning and maintenance of washrooms intelligently and in a structured way.

The environment

As well as Standards dealing with hygiene and health and safety, there is a growing onus on organisations to treat water management as an important ethical concern. The efficient use of water plays a key role in environmental policy and one that will grow in relevance as the impact of climate change grows.

Statistics from Envirowise have shown that offices alone waste around 310 million litres of water every working day, and Envirowise has estimated that a very few, simple and inexpensive measures could save industry over £300m a year. There are a number of such steps that building and facilities managers can take, both at the initial installation of washroom facilities and at retrofit, to ensure that water is used as efficiently as possible to both save money and help the environment.

These can range from the straightforward to the more sophisticated. At the most simple level, one of the most effective methods is a product such as an inexpensive water saving device which, when placed into a cistern, can save as much as a third of the water used by toilets. When you consider that around 86% of the 35 litres each office worker in the UK uses each day is simply flushed away, this can represent an enormous saving. Carelessly or maliciously left-on taps can also be a serious source of water wastage, as well as causing floods, so it can be important to use push-action taps that deliver a set amount of water for hand washing.

Automatic flushing urinals are extremely useful as a way of maintaining hygiene, and are usually designed to be as efficient as possible in terms of the amount of water they use. Nevertheless, they can also be extremely wasteful of water when the number of flushes is out of sync with the number of people using the facility. To help counter this, an intelligent water management system can regulate the flushing of the urinal to match the number of people using the washroom. DEFRA is able to make recommendations about the best systems. It may also be worthwhile asking for any local by-laws regarding water management.

Health regulations

Under the Control of Substances Hazardous to Health (COSHH) Regulations, owners and operators of all commercial premises, including schools, offices, hotels and leisure centres, have a statutory duty to control the risk of legionella bacteria that causes Legionnaires' Disease. Legionella is usually found in stagnant water, for example in showerheads or spray taps that have been allowed to get clogged up with limescale, air conditioning cooling systems or leaking pipes. Maintenance of washrooms and other areas is, therefore, key. Regularly cleaned toilets, with appropriate facilities to clean hands, is one of the key areas to focus on when looking to reduce infections.

Washroom allocation and the gender gap

The problem of allocating enough washrooms for people has always been with us. At the Great Exhibition of 1851, the engineer, George Jennings (who was also instrumental in introducing public conveniences to London and other cities and helped to develop the modern WC) was tasked with providing large scale 'elimination facilities' for the public for the first time. The toilets he introduced were reportedly used some 830,000 times over the 141 days of the exhibition, leading Jennings to acknowledge "the necessity of making similar provisions for the public whenever large numbers are congregated to alleviate the sufferings which must be endured by all, but more especially by females by account of the want of them".

The minimum number of facilities that organisations in the UK should provide is governed by the Workplace (Health, Safety and Welfare) Regulations 1992, supported by a Code of Practice that gives precise details regarding the numbers of toilets and hand basins and so on based on the numbers and sexes of employees. The Regulations try to ensure that, where the facilities are used by women, there is greater access to more toilet facilities than where toilets are used by men only (*see Tables 1 and 2*).

Number of people at work	Number of toilets	Number of washbasins
1-5	1	1
6-25	2	2
26-50	3	3
51-75	4	4
76-100	5	5

Table 1. Number of toilets and washbasins for mixed use (or women only).

Number of men at work	Number of toilets	Number of urinals
1-15	1	1
16-30	2	1
31-45	2	2
46-60	3	2
61-75	3	3
76-90	4	3
91-100	4	4

Table 2. Toilets used by men only.

Conclusion

The Regulations are a useful guide. It is always important that you look at the wider picture when you are looking into the provision of these facilities. It is essential that you understand the culture of the organisation or bar or restaurant or venue or wherever before you specify anything. You can often make a good case for providing more and better facilities than is simply the legal requirement. Less is definitely not more where washrooms are concerned.

The UK Regulations also touch on the unisex issue. The Regulations state that 'specific facilities for males and females may need to be provided, except where each toilet is in a separate room capable of being secured from the inside'. The demand for unisex toilets is obviously still largely based on pressure of space but it is also dependent on new cultural attitudes that allow us to view them as acceptable. The consensus seems to be that they have their place, but it is incredibly important to be aware of the sensitivities surrounding their use.

Growth in population and the changing requirements for space, changing attitudes, changing social norms, cleanliness and hygiene standards and the greater number of women in the workforce are continuing to have a profound effect on washrooms. People expect well cared-for facilities and they will often make a judgement about a firm or a venue based on the quality of the washroom design and maintenance.

See also: Accessible environments, p.18; Building Regulations, p.76; Water fittings, p.720; Water quality, p.725.

Sources of further information

L24 – Workplace health, safety and welfare, approved code of practice and guidance: www.hse.gov.uk/pubns/priced/l24.pdf

Trade unions

Pinsent Masons Employment Group

Key points

- A trade union is an organisation consisting of workers whose main purpose is the regulation of relations between workers and their employers.
- Employers are prevented from offering inducements to their employees not to be a member of a trade union, not to take part in the activities of a trade union, not to make use of the services of a trade union and not to give up the right to have their terms and conditions of employment determined by a collective agreement.
- Further protection is provided to ensure that employees should not suffer detrimental actions for being a union member or using a union's services.

Legislation

- Trade Union and Labour Relations (Consolidation) Act 1992.
- Trade Union Reform and Employment Rights Act 1993.
- Employment Relations Act 1999.
- Employment Relations Act 2004.

Trade unions and collective agreements

In some industries, negotiated collective agreements exist relating to pay and terms of employment, and in some circumstances those agreements can also form part of the workers' contracts of employment.

A collective agreement may not always be enforceable between the union and the employer. However, the terms of a collective agreement may become incorporated into an individual employee's contract of employment, and so themselves become terms and conditions of employment.

Collective agreements can be incorporated if the employment contract expressly says so, or if the custom and practice in the industry is that the collective agreements

are impliedly incorporated. However, some parts of collective agreements are not appropriate for incorporation.

Generally, once the terms of a collective agreement are incorporated into a contract of employment, they become terms of the contract and in some cases can remain in force even if the original collective agreement terminates.

Trade union recognition

The Employment Relations Act 1999 (ERA) created rules for trade unions to be recognised by employers on a statutory basis as long as certain conditions are fulfilled (*see below*).

In general terms, however, 'recognition' of a trade union is important in a number of ways. If a union is recognised, employers will have certain duties, for example:

- to consult with the union and its representatives on collective redundancy situations;
- to disclose information for collective bargaining purposes; and
- to allow time off to employees engaged in trade union activities or duties.

In addition, employers are under a duty to provide information to and consult with recognised trade unions concerning TUPE transfers.

Statutory recognition

In certain circumstances, even outside the provisions of the ERA, trade union recognition can take place voluntarily. An employer can voluntarily recognise a trade union, either expressly by stating so or by clear conduct that shows an implied agreement to recognise that union.

Accordingly, an employer that actually enters into negotiation with a trade union about terms and conditions of employment, conditions of work, employee discipline, trade union membership, etc., may be deemed to recognise the union voluntarily.

However, the statutory procedure also allows the trade union to apply for recognition so that it can conduct collective bargaining regarding pay, hours and holidays. The procedures are complex and are set out in Schedule A1 to the Trade Union and Labour Relations (Consolidation) Act 1992 (TUL(C)RA),

the legislation containing the recognition machinery introduced by Schedule 1 to ERA. To trigger the statutory procedure, the trade union must apply to the employer in respect of the workers who wish to constitute a bargaining unit (BU). A BU is determined by a number of factors that may result in a sector of the workforce being identified as a BU even though they may not have been the subject of separate negotiations in the past or even where the employer wishes to negotiate with the whole workforce.

The request to the employer must:

■ be in writing;
■ identify the relevant trade union and the BU; and
■ state that the request is made under paragraph eight of Schedule A1 to TUL(C)RA.

Further, the trade union must be independent and the employer must employ at least 21 workers.

Negotiation
The employer should, within ten working days of receiving the written request from

the trade union, accept the request, reject it, or offer to negotiate.

If the parties agree on the BU and that the trade union should be recognised in respect of the BU, that is the end of the statutory procedure.

However, if the employer rejects the trade union's request outright or fails to respond, the union can apply to the Central Arbitration Committee (CAC).

The employer or the trade union may request the Arbitration, Conciliation and Advisory Service (ACAS) to assist in conducting negotiations. If the employer proposes that ACAS assistance be requested, and the union fails to respond within ten working days of the proposal or rejects such a proposal, no application to the CAC can be made. This is provided that the proposal is made by the employer within ten working days of having informed the union of its willingness to negotiate.

The trade union may approach the CAC if no agreement is reached between the parties, and if the employer fails to respond to the request within the ten-working-day period. If the employer informs the union within ten working days that it does not accept the request but is willing to negotiate, then there is an additional 20 days for negotiation, starting the day after the first ten-day period ends. If no agreement is reached at the end of the additional 20-day period, or if the parties agree a BU, but do not agree that the trade union is to be recognised, the trade union may apply to the CAC. On a practical note, if the employer reaches an agreement with the trade union that a ballot on recognition can take place, it may wish to include an undertaking by the trade union that the latter will not make another request for recognition for a period of time.

The CAC may accept the request if the initial request for recognition was valid and is on the face of it 'admissible'. The CAC will normally decide within ten working days of it receiving the request whether it may accept the claim. The CAC decides if the application is admissible by asking whether the trade union has 10% membership and whether the majority is likely to be in favour of recognition.

If the CAC accepts an application, but the BU has not been agreed, the employer must, within five working days, supply certain information about his workforce to the union and the CAC. If the BU has not been agreed by the parties, the CAC will try to help the parties to agree a BU within 20 working days of it giving notice of its acceptance. If the claim is not accepted by the CAC, this is an end to the statutory procedure.

If an agreement is reached between the parties, or if the CAC determines the BU, and this BU is different from the one originally proposed, then the validity test must be applied again.

If the CAC is satisfied that more than 50% of the workers constituting the BU are members of the trade union, it must usually issue a declaration that 'the trade union is recognised as entitled to conduct collective bargaining on behalf of the workers constituting the BU', but may hold a ballot if any of the following three factors apply:

1. It is in the interests of good industrial relations; or
2. A significant number of the trade union members within the BU informs the CAC that they do not wish the trade union to conduct collective bargaining on their behalf; or
3. Evidence leads the CAC to doubt whether a significant number of trade union members really want the trade union to conduct collective bargaining on their behalf.

If any of these conditions apply, or if the CAC is not satisfied that the majority of the workers in the BU are members of the union, then the CAC must arrange to hold a secret recognition ballot in which the workers constituting the BU are asked whether they want the trade union to conduct collective bargaining on their behalf.

Within ten working days of receiving the CAC notice, the trade union, or the trade union and the employer together, may notify the CAC that they do not want a ballot to be held. If a ballot is held in any event or if no objection is made, the ballot will be conducted by a Qualified Independent Person (QIP), who is appointed by the CAC. A QIP can, for example, be a practising solicitor.

The ballot must take place within 20 working days from the day the QIP is appointed, or such longer period as the CAC may decide. It may be conducted at a workplace, by post, or by a combination of these two.

The CAC must inform the employer and the trade union of the result of the ballot as soon as it is reasonably practicable after it has itself been so informed by the QIP.

If a majority of the workers voting in the ballot, and at least 40% of the workers constituting the BU, vote in favour of recognition, the CAC must issue a declaration that the trade union is recognised as entitled to conduct collective bargaining on behalf of the BU.

The parties will then have a 30-working-day negotiation period in which they may negotiate with a view to agreeing a method by which they will conduct collective bargaining. If no agreement is reached, the employer or the trade union may apply to the CAC for assistance.

If an agreement still cannot be reached, the CAC must take a decision.

The Code of Practice, 'Access and Unfair Practices during Recognition and Derecognition Ballots' (see *Sources of further information*), gives guidance regarding the union's access to workers during the period of recognition ballots and the avoidance of unfair practices whilst campaigning during that period. Whilst the Code imposes no legal obligations, its provisions are admissible in evidence and will be taken into account by any Court, Tribunal or the CAC where relevant.

See also: Health and safety consultation, p.368.

Sources of further information

Central Arbitration Committee (CAC): www.cac.gov.uk/

ACAS: www.acas.org.uk

DTI: Code of Practice 'Access and Unfair Practices during Recognition and Derecognition Ballots': www.bis.gov.uk/files/file14418.pdf

Training

Lizzy Campbell, Anderson Strathern

Key points
- Well trained staff can help businesses retain a competitive edge.
- Training staff in health and safety will help reduce risk to the business.
- Eligible staff now have the right to request time off for training.

Legislation
- Health and Safety at Work etc. Act 1974.
- Corporate Manslaughter and Corporate Homicide Act 2007.
- Health and Safety (Offences) Act 2008.
- The Apprenticeships, Skills, Children and Learning Act 2009.

Why train your staff?

A business' most important resource is its employees. Giving employees the tools to enable them to excel is one way of making sure that they remain highly motivated and committed. Well-trained staff can be key to businesses achieving improved quality and increased productivity.

The Apprenticeships, Skills, Children and Learning Act 2009 received Royal Assent on 12 November 2009, bringing in a host of new measures to prepare for the UK's long-term economic and social needs.

This Act introduced a new apprenticeship structure that is applicable to England and Wales and a new 'right to train' for employees (subject to meeting certain conditions) which is also applicable to Scotland, under Section 40 of the Act. It has applied to businesses with 250 or more employees since 6 April 2010. Although it was due to be extended to employers of all sizes from April 2011, the Government has delayed the extension of its application to all employers in order to further assess the potential impact.

It is the Government's opinion that introducing a new right for employees to request time off work to undertake study or training will assist people in strengthening their skills and future employability. In helping individuals reach their full potential it is hoped that the working population will be better equipped to support the economic recovery of the nation and be more competitive on a global basis.

Well trained staff will give a business that competitive edge. Proper staff training will, of course, also help businesses manage risk. So, in terms of workplace law, what are the key areas where training will be of greatest benefit?

Health and safety

Every employer has a legal duty to ensure, so far as is reasonably practicable, the health, safety and welfare at work of all its employees as set out in the Health and Safety at Work etc. Act 1974 (HSWA). A breach of health and safety regulations can result in an employee being injured (with the possibility of a personal injury claim against the employer to follow).

Even if no one is injured, a breach may also expose the business, and/or its responsible officers, to the risk of criminal prosecution.

This duty towards an employee specifically extends (HSWA Section 2(2)(c)) to the provision of such information, instruction, training and supervision as is necessary to ensure the employee's heath and safety at work. A failure to provide such training etc. is an offence in its own right, and can of itself result in a criminal charge.

It is therefore essential that an employer trains its employees, where appropriate, in relation to matters such as lifting and manual handling and the proper operation of plant and machinery. And, of course, it follows that the provision of such training reduces risk for both employee and employer, all the more relevant since the coming into force of the Corporate Manslaughter and Corporate Homicide Act 2007, which created a new offence of corporate manslaughter (corporate homicide in Scotland), and the Health and Safety (Offences) Act 2008, which increased the penalties available to the courts for breach of the HSWA.

An employer's duty to safeguard the health, safety and welfare of its employees extends to mental, as well as physical, health. Unreasonable stress at work can adversely impact upon the health of employees. Ensuring that staff are equipped with the skills to enable them to carry out their jobs can help to reduce stress. However, managers also need to be trained to identify signs of stress and to find ways of reducing it.

See also: Corporate manslaughter, p.191; Display Screen Equipment, p.244; First aid, p.325; Manual handling, p.474; Risk assessments, p.624; Workplace health, safety and welfare, p.768.

Sources of further information

Workplace Law is one of the leading providers of IOSH training courses in the UK and has helped thousands of students get accredited. We have a commitment to providing the highest level of IOSH training available and offer the full portfolio of IOSH courses, including:

- IOSH Managing Safely
- IOSH Working Safely
- IOSH Directing Safely
- IOSH Safety for Senior Executives

If you are looking towards a future in health and safety, or have staff who need training, we have the course for you. Our IOSH courses offer training solutions across all levels of your organisation.

We run a busy schedule of public classroom based training courses or we can run face-to-face training in-house on your premises. We have also developed state of the art e-learning courses that prove popular for individuals or as multiple user licenses for larger organisations – this is a flexible and cost effective solution that is becoming ever more popular with our growing number of students.

Workplace Law IOSH courses are all about:

- *Fun* – our IOSH courses have been designed with our students in mind to help make the learning experience as fun and engaging as possible.
- *Getting involved* – Workplace Law is the fastest growing IOSH training provider in the UK with a deserved reputation for excellence. You will be joining an expanding number of UK professionals gaining accreditation the Workplace Law way.
- *Support* – our expert tutors provide students with real world experience and are on hand to guide you through our courses and ensure you get the accreditation you are looking for. With satisfaction rates averaging over 90% across our courses you can be sure that you are in safe hands.
- *Value for money* – we know that when you choose Workplace Law you are putting your trust in us. We firmly believe that the quality of our training – whether online or in the classroom – is at the highest level there is, ensuring that our courses are the best value for money in the market.

Visit www.workplacelaw.net/training for more information.

Trespassers and squatters

Melissa Thompson, Pinsent Masons Property Group

Key points

- A trespasser or squatter is a person who occupies land without the owner's consent. This extends to the airspace above land and to below ground level.
- The recommended remedy is for the landowner to obtain a possession order to require the trespasser to leave. Injunction proceedings may be instigated in addition to restrain any further trespass and obstruction / interference from existing trespassers.
- Self-help is not advisable as it is easy to commit an offence under the Criminal Law Act 1977. The police have rights to arrest trespassers in some cases, but rarely use them.
- A landowner owes a limited duty of care to a trespasser and may be liable in damages for a trespasser injured as a result of a hazard on the land. This is particularly the case in respect of children.
- Trespassers can obtain certain rights by the long-term occupation of land without the owner's permission.

Legislation

- Criminal Law Act 1977.
- Criminal Justice and Public Order Act 1994.
- Land Registration Act 2002.

Steps available to a landowner

Police

The police have powers to arrest and remove trespassers or squatters in certain circumstances:

- If there is evidence that a squatter has used force to break into a building – e.g. a witness has seen the squatter breaking a lock or window – then the police can arrest the squatter on suspicion of causing criminal damage. Once the squatters have been removed, the landowner can re-secure the premises.
- The police have powers to arrest a person for a breach of the peace and/or for the offence of aggravated trespass under Sections 68 and 69 of the Criminal Justice and Public Order

Act 1994. However, while the police may be prepared to use their powers in the case of disruptive trespassers who are seeking to prevent a lawful activity, they are less likely to do so in the more usual situation where land is occupied by passive squatters.

In practice, the police tend not to get involved until a court order has been obtained from the civil courts.

Self-help

A landowner is entitled to use reasonable force to prevent a person from committing an act of trespass on his land. However, the use of reasonable force is not usually appropriate because of the following factors:

- It is a criminal offence under Section 6(1) of the Criminal Law Act 1977 for a person to use or threaten violence to secure entry if there is someone present on the property who is opposed to entry. Violence

---

> **Case study**
>
> *University of Sussex v. Protestors* (2010)
>
> In this case, a student protest on a university campus escalated to the extent that key members of university personnel, located in an administrative building, were prevented from leaving their offices by the protestors who took occupation of the building. In light of these exceptional circumstances where there were legitimate concerns regarding the health and safety of the university staff and the risk of property damage, a claim was brought on the same day in the High Court by solicitors representing the university.
>
> As these were particularly serious circumstances, the judge granted an interim injunction immediately and agreed to abridge time for service of proceedings to 45 minutes. The possession proceedings were heard by the judge later that evening and the judge awarded the university landowner a possession order that extended to the whole of the university campus. This order prevented the protestors from moving from building to building across the campus once they had been evicted from their original protest site, as they had threatened to do.
>
> While this case has unusual and individual facts, it is useful to note how quickly it is possible to obtain injunctions and possession orders against trespassers in circumstances where there is a risk of property damage or to the health and safety of the landowners or their staff.

includes violence to property as well as to persons and so an owner who smashes a window or a lock to gain access where there are trespassers in the building can commit a criminal offence, even if he is the owner of the building. In practice, this means that a landowner cannot force entry to a building unless he is satisfied that the squatter is not actually in the building at that time.

- What constitutes reasonable force is a question of fact and degree and gives rise to considerable uncertainty.
- The use of any force can cause a situation to degenerate into violence.

Possession order

There is a special procedure to obtain a possession order against trespassers or squatters contained in Part 55 of the Civil Procedure Rules. Generally a claim must be started in the County Court for the District in which the land is situated, although in exceptional circumstances the claim can be brought in the High Court where, for example, there are serious concerns about health and safety, fear of property damage, etc. In order to prepare the claim, solicitors will generally require full details of property ownership and full details of the factual circumstances surrounding any trespass. Generally, a hearing date will be given when the original application is issued by the court.

The trespassers must be served, in the case of residential property, no fewer than five working days before the hearing date, and, in the case of all other land, no fewer than two working days before the hearing date. In particularly serious circumstances an application can be made to the court to abridge time for service.

Where trespassers are named, these individuals must be served personally. Where the names of the trespassers are not known, the claim will be brought against 'persons unknown' and service can be effected by attaching the claim form and any witness statements to the main door or some other part of the building where it is clearly visible, and by placing it through the letter box in a sealed transparent envelope addressed to the occupiers. Where the squatters are occupying open land rather than buildings, service may be effected by placing stakes (to which the relevant documents are attached) on the land at prominent positions, again in sealed transparent envelopes addressed to the occupiers.

At the possession hearing, assuming there is no defence to the trespass, a possession order should be made. This will cover the area actually occupied by the trespassers but can in some circumstances cover a wider, currently unoccupied area if there is a real risk of the trespassers moving there from the originally occupied area.

When a possession order has been obtained in the County Court, the landowner may have the matter transferred to the High Court for enforcement. This is by way of a High Court Enforcement Officer. This is more costly than enforcing through a County Court bailiff, but generally ensures that enforcement will take place in a matter of days rather than weeks.

Costs are not generally sought or ordered since the majority of trespassers will be unnamed and make costs recovery practically impossible. However, on occasion the courts are willing to make

trespassers give their names to court and make a costs order against these named individuals where they have no defence to an order for possession and they are seeking to delay matters intentionally to increase the Claimant's costs.

Interim possession order

There is also a procedure to obtain interim possession orders. The Court issues the claim form and the application for the interim possession order, setting a date for the hearing no fewer than three days after the date of issue. All documents have to be served within 24 hours of the issue of the application.

If an interim possession order is granted, the defendant will be required to vacate the premises within 24 hours, and it is a criminal offence if he does not. There are two potential difficulties with the interim possession proceedings. First, having made an interim possession order, the Court sets a date for a final possession hearing; hence there is a need for a second attendance at Court, which increases costs. Second, it is not possible to get a court bailiff to enforce an interim possession order. Accordingly, if the trespassers do not vacate the premises, and the police cannot be persuaded to assist, it will be necessary to wait until after the second Court hearing to obtain a final possession order that can be converted into a warrant for possession, which the Court bailiff will need to assist.

As standard possession proceedings can often be dealt with as quickly and do not require two hearings, these are often to be preferred over interim possession order proceedings.

Injunctions

Injunction proceedings can be issued in tandem with possession proceedings where there is a concern that further trespass will occur or that existing trespassers are likely to cause disruption or otherwise act obstructively. It is a contempt of court to breach the terms of an injunction, which prohibits further trespass or certain behaviour.

Generally, interim injunctions suffice and will often be applied for and obtained at a hearing that the trespassers are not notified of and that can take place very quickly, depending on the urgency of the situation. Any interim injunction ordered is usually served on the trespassers at the same time as the issued possession proceedings.

Although the injunction does not entitle the landowner to recover possession (and he must wait for a final possession order to do this) on occasion, the service of an interim injunction can lead the trespassers to leave prior to any possession order being made.

Theoretically, a Court can order a hearing to deal with the grant of a final injunction, but this is rare since interim injunctions, coupled with final possession orders, generally achieve the desired result of bringing an end to the trespass.

Law reform
It should be noted that the Ministry of Justice is currently conducting a consultation on the extent of problems caused by squatters and whether and how existing criminal and civil mechanisms should be strengthened to deal with them. Accordingly, the law in this area may well change in the future. The consultation closes in October 2011.

Adverse possession
Surprising though it may seem, trespassers can obtain certain rights by the long-term occupation of land without the owner's permission. These rights have changed following the coming into force of the Land Registration Act 2002 on 13 October 2003 (although it should be noted that this Act, and any changes it has introduced, only apply to registered land):

- A person who has occupied another person's land without permission and has excluded the landowner from that land (e.g. by enclosing the land with a fence) for a period of 12 years *prior* to 13 October 2003 can become the legal landowner of the land he has unlawfully occupied.
- In order to become the legal owner, it must be shown that the squatter has possessed the land by dealing with it exclusively as an occupying owner might have been expected to deal with it and intentionally possessing it to the exclusion of all others.
- *Following* 13 October 2003, the squatter only needs to prove possession of the land for ten years (unless the land in question is unregistered, in which case 12 years' possession is still required).
- However, despite reducing the amount of time a squatter needs to possess registered land from 12 years to ten, the Land Registration Act 2002 has introduced a new three-part test that the squatter must pass to claim legal ownership of the land. As this requires squatters to make formal applications to be registered as replacement legal owners of the land in place of the original owners, and the original owners of the land in question are given notice of such applications and can object to these, it is expected that this new procedure will make it easier for the landowner to prevent the squatter from replacing him as the legal owner of the land.

- Adverse possession most frequently happens when an adjoining landowner extends his garden. However, it does occasionally occur with much bigger parcels of land and buildings. Consequently, once evidence of an encroachment or squatter comes to light, a landowner should take prompt action.

The UK law in relation to adverse possession will not infringe human rights.

> *See also*: Landlord and tenant: possession issues, p.438; Occupiers' liability, p.524.

See also: Landlord and tenant: possession issues, p.438; Occupiers' liability, p.524.

Sources of further information

Civil Procedure Rules Part 55 – Possession Claims:
www.justice.gov.uk/civil/procrules_fin/contents/parts/part55.htm

Vehicles at work

Kathryn Gilbertson, Greenwoods Solicitors LLP

Key points

- There are many specific industries where vehicles are designed and used for specific workplace tasks. These are beyond the scope of this chapter.
- Approximately 60 deaths and 6,000 serious accidents each year are caused by vehicles in the workplace.
- Employers who need to provide vehicles as part of their safe systems of work must ensure that they put appropriate control measures in place.
- The safe system of work should include:
  - the workplace (vehicle routes, provision for pedestrians, signage);
  - vehicles (safety features, good maintenance);
  - employees (driver training, traffic hazard briefing, competence); and
  - vehicle activities (loading and unloading, refuelling or recharging, reversing, tipping, sheeting and unsheeting).

Legislation

- Health and Safety at Work etc. Act 1974.
- Workplace (Health, Safety and Welfare) Regulations 1992.
- Provision and Use of Work Equipment Regulations 1998.
- Management of Health and Safety at Work Regulations 1999.

Statutory requirements

An employer's principal legal duty is, so far as is reasonably practicable, to provide and maintain a safe system of work and to take all reasonably practicable precautions to ensure the health and safety of all the workers in the workplace.

The Management of Health and Safety at Work Regulations 1999 require that all employers assess the risks to the health and safety of their employees and of anyone who may be affected by their work activity. The risk assessment process should cover the following:

- Identification of all the hazards involving vehicles – moving vehicles,

causing injury or damage by driving, loading and unloading or refuelling or recharging vehicles, maintenance;
- Consideration of the risks involved;
- Identifying the people involved;
- Evaluating the existing control measures;
- Recommending new control measures; and
- Recording the assessment.

It is important that all the risks are addressed. Risks should be removed if possible or, if not, control measures that will reduce the risks to acceptable levels be put into place. In relation to vehicles, the risk control measures will include safe systems of work.

The Management of Health and Safety at Work Regulations also require that employers shall:

- in entrusting tasks to employees, take into account their capabilities as regards health and safety; and
- ensure that employees are provided with adequate health and safety

training on being recruited into the employer's undertaking and on being exposed to new or increased risks at the workplace.

The Workplace (Health, Safety and Welfare) Regulations place the following duties, in relation to vehicles, upon employers:

■ The workplace shall be organised in such a way that vehicles can circulate in a safe manner.
■ Traffic routes must be suitable for the vehicles using them.
■ All routes must be suitably indicated.
■ The workplace shall be maintained in an efficient state, in efficient working order and in good repair.
■ Every floor in a workplace and the surface of every traffic route shall be kept free of obstruction.

This should be under constant review to ensure compliance.

The Provision and Use of Work Equipment Regulations 1998 place the following duties, in relation to vehicles, upon employers:

■ To ensure that work equipment (which includes vehicles) is so constructed or adapted as to be suitable for the purpose for which it is used or provided.
■ In selecting the work equipment, every employer shall have regard to the working conditions and to the risks to the health and safety of persons which exist in the premises or undertaking in which that work equipment is to be used, and any additional risk posed by the use of that work equipment.
■ To ensure the work equipment is maintained in efficient working order and in good repair.

Common factors that will need consideration by employers in the majority of workplace situations include the following:

■ Vehicle routes meet the needs of pedestrians and vehicles.
■ Traffic routes are appropriate to the types and quantities of vehicles.

- Safety features are in place, e.g. signs and markings, barriers and humps.
- Reversing manoeuvres are minimised.
- Safe parking for all drivers.
- Loading and unloading procedures arranged for safety.
- Vehicles fitted with all necessary safety equipment and features.
- Vehicles maintained in good working order.
- Driver selection and training procedures sufficient to ensure that employees asked to drive at work are suitably experienced and competent, and remain so.
- Adequate briefing on workplace driving hazards.
- Adequate supervision and inspection of workplace driving activities to ensure safe systems of work are being followed.

It is strongly advised for employers to put in place a written policy covering driving at work, which takes into account the factors mentioned above, not forgetting private cars used by employees for business use, for which employers are to have responsibility. This should ensure checks such as whether the car is insured for business, if it has a valid MOT, and so on.

The Health Act 2006 and the Smoke Free (Exemptions and Vehicle) Regulations 2007 provide that a vehicle that is used by the public in the course of paid or unpaid work by more than one person must be smoke-free. See 'Driving at work' (p.250) and 'Smoking' (p.653).

> *See also*: Accident investigations, p.29; Driving at work, p.250; Mobile phones at work, p.496; Risk assessments, p.624; Smoking, p.653.

Sources of further information

INDG 199 – Workplace transport safety
www.hse.gov.uk/pubns/indg199.pdf

HSG 136 Workplace transport safety – an employer's guide:
www.hse.gov.uk/pubns/priced/hsg136.pdf

Workplace Law's **Driving at Work Policy and Management Guide v.5.0** helps you cover yourself and your staff and ensure that your employees keep to the highest standards of safe driving at work. This comprehensive new edition of the policy and management guide updates several elements of the original including the implications of recent legislation such as the Health Act 2006, the Road Safety Act 2007 and the Corporate Manslaughter and Corporate Homicide Act 2007. If your business hasn't already got a driving at work policy in place, or your current policy is not up-to-date, this is an essential publication. The policy highlights the issue of liability should prosecution occur following a driving at work accident, and who might face prosecution as a result. For more details visit www.workplacelaw.net.

Vibration

Craig Scott, Bureau Veritas UK Ltd

Key points

- The Control of Vibration Regulations 2005 came into force on 6 July 2005.
- The Regulations cover two aspects – Hand–Arm Vibration (HAV) and Whole-Body Vibration (WBV).
- There is a transitional period for exposure limits up to 2010, which applies to work equipment already in use before July 2007.
- Whole-Body Vibration exposure limits in the agriculture and forestry sectors are extended to 2014.
- Exposure limits may be exceeded during the transitional period, providing you have complied with all the other requirements of the Regulations and taken all reasonably practicable actions to reduce exposure as much as you can.
- The emphasis of the Regulations is on control and taking action, not on continual assessment.
- The Control of Vibration Regulations 2005 do not apply to work taking place in ships, boats or other vessels. Hand–Arm and Whole-Body Vibration is separately dealt with under the Maritime and Coastguard Agency Regulations, Statutory Instruments 2007 No. 3077.
- Be aware of Whole-Body Vibration risks where any commercial / industrial / construction vehicles are driven regularly for most of the day.

Legislation

- Health and Safety at Work etc. Act 1974.
- Social Security (Industrial Injuries) (Prescribed Diseases) Regulations 1985.
- Personal Protective Equipment at Work Regulations 1992.
- Supply of Machinery (Safety) Regulations 1992.
- Reporting of Injuries, Diseases and Dangerous Occurrences Regulations 1995.
- Provision and Use of Work Equipment Regulations 1998.
- Management of Health and Safety at Work Regulations 1999.
- Control of Noise at Work Regulations 2005.
- Control of Vibration Regulations 2005.

Whole-Body Vibration

- Exposure occurs when vibration is transmitted through the seat or feet.
- Regular long-term exposure is associated with back pain and other muscle fatigue complaints.

Hand–Arm Vibration

This is transmitted from work processes into workers' hands and arms. Common processes are:

- operating handheld or hand-guided power tools; and
- holding materials being processed by machines.

Ill health symptoms

- Tingling and numbness in the fingers.
- Not being able to feel things properly.
- Loss of strength in the hands.

Facts

- Around five million workers are exposed to Hand–Arm Vibration in the workplace.
- Two million of these workers are exposed to levels of vibration where there are clear risks of developing disease.
- Hand–Arm Vibration comes from the use of handheld power tools, and is the cause of significant ill health such as painful and disabling disorders of the blood vessels, nerves, joints and muscles of the hands and arms.
- Regular long-term exposure to Whole-Body Vibration is associated with back pain, alongside other factors such as poor posture and heavy lifting.

- Fingers going white (blanching) and becoming red and painful on recovery particularly in the cold and wet. Often referred to as dead finger, dead hand or white finger.

Legal duties

Employers must:

- assess the vibration risk;
- establish if the daily exposure action value (EAV) will be exceeded;
- establish if the daily exposure limit value (ELV) will be exceeded;
- eliminate the risk or reduce exposure to a level as is reasonably practicable;
- provide health surveillance to employees who continue to be regularly exposed above the action value;
- provide information and training;
- consult trade union safety representatives or employee representatives on your proposals;
- keep records of the risk assessment and controls;
- keep health records for employees under health surveillance; and
- carry out regular reviews and take actions to reduce exposure.

Exposure action value

The exposure action value (EAV) is a daily (eight hours) amount of vibration exposure above which employers are required to

take action to control exposure. Exposure is quantified in terms of the acceleration of the surface in contact with the hand. The acceleration of the surface is normally expressed in units of metres per second squared (m/s^2):

- For Whole-Body Vibration the EAV is a daily exposure of $0.5/s^2$ A(8).
- For Hand–Arm Vibration the EAV is a daily exposure of $2.5 \ m/s^2$ A(8).

Exposure limit value (ELV)

The exposure limit value (ELV) is the maximum amount of vibration an employee may be exposed to on any single day. It represents a high risk above which employees should not be exposed:

- For Whole-Body Vibration the ELV is a daily exposure of $1.15 m/s^2$ A(8).
- For Hand–Arm Vibration the ELV is a daily exposure of $5 m/s^2$ A(8).

Estimating exposure

Care should be exercised in using manufacturers' vibration data. Check if the information represents the way you use the equipment. Specialist equipment and a competent person will be required if you want to obtain vibration measurements for your own tools. Many factors affect the readings and the experience of the competent person is important in achieving realistic readings.

Vibration

Tool type	Lowest	Typical	Highest
Road breakers	5m/s²	12m/s²	20m/s²
Demolition hammers	8m/s²	15m/s²	25m/s²
Hammer drills / combi hammers	6m/s²	9m/s²	25m/s²
Needle scalers	5m/s²	-	18m/s²
Scabblers (hammer type)	-	-	40m/s²
Angle grinders	4m/s²	-	8m/s²
Clay spades / jigger picks	-	16m/s²	-
Chipping hammers (metal)	-	18m/s²	-
Stone-working hammers	10m/s²	-	30m/s²
Chainsaws	-	6m/s²	-
Brushcutters	2m/s²	4m/s²	-
Sanders (random orbital)	-	7-10m/s²	-

Table 1. How vibration level and duration affect exposure. (Source: HSE)

Tool vibration (m/s²)	3	4	5	6	7	10	12	15
Points per hour (approx)	20	30	50	70	100	200	300	450

Table 2. Exposure points. (Source: HSE)

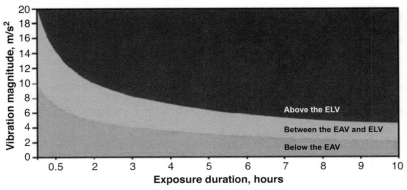

The HSE has produced an exposure calculator at www.hse.gov.uk/vibration, which can be used to assess each employee's daily exposure. Alternatively you can use the simple 'exposure points' system in Table 2.

Multiply the points assigned to the tool vibration by the number of hours of daily 'trigger time' for the tool(s) and then compare the total with the exposure action value (EAV) and exposure limit value (ELV) points.

- 100 points per day = exposure action value (EAV).
- 400 points per day = exposure limit value (ELV).

Example
Someone using an angle grinder with a vibration level of 7m/s² for five hours a day would reach the EAV (Exposure Action Value) level in one hour and the ELV (Exposure Limit Value) in four hours. In this example, if the grinder is used for five hours a day the ELV has been exceeded by one hour.

Control measures
- Try to find an alternative way of working that does not involve the use of vibrating tools or exposure to activities that create vibration.
- Ensure equipment is suitable and can do the work efficiently – if it is too small or not powerful enough it will extend the vibration exposure.
- Check the vibration data for equipment before purchase – choose equipment with the lowest vibration rating.
- Devices such as jigs reduce the need to grip items.
- Correct maintenance for equipment will lessen the likelihood of an increase in vibration, e.g. make sure cutting tools are kept sharp.
- Limit the time to exposure, plan work, rotate employees.
- Provide the correct PPE.

Information and training
Inform employees of the risk assessment and the control measures and provide any training that may be considered necessary. Information to employees can include:

- instruction in the use of equipment, viz. avoid gripping or forcing a tool or work piece more than you have to;
- keeping hands warm and dry;
- reducing smoking because smoking reduces the blood flow;
- massaging and exercising fingers during work breaks; and
- posture and manual handling training when taking into account Whole-Body Vibration exposure.

Health surveillance
You must provide health surveillance for all employees who, despite your action to control the risk, are likely to be regularly exposed above the EAV. This will provide you with useful feedback information on the effectiveness of your vibration control measures as well as monitoring those people at risk.

See also: Health surveillance, p.396; Occupational health, p.521.

Sources of further information

Vibration at work: www.hse.gov.uk/vibration

Violence at work

Kathryn Gilbertson, Greenwoods Solicitors LLP

Key points

- In January 2011, the HSE published its Violence at work report, which presented the findings of the 2009/10 British Crime Survey. The report identified that the risk of being a victim of actual or threatened violence at work is low – 1.4% of working adults were victims of one or more incidents at work.
- The police were most at risk of violence at work, with 9% having experienced violence, followed by health professionals (3.8%) and social welfare professionals (2.6%).
- Violence can be physical or emotional. It can be a single outburst, projected at a particular person, or a repeated series of attacks against a person or a particular group. Violence can and often does result in physical or emotional injury to its recipient and can have a significant effect on the lives of those around them.
- Many victims perceive that the violence is the result of many factors coming together and is often fuelled by anger, frustration and it is exasperated by alcohol and drug consumption.

Legislation

- Health and Safety at Work etc. Act 1974.
- Reporting of Injuries, Diseases and Dangerous Occurrences Regulations 1995.
- Management of Health and Safety at Work Regulations 1999.

Definitions and duties

The Health and Safety Executive (HSE) defines work-related violence as 'any incident in which a person is abused, threatened or assaulted in circumstances relating to their work'.

Section 2 of the Health and Safety at Work etc. Act 1974 requires employers, so far as is reasonably practicable, to ensure the health, safety and welfare of their employees. This duty includes providing a working environment that is safe and without risks to health. Thus, employers

are required to protect their staff from the risks associated with work-related violence.

The Management of Health and Safety at Work Regulations 1999 require employers to assess the risks to their employees and to take appropriate measures to prevent or reduce the risk.

Horseplay amongst employees may result in violence that falls within the scope of Section 7 of HSWA.

Additionally, the Reporting of Injuries, Diseases and Dangerous Occurrences Regulations 1995 (RIDDOR) require employers to notify an accident at work to any employee resulting in death, major injury, or incapacity to work three or more days. This includes accidents and injuries that occur as a result of any non-consensual physical violence done to a person at work.

Case study

The HSE will investigate and take action against incidents of workplace violence that come to their attention, such as an improvement notice issued against the employer.

Whilst somewhat aged (December 2003), the improvement notice against West Dorset General Hospitals NHS Trust shows the detailed responses expected by the HSE to combat workplace stress and violence. Other examples of enforcement action include an improvement notice against a Local Authority requiring the implementation of control measures for violence and aggression risks in home to school transport services and an improvement notice issued against an NHS Trust because it had not ensured:

- that staff at risk from violence and aggression were provided with adequate training in understanding and dealing with aggression in the workplace;
- that an adequate risk assessment was in place; and
- that a policy for managing violence to staff was in place.

Mental Health Matters Ltd, a North-East based charity, was fined £30,000 and ordered to pay £20,000 costs after one of its employees was killed by a service user. The victim was a support worker who was stabbed to death whilst visiting a patient at his home. The attack happened on her final day of her probation period.

The prosecution told the court that the patient's mental health was known to be deteriorating and that the company failed to respond to a number of warning signs. The charity also failed to provide the victim with the level of protection needed for this work.

The HSE commented, "if Mental Health Matters had carried out a risk assessment, it would have resulted in the visiting arrangements being reviewed".

Effects

The employer too can be affected by the consequences of workplace violence, financially and morally. Where workplace violence exists and is not managed, employers may well face low morale, high staff turnover, and difficulties in recruiting associated with an environment that is not conducive to healthy working and has a poor reputation in the public eye. Further financial impacts may occur from civil action and resulting compensation payments, prosecution and increased insurance premiums.

Risk assessments

Identify the hazards

The employer should find out if there are problems with violence at work by consult with employees, reviewing accident books and sickness records and ensure that suitable and sufficient risk assessments are carried out.

Key elements:

- What is the perception of violence amongst employees?

Facts

- The British Crime Survey (BCS) estimated that there were 677,000 incidents of violence at work in 2009/10.
- There were approximately 310,000 assaults and 366,000 threats of violence.
- 43% of people assaulted or threatened were repeat victims.
- 65% of cases involved strangers.
- Victims said that the offender was under the influence of alcohol in 38% of incidents.
- 37% of assaults resulted in injury.

- Is there evidence of verbal aggression in the workplace?
- Is there evidence of physical aggression in the workplace?
- Are there other factors influencing behaviours that can lead to violence and aggression, i.e. environmental factors, work patterns, training issues, organisational structure and job design?

Identify who is at risk and how they could be harmed

Many groups of people could be described as being 'at risk' of violence at work. Typical groups of at-risk individuals include those who are involved in providing services and are involved in direct communication with the public – for example, shop assistants, call centre operatives, nurses and doctors, couriers, teachers, security guards, etc.

It is important, however, to consider the less obvious 'at-risk' groups in your organisation – those representing authority such as line managers and HR managers; those who may be dealing with sensitive health issues such as occupational health professionals; and employees who travel off site or overseas where they may be at higher risk purely because of their nationality and/or ethnicity.

An employer will need to carefully examine the work and working practices of their identified 'at-risk' groups to determine how they could be harmed. For example, is it more likely to be verbal aggression rather than physical aggression, and are there common incidents amongst those people such as physical assault with weapons?

Controls

Consider what controls you have in place and whether those controls are effective.

For example, ensure that you have in place:

- a system for reporting incidents that encourages participation and provides appropriate confidentiality, along with analysis of those figures to provide critical information on trends and the efficacy of controls;
- a policy that specifically addresses violence in the workplace, reporting incidents, debriefing from incidents, post-incident help and support, external legal assistance;
- other policies that may indirectly prevent violence in the workplace, i.e. lone working policy, security systems, approach to organisational communication;
- good job design with employees provided with relevant information to carry out their work safely, i.e. pre-travel briefings for work overseas or in high risk areas;

- training for employees to enable them to identify potentially violent behaviours and diffuse 'hot' situations, use personal protection measures and tactics when appropriate and be able to adjust personal behaviours accordingly;
- other appropriate physical controls relating to building layout and ambience which may include alternative methods of communicating, i.e. visual display boards, and alarm systems, etc.; and
- an environment that is clean and comfortable for employees, visitors and clients alike.

It is essential that control measures are relevant to the organisation and the risks it faces. The findings of the risk assessment should be recorded and the employer should have in place a system for monitoring, reviewing and auditing of control measures on a regular basis.

The most effective systems are those that have involved employee (and where appropriate public) consultation, have created a culture in which employees and visitors alike feel comfortable and safe, and demonstrate a balanced approach to the prevention and management of violence.

> *See also*: Loneworking, p.467; Reporting of Injuries, Diseases and Dangerous Occurrences (RIDDOR), p.618; Risk assessments, p.624; Stress, p.659.

Sources of further information

INDG 69 Violence at work: A guide for employers
www.hse.gov.uk/pubns/indg69.pdf

HSE example policy on work related violence:
www.hse.gov.uk/violence/toolkit/examplepolicy.pdf

HSE report – Violence at work – findings of the 2009/10 British Crime Survey:
www.hse.gov.uk/statistics/causdis/violence/british-crime-survey2009-10.pdf

Visitor safety

Andrew Richardson, URS Scott Wilson

Key points

- Employers have a duty to provide a safe environment for visitors to their premises.
- Employers' emergency procedures should cover all visitors to the premises.
- Visitors' needs may be different from, and/or more onerous than, the needs of regular employees.
- Children are less careful than adults and need more controls.
- The duty extends to uninvited visitors such as trespassers, and to tenants.
- Employers are liable for the actions of their employees who injure visitors.
- Employees must help their employer to ensure a safe and healthy workplace.

Legislation

- Occupiers' Liability Acts 1957 and 1984.
- Health and Safety at Work etc. Act 1974.
- Workplace (Health, Safety and Welfare) Regulations 1992.
- Management of Health and Safety at Work Regulations 1999.

Occupiers' Liability Acts 1957 and 1984

From the employer's perspective, the Occupiers' Liability Act 1957 places upon the occupier a common law duty of care to all visitors, to take such care as is reasonable to see that visitors will be reasonably safe in using the premises for the purposes for which they were invited or permitted to be there. The Act also points out that children will inevitably be less careful than adults.

The Occupiers' Liability Act 1984 amended the 1957 Act slightly to include a duty to unlawful visitors (e.g. trespassers) such that if:

- the occupier is aware of the danger, and

- the occupier knows the person could put himself at risk, and
- the risk is one that the occupier could reasonably be expected to do something about,

then the same common law duty of care is owed as is owed to lawful visitors.

Employers are also liable for the actions of their employees and, in relation to visitors, have a vicarious liability for those actions. If an employee injures a visitor during the course of his work, then the employer is liable.

Health and Safety at Work etc. Act 1974

Under Section 3 of the Act an employer has a duty to 'conduct his undertaking in such a way as to ensure, so far as is reasonably practicable, that persons not in his employment (e.g. visitors) who may be affected thereby are not thereby exposed to risks to their health and safety'.

Section 4 requires that anyone in control of the premises or plant used by persons not in their employment should, so far as is reasonably practicable:

- ensure safe access and egress to the premises and plant; and
- ensure that plant or substances in the premises, or provided for their use, are safe and without risk to health.

These obligations are also transferred to any tenants of a building.

Section 7 places a general obligation upon all employees to take reasonable care of their own health and safety and that of others who may be affected by their own acts or omissions. They also have to cooperate with the employer so as to ensure that the employer can comply with all of the above statutory obligations.

Finally, under Section 8, no person should intentionally or recklessly misuse or interfere with anything provided under the Act and other legislation in the interests of health and safety.

Management of Health and Safety at Work Regulations 1999

These Regulations identify a number of general duties that, if followed, will allow the employer to meet those obligations detailed in the Acts mentioned above. These are principally to carry out risk assessments and to set up emergency procedures.

All employers must assess the risks to health and safety of their employees and of anyone who may be affected by their work activity; this will include visitors. The risk assessment process should:

- identify all the hazards;
- consider the risks involved and identify the people involved (employees, visitors, the public);
- evaluate the existing control measures and recommend any new control measures;
- record the assessment; and
- review and revise the assessment as necessary.

It is important to address all of the risks, eliminate them if possible and, if not, to put in place control measures that will reduce the risks to acceptable levels.

The Regulations also require employers to set up emergency procedures for serious and imminent dangers and to appoint competent persons to ensure compliance with those procedures. The procedures should cover all visitors to the premises – it should be borne in mind that visitors are likely to have different and/or greater needs than experienced employees.

Workplace (Health, Safety and Welfare) Regulations 1992

These Regulations define the various criteria for a safe and healthy workplace environment that an employer must provide for his employees. Occupiers have a duty of care towards visitors so, even though visitors may not be 'working', it is reasonable to conclude that workplace facilities that could impact on the health and safety of visitors, e.g. adequacy of lighting, reasonable temperature, non-slip floor surfaces, safe traffic routes. Etc. should be provided to meet the reasonably expected needs of visitors.

Issues for visitors

Particular issues of concern to visitors that need to be addressed include:

- they need information so they understand the risks and controls – in some cases they may need this in advance of arriving;
- if English is not their first language they may need additional briefings;
- they may need more supervision than employees, either because of their lack of familiarity with their surroundings or because of the special nature of their visit; and
- making the site more accessible and 'user-friendly' for them.

See also: Bomb alerts, p.56; CCTV monitoring, p.129; Children at work, p.143; Disability access and egress, p.238; Fire, means of escape, p.314; Health and safety at work, p.361; Slips, trips and falls, p.650; Trespassers and squatters, p.681.

Sources of further information

L21 Management of Health and Safety at Work Regulations 1999: www.hse.gov.uk/pubns/priced/l21.pdf

The HSE's *Workplace Transport Safety Information Sheet WPT17 – Preparing for visitors* provides guidance on a broad range of visitor management issues. It can be downloaded at www.hse.gov.uk/pubns/wpt17.pdf

Hazardous waste

Mark Webster, Bureau Veritas UK Ltd

Key points

- During 2005, Hazardous Waste Regulations (HWR) were made in England, Wales and Northern Ireland to revoke the respective Special Waste Regulations. The HWR provide a new system for the control of those wastes that are harmful to human health or the environment, or are difficult to handle. The Regulations provide a process for the control of such wastes from their production to their final disposal destination, i.e. a cradle-to-grave documentation trail for the movement of hazardous waste.
- Amendments to the Special Waste Regulations 1996 were made in Scotland in 2003 to meet the requirements of the Hazardous Waste Directive.
- Recently, these Regulations were amended by the Waste (England and Wales) (Amendment) Regulations 2009, which changed slightly the obligations placed upon producers and carriers of hazardous waste.
- Furthermore, two sets of Regulations that have recently come into place in the UK and Ireland deal with specific types of waste, both of which are generally considered hazardous, and thus merit discussion here. These are the Waste Electrical and Electronic Equipment Regulations 2006 (amended in 2007 and 2009), and the Batteries and Accumulators Regulations 2009.

Legislation

- Environmental Protection Act 1990.
- Special Waste (Scotland) Amendment Regulations 2003.
- Hazardous Waste (England and Wales) Regulations 2005.
- Hazardous Waste (Wales) Regulations 2005 (amended 2009).
- Hazardous Waste Regulations (Northern Ireland) 2005 (as amended).
- The List of Wastes (England) Regulations 2005.
- Waste Electrical and Electronic Equipment Regulations 2006 (amended in 2007 and 2009).
- The Batteries and Accumulators (Placing on the Market) Regulations 2008.
- The Batteries and Accumulators Regulations 2009.
- The Waste (England and Wales) (Amendment) Regulations 2009.
- Waste (England and Wales) Regulations 2011.
- Waste (Miscellaneous Provisions) (Wales) 2011 Regulations.

Hazardous Waste Regulations

- 'Special waste' is now termed 'hazardous waste'.
- All hazardous wastes must be characterised by codes from the associated List of Wastes Regulations 2005, using guidance issued by the Environment Agency (EA)/Scottish Environment Protection Agency (SEPA) for wastes with mirror entries in the List of Wastes to determine if such wastes are hazardous or not.
- An additional 200 types of waste are now classified as hazardous. Common office items such as fluorescent bulbs, computer monitors, Ni-Cd, lead or mercury batteries and televisions are examples of newly-classified hazardous wastes.

- Mixing of different categories of hazardous waste and the mixing of hazardous waste with non-hazardous waste is not permitted. It is important to note that *each specific hazardous waste type cannot be mixed or co-disposed* (i.e. batteries cannot be mixed with computer monitors).
- Producers of hazardous waste must undertake site premises notification with the EA and will receive a unique registration number (also known as a premises code) which is used in the issuing of consignment note numbers.
- Cessation of the requirement to pre-notify the EA prior to movement of hazardous waste in certain circumstances.
- Introduction of increased record keeping requirements – requirement on waste producers to maintain a register, and keep hazardous waste consignment notes for a minimum of three years.
- Must report on hazardous waste produced quarterly to the EA/SEPA.

The cross-border movement of hazardous waste within the UK has been accounted for within the respective Regulations. Should you need to arrange such a cross-border movement, it is recommended that you seek advice from an appropriately qualified source. Refer to the document entitled *Consignment Notes: Cross Border Movements: A Guide to the Hazardous Waste Regulations* (HWR05) published by the EA – see *Sources of further information*.

Implementation of the revised Waste Framework Directive, which is the primary European legislation providing a strategic framework for the management of waste, has brought some changes to the Hazardous Waste Regulations. These changes have been brought in by the Waste (England and Wales) Regulations 2011 and the Waste (Miscellaneous

Provisions) (Wales) 2011 Regulations and will affect those who produce or manage hazardous wastes. All the current requirements of the HWR will remain but there will be some changes to certain aspects of the regulations. In Wales the changes to the HWR will be the same, although they may be implemented through different regulations. The major changes to the way hazardous waste is managed are listed below and came into effect on 29 March 2011:

- Mixing of hazardous waste can only be carried out if you hold an appropriate permit allowing you to do this and the activity must comply with Best Available Techniques (BAT). Those operations authorised under IPPC already have this requirement but it will now be applied to all permitted mixing activities. This new provision will be applied to new permits issued after April 2011 and to existing authorisations on their first review.
- The method of classification and assessment of hazardous waste is generally unchanged but a new hazardous property (H13 sensitising) has been introduced and will need to be used when assessing hazardous waste.
- There are changes to the record keeping requirements, in particular for brokers.

The following changes came into effect on 28 September 2011:

- The consignment note has been amended and the multiple consignment procedure has been simplified with the requirement to use the modified standard consignment note. The same type of consignment note can then be used for multiple collection rounds as for single collections.

- The waste hierarchy must be considered and applied in a priority order of prevention, preparing for re-use, recycling, other recovery (for example, energy recovery) and disposal when hazardous waste is transferred. The waste hierarchy is expected to be implemented through amended Duty of Care requirements from the autumn of 2011. Hence, when completing a consignment note, from 28 September 2011 a consignor must sign the declaration in Part D to indicate they have considered the hierarchy before transferring waste.

Producer's responsibility (England and Wales)

Producers of hazardous wastes in England and Wales must ensure that their organisation (and individual locations) can fully demonstrate to regulators that they comply with all elements of the new Regulations. In order to ensure compliance, all producers must address the following actions:

- Register all premises that produce hazardous waste with the EA prior to any collection of hazardous waste. It is an offence to produce, remove or cause to be removed, or transport, hazardous wastes from non-notified premises. Re-notification must be undertaken on an annual basis. Some premises may be exempt from registering their premises with the EA, for instance if they produce less than 500kg of hazardous waste in a 12-month period (*see below*). Premises in Scotland or Northern Ireland are not required to register.
- Ensure that the consignee issues returns to the waste producer, notifying the receipt of their waste. This requirement aims to close the loop by demonstrating that the disposal point receives the waste that was intended for its site, thus

confirming that the waste followed its intended disposal route.
- Retain the quarterly returns provided to the waste producer by the waste management contractor for provision to the EA if requested. The information should include details of the wastes disposed of, the quantities, and the location of treatment / disposal for each consignment of waste removed.
- Undertake and permit authorities to undertake inspections of waste management premises, procedures and documentation.
- Control consignment notes and effectively manage records.
- Whilst waste management contractors can be valuable partners in achieving compliance, the responsibilities for premises notification and the provision and partial completion of the consignment notes rest with the waste producer.

Exemptions to premises notification (England and Wales)

There are exemptions to premises notification (fewer than recently) and the definition of 'premises' has changed such that all places where hazardous waste is produced are now considered to be 'premises'. Exclusions or exemptions from notification may include:

- Where mobile service operators (e.g. mechanics) produce all of your hazardous waste.
- Any premises where all hazardous waste is collected by either a registered carrier or exempt carrier *and* less than 500kg of hazardous waste is produced in any 12-month period. In the event that you plan to, or actually do produce, collect or remove 500kg or more, there is the requirement to notify such premises.
- Where hazardous waste is removed from a ship, the ship does not need to be registered. There is currently

no limit on the amount of hazardous waste that can be produced on, or removed from, a ship.

For further information refer to the EA document entitled *Do I need to notify my premises? A Guide to the Hazardous Waste Regulations* – see *Sources of further information*.

In addition, the EA will not pursue the need for a premises registration for the highways maintenance, highways spillage, railway tracks, waterways utility infrastructure or ambulances and other mobile healthcare providers. Nor will the EA pursue a consignment note for the removal of hazardous waste from these locations.

For more information on registering, refer to the appropriate section of the EA's website – see *Sources of further information*.

In addition to the above, if waste has been fly-tipped on to land (i.e. in contravention of Section 33 to the Environmental Protection Act 1990), it may be removed from those premises without requiring notification of those premises, but would require a waste consignment note prior to removal and to be removed/transported by a registered waste carrier.

Charges associated with the Regulations (England and Wales)

The notification of locations with the EA will result in one of the charges below:

- £28 for each premises notified in writing.
- £23 for each premises notified by telephone.
- £18 for each premises notified electronically (online).

Most waste management contractors will register premises for their waste producer customers, but it is the waste producer that is responsible for ensuring registration has occurred and for managing the use of the unique premises code on sequential consignment notes. In addition to the registration fees that a company will have to pay, there is also the potential for fines to be issued by the EA if the Regulations are not effectively implemented. Since April 2007 the EA has had powers under Regulation 70 of the HWR to issue Fixed Penalty Notices (FPNs) for certain environmental offences, such as:

- failure to register premises;
- failure to complete and store consignment notes; and
- failure to apply for reviews of existing permits.

The FPN is currently £300, payable within 28 days of issue of the notice, and if left unpaid the EA may resort to prosecution.

It should also be noted that hazardous waste disposal costs have increased due both to the decreased landfill disposal capacity following implementation of European Directive 1999/31/EC on the Landfill of Waste (Landfill Directive), increased waste treatment costs and also increasing Landfill Tax – a tax on the disposal of waste. There is therefore a real financial gain to be made in effectively reducing and managing the disposal of hazardous wastes.

Waste Electrical and Electronic Equipment Regulations 2006 (amended in 2007)

- All producers (manufacturers, those who sell EEE under their own brand, and those who import EEE) who put Waste Electrical and Electronic Equipment (WEEE) on the market are responsible for financing the costs of the collection, treatment, recovery and environmentally-sound disposal of WEEE from users and private households.

- Any producer who has any obligation under Regulation 8 (Financing: WEEE from private households) and Regulation 9 (Financing: WEEE from users other than private households) must join an approved compliance scheme.
- A producer must provide a declaration of compliance, together with supporting evidence, to the appropriate authority.
- A producer must mark EEE that he puts on the market with the crossed-out wheelie bin symbol, a producer identification mark and a date mark.
- A producer must also provide information on re-use and environmentally-sound treatment for each new type of EEE put on the market by that producer.
- An operator of a compliance scheme has certain obligations in relation to the re-use of whole appliances, treatment and recovery for any WEEE that he is responsible for.
- An operator of a compliance scheme has reporting, compliance and record-keeping obligations.
- The distributor is responsible for providing in-store take-back service for customers in relation to specified WEEE, unless he is a member of a distributor take-back scheme.

The 2007 Amendment Regulations prescribe the format for distributor records and the right for a final holder of WEEE from private households to return it into the system free of charge. For more information see 'Waste Electrical and Electronic Equipment: the WEEE and RoHS Regulations' (p.737).

The 2009 Amendment Regulations are intended to improve the Producer Compliance Scheme approval process and reduce the administrative burdens placed on business by simplifying the data reporting requirements and the evidence system. The amendments will produce an overall reduction in the amount of data to be submitted to the environment agencies; allow evidence to be issued on the receipt of separately collected waste electrical and electronic equipment at treatment facilities; remove the need for Producer Compliance Schemes to apply for approval every three compliance periods, although they will be required to submit 'rolling' three-year operational plans annually; and ensure all treatment facilities approved under the Regulations are able to meet the minimum standards set for recycling and recovery.

The Batteries and Accumulators (Placing on the Market) Regulations 2008

The Regulations deal with technical requirements affecting the manufacture, marketing and labelling of new batteries, and the design of certain battery-powered equipment. The objective of the Regulations is to aid persons placing batteries and accumulators, or certain electrical and electronic equipment that may contain or incorporate batteries and accumulators on the EU market for the first time on or after September 2008, including:

- facilitating the free movement of compliant batteries throughout the EU;
- reducing the quantities of heavy metals that batteries are allowed to contain, as an environmental protection measure; and
- introducing a labelling regime in preparation for 'producer responsibility' legislation aimed at batteries, with a view to achieving high collection and recycling rates.

The Regulations incorporate various requirements, including restrictions on the use of mercury and cadmium in battery or accumulator manufacture, with exceptions concerning usage in specific circumstances or, in the case of cadmium, industrial applications, application of the 'crossed out wheelie-bin' and the chemical symbols for lead, mercury or cadmium on batteries, where appropriate obligations upon manufacturers to design batteries in such a way as to facilitate the removal of waste batteries from certain appliances. For more information, see '*Batteries*' (p.46).

The subsequent Batteries and Accumulators Regulations 2009 relate to the collection, treatment and recycling of waste automotive, industrial and portable batteries. They include producer responsibility requirements for those placing new batteries on the UK market and requirements for those selling new portable batteries, or collecting, treating, recycling or exporting automotive,

industrial or portable batteries when they become waste. There are substantial differences in obligations depending on whether the batteries in question are automotive, industrial or portable and affected parties need to be clear which type or types they deal with in order to establish how the Regulations affect them.

> *See also*: Batteries, p.46;
> Packaging, p.541; Recycling,
> p.605; Waste management, p.707;
> Waste Electrical and Electronic
> Equipment: the WEEE and RoHS
> Regulations, p.737.

Sources of further information

Environment Agency – Dealing with hazardous waste:
www.environment-agency.gov.uk/business/topics/waste/32180.aspx

A considerable amount of information on waste management can be found on the EA's website: www.environment-agency.gov.uk

EA publications:

Consignment Notes: Cross Border Movements: A Guide to the Hazardous Waste Regulations (HWR03D):
http://publications.environment-agency.gov.uk/pdf/GEHO0507BMSL-e-e.pdf

Mobile Services: A Guide to the Hazardous Waste Regulations (HWR07):
http://publications.environment-agency.gov.uk/pdf/GEHO0409BPSM-e-e.pdf

The Department for Environment, Food and Rural Affairs (DEFRA) also offers much information on environmental, including waste management, topics at www.defra.gov.uk/

Do I need to notify my premises? A Guide to the Hazardous Waste Regulations:
http://publications.environment-agency.gov.uk/PDF/GEHO0710BSVQ-E-E.pdf

The DTI's Envirowise programme can supply information and assistance via its helpline on 0800 585 794, or website at www.envirowise.gov.uk

WRAP works in partnership to encourage and enable businesses and consumers to be more efficient in their use of materials and recycling at www.wrap.org.uk or contact the helpline on 0808 100 2040.

NetRegs is a partnership between the UK environmental regulators – the Environment Agency in England and Wales, SEPA in Scotland and the Environment and Heritage Service in Northern Ireland – and provides free environmental guidance for small and medium-sized businesses throughout the UK. Help and advice is available by business type, environmental legislation and environmental topic and has useful links to other websites. Go to www.netregs.gov.uk/netregs/ or telephone the EA National Customer Contact Centre on 08708 506 506.

National Industrial Symbiosis Programme (NISP) works directly with businesses of all sizes and sectors with the aim of improving cross-industry resource efficiency via commercial trading of materials, energy and water and sharing assets, logistics and expertise. It engages traditionally separate industries and other organisations in a collective approach to competitive advantage involving physical exchange of materials, energy, water and/or by-products, together with the shared use of assets, logistics and expertise. NISP is delivered at regional level across the UK. Go to www.nisp.org.uk or telephone 0121 433 2650.

Waste management

Mike Lachowicz, Bureau Veritas UK Ltd

Key points

- The UK currently produces around 288.6 million tonnes of waste annually according to the most recent DEFRA data from 2008 – a quarter of which is from households and business.

- The remainder derives from construction and demolition, sewage sludge, farm waste and spoils from mines and dredging of rivers. Businesses handling, storing, transporting, treating and disposing of waste must be aware of a wide range of waste-related legislation to ensure compliance with the law and avoid prosecution. However, businesses can also save around 1% of turnover each year through effective waste minimisation programmes.

- The movement and trans-shipment of waste is governed by tight requirements for documentary proof containing detailed descriptions. Failure to follow correct procedures can result in fines by magistrates of up to £50,000 and/or 12 months' imprisonment for each individual offence, and in the Crown Court it could result in unlimited fines or a five-year prison sentence. Since 2001 the scale and frequency of fines for illegal dumping and fly-tipping offences have both increased dramatically as case law from 1990s legislation is fine-tuned.

- The recent change in government has resulted in a series of new policies and initiatives such as the Government Review of Waste Policy in England 2011 announced in June 2011, which sets out its intention to continue its commitment to the waste hierarchy by increasing re-use and recycling; reduction of total waste and waste sent to landfill; and to encourage resource efficiency as well as partnerships such as 'voluntary responsibility deals', a comprehensive Waste Prevention Programme and 'good behaviour' by businesses, the civil society and individuals.

Legislation

Listed below is some of the most pertinent legislation related to waste management activities, which implement many European Directives into UK law, together with pollution prevention and control (PPC)/ environmental permitting, contaminated land and groundwater related legislation.

- Control of Pollution (Amendments) Act 1989.
- Environmental Protection Act (Part II) 1990.
- Environmental Protection (Duty of Care) Regulations 1991 as amended by the Environmental Protection (Duty of Care) (England) (Amendment) Regulations 2003.
- Controlled Waste Regulations 1992, plus subsequent amendments (1993) as replaced by the Waste Management (England and Wales) Regulations 2006 and Waste (Scotland) Regulations 2005, to be replaced by the Controlled Waste (England and Wales) Regulations 2011.
- Landfill Tax Regulations 1996 (amended 2011).
- The Producer Responsibility Obligations (Packaging Waste) (England and Wales) Regulations

1997 (as amended 2000, 2007 and 2008).

- The Landfill (England and Wales) Regulations 2002, as amended by the Landfill (England and Wales) Regulations 2004 and the Landfill (England and Wales) (Amendment) Regulations 2005.
- The End-of-Life Vehicles Regulations 2003 and (Amendment) Regulations 2010.
- End-of-Life Vehicles (Storage and Treatment) (Scotland) Regulations 2003.
- Landfill (Scotland) Amendment Regulations 2003.
- Packaging (Essential Requirements) Regulations 2003 (amended 2006 and 2010).
- The Special Waste Amendment (Scotland) Regulations 2004.
- End-of-Life Vehicles (Producer Responsibility) Regulations 2005.
- Hazardous Waste (England and Wales) Regulations 2005 (amended 2009).
- Hazardous Waste Regulations (Northern Ireland) 2005 (as amended 2009).
- List of Wastes (England) Regulations 2005 (as amended) and Explanatory Memorandum to the List of Wastes (England) Regulations 2005 No. 895 (Wales).
- Waste Electrical and Electronic Equipment Regulations 2006 and Waste Electrical and Electronic Equipment (Amendment) Regulations 2010.
- Environmental Permitting (England and Wales) Regulations 2007 (amended 2010).
- The Site Waste Management Plans Regulations 2008 (England) and the Site Waste Management Plans Regulations (Northern Ireland) 2011.
- The Waste Batteries and Accumulators Regulations 2009, Waste Batteries and Accumulators (Charges) Regulations (Northern Ireland) 2009.

- Waste Batteries and Accumulators (Treatment and Disposal) Regulations (Northern Ireland) 2009 and Waste Batteries (Scotland) Regulations 2009.
- Waste (England and Wales) Regulations 2011.
- Waste (Miscellaneous Provisions) (Wales) Regulations 2011.

General guidance

Steps to consider in the overall process of waste generation through to disposal are outlined below.

Step one – Identify the types and sources of waste from your business

Firstly, create a list of each of the waste streams originating from your facility. This might include:

- general waste (day-to-day food and drink items, food wrappers, etc.);
- waste packaging materials;
- sanitary waste from toilets or medical waste from a first aid room;
- medical waste from first aid or healthcare centres;
- building and other construction wastes; and
- waste oils from maintenance operations.

As a minimum, those responsible for waste management need to identify and separate out the different waste streams – inert, non-hazardous and hazardous, in particular the latter which includes asbestos, lead-acid batteries, electrical equipment containing hazardous components such as cathode ray tubes (e.g. televisions and computer monitors), oily sludges, solvents, fluorescent light tubes, chemical wastes and pesticides etc. – from other waste streams. The segregation of the differing waste streams needs to be at source as cross-contamination of waste streams must be avoided.

Facts

- At £11bn, employing over 120,000 people, with forecast growth of 3-4% per year for the next few years, the UK waste management sector is a significant industry in its own right. Approximately two-thirds of its operations are attributable to industry and commerce; one-third to local government for household collections.

- The direct cost of waste for the average enterprise amounts to around 1% of turnover, but the true costs for a business according to DEFRA (*The Further Benefits of Business Resource Efficiency,* March 2011) are estimated at closer to 4.5% of turnover or £22.6bn annual savings from low or no cost measures (over half of which could come from reducing waste).

- Non-hazardous business and commercial waste now costs over £100 per tonne in total for disposal with costs still increasing as the landfill tax is raised towards a floor of £80 per tonne by 2014/15.

- Waste production in the UK continues to decline, producing 288.6m tonnes in 2008 with the largest contribution being from the construction and demolition sector. The construction and demolition sector in England produced 86.9m tonnes, of which 62% was recycled or re-used and 26% sent to landfill.

- Domestic waste in the UK was a total 32.5m tonnes in 2009, of which 49% was landfilled and 42% had some value recycled or recovered.

- British industry already has a far more successful track record on recycling than British households, with most companies achieving over 50% recycling or re-use and many companies aiming for zero landfill. However, small and medium enterprises (SMEs), with between zero and 49 employees, produced 16.6 million tonnes of C&I waste in 2009, or 35% of total C&I waste.

- The latest available DEFRA statistics from its Waste Data Overview (June 2011) for England quote a total of 47.9m tonnes of industrial and commercial waste, of which 52% was recycled or re-used while 24% was sent to landfill.

- As a nation we consume around 650m tonnes of materials as raw material inputs at the front end of the economy. The weight of finished goods sold each year in the UK amounts to around 60m tonnes (c.10% of inputs) and the weight of total goods sold is approximately 50% food, 50% all other consumer goods.

- The 10:1 ratio of inputs to finished good outputs is approximately true for many business and commercial enterprises (including energy and construction) although the specific ratio varies according to the nature of the enterprise (see www.massbalance.org).

- A typical cost of waste for the average SME is over £1,000 per year – excluding costs for any hazardous waste 'extras' like disposal of light bulbs and electronic waste.

The European Waste Catalogue (EWC) lists all wastes, grouped according to generic industry or process. Each identified waste type is allocated a six-digit code. A waste is hazardous if it is classified as such in the EWC. Hazardous wastes are identified in the EWC with an asterisk (*).

Some wastes are classed as inert or hazardous outright, whereas some wastes require prior assessment to determine which category they fall into. For instance, inert wastes include glass, bricks, naturally occurring soil and stones (excluding topsoil and peat) whereas hazardous wastes include those that are explosive, flammable, corrosive or carcinogenic. The assessment is required to be undertaken by the 'owner' of the waste, or on their behalf by a third party and must be undertaken prior to recovery, re-use, recycling or disposal to account for relevant procedures to follow.

The Environment Agency (EA) has produced guidance to help waste categorisation in its document entitled *Guidance on Sampling and Testing of Wastes to Meet Landfill Waste Acceptance Procedures*. Useful information is also provided in the Landfill (England and Wales) Regulations 2002, and amendments (2004 and 2005) and in Technical Guidance document WM2 (updated in April 2011), which contains a consolidated version of the EWC and provides advice on the classification and assessment of hazardous waste. Also refer to the document entitled *What is Hazardous Waste*, published by the EA. See *Sources of further information*.

It is important to stress that it is worth considering options available in the event of waste streams being classified as non-hazardous and hazardous as this will have inherent cost implications for waste management. It may be possible to treat the waste in some way so it is re-categorised to a non-hazardous or even inert waste.

If you are unsure about whether particular waste streams or wastes are inert, non-hazardous or hazardous and are unsure of the most appropriate route for

re-use, recycling or disposal it is advisable to first liaise with your local EA office or call the BIS-sponsored WRAP / Envirowise programme (helpline 0800 585 794).

Step two – Register with the Environment Agency (if producing more than 500kg of hazardous waste per annum)

All businesses in England and Wales that produce hazardous waste have a legal duty to register their premises with the EA. The duty to notify premises rests with the producer of the waste. This would normally be regarded as the owner or occupier of the site. However, where waste is produced by a visiting mobile service, the duty may fall on the person operating that service. Hazardous waste producer registrations are valid for 12 months from the date of registration. It is essential to have a registration number, which your licensed hazardous waste contractor will insist on when collecting your waste. In the event of premises continuing to produce hazardous waste after the initial 12-month period, the registration must be renewed, up to one month in advance of the expiry date. Premises that fail to renew their registrations may be liable to enforcement action. The EA will not inform premises to renew their registration. Each location that is registered will be given a unique registration number or 'premises code'.

Some types of premises may be exempt from registering if they produce less than 500kg of hazardous waste in any 12-month period. However, if exemption applies, e.g. office and shop premises, dental and doctors' surgeries, some schools, charities and voluntary organisations, there is still a requirement to use a Hazardous Waste Consignment Note (HWCN) to accompany transfer of hazardous waste. In the event that it appears that the limit of 500kg will be exceeded, the EA must be notified immediately and before the limit is exceeded. As an example, 500kg equates to approximately 25 small TVs; 35 lead-acid batteries; 1,250 fluorescent tubes; 13 small domestic fridges.

Information required for registering as a hazardous waste producer includes the following:

Organisation:

- Name of organisation.
- Address and postcode of the applicant.
- Contact name of the applicant.
- Contact details for the applicant.

Sites to be registered:

- Name of organisation for which notification is required.
- Address and postcodes of all sites that require notification.
- Contact names and telephone number for contact person at each site to be registered.
- Previous registration number (if applicable).
- The Standard Industry Classification (SIC) code for the main activity that produces the waste (the 2007 UK SIC listing codes must be used and are obtainable from www.statistics.gov.uk/statbase/product.asp?vlnk=14012).
- The number of employees working at the premises for which notification is required.
- Customer reference.
- Proposed notification start date.

Sites producing hazardous waste are required to provide separate registrations, although multiple sites can be registered on the same notification (up to a maximum of 2,000 premises). Therefore, a head office could register all its sites centrally, but each site would have a separate unique registration number and require a separate fee.

Applications for registration can be made using the internet, on disk, via email, by phone or on a paper form, and each method attracts a charge of up to £28 – the amount paid depends on the method of registration. For more information on registering, refer to the appropriate section of the EA's website at www.environment-agency.gov.uk/business/topics/waste/32198.aspx, contact the EA on 08708 502 858 (weekdays 09.00 to 17.00) and refer to the documents entitled *Do I Need to Notify My Premises?* and *How to Register Your Premises, A Guide to the Hazardous Waste Regulations*. See *Sources of further information.*

Step three – Waste storage

Wastes should be appropriately and securely stored at all times to ensure compliance with the Duty of Care Regulations, and must be prevented from causing pollution or harm. It is not appropriate to burn, bury or pour away wastes or illegally dispose of wastes, for instance on another premises.

Fly-tipping is illegal and, under recent changes to Section 33 of the Environmental Protection Act (EPA) 1990 by the Clean Neighbourhoods and Environment Act 2005, can be punishable by fines up to £50,000 in Magistrates' Courts, or unlimited fines in higher courts, and 12 months' imprisonment. If prosecuted in the Crown Court, the fine may be unlimited or carry up to five years' imprisonment. Since April 2007 the EA has had powers under Regulation 70 of the Hazardous Waste Regulations 2005 to issue Fixed Penalty Notices (FPNs) for certain environmental offences, including failure to provide evidence of being a registered waste carrier, failing to comply with laws on the duty of care, and certain offences relating to managing hazardous waste as set out in the Hazardous Waste Regulations. The FPN is currently £300, payable within 28 days of issue of the notice, and if left unpaid the EA may resort to prosecution. Ensure your own waste isn't fly-tipped by checking that any waste contractor you use is operating legally. Call the Environment Agency on 0370 850 6506 and ask for a waste carrier registration check or check online.

Step four – Identify the most appropriate means of waste management

Regular reviews of waste management options should be undertaken – contractors' costs, escalating landfill tax and other measures make waste disposal expensive. For example, check whether wastes can be re-used, recycled or reclaimed or indeed if the processes undertaken can be modified to reduce the amount, or even types, of wastes generated. Not only is this a more environmentally friendly and sustainable option, but it may have cost benefits too, in the short- or long-term.

Other examples include monitoring the use of raw materials, chemicals, water, etc., with a view to reducing them, use of alternatives with low(er) environmental impact, use of oil interceptors and segregation of 'clean' and 'dirty' products to minimise the amounts of effluent or hazardous waste generated, or even introduction of compactors to reduce waste volume/bulk, which therefore requires fewer waste collections. Each of these may have an impact on waste management and associated disposal costs.

Step five – Select authorised waste carriers and disposal contractors

It is a legal requirement to transfer controlled waste by use of an authorised waste carrier such as a registered carrier or holder of an environmental permit (formerly a waste management licence), unless, for instance, you are carrying only your own waste to be disposed of or recovered, unless it is construction or demolition waste. Current exemptions include animal by-products, mines and quarries waste or agricultural waste, yet if these are mixed you will need to register as a waste carrier. However, this exemption is subject to change following a judgment in the European Court of Justice (CIWM, August 2007). Charities and voluntary organisations may collect or transport waste on a professional basis but must register with the EA as a waste transporter. Registration as an authorised carrier is still required even if waste is only occasionally transported.

Some commercial or trade wastes can be handled by your Local Authority's collection or disposal schemes, but check with them first. Ensure that waste contractors' authorisations are current and valid for your particular waste materials. Check actual copies of contractors' licences and, if unsure on any point, double-check with the EA. If you suspect any problems, suspend waste transfers and alert the EA.

Step six – Maintain comprehensive records

All controlled waste (i.e. generated from households, commerce or industry) transfers/movements, intermediate storage, recovery, or disposal must be accompanied by a waste transfer note (WTN). This should include appropriate information such as a written description of the waste and name of the person to whom the waste is being transferred.

A transfer note should be completed for each individual transfer or, alternatively, a 'season ticket' can be completed which covers a pre-agreed (multiple) sequence of collections of small amounts of waste from more than one premises collected on the same vehicle and being delivered to the same consignee. Waste carriers or contractors will often provide this as part of their overall service. In addition, all and each hazardous waste transfer must be accompanied by a consignment note (CN), even if from premises exempt from registration. The single movement HWCN is a three-part form and each part of the form should be completed. They are colour-coded and labelled as follows:

- Producer's / holder's / consignor's copy (White).
- Carrier's copy (Gold).
- Consignee's copy (Pink).

Each member of the chain has legal responsibility for ensuring that the procedures are followed and paperwork is all correct. Checking is not only in one direction. No one should accept waste from a source that seems to be in breach of the duty of care. Waste may only come either from the person who first produces or imports it or from someone who has received it.

All hazardous waste to be accepted at permitted waste management facilities must also meet minimum regulatory obligations relating to waste pre-acceptance, waste acceptance and waste storage. To find out how to meet the waste acceptance criteria of particular wastes, talk to your waste contractor or waste facility operator.

HWCNs must be kept on file for a minimum period of three years from the date on which the waste was transferred to another person or where it was disposed. It is the record-holder's responsibility to ensure that all records are kept securely and are readily retrievable at all times. In the case of hazardous waste, a consignee (receiver of waste) must also keep detailed records showing the location(s) of where wastes are kept or deposited and

provide returns to producers, holders or consignors.

For further details refer to the documents entitled *Consignment Notes: A Guide to the Hazardous Waste Regulations Version 1, April 2011; Consignment Notes: Standard Procedure, A Guide to the Hazardous Waste Regulations*; and *Record Keeping. A Guide to Hazardous Waste Regulations,* prepared by the EA – see *Sources of further information.*

Additional points for landlords or managing agents

A waste broker is an individual or company who arranges for the disposal or recovery of controlled waste on behalf of another party. Such arrangements will include those for the transfer of waste. Brokers do not handle the waste themselves or take it in their own physical possession, but control what happens to it.

Subject to various provisions, it is an offence for anyone to arrange on behalf of another person for the disposal or recovery of controlled waste if they are not registered as a broker. Anyone subject to the duty of care must ensure that, insofar as they use a broker when transferring waste, they use a registered broker or one exempt from the registration requirements.

Hence, where landlords (in the broadest meaning) or their managing agents, arrange for waste to be disposed of on behalf of their tenants, this may be a waste-brokering operation and the broker would have to register as a waste broker with the EA, prior to providing the service to the tenants. Speak to the EA in the first instance, if advice is required, or other specialists.

Additional considerations

As part of the bigger picture for waste management, it is important to understand the legal definition of waste and the implications of generating, handling, storing, transporting, treating and disposing of controlled waste. Cost considerations for each of these elements should be taken into account as waste disposal costs have increased over recent years due both to the decreased landfill disposal capacity following implementation of the Landfill Directive, increased waste treatment costs and also to the increased data return fees that waste management contractors have to pay the EA. There is therefore a real financial gain to be made in effectively reducing and managing the disposal of wastes and in particular hazardous waste types.

In response to the Waste Framework Directive (Directive 2008/98/EC) in England and Wales (WFD), the Government has brought into force two new sets of regulations in England and Wales in 2011. The Waste (England and Wales) Regulations 2011 were introduced on 29 March 2011 and require businesses to:

- confirm that they have applied the waste management hierarchy when transferring waste, and include a declaration on their waste transfer note or consignment note;
- introduce a two-tier system for waste carrier and broker registration, including a new concept of a waste dealer;
- make amendments to hazardous waste controls; and
- exclude some categories of waste from waste controls.

The Waste (Miscellaneous Provisions) (Wales) Regulations 2011 also came into effect at this time and are supplemental to the Waste (England and Wales) Regulations 2011 with minor changes to hazardous waste, landfill, list of wastes, planning and environmental damage.

DEFRA and the Welsh Assembly are proposing to replace the Controlled Waste Regulations 1992 with the Controlled Waste (England and Wales) Regulations 2011, expected to come into force in October 2011. The Controlled Waste Regulations 1992 list different types of household waste that local authorities can charge to collect but not to dispose of. This is a barrier to achieving the Government's plan for a zero waste economy – the Government wants organisations to pay the full costs of their waste disposal. The Government also aims to restructure the Controlled Waste Regulations to make them easier to use.

Draft regulations dealing with seizure of vehicles, to be called the Control of Waste (Authority to Transport Waste and Dealing with Seized Property) (England and Wales) Regulations 2010, are to be published shortly and are expected to come into force later in 2011.

The proposals will affect businesses that:
- produce waste;
- import waste;
- carry or transport waste;
- keep or store waste;
- treat waste;
- dispose of waste; and/or
- operate as waste brokers or dealers.

This means that the proposals will affect most businesses. In particular, they will affect the waste management industry and farmers.

See also: Batteries, p.46; Recycling, p.605; Hazardous waste, p.700; Waste Electrical and Electronic Equipment: the WEEE and RoHS Regulations, p.737.

Sources of further information

Much of the Government's environmental information resources, including those related to waste management, are currently being reorganised into fewer, 'one stop shop' agencies.

A considerable amount of information on waste management can be found on the EA's website: www.environment-agency.gov.uk/business/topics/waste/default.aspx

The Department for Environment Food and Rural Affairs (DEFRA) website also offers much information on environmental, including waste management: www.defra.gov.uk/environment/waste/index.htm

WRAP works in partnership to encourage and enable businesses and consumers to be more efficient in their use of materials and recycling: www.wrap.org.uk Helpline: 0808 100 2040.

The Envirowise programme has been merged with that of WRAP and can supply information and assistance via its helpline on 0800 585 794, or www.envirowise.wrap.org.uk/

NetRegs is a partnership between the UK environmental regulators – the Environment Agency in England and Wales, SEPA in Scotland and the Environment and Heritage Service in Northern Ireland – which provides free environmental guidance for small and medium-sized businesses throughout the UK. Help and advice is available by business type, environmental legislation and environmental topic and has useful links to other websites: www.netregs.gov.uk/netregs

EA National Customer Contact Centre: 08708 506 506; SEPA: 01786 457710; NIEA: netregs@doeni.gov.uk

The National Industrial Symbiosis Programme (NISP) works directly with businesses of all sizes and sectors with the aim of improving cross-industry resource efficiency via commercial trading of materials, energy and water and sharing assets, logistics and expertise. It engages traditionally separate industries and other organisations in a collective approach to competitive advantage involving physical exchange of materials, energy, water and/or by-products together with the shared use of assets, logistics and expertise. NISP is delivered at regional level across the UK: www.nisp.org.uk Telephone: 0845 094 9501.

The Environment Agency (EA) has produced guidance to help waste categorisation in its document entitled *Guidance on Sampling and Testing of Wastes to Meet Landfill Waste Acceptance Procedures.*

Sources of further information – *continued*

Technical Guidance document WM2 (updated in April 2011), which contains a consolidated version of the EWC and provides advice on the classification and assessment of hazardous waste.

EA publications

What is Hazardous Waste?:
http://publications.environment-agency.gov.uk/PDF/GEHO0411BTQZ-E-E.pdf

Do I need to notify my premises? A Guide to the Hazardous Waste Regulations:
http://publications.environment-agency.gov.uk/PDF/GEHO0710BSVQ-E-E.pdf

How to Register Your Premises, A Guide to the Hazardous Waste Regulations:
http://publications.environment-agency.gov.uk/PDF/GEHO1110BTGF-E-E.pdf

Consignment Notes: A Guide to the Hazardous Waste Regulations:
http://publications.environment-agency.gov.uk/PDF/GEHO0311BTPW-E-E.pdf

Consignment Notes: Standard Procedure, A Guide to the Hazardous Waste Regulations:
http://publications.environment-agency.gov.uk/PDF/GEHO0311BTPW-E-E.pdf

Record Keeping. A Guide to Hazardous Waste Regulations (HWR05):
http://publications.environment-agency.gov.uk/PDF/GEHO0611BTUV-E-E.pdf

Comment ...

The full force of the law
Workplace Law Environmental

Since 4 January 2011, the Environment Agency has started using new enforcement powers, known as civil sanctions, against companies that flout environmental laws. Heralded as a more flexible approach towards enforcement, what will these sanctions mean for employers?

Criminal law has not always allowed flexibility in enforcement of environmental legislation. In some instances, regulators have been forced to choose between issuing warning letters and cautions at one level, and taking criminal proceedings at another. This has often led to an imbalance in enforcement whereby criminal law has not been flexible enough to deal proportionally with offences, or ensure fines adequately cover the costs of non-compliance.

However, further to a recent review about the effectiveness of regulation, new civil sanctions regulations have been introduced to provide environmental regulators with a more flexible and balanced set of tools to deal with environmental offences. The Environment Agency in England and Wales has started to use new enforcement powers, called civil sanctions, since 4 January 2011.

Civil sanctions can be used against a business committing certain environmental offences, as an alternative to prosecution and criminal penalties of fines and imprisonment. They allow the Environment Agency (and other regulators) to take action that is proportionate to the offence

 **workplace law**
environmental

Workplace Law Environmental specialises in environmental management and compliance.

We are an established training provider, approved by the Institute of Environmental Management and Assessment (IEMA) to provide certified courses at foundation level. Our specialist consultancy work includes environmental auditing, advice on environmental management systems, including ISO 14001, BREAAM and energy efficiency assessments, and compliance with environmental regulations.

and the offender, and reflect the fact that most offences committed by businesses are unintentional.

The EA will still be able to use criminal punishments for serious offences, but the Government believes civil sanctions will make environmental law enforcement more flexible and effective for both regulators and businesses.

Dr Anna Willetts, Business Defence Solicitor at Greenwoods Solicitors, believes this to be a fairer approach: "Some environmental offences are of very low magnitude, such as forgetting to send in a quarterly return on time, which does not, of itself, harm or damage the environment. Under the current criminal law regime, such offences may be harshly punished, for example a fine up to £50,000 or a term of imprisonment, or even an unlimited fine, can be made under the Environmental Permitting Regulations 2010.

"As environmental offences are strict liability offences, even a minor breach of a Permit can result in a prosecution. Offences may be simply paperwork offences committed in ignorance, rather than purposely attempting to pollute or harm the environment, but can still result in a harsh penalty. The Civil Sanctions may provide a fairer approach to environmental crimes at the lower end of the scale. They would mean that an organisation, or individual, does not then have a criminal record. A small monetary punishment, such as a Fixed Penalty Notice of a few hundred pounds, for example, could be enough to remind the offender not to commit the offence again, especially if it was committed inadvertently."

It was identified some time ago that the current sanctioning framework for dealing with environmental offences was capable of improvement. Regulators often had to choose between issuing a warning letter or caution and taking criminal proceedings without easy access to proportionate intermediate sanctions that act as a deterrent, leading to a 'compliance deficit'. The current enforcement system therefore relies heavily on criminal sanctions and this is sometimes disproportionate. Also, fines do not always cover the costs to society of harm from non-compliance and therefore do not act as an appropriate deterrent.

> "The point of these sanctions is to allow regulators to distinguish more effectively between those with a good general approach to compliance and those who tend to disregard the law, and hopefully punish the latter more harshly."

There are many different ways in which a business can breach environmental law, ranging from the deliberate discharge of hazardous substances to a river, to forgetting to keep a record of a transfer note for the disposal of waste, and everything in between. There are clearly many ways of breaking the law, but historically not a corresponding suite of sanctions that could be applied for each level of offence. Increasing the range of sanctions available to more accurately match the offence is one of the main reasons these Regulations have come in, and with these sanctions the Environment Agency can look at each offence and apply a more proportionate response.

Before, the regulators had been forced to make difficult decisions on how best to apply sanctions in the most effective way, and the narrower range and more inflexible nature of criminal sanctions has not always enabled the most appropriate and desired outcome to be reached.

Willetts predicts that the new sanctions will lead to a reduced number of companies appearing in Court for environmental offences. She says: "The point of these sanctions is to allow regulators to distinguish more effectively between those with a good general approach to compliance and those who tend to disregard the law, and hopefully punish the latter more harshly. It should keep the minor and 'accidental' offenders out of the Courts, and save valuable Court time for the most serious and consistent offenders."

The guidance says quite explicitly that the civil sanctions aren't going to replace the existing framework for criminal sanctions. The idea is that the existing criminal sanctions and the civil sanctions will work side by side, resulting in more options for the regulatory authorities in terms of proportionate enforcement. If a business is deliberately causing pollution, the guidance suggests criminal sanctions will still be the best route for the regulators. However, if pollution is being caused by accident, or if the company causing the pollution thought they had put sufficient measures in place to prevent pollution, the option of a less severe sanction can be considered.

But it's not all good news for employers. Whilst environmental regulators may no longer take non-compliant companies to court for minor offences, they now have greater powers to deal with those offences proportionately. Instead of merely a slap on the wrist, or a nasty letter, companies may find themselves dealing with official sanctions that require them to pay for their offences.

Previously, either to their benefit or to their detriment, the range of sanctions applied to business' non-compliance were not always proportional to the offence committed. Sanctions weren't being applied in a fair and consistent manner.

Now that such 'action' sanctions (such as compliance and restoration notices) can be applied, there is more scope for

the damage that companies cause to the environment to be dealt with.

"I hope that the 'action penalties' will be used, as I feel this is the best way of protecting and preserving the environment if pollution or harm has been caused," says Willetts. "Slightly cynically, however, I wonder if the options of Fixed Monetary and Variable Monetary Penalties may be more frequently used as this will result in the immediate garnering of funds. These financial penalties are the most likely sanctions for paperwork and other administrative offences where no remedial action is needed."

So will civil sanctions lead to more awareness of environmental offences?

Anna Willetts believes so: "Potentially yes, although there is the competing thought that the more unscrupulous will think 'Well, it won't cost much so I'll just see if I get caught'. Although these people are more likely to be the persistent serious offenders, who end up in Court anyway! The majority of companies try to do the right thing and already comply with the law. Companies should use this as an opportunity to review their current procedures, take advice and document their environmental procedures more robustly. Whilst the civil sanctions will have their place for more minor offences, environmental breaches are still covered by the criminal law. A Regulator will not hesitate to instigate criminal proceedings where it has evidence of breach and where it believes it is in the public interest. The civil sanctions are only part of the armoury of tools for the Regulator to use – they are not obliged to do so and they will continue to prosecute. Penalties, whether imposed by the criminal court or via civil sanctions, may be considerable and businesses would be unwise to intentionally operate outside the law."

Water fittings

Steve Tuckwell, WRAS

Key points

- National regulations or by-laws apply to all plumbing systems in premises that have a connection to the public water supply.
- The regulations' purpose is to prevent waste, misuse, undue consumption, contamination and erroneous measurement of water supplied by a public water supplier.
- The installers and users of plumbing systems have a legal duty to comply with the regulations or by-laws.
- All plumbing fittings and water-using appliances must be designed, constructed, installed, used and maintained to meet the requirements of the regulations.
- In most circumstances, it is a criminal offence to begin the installation of water fittings or appliances, without the consent of the water supplier.
- Consent is gained by notifying the water supplier in advance of installation work.

Legislation

- Water Supply (Water Fittings) Regulations 1999 (for England and Wales).
- Scottish Water By-laws 2004.
- Water Supply (Water Fittings) Regulations (Northern Ireland) 2009.

The Water Supply (Water Fittings) Regulations came into force in England and Wales on 1 July 1999 to replace the suppliers' Water By-laws which had applied since 1989. Technically identical requirements were made for Scotland on 4 April 2000, which have been updated as the Scottish Water By-laws 2004. From August 2009, revised regulations in Northern Ireland were introduced, which in most regards are technically identical to the rest of the UK. In this chapter, 'Regulations' refers to the legislation covering all parts of the UK, with any significant differences indicated.

In all premises that have a public water supply connection, these Regulations control the design and installation of water systems, their maintenance and their disconnection, together with the use of the water that is supplied. They apply to pipework, including the underground supply pipe connecting premises to the water main from the point where the ownership of the pipe passes from the public water supplier to private ownership, to all fittings such as pipes, valves, taps, pumps and storage cisterns, and to all appliances, machines, sanitary ware and hosepipes that are connected to the plumbing system or receive water from it. The Regulations do not apply to premises that have no public water supply, e.g. those with only a privately owned borehole or well supply. However, the Regulations do provide a Code of Practice for plumbing systems used with private supplies, where prevention of contamination and waste of water are equally important.

Implications

What are the implications for workplace managers? It is important for your health and safety responsibilities that the Regulations are complied with to protect all those using your premises against contamination of water for domestic purposes (bathing, cooking, drinking). Compliance also ensures reliable and robust plumbing systems and the efficient use of water. Underlying all this, compliance is necessary to avoid criminal prosecution and possible claims for compensation from those affected by incidents.

The Regulations place legal duties on installers and users of plumbing systems as follows, but introduce a benefit too. The duties include:

- Giving advanced notification and having the Water Supplier's consent for proposed plumbing work (new plumbing installations in all premises and changes to the plumbing in existing non-domestic premises).
- All points of use must be adequately protected against contamination by backflow.
- WCs and urinals must be installed and maintained to consume no more than the permitted flush volumes (full flush, six litres; dual flush, less than two-thirds full flush.)
- Approved contractors are permitted to do certain types of work without the prior consent of the water supplier, so providing more flexibility in carrying out work. They also provide their clients with the reassurance of a certificate guaranteeing compliance of their work with the Regulations. These aspects are described more fully below.

Notification
The Regulations require prior notification to the water supplier of the intention to install any water fittings. The only exception is the extension of the water system of an existing house, which does not have

to be notified unless there are certain specific items to be installed (*see below*). A table included in the Regulations lists certain types of fittings that need specific approval.

Many of these relate to water conservation requirements, such as the need to notify the installation of a bath having a capacity of more than 230 litres; a pump drawing more than 12 litres per minute; a water treatment unit incorporating reverse osmosis, producing a waste water discharge or requiring the use of water for regeneration or cleaning; an automatic garden watering system; the laying of underground pipes outside the minimum depth of 750mm or the maximum of 1,350mm; and the construction of a swimming pool with a capacity greater than 10,000 litres being replenished automatically and from the water supplier's mains.

The remaining items requiring notification relate to contamination hazards such as the installation of a bidet with an ascending spray or flexible hose (and, in Northern Ireland, a WC with an ascending spray or flexible hose), and backflow protection devices designed to protect against the higher categories of risk.

Additionally, in Northern Ireland, notification is required for:

- greywater, recycled water, reclaimed water and rainwater harvesting systems;
- water systems for firefighting, including domestic sprinklers;
- a flexible shower hose or other flexible outlet for use in conjunction with a WC;
- a 'shower-toilet' or 'bidet-toilet,' where, either as part of the WC itself or as an addition or adaptation of it, a stream of water is provided from below the spillover level of the WC pan for personal cleansing.

The written notification must include plans of the relevant parts of the premises, plans of the plumbing layout and fittings to be installed, and details of the person making the notification and those to whom a consent should be sent. Within ten working days of receipt of the notification, the water supplier must either grant consent, with conditions if necessary, or refuse consent, otherwise consent is deemed to have been given and installation can start. A leaflet summarising notification requirements is available on the WRAS website (www.wras.co.uk).

Suitability of fittings

There is no need to remove, replace, alter, disconnect or stop using any fittings that were lawfully installed under the by-laws before the Regulations came into force. However, in all new installations or modifications or extensions of existing systems, water fittings must comply – by being of an appropriate quality and standard and suitable for their intended purpose. To demonstrate that they are compliant, fittings must meet one of the following criteria:

- Carry an appropriate CE mark;
- Conform with an appropriate European harmonised standard or European Technical Approval (ETA);
- Be manufactured in accordance with an appropriate British Standard or an equivalent standard of a state that is a member of the European Economic Area; or
- Comply with the performance specification approved by the 'regulator' (e.g. the Secretary of State).

Currently there are no relevant CE marks or harmonised standards available. Some water fittings with 'Kitemarks' are manufactured to relevant British Standards. The easiest way to demonstrate compliance of water fittings is to choose those that manufacturers

have voluntarily submitted for checking by WRAS. These fittings and materials are assessed by the water suppliers' representatives themselves and are listed as WRAS Approved Products in the online *Water Fittings and Materials Directory* (see *Sources of further information*).

The owner or occupier of the property is liable if installed fittings do not comply with the Regulations and if the installation has not been carried out 'in a workmanlike manner' – e.g. in accordance with an appropriate standard such as BS 6700: 2006 'Design, installation, testing and maintenance of services supplying water for domestic use within buildings and their curtilages – Specification for the plumbing installations for domestic purposes in buildings'.

Preventing contamination by backflow

If the usual direction of flow of water in pipes is reversed, there is a risk of contamination being drawn into the pipework from downstream appliances or processes and affecting drinking water supplies, either in the premises themselves or in adjacent premises. The Regulations define five levels of backflow risk ('fluid categories') according to the risk the contaminants pose to health and set out the acceptable devices for preventing backflow at each level of risk. When designing new plumbing systems or making changes to existing ones, a risk assessment is required to determine the level of risk and suitable protective devices must then be installed. More information is given about this in the WRAS *Water Regulations Guide – see Sources of further information*.

Requirements for WCs and urinals

The maximum flush volume for new WCs is six litres and European-style drop valves, flap valves, flushing valves and flushing cisterns are permitted, provided they have undergone testing in accordance with a performance specification approved by the Government. It is an offence to use valves and WCs that do not comply. Like-for-like replacements of already installed WCs exceeding six litres flush will still be permitted. WC cisterns no longer have to have an external warning pipe to indicate if the inlet valve is leaking. Instead, an internal arrangement is permitted, allowing water to run into the back of the WC pan via the flush pipe. Dual-flush WCs are reintroduced, but the smaller flush must not be greater than two-thirds of the larger flush.

Water saving

In June 2003 the regulator approved the retrofitting of devices into existing seven-and-a-half or nine-litre 'by-laws' WC cisterns, which were installed before July 1999, to modify them to provide dual-flush or interruptible flush. This offers significant water savings for older premises. Cisterns used to be required for flushing urinals, but in non-domestic premises the Regulations permit automatic urinal control with water direct from the mains supply, via suitable backflow protection. This is a major water conservation initiative and workplace managers should now give consideration to taking advantage of it.

In addition to water conservation arising from the use of new types of WCs and urinals, the legislation makes reference to recycled or 'grey' water and the need to mark pipes carrying such water to reduce the risk of dangerous cross-connection with wholesome water.

Approved contractors

The Regulations define 'approved contractors' and give them certain benefits and responsibilities. They can be accredited by the water suppliers and, except in Northern Ireland, by other organisations authorised by the Secretary of State. Among the water suppliers in England and Wales, Anglian, Severn Trent, and Thames Water operate their own schemes; the remainder support the Water Industry Approved Plumbers Scheme (WIAPS) administered by WRAS. Members of WIAPS have demonstrated their experience in plumbing and knowledge of the Regulations. Names and addresses of WIAPS members are given on the WRAS website. Other schemes are operated by the Chartered Institute of Plumbing and Heating Engineering (formerly the Institute of Plumbing), the Association of Plumbing and Heating Contractors, and the Scottish and Northern Ireland Plumbing Employers' Federation.

Approved plumbers must give their customers a certificate of compliance for their work, which the customers can use as a defence in the event of any prosecution for non-compliance associated with the installation work. Approved plumbers can also undertake some types of work without the need for prior consent, which can provide flexibility for the timing of projects.

Practical conclusions

Workplace managers need to be fully aware of the legal requirements that the Regulations place upon the owners and occupiers of premises. Not only does this make sense to avoid possible prosecution and remedial costs if contraventions are found by the water suppliers that enforce the Regulations, but it will prevent contamination of drinking water in the premises and ensure efficient use of water.

See also: Building Regulations, p.76; Legionella, p.442; Toilets, p.669; Water quality, p.725.

Sources of further information

The Water Regulations Advisory Scheme (WRAS) website provides downloads of advice leaflets and guidance notes, details of other publications, recent interpretations of the Regulations.

WRAS: www.wras.co.uk

WRAS: *Water Fittings and Materials Directory*: www.wras.co.uk/directory

WRAS: *Water Regulations Guide*: www.wras.co.uk/Regulations_guide.htm

The Water Industry Approved Plumbers Scheme (WIAPS):
www.wras.co.uk/WIAPS

Chartered Institute of Plumbing and Heating Engineering
www.iphe.org.uk

Association of Plumbing and Heating Contractors
www.competentpersonsscheme.co.uk

Scottish and Northern Ireland Plumbing Employers' Federation
www.snipef.org

Water quality

Colin Malcolm, Workplace Law

Key points

- There are many ways in which businesses can cause water pollution, and the number of incidents, whilst exhibiting an overall trend of decline, is still a cause for concern.
- Virtually all businesses have the potential to cause water pollution. This is principally because whilst water itself may not always be visible, a pathway will normally exist to transport polluting substances from a business premises to a water body.
- Managing compliance with water pollution can broadly be split up by the three main water receptors; the foul sewerage system, the surface water system and groundwater. Each receptor is, in different ways, very susceptible to sources of pollution.
- As an overarching rule, no substances except clean uncontaminated water should be discharged to any water receptor without prior discussion and approval with the appropriate regulatory authority. Discharges to foul sewer are regulated by the local sewerage company, while discharges to surface waters and groundwater are regulated by the Environment Agency.
- Specific legislation exists to either prohibit or control the admittance of polluting materials to each receptor and, as such, the most important first step in site water management is understanding the type of drainage system present on your site. If existing drainage plans and historical records are inadequate, specialist firms can be engaged to carry out drainage surveys and prepare up to date plans on your behalf.

Legislation

- Water Industry Act 1991.
- Water Resources Act 1991.
- The Anti-Pollution Works Regulations 1999.
- The Environmental Damage Regulations 2009.
- Environmental Permitting Regulations 2010.

Discharge to surface water

The Environmental Permitting Regulations 2010 provide the regulatory framework for preventing pollution of surface waters. Coastal waters, estuaries, rivers, lakes, streams, canals and reservoirs are all examples of surface waters.

Note: The Environmental Permitting Regulations 2010 have replaced the sections of the Water Resources Act 1991 that relate to the regulation of discharges to controlled waters. Previously issued discharge consents under the Water Resources Act 1991 will automatically become Environmental Permits from 6 April 2010.

The Regulations define 'water discharge activities', and there are several different activities defined. However, of particular relevance to most businesses is Paragraph 3(1)(a) of Schedule 21 to the Regulations, which identifies:

> **Facts**
>
> Putting to one side common examples of water pollutants such as oil or solvents, seemingly innocuous substances such as milk have a significant impact on the water environment. Milk has a high Biological Oxygen Demand (*a measure of the amount of oxygen required by biological organisms to deal with organic matter*) and as such it contributes to de-oxygenate water, to the detriment of many aquatic species that rely on available and stable levels of dissolved oxygen. Furthermore, dissolved oxygen content in water decreases with increasing temperatures; hence the temperature alone of any discharge can impact on water pollution, even before the chemical composition is factored in.

The discharge or entry to inland freshwaters, coastal waters or relevant territorial waters of any:

- poisonous, noxious or polluting matter;
- waste matter; or
- trade effluent or sewage effluent.

Only clean, uncontaminated water should be discharged to surface water. Any other discharge (that has not previously been authorised) should be stopped immediately as it may be an offence under the Environmental Permitting Regulations 2010. Some discharges to surface waters can be permitted, while there are also certain activities that are exempt from the requirements. Discussions should be held with the Environment Agency regarding applicability of the Regulations and, if applicable, permitting requirements for the discharge of anything other than clean uncontaminated water.

There are three key steps in deciding if and then what type of Environmental Permit will apply to your site if a discharge to surface waters is undertaken. Firstly, there is a narrow band of exempt activities that do not require an Environmental Permit. These are quite specific, for instance small discharges of domestic sewage from septic tanks and, as such,

are therefore unlikely to be relevant for the majority of businesses.

Secondly, well defined and low risk discharges are regulated under a system of 'standard permits'. These define specific conditions that must be met and, providing the operator can comply with the defined conditions, an Environmental Permit to discharge can be granted. Example activities where a standard permit can be issued are the discharge of secondary treated domestic sewage or cooling water. For all other activities, a bespoke Environmental Permit will be required, which entails completing and submitting an application to the Environment Agency. Subject to approval, a permit will be issued that contains detailed operating conditions specifying how the water discharge activity is to be managed.

Drainage systems in many businesses have connections to both surface water and foul sewer systems. Whilst the latter incorporate pre-treatment systems prior to final discharge, this is not always the case with surface water systems. It is important to be absolutely clear, however, that the presence of treatment plant in foul sewer systems should not in any way be seen as a green light for unauthorised discharges.

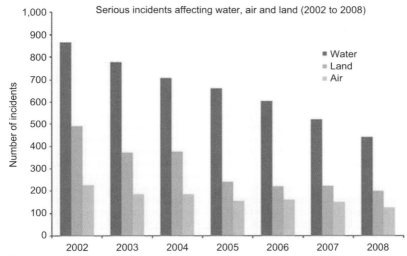

Serious incidents affecting water, air and land (2002 to 2008)

Source: Environment Agency.

Whilst it is the case that discharges to surface water drainage systems do not always pass through any form of pre-treatment, there are exceptions. Of particular relevance to many business are separators (oil interceptors), which are simply underground chambers connected to the drainage system designed to separate out and contain oils and some other polluting materials. It is important to enter into dialogue with the Environment Agency if an interceptor is planned or used at your site, as the type of system, pollutant removal efficiency and discharge point may need to be assessed.

Note: dialogue should also be sought from the relevant sewage undertaker if a separator exists when a foul sewer discharge is made.

The pollutant removal efficiency of separators is dependent upon several factors, including amongst others emptying frequency and regular maintenance. It

is therefore important that existing site operational controls such as planned preventative maintenance programmes include separators.

The majority of water pollution incidents that are brought before a court of law relate to unplanned discharges. Sitting behind this is nearly always a poor understanding and management of how a water pollution incident could occur, be it under normal, abnormal or emergency scenarios.

As a starting point, it is important to assess all areas of a site to identify potential situations whereby sources of pollution could enter surface waters. Typical areas to focus on include separators, raw material and waste storage areas (particularly where liquid or dusty materials are kept). Secondary containment provision, if relevant to storage, should also be subject to regular integrity and general suitability tests.

Once all sources of pollution have been identified, potential pathways for polluting material to reach surface water entry points (i.e. drains, direct run-off, topography) should be identified. The question of whether or how to obstruct pathways is subjective and will depend on the risk potential for a pollution incident to occur. For instance, a well designed bund (secondary containment) will provide an effective barrier should the primary container leak, but a surface water drain immediately adjacent to the bund should not immediately be ignored.

As mentioned earlier, an up to date drainage plan is of paramount importance. It is also highly beneficial to colour code drains and other drainage entrance points to visibly differentiate between surface water and foul sewerage drainage systems. Competence and training in both preventing and dealing with water pollution scenarios should be provided and up to date on sites where risk exists. For more complex sites, where a site risk assessment has identified significant potential for surface water pollution to occur, albeit under abnormal conditions, an emergency plan to manage pollution potential should be developed and routinely tested.

Contaminated rainwater is often overlooked but it can be a relevant potential source of surface water pollution, particularly in industrial or manufacturing sectors. Common areas to focus on include roof areas, particularly where particulate or other polluting emission deposition may build up and be washed away to surface waters by rainfall. Other open areas of a site where residual spills of chemical, oil or other polluting substances exist should also be cleaned regularly to avoid similar occurrences.

Discharge to groundwater

The Environmental Permitting Regulations 2010 provide the regulatory framework for preventing pollution of groundwater, which is defined as all water that is below the surface of the ground in the saturation zone and in direct contact with the ground or subsoil.

The Regulations prevent the input of hazardous substances to groundwater and limit the input of non-hazardous pollutants to groundwater. Authorisations granted under the now-revoked Groundwater Regulations 1998 and Groundwater (England and Wales) Regulations 2009 remain valid and will automatically become permits under the Environmental Permitting Regulations 2010 from 6 April 2010.

The terms 'hazardous substances' and 'non-hazardous pollutants' are used instead of List I and List II substances, both of which were key phrases under the Groundwater Regulations 1998. In both cases these new terms are slightly wider in scope than those they succeed. Hazardous substances are defined as substances or groups of substances that are toxic, persistent and liable to bio-accumulate, and other substances or groups of substances that give rise to an equivalent level of concern. All substances that are not determined to be hazardous are potentially non-hazardous pollutants.

Under the Regulations, the Environment Agency is required to publish substances that it considers to be hazardous substances. An up to date list of hazardous substances can be viewed on the website of the Joint Agencies Groundwater Directive Advisory Group (see *Sources of further information*). The Advisory Group has responsibility to peer review the Environment Agency substance assessment, ultimately advising whether substances are classified as hazardous substances or non-hazardous pollutants. Two important points need to be made about the classification process; firstly, the JADAG list of hazardous substances should not be taken to be absolute as a non-listed substance may be hazardous if it were to be assessed, and secondly, as a non-hazardous pollutant is any pollutant other than a hazardous substance, the scope is much wider than in previous groundwater regimes.

The Regulations introduce the term 'groundwater activity', and it is an offence to cause or knowingly permit a groundwater activity unless authorised by an Environmental Permit or an exemption. Groundwater activities include but are not limited to:

- the discharge of a pollutant that results in or might lead to a direct or indirect input to groundwater; and/or
- any other discharge that might lead to a direct or indirect input of a pollutant to groundwater.

It is important to be clear on the terms 'direct input', which is defined as the introduction of a pollutant to groundwater without percolation through soil or subsoil, and 'indirect input', which is defined as the introduction of a pollutant to groundwater after percolation through soil or subsoil. This is a key issue for business as, in the majority of cases, direct input will not be relevant as this typically relates to issues such as landfill, borehole or other deliberate geological intrusion. Of more relevance is the potential for indirect input to occur. This may become relevant should, for example, an unplanned spillage of polluting matter have potential to reach groundwater via a permeable surface or other pathway.

It is an offence under the Regulations to allow hazardous substances, which include pesticides, sheep dip, solvents, hydrocarbons, mercury, cadmium and cyanide, to pollute groundwater. Non-hazardous pollutants must be limited from entering groundwater so as to ensure that such inputs do not cause pollution of groundwater.

Discharge to foul sewer

Discharge to foul sewer that has not been previously authorised can cause damage to the sewerage system and treatment works, to employees engaged in sewerage system activities, and the general public.

The Water Industry Act 1991 provides the regulatory framework for the discharge of trade effluent to sewer. Trade effluent is defined as liquid waste discharged from premises being used for a business, trade or industry. Trade effluent can include any liquid wastes, regardless of quantity or how non-hazardous it may appear. The only liquid wastes that are not classed as trade effluent are domestic sewage and clean, uncontaminated surface water. Some examples of activities that would give rise to trade effluent include the following:

- Disposing of out of date liquid products down the toilet;
- Washing or cleaning products in the sink with detergents, solvents or other substances;
- Vehicle washing activities;
- Diluting chemicals with water prior to disposal in the foul sewer; and/or
- Tipping small amounts of liquid waste down the foul sewer drain.

Under the Regulations, it is an offence to discharge trade effluent to a public foul sewer or a private foul sewer that connects to a public sewer, without prior consent.

It is important for businesses to be aware that they may be in breach of the Regulations even if a discharge to foul sewer is made on an infrequent basis and in very small quantities. Prior authorisation must be obtained by the sewerage undertaker before discharge, and this will either be a trade effluent consent or agreement. Trade effluent consents contain specific conditions and prescriptive limits on the effluent composition, such as volume, temperature, pH and, if relevant, upper limits on individual or groups of substances that exhibit hazardous properties, such as heavy metals for example. An agreement is less detailed and is typically used for effluent that has reduced hazard potential to the sewerage system.

Other water regulations

The Anti-Pollution Works Regulations 1999 enable the Environment Agency to either require a person or carry out itself works to prevent or mitigate pollution caused to surface waters. The Environmental Damage Regulations 2009 similarly have the provision to require preventative works to be carried out to prevent water and groundwater pollution, and also have significant provisions should an incident occur.

For example, in addition to measures taken to return a watercourse to its previous condition, the Regulations also apply if the watercourse does not fully recover and for the loss of environmental resources and environmental services pending recovery.

The Environmental Damage Regulations 2009 operate without prejudice to existing legislation; that is, existing legislation continues to apply, and if it imposes additional obligations on operators, then those would need to be complied with. The Regulations also include damage to species, habitats and risk to human health from contaminated land and will only

apply in the most serious of incidents. In cases of water pollution, the Regulations are limited to a defined list of operators, principally those already under existing regulatory regimes.

Pollution prevention

Water prosecutions are often the result of an incident that can be prevented by relatively straightforward environmental management controls. Example issues for consideration to minimise water pollution risk include:

- Having an up to date drainage plan.
- Working knowledge of applicable water law and bespoke compliance requirements.
- Management controls to anticipate and integrate relevant new or amended water law.
- Understand all site entry points to water bodies.
- Ensuring an up to date inventory of substances is retained.
- Completing a risk assessment to understand all potential pathways for substances to reach a water body under normal, abnormal and emergency scenarios.
- Assessing the need for preventative measures, such as indoor storage, secondary containment, drain covers and other spill control measures.
- Assessing training and competency requirements.
- Regularly auditing and documenting the effectiveness of staff competence and preventative measures, including management responsibility for close out actions.
- Routine testing of spill management and emergency response procedures.
- Understanding the water pollution implications of contractors or other third parties working on site.
- Constant supervision of deliveries where water pollution risks are identified.

See also: Contaminated land, p.182; Environmental management systems, p.286; Environmental risk assessments, p.292; Legionella, p.442; Waste management, p.707; Water fittings, p.720.

Sources of further information

The Environment Agency has a number of useful Pollution Prevention Guidelines (on containment structures, SUDS etc.) available at www.environment-agency.gov.uk.

Joint Agencies Groundwater Directive Advisory Group: www.wfduk.org./jagdag

Weather

Sally Cummings, Kennedys

Key points
- Employers should consider weather conditions and their effect on the working environment when carrying out risk assessments.
- Air temperature is only one of many weather-related factors that should be considered.

Legislation
- Health and Safety at Work etc. Act 1974.
- Personal Protective Equipment at Work Regulations 1992.
- Workplace (Health, Safety and Welfare) Regulations 1992.
- Management of Health and Safety at Work Regulations 1999.

The Health and Safety at Work etc. Act 1974 (HSWA) requires employers to take all reasonably practicable steps to ensure the health, safety and welfare of employees at work.

A number of different health and safety regulations impose more specific duties that may require employers to take into account the effects of weather in order to comply with them.

The Workplace (Health, Safety and Welfare) Regulations 1992
Regulation 7 requires that the temperature of all indoor workplaces is 'reasonable' during work hours. The relevant ACoP advises that, subject to practicability, the temperature 'should provide reasonable comfort without the need for special clothing'. Where that is impractical because of hot or cold processes, all reasonable steps should be taken to achieve a temperature as close as possible to comfortable. The ACoP further advises

that the minimum temperature should be 16°C, or 13°C if much of the work is physical. The ACoP does not advise on a maximum temperature.

Regulation 11 provides that all outdoor workstations should, so far as is reasonably practicable, provide protection from adverse weather conditions.

The Personal Protective Equipment at Work Regulations 1992
Regulation 4 states that employees must be provided with 'suitable' PPE, which is appropriate for conditions in the workplace. In terms of suitability, the ACoP suggests that consideration be given to environmental factors such as the weather if working outside.

The Management of Health and Safety at Work Regulations 1999
Various Regulations, e.g. 3, 16 and 19, and the ACoP, refer to the need to pay attention to the particular risks posed to young people and new and expectant mothers by environmental conditions such as extremes of temperature.

All potentially adverse weather, resulting in working conditions that are hot, cold, windy, icy, or wet, should be taken into account when employers undertake risk assessments, and appropriate measures should be implemented.

Hot weather

Exposure to the sun can cause skin damage such as blistering and even skin cancer, which has become the commonest form of cancer in the UK. Exposure to heat can also cause symptoms of heat stress, from a mild inability to concentrate and increased irritability, to heat stroke. Productivity has been shown to decrease at temperatures of 25°C and above and the risks to workers' health increase as conditions deteriorate from those accepted as comfortable. Calls from organisations for an upper temperature limit have been made in the past.

In July 2009, and following pressure from USDAW (Union of Shop, Distributive and Allied Workers), the Government asked the HSE to review the Regulations on workplace temperature. John Hannett, USDAW General Secretary, had placed hope in a maximum workplace temperature being imposed. However, there is currently no such upper limit expressed in health and safety legislation or guidance. Instead the HSE has published detailed guidance on thermal comfort.

This guidance provides that where the temperature in a workroom would otherwise be uncomfortably high, for example because of hot processes or the design of the building, all reasonable steps should be taken to achieve a reasonably comfortable temperature, for example by:

- insulating hot plants or pipes;
- providing air-cooling plant;
- shading windows; and/or
- siting workstations away from places subject to radiant heat.

The guidance states that if the percentage of workers complaining about thermal discomfort exceeds the recommended figure (10% of employees in air conditioned offices, 15% of employees in naturally ventilated offices and 20% of employees in retail businesses, warehouses and other indoor environments that may not have air conditioning) the employer should carry out a risk assessment and act on the results of that assessment.

Although all workers should take steps to protect themselves, employers are also under a duty to consider implementing measures, outdoors or indoors, such as:

- shading employees from direct sunlight;
- the insulation of hot plant and pipes;
- the provision of fans or air-cooling plant;
- ensuring hats and other suitable clothing is worn;
- suitable rest breaks;
- scheduling work during a cooler time of day or year;
- additional supplies of drinking water; and/or
- educating workers.

In certain industries, such as glass manufacturing plants, mines, boiler rooms and bakeries, working in the heat may be the norm; nevertheless, employers should be aware of the impact of seasonal changes in the ambient temperature and provide advice on heat stress.

There is no legal obligation for employers to provide sun cream or sunglasses for outdoor workers but they should consider providing sun protection advice as part of health and safety training.

There was concern in 2005 when it was proposed that an EU Directive on optical radiation, originally intended to regulate the health and safety of workers regarding artificial radiation only, should be amended to cover natural radiation from the sun. The result would have been a specific requirement for employers to carry out risk assessments and devise measures to reduce the risks of sun exposure. In September 2005, however, after a vote by the European Parliament, the EU Commissioner for Employment, Social Affairs and Equal Opportunities, Vladimir Spidla, announced that the reference to sunlight would be removed from the Directive.

Cold weather

Just as for hot weather and hot conditions, employers should risk assess their employees' workplaces and activities with the effects of cold weather in mind.

The ACoP provides that temperature in workrooms should normally be at least 16°C unless much of the work involves severe physical effort in which case the temperature should be at least 13°C.

Possible risk mitigation measures include:

- the provision of mobile facilities such as heated cabins;
- warm drinks and sufficient breaks in which to have them and warm up;
- rescheduling work to warmer times of the year;
- systems of work to limit exposure, such as job rotation; and
- appropriate PPE for cold environments.

As far as the last of these measures is concerned, the House of Lords' decision in the case of *Fytche v. Wincanton Logistics plc* (2004) provides some useful guidance. A lorry driver employed to collect milk from farms was provided by his employer with steel-capped safety boots to protect his toes from falling objects. In exceptionally wintry weather, the driver's tanker became stuck on a remote snow-covered road. Although the employer's standing instruction in these circumstances was for drivers to use the cab telephone to

call for help and wait to be rescued, this driver decided to dig himself out. One of his boots had a tiny hole in it and, as a result, freezing water penetrated and caused mild frostbite to the driver's little toe. He subsequently sued his employer on the basis that it had failed to maintain his boots in an efficient state, efficient working order and good repair as required by Regulation 7 of the PPE at Work Regulations 1992.

Having failed at first instance and on appeal, the driver's case finally ended up in the House of Lords, who confirmed that the requirement to maintain was restricted to ensuring PPE was efficient for the purpose of protecting against the relevant risk. The boots had been adequate for the driver's ordinary conditions of work, the tiny hole did not constitute a breach of the employer's duty to maintain, so the driver's appeal was dismissed.

Travelling to and from work in adverse weather

What happens if employees are unable to get into work because of adverse weather conditions? Much will depend on whether or not the reason for non-attendance is reasonable, i.e. whether the employee has made a reasonable effort to get to work. An employee who fails to attend without a reasonable excuse is absent without leave and, subject to the terms of the employment contract, the employer may take disciplinary action, which might include withholding pay. If the excuse for non-attendance is reasonable, and the employment contract allows it, the employer may still withhold pay or, if the employee enjoys more than the 28-day statutory holiday entitlement, require holiday to be taken. However, employers who allow time off with pay in these circumstances will often benefit from improved employee morale, attitude and loyalty, and a common sense approach should be adopted wherever possible. From a safety perspective, employers should be wary of exerting undue pressure on employees to drive to work in dangerous conditions.

Gritting around premises

In early January 2010, the *Telegraph* ran a story headed 'Health and safety experts warn: don't clear icy pavements, you could get sued', in which it reported on what it said was the contrasting advice of RoSPA and IOSH on the gritting of icy pavements against a background of record numbers of hospital admissions for broken bones during the preceding cold snap and the attendant risk of litigation. While, the *Telegraph* reported, RoSPA encouraged individuals and organisations to demonstrate a 'good attitude' by dealing with icy paths even if they were not within the bounds of their own property, IOSH counselled, 'it is probably worth stopping at the boundaries of the property under your control', apparently on the basis that to do more would risk criticism and potential injury claim if it was not done perfectly.

IOSH's media team hit back the next day with its own criticism of the *Telegraph*'s "completely irresponsible" reporting, pointing out that the advice attributed to it had actually formed part of an independent contribution to *SHP Magazine*. IOSH sought to set the record straight by putting forward its true position on the subject:

"Deciding whether to grit beyond the boundaries of their property needs to be carefully considered by companies. If access to the premises is covered in ice, companies may choose to grit the access to help their staff and visitors arrive and leave safely, even though it's not their property. However, in this instance, if they failed to grit the surface properly and someone had an accident as a result, then they could incur some liability.

"As a general rule, though, it's sensible for firms to consider the risks and take reasonable steps to prevent accidents from happening. If this means gritting outside the boundaries of your workplace, then it's better to do that than to have people slipping over or involved in car crashes on your doorstep."

The most pithy comment, however, came from the past president of the British Orthopaedic Association: "People need a bit of grit, in both senses".

Conclusion

Although employers may often feel as though the law is stacked against them, requiring them to cross every single 't' and dot every single 'i' in the pursuit of absolute safety, the weather-related 1999 High Court case of *Groody v. West Sussex County Council* (1999) might give some comfort. An office worker employed in the County Hall offices in Chichester claimed she fell and was injured when she was dazzled by bright sunlight reflected off a photocopier and failed to see a step to a low platform. Despite the fact that she had successfully negotiated the step on a daily basis and again just prior to her accident and it had a fixed right-angled white edging, the worker alleged her employer was negligent in failing to (a) post a warning notice, (b) carpet the platform in a different colour to the rest of the floor, and (c) shade the windows to prevent sunlight from entering and reflecting off the photocopier. In deciding against the worker, the Court held that the key question was whether it was reasonably foreseeable that there was a risk in clear weather, at certain times of the day and year, that the sun would shine directly into the office and, either directly or indirectly, by means of reflection, so distract the people inside from their activities as to expose them to harm. In the Court's view, to state the worker's proposition revealed its absurdity.

See also: Occupational cancers, p.512; Outdoor workers, p.528; Personal Protective Equipment, p.576; Radiation, p.602; Temperature and ventilation, p.666.

Sources of further information

HSE – thermal comfort: www.hse.gov.uk/temperature/thermal/index.htm

HSE – Sun protection: advice for employers of outdoor workers: www.hse.gov.uk/pubns/indg337.pdf

Waste Electrical and Electronic Equipment: the WEEE and RoHS Regulations

Phil Conran, 360 Environmental

Key points

■ WEEE stands for Waste Electrical and Electronic Equipment.

■ RoHS stands for Restriction of the use of certain Hazardous Substances (in waste electronic and electrical equipment).

■ The UK produces around one million tonnes of waste electrical and electronic equipment (EEE) every year.

■ The WEEE and RoHS Directives (2003/108/EC and 2002/95/EC) aim to minimise the impact of EEE on the environment. (RoHS has now been revised by the EC.)

■ The Waste Electrical and Electronic Equipment Regulations 2006 (the WEEE Regulations) came into force on 2 January 2007 and were implemented on 1 July that year. The RoHS Directive and the UK RoHS Regulations came into force on 1 July 2006.

Legislation

■ Waste Electrical and Electronic Equipment Regulations 2006.

■ Restriction of the Use of Certain Hazardous Substances in Electrical and Electronic Equipment Regulations 2008.

Overview

The law on WEEE requires certain companies to provide and pay for the costs of the collection, treatment and recycling of EEE for which they are responsible at the end of its useful life. This applies to any company that manufactures EEE, re-sells EEE produced by others under its own brand, or imports or exports EEE to an EU Member State. More limited obligations apply to distributors (including retailers) of EEE and, in certain circumstances, final business users.

The RoHS Regulations implement EU Directive 2002/95, which bans the placing on the EU market of new electrical and electronic equipment containing more than agreed levels of lead, cadmium, mercury, hexavalent chromium, polybrominated biphenyl (PBB) and polybrominated diphenyl ether (PBDE) flame retardants.

Key features of the WEEE Regulations

■ System is overseen and regulated by the environmental agencies (EA in England and Wales, SEPA in Scotland and NIEA in Northern Ireland).

■ Single national Distributor Takeback Scheme (DTS) provides a national network of Designated Collection Facilities (DCFs) to collect household WEEE.

■ Direct registration with the Agencies is not permitted. All 'producers' of EEE, regardless of size, must join a Government-approved producer compliance scheme (PCS).

- Each PCS will:
  - register members with the Agency;
  - provide all required data and reports to the Agency including the amount of EEE by category that its members placed on the market in the preceding quarter and quarterly reports on quantities of WEEE collected, re-used, treated, recycled and recovered;
  - arrange on behalf of members the collection, transportation, treatment and reprocessing of the required amount of WEEE deposited at DCFs to meet their obligations;
  - arrange settlement at the end of each compliance period to balance evidence against market share where under or over collection has occurred; and
  - arrange for the collection of any non-household WEEE in relation to their member's obligations.
- Each individual PCS to contractually agree with its members how the members share the costs of dealing with their aggregate WEEE obligations.
- A Code of Practice governs collection of WEEE from DCFs and the relationship between PCSs and Local Authorities.
- Compliance periods are annual calendar years.
- Schemes must supply data to the Agencies each quarter showing the amount of household EEE placed on to the market by their members and annually showing the amount of non-household WEEE placed on the market.
- Schemes and AATF/AEs (Approved Authorised Treatment Facilities and Approved Exporters) must also provide quarterly data for the amount of WEEE collected and treated.
- At the end of each year, the Environment Agency calculates scheme obligations based on actual amounts of household EEE placed on the market and the amounts of household WEEE collected nationally. Schemes are then notified of their final obligations and will have to show how they have discharged them by 30 April.
- The financing obligation is met by a scheme collecting sufficient household WEEE to meet its market share. Evidence that this WEEE has been treated and recycled has to be placed by AATFs / AEs on to an electronic 'Settlement Centre' that is administered by BIS.
- AATFs and AEs of WEEE issue 'Evidence Notes' through an electronic Settlement Centre administered by BIS confirming treatment, recycling and recovery of quantities of WEEE. These are then used by PCSs to demonstrate compliance with the environmental agencies using the Settlement Centre to verify that each scheme has met its market share obligations.
- PCSs may trade Evidence Notes with other schemes but as a condition of their approval, must be able to demonstrate contractual agreements for the collection of their expected evidence requirements.

Scope of the WEEE Regulations

Products covered
The EU Directive lists ten categories. Because of the way WEEE is collected, the UK has separated out cooling equipment, display equipment and gas discharge lamps from categories 1, 3, 4 and 5 and created three new categories:

1. Large household appliances (e.g. white goods).
2. Small household appliances (e.g. vacuums, irons, toasters, etc.).
3. IT and telecoms equipment (e.g. computers, printers, calculators, phones, answer machines, etc.).

4. Consumer equipment (e.g. radios, TVs, hi-fi equipment, electronic musical instruments, etc.).
5. Lighting equipment.
6. Electrical and electronic tools (e.g. drills, saws, sewing machines etc., but excluding large stationary industrial tools).
7. Toys and leisure and sports equipment (e.g. train sets, video games, coin slot machines, etc.).
8. Medical devices (e.g. dialysis machines, ventilators etc.). (*Note that this category is not covered by the RoHS Directive at present – see below.*)
9. Monitoring and Control instruments (e.g. smoke detectors, thermostats, etc.). (*Note that this category is also not covered by the RoHS Directive at present.*)
10. Automatic dispensers (e.g. ATMs, vending machines, etc.).
11. Display equipment (e.g. TVs and monitors, regardless of type).
12. Cooling equipment (e.g. fridges, freezers, air conditioning equipment).
13. Gas discharge lamps (e.g. fluorescent tubes, compact energy saving bulbs).

It should be noted that even though a category may include the word 'household' it applies equally to non-household equipment. Therefore, a commercial electric cooker would fall into category 1.

Are you affected?

For a preliminary guide, consider the questions below (definitive legal advice should be sought in every case).

1. Does your business:
   - manufacture; or
   - sell under its own brand (even if manufactured by another); or
   - distribute (which includes retailing and distance selling, e.g. via the internet); or
   - import into the UK (for WEEE) or EU (for RoHS); or
   - export into the UK (for WEEE),

any electrical equipment not exceeding 1,000 volts AC and 1,500 volts DC?

If so go to Question 2. If not, you are unlikely to be obligated.

2. Does that EEE fall into one of the above broad product categories? These categories, taken from Annex IB to the WEEE Directive, are not exhaustive.

If so, go to Question 3. If not, you unlikely to be obligated.

3. Does your equipment fall into any of the exempt categories (described below)?

Exempt products:

- Part of another type of equipment that does not fall within the scope of the WEEE Directive (for example a car radio, vehicles being excluded), or part of a fixed installation;
- Intended solely for national security/military use;
- Large-scale, stationary industrial tools;
- Luminaires (such as lamp holders and sockets) in households (this exception applies to WEEE only, not RoHS);
- Implanted or infected medical products; and
- Filament light bulbs (this exception also applies only to WEEE, not RoHS).

If not, go to Question 4. If so, the new laws probably do not apply to your business.

Note: there are other specific exceptions to the RoHS directive that may be relevant but are too detailed to be included in this chapter.

4. Is the main power source of your product electricity, whether mains or battery?

If so, go to Question 5. If not, the new laws probably do not apply to your business.

5. Is electricity needed for primary function of the product?

If so, the product in question may well be covered by these new laws.

If you suspect the new laws may be applicable to your business, you need to act now.

Practical implications

It will be a criminal offence not to comply with these various obligations. There are no exemptions for small companies or persons handling only small quantities of WEEE.

Summary of key impacts

Producers

If you are a 'producer' the following are the key impacts upon you.

Compliance scheme membership

You will need to be a member of a PCS and that PCS will need to register you with one of the Environmental Agencies if you wish to continue lawfully to place EEE on the UK market. PCSs must re-register their members with their Agency by 30 November each year.

Household WEEE financing

Since 1 July 2007, producers (through their schemes) have had to pay for the cost of the collection, treatment, recovery and environmentally sound disposal of household WEEE deposited at designated collection facilities on a market share basis. Slightly more complex obligations apply in relation to B2B WEEE and these are addressed below.

Business WEEE financing

The rules regarding responsibility for business WEEE depend on whether the WEEE in question is historical business EEE (business EEE placed on the market before 13 August 2005 that has become WEEE) or new business EEE (business EEE placed on the market after 13 August 2005 that has since become WEEE).

If a producer has supplied or supplies new EEE to business users after 13 August 2005, it will be required to finance its treatment, recovery and disposal when that new EEE becomes waste in just the same way as for household WEEE, unless it contractually agrees otherwise with the business user in question. Subject to the provision below, it will not, however, have obligations in relation to historical (pre-13 August 2005) business EEE.

If a producer supplies new EEE to a business user after 13 August 2005, which replaces on a like-for-like basis historical EEE currently held by that business user, it will be required to finance the treatment, recovery and disposal of the original, replaced EEE – again unless it contractually agrees otherwise with the business user in question. This is so even though the same producer may not have supplied the original EEE at all.

'Like-for-like' purchases mean products that are intended to replace the relevant WEEE and are of an equivalent type or fulfil the same function on a common sense approach, even though not an identical replacement.

Buyers and sellers of business EEE should therefore be considering the position they intend to take on contractual allocation of liability for WEEE compliance and what language may therefore need to be introduced into standard terms of business for the sale / purchase of EEE.

If historical business EEE is simply disposed of without 'like-for-like' replacement, the business end-user bears the financing obligation (*see below*).

Product marking

Producers must also ensure that they comply with WEEE information and marking requirements. These require producers to mark all EEE they place on the market after 1 April 2007 with:

- the designated symbol of a crossed-out wheelie bin, the producer's identity; and
- an indication of placement on the market after 13 August 2005. This is generally satisfied by a small bar beneath the crossed-out wheelie bin.

Distributors

If you are a retailer of, or otherwise provide, new EEE of any of the kinds described above to household users on a commercial basis, then the WEEE Regulations impose two key obligations described below.

Free take-back of WEEE

- Retailers of electrical equipment have the option of either offering their customers free take-back of their WEEE when they make a 'like-for-like' purchase of new equipment (e.g. by in-store take-back or by collection-on-delivery, provided it is free of charge) or by participating in a Government-approved retailer compliance scheme (the Distributor Take-back Scheme or DTS) which helps to fund Local Authority take back at Civic Amenity sites. In practice, the latter is likely to be the most convenient route for many retailers, particularly distance sellers.
- WEEE thus collected will then have to be properly disposed of by the retailer a free disposal point operated by a

Compliance Scheme. This distributor take-back obligation does not apply to sales to business users.

Provision of information to customers

Retailers and other distributors must take adequate steps to ensure that private householders are informed of:

- the WEEE take-back facilities available to them (either from the retailer / distributor itself or other facilities);
- the meaning of the crossed-out wheelie bin symbol on products covered by the WEEE Directive;
- the role of users in private households in contributing to re-use, recycling and other forms of recovery of WEEE; and
- the potential effects on the environment and human health of WEEE.

It will be an offence for any affected retailer / distributor not to comply with these duties.

Distributors have no obligations under the RoHS Directive, which applies only to producers, and no obligations in relation to WEEE beyond the two identified above.

Business end-users

There are circumstances in which businesses can have obligations under the WEEE Directive, even though they do not themselves produce or sell EEE.

Duties of business end-users of EEE

The end-user will assume full responsibility for financing the costs of the treatment, recovery and sound disposal of the WEEE only in the following cases:

- Where a business has purchased EEE before 13 August 2005 and does not buy 'like-for-like' replacement goods when it reaches the end of its useful life and is disposed of;

- Where a business has purchased EEE before 13 August 2005 and does replace it with 'like-for-like' goods, but agrees with its supplier to accept responsibility for the WEEE being replaced; and
- Where a business purchases EEE after 13 August 2005 and agrees with its supplier to accept responsibility for it when it becomes WEEE.

Where non-household WEEE is disposed of by a producer under the obligations of the Regulations, e.g. the producer arranges collection from the end user and manages the disposal, the WEEE must be taken to an AATF/AE and the producer should ensure that the AATF/AE places evidence on to the Settlement Centre for that WEEE.

RoHS compliance: restrictions on use of certain hazardous substances

Since 1 July 2006, subject to certain limited exemptions, producers must ensure and be able to demonstrate that their EEE products placed on the EU market, and the components in their EEE products (even if supplied by others), do not contain more than the prescribed amounts of lead, mercury, cadmium, hexavalent chromium, polybrominated biphenyls (PBBs) or polybrominated diphenyl ethers (PBDEs). Those prescribed amounts are shown below.

RoHS maximum concentration values:

- Up to 0.1% by weight in homogenous materials for lead, mercury, hexavalent chromium, PBB and PBDE; and
- Up to 0.01% by weight in homogenous materials for cadmium.

The EC now published revised RoHS Regulations that member states must transpose by 2 January 2013.

RoHS exemptions

While the RoHS Directive essentially applies to all of the same products as WEEE, there are some very important additional exceptions. Some have already been specified. In addition you should note those below:

- For the time being, the RoHS requirements do not apply to products within categories 8 or 9 of Annex IB of the WEEE directive, namely Medical Devices (category 8) and Monitoring and Control Instruments (category 9). However, the amended Directive removes these exemptions. Specific applications of the prescribed substances are exempted, as specified in the Annex to the RoHS Directive itself as amended (for example, lead in the glass of cathode ray tubes, electronic components and fluorescent tubes). Various further exemptions are being considered at present.
- The requirements of RoHS do not apply to spare parts for the repair of, or to the re-use of, EEE put on the market before 1 July 2006.

RoHS compliance and supply chain auditing

The UK has adopted a self-declaration approach to compliance. However, this is bolstered by risk-based market surveillance by the National Weights and Measures Laboratory. Technical documents and other information sufficient to demonstrate product compliance must be kept by producers for a minimum of four years and produced on request.

This will change under the revised RoHS requirements and will require all RoHS affected equipment to be CE kite-marked.

If your product contains materials supplied by third parties you should be carrying out adequate RoHS compliance auditing of your supply chain.

If you market your product in other EU Member States, you will need to find out what approach to implementation and compliance is being adopted in each state and ensure that your product is compliant there as well.

See also: Batteries, p.46; Recycling, p.605; Waste management, p.707.

RoHS due diligence defence

A defence of due diligence is available in the UK to prosecution under the RoHS legislation, provided the business can show that it took 'all reasonable steps and exercised all due diligence' to avoid committing the offence.

Sources of further information

Environment Agency: www.environment-agency.gov.uk

RoHS: www.rohs.gov.uk/

Research has shown that many businesses are unaware of their duties under the Regulations, and with this in mind Workplace Law has produced the **Guide to WEEE 2007**, a downloadable publication, fully updated this year, that explains the varying responsibilities and roles under the new legislation. For more information visit www.workplacelaw.net.

Wheelclamping

Dale Collins, Bond Pearce LLP

Key points

- Currently land owners and occupiers who allow their land to be used as a car park have the right to restrict that right to specified persons, e.g. those who have purchased and are using a valid ticket. Running with that right is the right to demand a payment for unlawful use and the ability to prevent the removal of unauthorised vehicles until such time as a payment is made. However, there is currently before Parliament a Bill (the Protection of Freedoms Bill) which, if enacted, will make it a criminal offence to clamp (immobilise) a vehicle on private land except where one has the lawful authority to do so (e.g. on behalf of a local authority, the DVLA or the police).
- Where such a system is in place, however, the land owner or occupier must ensure that the person using the car park is aware of that fact to prevent actions for trespass and criminal damage being pursued.
- In accordance with the Private Security Industry Act 2001, since 3 May 2005 the wheelclamper must be licensed with the Security Industry Authority, failing which criminal offences are committed with, potentially, large fines being imposed.
- The Crime and Security Act 2010, which received Royal Assent on 8 April 2010, made provision for the licensing of businesses that undertake vehicle immobilisation activities. However, these provisions were essentially enabling provisions – i.e. they required Regulations to bring them into force, and this did not occur before the 2010 General Election was called. The Protection of Freedoms Bill aims to abolish this proposed licensing scheme.

Legislation

- Private Security Industry Act 2001.
- Crime and Security Act 2010.
- Protection of Freedoms Bill.

Cases

- *Arthur v. Anker* (1996).
- *Vine v. Waltham Forest London Borough Council* (2000).
- Rowencroft Immobilisers.

The current law

A private landowner has a right to take reasonable steps to protect his land, and any interest in that land, from harm. Anything placed on that land without his consent is a trespass. A trespass is an actionable tort (civil wrong), which gives rise to a claim in damages for compensation. It also gives rise to the right of 'self-help' – in other words, the ability to take action oneself to remove the trespassing article.

Clearly, vehicles parked unlawfully on private land are trespassing, and, although there may be no physical damage being caused by their being parked on the land, a claim can arise as damage does not need to be physical – it can arise from the landowner being unable to use that space. The car park owner or manager will want the car removed and the ability to claim immediate damages to represent the loss. He does not want to have to issue proceedings in court. The only way to do this is to ensure that the car is not removed without such payment being made.

Case studies

There are still rogue clampers out there, but they are being caught and punished. In May 2008, Rebecca Meakin was sentenced to serve four years' imprisonment after being found guilty to charges of blackmail. She was the proprietor of Rowencroft Immobilisers, a clamping and towing service to private landowners in Staffordshire, West Midlands and Wiltshire.

In February 2006 a series of files were presented to the CPS for evidential review relating to incidents centred on two public houses in Staffordshire. At both locations, Rowancroft Immobilisers were operating and incidents included allegations of assault, intimidation, theft and threatening behaviour.

The report review regularly showed an increase from the advertised £95 release fee to the £295 towing fee once the recovery truck had been called. This happened within minutes of the motorist parking because the recovery truck was already parked close by.

The company practice of only accepting cash or postal orders increased tension with motorists. When Rowancroft towed away vehicles they refused to state where they had been taken – significant evidence became available to show they were being abandoned on uncontrolled car parks or waste land.

The company had a P.O. Box in Worcester as their business address and it used mobile telephone numbers for contact with victims. It was not registered for VAT, had no business bank account and was not a registered company. Records of income and expenditure were not maintained.

Rowancroft had a significant impact on police resources as members of the public were calling for assistance and officers were required to attend, sometimes leading to specific allegations, statements and arrests.

The basis of the blackmail enquiry was that it appeared that Rowencroft was making demands for money and reinforcing the demand with menaces:

- The motorists were prepared to pay £95 for the release of the clamp, but £295 was demanded.
- In some cases the motorists were only on the car park for minutes or even seconds and were boxed in and not given the opportunity to drive away when the driver realised there were parking restrictions.
- Some victims parked, left the car park to call into a nearby shop and returned within a couple of minutes. It was considered that to clamp and demand £95 in these circumstances was not reasonable. When the motorist argued there should not be a charge, the fee would be increased to £295 on the basis that the tow truck was coming.

<hr />

Case studies – *continued*

- There were never any allowances made for being elderly / disabled / a doctor on call / mothers with babies / other vulnerable people.
- The car was still on site, not hooked up or removed but a tow fee became payable at the discretion of the clamper.

The menaces were:

- the threat that the car would be taken away;
- the location that the car was to be taken to was never disclosed; and
- veiled threats of violence by the presence of 'heavies'.

<hr />

The way to do this is through the use of wheelclamps.

However, the placing of a wheelclamp on another person's vehicle is itself a trespass, allowing the car owner to bring a claim for damages. In addition, should the installation of the clamp (or its removal) cause damage, the clamper could be liable for that damage and face a charge of criminal damage in the criminal courts.

Also, the wheelclamping itself does not remove the problem, i.e. the loss of car-parking space. In fact, it exacerbates the problem by potentially keeping the vehicle *in situ*.

Thus, in addition to the clamping, which is essentially a detention to secure payment of the fee, it is also necessary to have the ability to remove the obstructing vehicle. This removal can again cause damage.

So, from where do wheelclampers derive their authority preventing them being pursued or prosecuted? It is all a question of consent. Where it can be proved that the driver either saw or should have seen a sign prohibiting parking and warning that a wheelclamping regime was in place, the wheelclamping is lawful as the driver is said to accept, either explicitly or

by implication, the consequences of his action.

This is best seen in the cases of *Arthur v. Anker* (1996) and *Vine v. Waltham Forest London Borough Council* (2000). In the former, the wheelclamper's actions were held to be lawful as the driver had seen the sign but decided to ignore it. In the latter, as the driver had not seen the sign prohibiting parking, she was found not to have consented to the wheelclamping and thus could recover damages.

In the *Vine* case, Lord Justice May stated:

"A motorist who appreciates that there are warning signs obviously intended to affect the use of private property for parking vehicles, but who does not read the detailed warning, might, depending on the facts, be held to have consented to, or willingly assumed, the risk of a vehicle being clamped, if the unread warning sign in fact gives sufficient warning that trespassing vehicles would be clamped."

In other words, if the notices are such that it could not reasonably be argued that they were not seen, and if there were signs that made it abundantly clear what would happen if unauthorised parking occurred, the court would infer consent.

Fees

In the *Arthur* case, Master of the Rolls Bingham in his judgment stated:

"I would not accept that the clamper could exact any unreasonable or exorbitant charge for releasing the car, and the court would be very slow to find implied acceptance of such a charge."

In the *Vine* case the fee was £105, and in the *Arthur* case it was £40 to release the clamp and £90 plus storage to return the car. In both of those cases, these fees were considered reasonable. It is clear that the recovery of the costs associated with the wheelclamping scheme is acceptable, but if an attempt was made to make a large profit that would not be acceptable.

Removal

The problem with simply wheelclamping a vehicle that is illegally parked is that it does not free up the space – in other words, it does not solve the problem. The owners of a private car park are entitled, therefore, to reduce their losses by removing the vehicle, provided that removal is specified on the notice.

The additional difficulty with removal, however, is that there is an increased danger of damage being caused during the removal process. However, whether the trespassing owner will be entitled to claim for such damage will again be down to a question of consent. By parking unlawfully in the full knowledge that removal is a possibility, the driver is accepting that there may be some damage incurred, and, provided that damage is not unreasonable or unexpected, there should be no claim.

Signage

Signage is vitally important in this area. Signs must be prominently displayed where they can be read and where they will be visible to anyone using that car park, and must include, as a minimum, the following information:

- The notice must specify that wheelclamping is in operation, the consequences of parking unlawfully and the cost of removal. (It may be worth considering that as well as written warnings there should be some type of pictogram making it clear that there is no parking except by permit.)
- The notice must contain the contact details for the removal of the wheelclamp (and that removal must be able to take place within a reasonable time).
- If the vehicle is to be removed that fact must be stated.

Guidance

The British Parking Association (BPA) launched its Code of Practice for *Parking Enforcement on Private Land and Unregulated Public Car Parks, Part 1: Vehicle Immobilisation or Removal*, in April 2006 (see *Sources of further information*).

The Code provides a model of best practice for individuals or organisations undertaking vehicle immobilisation or removal of vehicles on private land. Its main objective is to help raise standards in the parking sector and ensure that vehicle immobilisation and removal is undertaken in a responsible, effective and efficient manner. It also suggests 'best practice' for fees, providing £125 for the removal of a clamp on a private vehicle and £250 for the removal and return of a private vehicle that has been towed.

The RAC Foundation for Motoring states:

'The RAC Foundation for Motoring wholeheartedly supports the BPA wheelclamping Code of Practice. It is concise, fair, workable and, above all, responsible, and represents a genuine attempt to introduce a welcome element of reassurance into the sphere of off-street wheel clamping.

'Whilst the Code is primarily directed at BPA members, themselves responsible organisations, it could and should be considered by the Security Industry Authority (SIA) as a model of good practice and be incorporated into the SIA Approved Contractors Scheme.'

The document provides what is effectively good practice, including information on the types of vehicle that should not be clamped, expected release times to be achieved (usually no more than two hours and certainly never beyond four) and the types of sign to be used in car parks and to be placed on a vehicle following the clamping. Other practical issues to be considered include the following:

- Photographing the car in relation to the sign (useful evidence).
- Checking the insurance position with regard to public liability.
- Training for those carrying out the wheelclamping and risk-assessing their tasks.

The requirement for a licence

The Private Security Industry Act 2001 has been effective from 1 April 2003. The Act created the Security Industry Authority (SIA), whose job it is to regulate and oversee the private security industry, which includes those undertaking wheelclamping or vehicle immobilisation.

The 'Regulation' is by way of a licensing regime for those involved and the development of criteria for their professional training. Not only do those working for wheelclamping firms have to be licensed; so, too, do in-house clampers.

There are two types of licence – the Front-line licence (for the clampers / immobilisers) and the Non front-line

licence (for the managers and supervisors of the clampers / immobilisers). The front-line licence costs £245 and is valid for one year. A non front-line licence also costs £245, but is valid for three years. Before such a licence is issued, the individual will be checked out by the SIA (criminal record checks and so on) to ensure he is fit to hold a licence.

Undertaking clamping without a licence is a criminal offence for the individual, giving rise to a fine of £5,000 and/or six months' imprisonment. In addition, the occupier of the land on which the wheelclamping takes place also commits an offence if an unlicensed clamper is working on his site. The only defence is that either he did not know that the person did not hold a licence or he took all reasonable steps to prevent such an unlicensed person from clamping on the land.

In the Magistrates' Court the penalty is the same as it is for the unlicensed clamper, but the matter may be heard in the Crown Court, where there is an unlimited fine and up to five years' imprisonment. Where the occupier is a company, not only can the company be held liable, but, where it can be shown that a director, manager or similar officer consented to, connived in, or was negligent as to, the offence, then that individual can also be personally prosecuted.

The SIA has issued a guidance entitled *Get Licensed*, and this is available on its website (www.the-sia.org.uk). The SIA has the authority to withdraw (revoke) a licence at any time if the licence holder fails to meet the licensing requirements. It will withdraw a licence if the licence holder:

- is not the person to whom the named licence should have been issued;
- does not have the training qualifications that were claimed on application;

receives a conviction, caution or warning for a relevant offence; and/or

loses, or did not have when they applied, the right to remain or work in the UK.

It may also consider taking away a licence if:

the licence-holder breaks the conditions on which the licence was issued;

the SIA receives non-conviction information suggesting that there is a case for having the licence withdrawn; and/or

the licence-holder becomes subject to detention because of mental disorder.

There is also a power to suspend a licence, with such suspensions having immediate effect. The SIA states that it will consider suspension only where it is reasonably satisfied that a clear threat to public safety could exist if it did not suspend the licence. This usually means that a serious offence has allegedly taken place, where the licence holder has been charged but bailed. It will suspend a licence in other circumstances if it is in the public interest to do so. Since the regime became effective there have been ten revocations of vehicle immobiliser licences.

The SIA published, in December 2009, its revised Enforcement Policy Code of Practice, which details its powers and how it will use those powers to ensure proactive and effective enforcement; this is also available on its website (www.the-sia.org.uk).

The following illustrates the enforcement activity undertaken by the SIA in the year ending 31 March 2011:

Written warnings issued: 307.

Improvement notices issued: 43.

Licence revocations: 3,662.

Recent developments

Although the licensing regime has gone some way towards legitimising the vehicle immobilisation industry, and removing those who sought to take advantage of loopholes in the law, there remains a perception (which is to an extent justified) that there are still 'rogues' using wheelclamping as a way of extorting money from drivers.

This led to the Home Office publishing, on 30 April 2009, a consultation paper entitled *Licensing of Vehicle Immobilisation Businesses*, which sought the views of stakeholders and others on further controls of the industry.

As Alan Campbell MP stated in the foreword to the Paper:

'The licensing of individual vehicle immobilisers has gone a long way to reducing criminality and improving standards in the industry, but it has become clear that the existing licensing scheme does not address all the concerns the public have. There are clearly a minority of businesses indulging in unacceptable behaviour, including unclear signage and excessive fees.

'The Government intends to act to prevent abuses by some vehicle immobilisation businesses and their employees. This consultation paper is a vital step towards this. It seeks views on a range of options for controls on vehicle immobilisation businesses.'

The Paper identified the main areas of complaint as:

inadequate signage;

the size of release fees;

immediate clamping and/or towing away;

operatives refusing to identify themselves;

Wheelclamping

- operatives luring people to park, or intimidating them;
- SIA's lack of powers against these types of conduct; and
- operatives clamping blue-badged disabled users' vehicles.

The Paper proposed four options for consideration:

1. *Status quo*. No change to current requirements.
2. *A voluntary code of practice*. This would reflect the current code issued by the British Parking Association mentioned above.
3. *Compulsory membership of a Business Licensing Scheme*. This would require compliance with a compulsory code of practice through an accredited third party. This third party would decide on whether each vehicle immobilisation business met the requirements for qualification to the scheme and would be responsible for monitoring compliance. The SIA would, however, remain as the regulator.
4. *Compulsory Approved Contractor Scheme*. This would make the current voluntary scheme compulsory for all businesses in the vehicle immobilisation sector.

The Government's preferred option in the Paper was option three. The consultation period ended on 23 July 2009, and on 20 November 2009, during its last legislative session before the general election, the Labour Government published the Crime and Security Bill 2009–10. A Home Office press notice set out the wheel clamping provisions in the bill:

'Proposals within the Bill will make it mandatory for all wheelclamping businesses to be licensed under the terms of a strict code of conduct. The code will include a cap on fines, time limits on towing cars unreasonably

quickly after being clamped and set out clear instructions for putting up signs warning drivers that clamping takes place. Ministers are also looking to introduce an independent appeals process for motorists who feel unfairly penalised by firms and their employees.

'Any company which breaches the terms of their licence could lose their right to practise and face up to five years in prison or a substantial fine.'

The Crime and Security Act 2010 received Royal Assent on 8 April 2010. Sections 42–44 and Schedule 1 to the Act provide for the licensing of businesses that undertake vehicle immobilisation activities. The assumption was that once businesses were subject to a licence they could then be obligated to abide by a proper appeals process and follow a statutory Code of Practice which would set standards for signage and maximum charges. Although the Act received Royal Assent, the provisions in the Act were essentially enabling provisions – i.e. they require Regulations to bring them into force and this did not occur before the 2010 General Election was called.

The Conservative–Liberal Democrat Coalition Government that took office in May 2010 stated in its Coalition Agreement that one of the Government's transport priorities was to 'tackle rogue private sector wheelclampers'. However, it was initially not clear whether they intended to bring the provisions of the 2010 Act into force or tackle the problem in some other way. In response to an adjournment debate on 15 June 2010, the Home Office Minister, Lynne Featherstone, reiterated the Government's commitment to 'tackle the menace of rogue private sector wheelclampers'. On 17 August 2010, Ms Featherstone announced the Government's intention to introduce measures in the 'Freedom Bill' to provide for an outright ban of clamping on private land, where it is carried out by private companies. The provisions of the 2010 Act introduced by the previous government would therefore not be brought into force.

Clause 54 of the Protection of Freedoms Bill, currently going through Parliament, will effectively make it a criminal offence to clamp (immobilise) a vehicle on private land except where one has the lawful authority to do so (e.g. on behalf of a local authority, the DVLA or the police). The Explanatory Notes to the Bill have a good explanation of how this will work in practice.

The maximum penalty will be a fine of £5,000 on summary conviction or an unlimited fine on indictment. Schedule 7, Part 3 abolishes the licensing regime set up under the previous government and the uncommenced provisions of the 2010 Act (*see above*). The Bill went through the Committee Stage in the House of Commons on 17 May 2011, and currently awaits a date for the Report stage before the House of Commons.

Conclusion

There is little doubt that the regulation of those who undertake wheelclamping (or, more strictly, vehicle immobilisation) on

private land is needed. Examples of the boorish behaviour of some 'clampers' are many and varied, and it is such behaviour that the new regime was created to prevent.

- A man who broke down on a busy road pulled into the car park at a local pub. He went to find a phone box so that he could call out the RAC, leaving his 82-year-old disabled wife in the vehicle. Clampers appeared immediately and demanded that the woman move the car. Her disability prevented her from doing this, so they clamped the vehicle and demanded £80 for its removal.
- A clamper in London impounded a car without notifying the owner and then gave it to his daughter to drive.
- A hearse was clamped with a dead body in the back.
- Some clampers have demanded wedding rings, gold teeth or even sexual favours in lieu of payment.
- Clampers in Doncaster threatened to hold a mother's three-year-old daughter ransom.

See also: Parking, p.547; Vehicles at work, p.686.

Sources of further information

British Parking Association: www.britishparking.co.uk

Security Industry Authority: www.the-sia.org.uk

British Parking Association – Code of Practice for Parking Enforcement on Private Land and Unregulated Public Car Parks, Part 1: Vehicle Immobilisation or Removal: www.britishparking.co.uk/write/BPA_CodeofPractice_v8.pdf

Work equipment

Nick Godwin, Fentons Solicitors LLP

Key points

- The law relating to work equipment is found in specific statutory regulations, which impose duties on all employers. These duties relate to the suitability, selection, maintenance and use of work equipment, as well as training to be provided for those using work equipment.
- In addition to statutory regulations, employers have common law duties in respect of work equipment.

Legislation

- Employers' Liability (Defective Equipment) Act 1969.
- Personal Protective Equipment at Work Regulations 1992.
- The Workplace (Health, Safety and Welfare) Regulations 1992.
- The Lifting Operations and Lifting Equipment Regulations 1998.
- Provision and Use of Work Equipment Regulations 1998 (PUWER).
- Management of Health and Safety at Work Regulations 1999.

Provision and Use of Work Equipment Regulations 1998 (PUWER)

These Regulations form the core obligations concerning work equipment. PUWER is intended to cover all work equipment (except a ship's work equipment, which is, generally, excluded from the Regulations) and to impose general duties. HSE guidance makes it clear that PUWER must be considered in conjunction with the Management of Health and Safety at Work Regulations 1999. These Regulations impose, in particular, a requirement for employers to assess risks to the health and safety of employees and others who are affected by work carried out on behalf of the employer.

To whom do the Regulations apply?

The Regulations apply to all employers. This includes the self-employed, for the purposes of equipment that they use at work. The Regulations also apply to anyone who has control of work equipment, control of any person who uses or supervises the use of work equipment or control over the way such equipment is used.

What is work equipment?

This is defined as any machinery, appliance, apparatus, tool or installation for use at work (whether exclusively or not). This is a wide definition and extends to almost all equipment provided for use or used at work, including, for example, parts of machinery, desks, chairs, pens, trolleys and large machine presses. It may also include tools, etc., provided by the employee himself, provided they are used at work and provided the employer expressly or impliedly permits the use or is deemed to have permitted the use (see *Couzens v. T McGee & Co Ltd*). This might include knowing about but turning a blind eye to an item of equipment being used.

However, a distinction does need to be drawn between equipment used at work

Case study

Smith v. Northamptonshire County Council (2008)

The claimant was employed by the defendant council as a carer. She regularly collected a client, who was a wheelchair user, from her home to take her for day care. Wheelchair access to the client's property was by way of a wooden ramp installed by the NHS. The claimant was injured when the edge of the ramp gave way while she was pushing the wheelchair along the ramp. She sought damages for personal injury from the council, alleging, inter alia, that the council was in breach of Regulation 5(1) of the Provision and Use of Work Equipment Regulations 1998 by failing to maintain the ramp in an efficient state or in good repair.

The judge at first instance found that the ramp was 'work equipment' and equipment 'used by an employee at work' within Regulations 2(1) and 3(2) and that there had been a breach of Regulation 5(1).

The council appealed successfully to the Court of Appeal before the House of Lords upheld the decision upon further appeal. Although it was conceded that the ramp was work equipment as defined by the Regulations, the House of Lords held that work equipment must be incorporated into and adopted as part of the employer's undertaking before a breach of the Regulations could be found.

Each case turns on its own facts, and, in the present case, where the ramp was part of someone else's property, had been installed by a third party as a permanent fixture and was used most of the time by persons other than the council's employees, the council had no control over the ramp sufficient to enable it to perform the obligation as to its maintenance imposed by Regulation 5(1).

Accordingly, whilst the ramp was 'work equipment' it was not 'work equipment' used by the claimant at work for the purposes of the Regulations and there was no breach of the statutory duty by the council.

over which the employer has control and that over which it has little or no control. This was considered by the House of Lords, who upheld the Court of Appeal's decision in *Smith v. Northamptonshire County Council* (2009) (see *case study*) that a Local Authority was not liable for failing to maintain a ramp used by its employee at a person's house. For the Authority to be caught by the Regulations the ramp had to be incorporated into and adopted as part of its undertaking. It was not, so the ramp was not caught by the Regulations.

General obligations

Regulation 4 – Suitability

Work equipment must be suitable (whether in its original state or when adapted) for the purpose for which it is used or provided. Suitability is interpreted in the light of risks to health and safety of those using the equipment, not merely from the point of view of acceptability for the job. Although a heavy burden for employers, it is not absolute. By way of example, work equipment that is ostensibly suitable but only rendered unsuitable because of the manner in which it is used by an employee is unlikely to be deemed unsuitable for the purposes of the Regulations. See *Mason v. Satelcom Ltd and East Potential Ltd* (2008). However, depending on the circumstances, the employer might be in breach of Regulations 8 and/or 9, discussed further below.

Regulation 5 – Maintenance

This places an absolute obligation on employers to ensure that all work equipment is maintained in an efficient state, in efficient working order and in good repair. The significance of the absolute obligation should be understood. An employer is likely to be liable for defects in machinery or equipment, even though examination would not have revealed the defect. This underlines the importance of ensuring equipment is safe. A useful illustration is the case of *Stark v. Post Office* (2000). Mr Stark was a postman who had been provided with a bicycle to use when delivering mail. Part of the bicycle's front brake broke and lodged in the front wheel, which then locked. Mr Stark was thrown over the handlebars and injured. The Court accepted that even a 'perfectly rigorous' inspection would not have found the defect. Even so, the correct interpretation of the relevant regulation meant that the employer was still at fault. Equipment should, nevertheless, be inspected regularly and results of such inspections recorded. If machinery has a maintenance log, that log should be kept up to date.

Although the duty to maintain is absolute, it does not extend to removing transient or temporary conditions, e.g. rainwater accumulating on the step of a bus where the step is considered to be work equipment.

Regulation 6 – Inspection

When the safety of work equipment depends on installation conditions, the employer must inspect after installation and before equipment is put into use for the first time, or after assembly at a new site or location. Similarly, if the equipment is exposed to conditions that may lead to a deterioration, which could in turn lead to a dangerous situation, there must be regular inspections, as well as inspections each time such conditions occur. Results of inspections must be logged and kept until the next inspection is recorded.

Regulation 7 – Specific risks

Where use of work equipment is likely to involve a specific risk to health and safety, Regulation 7 provides that use of that equipment should be restricted to those given the task of using it and that any repairs, modifications, maintenance or servicing be restricted to those specially selected and trained to do this. It envisages removing such dangerous equipment from the general work population, reserving it for those individuals who are trained to use it.

Regulation 8 – Information and instructions

Employers must give adequate health and safety information to users of work equipment and, where appropriate, written instructions. It is the employer's responsibility to decide on which is the appropriate method of giving the information.

Regulation 9 – Training

This Regulation imposes a duty on employers to ensure that those using work equipment, and also supervisors / managers, have received adequate training for the purposes of health and safety. Whether given orally or in writing, the guidance should be easily understood by the person to whom it is being given. It should be noted that training is required for all persons who use work equipment, not just employees. Further, the requirement to train is ongoing and does not just apply on recruitment.

Specific requirements

Regulation 10 – Conformity with Community requirements
Work equipment must comply with EC requirements relating to health and safety information, instructions and kite marking.

Regulation 11 – Dangerous parts
There is an absolute obligation to ensure that effective measures are taken, in accordance with PUWER, to prevent access to dangerous parts of machinery. This Regulation is subject to the general requirement in Regulation 4, that work equipment must be suitable, but also deals in detail with requirements for guards and protection devices. The measures that employers must take are placed in a hierarchy comprising of four levels. The 'best' protection is a fixed enclosing guard; next, other guards or protection devices; followed by protection appliances, e.g. pushsticks; and, finally, the provision of training, guidance and supervision.

Regulation 12 – Protection against specified hazards
Employers must take measures to prevent or, where not reasonably practicable, adequately control, exposure of those using work equipment to risks to health and safety from various hazards, including articles falling or being ejected from work equipment, work equipment catching fire, and the unintended or premature explosion of work equipment. Regulation 12 does not apply in certain high-risk work areas where other specific regulations apply instead, e.g. Control of Asbestos at Work Regulations 1987; Noise at Work Regulations 1989; Control of Lead at Work Regulations 1998.

Risk assessments should identify hazards and assess risks likely to arise from them. Steps should then be taken to prevent the hazard arising. If this is not reasonably practicable, steps should be taken to control exposure to these hazards.

Regulation 13 – High or very low temperature
There is an absolute obligation to ensure work equipment (which includes any article or substance produced, used or stored in the work equipment) has appropriate protection to prevent injury by burn, scald or sear if it is used at a very high or very low temperature. It is anticipated that the risk should be reduced by practical engineering means, i.e. insulation or shielding. If this cannot be done, consideration needs to be given to personal protective equipment.

Regulations 14-18 – Controls
These Regulations set out detailed obligations for the use of and provision of controls of work equipment. There are separate provisions for controls that start work equipment (Regulation 14), stop work equipment (Regulation 15) and emergency stop controls (Regulation 16). Regulation 17 imposes an absolute obligation on employers to ensure that all controls for work equipment are clearly visible and identifiable and Regulation 18 deals with an employer's obligations to ensure the safety of control systems.

Regulation 19 – Isolation from sources of energy
This sets out an employer's duty to isolate appropriate work equipment from power sources to prevent risks to employees' health and safety.

Regulation 20 – Stability
Work equipment or parts of such equipment must be stabilised as necessary, for the purposes of health and safety. Stabilisation

can be effected by a number of means, e.g. fastening or clamping.

Regulation 21 – Lighting
Suitable and sufficient lighting, taking into account the work being carried out, must be provided. This means the light provided must suit the job being carried out – this may mean special or additional lighting.

Regulation 22 – Maintenance operations
Appropriate measures must be taken to ensure that work equipment can, so far as is reasonably practicable, be maintained either when shut down or in such a way as to prevent risk to the health and safety of persons carrying out such maintenance.

Regulation 23 – Markings
All work equipment must be marked in a clearly visible way, with any markings needed for health and safety reasons. This includes clear identification of controls, safe working loads and the contents of storage vessels. The purpose of this Regulation is to ensure workers know what they are dealing with.

Regulation 24 – Warnings
All work equipment must carry any warnings or warning devices appropriate for health and safety. Such warnings must be clear and easily understood.

Regulations 25–30
These deal with requirements relating to mobile work equipment, including forklift trucks and self-propelled work equipment and are discussed separately below. Regulation 25 requires employers to ensure no employee is carried on mobile work equipment unless it is suitable for carrying persons and has incorporated safety features. Regulation 26 imposes specific requirements relating to the need to stabilise work equipment ridden on by employees.

Regulation 27 imposes a requirement relating to reducing the risk of forklift trucks rolling or overturning.

Regulation 28 deals with safeguards that self-propelled work equipment should have to protect the safety of employees using it whilst it is in motion. Non-exhaustive examples include a requirement for the equipment to have a device for braking and stopping and facilities for preventing it being started by an unauthorised person. Other safeguards that may be required depend on the circumstances and environment in which the work equipment is being used and are set out in full in the Regulations.

Regulation 29 deals with safeguards that apply to remote-controlled self-propelled work equipment. Where there is a risk to a person's safety when in motion, the equipment must stop automatically once it leaves control range and incorporate features guarding against crushing or impact. Regulation 30 imposes requirements relating to minimising risks from seizure of drive shafts.

Regulations 31–35
These apply to some power presses only and impose requirements in respect of guarding, examination, reporting and keeping of information. These requirements are set out in detail in the Regulations. Any breach of the Regulations may give rise to both civil and criminal liability.

Guidance
The Regulations are accompanied by guidance. This contains some Approved Code of Practice material, as well as general information.

Other legislation
Although much of the law relating to work equipment is contained within PUWER,

there are other statutory provisions that may apply to specific types of work equipment, which may need to be considered in conjunction with PUWER. These include:

- *The Personal Protective Equipment at Work Regulations 1992.* These govern equipment worn or held by employees and provided for safety.
- *The Lifting Operations and Lifting Equipment Regulations 1998.* These Regulations deal specifically with lifting equipment.
- *The Workplace (Health, Safety and Welfare) Regulations 1992.* In some circumstances it may not be clear whether 'equipment' is, in fact, work equipment – for the purposes of PUWER – or whether it forms part of the workplace itself. In *PRP Architects v. Reid* (2007) a lift was work equipment but it was suggested that fire doors were more likely part of the workplace whilst in other cases doors and a door closure were both considered to be work equipment. If in doubt, reference should be made to these Regulations, which impose upon employers specific duties in relation to the safety of the workplace – particularly in respect of maintenance.
- *Employers' Liability (Defective Equipment) Act 1969.* This legislation imposes liability on an employer for injury to employees caused by any defect in equipment provided by the employer for use at work. If an employee suffers personal injury as a result of a defect in equipment and that defect is attributable to a third party, the injury will be deemed also to be attributable to negligence on the part of the employer (even if the defect could not have been discovered by proper examination). Although primarily liable to the employee, the employer may be able to pass on some of the costs of liability to the supplier of the equipment.

Common law

An employer owes duties to his employees to provide a safe system of work; a safe place of work; safe work equipment and safe co-workers. They are non-delegable – employers cannot absolve themselves of these duties by delegating responsibility to others (e.g. contractors).

See also: Construction site health and safety, p.177; Display Screen Equipment, p.244; Driving at work, p.250; Electricity and electrical equipment, p.256; Head protection, p.355; Health and safety at work, p.361; Ladders, p.425; Lifting equipment, p.448; Personal Protective Equipment, p.576.

Sources of further information

L113 Safe use of lifting equipment: www.hse.gov.uk/pubns/priced/l113.pdf

INDG 291 Simple guide to the Provision and Use of Work Equipment Regulations 1998: www.hse.gov.uk/pubns/indg291.pdf

L22 Safe use of work equipment: www.hse.gov.uk/pubns/priced/l22.pdf

Working time

Pinsent Masons Employment Group

Key points

- The Working Time Regulations 1998 (which implement the EC Working Time Directive into UK law) limit working hours and provide for rest breaks and holidays.
- The Business Link website (see *sources of further information*) provides further information on the Regulations.
- Employees can opt out of the 48-hour week, and other rights can be softened or extended in 'special cases' or by agreement.
- The Regulations do not apply to some sectors, or to time that is not 'working time'.

Legislation

The Working Time Regulations 1998 came into force on 1 October 1998.

The Regulations limit working hours and provide for rest breaks and minimum paid holiday rights. Night workers have special rights that are covered in the Regulations – see *'Night Working'* (p.502).

Workers

The Regulations apply to 'workers' – this includes employees, temporary workers and freelancers, but not individuals who are genuinely self-employed. Young workers are also covered by the Regulations and are protected by special rights such as greater rest break entitlements – see *'Young Persons'* (p.774).

Some types of workers are excluded from the Regulations altogether, and some are subject to separate regulation or partial exemption. The Regulations were amended from 1 August 2003 to extend working time measures in full to all workers in road transport (other than those covered by the Road Transport Directive), non-mobile workers in road, sea, inland waterways or lake transport, to workers in the railway and offshore sectors and to all workers in aviation who are not covered by the Aviation Directive.

Since 1 August 2004 the Regulations have also applied to junior doctors, with some exceptions and special rules.

Working time

'Working time' is defined as any period during which a worker is 'working, at his employer's disposal and carrying out his activity or duties'; any period during which the worker is receiving 'relevant training'; or any additional period that is agreed in a relevant agreement to be 'working time'. This can lead to uncertainty in some cases, but it is clear that working time will not usually include time spent travelling to and from the workplace and time during rest breaks. Time spent on call has been the subject of much debate in case law, which has concluded that 'on call' time constitutes working time if the employee is required to be in the workplace rather than at home, even if the worker is asleep for some or all of that time.

Unmeasured working time

The provisions relating to the 48-hour week (*see below*), night work, and minimum rest

Case study

Rawlings v. The Direct Garage Door Company Ltd (2009)

The non-payment of holiday pay to a worker who has been on sick leave can amount to an unlawful deduction from wages even where the worker has not actually requested to take the leave in question.

The claimant was employed at a company where the holiday year ran from 1 January to 31 December. He was absent on sick leave throughout the whole of 2004 and remained on sick leave until his employment was terminated on 5 April 2006. In 2004 he exercised his right to take his statutory holiday, despite being on sick leave, and received holiday pay for the 2004 period, but he failed to follow the same process for 2005 and 2006.

The claimant commenced proceedings under the Working Time Regulations (WTR), seeking holiday pay for 2005 and for 2006 (up to his resignation in April). He brought the claim as an unlawful deduction from wages claim.

The Employment Tribunal relied on the ECJ judgment in *Stringer* (*see below*) to uphold the claimant's unlawful deduction claim for holiday pay for those periods. The Tribunal found that the claimant had been unable to take annual leave during 2005 and 2006 due to illness and he was entitled to be paid holiday pay, as there had been a series of unlawful deductions from wages.

This case is one of the first Employment Tribunal decisions that has tried to deal with the conflict between the WTR, which provide that statutory holiday may only be taken in the holiday year in which it accrues, and ECJ case law, which provides that workers who are off sick and cannot take their holiday in any holiday year must be allowed to carry it over to a later date. Other cases have the opposite view. Although Employment Tribunals decisions are not binding, the Tribunal in this case decided that the employee could not take his 2005 statutory holiday entitlement as he was off sick. On this basis it allowed him, in effect, to carry it over into 2006 and to be paid in lieu on termination. There have been no Employment Appeal Tribunal decisions dealing with this issue.

periods will not apply where a worker's work is not measured or predetermined or can otherwise be determined by the worker himself.

Examples are managing executives or other persons who have discretion over whether to work or not on a given day without needing to consult the employer.

48-hour week

An employer is expected to take all reasonable steps in keeping with the need to protect health and safety, to ensure that in principle each worker works no more than 48 hours on average in each working week. Young workers may not work more than eight hours a day or 40 hours a week and, unlike the working hours of adult workers, there are no averaging provisions for young workers.

The average working time is calculated across a 17-week rolling reference period immediately prior to the calculation date. This reference period can be extended in certain circumstances such as where the worker is a new starter; where there is a relevant agreement replacing the rolling 17-week periods with fixed successive periods of 17 weeks; under a collective or workforce agreement; and where special case provisions apply to the worker.

The time spent working for any employer is included as working time, so care is needed if an employer knows or should know that an employee has more than one job. A worker cannot be forced to work more than these hours if the hours constitute 'working time'.

The Regulations allow a worker to opt out of the 48-hour week restriction by written agreement in a number of ways, including by way of an amendment to the individual's contract of employment, but it must be in writing. The opt-out agreement can last for a fixed period or indefinitely. Any opted out worker can cancel the opt-out by giving at least seven days' notice, unless the opt-out agreement provides for longer notice, which cannot exceed three months in any event. Even if a worker has agreed to opt out, he cannot be required to work excessively long hours if this creates a reasonably foreseeable risk to health and safety.

Where a worker has contracted out of the 48-hour week, the employer no longer needs to keep records showing the number of hours actually worked by the opted-out individual.

In these circumstances only a list of those who have opted out is necessary.

See below for the current position on the continued availability of the opt-out.

Rest periods

Under the Regulations, workers are entitled to regular breaks in the working day and rest periods between working days.

Employers must provide that rest periods can be taken, but there is no need to ensure they are actually taken. The rest period is in addition to annual leave and can be paid or unpaid.

The provisions can be summarised as follows:

- There should be a minimum rest period of 11 uninterrupted hours between each working day.
- Young workers are entitled to 12 hours' uninterrupted rest in each 24-hour period.
- There should be a minimum weekly rest period of not less than 24 uninterrupted hours in each seven-day period.
- Days off can be averaged over a two-week period.
- Workers who work for six hours are entitled to a 20-minute break.
- There should be adequate rest breaks where monotonous work places the worker at risk.

Special cases

Workers can be asked to work without breaks in a number of 'special cases'.

The basis of the special cases is that, if they exist, there is a reasonable need

for work to be carried out quickly in a confined period. If because of one of the 'special cases' a worker is not able to take a rest break when he would ordinarily be entitled to do so, he should be allowed to take an equivalent rest break as soon as reasonably practicable thereafter.

Also, where special cases exist, the 17-week average period for the 48-hour week can be extended to 26 weeks. These include:

- where there is a 'foreseeable surge of activity';
- where 'unusual and unforeseeable circumstances beyond the control of the worker's employer' exist;
- where continuity of service or production is needed (e.g. hospital care, prisons, media, refuse and where a need exists to keep machines running);
- where permanent presence is needed (e.g. security and surveillance); and
- where there is great distance between the workplace and an employee's home, or between different places of work.

Annual leave

Subject to certain exceptions, workers have a statutory right to a minimum of 5.6 weeks' paid annual leave. This rose from 4.8 weeks on 1 April 2009. A part-time worker is entitled to 5.6 weeks' holiday reduced pro-rata according to the amount of days they work. Where a worker begins employment part-way through a leave year, he is entitled in that leave year to the proportion of the 5.6 weeks' annual leave that is equal to the proportion of the leave year for which he is employed.

There is no statutory right under the Regulations to time off on bank holidays. The October 2007 and April 2009 increases in annual leave gave full-time workers a further eight days' holiday to address the fact that because the Regulations did not provide an entitlement to bank holidays, many employers were counting the eight public holidays against the then four-week annual leave entitlement. Note, however, there is no right to time off to be taken on bank holidays, and whether a worker can be required to work on a bank holiday is a matter for the employment contract or managerial prerogative.

Contractual holiday provisions should be checked to ensure enough holiday is given, and can also be used to fill in gaps in the Regulations, including, for example, in relation to the clawback of overpaid holiday pay when an employee leaves.

For the purposes of the statutory leave entitlement, workers are entitled to be paid a 'week's pay' for each week of annual leave. The exact way in which payment is calculated depends on a number of factors but, effectively, where a worker is paid an annual, monthly or weekly amount to which he is contractually entitled, his holiday pay will be the weekly equivalent of that amount. However, where a worker receives a varying amount of pay each week, which is not contractually provided for or agreed, a 'week's pay' must be calculated in accordance with the average amount of pay the worker received in the 12-week period prior to the date of payment. Specific pro rata rules apply to untaken holiday when an employee leaves.

The Regulations provide a right to stipulate when a worker can take his leave entitlement, including notice provisions for the employer and the employee.

Annual leave and sickness
The interaction of holiday rights and sick leave was in a state of uncertainty for some time. The issues were resolved (in part) by the ECJ decision in *Stringer v.*

HMRC. Very generally, the position now is that:

- a worker on sick leave accrues statutory holiday;
- a worker can take statutory paid holiday during sick leave; and
- if a worker falls sick during a period of pre-booked statutory holiday, they can choose whether to take the period when they are sick as sick leave or as statutory holiday.

This is a very complex area and the law is in a state of flux in certain areas.

The Government has launched a consultation to amend the Working Time Regulations to ensure that UK legislation is consistent with the Working Time Directive, as interpreted in a number of judgments of the ECJ, relating to the interaction of annual leave with sick pay, maternity pay and parental leave.

Agreements
Various parts of the Regulations can be disapplied or softened by specific agreements. This can be done by using a 'collective agreement' between an employer and trade union, a 'workforce agreement' between an employer and its workers or trade union, or 'individual agreements' between an employer and a worker.

Records
Employers must keep adequate records to show in particular whether the limits in the Regulations dealing with the 48-hour week and night work are being complied with.

The courts will expect employers to be able to show they are complying with the Regulations and policing working time.

Officers of the HSE are entitled to investigate an employer's working time practices, and can demand to see copies of its records.

Employers are not required to keep records of annual leave taken by workers and therefore there is no penalty if records are not retained. However, it is advisable for employers to keep records to ensure that there are no disputes over a particular worker's entitlement.

Enforcement

The Regulations provide a wide range of sanctions, depending on the breach in question. The limits on working time and the record-keeping requirements are enforced by the HSE or Local Authority Environmental Health Departments. They can issue 'improvement' or 'prohibition' notices, which attract unlimited fines and up to two years' imprisonment for directors if such a notice is not complied with.

Workers may present a complaint to an Employment Tribunal in connection with any failure by their employer to provide them with the relevant protections afforded by the Regulations. Where a worker is also an employee, and is dismissed as a result of exercising a right under the Regulations, his dismissal will be deemed to be automatically unfair. Employees may present a claim to an Employment Tribunal regardless of length of service.

EC proposals for change

The European Commission was obliged to review certain aspects of the European Working Time Directive, which the Working Time Regulations implemented in the UK. In June 2008, the EU Council published proposed wording for a directive to amend the Working Time Directive. The agreed position was that the opt-out would remain, but the following restrictions would apply:

- Workers will have to renew the opt-out in writing annually.
- Workers will be able to opt back in with immediate effect during the first six months of employment or up to

three months after the end of any probationary period, whichever is longer. This means that the notice period for opting back in will be two months rather than the current three months.

- An opt-out will be void if signed at the same time as the employment contract.
- An opt-out will be void if signed within four weeks of starting work. (This provision will not apply to workers who work for an employer for fewer than ten weeks in a 12-month period.)
- No worker can work for more than 60 hours a week, averaged over three months, unless permitted in a collective agreement or agreement. (This provision will not apply to workers who work for an employer for fewer than ten weeks in a 12-month period.)
- Working time plus inactive on-call time cannot exceed 65 hours a week, averaged over three months, unless permitted in a collective agreement. (This provision will not apply to workers who work for an employer for fewer than ten weeks in a 12-month period.)

In addition:

- The reference period for calculating the 48-hour week may be extended to six months, 'for objective or technical reasons, or reasons concerning the organisation of work'.
- There will be a new category of time called 'inactive part of on-call time', which counts as neither working time nor a rest period.
- Compensatory rest may be given after a 'reasonable period', rather than straight after the shift to which it relates.

However, the amended Directive is not yet in force because agreement could not be reached. In particular, several member states wanted to end the opt-out altogether, which faced fierce opposition from other member states, including the UK. In April 2009, negotiations in Europe came to an end without any agreement having been reached on the proposals. The Commission effectively recommenced the process in March 2010 by launching a fresh consultation on proposals to comprehensively review the legislation. The legal framework in this area therefore remains largely unchanged and the opt-out system remains. However, talks are expected to resume at some point in the future.

See also: Night working, p.502; Stress, p.659; Young persons, p.774.

Sources of further information

Business Link – Working Time: www.businesslink.gov.uk/bdotg/action/layer?r.l1=1073858787&topicId=1073858926&r.lc=en&r.s=tl

Comment ...

Getting better

Mark Eltringham

The quest for a proper understanding of the links between the places in which we work and our wellbeing and productivity has been ongoing for a very long time. It predates the health and safety debate as we now know it, with roots in research such as that carried out at the Hawthorne Works in Chicago in the late 1920s. The Hawthorne research has become seminal, not only in the study of productivity and wellbeing, but also in wider management thinking. When it was discovered that productivity fell back to some degree at the end of the experiments, a second interpretation was postulated; namely that the workers were not merely responding to better conditions but also to the experiment itself. They welcomed the focus on their wellbeing.

According to Koray Malhan of furniture designers Koleksiyon, this same complex relationship between design and management is still the underlying principle of health and safety in offices.

"Obviously there are elements of design in the workplace that address the safety and wellbeing of the people who work in it. At a basic level this will meet the relevant legislation for lighting, ergonomics or air quality and so on, but people like to know that their employers are paying attention to their wellbeing. Meeting legislative obligations is best seen as the foundation for a more enlightened approach to worker wellbeing.

"For example, suppliers may say that such and such a product is 'ergonomic'

Mark Eltringham is a journalist, marketeer, consultant and speaker. He has worked in the workplace and facilities management sector for 20 years in a number of roles. He lives in Cheshire, has three children and a dog. He supports Stoke City, which was seen as noble when they were completely useless, and is now seen as disreputable now that they have appeared in a Cup Final and routinely beat Arsenal.

and meets the demands of the Health and Safety and Work etc. Act and the Display Screen Equipment Regulations," says Koray. "That may be true but really the claims are meaningless unless you look at things in a design and cultural context. Ergonomics is defined as the relationship between people and their environment. That relationship is inherently a two-way thing. At the heart of it must be the belief that you are looking after people for the right reasons. It's no longer enough to try to minimise the risk of harm; you have to look at improving wellbeing and productivity."

A similar debate is apparent in the area of lighting design. According to Lee Kilminster of RS Components, there is an opportunity to look at the issue in a more sophisticated way. "The wrong approach and a tight budget can lead people to focus on accepting the norm to meet their legislative obligations," says Lee. "That's great on one level but it is the human issues that must underlie decisions about light. The office has become more of a social space, more domestic, and that has to be reflected in lighting systems. The most important thing is to create something more humane than the harsh, cold over-illumination you get in a lot of offices as a result of adopting the lowest common denominator approach. You can use light to deal with specific workplace issues not normally associated with light. For example, in open plans, one of the reasons why there can be a lot of background noise is that people are talking too loudly, as harsh direct downward lighting creates shadows across people's mouths, distorting the visual clues you get about what someone is saying and causing everybody to speak more loudly."

This problem of acoustics has become particularly acute in recent years as more and more sources of noise have been introduced to workplaces, and as space standards have changed. According to the British Council for Offices, most people in the UK work in open plan offices at workstations that are on average about 20% smaller than they were ten years ago.

> "Certain organisations, like certain people, are inherently louder than others. In many cases, noise is important for creating a buzz. Turning the workplace into a soundless crypt will be counterproductive. So it's important that the physical environment reflects culture."

We're closer to our neighbours, so we are more likely to hear them.

The problem is highlighted in a report from the Commission for Architecture and the Built Environment (CABE), which claims people enjoy a 38% improvement in their ability to perform many tasks if they work in a workplace where acoustic conditions have been optimised. The same survey also reported that people perform 16% better in memory tests, and 40% better in mental arithmetic tests, when they aren't disrupted by undue noise. Other reports go even further. A study published in the *British Journal of Psychology* has highlighted the role that 'irrelevant noise' plays, not only in disrupting work, but also in increasing stress levels and decreasing job satisfaction.

"Although we are aware of the harmful effects of noise, and much as many people claim they would like to work in enclosed offices, the cost of space and the contemporary focus on team working dictate that the open plan is here to stay as the norm for most of them," says Ann Clarke of Claremont Group Interiors. "Fortunately it is possible to reach some sort of balance between the often conflicting need for us to work in privacy but also communicate as part of teams."

"Problems and solution arise first at an architectural level," she says. "Sound is prone to bounce off ceilings and follow sight lines, so the way a building is designed can have a significant impact on noise levels in its interior. The type and shape of a building is often beyond the control of the organisations that inhabit them so, regardless of its architecture, there are several basic elements to address to deal with problems of noise in a building, including ceiling systems, sound masking systems, systems furniture, flooring and interior design.

"Design is becoming increasingly important as firms reduce the amount of space they allocate to people in the office. Something as basic as the move from bulky CRT monitors to flat screens and laptops has helped the average workstation to shrink by over 20% in the last few years, saving space and money, but with the potential for counter-productive cramming.

"And finally, there's the issue of culture," Ann continues. "Certain organisations, like certain people, are inherently louder than others. In many cases, noise is important for creating a buzz. Turning the workplace into a soundless crypt will be counterproductive. So it's important that the physical environment reflects culture. What is important is not to get rid of noise, but irrelevant noise. A good start may be to ban certain ring tones."

Changing approaches to office design are also having an impact on what we breathe. It is estimated that even in a typical office each person and their computer equipment will generate some 1,500W of energy per hour, the equivalent of a fan heater. The problem is significantly more acute in certain environments such as dealer rooms where people typically use multiple flat screens in large open plan areas with occupancy densities of above 7m^2

per person and where work can carry on around the clock.

According to Kevin Ager of DAS Business Furniture, in these circumstances the challenge is how to balance sustainable policies with the need to continue to address traditional air conditioning-related issues such as employee comfort, acceptable workplace temperature standards and changing seasonal usage patterns.

Although there have been several calls for a maximum workplace temperature standard to be officially introduced, there is – at present – no legal standard in Britain that limits the upper levels of workplace temperatures. However, according to the Workplace (Health, Safety and Welfare) Regulations 1992, the upper allowable temperature in workplaces during working hours should be 'reasonable'.

"Around half of workplaces now feature air conditioning compared with just 10% in 1994," says Ager, "but they are coming under increased pressure from changing working patterns and technology and many are still not adequately equipped to provide relief on the hottest days.

"This isn't solely an issue of changing technology and space standards," he continues. "The UK has seen a shift in HVAC energy consumption patterns from a winter peak to a summer peak. People are more concerned about being too hot in the workplace than too cold. Businesses too are worried about the impact of heat. The risks to workers include dehydration, fatigue, increased heart beat and risk of infection, dizziness, fainting; and heat cramps due to loss of water and salt. So there is a clear need to keep workers cool enough, even if it isn't yet a clearly defined legal requirement."

Workplace health, safety and welfare

Kathryn Gilbertson, Greenwoods Solicitors LLP

Key points

- The Workplace (Health, Safety and Welfare) Regulations 1992 impose duties on an employer in respect of the health, safety and welfare of the persons in a workplace.
- They do not apply to construction sites, temporary work sites and certain agricultural undertakings.
- They specify the minimum standards to be attained and employers are encouraged to exceed these wherever possible.
- All workplaces must be smoke-free.

Legislation

- Health and Safety at Work etc. Act 1974.
- Workplace (Health, Safety and Welfare) Regulations 1992.
- Health Act 2006.
- Smoke Free (Premises and Enforcement) Regulations 2006.

Workplace (Health, Safety and welfare) Regulations 1992

Regulation 5 – Maintenance of workplace and of equipment, devices and systems

These should be maintained in an efficient state, in efficient working order and in good repair. The Regulation creates an absolute duty that is not limited to what is reasonably practicable. Thus, it is no defence to an employer to show that the cause of a lift failure could not be discovered. 'Efficient' is considered from the view of health, safety and welfare and not productivity or economy. Thus steps should be taken to ensure that repair and maintenance work is carried out properly.

Regulation 6 – Ventilation

Effective and suitable ventilation shall be provided to ensure that the workplace has a sufficient quantity of fresh or purified air. Enclosed workplaces should be sufficiently well ventilated so that stale air and air which is hot or humid is replaced at a reasonable rate. Whilst it may not always be possible to remove smells coming in from outside, reasonable steps should be taken to minimise these. Air that is being introduced should be free from any impurity that is likely to give offence or cause ill health.

Regulation 7 – Temperature in indoor places

During working hours, the temperature in workplaces inside buildings shall be reasonable. Thus workers should be comfortable without the need for special clothing. Normally the temperature should be at least 16°C unless much of the work involves severe physical effort, in which case the temperature should be at least 13°C. Where a reasonably comfortable temperature cannot be achieved throughout a work room, local heating or cooling should be provided. Thermometers should be available to enable persons to check the temperature of the workplace. These do not need to be provided in each work room. Site-specific risk

Case studies

A Cheshire recycling company was fined £10,000 after a worker lost part of his leg when he was crushed by an 18-tonne truck. The victim was working in the tipping bay when he was struck by an articulated shovel loader.

The HSE prosecuted WSR Recycling Ltd after the incident, which led to the worker's left leg being amputated below the knee. The company admitted that it did not ensure that pedestrians and vehicles could move around in the bay safely.

It noted that it was foreseeable that pedestrians would be working in the same area as trucks, and so measures should have been taken to manage the risks. The site should have been properly supervised so that workers were kept away from moving vehicles.

---

assessments should be undertaken where heat stress or cold stress has been identified.

Regulation 8 – Lighting

Every workplace shall have suitable and sufficient lighting which should be, so far as is reasonably practicable, through natural lighting. Lighting levels should be sufficient to enable people to work without experiencing eye strain. Staircases in particular should be well lit, such that shadows are not cast over the main parts of the treads. Windows and sky lights should be cleaned regularly and kept free from unnecessary obstructions to admit maximum daylight.

Regulation 9 – Cleanliness and waste materials

Workplaces, including furniture, furnishings and fittings, shall be kept sufficiently clean. Waste materials should not be allowed to accumulate in the workplace except in suitable bins. Floors and indoor traffic routes should be cleaned at least once a week. Refuse should be removed at least daily.

Regulation 10 – Room dimensions and space

Every room shall have sufficient floor area, height and unoccupied space for the purposes of health, safety and welfare. Work rooms should allow enough free space for people to get to and from workstations and to move around within the room with ease. Eleven cubic metres per person is a minimum and this may be insufficient if the room is taken up with furniture etc. However, this figure does not apply to retain sales kiosks, attendance shelters etc.

Regulation 11 – Workstations and seating

A workstation must be suitable for the work being carried out in it and the person working within it. It shall provide protection from adverse weather, be arranged such that a person can leave swiftly if there is an emergency, and be provided with a suitable seat for them to carry out work whilst sitting. The workstation should allow the freedom of movement and the ability for the person to stand upright.

Seating should support the lower back and a footrest should be provided for any worker who cannot comfortably place their feet flat on the floor. Where visual display units (VDUs) are provided then employers should also refer to the Health and Safety (Display Screen Equipment) Regulations 1992.

Regulation 12 – Condition of floors and traffic routes

Floors should be of a sound construction without holes, slopes or an uneven or slippery surface so as to expose any person to a risk to their health or safety. Floors and traffic routes should be kept free from obstruction that may cause people to slip, trip or fall. Handrails should be provided on staircases except where they may cause obstruction.

Arrangements should be made to minimise the risk from snow and ice – which may involve gritting, snow clearing or the closure of some routes.

Regulation 13 – Falls or falling objects

Where there is a risk of a person falling a distance likely to cause personal injury or a person is likely to be struck by a falling object, then secure fencing or coverings should be provided. Fencing should be provided where a person may fall two metres or more. Such fencing should consist of two guard rails at suitable heights. Fencing should be strong and stable to restrain any person or object falling on to or against it. Particular care should be undertaken where roof work is being carried out.

Materials should be stored and stacked in such a way that they are unlikely to fall or cause injury. This could be achieved through palletisation, or by setting limits for the height of stacks to maintain stability.

Regulations 14, 15 and 16 – Windows

Transparent or translucent surfaces (windows, skylights, partitions, doors, etc.) shall be made of safety material. Since there is a danger of breakage, glass doors and partitions should be marked to make them obvious. Safety glass that is laminated or toughened is recommended.

All windows and skylights must be able to be cleaned safely and consideration must also be given to any equipment that needs to be used when undertaking this cleaning.

Regulation 17 – Traffic routes

Workplaces should be organised in such a way that pedestrians and vehicles can move in a safe manner. Pedestrians and vehicles should be separated to enable people and traffic to pass near each other without any collision. This is particularly important where vehicles have to reverse into service yards, on to loading docks, or where pedestrian and vehicle routes cross. The use of signage, marked routes and segregation must be carefully managed.

Regulation 18 – Doors and gates

These should be properly constructed with devices to prevent them coming off their tracks or hinges during use. Thus, any powered door should be able to be operated manually if the power fails. Doors and gates which swing in both directions should have a transparent panel positioned to enable a person in a wheelchair to be seen from either side.

Regulation 19 – Escalators and moving walkways

These should work safely, and be fitted with emergency stop controls that are readily accessible and easily identifiable.

Regulation 20 and 21 – Sanitary conveniences and washing facilities

Sufficient facilities should be provided to enable everyone at work to use them without undue delay. They should be readily accessible, kept clean, adequately ventilated and well lit. Washing facilities should have hot and cold running water, soap and a means of hand drying.

The minimum number of facilities required is:

- Up to five people – one toilet and one wash station.
- Six to 25 people – two toilets and two wash stations.
- One extra toilet and wash station for every subsequent 25 people.

For men, a mixture of toilets and urinals can be provided. See '*Toilets*' (p.669).

Regulation 22 – Drinking water
Wholesome drinking water should be provided. It should be readily accessible and clearly marked that it is suitable for drinking purposes. It should not be installed in sanitary accommodation but be readily accessible in another location. Cups or glasses must be provided unless a water fountain is used.

Regulation 23 and 24 – Clothing
Employers should provide facilities for employees to change their clothing. This would include storage facilities for personal clothing not being worn at work, as well as special clothing to be worn at work that is not taken home. As a minimum this could be a peg or hook for each worker in a changing room. The accommodation should allow clothing to dry and should be secure.

Regulation 25 – Facilities for rest and to eat meals
Rest areas should be large enough and have sufficient seats with back rests and tables for a number of workers likely to use them at any one time. The rest room shall include facilities to prepare or obtain a hot drink (such as a kettle, vending machine or canteen) as well as providing a means for heating their own food if hot food cannot be obtained in the workplace or reasonably near to it.

Health Act 2006
All workplaces must be smoke-free. See '*Smoking*' (p.653).

> *See also*: Furniture, p.340;
> Lighting, p.451; Smoking, p.653;
> Temperature and ventilation, p.666;
> Toilets, p.669.

Sources of further information

L24 Workplace, Health, Safety and Welfare: www.hse.gov.uk/pubns/priced/l24.pdf

INDG 244 Workplace, Health, Safety and Welfare: a short guide for Managers: www.hse.gov.uk/pubns/indg244.pdf

Work-related upper limb disorders

Andrew Richardson, URS Scott Wilson

Key points

- Musculoskeletal problems of the arm, hand, shoulder and neck can be found across a broad range of work activities. These are properly referred to as work-related upper limb disorders (WRULDs). Often, especially where pain in the arm occurs among computer users, some forms of WRULD are referred to as repetitive strain injury (RSI). This can be confusing because RSI is not a medical diagnosis. For clarity and to avoid confusion, only WRULDs will be referred to in this chapter.

- WRULDs is a generic term for a group of musculoskeletal injuries that affect the muscles, tendons, joints and bones, usually in the hand, arm or shoulder, which are generally caused by frequent or repetitive movement of the arms, wrists and fingers.

- ULDs can be caused by non-work activities. Employers need to ensure that the tasks they allocate to workers do not make the injury any worse.

- WRULDs can be avoided by ergonomic improvements in the workplace, i.e. improving the interface between person and machine. The job must be matched to the person.

- Employers need to make adjustments to the task, workstation, work environment and/or work organisation.

Legislation

- Health and Safety at Work etc. Act 1974.
- Health and Safety (Display Screen Equipment) Regulations 1992 (as amended by the Health and Safety (Miscellaneous Amendments) Regulations 2002).
- Manual Handling Operations Regulations 1992 (amended 2007).
- Personal Protective Equipment at Work Regulations 1992.
- Reporting of Injuries, Diseases and Dangerous Occurrences Regulations 1995.
- Provision and Use of Work Equipment Regulations 1998.
- Management of Health and Safety at Work Regulations 1999.

Dealing with WRULDs

WRULDs are musculoskeletal disorders of the arm, hand, shoulder and neck.

They can range from temporary fatigue or soreness of the limbs through to chronic soft tissue disorders such as tendonitis and carpal tunnel syndrome. Personnel can also suffer from occupational cramp. Symptoms of WRULDs include tenderness, aches, pains, stiffness, weakness, tingling, numbness, cramp and swelling.

WRULDs can be caused by forceful or repetitive activities or poor posture. They are widespread across a range of industries and jobs, but are particularly associated with the work of computer users and assembly workers.

The way that the workplace is arranged and managed can cause WRULDs or make existing medical conditions worse.

Employers have a legal duty under the Health and Safety at Work etc. Act 1974 and the Management of Health and Safety at Work Regulations 1999 to carry out risk assessments and put into place measures to prevent WRULDs and/or stop any existing medical conditions becoming worse.

The HSE advocates the following management framework to all employers for dealing with WRULDs:

- *Understand the issues and commit to action*. Both the employer and employee should have an understanding of WRULDs and should be committed to carrying out actions to prevent them. Positive leadership, with a policy on WRULDs and the necessary systems in place, will help promote a positive safety culture.
- *Create the right organisational environment*. This should foster active employee participation and involvement, establish clear lines of communication and encourage employer–employee partnerships in carrying out the following framework steps:
  - Assess the risk of WRULDs in your workplace. Managers and workers should carry out assessments in a systematic way to identify the risks and prioritise them for action.
  - Reduce the risks of WRULDs. A process of risk reduction should be undertaken using an ergonomic approach. Where possible, risks should be eliminated or reduced at source. Implementation should include the

workforce, as this makes it more effective.
- Educate and inform your workforce. The provision of education and information is vital. Training will support all aspects of this framework and should be an ongoing activity, not a one-off task.
- Manage any incidence of WRULDs. Employees should be encouraged to identify symptoms and to report them as soon as possible before they become persistent. Employers should respond quickly by reviewing the risks and introducing more effective controls. Employers must reassure employees that reporting symptoms will not prejudice their job or position. Early medical detection can stop further deterioration and help aid return to work.
- Carry out regular checks on programme effectiveness. This will ensure the framework remains effective and will improve its effectiveness.

Where workers use computer equipment, employers must comply with the requirements of the Display Screen Equipment Regulations – see '*Display Screen Equipment*' (p.244).

See also: Display Screen Equipment, p.244; Health surveillance, p.396; Workplace health, safety and welfare, p.768.

Sources of further information

INDG 171 Aching arms (or RSI) in small businesses: www.hse.gov.uk/pubns/indg171.pdf

Young persons

Pinsent Masons Employment Group

Key points

- For health and safety purposes, a young person is under the age of 18 years. A child is a person not over compulsory school age, currently 16 years.
- Employers should carry out risk assessments before employing young persons.
- General health and safety duties prevent young workers being used for work beyond their capabilities.
- Young persons should receive appropriate health and safety training.
- Young workers have particular rights under the Working Time Regulations 1998, particularly relating to rest breaks and night work assessments.
- Particular hourly rates apply to young workers for minimum wage purposes. There are different definitions of young workers for these purposes.

Legislation

- Children and Young Persons Act 1933 (as amended).
- Health and Safety at Work etc Act 1974.
- Management of Health and Safety at Work Regulations 1999.

General health and safety issues

Every employer must ensure, so far as is reasonably practicable, the health and safety at work of its employees. In particular, every employer should ensure that young persons employed by him are protected at work from any risks to their health or safety that are a consequence of:

- their lack of training and experience;
- their lack of awareness of existing or potential risk;
- their physical / psychological capacity;
- the pace of work;
- temperature extremes, noise, vibration, or radiation;
- compressed air and diving; and
- hazardous substances.

Failure to do so may result in civil liability for a breach of statutory duty on the part of the employer.

Risk assessments

An employer should not employ a young person unless he has carried out a risk assessment to ensure that all relevant hazards and consequent risks have been identified. The risk assessment must be carried out before the young person starts work.

In particular, an employer should consider:

- the fitting-out and layout of the workplace where the young person will work;
- the nature of any physical, biological and chemical agents they could be exposed to, how long they could be exposed and to what extent;
- what types of equipment they will be required to use;
- how the work and processes involved are organised;
- the need to assess and provide health and safety training; and
- risks from the particular agents, processes and work.

It should then identify the measures it needs to take to control or eliminate health and safety risks.

There is no need for a new risk assessment to be carried out every time a young person is employed, as long as the current risk assessment takes account of the characteristics of young people and activities that present significant risks to their health and safety. In all cases, the risk assessment will need to be reviewed if the nature of the work changes or if the employer has reason to believe that it is no longer valid. There are additional obligations that apply in respect of children. Employers should let the parents / carers of any children know the key findings of the risk assessment and control measures introduced before the child starts work.

Restrictions

Except in special circumstances, employers should not employ young people to do work that:

- is beyond their physical or psychological capacity;
- exposes them to substances chronically harmful to human health, e.g. toxic or carcinogenic substances;
- exposes them to radiation;
- involves a risk of accidents that they are unlikely to recognise because of a lack of experience, training or attention to safety; or
- involves a risk to their health from extreme heat, noise or vibration.

These restrictions do not apply in special circumstances where young people over compulsory school leaving age are doing work necessary for their training, under proper supervision from a competent person and provided the risks are reduced to the lowest level, so far as is reasonably practicable.

Training and supervision

Young persons should receive training when they first start a job to ensure they can do the work without putting themselves and others at risk. Employers must ensure that training is undertaken and check that key messages have been understood. Young persons should also receive training and instruction on the hazards and risks present in the workplace, and on the preventative and control measures put in place to protect their health and safety. This should include a basic introduction to health and safety e.g. first aid, fire evacuation procedures, etc.

Employers should also ensure that young people receive effective supervision.

See also: Children at work, p.143; Health surveillance, p.396; Night working, p.502.

Sources of further information

The Workplace Law website has been one of the UK's leading legal information sites since its launch in 2002. As well as providing free news and forums, our Information Centre provides you with a 'one-stop' shop' where you will find all you need to know to manage your workplace and fulfil your legal obligations. Content is added and updated regularly by our editorial team who utilise a wealth of in-house experts and legal consultants. Visit www.workplacelaw.net for more information.

workplace law
here's to the future™

The contributors

Maria Anderson, Health and Safety Consultant
maria.anderson@workplacelaw.net

Maria Anderson graduated as a Chemical Engineer and has over ten years' experience in manufacturing and quality systems, having worked in a wide range of industries. She gained her NEBOSH diploma in 2007 and has since been intensively involved in improving the health and safety culture of organisations and the development and implementation of management systems such as ISO 9001, ISO 14001 and OHSAS 18001. Maria is also an Associate Member of the Project Safety and holds both the NEBOSH Fire and NEBOSH Construction certificates.

Rob Castledine, Associate Director, Health and Safety
rob.castledine@workplacelaw.net

Rob Castledine is a qualified Environmental Health Officer with over 20 years' experience in occupational safety and health. Rob now focuses on health and safety management, including development of policy and procedures, safety auditing and assisting clients to develop their training requirements at all levels. Rob takes a lead on our training programmes, including the Corporate Health and Safety Briefing, IOSH Safety for Senior Executives and IOSH Directing Safely.

Kate Gardner, Business Manager Health, Safety and FM
kate.gardner@workplacelaw.net

Kate Gardner has more than 15 years' practical experience in both health and safety and facilities management in the pharmaceutical and biotechnology sectors. A member of IOSH, the International Institute of Risk and Safety Managers (IIRSM), Kate also sits on the East Region Committee for the British Institute of Facilities Managers, and is a qualified Display Screen Equipment assessor and NEBOSH tutor.

Colin Malcolm, Principal Consultant – Environment
colin.malcolm@workplacelaw.net

Colin Malcolm has an MA in Environmental Management, is a Full Member of the Institute of Environmental Management and Assessment (IEMA) and sits on the IEMA Full Membership assessment panel. He is experienced in supporting industrial and commercial sector clients in a breadth of environmental management and sustainability issues.

Heidi Thompson, HR Consultant
heidi.thompson@workplacelaw.net

Heidi Thompson is a Chartered Member of the CIPD, with a Masters Degree in HR Management with over 17 years' experience working in both the private and public sector for a variety of different companies, Government bodies, schools and healthcare. Heidi advises clients at strategic and operational level covering a broad range of HR issues.

Simon Toseland, Head of Health and Safety
simon.toseland@workplacelaw.net

Simon Toseland has over ten years' experience of delivering health and safety consultancy and training, is a Chartered Member of IOSH, a Registered Member of the Association for Project Safety, and a Graduate Member of the Institute of Fire Engineers. In January 2011 Simon became approved on the Occupational Safety and Health Consultants Register.

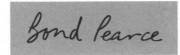

Bond Pearce LLP is a leading national business law firm providing commercial, corporate, real estate, dispute resolution and employment services to some of the UK's pre-eminent corporate and public sector organisations. The firm advises in excess of 40 FTSE 350 companies, making it one of the leading FTSE advisors outside of London.

Bond Pearce is a national leader in energy, retail, real estate and insurance and is also recognised for having one of the largest dispute resolution practices in the country, a heavyweight property, planning and environment team and full service corporate, finance and commercial capability.

The firm's growth has been based on forging strong client relationships, delivering effective business solutions and recruiting high-calibre people across the firm. Clients have access to more than 440 legal professionals, 72 of whom are recognised as 'Leaders in their Field' in the 2012 edition of *Chambers* with best of the UK status in areas such as Retail, Insurance, Energy, Local Government and Health and Safety specialisms.

www.bondpearce.com

Dale Collins, Solicitor-Advocate
dale.collins@bondpearce.com

Dale Collins is a Solicitor-Advocate and is recognised in the legal directories for his experience and expertise in the field of health and safety law. He has been a criminal advocate for over 22 years and has extensive advocacy experience in the criminal courts and before other tribunals, both from a prosecution and defence perspective, having dealt with everything from pollution of watercourses to corporate manslaughter. Dale also has an MA in Environmental Law and is an experienced lecturer.

Jagdeep Tiwana, Partner
jagdeep.tiwana@bondpearce.com

Jagdeep Tiwana is a Partner in the Regulatory Team at Bond Pearce. She has extensive regulatory experience, having successfully defended many clients facing prosecution for health and safety offences. She has also acted as a prosecutor for both HSE and Local Authorities. Jagdeep has particular expertise in dealing with serious and fatal accidents and advising companies, directors and senior employees on their potential liability. She regularly represents clients throughout the investigation, inquest and prosecution process.

BUREAU VERITAS

Bureau Veritas UK Ltd provides a broad range of consulting and laboratory services to companies operating within the built and natural environments.

Our services are deployed to a wide range of clients and sectors, including major blue chip companies, offshore operators, consumer products, and local authorities.

The built environment division closely monitors and evaluates risks relating to people working within the workplace and any other type of building. The built environment has a major influence on health, welfare, morale and productivity of employees, contractors and visitors alike. Managing these risk factors is a key issue for all organisations and new legal precedents are being set to ensure statutory requirements are met.

Bureau Veritas UK Ltd has the expert knowledge, resource and expertise to support organisations in providing a safe and healthy workplace, providing a comprehensive management solution.

www.uk.bureauveritas.co.uk

Stephen Day, Head of Fire Science
stephen.day@uk.bureauveritas.com

Stephen Day heads Fire Science at the London laboratories of Bureau Veritas UK & Ireland Consulting, having over 20 years' experience in front line fire, safety, and radiological activities. His role has given him experience in a wide range of incidents and emergencies and includes commissioning and running a UKAS accredited laboratory and health and safety auditing. He has investigated many sites contaminated with radioactivity throughout the UK as well as in Russia, advised on remediation options, and acted as technical advisor to LARRMACC.

Dave Fray, Engineering Specialist – Cranes and lifting equipment
dave.fray@uk.bureauveritas.com

Dave Fray joined Bureau Veritas UK Ltd as a Crane Engineer surveyor in September 1991. Dave is now responsible for the training and development of our Crane Engineer Surveyors and has specialist knowledge of CE marking in the cranes and lifting equipment industry. Dave works with Bureau Veritas China on a number of these projects for customers across Europe.

Sally Goodman, Technical Director
sally.goodman@uk.bureauveritas.com

Sally Goodman is Technical Director, Sustainability Services at Bureau Veritas UK Limited and is a management systems and auditing expert. Sally is involved in training, sustainable development strategy, corporate social responsibility, environmental management systems and environmental supply chain management.

Mike Lachowicz, Senior Consultant
mike.lachowicz@uk.bureauveritas.com

Mike Lachowicz joined Bureau Veritas' Sustainability Services team as a Senior Consultant in 2006, since when he has worked with assurance projects, energy and HSE management system development, implementation, training and auditing for a variety of clients. He currently specialises in the Facilities and Property Management sectors.

Craig Scott, Senior Consultant
craig.scott@uk.bureauveritas.com

Craig Scott has over ten years' experience in a wide range of noise and vibration investigations, which have included monitoring, prediction and assessment. Craig has specialist expertise in Environmental Impact Assessment, workplace noise and vibration exposure and building acoustics. Craig's role within Bureau Veritas UK Ltd also includes the provision of training and awareness courses in workplace noise and vibration exposure.

James Tiernan, Senior Sustainability Consultant
james.tiernan@uk.bureauveritas.com

James Tiernan has an MSc in Sustainable Energy and Environment, is a member of CIBSE and an experienced BREEAM, Code for Sustainable Homes and Green Rating assessor. He has also worked on numerous renewable energy studies and sustainability strategies for new and existing developments, as well as conducting carbon footprinting studies, integrated energy strategies and building energy modelling using dynamic simulation software.

Mark Webster, Principal Consultant
mark.webster@uk.bureauveritas.com

Mark Webster provides health, safety and environment support to a diverse range of clients, across differing sectors, including waste management. He has been involved with the waste management sector for over 20 years, typically for independent third party assessment of compliance and/or environmental liability.

Gillies Associates Limited provides multi-disciplinary occupational hygiene, health and safety and environmental management services to businesses in the manufacturing and service sectors, including public and private sector organisations. Our comprehensive range of services includes evaluation of legal compliance, risk assessments, design and implementation of management systems, measuring exposure (dust, fumes, noise, etc.), tailored staff training courses, and REACH business health checks and compliance services.

We pride ourselves on being a dynamic and responsive company with a practical approach, providing a cost-effective and high quality service in all aspects. We aim to work in tandem with clients to help them manage HS and E issues. Our qualified, experienced consultants offer technical expertise with a sensible approach to cost-effective management and control of risks to health and the environment.

www.gilliesassociates.co.uk

Andy Gillies, Principal Consultant
andy@gilliesassociates.co.uk

Andy Gillies is a Senior Occupational Hygienist and Environmental Consultant, with over 30 years' practical experience of EH and S, working in the steel-making and chemical industries and, since 1995 when he founded Gillies Associates Ltd, the consultancy field. Andy has MScs in both Occupational Hygiene and Environmental Management and was President of the British Occupational Hygiene Society (BOHS) from 2006-07. He is a Fellow of the Faculty of Occupational Hygiene and an Associate member and registered environmental auditor with the Institute of Environmental Management and Assessment (IEMA).

GREENWOODS
SOLICITORS LLP

We provide top quality legal advice and pragmatic solutions to our local, national and international clients.

We are proactive, we look for solutions (rather than just problems) and adopt a 'can-do' attitude. We are committed to drafting legal documents in plain English and to communicating with our clients in a straightforward manner.

To ensure our lawyers remain at the forefront of the law we run an active training programme which includes the development of legal skills, business acumen and client relationship skills.

www.greenwoods.co.uk

Robert Dillarstone, Director
rdillarstone@greenwoods.co.uk

Robert Dillarstone is Greenwoods' Managing Director. He has acted in cases both before the Court of Appeal and the House of Lords. His strength lies in his commercial, practical and personable approach to delivering solutions to clients. Robert has received consistently strong reviews from the legal directories over many years which independently analyse legal service provision.

Past reviews have referred to the team's 'excellent reputation as far afield as the City' (*Legal 500*) and to Robert as 'talented and effective' and 'never ducking an issue' (Chambers and Partners' *A Client's Guide to the Legal Profession*). The 2009 edition of *The Legal 500* reports, "'I trust them 100%', say clients of Greenwoods Solicitors LLP" and refers to Robert as an 'outstanding employment lawyer'.

Kathryn Gilbertson, Director
kgilbertson@greenwoods.co.uk

Kathryn Gilbertson heads Greenwoods Solicitors LLP's nationally renowned Business Defence team advising on all aspects of regulatory law, including corporate manslaughter, health and safety issues, food safety and trading law matters.

Kathryn, with the support of her team, has represented numerous companies in trials and inquests concerning fatal accidents, major workplace incidents and safety issues. A recognised leader in her field, she has a reputation for an innovative approach to regulatory compliance issues and for her astute commercial awareness.

Lisa Jinks, Associate
ljinks@greenwoods.co.uk

Lisa Jinks has substantial experience in all aspects of employment law, having qualified in 1992. Her all-round expertise includes corporate immigration advice, an area that Lisa built up since joining Greenwoods Solicitors LLP.

One of Lisa's key responsibilities is employmentlaw@work, Greenwoods' up-to-the minute email update service which has been a great success since its launch in 2002.

John Macaulay, Director, Head of Employment and Employee Benefits
jmacaulay@greenwoods.co.uk

John Macaulay has a wealth of experience in advising on all aspects of employment law and has spent significant time 'at the sharp end', arguing clients' cases at Tribunal.

John leads the Department's training programme, delivering seminars and workshops, both regionally and nationally, on a wide range of topics such as discrimination, data protection, workplace consultation, stress at work and cyber liability, to diverse audiences including HR professionals, owners / managers, in-house lawyers, doctors and occupational health professionals.

John recognises the importance of the human element in any employment matter; he has a degree in Psychology and is a trained counsellor, specialising in the 'person-centred' approach. His skill is therefore to deliver practical solutions that will actually work in the highly complex field of human relationships.

Dr Anna Willetts, Solicitor
aewilletts@greenwoods.co.uk

Dr Anna Willetts is dual qualified – being both a solicitor and an environmental expert. Anna had a previous career as an environmental consultant, working for environmental consultancy, White Young Green, and engineering firm, Atkins, after completing her PhD in waste management and geochemistry. As a consultant, Anna advised clients on a broad range of environmental and waste matters, including preparation of PPC permits for landfill sites and river dredging, exemptions from waste licensing regulations, surrendering site licences following contaminated land remediation, as well as groundwater, landfill gas and contaminated land risk assessments.

She also undertook technical due diligence assessments, including one for the purchase of a hazardous landfill site. Anna has prepared expert witness reports for landfill site construction appeals, hydrogeological risk assessments and planning applications.

Esther Woodhouse, Executive
ejwoodhouse@greenwoods.co.uk

Esther Woodhouse joined the Business Defence Team in January 2009 as an Executive and is currently training to be a qualified Legal Executive Lawyer.

She is a qualified Accredited Representative with over seven years' experience in representing clients at the police station who face being interviewed under caution, in often very stressful situations. In addition, she has experience attending voluntary interviews involving the HSE and other governing bodies.

Kennedys
Legal advice in black and white

Having acted for individuals and organisations involved in many of the major cases in this area, Kennedys' Health and Safety team is widely recognised as a leader in its field. The team is ranked number one in the UK for health and safety in the current edition of the *Legal 500*, with a number of its lawyers singled out for commendation. The wider Kennedys firm has traditionally been known as a leading insurance-driven commercial litigation practice, but is also recognised for skills in the non-contentious commercial field, particularly within the insurance, construction and transport industries. In addition, it has a fast-growing reputation for its work in employment law and the healthcare and insolvency sectors.

www.kennedys-law.com

Susan Cha, Solicitor
s.cha@kennedys-law.com

Susan Cha advises and defends companies and individuals in police and HSE investigations and prosecutions arising out of work-related accidents and incidents. She also has extensive experience of public inquiry work.

Sally Cummins, Solicitor
s.cummins@kennedys-law.com

Sally is a Solicitor in the Health and Safety Group. She has acted for public and private bodies on contentious and non-contentious health and safety and regulatory matters. She has experience of advising on compliance with health and safety legislation, preparing for administrative and statutory appeals, defending companies in regulatory investigations and attending coroners' inquests. Sally also has experience in conducting health and safety reviews.

Daniel McShee, Partner
d.mcshee@kennedys-law.com

Daniel McShee specialises in disaster litigation, public inquiries, criminal law in industry and health and safety prosecutions and enforcement actions. He has a particular interest in the law of corporate manslaughter, having been involved in a number of such actions, and writes and presents regularly on this and other related topics.

Karen Patterson, Solicitor
k.patterson@kennedys-law.com

Karen Patterson assists in defending companies and individuals in regulatory investigations and prosecutions. She also advises and assists companies with various commercial and employment litigation matters.

David Wright, Partner
d.wright@kennedys-law.com

David Wright specialises in advising organisations and individuals who are under investigation or being prosecuted by the authorities in relation to work-related accidents. He presents and writes regularly on health and safety and other regulatory issues and also acts for companies and individuals in relation to commercial and employment-related disputes.

Pinsent Masons

Pinsent Masons LLP is one of the most highly regarded property law firms. The firm's strategy is to be recognised as the pre-eminent legal advisor serving the needs of the energy and utilities, financial services and insurance, government, infrastructure and construction, manufacturing and engineering, real estate, technology and facilities management industries. The team was a finalist in the category of Real Estate Team of the Year at The Lawyer Awards 2008. The firm ranks in the UK top 15 law firms and is one of Europe's leaders. Clients benefit from the depth of industry knowledge of the firm's teams of specialist lawyers who provide a comprehensive service in health and safety, employment, PPP/PFI projects, dispute resolution (property and commercial), commercial property, planning, environment and sustainability, corporate and commercial, data protection, pensions, insolvency and taxation.

www.pinsentmasons.com

Leyton Bell, Solicitor
leyton.bell@pinsentmasons.com

Leyton Bell is a Solicitor in the Property Litigation team and advises upon a wide range of contentious property matters. These include landlord and tenant disputes, freehold issues, possession proceedings and lease renewals. He has acted for corporate investors, large land owners, Local Authorities and educational institutions.

Kevin Boa, Partner
kevin.boa@pinsentmasons.com

Kevin Boa is a Partner in the Property Group and is qualified in England and Scotland. He specialises in advice on commercial property transactions and projects. He has particular experience in development, regeneration and energy work and he also advises landlords, tenants, buyers and sellers of commercial property.

Anna Cartledge, Associate
anna.cartledge@pinsentmasons.com

Anna Cartledge is an Associate in the Pinsent Masons LLP Planning and Environment Team (one of the UK's largest P&E practices) and is based in Birmingham. Anna specialises in planning, highways and compulsory purchase work for a range of commercial clients including retailers, developers, landowners, banks, local authorities, housing associations and government agencies. She advises on all aspects of CPO, highways stopping-up procedures, drafting /negotiating Section 106/38/278 planning / highways agreements and planning conditions, affordable housing, listed buildings, planning appeals and judicial review.

Helen Nicholson, Solicitor
helen.nicholson@pinsentmasons.com

Helen Nicholson is a Solicitor in the firm's Planning and Environmental team. She studied at Wadham College, Oxford before training at a leading national firm and now specialises in both planning and environmental matters for both the public and private sectors.

Jonathan Riley, Partner
jonathan.riley@pinsentmasons.com

Jonathan Riley is a Partner in the firm's Planning and Environmental Unit. He trained at a leading city firm and has extensive experience as a specialist planning solicitor acting for both the private and public sectors. He formerly combined private practice with being a lecturer in law at Keble College, Oxford.

Michael Smith, Senior Associate
michael.smith@pinsentmasons.com

Michael Smith is a Senior Associate in the firm's specialist Property Litigation team, forming part of the Property Group. He advises on all aspects of property law with an emphasis on dilapidations claims, opposed and unopposed lease renewals, telecommunications-related property matters, forfeiture and possession actions, as well as advising on contentious issues arising in all types of commercial transactions. He currently acts for a wide variety of clients, including government departments, listed companies, property developers, property investments companies and banks.

Melissa Thompson, Senior Associate
melissa.thompson@pinsentmasons.com

Melissa Thompson is a Senior Associate in the Property Litigation team at Pinsent Masons. She regularly advises in all areas of property litigation, including landlord and tenant matters, trespass and possession. She acts for retailers, banks, developers and property companies. Melissa qualified as a Solicitor Advocate in June 2006.

Stuart Wortley, Partner
stuart.wortley@pinsentmasons.com

Stuart Wortley leads the firm's national Property Litigation team comprising lawyers in Birmingham, Leeds and London. He worked from the Leeds office for nine years but is now based in London. He advises on all aspects of property law (in particular cases involving landlord and tenant issues) but concentrates on resolving disputes and finding solutions to property-related problems. He acts for government departments, banks, property developers, property investment companies and corporate clients.

Scott Wilson is part of URS Corporation, a worldwide leading provider of engineering, construction and technical services for public agencies and private sector companies around the world. URS Scott Wilson offers a comprehensive portfolio of professional services, access to a broad range of professional and technical resources, and a network of offices throughout the world. As a leading provider of multi-disciplinary consultancy services worldwide, URS Scott Wilson offers comprehensive advice and support on all building related topics. By providing asset management advice such as strategy planning, feasibility studies, space planning, project management and health and safety advice, URS Scott Wilson enables the client to manage their assets in the most economical and beneficial way for their business. URS Scott Wilson has extensive experience of providing services under the CDM Regulations, as well as general health and safety advice to clients such as audits, policies, procedures, asbestos management plans and workplace inspections. URS Scott Wilson's wide range of services also includes considerable experience in project management where the company has been involved in projects ranging from minor works to significant new developments such as Portsmouth's Spinnaker Tower and London's Crossrail, the largest infrastructure project in Europe.

www.URS-scottwilson.com

Andrew Richardson, Associate, Health and Safety Services
andrew.richardson@scottwilson.com

Andrew Richardson is an Associate in Health and Safety Services for URS Scott Wilson. He is a Chartered Civil Engineer with almost 40 years' experience of designing and project managing an extensive range of projects including commercial, industrial, military and historic buildings and infrastructure schemes such as highways and railways. On moving into health and safety ten years ago, he carried out the role of Planning Supervisor under the CDM Regulations. He holds a NEBOSH General Certificate in Occupational Safety and Health and is a Technician Member of IOSH. At URS Scott Wilson, as well as providing services under the CDM Regulations, he is currently the Health and Safety Manager for the company's 150-strong team working on the Crossrail project in London. He has also developed and presented a number of safety-related courses, including CDM and Working at Height. He provides health and safety consultancy advice to many respected companies and organisations.

shoosmiths

Shoosmiths is amongst the UK's largest national law firms offering you great value and a readily accessible, personal service through a network of nine offices. We can help you with advice on legal areas including construction, health and safety, environment, crisis management, employment, pensions, property, dispute resolution, corporate, commercial, insolvency and taxation. Our commitment to safety is evidenced by our winning the RoSPA award for health and safety performance in the Commercial and Business Services Industry Sector for the third year in succession.

www.shoosmiths.com

Hayley Saunders, Associate
hayley.saunders@shoosmiths.co.uk

Hayley Saunders is an Associate in the Shoosmiths Regulatory team and is experienced in all types of regulatory law, including health and safety / manslaughter offences, trading standards, environmental, food and transport law. She advises and represents companies and individuals facing prosecution, as well as advising clients on drafting business compliance documents. Hayley also conducts appeals against improvement notices, advises clients during HSE and PACE interviews and attends inquests. She is also a member of the Health and Safety Lawyers Association (HSLA).

Sophie Wilkinson, Associate
sophie.wilkinson@shoosmiths.co.uk

Sophie Wilkinson is an Associate at Shoosmiths. She specialises in advice on environmental aspects of property and corporate transactions, waste management licensing and compliance with waste legislation, water law, REACH and other specific environmental legislation. Her recent experience includes carrying out environmental due diligence on the acquisition of an environmentally sensitive business, advising on environmental liability transfer clauses and remediation provisions on the sale of a contaminated site and advising on whether materials under a supply contract constituted waste and the consequences of the requirements under EU and UK law. She has also got experience in relation to compliance with asbestos legislation, energy performance certificates, water law and regulation. Sophie is a member of the United Kingdom Environmental Law Association.

Guy Willetts, Partner
guy.willetts@shoosmiths.co.uk

Guy Willetts is a Partner in the Property Litigation team at Shoosmiths. He qualified in 1988 and spent 13 years in practices in the City of London, seven of which were spent as Head of Property and Environmental Litigation at Theodore Goddard, before joining Shoosmiths as a Partner in 2001.

He has been involved in high profile environmental law cases concerning dealings with regulatory authorities as well as civil nuisance claims contamination claims and toxic torts. He has also been involved in the regulatory side of significant corporate deals involving potentially contaminating processes such as the acquisition of coal fired power stations.

Visor Consultants (UK) Ltd.
180 Piccadilly, London, W1J 9HF
E: info@visorconsultants.com
W: www.visorconsultants.com

Visor Consultants are specialists in Crisis management (CM), Business Continuity (BC) and Operational Risk Management (ORM). We were created in 1995 and since then have delivered many highly cost saving and successful projects, workshops, exercises and thought leadership sessions. In particular, we have supplied key members to the following groups:

- BC Standard Working Party (BS 25999).

- UK Strategic Defence Review (IPPR).

- UK Cabinet Office / BSI Steering Group for CM guidance (PAS 200 - launched 29 September 2011).

Visor Consultants have been able to support many organisations to prevent chaos in a crisis and increase their overall resilience. We can quantify your various risks and hazards; advise you on which pose the greatest threats; assist you to eliminate - or at least contain them and help develop your CM / BC skills. We can also guide you on running exercises, speaking to the Media when a crisis rapidly unfolds, dealing with stress & trauma, risk mitigation, coaching your CM(s) and help to design your resilience overall strategies to make you more competitive. Our consultants have, so far, won four prestigious BC awards.

Our clients range from departments in the UK Government to some of the most famous names in global business. We also specialise in bringing new ideas, structures and procedures to a range of medium to small organisations in all business sectors.

www.visorconsultants.com

Other contributors

Anderson Strathern LLP: Lizzy Campbell
www.andersonstrathern.co.uk

Anderson Strathern LLP is a dynamic and progressive full service law firm. The firm is focused on its clients' needs and achieving excellence in terms of the quality of advice and its overall service. Anderson Strathern's impressive client base includes commercial, heritage and public sector clients. The firm has recognised strengths in all aspects of property, corporate services, dispute resolution, employment, discrimination and private client management.

Its parliamentary and public law team provides a unique service and is relevant to many areas of business. The firm has the largest number of Accredited Specialists of any Scottish law firm and was the only new appointment to the Scottish Government Legal Framework Agreement Panel.

Anderson Strathern is a member firm of the Association of European Lawyers. While much of its client work has a Scottish focus, a significant number of the firm's solicitors are dual qualified in England as well as Scotland and the firm acts on behalf of clients in transactions throughout the UK, while the firm's Employment Unit represents employer clients in Employment Tribunals in Scotland, England and Northern Ireland.

Lizzy Campbell is a Senior Solicitor in the Employment Unit. Lizzy advises both employer and employee clients on a wide range of employment issues including unfair dismissal, TUPE and performance management as well as negotiating Compromise Agreements.

Lizzy is also involved in representing clients in the Employment Tribunal. Lizzy regularly delivers training on employment law topics, including, most recently, sessions on performance management and managing disciplinary / grievance investigations. Lizzy recently obtained the CIPD Certificate in Training Practice.

Berwin Leighton Paisner: Marc Hanson and Jackie Thomas
www.blplaw.com

Berwin Leighton Paisner is a leading law firm based in the City of London. We pride ourselves on our superb client base including many leading companies and financial institutions. We strive to lead the market in the excellence of our service delivery.

We are well known for our extraordinary success in developing market leadership positions within the real estate, corporate and finance areas. We distinguish ourselves by working with our clients in creative and innovative ways to achieve commercial solutions.

Marc Hanson is Head of the Commercial Construction Team, specialising in all aspects of facilities management law, from contract drafting to dispute resolution. He has extensive experience of advising property owners, contractors, institutions and public authorities in connection with major domestic and international projects.

Jackie Thomas specialises in all aspects of employment law, and advises on both contentious and non-contentious employment law issues, including executive service contracts, discrimination and issues arising from the termination of employment.

BIFM: Ian R. Fielder
www.bifm.org.uk

Ian R. Fielder joined the British Institute of Facilities Management (BIFM) as CEO in April 2004. His career in workplace management spans over 31 years and covers both the private and public sectors. His involvement in the industry has included 13 years' working in the NHS, eight years with IBM and then over ten years with Procord / Johnson Controls.

Ian is an active board member of Global FM, an association of associations, set up in 2006 to promote facilities management internationally. Ian is a Fellow of the Royal Society of Arts (FRSA) and is also a Board Adviser of the Asset Skills Council, representing the FM industry.

Blake Lapthorn Tarlo Lyons: Michael Brandman
www.bllaw.co.uk

Blake Lapthorn Tarlo Lyons provides the full range of legal services needed by companies and organisations, with an industry-focused approach through specialist industry sector groups. Combining specialist expertise with a strong commercial approach, the firm delivers results for clients, whether they are multinational companies, owner-managed businesses, government agencies or not-for-profit organisations.

The firm's size and locations gives it the breadth of experience and the depth of resources to provide the highest levels of service and the capacity to handle large and complex projects at very competitive rates. As well as the full range of corporate and commercial services, the industry sector group draws together leading advisors from different legal disciplines, to provide targeted advice built on an understanding of the issues faced by a particular sector.

Michael Brandman joined Blake Lapthorn Tarlo Lyons as a Partner in 1990.

He specialises in property litigation:

- advice and implementation of landlord's remedies for tenant default, including distress for rent, forfeiture and relief against forfeiture, statutory notices and fixed charge demands under the Landlord & tenant (Covenant) Act, section 146 Law of Property 1925, statutory demands under the Insolvency Act, applications for restraining orders; and

- advice on tenant insolvency, dealing with liquidators and receivers.

BPE: Lisa Gettins, Sarah Lee and Heyma Vij
www.bpe.co.uk

BPE is a client-focused commercial law firm. We offer a full range of services including Corporate, Commercial Property, Commercial Litigation, Commercial, Science and Technology, Employment, Construction and Engineering, Lender Services, Private Client and Personal Injury.

Lisa Gettins has 14 years' experience specialising in employment law, and advises on all aspects of contentious and non-contentious work. During this time, Lisa has worked in-house in manufacturing, for a firm specialising in public sector work and for an international law firm. She has extensive experience advising private and public sector employers, particularly in relation to TUPE, defending high value discrimination claims, strategic advice on business restructuring, and trade union issues, ranging from recognition to strike action.

She is an experienced Tribunal advocate in both Tribunal and the EAT and has also dealt with a number of Court of Appeal cases, including reported cases in the fields of whistleblowing, constructive dismissal, sex discrimination and TUPE. Lisa is also CIPD qualified, regularly dealing with policy and consultation papers on behalf of CIPD.

Before joining BPE Solicitors LLP, Sarah Lee worked for an international law firm based in Birmingham. She is a very down-to-earth lawyer, who quickly gains the confidence of senior executives with her excellent communication and client handling skills. Sarah is well versed in all aspects of employment law and has successfully defended national clients in a wide range of Tribunal claims, including complex whistleblowing, discrimination and unfair dismissal. Her first class negotiation skills have also facilitated a number of successful commercial settlements. Sarah is equally at home providing practical advice to clients on day-to-day employment issues and more strategic advice, such as redundancy and restructuring.

Heyma Vij qualified with BPE Solicitors LLP having gained extensive legal experience with a city firm, and now advises upon both contentious and non-contentious matters. Having dealt with a number of complex Employment Tribunal claims, Heyma has developed a specialism in dealing with national minimum wage and working time issues. Heyma is also involved with advising on the implications of outsourcing and TUPE queries whilst working closely with the corporate team during acquisitions and disposals to provide reliable, effective support.

British Parking Association: Kelvin Reynolds
www.britishparking.co.uk

The British Parking Association (BPA) is the largest Professional Association in Europe representing organisations in the Parking and Traffic Management Industry. As the recognised authority within the parking industry, the BPA represents, promotes and influences the best interests of the parking and traffic management sectors throughout the UK and Europe. It provides its members with a range of benefits all aimed at helping the professional in their day-to-day work.

Kelvin Reynolds is Director of Operations and Technical Services at the BPA.

Kelvin joined the BPA in March 2004 where he took responsibility for managing the Safer Parking Scheme as well as the Association's development of technical services to its expanding membership. Kelvin has directed the BPA's involvement with the Government's development of Civil Parking Enforcement and the review of the Blue Badge Scheme for disabled people. He is also leading the development of the BPA's Approved Operator Scheme and its Codes of Practice and the BPA relationship with the DVLA. He is currently a member of the DfT Traffic Signs Review –Traffic Signs and Enforcement Group, and contributes to the DfT Blue Badge Reform Programme. He also leads on the development of The Parking Forum, a BPA inspired multi-agency 'think-tank' which develops collaborative statements about parking policy and provision in the UK.

From Parking Services Manager at the City of London in the 1980s and 90s, through to Transport and Infrastructure Manager at Bluewater, Europe's largest retail and leisure destination, with 13,000 parking spaces of its own, Kelvin brings a wealth of knowledge and experience to his role. He was a member of the BPA Executive Council in the 1990s too and has been a major contributor to the successful development of the BPA's Parkex International. He is a Judge in the annual British Parking Award.

Building Compliance Associates: Gavin Miller
www.easycompliance.co.uk

Gavin Miller is the owner of Building Compliance Associates Ltd, with offices in London and Edinburgh. He has over 17 years' practical experience in industrial and facilities management, energy and sustainability consultancy, health and safety management and compliance services.

A Chartered Energy Engineer and Chartered Environmentalist, Gavin provides consultancy support in all elements of building compliance legislation to clients in all sectors, with clients including the MoD, property management companies, retail organisations and large industrial sites. Gavin's breadth of experience means he enjoys working with clients at all levels, from boardroom presentations to practical site work.

· ·

Butler & Young Group: Dave Allen and Steve Cooper
www.byl.co.uk

Dave Allen is Director at Butler & Young Group and is responsible for setting technical policy in the group and ensuring a quality service is maintained through technical and procedural auditing. He has 25 years' experience in Building Control, dealing with projects as diverse as towers in Canary Wharf and listed barns in Herefordshire. He is actively involved in national building control working groups and has recently been involved in the implementation of a new IT system at Butler & Young.

Steve Cooper is Standards and Warranty Manager for Butler and Young Ltd and is a member of the Association of Building Engineers. He has over 25 years' experience in the construction industry, having spent nearly all of that time in Building Control. During this time he has been employed in both the private and public sector and has experience in dealing with the full spectrum of building work, from the refurbishment of historic buildings to modern methods of construction.

Charles Russell: Nicola McMahon
www.cr-law.co.uk

Over 100 years of Charles Russell's diligent client service has given it a unique position as a firm which successfully combines traditional values with the progressive practices of a modern commercial firm. The result is a firm which enjoys a refreshing diversity of work. Through the quality, intelligence and commitment of our staff, we have achieved a balance which offers real benefits to our clients whether they are seeking legal advice for themselves, their family, or on behalf of their business. We increasingly find that clients require a full range of legal services and so we are committed to providing legal services to both commercial and private clients. After all, where there is commerce there are people.

Having qualified into the Charles Russell Employment and Pensions Service Group in 2008, Nicola McMahon advises on all aspects of employment law, both contentious and non contentious, and corporate immigration matters.

CMS Cameron McKenna: Jan Burgess
www.cms-cmck.com

CMS Cameron McKenna is a full service international law firm, advising businesses, financial institutions, governments and other public sector bodies. The firm has about 200 partners and employs over 1,600 people. CMS Cameron McKenna has a strong network of offices throughout the UK, CEE and Western Europe. Our lawyers have strong expertise in many legal areas including facilities management, construction, health and safety, projects and project finance, real estate, environment, financial services, corporate, energy and natural resources, insurance and reinsurance, technology, life sciences, intellectual property, human resources, pensions, competition, European law, arbitration and litigation.

Jan Burgess is the Partner for Health and Safety for the firm, and a Solicitor Advocate practising health and safety law in England, Wales and Scotland. Her client base consists of a large number of multinational companies, many of them engaged in the highly regulated oil and gas industry in the North Sea. Her recent experience includes a number of high profile matters – including fatalities both onshore and offshore. In addition to representing companies in health and safety prosecutions and fatal accident inquiries / coroners' inquests she and her large team of dedicated health and safety lawyers provides regular advice on current and forthcoming legislation in this area, along with assisting with health and safety regulatory audits, drafting of policy and procedures and assisting multinationals with rolling out global health and safety standards.

Dawson Asbestos Consulting Ltd: Mick Dawson
www.dac-asbestos.co.uk

For over 20 years Mick Dawson worked for some of the larger asbestos consultancies in the UK. At the beginning of 2010 he formed his own consultancy, offering bespoke advice and techncial expertise to the consulting, contracting and dutyholder sectors.

He is a Chartered Environmentalist and holds the BOHS Certificate of Competence in Asbestos, as well as being a tutor and examiner for the BOHS proficiency modules and S301 oral exam.

DLA Piper: Jonathan Exten-Wright
www.dlapiper.com

Jonathan Exten-Wright practises employment law on behalf of public and private companies. He is experienced in senior executive issues and boardroom disputes; injunctive proceedings; in central and local government transactions; contracting-out / outsourcing; private finance initiatives, mergers and acquisitions, and sales of businesses; trade union recognition and negotiations; works councils and collective consultation; redundancy and change programmes; contract variation; and discrimination. In addition Jonathan advices on partnership and LLP disputes. Jonathan is also involved in pan-European labour law support. Clients include various well known financial service organisations, hoteliers, telecom and media organisations and public sector bodies.

Dundas & Wilson LLP: Mandy Laurie
www.dundas-wilson.com

Dundas & Wilson is a commercial law firm highly rated by clients as well as by *Chambers* and *Legal 500* directories. Our clients value our determined focus. They rely on the quality of our advice and instruct us to oppose, work with or on panels alongside major London firms. They choose us time and again because we work tirelessly to deliver the right results at the right value.

Mandy Laurie is a Partner in Dundas & Wilson's employment team and has considerable experience in representing both private and public sector clients throughout the UK. She is described as an inspirational solicitor, a unique 'all-rounder', who has an incredibly engaging manner with clients. Excellent in advocacy, she is viewed as being very practical and commercially focused. Mandy regularly gives advice on potentially contentious issues and problems with an international aspect, and has a particular interest in discrimination and equality issues.

Fentons LLP: Nick Godwin
www.fentons.co.uk

Fentons Solicitors LLP is one of the country's leading claimant personal injury firms handling complex and substantial litigation for accident victims nationwide.

Fentons employs 150 lawyers dedicated to pursuing compensation claims on behalf of victims who suffer injuries in all forms of accident. The firm has a specialist team of highly qualified lawyers to investigate and pursue claims arising from accidents in the workplace, with long-standing experience in advising on all aspects of health and safety issues.

Fentons plays an active role in drafting codes of best practice and provides training in the public and private sectors. The firm enjoys a national profile and campaigns for improvement of health and safety standards in the workplace whilst actively supporting numerous charities dedicated to victims of injury.

Nick Godwin is an Associate Solicitor in Fentons' Serious Injury team. He has many years' experience advising individuals and companies on health and safety issues, with particular expertise in all aspects of employers' liability, including workplace accidents and industrial disease matters.

Furniture Industry Research Association (FIRA): Phil Reynolds
www.fira.co.uk

For over half a century, the Furniture Industry Research Association (FIRA) has driven the need for higher standards through testing, research and innovation for the furniture and allied industries. New and better materials, improved processes and appropriate standards have been developed to enhance the quality of furniture and assist manufacturers and retailers to become more competitive. Information on our members' products can be found on the FIRA website. A non-government-funded organisation, FIRA is supported by all sections of the furniture industry, ensuring ongoing research programmes that bring benefits to all and at the same time providing the Association with the influence and capability to help shape legislation and regulations.

Phil Reynolds is the Principal Technical Manager at FIRA. Phil is the Convenor of both CEN TC207 WG 5 Contract Furniture, and WG9 Common test methods for furniture, two of the European committee developing standards for furniture, and Chairman of BSI FW0/2, the equivalent British Standards Committee. Phil is also a Director of FIRA-CMA, offering furniture testing services in Hong Kong and the far east.

Fork Lift Truck Association: David Ellison
www.fork-truck.org.uk

The Fork Lift Truck Association is the UK's independent authority on forklift trucks and your first port of call for information and advice on forklift trucks and materials handling issues.

The FLTA Membership comprises the UK's most respected forklift truck dealers, manufacturers, suppliers and training companies. Our Members adhere to a strict Code of Practice and are vetted to ensure the highest standards of professionalism, safety and customer service.

Alternatively, if you own or operate a forklift truck, membership of the Associations Safe User Group (SUG) provides you with all the relevant information you need to keep your business operating safely, legally and productively, in clear, concise terms.

David Ellison served for 25 years as a logistics officer with the British Army. A specialist in road transport and distribution technology, he saw service in Europe, North, South and Central America, Africa and the Middle East. This was followed by a short spell as a professional trainer before joining the Fork Lift Truck Association in 2000. As Chief Executive, he has been heavily involved in all aspects of this expanding trade association, providing guidance and advice to its many members as well as government officials, business and the public.

Greta Thornbory, Occupational Health Consultant
www.gtenterprises-uk.com

Greta Thornbory is an Occupational Health and Educational Consultant with over 30 years' experience in OH practice and teaching. During that time she has worked with government departments, professional bodies, pharmaceutical, educational and other companies, including several multi-nationals, on a variety of occupational health and safety projects. She also worked for the Royal College of Nursing for 12 years as a senior lecturer and programme director of both occupational health and continuing professional development, during which time she was responsible for many of the 95 Nursing Update programmes for the learning zone on BBC TV and was the RCN representative on the UN Environment and Development Round Table on Health and Environment.

She is now Consulting and CPD Editor of *Occupational Health* and is responsible for commissioning and editing the monthly multidisciplinary CPD articles and resources for professional updating purposes. She is the co-author with Joan Lewis of *Employment Law and Occupational health: a practical handbook* first published in 2006 (2nd ed. 2010) and edited *Public Health Nursing*: a textbook for health visitors, school nurses and occupational health nurses, published in July 2009.

GVA: Alex Stevens
www.gva.co.uk

GVA is a multi-disciplinary practice of Chartered Surveyors with a consultancy bias. There are approximately 1,200 personnel and we have offices throughout the UK. Our strength is the size and reach of our regional offices, which is unique among the larger surveying practices.

Alex Stevens is a Senior Director in the Lease and Rating Consultancy Department of GVA. He has over 30 years' of rating experience, written a number of articles for the Estates Gazette and other professional magazines, amd has successfully appeared as an Expert Witness before the Land's Tribunal on several occasions. He specialises in Central London offices, data centres and utilities.

Heating and Ventilating Contractors' Association: Bob Towse
www.hvca.org.uk

The HVCA is the UK's leading trade association for the hvacr industry, and was established in 1904. Bob Towse is Head of Technical and Safety for the Heating and Ventilating Contractors' Association (HVCA). He took on this role in 2001 and is responsible for technical and health and safety policy for the Association. Prior to joining the HVCA, Bob was technical manager and field operations manager for CORGI, the Council for Registered Gas Installers. Bob is a member of the Council of the Institution of Gas Engineers and Managers (IGEM).

Hogan Lovells: Hugh Merritt
www.hoganlovells.com

Formed through the combination of two top international law firms, Hogan Lovells has more than 40 offices around the world, including associated offices. With a presence in the world's major financial and commercial markets, we are well placed to provide excellent business-oriented advice to our clients locally and internationally.

Hugh Merritt is an Associate in the Financial Services Team at Hogan Lovells International LLP. He advises on all aspects of financial services law and regulation, including products, day-to-day compliance issues and commercial and corporate transactions in the financial services sector.

IBB Solicitors: Andrew Olins
www.ibblaw.co.uk

Andrew Olins is a Partner and Group Head of the commercial property team at IBB Solicitors. He specialises in resolving disputes in relation to the buying and selling of development sites where overage provisions, covenants and easements are in issue. He also resolves landlord and tenant disputes, including lease renewal proceedings, dilapidations claims, rent review disputes and other issues arising from substantial property portfolios.

He is widely published, particularly on landlord and tenant matters. He is a Professor of Land Law at Brunel University where he tutors third-year law undergraduates part-time. He is an ADR Group Accredited Mediator and is also a member of the Property Litigation Association.

Loch Associates: Sophie Applewhite, Chloe Harrold and Pam Loch
www.lochassociates.co.uk

Loch Associates Employment Lawyers provides commercial employment legal advice and HR services to employers and executive employees. Based in the South East with offices in London and Tunbridge Wells, we act for clients right across the UK in a variety of sectors.Our specialist knowledge and skills in this complex legal area and our commercial approach provide our clients with a valuable insight to make the right decisions. We provide advice to companies and executive employees in a wide range of sectors, including professional services, manufacturing, outsourcing, leisure, PR and Marketing, financial services and IT, acting for small businesses and charities through to companies employing many thousands of employees.

Sophie Applewhite is an Associate Employment Lawyer. She advises on both contentious and non-contentious employment law issues and represents clients at Employment Tribunals. Her experience includes advising both employers and employees on unfair dismissal, discipline and grievance issues, redundancy, compromise agreements and employment contracts.

Chloe Harrold is a dual qualified English Employment Lawyer and New York Attorney. Chloe's employment law experience includes advising both employers and employees on a variety of employment law areas such as redundancy, restructuring, unfair dismissal and discrimination and also includes advising employers in relation to incentive and benefit arrangements.

Founder of Loch Associates Employment Lawyers, Pam Loch is a dual qualified Scottish and English Employment Lawyer with extensive experience in contentious and non-contentious employment matters, having acted for employers and employees advising on all aspects of employment law in England and Scotland. She advises plcs, private companies, charities and individuals. Her experience is extensive but includes acting for law firms, football clubs and in sectors including manufacturing, media, advertising, financial institutions, technology and leisure.

MacRoberts: David Flint and Valerie Surgenor
www.macroberts.com

MacRoberts LLP, one of Scotland's leading commercial law firms, prides itself on being highly-attuned to clients' needs. Over many years, a huge range of leading British and international businesses, banks and other financial institutions have continued to trust MacRoberts' lawyers to lend a clear insight to the commercial and legal issues that face them. In an increasingly frantic world there is no substitute for a commercial partner on whom you can rely. Our full service offices in Glasgow and Edinburgh, combined with the extensive experience of over 40 partners, means that we are able to offer clients innovative and cost effective solutions based solidly in commercial practice. We provide a comprehensive range of legal services across all sectors and our commitment to the success of our clients remains at the core of everything we do.

David Flint specialises in all aspects of non-contentious intellectual property, with particular emphasis on computer-related contracts and issues. He is acknowledged as a leading expert in intellectual property law, computer / IT law, and European law. He has also specialised in corporate insolvency for almost 30 years, is a licensed insolvency practitioner and has an extensive practice in competition law.

Primarily practising in non-contentious intellectual property, information technology and compliance and regulatory matters, Valerie Surgenor has a keen interest in the areas of information management and data security, having particular experience in the carrying out of European- and UK-wide Data Protection Compliance Audits. As a member of the MacRoberts Compliance and Regulatory Group, Valerie advises on the areas of anti-corruption and the new UK Bribery Act and related compliance issues.

Metis Law: David Sinclair
www.metislaw.com

Metis Law LLP is a new style of regional law firm with a strong property and construction base. Metis acts across the public and private sector, providing businesses with tailored services and innovative products. The firm has a reputation for delivering solutions to a range of property, facilities and construction organisations.

David Sinclair is one of a few dual-qualified Solicitors and safety practitioners in the UK, who have achieved Chartered Health and Safety Practitioner status. Trained originally as a mining engineer, David has a BSc Honours degree in Occupational Health and Safety and a Durham University postgraduate diploma in Environmental Management.

Ortalan: Louise Smail
www.ortalan.com

Louise Smail has extensive experience in the field of risk management with a wide range of organisations, including local government, emergency services, railways and process industries. Louise acted as Advanced Works Safety Manager for Union Railways Limited, looking at the issues of environment and consents associated with the works. Louise now runs her own consultancy, Ortalan, where she works on high profile projects looking at business risk evaluation, business continuity and performance and the impact of changes in legislation in both the UK and Europe.

PHS Compliance: Paul Caddick
www.phscompliance.co.uk

PHS Compliance is a specialist service provider in compliance for electricity, gas, fire and emergency lighting. The company is part of the PHS Group and operates nationally, with a network of regional offices. The services offered focus on compliance with the UK's Regulatory Standards and ACoPs (Approved Codes of Practice). Facets within this complete compliance approach include safety inspection, testing, remedial repair work, planned maintenance and emergency repairs. PHS Compliance has a history of technical expertise that extends over 20 years and holds the UKAS Inspection accreditation in recognition of its service standards.

Paul Caddick is Managing Director of PHS Compliance. He is registered as a chartered engineer and qualified in electrical engineering; nationally recognised as a technical expert within the field of electrical safety. Paul's career began with the NICEIC, working on a special project to analyse electrical accidents, predominantly in the domestic market. From this involvement in safety and its management, Paul joined the HSE working as an inspector, conducting both routine inspections and also investigating accidents and incidents. As a highly regarded technical expert, Paul has maintained his association with the HSE and the Executive Forum in Scotland, assisting with investigation work on safety incidents. His opinion is sought and respected within Government, the NHS and private organisations.

PHS Washrooms: Hayley O'Donovan
www.phs.co.uk/washrooms

PHS Washrooms is the UK's leading provider of away-from-home washroom services. They offer a wide range of products and services to enhance your washroom, tackle legislative issues and to help you meet your sustainability targets.

Hayley O'Donovan is the Marketing Executive for PHS Washrooms. With full nationwide coverage, over 45 years' experience and a team of qualified, friendly service technicians offering unrivalled standards of service, PHS Washrooms is in a great position to meet your washroom requirements, whatever they might be.

PPL: Christine Geissmar
www.ppluk.com

PPL is the UK-based music licensing company that licenses recorded music for broadcast, online and public performance use. Established in 1934, PPL carries out this role on behalf of thousands of record company and performer members. If recorded music is played in public – including offices, factories and general workplaces – music licences are legally required.

PRS for Music: Barney Hooper
www.prsformusic.com

PRS for Music represents 80,000 songwriters, composers and music publishers in the UK. As a not-for-profit organisation it ensures creators are paid whenever their music is played, performed or reproduced; championing the importance of copyright to protect and support the UK music industry. The UK has a proud tradition of creating wonderful music that is enjoyed the world over and PRS for Music has been supporting the creators of that music since 1914.

PRS for Music provides business and community groups with easy access to over ten million songs through its music licences. Barney Hooper is Head of PR and Corporate Communications at PRS for Music.

Rentokil: David Cross
www.rentokil.co.uk

Rentokil is the UK's largest supplier of pest control services and products to businesses and homes. Established more than 80 years ago, Rentokil Pest Control is part of Rentokil Initial plc, one of the largest global providers of business support services. Rentokil Pest Control has over 600 technicians throughout the UK, fully trained to prevent pest problems, and to quickly resolve any issues that do occur. The technicians are supported by the most advanced range of technology in the industry, much of which is developed in-house by the European Technical Centre in Horsham. Rentokil is experienced in working with all types of business, and develops tailored programmes to meet individual requirements.

RICS
www.rics.org/uk

RICS is the world's leading professional body for qualifications and standards in land, property and construction.

As people, governments, banks and commercial organisations continue to demand more assurance of certified standards and ethics, attaining RICS status is the recognised mark of property professionalism.

Rollits: Chris Platts
www.rollits.com

Chris Platts is a Regulatory lawyer and Head of Dispute Resolution Department at Rollits LLP. Chris has been involved in health and safety law for 27 years. During that time he has represented a whole host of clients ranging from individuals to major enterprises. His present clients number among them major plcs, small and medium-sized businesses, together with individuals.

Chris advises clients on strategies following accidents, particularly with regard to investigation and appropriate responses to enforcing authorities, in an effort to avoid prosecutions or other enforcement action.

As well as assisting clients with risk management advice, Chris deals with criminal offences and misconduct, defending organisations and individuals under investigation or being prosecuted. Chris has extensive experience in all types of criminal investigations and prosecutions brought by regulators, including the Police, HSE, Environment Agency, Local Authorities, and others. Chris also has wide experience of attending coroners' inquests.

Scafftag: Alex Foulkes
www.scafftag.com

In 1983, Scafftag introduced the innovative visible status tagging system for scaffolding and the name Scafftag has since become synonymous with scaffold inspection tagging.

Without accurate, up to date and timely information, people are prone to make assumptions. And assumptions can lead to potentially damaging decisions for both them and your business performance. Tracking the identity, usage, safety, maintenance and inspection status of each and every item of equipment is therefore critical. Scafftag helps you manage and communicate the latest inspection / test status of equipment in terms of health and safety, maintenance and identity to improve safety compliance and business performance. Visual tags, inspection management software and records provide a complete, customisable inspections solution ensuring the latest equipment status is clearly visible at the point of use. This allows you and your employees to make informed decisions and improve health and safety at work.

Society of Light and Lighting: Liz Peck
www.cibse.org

The Society of Light and Lighting acts as the professional body for lighting. It has over 2,000 members in the UK and worldwide, and carries out a full range of activities. The Society is a company limited by guarantee, which is controlled by CIBSE.

Liz Peck studied Business and Finance at Sheffield Hallam University. She joined Concord Lighting in its customer service department in early 1999. When Concord merged with Marlin Lighting to form Concord:Marlin, she joined its Lighting Design department. Liz completed the LIF (Lighting Industry Federation) Certificate and then went on to complete an MSc in Light and Lighting at the Bartlett School of Architecture in London, as part of her continuing lighting education. She spent a further two years with Concord:Marlin before joining Philips Lighting as Senior Lighting Designer. Liz has been an active Member of the Society for a number of years, sitting on the Newsletter and Communications Committee and both CIBSE and SLL Council. She took up the role as Secretary of the Society in February 2008.

Steeles Law: Elizabeth Stevens
www.steeleslaw.co.uk

Steeles Law is a medium-sized, multi-disciplinary law firm with offices in London, Norwich (HQ) and Diss, providing legal services to a diverse mix of clients, from corporate organisations and institutions to public sector bodies, charities and individuals. The firm is particularly recognised for the strength of its specialist employment practitioners, who provide practical and commercially sensible advice across the full range of employment law and HR issues and who have experience of handling the most complex and sensitive workplace disputes.

Elizabeth Stevens is a Professional Support Lawyer in the employment team of Steeles Law. She regularly writes articles on employment law for publication, as well as designing and delivering seminars and training programmes for clients.

Taylor Wessing: Rachel Farr and Lorraine Smith
www.taylorwessing.com

Taylor Wessing is a powerful source of legal support for organisations doing business in or with Europe and is based primarily in the UK, France and Germany, with further offices in Brussels, Alicante, Dubai and Shanghai. A market leader in advising IP and technology-rich industries, Taylor Wessing boasts a strong reputation in the corporate, finance and real estate sectors alongside indepth experience across the full range of legal services including tax, commercial disputes and employment.

Rachel Farr is a Professional Support Lawyer in the Employment and Pensions Group, focusing on UK employment law. Her role includes monitoring legal developments and assessing their impact on the market, writing client updates, and training colleagues. She was previously an Associate in the Group, advising both employers and employees on both contentious and non-contentious employment issues, including the recruitment, rewarding and termination of directors and senior executives. Rachel joined the firm in 2003.

Lorraine Smith is a Professional Support Lawyer. She focuses on analysing the latest legal and market practice developments in corporate law and their impact for our clients and the transactions we advise on. She also organises and updates Taylor Wessing's corporate legal know-how resources, prepares bulletins for clients and provides internal legal training for our lawyers. She trained at Herbert Smith, and came to Taylor Wessing after a period at Lovells (now Hogan Lovells). She draws on transactional and advisory experience in private M&A, group reorganisations and corporate governance.

Tetra Consulting: Giles Green
www.tetraconsulting.co.uk

Giles Green joined Tetra Consulting in 2009 with 24 years' experience in legionella control; four as a water treatment service provider and then 20 as a consultant. In addition to a career as a hands-on consultant, he has developed policies and procedures for water hygiene management in buildings, has developed training courses and has contributed to guidance and standards, investigated outbreaks of legionnaires' disease and given evidence in Crown Court in the capacity of an expert witness.

TripleAconsult: Elspeth Grant
www.tripleaconsult.co.uk

TripleAconsult offers a highly professional, cost effective service to clients seeking to reduce the costs of meeting the increasing legislative burden in both the public and private sectors. This is achieved through the delivery of integrated solutions in the areas of health and safety, fire safety and disability discrimination, using mobile data solutions and online web-enabled databases to assist those managing distributed workplaces. Our consultants' approach is imaginative and proactive, recognising the need to understand and appreciate an organisation's culture within which they are working.

Elspeth Grant is a Director and co-founder of TripleAconsult. Elspeth is an Associate Member of the Institution of Fire Safety Managers, is an Associate Trainer with Workplace Law, and regularly writes in the FM press. Over the past seven years, she has worked closely with design teams and clients to ensure the implementation of effective accessible workplaces and specialises in the evacuation of disabled people from complex environments such as high rise residential accommodation.

UK CEED
www.ukceed.org

The UK Centre for Economic and Environmental Development (UK CEED) is an independent not-for-profit company that promotes eco-innovation and the economic benefits of sound environmental practice. UK CEED supports the growth and development of the low carbon and environmental goods and services sector, whose innovative solutions are needed to create an economically and environmentally sustainable future in the UK. The Centre was founded in 1984 and has played an important role in demonstrating how environmental protection and economic development priorities can be brought together.

Water Regulations Advisory Scheme (WRAS): Steve Tuckwell
www.wras.co.uk

The Water Regulations Advisory Scheme Ltd (WRAS) promotes the Water Fittings Regulations and Water By-laws throughout the UK and encourages their consistent interpretation and enforcement to prevent waste and contamination of public water supplies. Supported by the UK Water Suppliers, WRAS offers a free Regulations enquiry service for anyone to use, approves water fittings for their compliance with the Regulations, and provides a register of plumbers who are approved for their competence in the Water Fittings Regulations.

Woodfines: Claire Morrissey
www.woodfines.co.uk

With offices across Bedfordshire, Buckinghamshire and Cambridgeshire Woodfines Solicitors are at the heart of the community, offering a wide range of legal advice, delivered by a team of lawyers in a friendly and approachable manner. Our objective is to deliver pragmatic solutions whilst maintaining our 'client first' philosophy.

Although our origins can be traced back to the 1800s we are well known for our progressive and forward-thinking approach to legal advice as well as for adopting a holistic approach to the needs of our clients. Our team of lawyers are regularly called upon for public comment and articles relating to legislative changes and to provide informed opinion.

Claire Morrissey practices in legal aid and private criminal work and regulatory work including health and safety, environmental, trading standards, licensing and road transport law. Claire also conducts prosecution work on behalf of the Vehicle Operator and Services Agency. Claire trained at Eversheds in Birmingham between 2007 and 2009 and qualified as a Solicitor in September 2009.

360 Environmental: Phil Conran
www.360environmental.co.uk

Phil Conran is a Director at 360 Environmental Ltd, a consultancy that specialises in waste-related legislation. Phil worked at Biffa until the end of 2008 and was responsible for the development of the Biffpack and Transform compliance schemes. He also sat on various Government Advisory bodies including the WEEE Advisory Board and was involved in much of the development of the WEEE Regulations and now provides support to companies that have to be registered under the various areas of Producer Responsibility. 360 Environmental has a website that provides easy to understand explanations of waste legislation.

Directory of information sources

Access Association
01922 652010

www.accessassociation.co.uk

Acoustic Safety Programme
info@acousticsafety.org
www.acousticshock.org

Action on Smoking and Health (ASH)
020 7739 5902
www.ash.org.uk

Alcohol Concern
020 7264 0510
www.alcoholconcern.org.uk

Asbestos Removal Contractors
Association (ARCA)
01283 566467
www.arca.org.uk

Association for Project Safety (APS)
08456 121 290
www.aps.org.uk

Association for Specialist Fire Protection
(ASFP)
01252 357 832
www.asfp.org.uk

Association of British Insurers (ABI)
020 7600 3333
www.abi.org.uk

Association of Building Engineers (ABE)
0845 126 1058
www.abe.org.uk

Association of Chief Police Officers (ACPO)
020 7084 8950
www.acpo.police.uk

Association of Consultant Approved
Inspectors (ACAI)
020 7491 1914
www.acai.org.uk

Association of Consultant Architects (ACA)
020 8466 9079
www.acarchitects.co.uk

Association of Consulting and Engineers
(ACE)
020 7222 6557
www.acenet.co.uk

Association of Industrial Road Safety
Officers (AIRSO)
01903 506095
www.airso.org.uk

Association of Plumbing and Heating
Contractors (APHC)
01217 115030
www.competentpersonscheme.co.uk

Association of Security Consultants (ASC)
07071 224865
www.securityconsultants.org.uk

Association of Sustainability Practitioners
www.asp-online.org

Association of Technical Lighting and Access
Specialists
0115 955 8818
www.atlas.org.uk

Automobile Association (AA)
0800 085 2721
www.theaa.com

Bathroom Manufacturers Association (BMA)
01782 631619
www.bathroom-association.org

BioIndustry Association (BIA)
020 7565 7190
www.bioindustry.org

Blind in Business
020 7630 2180
www.blindinbusiness.co.uk

BRE Certification Ltd
01923 664000
www.bre.co.uk

British Approval for Fire Equipment (BAFE)
0844 335 0897
www.bafe.org.uk

British Association for Chemical Specialities
(BACS)
01423 700 249
www.bacsnet.org

British Association of Removers (BAR)
01923 699 480
www.bar.co.uk

British Automatic Fire Sprinkler Association
(BAFSA)
01353 659187
www.bafsa.org.uk

British Cement Association (BCA)
01276 608700
www.cementindustry.co.uk

British Cleaning Council (BCC)
01562 851129
www.britishcleaningcouncil.org

British Fire Consortium (BFC)
033 31235306
www.britishfireconsortium.org.uk

British Institute of Cleaning Science (BICSc)
01604 678710
www.bics.org.uk

British Institute of Facilities Management (BIFM)
0845 058 1356
www.bifm.org.uk

British Occupational Health Research
Foundation (BOHRF)
020 8449 7218
www.bohrf.org.uk

British Occupational Hygiene Society
(BOHS)
01332 298101
www.bohs.org

British Parking Association (BPA)
01444 447300
www.britishparking.co.uk

British Pest Control Association (BPCA)
01332 294 288
www.bpca.org.uk

British Plastics Federation (BPF)
020 7457 5000
www.bpf.co.uk

British Property Federation (BPF)
020 7828 0111
www.bpf.org.uk

British Red Cross Society
0844 871 1111
www.redcross.org.uk

British Retail Consortium (BRC)
020 7854 8900
www.brc.org.uk

British Safety Council
020 8741 1231
www.britsafe.org

British Security Industry Association (BSIA)
0845 389 3889
www.bsia.co.uk

British Standards Institution (BSI)
020 8996 9001
www.bsi-global.com

British Woodworking Federation (BWF)
0870 458 6939
www.bwf.org.uk

Building Cost Information Service Ltd (BCIS)
020 7695 1500
www.bcis.co.uk

Building Research Establishment Ltd (BRE)
01923 664 000
www.bre.co.uk

Building Services Research and Information
Association (BSRIA)
01344 465 600
www.bsria.co.uk

Directory

Business Continuity Institute (BCI)
0118 947 8215
www.thebci.org

Business Link
0845 600 9 006
www.businesslink.gov.uk

Cadw
01443 336 000
www.cadw.wales.gov.uk

Carbon Trust
0800 085 2005
www.carbontrust.co.uk

CCA – Customer Contact Association
0141 564 9010
www.cca-global.com

Centre for Accessible Environments (CAE)
020 7840 0125
www.cae.org.uk

Chartered Institute of Architectural
Technologists (CIAT)
020 7278 2206
www.ciat.org.uk

Chartered Institute of Building (CIOB)
01344 630700
www.ciob.org.uk

Chartered Institute of Environmental Health
(CIEH)
020 7928 6006
www.cieh.org

Chartered Institute of Plumbing and Heating
Engineering
01708 472791
www.ciphe.org.uk

Chartered Institute of Purchasing and
Supply (CIPS)
01780 756777
www.cips.org

Chartered Institution of Waste Management
(CIWM)
01604 620426
www.ciwm.co.uk

Chartered Institution of Building Services
Engineers (CIBSE)
020 8675 5211
www.cibse.org

Chartered Society of Physiotherapy (CSP)
020 7306 6666
www.csp.org.uk

Chemical Hazards Communication Society
0844 636 2427
www.chcs.org.uk

Chemical Industries Association (CIA)
020 7963 6772
www.cia.org.uk

Chief Fire Officers' Association (CFOA)
01827 302 300
www.cfoa.org.uk

Chubb UK
0800 321 666 (Chubb Fire)
0800 282494 (Chubb Electronic Security)
www.chubb.co.uk

CIRIA
020 7549 3300
www.ciria.org.uk

Civil Contingencies Secretariat (Cabinet
Office)
020 7276 3000
www.cabinetoffice.gov.uk

Clay Pipe Development Association (CPDA)
01494 791456
www.cpda.co.uk

College of Occupational Therapists (COT)
020 7357 6480
www.cot.co.uk

Commercial Occupational Health Providers
Association (COHPA)
01933 232 373
www.cohpa.co.uk

Commission for Architecture and the Built
Environment (CABE)
020 7070 6700
www.cabe.org.uk

Concrete Society
01276 607140
www.concrete.org.uk

Confederation of British Industry (CBI)
020 7379 7400
www.cbi.org.uk

Confederation of Paper Industries (CPI)
01793 889600
www.paper.org.uk

Consortium of European Building Control
(CEBC)
01473 748 182
www.cebc.eu

Constructing Excellence
0845 605 5556
www.constructingexcellence.org.uk

Construction Health and Safety Group
(CHSG)
01932 561 871 / 563 121
www.chsg.co.uk

Construction Industry Council (CIC)
020 7399 7400
www.cic.org.uk

CITB – Construction Skills
01485 577577
www.cskills.org

Contract Flooring Association (CFA)
0115 941 1126
www.cfa.org.uk

Contractors Health and Safety Assessment
Scheme (CHAS)
www.chas.gov.uk

Council of Registered Gas Installers
(CORGI)
0800 915 0485
www.trustcorgi.com

Cranfield Institute for Safety, Risk and
Reliability
01234 750 111
www.cranfield.ac.uk/safety

Criminal Records Bureau (CRB)
0870 9090811
www.crb.gov.uk

Crown Prosecution Service (CPS)
020 7796 8000
www.cps.gov.uk

Department for Environment, Food and Rural
Affairs (DEFRA)
0845 933 5577
www.defra.gov.uk

Department for Transport (DfT)
020 7944 8300
www.dft.gov.uk

Department of Health (DH)
020 7210 4850
www.dh.gov.uk

Design Council
020 7420 5200
www.designcouncil.org.uk

Direct Gov
www.direct.gov.uk

ELECSA Ltd
0845 634 9043
www.elecsa.org.uk

Electrical Contractors' Association (ECA)
020 7313 4800
www.eca.co.uk

Emergency Planning Society
0845 600 9587
www.the-eps.org

Energy Institute
020 7467 7100
www.energyinst.org.uk

Engineering and Construction Industry
Association (ECIA)
020 7799 2000
www.ecia.co.uk

Engineering Council UK (ECUK)
020 3206 0500
www.engc.org.uk

Directory

Engineering Employers Federation (EEF)
0800 458 1500
www.eef.org.uk

English Heritage
0870 333 1181
www.english-heritage.org.uk

Environment Agency
08708 506 506
www.environment-agency.gov.uk

Environment Council
020 7836 2626
www.the-environment-council.org.uk

Environment and Heritage Service (Northern Ireland) (EHSNI)
0845 302 0008
www.ni-environment.gov.uk

Environmental Services Association (ESA)
www.esauk.org

Envirowise
0800 585 794
www.envirowise.gov.uk

Ergonomics Society
01509 234904
www.ergonomics.org.uk

European Agency for Safety and Health at Work
+34 944 794 360
www.agency.osha.eu.int

Facilities Management Association (FMA)
07960 428 146
www.fmassociation.org.uk

Fall Arrest Safety Equipment Training (FASET)
01948 780 652
www.faset.org.uk

Federation of Environmental Trade Associations
0118 940 3416
www.feta.co.uk

Federation of Master Builders (FMB)
020 7242 7583
www.fmb.org.uk

Federation of Window Cleaners (FWC)
0161 432 8754
www.nfmwgc.com

Fire Industry Association
020 8549 5855
www.fia.uk.com

FireNet International
www.fire.org.uk

Fire Protection Association (FPA)
01608 812500
www.thefpa.co.uk

Fire Service College
www.fireservicecollege.ac.uk

Food Standards Agency (FSA)
020 7276 8829
www.food.gov.uk

Fork Lift Truck Association (FLTA)
01256 381441
www.fork-truck.org.uk

Friends of the Earth
020 7490 1555
www.foe.co.uk

Furniture Industry Research Association (FIRA)
01438 777700
www.fira.co.uk

Gangmasters Licensing Authority
0845 602 5020
www.gla.gov.uk

Gas Safe Register
0800 408 5500
www.gassaferegister.co.uk

Greenpeace
020 7865 8100
www.greenpeace.org.uk

Hazards Forum
020 7665 2230
www.hazardsforum.co.uk

Health and Safety Executive (HSE)
www.hse.gov.uk

Health and Safety Executive for Northern Ireland (HSENI)
www.hseni.gov.uk

Health and Safety Laboratory (HSL)
01298 218 000
www.hsl.gov.uk

Health Facilities Management Association
www.hefma.org.uk

Healthy Workplace Initiative (HWI)
www.newworkplaceinstitute.org

Heating and Ventilating Contractors' Association (HVCA)
020 7313 4900
www.hvca.org.uk

Historic Scotland
0131 668 8600
www.historic-scotland.gov.uk

HSE Books
01787 881165
www.hsebooks.com

Incident Contact Centre Website
www.riddor.gov.uk

Independent Safety Consultants Association
01621 874938
www.isca.org.uk

Industrial Rope Access Trade Association
01252 357 839
www.irata.org

Industry Committee for Emergency Lighting
www.icel.co.uk

Institute of Acoustics (IOA)
01727 848195
www.ioa.org.uk

Institute of Alcohol Studies (IAS)
01480 466 766
www.ias.org.uk

Institute of Environmental Management and Assessment (IEMA)
01522 540069
www.iema.net

Institute of Food Research (IFR)
01603 255000
www.ifr.ac.uk

Institute of Hospitality
020 8661 4900
www.instituteofhospitality.org

Institute of Risk Management (IRM)
020 7709 9808
www.theirm.org

Institution of Civil Engineers (ICE)
020 7222 7722
www.ice.org.uk

Institution of Engineering and Technology
01483 313 311
www.theiet.org

Institution of Fire Engineers (IFE)
01608 812580
www.ife.org.uk

Institution of Gas Engineers and Managers (IGEM)
0844 3754436
www.igem.org.uk

Institution of Lighting Professionals (ILP)
01788 576492
www.ilp.org.uk

Institution of Occupational Safety and Health (IOSH)
0116 257 3100
www.iosh.co.uk

Institution of Structural Engineers (IStructE)
020 7235 4535
www.istructe.org

International Facilities Management Association (USA)
+1 713 623 4362
www.ifma.org

International Institute for Environment and Development (IIED)
020 7388 2117
www.iied.org

Directory

International Institute of Risk and Safety
Management (IIRSM)
020 8741 9100
www.iirsm.org

International Lead Association
020 7935 6146
www.ila-lead.org

Joint Industry Board for the Electrical
Contracting Industry (JIB)
020 8302 0031
www.jib.org.uk

Land Registry
020 7917 8888
www.landreg.gov.uk

Legislation.gov.uk
www.legislation.gov.uk

Lift and Escalator Industry Association (LEIA)
020 7935 3013
www.leia.co.uk

Local Authority Building Control (LABC)
0844 561 6136
www.labc.co.uk

London Fire Brigade
020 8555 1200
www.london-fire.gov.uk

Mastic Asphalt Council (MAC)
01424 814400
www.masticasphaltcouncil.co.uk

Mind
0845 7660163
www.mind.org.uk

Mobile Operators Association (MOA)
020 7331 2015
www.mobilemastinfo.com

Motability
0845 456 4566
www.motability.co.uk

Motor Insurers' Information Centre (MIIC)
0845 165 2800
www.miic.org.uk

National Access and Scaffolding
Confederation (NASC)
020 7822 7400
www.nasc.org.uk

National Examination Board in Occupational
Safety and Health (NEBOSH)
0116 263 4700
www.nebosh.org.uk

National Federation of Builders (NFB)
0870 8989 091
www.builders.org.uk

National Federation of Demolition
Contractors
01442 217144
www.demolition-nfdc.com

National Inspection Council for Electrical
Installation Contracting (NICEIC)
0870 013 0382
www.niceic.org.uk

National Institute for Health and Clinical
Excellence
0845 003 7780
www.nice.org.uk

National Pest Technicians Association (NPTA)
www.npta.org.uk

National Quality Assurance (NQA)
01582 539000
www.nqa.com

National Security Inspectorate (NSI)
0845 006 3003
www.nsi.org.uk

National Trust
0845 800 1895
www.nationaltrust.org.uk

Natural England
0845 600 3078
www.naturalengland.org.uk

NHS Plus
www.nhsplus.nhs.uk

Noise Abatement Society (NAS)
01273 823 850
www.noiseabatementsociety.com

Occupational Safety and Health Consultants
Register
www.oshcr.org

Occupational Road Safety Alliance
(ORSA)
www.orsa.org.uk

Office Furniture Advisory Service (OFAS)
01344 779438
www.ofas.org.uk

Office of Gas and Electricity Markets
(Ofgem)
020 7901 7000
www.ofgem.gov.uk

Office of Water Services (Ofwat)
0121 625 1300
www.ofwat.gov.uk

Painting and Decorating Association
(PDA)
024 7635 3776
www.paintingdecoratingassociation.co.uk

Planning Inspectorate
0117 372 6372 (England)/
029 2082 3866 (Wales)
www.planning-inspectorate.gov.uk

Rentokil
0800 917 1989
www.rentokil.co.uk

Remploy
0800 155 2700
www.remploy.co.uk

Repetitive Strain Injuries Association
(RSIA)
023 8029 4500
www.rsi.org.uk

RIBA Bookshops
020 7256 7222
www.ribabookshops.com

RICS Books
0870 333 1600
www.ricsbooks.com

Robust Details Ltd
www.robustdetails.com

Royal Automobile Club (RAC)
01922 727 313
www.rac.co.uk

Royal Environmental Health Institute of
Scotland
0131 225 6999
www.rehis.org

Royal Incorporation of Architects in Scotland
(RIAS)
0131 229 7545
www.rias.org.uk

Royal Institute of British Architects (RIBA)
020 7580 5533
www.architecture.com

Royal Institution of Chartered Surveyors
(RICS)
0870 333 1600
www.rics.org

Royal National Institute of the Blind (RNIB)
020 7388 1266
www.rnib.org.uk

Royal National Institute for Deaf People
(RNID)
020 7296 8000
www.rnid.org.uk

Royal Society for the Prevention of Accidents
(RoSPA)
0121 248 2000
www.rospa.org.uk

Royal Society for Public Health (RSPH)
020 3177 1600
www.rsph.org

Safety and Reliability Society (SaRS)
0161 228 7824
www.sars.org.uk

Safety Assessment Federation (SAFed)
020 7582 3208
www.safed.co.uk

St John Ambulance
087000 104 950
www.sja.org.uk

Scafftag
www.scafftag.com
0845 601 0329

Scottish and Northern Ireland Plumbing
Employers' Federation (SNIPEF)
0131 225 2255
www.snipef.org

Scottish Association of Building Standards
Managers (SABSM)
www.sabsm.co.uk

Scottish Building Standards Agency
01506 600 400
www.sbsa.gov.uk

Scottish Environment Protection Agency (SEPA)
01786 457700
www.sepa.org.uk

Scottish Executive
08457 741 741
www.scotland.gov.uk/Home

Security Industry Authority (SIA)
0844 892 1025
www.the-sia.org.uk

Security Institute
08453 707 717
www.security-institute.org

Sign Design Society
020 8776 8866
www.signdesignsociety.co.uk

Society of Light and Lighting (SLL)
020 8675 5211
www.cibse.org

Society of Occupational Medicine (SOM)
020 7486 2641
www.som.org.uk

Stress Management Society
08701 999 235
www.stress.org.uk

Suzy Lamplugh Trust
020 7091 0014
www.suzylamplugh.org

The Stationery Office
0870 600 5522
www.tso.co.uk

Transco
www.nationalgrid.com/uk

United Kingdom Accreditation Service
(UKAS)
020 8917 8400
www.ukas.com

UK Centre for Economic & Environmental
Development (UK CEED)
01733 311644
www.ukceed.org

Valuation Office Agency (VOA)
020 7506 1700
www.voa.gov.uk

Wales Trades Union Congress (TUC Cymru)
020 7636 4030
www.tuc.org.uk

Water Regulations Advisory Scheme (WRAS)
01495 248454
www.wras.co.uk

Water Research Centre (WRc plc)
01793 865000
www.wrcplc.co.uk

Water UK
020 7344 1844
www.water.org.uk

The Welding Institute (TWI)
01223 899 000
www.twi.co.uk

Workplace Law
0871 777 8881
www.workplacelaw.net

Index